About IFPRI

The International Food Policy Research Institute (IFPRI), a CGIAR Research Center established in 1975, provides research-based policy solutions to sustainably reduce poverty and end hunger and malnutrition. IFPRI's strategic research aims to foster a climate-resilient and sustainable food supply; promote healthy diets and nutrition for all; build inclusive and efficient markets, trade systems, and food industries; transform agricultural and rural economies; and strengthen institutions and governance. Gender is integrated in all the Institute's work. Partnerships, communications, capacity strengthening, and data and knowledge management are essential components to translate IFPRI's research from action to impact. The Institute's regional and country programs play a critical role in responding to demand for food policy research and in delivering holistic support for country-led development. IFPRI collaborates with partners around the world.

About IFPRI's Peer-Review Process

IFPRI books are policy-relevant publications based on original and innovative research conducted at IFPRI. All manuscripts submitted for publication as IFPRI books undergo an extensive review procedure that is managed by IFPRI's Publications Review Committee (PRC). Upon submission to the PRC, the manuscript is reviewed by a PRC member. Once the manuscript is considered ready for external review, the PRC submits it to at least two external reviewers who are chosen for their familiarity with the subject matter and the country setting. Upon receipt of these blind external peer reviews, the PRC provides the author with an editorial decision and, when necessary, instructions for revision based on the external reviews. The PRC reassesses the revised manuscript and makes a recommendation regarding publication to the director general of IFPRI. With the director general's approval, the manuscript enters the editorial and production phase to become an IFPRI book.

Agricultural Development: New Perspectives in a Changing World

Edited by Keijiro Otsuka and Shenggen Fan

A Peer-Reviewed Publication

International Food Policy Research Institute
Washington, DC

Recommended citation: Otsuka, K., and S. Fan, eds. 2021. *Agricultural Development: New Perspectives in a Changing World*. Washington, DC: International Food Policy Research Institute. https://doi.org/10.2499/9780896293830

International Food Policy Research Institute, 1201 Eye Street, NW, 12th floor, Washington, DC 20005 USA, Telephone: +1-202-862-5600, www.ifpri.org

ISBN: 978-0-89629-383-0
DOI: https://doi.org/10.2499/9780896293830

Library of Congress Cataloging-in-Publication Data may be found on page viii.

CONTENTS

Library of Congress Cataloging-in-Publication Data

Names: Otsuka, Keijiro, editor. | Fan, Shenggen, editor.
Title: Agricultural development : new perspectives in a changing world / Keijiro Otsuka, Shenggen Fan.
Description: Washington, DC : International Food Policy Research Institute, [2021] | Includes bibliographical references and index.
Identifiers: LCCN 2020045410 (print) | LCCN 2020045411 (ebook) | ISBN 9780896293830 (paperback) | ISBN 9780896293847 (epub)
Subjects: LCSH: Agriculture—Developing countries. | Agricultural productivity—Developing countries. | Agriculture and state—Developing countries. | Sustainable agriculture—Developing countries.
Classification: LCC HD1417 . A36 2021 (print) | LCC HD1417 (ebook) | DDC 338.109172/4—dc23
LC record available at https://lccn.loc.gov/2020045410
LC ebook record available at https://lccn.loc.gov/2020045411

LIST OF TABLES

LIST OF FIGURES

FOREWORD

Agricultural development is a key catalyst for economic development in many low- and middle-income countries around the world and has been throughout history. Yet, understanding agricultural development in the 21st century requires an integrated framework, both conceptually and empirically. Advances in theory and empirical methodologies, new data, and developments in other scientific disciplines have contributed new insights. A new range of global challenges—including expanding global markets, urbanization and rural transformation, natural resource degradation, global warming, rising obesity, and the threat of pandemics—have brought a new urgency to advancing the field. Understanding agricultural development implies more than simply ascertaining what drives agricultural productivity growth; it also requires incorporating knowledge of household behavior and gender roles, nutrition, natural resource management, climate change, agrifood value chains, equity, public policies, and political economy issues. Agricultural development is essential not only to produce enough food, but also to reduce poverty, improve nutrition, promote equity, and support rural and food systems transformation.

This ambitious volume provides a comprehensive textbook on the current state of agricultural development. Contributions by top experts, including many IFPRI researchers, examine different development patterns around the world, the socioeconomic and institutional context for agricultural development, and new opportunities and challenges that will continue to reshape agrifood systems and how we study them. Together, these in-depth explorations of current research point to proven and promising policy options and strategies for developing countries.

The International Food Policy Research Institute is proud to be a leader in many of these advances in agricultural development, particularly in translating a growing understanding into viable and effective policy options for developing countries. We are pleased to publish *Agricultural Development: New Perspectives in a Changing World,* which, with its holistic approach, should prove invaluable to students, practitioners, researchers, and policymakers making decisions in today's changing world.

Johan Swinnen
Director General, IFPRI

PREFACE

After several major famines in the 1950s and 1960s, the main focus of agricultural development during the following two decades was on achieving food security and preventing hunger in developing countries. These challenges spurred the Green Revolution, characterized by the adoption of high-yielding wheat and rice varieties, which doubled or even tripled crop yields in a matter of 20 years. Without the Green Revolution, millions of people would have died of hunger and undernourishment. Policy played a critical role, creating incentives for farmers and supporting them with investments in irrigation, market infrastructure, and storage facilities.

The world has been changing dramatically since then, and so has the landscape of agriculture. In addition to hunger, poverty and undernutrition have become major global concerns. Rapid income growth and urbanization since the 1990s have led to increased demand for high-value agricultural products, such as fresh fruits and vegetables and animal-sourced foods. This demand, in turn, has both stimulated and been fueled by the emergence of modern food value chains led by giant supermarkets in both developed and developing countries. Owing in part to international trade liberalization, cross-border trade of agricultural products has increased enormously, and the political equilibrium has shifted to less distorted trade markets for agricultural products. Meanwhile, gender equity and women's empowerment are recognized as major global development goals. At the same time, there have been key innovations in credit and insurance markets, including microfinance and weather-based insurance programs, respectively. Above all, human beings face increasingly serious challenges from climate change, of which agriculture is both a perpetrator and a victim. In the face of climate change, natural resource management, including forest, biodiversity, soil, and water management, has come to the fore of agricultural development issues. Critical now is the role of agricultural research, which is expected to play a key role in climate change adaptation and mitigation efforts while also continuing to focus on increasing both staple and more healthy food production.

The year 2000 was a landmark year in the fight against global poverty. The United Nations Millennium Declaration, signed that September, committed world leaders to halving poverty by 2015. With this goal in mind, policy analysts began to focus on the definition and measures of poverty and to assess the impact of poverty reduction programs and policies. Since 2010, nutrition has risen to the top of the food policy research agenda. Agricultural economists and policy analysts are therefore focusing increasingly on linkages between agriculture and nutrition through diets, behavior change, water, sanitation, and hygiene (WASH), and woman's empowerment in agriculture.

In 2015, more than 190 countries endorsed the Sustainable Development Goals (SDGs) during the UN General Assembly in New York, marking another milestone. Since then, the food systems approach, first introduced in the 1980s, has regained momentum with new meanings and goals. Food policy researchers have been working on data, methods, and policy options to redesign food systems to deliver the outcomes required by the SDGs.

Currently, the study of agricultural development suffers from a lack of textbooks that address the diversity of agricultural issues in an integrated fashion. While the demand for this type of textbook is huge, it is unrealistic to think that a single or small group of researchers could deal with such a multitude of issues comprehensively and publish an integrated textbook successfully. We therefore decided to put together this volume on agricultural development, with chapters written primarily by researchers at the International Food Policy Research Institute (IFPRI), given their expertise in an array of specialized areas. In addition to current IFPRI researchers, we also invited several former IFPRI staff members and other outside researchers to contribute. Happily, nearly all accepted our invitation. Although the contributors have never met in seminars or conferences related to the book, apart from a symposium held at the International Conference of Agricultural Economists in Vancouver in 2018, we all made every effort to coordinate our various chapters. This reflects the common recognition that our profession needs a state-of-the art textbook.

We received enormous input from the two anonymous reviewers designated by IFPRI's Publications Review Committee and the committee's chair, Professor Gerald Shively, for which we are wholeheartedly grateful and which improved the quality of this volume. Since there are more than 40 contributors and they all certainly received useful comments and research assistance from a number of people, we cannot express our appreciation to each and every one of them. We are particularly grateful, however, to Emily EunYoung Cho for her skillful assistance to both editors throughout the entire process of preparing the manuscript. We would also like to thank Fumiyo Aburatani for her editorial assistance. We are grateful to Alejandro Nin-Pratt for preparing tables related to inputs, outputs, land and labor productivities, and total factor productivities used in many chapters.

We are happy to inform readers that this is an open-access book. Thus, anyone who would like to study agricultural development can download it free of charge. It is our hope that this edited volume contributes to the study of agricultural development and benefits both current and future researchers alike.

Keijiro Otsuka and Shenggen Fan

PART I

A Global Overview of Agriculture

Chapter 1

AGRICULTURAL DEVELOPMENT IN A CHANGING WORLD

Shenggen Fan and Keijiro Otsuka

The world has been changing rapidly, and major issues surrounding agriculture have evolved as well. In fact, over the last several decades major shifts have occurred in the thinking on and practice of agricultural development. Accordingly, agricultural development goals have moved far beyond traditional ones such as food production and availability, agricultural productivity, farmers' incomes (particularly those of smallholders), and employment. The set of new goals includes poverty reduction, adequate nutrition, functioning food value chains (FVCs), environmental sustainability, climate adaptation and mitigation, and gender equality and equity. Looking forward, agriculture will face new challenges and will have to be positioned to deliver broader development outcomes such as those mandated under the Sustainable Development Goals (SDGs).

It is therefore timely to publish a new volume that reflects the latest developments and new perspectives on agricultural development. Such a book will be useful to policymakers, students, and development economists alike. It will also fill an important gap, as few if any such comprehensive volumes exist in the field of agricultural development.

First, this is an edited volume that benefits from the expertise of numerous top scholars in different fields. This allows for coverage of a broad array of emerging and complex issues such as rapid urbanization and agricultural transformation, nutrition and health, and natural resource management and climate change.

Second, this book covers most developing countries and regions, providing a global perspective. Such a perspective is extremely useful because many relevant issues are global in nature, often crossing national borders, in part due to increasingly interconnected agricultural systems in a globalizing world. The global perspective is also useful because it can reflect the many commonalities and differences in major issues across regions.

Third, we follow developments since the framework of induced technological and institutional innovations was developed by Yujiro Hayami and

Vernon W. Ruttan (1985), particularly in Part II, where we review technological change, productivity growth, and beyond in major regions. Innovations are seen in new institutions, including those related to international trade, land and water rights, producer cooperatives for rural industrialization and contract farming, and social norms on women's status, as well as a variety of institutions to prevent or mitigate climate change. Technological and institutional innovations in agriculture have also been increasingly influenced by innovations outside of agriculture, for example, rapid development of information technologies and biotechnologies.

Fourth, this book departs from the Hayami-Ruttan and other conventional theses in several significant ways. Increasing production of high-value products, development of modern FVCs, nutrition and health, gender and intrahousehold resource allocation, insurance and credit, property rights and natural resource management, and climate change are newly emerging and interrelated issues. This book is a first attempt to analyze them in an integrated manner. It also reviews the literature that uses applied microeconomics to study development, including studies based on randomized controlled trials (RCTs) and natural and quasi experiments. Such studies are a critical step toward evidence-based agricultural policymaking.

Finally, the book will point out clear policy implications for key institutional innovations and explore policy reforms conducive to agricultural development. In particular, we will attempt to identify effective strategies for developing sustainable agriculture and reducing food insecurity and malnutrition, which are missing in the existing literature.

This volume is divided into four parts. Part I (A Global Overview of Agriculture) introduces the aims of the book, identifies major global trends and emerging global issues in agricultural development, and explains the structure of the book, including how the various chapters are interrelated.

Part II (Regional Issues in Agricultural Development) provides overviews of technological innovations, agricultural research, agricultural development, and economic transformation by major region—East Asia, South Asia, Africa south of the Sahara, Latin America and the Caribbean, and Eastern Europe and Central Asia. Specifically, chapters in Part II examine long-term changes in cultivation area, land-labor ratios, composition of agricultural products (for example, staple foods vs. high-value products, including livestock products), adoption of land-saving or yield-enhancing technologies (adoption of improved seeds and the use of chemical fertilizer) and labor-saving or mechanical technologies, total factor productivity, farm size, and land tenancy institutions. These chapters also identify newly emerging important issues in various regions, such as increased overweight and obesity, which are examined in depth by a series of thematic chapters in Part III.

Part III (Context for Agricultural Development) focuses on context for agricultural development, with particular attention to the role of (1) urbanization, rural-to-urban migration, poverty reduction, food security, and nutritional transitions; (2) emerging FVCs, development of nonfarm sectors, microfinance, weather-based crop insurance, and land markets in transforming agricultural and rural economies; (3) community organizations (for forest and irrigation management, development of rural industries, and production of high-value products) and land rights in facilitating land transactions, investment in land improvement, and management of common property resources; (4) social norms and women's ownership and control of land and other assets in the transformation of women's status; and (5) international agreements on international trade and agricultural policies in agricultural development. These issues are highly interrelated. For example, new FVCs emerge in response to increases in market demand for safe and high-quality products associated with the development of nonfarm sectors and urbanization, to globalized trade of high-value agricultural products, and to the increasing consciousness of health and nutrition among urban consumers.

Part IV (Emerging Challenges and Opportunities in Agricultural Development) delves into the emerging challenges and opportunities in agricultural development, particularly those associated with changes in food systems and climate change, by examining the growing scarcity of water, possible effects of climate change, and the future of agricultural research. This is followed by the concluding analysis of how to reshape and transform agrifood systems for environmentally sustainable and inclusive development toward achieving the SDGs.

In the remainder of this chapter, we provide an overview of agricultural growth by reviewing the evolution of agricultural development thinking through macro- and micro-lenses. We briefly review the literature influencing major shifts in agricultural development below; a comprehensive review of the literature on the contemporary and emerging issues can be found in Chapter 2. We conclude with a summary of the regional overview and key themes discussed in the rest of the book.

Evolution of Macroeconomics Literature on Agricultural Development

The theory of agricultural development has evolved over the years, as have its contexts, practices, and goals. Early development theories, most prominently, Lewis's dual economy theory in the 1950s, viewed the agriculture sector as a supplier of surplus labor whose share in the labor force and economy declined through the course of development (Lewis 1954). Surplus labor from the traditional, rural agriculture sector has a negligible or zero marginal product. The modern, urban, nonagricultural industrial sector has a higher marginal product and absorbs this surplus labor by creating jobs, thereby increasing aggregate output and incomes and stimulating economic growth. Ranis and Fei (1961) built upon this model to assert that without agricultural growth and sufficient food output, development of the industrial sector will be constrained. In these models, agriculture plays a critical but passive role in economic development and transformation.

In the 1960s, a new paradigm extended the duality model to view agricultural development as an engine for industrialization and economic growth. Conceptually, agriculture contributes to development of industry by raising farmers' incomes and earning foreign exchange and by generating surplus capital and labor (Johnston and Mellor 1961). Higher farmer incomes from higher output increase demand for farm inputs and value-added services, and higher incomes for farmers and laborers increase the demand for food and nonagricultural goods and services. Agricultural growth then has a multiplier effect on other sectors, jump-starting transformation and accelerating growth.

Schultz (1964) argued that poor smallholder farmers are rational, as they respond to price incentives and will adopt new profitable technologies, but are constrained by the absence of such technologies. Schultz's "efficient but poor" hypothesis inspired subsequent research by Hayami and Ruttan (1971, 1985), who formalized the theory of agricultural growth by highlighting the importance of technological innovations. They state that technological and institutional changes are induced through responses of actors to changing resource endowments. In other words, as scarcity of a factor of production increases, technology that saves the use of the factor is induced to develop, along with supportive institutions, such as property rights systems, public-sector research and extension systems, and marketing institutions. Binswanger (1974) later demonstrated that changes in product prices also play an important role in stimulating innovations.

With the rapid economic growth in developing countries in the 1990s, especially in Asia, Timmer (1988) marked a shift in the theory of agricultural development toward structural transformation. The theory noted that agriculture enhances labor productivity in the rural economy, increasing wages and driving urbanization and industrialization. As a result, the share of agriculture in economic output and employment falls, and urban economic activity in industry and services grows. At the same time, rural workers migrate to urban areas and the overall population undergoes a demographic transition (Timmer 1988, 2017).

Initially, empirical studies focused on the importance of agriculture for nonfarm economic growth as well as spillovers of growth across sectors, as seen in Gemmell, Lloyd, and Mathew (2000) and Tiffin and Irz (2006), among others. Agriculture was considered in the context of a broader development agenda, highlighting the role of agriculture in the formulation of development strategies. Studies also looked at agriculture's impact on poverty reduction—in particular, its role in driving overall economic growth and, indirectly, reductions in food prices, which especially benefit the poor. For instance, Ravallion and Chen (2007) found the impact of agricultural growth on poverty in China to be four times greater than that of nonagriculture. Christiaensen, Demery, and Kuhl (2011) found that the poverty reduction impact of agricultural growth in Africa south of the Sahara was two to three times greater than that of growth in nonagriculture, but also noted substantial heterogeneity depending on country characteristics like natural resources and initial economic conditions (Dercon and Gollin 2014).

Computable general equilibrium (CGE) modeling was later added as one of the tools for analyzing the role of agriculture in overall economic growth and poverty reduction (for example, in Dorosh and Haggblade 2003; Diao and Dorosh 2007; Dorosh and Thurlow 2012). Much of the CGE modeling focused on African economies found that agricultural productivity growth generates positive impacts on overall growth and positive poverty impacts as well. However, it should be noted that much of this literature was based on assumptions that the model economies are closed and must meet their food needs through domestic production (Dercon and Gollin 2014).

Together with agricultural development, the food value chain evolved over time as well. With rising incomes and growing urbanization, the FVC became spatially longer, stretching across rural and urban areas. This transition also led to the emergence of food industries such as milling and food processing to add value and transport food. Reardon et al. (2003) and others added to the notion of the modern FVC, focusing on the rapid rise of supermarkets, which further transformed and integrated food markets, driven by urbanization, economic growth, and improved infrastructure. Fan, Yosef, and Pandya-Lorch

(2019) further analyzed how agriculture can contribute to nutrition through FVCs, nutrition-sensitive programs, government policies, and private-sector investments.

Triggered by the 2008 food price crisis, Díaz-Bonilla (2015) systematically analyzed how macroeconomic policies like fiscal policies, monetary and financial policies, exchange rate policies, and trade policies can fundamentally affect the agricultural sector. He argues that a macroeconomic policy framework is needed to maintain fiscal balances and avoid the overvaluation of the exchange rate, thereby reducing significantly the possibility of financial/fiscal crises and ensuring that tradable products (particularly agricultural and food products) are not disadvantaged. A monetary policy that maintains low inflation levels is also needed. On credit, Díaz-Bonilla suggests that agriculture needs specific credit programs, institutions, and instruments that emphasize a variety of instruments and approaches for financial inclusion for the poor.

Evolution of Microeconomics Literature on Agricultural Development

Microeconomics in agricultural development mainly considers the behavior of rural households, namely, the constraints they face and the determinants of their decision-making. It has evolved over time from the analysis of resource allocation, particularly labor, land, and capital market failures, to technology adoption and extension services, property rights and externalities, intrahousehold bargaining, and risk management. While the macroeconomics literature essentially focuses on the role of agriculture in economic development and enabling macroeconomic policy environments, the microeconomics literature covers a diverse range of issues related to the behavior of rural households, justifying a much longer review.

Schultz's (1964) "efficient but poor" hypothesis fundamentally affected the microeconomics literature on agricultural development because if farmers are rational maximizers, a profit or utility maximization model can be applied to the analysis of farmers' behaviors in developing countries. Thus, Schultz stimulated the subsequent analysis of farm household behavior. Deeper understanding of farm household behaviors improves macroeconomic models and modeling of general equilibrium effects.

Singh, Squire, and Strauss (1986) pioneered the analysis of agricultural household behavior, which led to the debate in the late 1980s about nonseparability between production and consumption decisions of rural households and the relationship between productivity and farm size. These authors clarify the "non-separable behavior" of households in using resources, fundamentally as a result of missing or imperfect markets, particularly labor markets, implying that households are an integrated production and consumption unit.

Labor markets are widely assumed to fail, because agency costs make the productivity of hired labor less than that of family labor. While households with land endowments too small to hire labor behave according to predictions in the separable model, households with more land will begin to hire labor but face a growing monitoring cost of labor, creating the inverse farm productivity–farm size relationship. Carter and Yao (2002) call this "local non-separability." De Janvry, Fafchamps, and Sadoulet (1991) showed that transaction costs drive a wedge between producer and consumer prices, where some households do not purchase or sell the goods they produce and, hence, have limited response to price incentives. Similarly, Key, Sadoulet, and de Janvry (2000) and Bellemare and Barrett (2006) explore the endogeneity of market participation decisions and related econometric implications. In a recent paper, Foster and Rosenzweig (2017) observed a **U**-shaped relationship between farm productivity and farm scale—the initial fall in productivity as farm size increases from its lowest levels and the continuous upward trajectory as scale increases after a threshold—in low-income countries across the world. They show that the existence of labor-market transaction costs can explain why the smallest farms, which rely on family labor, are most efficient; slightly larger farms are least efficient because they employ a host of hired workers; and larger farms are as efficient as the smallest farms because they adopt large-scale machines.

The microeconomics literature on agricultural development pays special attention to transaction costs in land markets, which lead to market failures. Eswaran and Kotwal (1984) studied how productivity is influenced by distribution of land among rural households under land, labor, and capital market imperfections. The authors noted that an economy with high inequality of land distribution will produce less than an economy with more equal land distribution, because farmers endowed with less land have less favorable access to credit. Consequently, credit market reforms that equalize access to capital across farm households can also have effects similar to those of land redistribution reforms. Otsuka, Chuma, and Hayami (1992) applied the principal-agent theory to share tenancy issues and argued that share tenancy is chosen partly because it has risk-sharing advantages and partly because labor contracting is less efficient due to higher monitoring costs.

Market failures also occur due to difficulty in establishing land rights, the lack of which reduces incentives to invest in land improvement (Besley

1995). Feder et al. (1988) found that enhanced formal land-tenure security in Thailand offered substantial payoffs in increased investment in land improvement and enhanced productivity. In Latin America, Carter and Olinto (2003) found that the investment demand effects of property rights reform applied to everyone, but that credit supply expanded only for medium- and larger-scale farmers. In other words, property rights reform is shown to have substantial impacts for only relatively advantaged farmers.

The lack of secure property rights for community-owned resources leads to market failures because individual users do not pay attention to the negative impact of their resource extraction on the productivity of other community members. For instance, Robalino and Pfaff (2012) find in Costa Rica that individuals are more likely to deforest when their neighbors deforest. Internalization of resource-related externalities requires collective action among users (Ostrom 1990), but the likelihood of collective action to resolve natural-resource-related externalities depends on the costs of cooperation, which vary depending on numerous environmental factors, existing norms, and users (Pender and Scherr 2002; Godquin and Quisumbing 2008). Lawry et al. (2014) suggest that secure land property rights contribute to welfare through greater perceived security of ownership and consequent long-term investments. Land ownership can help incentivize farmers in terms of land security, which has a positive effect on farmers' decisions to adapt to climate change (Yegbemey et al. 2013).

The Green Revolution revealed the importance of purchased inputs and small farmers' access to credit. Feder (1985) showed that if capital access improved with land endowments, the relationship between farm productivity and size could become positive even with labor market failures. Monitoring costs due to poor contract enforcement lead to different lending and deposit rates in financial markets, meaning that the wealthy will tend to invest more than the poor. Poor farmers who lack assets may not be able to offer the collateral necessary to access credit (Banerjee 2006), and if poor farmers lack access to insurance markets, loan terms may become too risky to borrow (Boucher, Carter, and Guirkinger 2008). Certain technologies, for instance, are more difficult for smallholders to finance, especially when the need for access to credit is greatest in order to meet high up-front costs. Despite enthusiasm over the last 10 to 15 years about the potential of microfinance as a major driver of poverty reduction in developing countries, recent studies have pointed to the lack of evidence and mixed long-term results. For example, Banerjee et al. (2015) find that while a microcredit program in India helped small business investment and profits of preexisting businesses to increase, consumption did not see a significant increase. Durable goods expenditure increased, and "temptation goods" expenditure declined, but there were no significant changes in health, education, or women's empowerment. Very few significant differences were found between treatment and control groups two years later.

Technology adoption has been studied by many researchers since the 1970s, with focus on the spillover effect of technological information. Early adopters of technologies provide information for others on benefits and correct use while disproportionately bearing the cost of the learning process. Thus, there is incentive to strategically delay adoption and to free ride when information is more readily available (Foster and Rosenzweig 1995; Bandiera and Rasul 2006; Maertens 2017). Any type of positive spillover in an area or network creates incentive to postpone adoption, as seen also in adoption of health products that generate immunity benefits for others (Kremer and Miguel 2007). Unreliable supply and high prices of fertilizer and other inputs are often noted as major barriers to adoption of new technologies. Mitra et al. (2013) study the importance of information asymmetries in the price gap between farmgate and market. Farmers' organizations have the potential to address many of these constraints by improving farmers' bargaining power, aggregating demand, and reducing individual risk, while also enhancing smallholder competitiveness in markets (World Bank 2007). However, inequality of asset ownership affects the degree of profit extracted by members, and smallholders may benefit less than larger landholders (Banerjee et al. 2001; Bernard, de Janvry, and Sadoulet 2010).

More recently, the role of farmer-to-farmer extension, which is likely to be complementary to public-sector extension, has received increased attention, and many RCT-based studies are conducted to identify the desirable characteristics of farmer-trainers (see Takahashi, Muraoka, and Otsuka 2020 for a recent review of the literature). Further, recent advancements in information and communications technologies (ICTs) have been evaluated by many studies. A study in India finds that mobile-phone-based agricultural extension alters management practices, increasing adoption of more effective pesticides and levels of farmer education (Cole and Fernando 2012). At the same time, other studies have pointed to more mixed results. Information on markets and weather provided through mobile phones was found to have no significant impact on prices received by farmers, crop value-added, crop losses, or likelihood of changing crop varieties or cultivation practices (Fafchamps and Minten 2012). Similarly, Nakasone, Torero, and Minten (2014) find that while access to mobile phones has generally improved agricultural market performance at the macro level, impacts at the micro level are mixed.

The study of the impact of risk on household decision-making was pioneered by Binswanger and Sillers (1983), who saw that risk mattered most due to differences in access to credit and other financial markets that could be used to mediate risk. Binswanger and Rosenzweig (1986) argue that covariate risk in agriculture further suppresses development of agricultural loan markets in risk-prone, low-income, rainfed agricultural regions, where all farmers in the neighborhood suffer from the same risk. Deaton (1991) shows that a modest amount of risk can be managed by maintaining some savings that could be used to smooth consumption in the face of income fluctuations or credit constraints. Further, Carter and Barrett (2006) describe several types of dynamic models where risk and capital constraints create a poverty trap. Unequal asset distribution that leaves large numbers of households below a critical asset threshold results in stagnation of productivity with persistent poverty. Also, uninsured risk in the face of a poverty trap has far higher costs for households. Findings often show better-off households are better at smoothing consumption than are poorer ones (Morduch 1995; Barrett, Carter, and Timmer 2010). Risk considerations also factor into household portfolio choices about assets and activities, whereby if risk preferences are related to ex ante wealth, portfolio choices may reinforce pre-existing unequal wealth distribution. Rosenzweig and Binswanger (1993) find wealthier households in rural India hold higher risk and higher expected return portfolios, which lead to different growth rates and increasing inequality over time, also contributing to a poverty trap among less well-endowed agricultural households. In considering the role of assets to address household risk, studies have evaluated the role of livestock and its wide spectrum of benefits, such as cash income, food, savings and insurance, and social capital. A study of microeconomic data from 12 developing countries (Pica-Ciamarra et al. 2015) finds that the majority of rural households keep livestock, and less well-off households are more likely than better-off households to keep livestock. However, poorer households often lack the resources to invest in small animals. The study also suggests there are many factors at play, meaning that policies need to account for farming systems, species, uses of livestock, and different wealth groups. Livestock ownership has also been shown to be associated with additional benefits like animal food consumption and relative nutritional benefits (Kim et al. 2019).

Informal mechanisms of coping with risk include risk pooling among households in a community, including through community insurance schemes to protect against idiosyncratic shocks. Social networks have been found to encourage insurance adoption. For example, one study finds that knowledge diffusion through peer effects (i.e., social networks) among farmers in rural China resulted in a premium reduction of up to 13 percentage points (Cai, de Janvry, and Sadoulet 2015). In addition to knowledge diffusion, informal risk-sharing arrangements can also help to manage basis risk of formal insurance, especially when the main source of basis risk is an idiosyncratic risk, and help to avoid free-riding and coordination problems (de Janvry et al. 2014; Dercon and Gollin 2014; Geng et al. 2018).

Studies show that other informal mechanisms, such as self-help groups (SHGs), have broader benefits beyond addressing household risk. SHGs tied to microcredit in India have been shown to improve women's empowerment and nutritional intake, though the impact on asset formation or income was not significant (Deininger and Liu 2013). SHG skills-training components have been found to facilitate empowerment through development of financial skills and access to household decision-making (Brody et al. 2015). SHGs have also been found to increase women's access to information and their participation in some agricultural decisions, and participating women are more likely to take advantage of a greater number of public entitlement schemes (Kumar et al. 2019; Raghunathan, Kannan, and Quisumbing 2019).

Conventionally, each household is assumed to possess a single utility function; this is known as the *unitary model of household behavior*. In the 1990s, however, it was recognized that household members have different preferences. This is important because intrahousehold inequality is a nontrivial factor impacting distribution of goods among household members. Several studies analyze household bargaining functions based on individual utility functions and threat points based on assets that individuals can carry away from the household, as well as the external legal and social environments that shape individuals' ability to use those assets. Interventions influencing the exit option of one household member can affect the intrahousehold distribution of goods. For instance, interventions that enhance men's legal and economic control over land resources may weaken women's bargaining power and decrease their and their children's well-being (Haddad, Hoddinott, and Alderman 1997; Quisumbing 2003). Bargaining matters for household expenditure priorities, and this continues to shape agricultural resource allocations. Ownership and control of assets by women is shown to be important for poverty reduction and has positive development outcomes at both household and individual levels (Johnson et al. 2016).

The microdevelopment literature from the early 2000s has been linked to the rise of decision science and behavioral economics. For agriculture, this has been centered around questions regarding the ability of farmers to achieve efficient or optimal outcomes in the face of problems in decision-making

(Duflo 2006). Behavioral economics has shown that individuals may be motivated by factors other than profit maximization. Research indicates that default options can play an important role in coordinating behavior where needed (Choi et al. 2003). For example, agricultural technologies like pest-resistant seeds require coordination to generate sufficient demand to improve input markets. Farmers' organizations may be influential in this regard, by providing pest-resistant seeds as a default input to their members, to help facilitate adoption and demand. Savings commitment devices have also been the subject of several recent studies that show there is an unmet demand for commitment products among the very poor. In the Philippines, the introduction of a commitment savings product increased savings significantly after one year (Ashraf, Karlan, and Yin 2006). Similarly in Kenya, when testing free delivery and price subsidies to purchase fertilizer, a savings commitment brought in new adopters instead of subsidizing those who would have adopted anyway (Duflo, Kremer, and Robinson 2009).

Other behavioral issues, like mental accounting or separate household accounts, are linked to inefficient allocations of expenditures and savings behavior if individuals associate certain funds with different expenditures. Duflo and Udry (2004) found that rainfall shocks that increase the output of crops cultivated individually by either husbands or wives are associated with strong expenditure shifts toward adult private goods like jewelry. In contrast, shocks that increase the output of crops predominantly cultivated by wives shift expenditures toward food consumption, but shocks affecting cash crops cultivated by men have no effect on the purchases of food. The literature also suggests that individual decisions made under risk and uncertainty are subject to irrational biases (Kahneman 2003). For instance, individuals tend to weigh the value of losses more than the value of gains and may give undue weight to small probabilities, which negatively affects adoption of new agricultural technologies (Liu 2013). Bryan (2010) also suggests that more ambiguity-averse households demand less insurance, and Ross, Santos, and Capon (2012) show that more ambiguity-averse individuals are more likely to adopt improved varieties. Studies have also evaluated interventions that directly incorporate behavior change communication components. For instance, Olney et al. (2015) found that integration of two programs, an agriculture homestead food production program and a nutrition and health behavior change communication program targeted to women and children, helped improve several child health outcomes, including wasting, diarrhea prevalence, hemoglobin levels, and anemia.

Three prominent economists, Michael Kremer, Abhijit Banerjee, and Esther Duflo, received the 2019 Nobel Prize in Economic Sciences for their seminal work on evaluating the impact of poverty reduction programs. They reshaped economics research by designing a new approach to alleviating global poverty. Utilizing RCTs, they sought to answer key questions on global poverty at individual or group levels through specially designed field experiments. In just the past 20 years, this new research has contributed key evidence to inform global development efforts. In 2010, Barret and Carter (2010) reflected on the power and pitfalls of RCTs and after a decade of work on the subject identified three more key issues to consider: First, ethical risks continue to exist and have not been dealt with. Second, interventions often have extremely heterogeneous impacts, which could mean that in some cases other research methods would be more suitable. Finally, nonclassical measurement errors resulting from the use of an RCT could weaken a study's statistical power (Barrett and Carter 2020). Quisumbing et al. (2020) proposed to address such concerns by identifying and assessing programmatic pathways to impact with quantitative and qualitative methods; by studying similar programs implemented by different organizations across various settings; and by working closely with implementing partners on the design, research, and dissemination processes to inform adaptation and scaling-up of programs and policies.

Field experiments or lab-like field experiments (LFEs) can be used in analyzing policy issues in agricultural and rural development to address the above concerns and complement other methods. Viceisza (2015) highlighted four main purposes of LFEs. The first is to test theories or heuristic principles; the second is to identify and estimate parameters associated with various characteristics; the third is to explore the structural nature of parameters derived from empirical methods, including other types of experiments; and the fourth is to assess methodological difficulties associated with LFEs and how these can impact parameter estimates. He also emphasized the importance of generalizability for LFEs that are intended to inform policymaking, and in the process, he proposed basic principles for conducting LFEs and suggested directions for future research.

Key Themes Addressed in This Book

In this section, we briefly sketch the purposes and key themes addressed in subsequent chapters and lay out the structure of this book.

Agricultural Growth across Time and Regions

We look to recent years and across regions to examine recent technological innovations in and outside of agriculture (such as ICTs), changing contexts for agriculture (such as rapid urbanization), and emerging challenges (including climate change) that impact agricultural productivity and efficiency.

Chapters 2 to 7 build upon the Hayami and Ruttan hypothesis, based on technological change in agriculture as an engine of agricultural growth, and compare patterns of agricultural growth across seven major regions.[1] Table 1.1 compares growth rates of land productivity (*Y*/*A*), labor productivity (*Y*/*L*), and the land-labor ratio (*A*/*L*) across major regions over time.[2] As can be seen from Figure 1.1, enormous differences exist and enormous changes in labor productivity have occurred among the seven major regions over the last several decades. Such large differences in labor productivity can be partly attributed to the difference in factor endowment represented by the land-labor ratio. The close relationship between the land-labor ratio and mechanization is clearly reported in Chapters 3 to 7. It is equally important to observe the changes in land and labor productivity over time. Table 1.1 clearly shows that in the 1960s, 1970s, and 1980s, many regions experienced a decline in the land-labor ratio; this was due to rapid population growth. Figure 1.2 shows the changing relationship between labor and land productivity, with the logarithm of labor productivity on the horizontal axis and that of land productivity on the vertical axis. To examine the importance of technological change in affecting different growth paths, Table 1.2 shows the annual growth rate of total factor productivity (TFP) from the 1970s to the early 2010s. Five regional chapters show that while some regions fit the Hayami and Ruttan hypothesis, other regions' growth paths have differed over time. It is clear that East Asia took the high-land-productivity development path, which looks like a concave production function. An interesting finding from Figure 1.1 is that not only South Asia but also Africa south of the Sahara (SSA) seem to have been following the East Asian path. Indeed, it seems that South Asia is

---

1 The five chapters in Part II deal with (1) East Asia, (2) South Asia, (3) Africa south of the Sahara, (4) Latin America and the Caribbean, and (5) Eastern Europe and Central Asia. Latin America is divided into two areas—Mexico and Central America, and South America; Eastern Europe and Central Asia are divided into two areas—the former Soviet Union (FSU) excluding the Baltics, and Eastern Europe (that is, the Baltics, Central Europe, and Balkans)—because of the large differences in factor endowments between subregions, among other things.

2 Output (*Y*) is the gross production value in constant international US$ 2004–2006, labor (*L*) is the number of economically active adults in agriculture, and land (*A*) is the sum of arable land and land for permanent crops and pasture. Data are from FAOSTAT, supplemented by data from USDA Economic Research Service on China and from Eurostat on Eastern Europe and Central Asia.

TABLE 1.1 Growth rates in land productivity, labor productivity, and land per labor, over decades, by region

Region	Growth rates	1960s	1970s	1980s	1990s	2000s	2010s
East Asia	Land productivity	0.300	0.194	0.361	0.363	0.278	0.063
	Labor productivity	0.084	0.290	0.255	0.521	0.571	0.246
	Land per labor	−0.166	0.081	−0.077	0.116	0.230	0.172
South Asia	Land productivity	0.098	0.104	0.336	0.220	0.215	0.121
	Labor productivity	0.046	0.028	0.225	0.174	0.145	0.101
	Land per labor	−0.047	−0.069	−0.083	−0.038	−0.058	−0.018
Eastern Europe	Land productivity	0.224	0.230	0.116	−0.148	0.174	0.113
	Labor productivity	0.403	0.674	0.401	−0.003	0.496	0.192
	Land per labor	0.146	0.361	0.255	0.170	0.275	0.071
Former Soviet Union	Land productivity	0.259	0.096	0.135	−0.411	0.339	0.216
	Labor productivity	0.273	0.186	0.185	−0.437	0.531	0.377
	Land per labor	0.011	0.082	0.044	−0.044	0.144	0.133
Africa south of the Sahara	Land productivity	0.091	0.154	0.056	0.164	0.124	−0.016
	Labor productivity	0.114	−0.061	0.046	0.069	0.070	0.009
	Land per labor	0.021	−0.186	−0.009	−0.082	−0.048	0.025
Mexico and Central America	Land productivity	0.460	0.298	−0.026	0.184	0.183	0.131
	Labor productivity	0.324	0.175	0.021	0.248	0.257	0.138
	Land per labor	−0.093	−0.095	0.048	0.053	0.063	0.006
South America	Land productivity	0.020	0.131	0.220	0.254	0.157	0.018
	Labor productivity	0.165	0.188	0.307	0.329	0.422	0.140
	Land per labor	0.143	0.050	0.072	0.059	0.230	0.120
Caribbean	Land productivity	−0.149	−0.076	0.069	−0.187	0.093	0.157
	Labor productivity	0.011	0.060	0.109	−0.141	0.044	0.111
	Land per labor	0.189	0.148	0.037	0.056	−0.045	−0.039
West Asia and North Africa	Land productivity	0.214	0.255	0.218	0.216	0.287	0.041
	Labor productivity	0.222	0.256	0.156	0.226	0.251	0.073
	Land per labor	0.007	0.001	−0.051	0.008	−0.028	0.030
High Income	Land productivity	0.128	0.212	0.072	0.173	0.108	0.032
	Labor productivity	0.559	0.434	0.334	0.478	0.311	0.176
	Land per labor	0.382	0.183	0.245	0.260	0.183	0.139
Other	Land productivity	0.278	0.159	−0.092	0.284	0.267	0.064
	Labor productivity	0.182	0.076	−0.001	−0.045	0.010	−0.021
	Land per labor	−0.075	−0.072	0.100	−0.256	−0.203	−0.081

Source: Calculated from USDA (2019) except for land and output, which were obtained from FAO (2019).

Note: 1960s is limited to 1961–1969; 2010s is limited to 2010–2014.

FIGURE 1.1 Changes in labor productivity by major region, in US$1,000 (2004–2006) per person

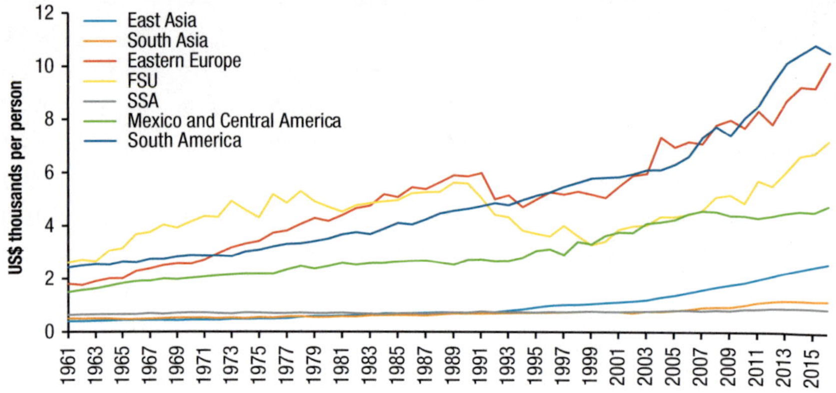

Source: Estimated from USDA (2019) and FAO (2019).
Note: FSU = former Soviet Union; SSA = Africa south of the Sahara.

FIGURE 1.2 Changes and differences in relationship between labor and land productivity by major region, double-log scale

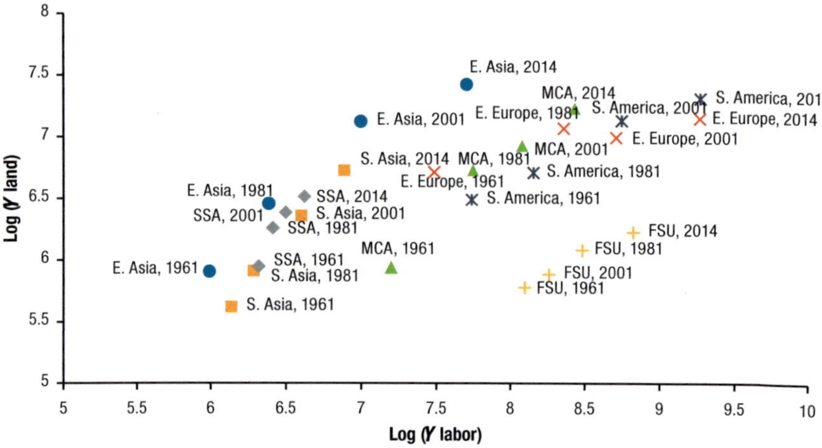

Source: Calculated from USDA (2019) except for land and output, which were obtained from FAO (2019).
Note: Y = productivity; FSU = former Soviet Union; SSA = Africa south of the Sahara; MCA = Mexico and Central America.

TABLE 1.2 Average total factor productivity growth rates, over decades, by region

Region	1970s	1980s	1990s	2000s	2010s
East Asia	−0.740	0.983	1.783	2.372	1.869
South Asia	−1.277	0.101	0.528	0.556	1.780
Eastern Europe	−0.115	0.578	1.098	1.067	1.646
Former Soviet Union	−0.263	−0.650	0.281	2.220	2.594
Africa south of the Sahara	−1.403	−0.297	1.345	1.199	0.759
Mexico and Central America	0.363	0.263	0.693	2.039	1.986
South America	−0.775	0.968	1.326	2.011	1.963
Caribbean	−0.736	0.190	−1.056	1.320	1.321
West Asia and North Africa	−0.153	0.661	1.929	1.740	1.742
High Income	0.993	1.188	1.493	1.354	1.593
Other	−2.051	0.552	1.719	−0.278	−0.006

Source: Estimated from USDA (2019) except for land and output, which were obtained from FAO (2019).
Note: 2010s is limited to 2010–2014.

attempting to catch up with East Asia (see Chapter 4) and that SSA also follows the Asian path so long as smallholder-based agriculture is maintained (Chapter 5). While the growth paths of Mexico and Central America, Eastern Europe, and South America may look similar, a closer examination reveals that Eastern Europe primarily improved its labor productivity, whereas South America and Mexico and Central America improved both labor and land productivity simultaneously.

On the whole, East Asia has performed well since the early 1980s, which is consistent with its frontier position in Figure 1.2. Growth rates of TFP in South Asia and SSA are lower than in East Asia, suggesting that their catch-up process has been taking place only slowly. However, it is remarkable that SSA achieved a TFP growth rate of more than 1 percent per year from the early 1990s to the mid-2000s, indicating that major technological changes have been taking place on this continent. As argued in Chapter 11, this may be a result of the Green Revolution in SSA. Indeed, land productivity began increasing in SSA in the 1990s (Chapter 5). A noteworthy trend is the relatively high and steady growth rate of TFP in South America, consistent with the rapid growth of both labor and land productivity as well as the region's active agricultural research activities (Chapter 6). Another conspicuous trend is the recent rise of TFP growth in the former Soviet Union (FSU), which may give rise to the potentially high production and export growth of this region (Chapter 7).

The five regional chapters examine the extent to which variations in the land-labor ratio can be explained by mechanization and variations in land productivity can be explained by fertilizer use per unit of land. They also explore new contexts and development imperatives for agriculture, including urbanization, nutrition transition, nonfarm employment, poverty and food security, and FVCs, among other things. Chapter 8 provides a summary view of regional experiences.

Nonfarm Activities and Urbanization

The nonfarm economy grows faster than agriculture, partly because of the low and decreasing income elasticity of demand for agricultural products and partly because of the ease of transfer of technologies in nonfarm sectors from developed to developing countries compared with agricultural technology, whose transfer is often constrained by climate factors. Thus, agriculture sectors tend to shrink at least relatively. The faster growth of nonfarm sectors creates an income gap between rural and urban areas. Consequently, economic transformation begins, which accompanies urbanization and rural-to-urban migration. In this process, people shift from low-income jobs in rural areas to high-income jobs in urban areas, thereby reducing the income gap and poverty.

As shown in Figure 1.3, the share of rural labor force has been declining in all seven regions under study.[3] The shares were high and declined slowly in SSA and South Asia because of the slow growth of nonfarm sectors in these regions. In contrast, labor share declined most rapidly in East Asia because of the successful development of nonfarm sectors in this region. The rural labor share in East Asia, however, was still higher than in Eastern Europe, FSU, Mexico and Central America, and South America in the 2010s.

An example of the close linkage between agriculture and urban centers is contract farming, which is designed to promote the production of new high-value products, such as fresh vegetables and fruits and livestock products, in response to increasing urban and foreign demand for such products (see Chapter 11). Obviously, farmers in developing countries must take advantage of the emergence of new production opportunities to increase their incomes and reduce poverty (World Bank 2007) (Table 1.3). Another major way to increase employment opportunities in rural areas is rural industrialization (Chapter 11). In general, rural industries are clustered because of the benefit of agglomeration and often linked with the processing of high-value products.

---

3 As is cogently pointed out in Chapter 8, the distinction between urban and rural areas is often arbitrary.

FIGURE 1.3 Declining share of rural labor force (%) by major region

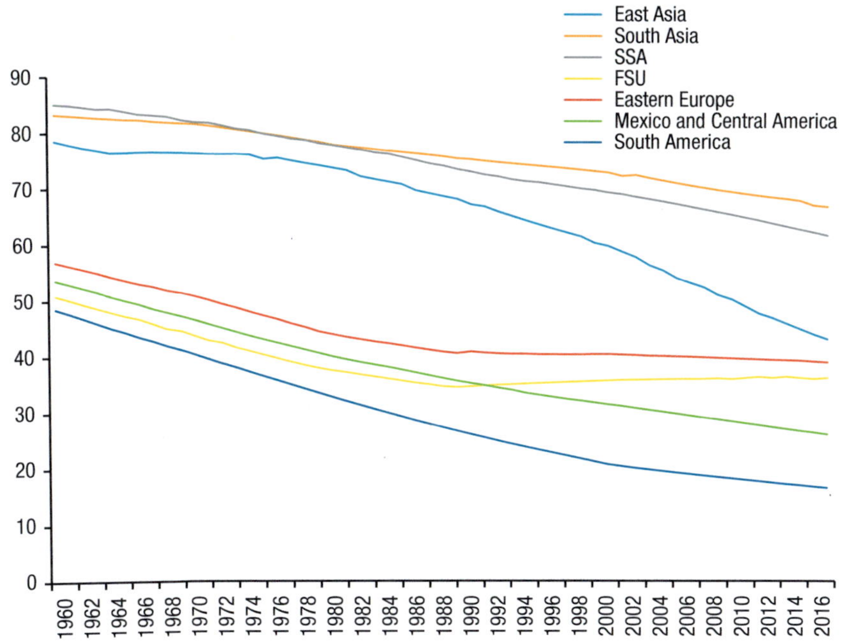

Source: Calculated from USDA (2019) and FAO (2019).

Note: FSU = former Soviet Union; SSA = Africa south of the Sahara

TABLE 1.3 Poverty headcount ratio at $1.90 a day (2011 PPP), percent of population, by region

Region	1981	1984	1987	1990	1993	1996	1999	2002	2005	2008	2011	2013	2015
South Asia	55.7	53.0	50.1	47.3	44.9	40.3	—	38.6	33.7	29.5	19.8	16.1	—
East Asia and Pacific	80.5	70.1	59.2	61.3	53.7	40.9	38.5	29.7	18.9	15.3	8.6	3.6	2.3
Europe and Central Asia	—	—	—	—	5.2	7.3	7.9	6.0	4.9	2.8	2.1	1.6	1.5
Latin America and Caribbean	13.5	16.5	13.5	14.8	14.0	13.7	13.5	11.8	9.9	6.9	5.7	4.6	3.9
Middle East and North Africa	—	8.9	8.1	6.2	7.0	6.2	3.8	3.4	3.1	2.7	2.7	2.6	4.2
Africa south of the Sahara	—	—	—	54.7	59.6	58.9	58.3	55.3	50.8	48.0	45.0	42.4	41.0

Source: Data from World Bank (2019).

Note: PPP = purchasing power parity; — = data not available.

Together with rural-to-urban migration, stimulating contract farming and developing rural industries are major means to facilitate rural transformation. As a result, poverty has declined significantly, particularly in East Asia (Table 1.3).

Urbanization and economic transformation are closely linked with diversification of diets caused by shifting consumer demand. Particularly after the classic "food problem" is solved, adequate nutrition and health become major "agricultural problems." Thus, Chapter 10 explores the linkages between agricultural development and nutrition, including increasing incidence of obesity. This is a critically important issue in developing countries because not only malnutrition but also stunting and anemia are highly prevalent among children, as can be clearly seen from Tables 1.4 and 1.5.

Changes in Markets and Food Value Chains

While Chapters 9 to 11 analyze the roles of agriculture in supply of staple foods, employment generation, poverty reduction, and nutritional transition, essentially from a sectoral point of view, Chapter 12 (Food Value Chain Transformation in Developing Regions), Chapter 13 (Agricultural Development and International Trade), and Chapter 14 (The Political Economy of Agricultural and Food Policies) deal with diverse issues of markets and international trade of agricultural products.

The continuous evolution of agricultural development can be seen through the remarkable transformation of the FVC in developing countries, particularly in the past 25 years. The share of grains and other staples in the food economy declined and was replaced by animal and horticultural products. Rapid globalization fundamentally changed FVCs in developing countries at a much faster pace than that of the transformation that occurred much earlier in developed countries. Chapter 12 describes this dramatic evolution and transformation of FVCs.

As discussed in Chapter 13, international food trade increased appreciably, driven by liberalization of trade and foreign direct investment (FDI) and increased demand for diverse processed foods and safe and high-quality fresh fruits and vegetables. On the supply side, FDI by European, US, and Japanese chains played a crucial role in the FVC transformation (Reardon et al. 2003). This rapid transformation of FVCs is reflected in increasing international trade of agricultural commodities. The amount of total agricultural trade as well as that of cereals, oilseeds, and high-value products (fruits, vegetables, meats, and dairy products) increased in recent years, as shown in Figure 1.4 (see also Chapter 13).

TABLE 1.4 Prevalence of undernourishment, by region, percent of population (3-year average)

Region	2000–2002	2002–2004	2004–2006	2006–2008	2008–2010	2010–2012	2012–2014	2014–2016	2016–2018
Caribbean	23.7	24.3	23.4	22.1	20.8	19.5	19.0	18.3	18.1
Central America	7.9	8.3	8.3	7.5	7.3	7.2	6.9	6.3	6.1
South America	11.5	9.7	7.9	6.4	5.6	5.0	4.7	5.0	5.4
Central Asia	13.8	13.6	11.0	8.8	7.8	6.7	5.8	5.5	5.6
Northeast Asia	14.5	14.3	14.0	13.2	11.9	10.6	9.4	8.5	8.4
Southeast Asia	21.5	19.9	18.4	16.7	14.3	11.7	10.5	9.8	9.4
South Asia	19.6	21.6	21.1	18.6	17.3	17.2	16.8	15.7	14.9
West Asia	10.1	9.8	9.5	9.2	8.9	9.2	9.9	11.0	12.1
Eastern Africa	38.3	36.7	34.4	33.1	32.1	30.6	30.3	30.3	30.9
Middle Africa	38.1	35.5	32.5	30.5	28.8	27.0	25.3	25.1	26.3
Northern Africa	6.5	6.2	6.1	5.6	5.2	4.9	7.6	7.0	7.1
Southern Africa	7.0	6.6	6.5	7.0	7.2	7.0	7.2	7.9	8.3
Western Africa	15.2	14.1	12.3	11.1	10.5	10.4	10.9	11.7	13.9
Oceania	5.4	5.5	5.5	5.2	5.1	5.3	5.7	5.9	6.1

Source: Data from FAO (2019).

TABLE 1.5 Prevalence of stunting in children under age 5, by region, percent of population

Region	1990	1995	2000	2005	2010	2015	2018
Caribbean	20.8	17.8	15.2	12.9	11.0	9.2	8.3
Central America	31.8	27.6	23.7	20.2	17.1	14.4	12.9
South America	19.1	16.2	13.6	11.4	9.5	7.9	7.1
Central Asia	—	34.9	28.0	22.0	17.0	12.9	10.9
Northeast Asia	35.6	26.6	19.2	13.4	9.2	6.2	4.9
South Asia	59.4	54.6	49.7	44.8	40.0	35.4	32.7
Southeast Asia	46.9	42.6	38.4	34.4	30.6	27.0	25.0
West Asia	28.7	25.8	23.1	20.6	18.3	16.2	15.1
Eastern Africa	51.9	48.8	45.8	42.8	39.8	36.9	35.2
Middle Africa	44.3	42.0	39.8	37.6	35.4	33.3	32.1
Northern Africa	28.0	25.8	23.7	21.7	19.9	18.2	17.2
Southern Africa	35.0	34.0	32.9	31.9	30.9	29.9	29.3
Western Africa	40.7	38.5	36.4	34.3	32.3	30.3	29.2
Oceania	36.0	36.4	36.8	37.2	37.6	37.9	38.2

Source: Data from UNICEF-WHO-World Bank Group (2019).
Note: — = data not available.

FIGURE 1.4 Changes in real value of net export of selected agricultural commodities by major region, billion 2011 US dollars

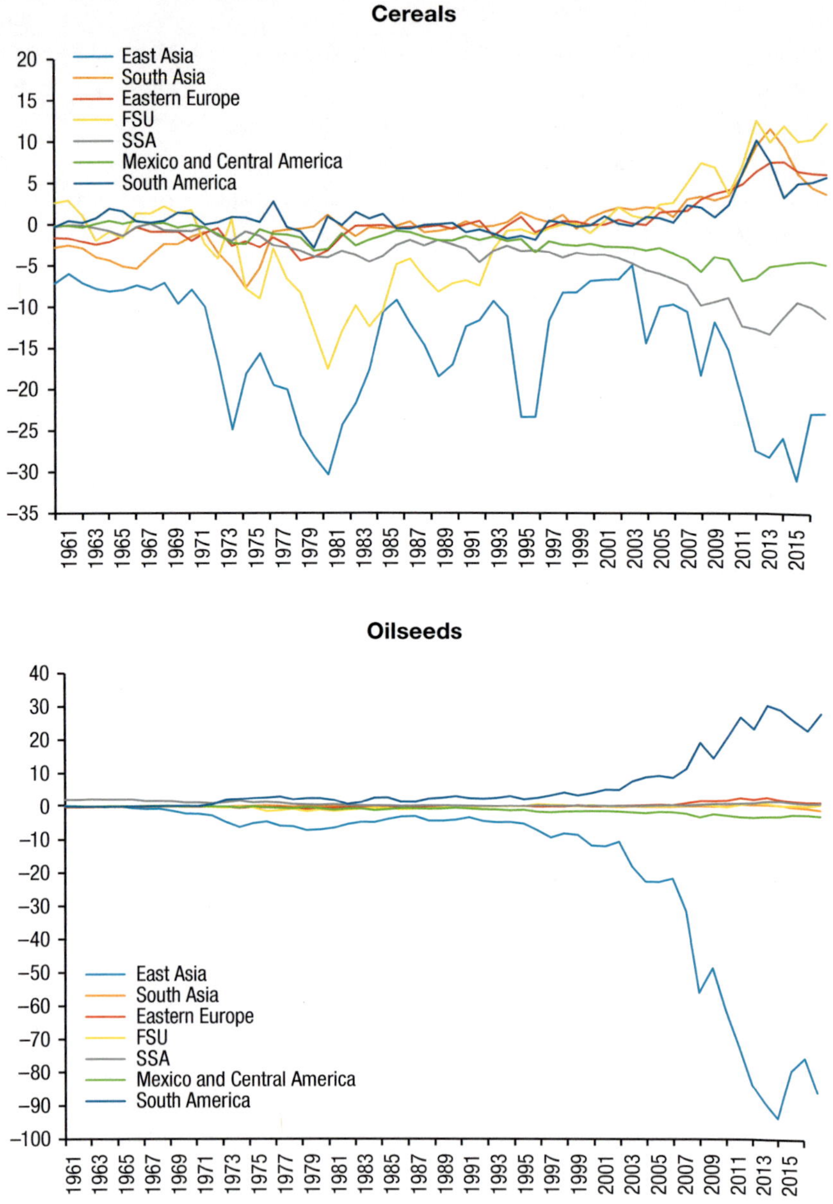

High-value products (fruits, vegetables, meats, and dairy)

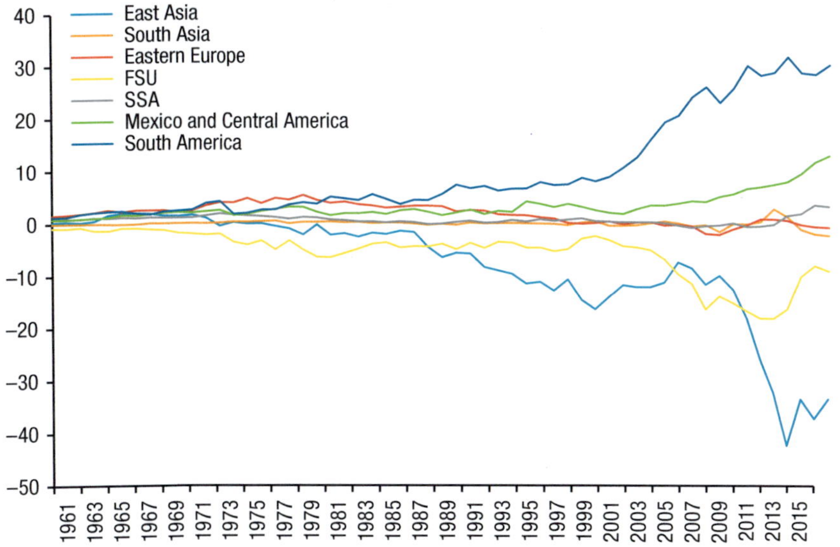

Source: Calculated from USDA (2019) and FAO (2019).
Note: FSU = former Soviet Union; SSA = Africa south of the Sahara.

Undistorted open markets maximize the net benefit that international trade can offer to boost global food security (Chapter 13). Yet agriculture in developing countries was traditionally discriminated against by export taxes and restricted import of agricultural inputs, whereas agriculture in developed countries was heavily protected by subsidies, tariffs, and restricted imports of agricultural products. In other words, farmers are taxed when they are the majority of the population but subsidized when they are the minority. This is referred to as the "development paradox." The question is why agricultural markets are so often distorted throughout much of history and across the globe. This is precisely the issue of political economy addressed in Chapter 14.

The Role of Gender, Credit, and Insurance in Household Decision-Making

Women are often handicapped in ownership and control of agricultural land and other assets in developing countries, as shown in Chapter 15. Since women tend to spend more resources on children's education and health and nutrition, their handicapped access to productive resources may lead to underinvestment in these and other areas. Understanding the roles of men and women, their decision-making powers, and the level of their cooperation within a household is essential for the analysis of household behaviors. Chapter 15 seeks deeper understanding of household behaviors in order to facilitate gender equity and investments in child schooling, nutrition, and health.

Access to institutional credit is critically important for smoothing consumption over time under risky agricultural production environments, financing purchase of modern inputs such as improved seeds and inorganic fertilizer, introducing new high-value products, and investing in land improvement to increase long-term productivity. Yet only a small fraction of farmers borrow from formal institutions. Furthermore, borrowing is used primarily for smoothing consumption, but not for raising agricultural productivity. While borrowing from microfinance institutions, which does not require collateral, is encouraged by many governments in developing countries and international organizations, these institutions are designed to provide small-scale and short-term lending primarily for nonfarm businesses. Thus, Chapter 16 seeks new effective institutions to provide credit for smallholders to accelerate agricultural development.

Also difficult is the design of efficient, effective, and widely acceptable insurance programs (Chapter 17). Traditional crop insurance programs, designed to insure farm-specific losses, are bound to fail, partly because of moral hazard (whereby an insured farmer does not exert optimum effort to reduce risk or mitigate its impact) and partly because of adverse selection (whereby only the most risk-prone farmers purchase insurance). Recently, index insurance programs have been promoted as a low-cost alternative to traditional insurance programs. In the case of index insurance products, payments are based on an independent measure, for example, rainfall, which is highly correlated with farm-level yield or revenue outcomes. Since rainfall is exogenous to the individual insurance policyholder, there is no incentive for a farmer to commit moral hazard. Despite high expectations for the wide adoption of index insurance, however, most index insurance programs have experienced low take-up rates. Thus, various solutions are sought to make insurance programs work in developing countries, as discussed in Chapter 17.

Climate Change and Sustainable Development of Agrifood Systems

Achieving sustainable agricultural development without sacrificing environmental quality is one of the most important challenges the world will need to meet, especially if we are to achieve poverty reduction, food security, and better nutrition and health. The global environment, however, has been deteriorating, and it is predicted that climate change will result in substantial reduction in agricultural production unless adequate investment in productivity-enhancing technological development is made (see Chapter 19). Efforts to expand agricultural areas in developing countries have resulted in deforestation, a major source of greenhouse gas emissions. Communal natural resources, such as forests, grazing land, and irrigation systems, also suffer from deterioration (see Chapter 18). In total, agriculture, forestry, and other land uses account for 24 percent of greenhouse gas emissions (Chapter 19). Also worrisome is increasing water scarcity due to rising demand for water in industrialized and urban areas, dietary changes toward more water-intensive foods, and frequent droughts associated with climate change (see Chapter 20).

It is important to strengthen both our ability to mitigate or reduce deterioration of global environments and the inefficient use of natural resources and our ability to adapt to a deteriorated global climate and environments. For adaptation to climate change, agricultural research plays a critical role (see Chapter 21). Adaptation can benefit from the use of biotechnologies, for example, gene editing, to develop crop varieties that are tolerant to heat, drought, salinity, and submergence and from the capacity to develop agronomic practices to use new varieties effectively. Efficient use of water is also an integral part of adaptation strategies, because the availability of irrigation water enhances resilience to frequent droughts. At present, the irrigation ratio is particularly low in low-income regions (Figure 1.5). For more effective adaptation to climate change, further irrigation investments will be required.

For sustainable resource management, it is essential to reduce the cost of measuring, monitoring, and verifying the use of natural resources or the emissions of greenhouse gases. In terms of global environmental issues, an international system is needed to determine the penalty amounts for resource use and greenhouse gas emissions and to enforce payments. Increasing the cost of natural resource use should contribute not only to the efficient allocation of such resources in the short term, but also to the development of natural-resource-saving technologies in the long term. If agriculture fails to contribute to solving climate change by not constructing the proper incentive systems in natural resource use and inducing the development of natural-resource-saving technologies, sustainable agricultural development may remain an unachievable dream and agriculture may become a culprit of global climate deterioration.

The challenge is how to mitigate climate change and preserve or improve ecosystems while delivering healthy foods to consumers. For these purposes, it will be critical to build a food system that is simultaneously nutrition- and health-driven, productive and efficient, environmentally sustainable and

FIGURE 1.5 Changes in irrigation ratio (%) by region

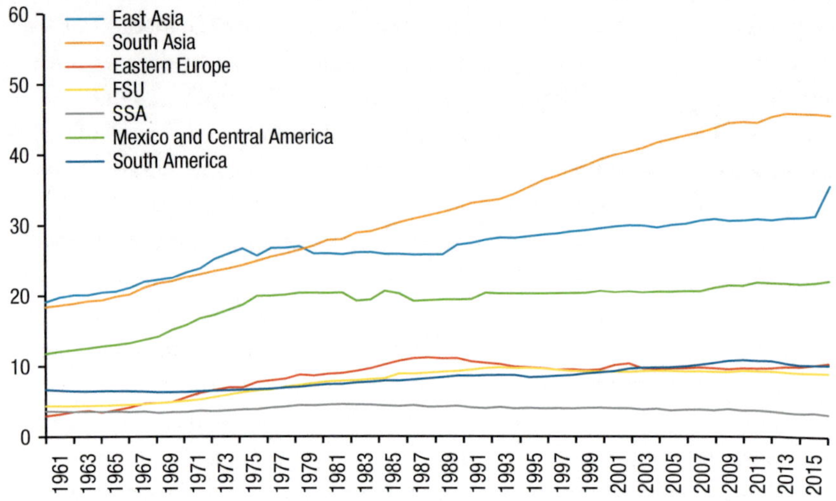

Source: Calculated from USDA (2019) and FAO (2019).
Note: FSU = former Soviet Union; SSA = Africa south of the Sahara.

climate smart, inclusive, and business friendly (Fan 2016). Chapter 22, the final chapter of the book, focuses on these critical and burning issues.

References

Ashraf, N., D. Karlan, and W. Yin. 2006. "Tying Odysseus to the Mast: Evidence from a Commitment Savings Product in the Philippines." *Quarterly Journal of Economics* 121 (2): 635–672.

Bandiera, O., and I. Rasul. 2006. "Social Networks and Technology Adoption in Northern Mozambique." *Economic Journal* 116 (514): 869–902.

Banerjee, A. 2006. "Inequality and Investment." Prepared for the World Bank's World Development Report.

Banerjee, A., E. Duflo, R. Glennerster, and C. Kinnan. 2015. "The Miracle of Microfinance? Evidence from a Randomized Evaluation." *American Economic Journal: Applied Economics* 7 (1): 22–53.

Banerjee, A., D. Mookherjee, K. Munshi, and D. Ray. 2001. "Inequality, Control Rights, and Rent Seeking: Sugar Cooperatives in Maharashtra." *Journal of Political Economy* 109 (1): 138–190.

Barret, C. B., and M. R. Carter. 2010. "The Power and Pitfalls of Experiments in Development Economics: Some Non-Random Reflections." *Applied Economic Perspectives and Policy* 32 (4), 515–548.

Barrett, C. B., and M. R. Carter. 2020. "Finding Our Balance? Revisiting the Randomization Revolution in Development Economics Ten Years Further On." *World Development* 127 (March 2020): 104789.

Barrett, C. B., M. R. Carter, and C. P. Timmer. 2010. "A Century-Long Perspective on Agricultural Development." *American Journal of Agricultural Economics* 92 (2): 447–468.

Bellemare, M. F., and C. B. Barrett. 2006. "An Ordered Tobit Model of Market Participation: Evidence from Kenya and Ethiopia." *American Journal of Agricultural Economics* 88: 324–337.

Bernard, T., A. de Janvry, and E. Sadoulet. 2010. "When Does Community Conservatism Constrain Village Organizations?" *Economic Development and Cultural Change* 58 (4): 609–641.

Besley, T. 1995. "Property Rights and Investment Incentives." *Journal of Political Economy* 103 (5): 903–937.

Binswanger, H. P. 1974. "The Measurement of Technical Change Biases with Many Factors of Production." *American Economic Review* 64 (6): 964–976.

Binswanger, H. P., and M. Rosenzweig. 1986. "Behavioural and Material Determinants of Production Relations in Agriculture." *Journal of Development Studies* 22: 503–539.

Binswanger, H. P., and D. Sillers. 1983. "Risk Aversion and Credit Constraints in Farmers' Decision-Making: A Reinterpretation." *Journal of Development Studies* 20 (1): 5–19.

Boucher, S. R., M. R. Carter, and C. Guirkinger. 2008. "Risk Rationing and Wealth Effects in Credit Markets: Theory and Implications for Agricultural Development." *American Journal of Agricultural Economics* 90 (2): 409–423.

Brody, C., T. de Hoop, M. Vojtkova, R. Warnock, M. Dunbar, P. Murthy, and S. L. Dworkin. 2015. "Economic Self-Help Group Programs for Improving Women's Empowerment: A Systematic Review." *Campbell Systematic Reviews* 11 (1): 1–182.

Bryan, G. 2010. *Ambiguity and Insurance*. Working Paper.

Cai, J., A. de Janvry, and E. Sadoulet. 2015. "Social Networks and the Decision to Insure." *American Economic Journal: Applied Economics* 7 (2): 81–108.

Carter, M. R., and C. B. Barrett. 2006. "The Economics of Poverty Traps and Persistent Poverty: An Asset Based Approach." *Journal of Development Studies* 42: 178–199.

Carter, M. R., and P. Olinto. 2003. "Getting Institutions Right for Whom? Credit Constraints and the Impact of Property Rights on the Quantity and Composition of Investment." *American Journal of Agricultural Economics* 85: 173–186.

Carter, M. R., and Y. Yao. 2002. "Local versus Global Separability in Agricultural Household Models: The Factor Price Equalization Effect of Land Transfer Rights in China." *American Journal of Agricultural Economics* 84: 702–715.

Choi, J. J., D. Laibson, B. Madrian, and A. Metrick. 2003. "Optimal Defaults." *American Economic Review* 93 (2): 180–185.

Christiaensen, L., L. Demery, and J. Kuhl. 2011. "The (Evolving) Role of Agriculture in Poverty Reduction: An Empirical Perspective." *Journal of Development Economics* 96 (2): 239–254.

Cole, S. A., and A. N. Fernando. 2012. *The Value of Advice: Evidence from Mobile Phone-Based Agricultural Extension*. Harvard Business School Working Paper 13–047.

de Janvry, A., V. Dequiedt, and E. Sadoulet. 2014. "The Demand for Insurance against Common Shocks." *Journal of Development Economics* 106 (January): 227–238.

de Janvry, A., M. Fafchamps, and E. Sadoulet. 1991. "Peasant Household Behavior with Missing Markets: Some Paradoxes Explained." *Economic Journal* 101: 1400–1417.

Deaton, A. 1991. "Savings and Liquidity Constraints." *Econometrica* 59: 1221–1248.

Deininger, K., and Y. Liu. 2013. "Economic and Social Impacts of an Innovative Self-Help Group Model in India." *World Development* 43: 149–163.

Dercon, S., and Gollin, D. 2014. "Agriculture in African Development: Theories and Strategies." *Annual Review of Resource Economics* 6 (1): 471–492.

Diao, X., and P. Dorosh. 2007. "Demand Constraints on Agricultural Growth in East and Southern Africa: A General Equilibrium Analysis." *Development Policy Review* 25 (3): 275–292.

Díaz-Bonilla, E. 2015. *Macroeconomics, Agriculture, and Food Security: A Guide to Policy Making in Developing Countries*. Washington, DC: International Food Policy Research Institute.

Dorosh, P., and S. Haggblade. 2003. "Growth Linkages, Price Effects and Income Distribution in Sub-Saharan Africa." *Journal of African Economies* 12 (2): 207–235.

Dorosh, P., and J. Thurlow. 2012. "Agglomeration, Growth and Regional Equity: An Analysis of Agriculture- versus Urban-Led Development in Uganda." *Journal of African Economies* 21 (1): 94–123.

Duflo, E. 2006. "Field Experiments in Development Economics." In *Advances in Economics and Econometrics: Theory and Applications, Ninth World Congress: Volume 2,* edited by Richard Blundell, Whitney Newey, and Torsten Persson. Econometric Society Monographs. Cambridge University Press.

Duflo, E., M. Kremer, and J. Robinson. 2009. *Nudging Farmers to Use Fertilizer: Theory and Experimental Evidence from Kenya*. NBER Working Paper No. 15131. Stanford, CA: National Bureau of Economic Research.

Duflo, E., and C. Udry. 2004. *Intrahousehold Resource Allocation in Cote d'Ivoire: Social Norms, Separate Accounts and Consumption Choices*. NBER Working Paper No. 10498. Stanford, CA: National Bureau of Economic Research.

Eswaran, M., and A. Kotwal. 1984. "Access to Capital and Agrarian Production Organization." *Economic Journal* 96: 482–498.

Fafchamps, M., and B. Minten. 2012. "Impact of SMS-Based Agricultural Information on Indian Farmers." *World Bank Economic Review* 26 (3): 383–414.

Fan, S. 2016. "Food Policy in 2015–2016: Reshaping the Global Food System for Sustainable Development." In *2016 Global Food Policy Report*, 1–11. Washington, DC: International Food Policy Research Institute.

Fan, S., S. Yosef., and R. Pandya-Lorch, eds. 2019. *Agriculture for Improved Nutrition: Seizing the Momentum*. Wallingford, UK: International Food Policy Research Institute and Centre for Agriculture and Bioscience International.

FAO (Food and Agriculture Organization of the United Nations). 2019. FAOSTAT database. http://faostat.fao.org.

Feder, G. 1985. "The Relation between Farm Size and Farm Productivity: The Role of Family Labor, Supervision and Credit Constraints." *Journal of Development Economics* 18: 85–101.

Feder, G., T. Oncham, Y. Chalamwond, and C. Hongladaron. 1988. *Land Policies and Farm Productivity in Thailand*. Baltimore: Johns Hopkins University Press.

Foster, A., and M. Rosenzweig. 1995. "Learning by Doing and from Others: Human Capital and Technological Change in Agriculture." *Journal of Political Economy* 103: 1176–1209.

———. 2017. "Are There Too Many Farms in the World? Labor-Market Transaction Costs, Machine Capacities and Optimal Farm Size." NBER Working Paper w23909. Stanford, CA: National Bureau of Economic Research.

Gemmell, N., T. Lloyd, and M. Mathew. 2000. "Agricultural Growth and Intersectoral Linkages in a Developing Economy." *Journal of Agricultural Economics* 51 (3): 353–370.

Geng, X., W. Janssens, B. Kramer, and M. van der List. 2018. "Health Insurance, a Friend in Need? Impacts of Formal Insurance and Crowding Out of Informal Insurance." *World Development* 111 (November): 196–210.

Godquin, M., and A. Quisumbing. 2008. "Separate but Equal? The Gendered Nature of Social Capital in Rural Philippine Communities." *Journal of International Development* 20 (1): 13–33.

Haddad, L., J. Hoddinott, and H. Alderman. 1997. *Intrahousehold Resource Allocation in Developing Countries: Methods, Models and Policies*. Baltimore: Johns Hopkins University Press.

Hayami, Y., and V. W. Ruttan. 1971. *Agricultural Development: An International Perspective.* Baltimore: Johns Hopkins University Press.

Hayami, Y., and V. W. Ruttan. 1985. *Agricultural Development: An International Perspective.* Rev. ed. Baltimore: Johns Hopkins University Press.

Johnson, N. L., C. Kovarik, R. Meinzen-Dick, J. Njuki, and A. Quisumbing. 2016. "Gender, Assets, and Agricultural Development: Lessons from Eight Projects." *World Development* 83 (July): 295–311.

Johnston, B., and J. Mellor. 1961. "The Role of Agriculture in Economic Development." *American Economic Review* 51 (4): 566–593.

Kahneman, D. 2003. "Maps of Bounded Rationality: Psychology for Behavioral Economics." *American Economic Review* 93 (5): 1449–1475.

Key, N., E. Sadoulet, and A. de Janvry. 2000. "Transactions Costs and Agricultural Household Supply Response." *American Journal of Agricultural Economics* 82: 245–259.

Kim, S. S., P. H. Nguyen, L. M. Tran, Y. Abebe, Y. Asrat, M. Tharaney, and P. Menon. 2019. "Maternal Behavioral Determinants and Livestock Ownership Are Associated with Animal Source Food Consumption among Young Children during Fasting in Rural Ethiopia." *Maternal and Child Nutrition* 15 (2): e12695.

Kremer, M., and E. Miguel. 2007. "The Illusion of Sustainability." *Quarterly Journal of Economics* 122 (3): 1007–1065.

Kumar, N., K. Raghunathan, A. Jilani, A. H. Arrieta, S. Chakrabarti, P. Menon, and A. R. Quisumbing. 2019. "Social Networks, Mobility, and Political Participation: The Potential for Women's Self-Help Groups to Improve Access and Use of Public Entitlement Schemes in India." *World Development* 114: 28–41.

Lawry, S., C. Samii, R. Hall, A. Leopold, D. Hornby, and F. Mtero. 2014. "The Impact of Land Property Rights Interventions on Investment and Agricultural Productivity in Developing Countries: A Systematic Review." *Campbell Systematic Reviews* 10: 1–104.

Lewis, W. A. 1954. "Economic Development with Unlimited Supplies of Labour." *Manchester School* 22 (2): 139–191.

Liu, E. 2013. "Time to Change What to Sow: Risk Preferences and Technology Adoption Decisions of Cotton Farmers in China." *Review of Economics and Statistics* 95 (4): 1386–1403.

Maertens, A. 2017. "Who Cares What Others Think (or Do)? Social Learning and Social Pressures in Cotton Farming in India." *American Journal of Agricultural Economics* 99 (4): 988–1007.

Mitra, S., D. Mookherjee, M. Torero, and S. Visaria. 2013. *Asymmetric Information and Middleman Margins: An Experiment with West Bengal Potato Farmers.* BREAD Working Paper No. 401 (October). Bureau for Research and Economic Analysis of Development.

Morduch, J. 1995. "Income Smoothing and Consumption Smoothing." *Journal of Economic Perspectives* 9 (3): 103–114.

Nakasone, E., M. Torero, and B. Minten. 2014. "The Power of Information: The ICT Revolution in Agricultural Development." *Annual Review of Resource Economics* 6 (1): 533–550.

Olney, D. K., A. Pedehombga, M. T. Ruel, and A. Dillon. 2015. "A 2-Year Integrated Agriculture and Nutrition and Health Behavior Change Communication Program Targeted to Women in Burkina Faso Reduces Anemia, Wasting, and Diarrhea in Children 3–12.9 Months of Age at Baseline: A Cluster-Randomized Controlled Trial." *Journal of Nutrition* 145 (6): 1317–1324.

Ostrom, E. 1990. *Governing the Commons: The Evolution of Institutions for Collective Action.* Cambridge: Cambridge University Press.

Otsuka, K., H. Chuma, and Y. Hayami. 1992. "Land and Labor Contracts in Agrarian Economies: Theories and Facts." *Journal of Economic Literature* 30 (4): 1965–2018.

Pender, J., and S. Scherr. 2002. "Organizational Development and Natural Resource Management: Evidence from Central Honduras." In *Innovation in Natural Resource Management: The Role of Property Rights and Collective Action in Developing Countries*, edited by R. Meinzen-Dick, A. Knox, B. Swallow, and F. Place. Baltimore: Johns Hopkins University Press.

Pica-Ciamarra, U., L. Tasciotti, J. Otte, and A. Zezza. 2015. "Livestock in the Household Economy: Cross-Country Evidence from Microeconomic Data." *Development Policy Review* 33: 61–81.

Quisumbing, A. 2003. *Household Decisions, Gender and Development: A Synthesis of Recent Research*. Baltimore: Johns Hopkins University Press.

Quisumbing, A., A. Ahmed, D. Gilligan et al. 2020. "Randomized Controlled Trials of Multi-Sectoral Programs: Lessons from Development Research." *World Development* 127 (March): 104822.

Raghunathan, K., S. Kannan, and A. Quisumbing. 2019. "Can Women's Self-Help Groups Improve Access to Information, Decision-Making, and Agricultural Practices? The Indian Case." *Agricultural Economics* 50 (5): 567–580.

Ranis, G., and J. C. H. Fei. 1961. "A Theory of Economic Development." *American Economic Review* 51 (4): 533–565.

Ravallion, M., and S. Chen. 2007. "China's (Uneven) Progress against Poverty." *Journal of Development Economics* 82 (1): 1–42.

Reardon, T., C. P. Timmer, C. B. Barrett, and J. Berdegué. 2003. "The Rise of Supermarkets in Africa, Asia, and Latin America." *American Journal of Agricultural Economics* 85 (5): 1140–1146.

Robalino, J., and A. Pfaff. 2012. "Contagious Development: Neighbor Interactions in Deforestation." *Journal of Development Economics* 97 (2): 427–436.

Rosenzweig, M., and H. P. Binswanger. 1993. "Wealth, Weather Risk and the Composition and Profitability of Agricultural Investments." *Economic Journal* 103 (416): 56–78.

Ross, N., P. Santos, and T. Capon. 2012. "Risk, Ambiguity and the Adoption of New Technologies: Experimental Evidence from a Developing Economy." 2012 Conference of the International Association of Agricultural Economists.

Schultz, T. W. 1964. *Transforming Traditional Agriculture*. New Haven, CT: Yale University Press.

Singh, I., L. Squire, and J. Strauss. 1986. *Agricultural Household Models*. Baltimore: Johns Hopkins University Press.

Takahashi, K., R. Muraoka, and K. Otsuka. 2020. "Technology Adoption, Impact, and Extension in Developing Country Agriculture: A Review of the Recent Literature." *Agricultural Economics*, 51 (1): 31–45.

Tiffin, R., and X. Irz. 2006. "Is Agriculture the Engine of Growth?" *Agricultural Economics* 35 (1): 79–89.

Timmer, C. P. 1988. "The Agricultural Transformation." In *Handbook of Development Economics*. Vol. 1, edited by H. Chenery and T. N. Srinivasan, 275–331. Amsterdam: Elsevier.

———. 2017. "Food Security, Structural Transformation, Markets and Government Policy." *Asia and the Pacific Policy Studies* 4: 4–19.

UNICEF-WHO-World Bank Group. 2019. Joint Child Malnutrition Estimates: Levels and Trends, 2019 edition. Accessed 2019. www.who.int/nutgrowthdb/estimates2018/en/.

USDA (US Department of Agriculture). 2019. Economic Research Service. www.ers.usda.gov.

Viceisza, A. 2015. "Creating a Lab in the Field: Economics Experiments for Policymaking." *Journal of Economic Surveys* 30 (5): 835–854.

World Bank. 2007. *World Development Report 2008: Agriculture for Development*. Washington, DC: World Bank.

———. 2019. World Development Indicators database. https://datacatalog.worldbank.org/dataset/world-development-indicators.

Yegbemey, R. N., J. A. Yabi, S. D. Tovignan, G. Gantoli, and S. E. H. Kokoye. 2013. "Farmers' Decisions to Adapt to Climate Change under Various Property Rights: A Case Study of Maize Farming in Northern Benin (West Africa)." *Land Use Policy* 34 (September): 168–175.

Chapter 2

GLOBAL ISSUES IN AGRICULTURAL DEVELOPMENT

Mark W. Rosegrant, Shenggen Fan, and Keijiro Otsuka

Chapter 1 reviewed dynamically changing global trends in agricultural development and identified emerging and diverse issues associated with the process of global agricultural development. This chapter reviews the literature on these topics in order to clarify the context under which regional issues (Part II), newly arising critical issues (Part III), and future challenges and opportunities (Part IV) are discussed. More specifically, this chapter addresses the major debates around agricultural development, policies, and technological changes to reduce or eliminate food insecurity and attain sustainable development.[1]

The first section addresses the potential for agricultural growth and structural transformation, including transformation of the rural areas and the dietary transition. The second section discusses markets, food value chains, and international trade, assessing the rapidly changing dynamics of delivering food from the farm to the table. Key topics include the impact of the rise of supermarkets in the developing world, contract farming, and food safety and food quality. The role of international trade and food price stabilization policy is also reviewed. The third section provides an overview of agricultural support policies, including agricultural subsidies, agricultural credit, and crop insurance. The fourth section assesses the potential for sustainable agricultural growth as a foundation for agricultural development by examining the role of technology, including biotechnology, sustainable water and land use, and property rights and climate change. The debate on future sources of sustainable growth is also reviewed.

---

1 Some sections of this chapter draw upon Rosegrant, M. W. 2014. *Food Security*. Thousand Oaks, CA, US: SAGE. http://ifpri.worldcat.org/title/food-security/oclc/871340599. Used with permission.

Agricultural Development and Transformation

Agricultural Growth and Poverty Reduction

The evolution of agricultural development thinking described in Chapter 1 provides an overview of structural transformation with agriculture-led economic growth, building on the Hayami and Ruttan (1971, 1985) model of structural transformation driven by induced technical change. Scientific progress continues, and the changing nature of technology may drive a new agricultural transformation. In addition to technological change in agriculture, rapid technological change outside of agriculture is creating new potential for technological change in agriculture and for new employment opportunities in rural towns and small cities. New science-based technologies are likely to be strongly laborsaving and can give a big technological advantage to agriculture in developed and middle-income countries. These new technologies are also likely to create economies of scale in agriculture that could create pressures to consolidate land ownership or operational size of farms, particularly in high-performing countries in East Asia. Food value chains will also be fundamentally influenced by these advanced technologies, with the potential for significant reduction in postharvest losses, and fundamental changes in contract farming. How these technologies play out will also influence employment, rural–urban migration, and the future of the structural transformation, as well as poverty reduction. These issues are addressed further in Chapters 3, 9, and 11.

Historically, the major sources of growth in agriculture have been increased input use and productivity growth. In addition to land and labor, increased use and productivity of inputs—including modern inputs such as fertilizer, improved seeds, and water—play a key role for agricultural productivity (Nin Pratt 2015). Other factors have also contributed to agricultural production growth. Craig, Pardey, and Roseboom (1997) demonstrate that improvements in input quality, such as better-educated labor and improved tractors, explain a significant part of the differences in agricultural productivity across countries. They also show that public goods, represented by agricultural research and road density, as well as life expectancy, are important in explaining between-country differences in agricultural productivity.

Productivity growth from technological innovations and institutional changes also contributes to agricultural growth. Key factors affecting agricultural productivity growth include climatic, environmental, technological, institutional, economic, and policy conditions. Measures that increase crop yields include institutional reforms, such as the household responsibility system in China and Viet Nam (Lin 1991; Rozelle and Swinnen 2004; Jin, Huang, Rozelle 2010). In addition, the development of new varieties, technological diffusion, land improvements, adoption of conservation tillage techniques, denser and earlier planting, irrigation, pest control, and weed control also contribute to enhanced yield. The factors that affect crop yields negatively include land degradation, adverse climate conditions, limited resource conditions, poor farm management, and disincentives like those in the former Soviet Union.

While major drivers of productivity growth vary by country and region, there is an extensive literature focused on understanding the determinants of the gains in productivity in agriculture that shows that public and private research and development investments have played a crucial role in realizing productivity growth in the agricultural sector (Hayami and Ruttan 1985; Alston et al. 2000, 2010; Evenson 2001; Fuglie and Schimmelpfennig 2010; Colman 2013). In developing countries, agricultural research and development has been mainly undertaken by public institutions such as government laboratories and universities, as well as nonprofit institutions such as the CGIAR centers, as is discussed by Evenson and Gollin (2003a, 2003b). However, private sector research has been increasing rapidly, particularly in relatively fast-growing countries such as Brazil and India.

Gollin, Hansen, and Wingender (2018) evaluate the impact of adopting high-yielding crop varieties (HYVs) in developing countries, focusing on the Green Revolution period from 1960 to 2000. They estimate that a 10 percentage point increase in HYV adoption increases gross domestic product (GDP) per capita by about 15 percent, by the direct effect on crop yields, factor adjustment, and structural transformation. The study also finds that HYV adoption reduced both fertility and mortality. Hazell (2009) further shows that Green Revolution investments achieved a high rate of economic return. But policies need to address concerns about the sustainability of intensive cultivation, including the possibility of soil degradation, chemical pollution, aquifer depletion, and soil salinity. Pingali and Rosegrant (2000) show that inappropriate management of modern inputs was the primary cause of these Green Revolution problems, driven mainly by inadequate extension and training, ineffective regulation of water quality, and input pricing and subsidy policies that made modern inputs too cheap and encouraged excessive use.

However, the Green Revolution has had limited impacts in Africa. Reasons for this include the lack of a dominant farming system (such as rice and wheat); predominance of rainfed rather than irrigated agriculture; lack of an effective extension system; poor soil fertility; underinvestment in agricultural research and development and rural infrastructure; lack of competitive markets and conducive enabling environments; the negative impact of poor human health on agriculture; minimal mechanization; and the predominance of customary land tenure (InterAcademy Council 2004). Recent literature focuses on the role of farmer-to-farmer extension, which is complementary to conventional public-sector extension, and discusses how to select and incentivize the farmer-trainers (see Takahashi, Muraoka, and Otsuka 2020 for a review). Prior to the implementation of structural adjustment programs, macro, trade, and taxation policies heavily penalized the agricultural sector, resulting in high costs and limited access to inputs and low output prices for farmers. Implementation of structural adjustment programs and agricultural market liberalization programs in the 1990s started reducing these penalties (Hazell 2009).

Agricultural growth can be a key driver of poverty reduction, more so than nonagricultural growth in some instances. Since poverty is largely a rural phenomenon and since many of the poor depend, directly or indirectly, on the farm sector for their incomes, growth that raises agricultural productivity and the incomes of small-scale farmers and landless laborers is particularly important in reducing poverty and improving food security (Rosegrant and Hazell 2000; Diao et al. 2014; Christiaensen and Martin 2018). In addition to driving agricultural development and transformation, productivity growth has far-reaching impacts beyond agricultural growth. Recent studies confirm the significant impact of agricultural growth on poverty reduction. Growth in agriculture is shown for many countries to be generally more effective at reducing poverty, in comparison to nonagricultural growth, by two or three times, and this advantage is more defined for the poorest populations (Thirtle, Lin, Piesse 2003; Christianesen and Martin 2018). Yet, Dercon and Gollin (2014) note that heterogeneity in the agricultural sector at all levels—in countries, subnational regions, villages/communities, individuals—affects potential pathways for agricultural development and impact on poverty reduction. Heterogeneity can be found in many factors, including climate and soils, policies, institutions, access to markets, and more, which is discussed further in Giller et al. (2011). Further, the authors indicate there is room for improvement in the empirical studies on agricultural growth considering agriculture's impact on poverty reduction. For example, more rigorous assessment of the costs and the opportunity costs of boosting growth in agriculture is needed. Studies should account for the relative cost-effectiveness of obtaining growth in agriculture compared with obtaining growth in other sectors and avoid assuming a straightforward path from investment to agricultural growth to poverty reduction. Overall balance is another important factor noted by the authors, as agricultural growth will not come in isolation but rather from its interaction with the rest of the economy.

Research-led technological change in agriculture in Africa and Asia has substantially reduced poverty, whereas productivity growth in industry and services has had a relatively modest impact except in East Asia (that is, Northeast and Southeast Asia) (Thirtle, Lin, and Piesse 2003). Fan et al. (2000) investigate the causes of the decline in rural poverty in India and the role that government investments have played. They show that government expenditures have both direct and indirect effects on poverty reduction. Such expenditures directly generate income through employment generation. Much larger indirect benefits are due to investments generating long-term agricultural and nonagricultural growth. Investments in roads and agricultural research and development have the largest impacts on poverty reduction and growth in agricultural productivity, followed by government spending on education, which may generate benefits over extended periods.

Nevertheless, nonagricultural growth also plays an important role in poverty reduction, including through the reallocation of labor from agriculture to nonagricultural sectors, which are often more skills intensive and labor-productivity enhancing. Nonagricultural growth can also be poverty reducing by contributing to agriculture. While the degrees of impact depend on the national subsectors involved, growth in some sectors, such as transport services and manufacturing (especially agro-processing) in Africa, can be as effective as growth in agriculture (Dorosh and Thurlow 2018). Increased efficiency of trade and transport also saw reduced transaction costs for all marketed products, particularly those with high cost margins, such as agriculture and food (Christiaensen and Martin 2018).

Masters (2005) assesses trends in poverty and undernutrition in Africa; the availability of food, rural population, and demography; farm output and input use; and determinants of input use and agricultural research effort. He points out that African farmers have adopted profitable new varieties when they have been available but that research and seed delivery systems have often not effectively developed suitable varieties. A Green Revolution in Africa will require the introduction of agronomic and other innovations for many crops and many environments, supported by investments to develop and spread new seed varieties. More appropriate and better-targeted research and development and effective extension systems for agriculture are essential for the economic future of the African continent (Otsuka and Larson 2013, 2016). Sources of agricultural growth will be further analyzed in five regional chapters in Part II (Chapters 3 to 7).

Meanwhile, the potential impact of the decreasing share of the agricultural sector on poverty reduction and other development outcomes should be taken into account when considering future investment priorities. Human capital investment that drives long-term development is also fundamentally influenced by agricultural growth and access to food. Alderman, Hoddinott, and Kinsey (2006) show that even short-term shocks such as drought and war early in life can have long-term negative effects on individuals' nutritional status. When such events generate reductions in consumption because households are unable to protect against the shock, they lead to long-term reductions in height and schooling achievement. Improvements in preschool nutrition and health—and prevention of losses in preschool nutritional and health status due to shocks—not only have immediate benefits, but also have long-term benefits in improved height and educational attainment.

The Dietary Transition

The structural transformation literature emphasizes the importance of agricultural growth as a means of reducing poverty and increasing food security through greater availability of food and lowering food prices. In his classic writing in the early 1980s, Amartya Sen introduced the *entitlement* approach, which contrasts with the more usual food availability approach, concentrating instead on the ability of the people to afford food through use of production possibilities, income, trade opportunities, and entitlements with respect to the state (Sen 1981). This shift in focus from food supply to individual access to food highlighted the important role of demand failure due to unemployment, adverse movement in the terms of trade, production failure, termination of transfers, or other forms of entitlement failure (Barrett, Carter, Timmer 2010). The past paradigm of agricultural development focused on growth, productivity, and poverty reduction. However, the current global context warrants a broader perspective toward agriculture, namely its contribution to improved nutrition, health, and human well-being (Fan and Pandya-Lorch 2012). In this regard, diets are key in our approach to modern agrifood systems.

This is ever more critical, as populations in those countries undergoing rapid economic growth and transformation are experiencing the most rapid nutritional transition. Beginning in Asia, but increasingly for the rest of the developing world, diets are shifting away from staples and toward livestock and dairy products, vegetables and fruits, and fats and oils. As income in both rural and urban areas increases, demand for more diversified foods is increasing and agricultural production in turn is driven by the demand to further diversify its products, which contribute to more-balanced diets (Fan, Yosef, and Pandya-Lorch 2019a).

At the same time, current food consumption patterns in much of the developing world are showing signs of convergence toward a Western diet. The diet transition is characterized by increased consumption of wheat, temperate fruits and vegetables, and high-protein and energy-dense food. Globalization and the consequent global interconnectedness of the urban middle class is the driving force behind the convergence of diets. The rapid spread of global supermarket chains and fast-food restaurants is reinforcing these trends (Hawkes, Harris, and Gillespie 2017). Growth in incomes in developing countries is driving particularly strong growth in per capita and total meat consumption, which in turn induces strong growth in feed consumption of cereals, particularly maize (Pingali 2006; Rosegrant, Leach, and Gerpacio 1999). The change in consumption trends has been slower in Africa and in some developing countries, especially in Africa south of the Sahara (Kearney 2010). The rapid urbanization of developing countries is another important factor, due to urban sedentary lifestyles and the rapid spread of food outlets for fast and energy-dense foods, as well as the increasing dependency on prepared food, usually consumed away from home (Caballero 2007; Ruel, Garrett, and Yosef 2017).

Globally, these changes in dietary patterns are particularly important in terms of the rise in overnutrition (that is, diet-related chronic disease) in many developing countries alongside persistent undernutrition (Kearney 2010; Hawkes, Harris, and Gillespie 2017). The dietary transition has in fact thrown a light on the new types of food insecurity that ought to influence the desirable path of agricultural development. In addition to rising overweight and obesity as described, the problem of undernourishment caused by insufficient intake of dietary energy and protein, as well as micronutrient deficiencies, all coexist. These three dimensions represent the "triple burden" of malnutrition (Gómez et al. 2013).

Overweight and obesity are the result of excessive dietary energy intake and are often associated with increased risk of diabetes and cardiovascular disease. These conditions are increasingly common in developing countries, with obesity reaching a global prevalence of 12 percent in 2008 (Gómez et al. 2013). According to the 2016 Global Burden of Disease (GBD) study, 72 percent of all cause-specific deaths worldwide were from noncommunicable diseases (NCDs), and most of the global burden of deaths and disabilities from these NCDs occur in low- and middle-income countries (Ecker 2019). Caballero

(2005) notes the increasing frequency in developing countries of underweight in children and overweight in adults, often within the same family. This phenomenon in developing countries occurs due to the changes in diet, food availability, and lifestyle in countries experiencing a socioeconomic and demographic transition. Moreover, evidence suggests that fetal undernutrition increases the risk of obesity and associated NCDs later in life (Ecker 2019).

Undernourishment contributes to negative health outcomes such as the prevalence of underweight (low weight for age), wasting (low weight for height), or stunting (low height for age), especially among children. Stunting measures the long-term negative effects of food intake deficiency and disease, and it afflicted 165 million children globally in 2011, for a prevalence of 26 percent. Stunting is caused primarily by maternal undernutrition and by poor nutrition and repeated infections in the first two years of a child's life. Wasting captures shorter-term, acute episodes of malnutrition. Underweight is a product of the other two indicators (Gómez et al. 2013).

Micronutrient malnutrition includes iodine, iron, and vitamin A deficiencies and can coexist with excessive consumption of macronutrients (that is, overweight and obesity). Vitamin A deficiency is the leading cause of blindness in children and harms proper growth and reproduction, leaving the body more vulnerable to infections (Gómez et al. 2013). Inadequate dietary intake and infectious disease are the most direct causes of undernutrition. Inadequate dietary intake weakens the immune system and increases susceptibility to disease. Infectious disease increases nutrient requirements and weakens the immune system by impairing absorption of essential nutrients and increasing nutritional requirements and decreasing purchasing power because of sickness, absenteeism, the inability to do work, and unemployment, as well as the increased time and money spent on treatment and care. Broader factors also influence nutrition, including economic, political, and social factors (UN Standing Committee on Nutrition 2010; Gómez et al. 2013). Particularly noteworthy is the empowerment of women within a household, as women are more conscious about nutrition and health than men.

Increasing productivity growth to expand food availability remains a high priority in many developing countries where undernourishment is still the dominant nutritional problem. Increases in food production are also essential to generate higher incomes in developing countries. But the broadening of malnutrition problems, together with the continued transformation of food systems in developing countries, calls for wider-ranging approaches and interventions to improving nutritional outcomes than have been used historically. Reducing the impact of these factors would require major changes in urban planning, transportation, public safety, and food production and marketing (Caballero 2007; Ruel, Garrett, and Yosef 2017). Interventions and policies should take account of the need for more sustainable diets that would include a sufficient supply of micronutrient-rich foods without excessive consumption of energy-dense, nutrient-poor foods (Kearney 2010). In promoting nutrition- and health-driven policies, targeting those most in need, particularly children under age two and marginalized populations underserved by essential health services, will be important. Furthermore, filling the knowledge gaps through research, scaling innovation solutions, and fueling partnerships between health, nutrition, and agriculture will be crucial (Fan, Yosef, and Pandya-Lorch 2019b). These issues are discussed further in Chapter 10 as well as Chapter 9.

Markets, Food Value Chains, and International Trade

The Dynamics of Markets and Food Value Chains

Effective agricultural markets benefit both producers and consumers and offer greater income opportunities by integrating small-scale producers into upgraded value chains, spurring agricultural development. Establishing competitive and efficient markets is an important challenge in achieving agricultural growth. Among the most important constraints are those related to costs in the value chain, including transport, communication, and transaction costs. The high costs to farmers and other actors due to poor infrastructure, lack of information, insufficient credit, and policy distortions reduce the efficiency of value chains and impede producers' ability to connect to market systems. An agricultural value chain is the series of activities by agents and markets from farm to final consumption that build product value. Millions of low-income people participate in agricultural value chains as consumers, producers, small-scale traders, processors, and retailers. Improvement in the performance of value chains would benefit large numbers of people (Beamon 1999; Lohman, Fortuin, and Wouters 2004; IFPRI 2011).

Revolutionary changes in processing, wholesale, and retail segments of the value chain are taking place. This process has been moving quickly in developing countries during the past three decades, beginning in Latin America, then in Asia, and later and more slowly in Africa. The transformation includes consolidation of value-chain operations, rapid institutional and organizational change, and modernization of the procurement system. Among the changes are the rapid rise of supermarkets, large and modern food processors, and logistics and wholesale firms. Reardon and Timmer (2012) show that "these changes have mainly been driven by the private sector, through both domestic and foreign direct investment, with public sector investment in infrastructure providing the underlying conditions."

Information and communications technologies (ICTs) have substantial potential to improve value-chain performance. Sensors linked to digital information systems can improve links between farmers and processors; reduce losses (with consequent reduction in water used in producing final food products, because less water-using primary food production is required to generate the final product) with digitally enabled harvest loans and digital warehouse receipts; monitor storage conditions and track provenance to allow grading and inform consumers; reduce costs of transport; increase choice of different types of transport for farmers; and increase access to timely information so that farmers know if and when transport is arriving (USAID 2017). The use of ICT is also expanding on the farm input side of the value chain. Private- and government-run management information systems provide knowledge related to farm inputs and farm management practices such as planting methods, seeds to use, sowing time and application of fertilizers, and current market prices. ICT-enabled applications allow agricultural input supply companies to improve operations and build capacity to expand outreach and meet farmer needs and provide real-time weather information and forecasts and drought early warning systems to farmers. Farmers can buy inputs (seed, fertilizer, etc.) using cell phones, reducing transaction costs.

Progress is being made, therefore, to turn the potential of digital information systems into reality for agricultural value chains. Yet major efforts are still required to achieve large-scale adoption in developing countries. The prices of many ICTs remain high for most developing country applications. Both the market and donors have a role to play in speeding up cost reductions for sensors and related technologies and in supporting local development partners in testing and refining technologies for context-specific applications. Innovation is especially important to integrate sensor technology and data applications into locally appropriate products and services that address problems affecting smallholders. Governments, entrepreneurs, and private-sector actors all have roles to play in promoting adoption by creating and disseminating actionable information to smallholders and others along the agricultural value chains (Feed the Future 2016).

The food-value-chain transformation appears to be improving food security for cities, by reducing marketing margins, offering lower consumer prices, and increasing the quality and diversity of food (Reardon et al. 2014). But evidence is mixed regarding the impact on farmers. Available evidence indicates that larger farmers with higher assets have higher participation rates in the transformation, even though smallholders increasingly participate in the production of high-value products (Otsuka, Nakano, and Takahashi 2016). Farmers who participate in transformed systems experience a net income gain and risk reduction, relative to those in only traditional markets. Gains can come both from rewards for quality differentiation and from a premium controlling for quality (Reardon and Timmer 2012).

Barrett et al. (2012) note that historically market sales of food have been heavily concentrated in the hands of a small number of large producers, even in regions and countries with broad-based market participation. The relatively high upfront investment required to participate in modern markets and the presence of fixed costs in making contracts are challenges to participation of smallholders, meaning major benefits of food-value-chain transformation may be concentrated in a relatively small share of farmers (Barrett et al. 2012). More recently, Ton et al. (2018) showed that smallholders can benefit from contract farming, yet the poorest farmers are rarely included, and contract farmers had significantly larger landholdings or more assets than the average farmers in the region. The authors also note that any broad assessment of the effect of contract farming on income is upwardly biased due to mechanically excluding contract farming arrangements that failed. Also, better specification of the services provided by the firm to the farmer is recommended to increase the relevance of the research for development practitioners and governments. Arouna et al. (2019) further provide insights from Benin that find that contracts that only included an agreement on price had nearly as large an impact as did contracts with additional attributes, suggesting that once price uncertainty is resolved, farmers are able to address other constraints on their own. More generally, Bellemare and Bloem (2018) review recent literature on contract farming, finding considerable heterogeneity across contexts that limit external validity, as well as issues relating to the credible causal identification of treatment effects that limit internal validity. Further, many findings almost always apply narrowly to a handful of crops, to a restricted geographical area, or to a single year, and the authors concluded that it is difficult to draw any broad policy-relevant conclusions from the literature on contract farming.

Minten et al. (2009) note that some literature points to modern retailing companies becoming increasingly dominant in international and local markets in fruits and vegetables, with mostly negative implications for small farmers in Africa, under the right circumstances. But they also show that, "given the right incentives and contracting systems, small farmers in developing countries—and in Africa in particular—can participate successfully in these emerging value chains." Benefits are generated from improved access to inputs, credit, extension services, and technology adoption and from productivity spillover effects on other crops and enhanced income stability. A review of the literature by Otsuka, Nakano, and Takahashi (2016) also suggests that producer cooperatives spontaneously formed by smallholders attempt to reduce transaction costs between the buyers and a large number of smallholders. Earlier, Michelson (2013) found that farmers who supply supermarkets experience an increase in their stock of household assets of about 16 percent on average in Nicaragua. However, the location of supermarket procurement centers and the rate of household participation are strongly impacted by access to roads, markets, and sufficient water for crop production.

An important innovation linking small farmers to markets is the rapid spread of mobile phones and other information technology. As Aker and Mbiti (2010) note, access to mobile phones in Africa south of the Sahara has increased dramatically; 60 percent of the population had mobile phone coverage by 2010. Mobile phones can improve access to and use of information, thereby reducing marketing costs, improving coordination, and increasing market efficiency while also facilitating communication in response to shocks to reduce exposure to risk.

For example, Suri and Jack (2016) find that the Kenyan mobile money system, M-Pesa, has increased per capita consumption levels and lifted households out of poverty. The impacts, which are more pronounced for female-headed households, appear to be driven by changes in financial behavior and labor market outcomes, such as occupational choice, especially for women who moved out of farming and into business. Mobile money has therefore increased the efficiency of the allocation of consumption over time while allowing a more efficient allocation of labor, resulting in a meaningful reduction of poverty in Kenya (Suri and Jack 2016). Although M-Pesa has been very successful and mobile money is expanding several other economies, its use remains mostly limited to person-to-person transactions that take place over long distances and those that are in places where holding cash is risky. Outside these applications, there has been less success, and financial innovation around mobile money is still in its early stages. Expansion in the scope and impact of mobile money will require innovation in financial systems and regulations and payment modalities and growing knowledge and acceptance from consumers (Suri 2017).

The issue of dynamic changes in food value chains will be discussed in Chapter 12 as well as in Chapter 11.

International Trade and Price Stabilization

Trade policies are important determinants of agricultural development. Broad economic growth, based on stable macroeconomic environments, can help create trade and domestic markets for agricultural commodities and the generation of capital for investment in agriculture. In addition to setting the economic environment for overall economic growth, these policies can also have a profound impact on the performance of individual sectors of the economy such as agriculture. Trade policies can affect agriculture by either taxing (dis-protecting) or subsidizing (protecting) the sector. Trade restrictions have a direct effect on the domestic prices of tradable (often agricultural) goods and have an impact on the real exchange rate, which in turn affects the domestic prices of tradable goods in relation to locally produced goods. For example, import duties and quotas raise the domestic price of import-competing products in relation to exportables (including many agricultural commodities) and therefore encourage a shift away from export production. The same policy instruments reduce the demand for imports, which lowers the price of foreign exchange so that the domestic prices of tradable goods fall in relation to non-tradable goods and hence indirectly bias production incentives against both import-competing and export goods such as agricultural products.

Alternatively, trade policies can protect agriculture or specific commodities within the agricultural sector. Agricultural protection, by raising domestic food and agricultural prices above world prices, penalizes consumers and introduces inefficiency by attracting excess resources to production of the protected commodity or sector and by rendering unprotected sectors less competitive. Protectionist policies require large government fiscal outlays to farmers to pay price supports and subsidies and may also encourage excessive use of agricultural chemicals, thus damaging the environment. Despite these negative impacts of interventions, the agricultural and food sector has been subjected to some of the heaviest governmental interventions over most of the world and much of history, accounting for an estimated 70 percent of the global welfare cost of all merchandise trade distortions in 2004, even though the agricultural sector contributes only 6 percent of global trade and 3 percent of global gross domestic product (Anderson, Rausser, and Swinnen 2013).

For developed countries, the main reason for restricting food trade has been to protect domestic producers from import competition, but these measures harm not only domestic consumers and exporters of other products but also foreign producers and traders of food products. Agricultural protection and subsidies in developed countries have depressed international prices of farm products, reducing agricultural income in developing countries.

"Protectionist policies of developed countries have thus been partly responsible for international income inequality and poverty in developing countries," say Anderson et al. (2013). During the past decade, economic growth, changes in political economy induced by such growth, and changes in governance and media structures have driven some reduction of antiagricultural policies in developing countries. Reductions in the protection of agriculture have also occurred in some developed countries, especially in the European Union (Anderson et al. 2013).

Because of these negative effects of taxation and protection through biased trade policy, removal of these biased policies can generate large economic and food security benefits. Winters, McCulloch, and McKay (2004) provide a review of the evidence on trade liberalization and poverty, vulnerability, productivity, and governance revenue. Their review of the empirical evidence shows that trade liberalization will alleviate poverty in the long run and lends no support to the position that trade liberalization generally has an adverse impact on poverty. Anderson, Cockburn, and Martin (2011) provide more support in a recent review of global trade liberalization, with all but 2 of the 32 cases analyzed indicating that overall global trade reform decreases poverty.

Winters et al. (2004) also note that trade liberalization necessarily implies changes in the distribution of income; it may well reduce the well-being of some people in the short term, and some of these people may be poor. Supporting policies will sometimes be needed to reduce the adverse impacts of trade liberalization, including the impact on poverty. Poorer households are less able than richer ones to protect themselves against negative effects or to take advantage of the opportunities created by policy reform, increasing the desirability of supporting policies such as social protection. Winters et al. (2004) conclude that, "although trade liberalization may not be the most powerful or direct mechanism for addressing poverty in a country, it is one of the easiest to change. Thus, trade reform may be one of the most cost-effective anti-poverty policies available to governments."

Food price risks and instability have contributed to the reluctance of countries to liberalize food markets because of fears about the impacts on food price instability, or out of the belief that food prices have become more unstable in countries that have already liberalized food markets. Byerlee, Jayne, and Myers (2006) explain that policy that targets price stability, especially using public grain reserves, has often had poor results, with consumers facing greater instability in food prices and availability when strategic reserves are used. Policies should instead focus on long-run investments in sustainable market development and productivity growth and the use of market-based instruments and targeted safety nets to manage the risks of adverse food market outcomes. "Such an approach can provide short-term relief from market shocks for the poor, while maintaining efficient long-run responses to market signals, thereby remaining compatible with longer-run market development," say Byerlee et al. (2006).

Wright (2012) further addresses the role of grain reserves and related policies in managing grain market volatility. Sharply rising food prices in 2007 and 2008 and again in 2010 and 2011 were caused by several factors, including rapid growth in demand for biofuels; rising energy, fertilizer, and fuel prices; bad weather; and commodity speculation. Also in play were long-run trends in supply-and-demand fundamentals, including low investment in agricultural research and slow crop yield growth (Rosegrant, Tokgoz, and Bhandary 2013). Further exacerbating the increases in rice prices in 2008 was the activation of export controls by a few exporters, including India (November 2007), Viet Nam and Egypt (January 2008), China (January 2008), and Cambodia (March 2008), limiting the availability of rice in the world market and increasing the level and volatility of prices (Headey, Malaiyandi, and Fan 2009). Key wheat suppliers also imposed export bans or taxes (Wright 2012).

Despite the contributions of trade restrictions to increased levels and volatility in prices, the 2007–2011 food price increases led to increased interest in many countries in implementing price stabilization policies. As Wright (2012) notes, "many different policy interventions have been used in attempts to reduce grain price volatility or support price levels. These include controls or sanctions on private 'hoarding' or 'speculation,' buffer stocks, buffer funds, strategic reserves, use of options and futures, marketing boards, and price floors, all of which obviously affect storage incentives. Other measures that can also affect storage are trade barriers, export taxes, interest rate policies, and production controls." In the past, a favored policy has been market stabilization using a price band bounded by the floor and ceiling prices to reduce the spikes and troughs typical of commodity prices. However, price-band schemes are unsustainable and expensive and can be extremely destabilizing when they fail (Wright 2012).

The demonstrated failure of many grain exporters to maintain uninterrupted market access to their supplies highlights the desirability of commitment-reinforcing mechanisms for international grain market participants. Among the proposals are an internationally coordinated global food reserve and a global virtual grain reserve. But Wright (2012) explains that such reserve schemes "are highly unlikely to have superior forecasting ability compared to the market as a whole, so will lose money on average, and eventually exhaust its budget, as did similar schemes in the past." The critical role of undistorted international trade in improving poverty and boosting agricultural growth will be discussed in Chapter 13, followed by the analysis of political economy factors, which lead to the distortion of agricultural markets, in Chapter 14.

Agricultural Support Policies

Policies related to agricultural development, including public investment in agricultural research and development, rural infrastructure, and irrigation, as well as trade and price-stabilization policies were discussed above. This section focuses in more detail on several key policies, including input subsidy, agricultural credit, and crop insurance.

Input Use and Input Subsidies

Government subsidies are often justified as providing public goods or counteracting the impact of market failures. But instead, governments often intervene to provide large subsidies to private goods (such as fertilizer and credit), displacing the supply of public goods (research, roads, and education). Together with reducing the supply of infrastructure and other public goods, public expenditures on subsidies often result in underinvestment in research and development and in inadequate sanitary and environmental protection. López and Galinato (2007) show that "reducing the share of subsidies to private goods in the government's budget therefore has a large and significant positive impact on rural per capita income, reduces certain undesirable environmental effects associated with output growth, and contributes to poverty reduction."

Fertilizer subsidies have been among the most common and most expensive subsidies in agriculture, intended to increase fertilizer use, crop production, and income. Ricker-Gilbert, Jayne, and Shively (2013) review the evidence on fertilizer subsidies in Africa, noting that "policy recommendations about returns to alternative investments must compare the benefits and costs of subsidizing fertilizer with other alternative public investments and policies capable of promoting smallholder food security and poverty reduction." They find some evidence of positive increases in maize production from fertilizer subsidies in Malawi, but these are small relative to the size and scope of Malawi's subsidy program. Moreover, the impacts of greater maize production must be weighed against the unintended effects of the subsidies, including the displacement of other crops by maize, crowding out of the private seed and fertilizer sectors, and the impact of reduced crop diversification on agricultural sustainability, smallholder vulnerability, and nutritional balance (Ricker-Gilbert, Jayne, and Shively 2013).

Given the negative effects of subsidies, are there appropriate uses for them? Small-scale "smart" fertilizer subsidies to farmers may be cost-effective in stimulating farmers to adopt and utilize fertilizer appropriately together with new production technology. Temporary subsidies during the early stage of fertilizer adoption may be effective in overcoming the fixed costs related to adoption of new technology and in inducing farmer experimentation and learning during periods of rapidly changing technological potential. Such temporary subsidies should be phased out as adoption and appropriate use of fertilizer become widespread. But the phaseout of subsidies becomes difficult once they are in place and develop political support.

Duflo, Kremer, and Robinson (2011) suggest an innovative variation on a smart fertilizer subsidy, based on some farmers being biased toward the present and tending to procrastinate on decisions, postponing fertilizer purchases until later periods. Under these conditions many farmers in Western Kenya fail to make apparently profitable fertilizer investments, but they do buy fertilizer in response to small, time-limited discounts on the cost of buying fertilizer just after harvest. Results suggest that this policy can yield higher welfare than either no subsidies or heavy subsidies. Small, time-limited reductions in the cost of purchasing fertilizer at the time of harvest induce substantial increases in fertilizer use, comparable to those induced by much larger price reductions later in the season (Duflo, Kremer, and Robinson 2011).

This approach may provide a design for effective small-scale "smart subsidies," but smart subsidies have often grown into costly large-scale input subsidy programs. Policymakers must consider whether input subsidy programs can be sustainable given their high direct and opportunity costs (Ricker-Gilbert, Jayne, and Shively 2013). A more appropriate role for the government in enhancing fertilizer use would be to provide public goods, including extension services to advise farmers on the appropriate quantity, quality, and timing of fertilizer applications. Several studies have concluded that commercial farmers should pay for extension advice, while the government should provide extension services to small producers free of charge (Rivera 2001). Extension policies and strategies need to define an effective division of labor between public extension and farmer-to-farmer extension (Takahashi, Muraoka, and Otsuka 2020) and identify overall objectives for public sector involvement in extension (Rivera 2001). Another challenge to privatizing extension services is the lack of private providers, especially in remote areas (Jagger and Pender

2003). But in some countries that have privatized provision of advisory services, many service providers have emerged. For example, in Mali and Uganda, there are many nongovernmental organizations (NGOs), private companies, and semiautonomous bodies engaged in delivering extension advice to farmers (Qamar 2002). Improved coordination of multiple extension service providers is also required. Moreover, governments should upgrade infrastructure affecting supply and distribution, such as roads and ports. Instead of relying on subsidies, improved fertilizer policy should be accompanied by integrated soil fertility management based on both organic and inorganic fertilizer application and intercropping of staple crops with beans for sustainable productivity growth (Otsuka and Muraoka 2017).

Mechanization is another private good that often has received public policy support. Binswanger (1986) shows that "mechanization is most profitable and contributes most to growth where land is abundant, where labor is scarce relative to land, and where labor is moving rapidly off the land." Mechanization facilitates the trend toward larger farms, and large farms adopt new forms of machinery considerably faster than small farms. However, when rental or contract operations are developed, mechanization can also spread to smaller farms (Binswanger 1986). Evidence shows that, in recent years, contracting of machinery has become more widespread in Asia, as is consolidation of farm management into larger units while ownership of individual smallholdings is often maintained (see, for example, Yang et al. 2013).

While mechanization rarely leads directly to increase in yields, it can stimulate growth when it makes a new method of crop production profitable, as in the case of irrigated farming using pumps (Binswanger 1986). Pump irrigation has been a major driver of agricultural growth in Asia, including at least four very different situations: (1) the use of tube wells to tap deep aquifers for irrigation, as are commonly found in the semiarid regions of Punjab in India, and in the North Central Plain of China, and to supply water to urban areas; (2) the use of low lift pumps to tap shallow alluvial aquifers that are usually replenished every year; (3) pumping of water (either for irrigation or for drainage) in major river deltas such as the Chao Phraya and the Mekong; and (4) other situations where pumps are used to recycle water from rivers and drains (Barker and Rosegrant 2007).

Government interventions to push mechanization have had little success and have been very costly. Pakistan subsidized big tractors and prohibited imports of all but a few brands, making it almost impossible for innovating firms to import smaller foreign designs for local adaptation. Thailand's contrasting laissez-faire policy "resulted in the development of indigenous power tillers and small tractors, a wider choice of machinery, and few adverse social consequences" (Binswanger 1986). Only when private tube well imports and markets were deregulated did the small-scale tube well revolution take off in Bangladesh (Barker and Rosegrant 2007). Similarly, state-led mechanization pushes in Africa often failed due to lack of understanding of the nature of demand for mechanization technologies among farmers and insufficient knowledge of private-sector functions (Diao, Silver, and Takeshima 2016). Until recently in Africa south of the Sahara, power tillers and small tractors were seldom used. However there has been increasing demand for mechanization there with increasing rural wages and urbanization that has increased demand for labor-intensive crops. The rising demand for mechanization and the increased number of medium-scale farmers owning tractors in some countries have created a private tractor hiring market in some African countries (Diao et al. 2016). Governments can support research and development to manufacture smaller machinery suitable to local conditions and demonstrate its advantages with training programs and reduce tariffs for imports of machinery and spare parts but should not repeat the history of failures of government-run or government-subsidized promotion of mechanization. Rather than subsidies, Merrey and Lefore (2018), based on a review of recent pilot studies, find that the most promising models to support small farmer adoption of irrigation are "partnerships among microfinance and other institutions (e.g., manufacturers, retailers, agricultural advisors); 'pay-as-you-go' or rent-to-own models that spread out payments and enable farmers to begin benefiting immediately, while minimizing risk because the pump is the collateral; and contractor or utility models where entrepreneurs offer irrigation services rather than selling equipment."

Remote sensing and precision agriculture coupled with ICT have potential to boost agricultural productivity and income through improved input markets and input-use efficiency. Precision agriculture has mainly been for large-scale farming, and these and other advanced technologies coming on stream could be characterized by economies of scale or barriers to entry based on expertise and financial start-up costs, which can make advanced and larger farmers more efficient than small farmers. Thus, it is essential to develop precision technologies suited to small farmers, as well as enabling conditions such as credit and contracting services for precision equipment. Promising applications are seeking to bridge this gap for small farmers. Hello Tractor is an example of a custom tractor company that seeks to fill the demand for tractor hire services mentioned above. It is a business platform of entrepreneurs operating in Africa and Central America, in which small farmers use their mobile phones to contract with Hello Tractor for tractors to plow and harvest their fields, track when they will arrive, and make mobile money payments. Hello Tractor uses smart tractors linked to the cloud with a GPS antenna and international SIM card for remote monitoring (Davies and Garrett 2018).

Ekekwe (2017), summarized in Rosegrant (2019), identifies some promising remote-sensing applications for small farmers. Aerial images from satellites or drones, weather forecasts, and soil sensors are making it possible to manage crop growth in real time. Zenvus, a Nigerian precision-farming start-up, measures and analyzes soil data like temperature, nutrients, and vegetative health to help farmers apply the right fertilizer and optimally irrigate their farms. The system can improve farm productivity and reduce input waste by using analytics to facilitate data-driven farming practices for small-scale farmers. UjuziKilimo, a Kenyan start-up, seeks to use big data and analytic capabilities to transform farmers into a knowledge-based community with the goal of improving productivity by adjusting irrigation. SunCulture, which sells drip irrigation kits that use solar energy to pump water from any source, can make irrigation affordable.

One example of precision agriculture for irrigation in Africa is FruitLook, used by farmers in the Western Cape in South Africa, which is a state-of-the-art information technology that helps deciduous fruit and grape farmers become water efficient and climate smart. A second example is the Chameleon and Wetting Front Detector sensors that have enabled small-scale farmers in Mozambique, Tanzania, and Zimbabwe to cut irrigation frequency and improve productivity (Ncube et al. 2018).

While these new ICT-enabled technologies show promise, they would still need to be upscaled to achieve broader impact. Aker et al. (2016) state that, although there is substantial potential for ICT-based services to address farmers' and traders' information and credit market constraints, economic research suggests that the impacts of such services on agricultural adoption, behavior, and welfare are mixed. To improve their effectiveness, information services need to be high quality and preferably from a trusted source. Services are best "delivered via platforms that build upon local ICT access and usage, paying particular attention to the gender digital divide" (Aker et al. 2016). IFPRI, CCAP, and Asian Development Bank (2019) show that government policies can facilitate the effective application of ICT, including e-commerce, in rural China: investing in storage and transportation for the development of e-commerce in agriculture; investing in farmers' capacity building through practical training; providing financial and credit supports for farmers, particularly to small farmers; facilitating farmers working together in e-commerce; and improving market regulations and providing a favorable market environment for the development of agricultural e-commerce.

Agricultural Credit and Crop Insurance

Simultaneous reform to liberalize and integrate domestic financial capital markets would reduce the costs of increased price variability through risk pooling on an economywide basis. Financial integration for risk spreading and asset accumulation is critical at the rural household level as well. To exploit the income-enhancing potential of the commercialization of agriculture, financial markets must accommodate the increased ability of households to save and build productive asset bases and improve human resources. The process of commercialization itself can provide the critical market size required for efficient, unsubsidized rural banking with low overhead costs. Effective rural financial institutions can in turn assist in the spread of the benefits of commercialization more widely across the community and region.

Farmers' financial needs are becoming more complex and diverse and include access to deposit and savings accounts and possibly also investment loans for nonfarm business activity as well as for agriculture. There is also increasing demand for financial services by many small-scale and part-time nonfarm businesses, especially in the service sector, as witnessed by the recent explosion in microfinance in rural areas. More flexible and customer-oriented financial services are required to meet these needs. In the face of these growing needs, government intervention has often had negative impacts on rural financial markets, limiting their ability to serve not just the rural nonfarm economy but even farmers themselves. Government interventions have included lending requirements imposed on banks, refinance schemes, loans at preferential interest rates, credit guarantees, and lending by government-operated development finance institutions. But these interventions have in most cases had limited impact on the adoption of new technology or on agricultural production, while seriously impairing the banks, cooperatives, and specialized agricultural development banks that have tried to implement them. Moreover, government interventions in rural financial markets have often failed to provide savings and other financial services demanded by farmers. The general failure of directed and subsidized credit calls for a new approach that limits the role of financial markets to financial intermediation rather than serving as a tool to stimulate production, compensating for distortions in other markets, and alleviating poverty. The appropriate roles for government are to help the poor while recognizing high fixed costs of lending to the poor and to create an environment in which misallocation of financial resources by financial institutions, arising not only from moral hazard and adverse selection but also from other failures, can be minimized. Among other things, this means macroeconomic stability, reasonably low levels of inflation, procedures to enforce contracts, the protection of property rights, and a regulatory and supervisory system that can ensure prudent financial operations (Rosegrant and Hazell 2000).

More recently, careful development of products, policies, institutions, and supportive infrastructure has led to greater success than the old paradigm of subsidized credit to farmers. Meyer, Zander, and Fritsche (2011) note that analysis of the past use of credit subsidies has led to guidelines for "smart" or "market-friendly" subsidies. These guidelines include the following: invest in the creation of public goods that benefit the entire financial sector; subsidize institutions rather than borrowers to reduce distortions, particularly where there is natural spillover to nonsubsidized institutions; but require quantitative performance measures to maintain incentives for high performance matching contributions from grant recipients to demonstrate commitment.

Nevertheless, for poor households, private financial markets may not work. Santos and Barrett (2011) examine social networks and loans to pastoralists in Ethiopia and find that the persistently poor are often excluded from the social networks needed to obtain loans. This social exclusion leaves the poor vulnerable to shocks and largely without credit networks to fall back on in times of need. In this context, worries about the crowding-out effect of public interventions targeted to the poorest appear misplaced, because there is a high probability of poorer members being left uninsured. Empirical results suggest that, up to some wealth level, public transfers may even increase private transfers (Santos and Barrett 2011).

Microfinance programs to assist very small farms and part-time nonfarm activity (especially among women and the poor) have expanded significantly in recent decades and are serving a segment of the rural nonfarm economy that has not been widely reached before. Yet, the ultra poor, who cannot afford to repay the loans, are usually denied access to microfinance. Even if this is the case, microfinance programs may well lead to favorable poverty-reduction effects for the direct beneficiaries. However, whether they lead to growth in total nonfarm income and employment has been a difficult question to answer because microfinance clients are self-selected, which can invalidate comparisons over time between clients and nonclients. Thus, Banerjee et al. (2013) and Banerjee, Karlan, and Zinman (2015) include rigorous impact evaluation case studies of six microcredit programs in selected countries. These studies find little evidence of transformative effects, lacking clear evidence of reductions in poverty, substantial improvements in living standards, or improvements in social indicators. However, there is stronger evidence that businesses expand, though to a limited extent, with some increases in profits. Authors also find some evidence of effects on occupational choice, business scale, consumption choice, female decision power, and improved risk management. At the same time, studies find little evidence of harmful effects, even with individual lending (in Bosnia and Mongolia) and even at a high real interest rate (in Mexico). Microfinance is discussed further in Chapter 16.

Smith and Glauber (2012) show that the range and scope of crop insurance products have rapidly expanded over the past 50 years, particularly in developed countries. This expansion has been driven by a wide range of government support, including subsidized premiums and delivery and public provision of reinsurance services. Government support made multiple-peril or all-risk insurance widely available for crops and attractive to farmers as an income-support mechanism. However, willingness-to-pay and demand studies show that few farmers are willing to pay the full commercial cost of crop insurance, most likely because they have many other, cheaper ways of managing risk. Some developing countries, such as India and Mexico, have established large-scale, highly subsidized insurance programs like those in developed countries. But the heavy subsidies paid by agricultural insurance programs for multiple-peril insurance in developed countries are usually not fiscally sustainable for governments in developing countries with limited financial resources.

The alternative insurance instrument most often proposed is index-based insurance. Miranda and Farrin (2012) provide a succinct description of index insurance: "Unlike conventional insurance, index insurance indemnifies the insured based on the observed value of a specified 'index.' Ideally, an index is a random variable that is objectively observable, reliably measurable, and highly correlated with the losses of the insured, and additionally cannot be influenced by the actions of the insured. The most widely used in index insurance contract designs is rainfall." Because index insurance relies on publicly available information, is standardized and transparent, and cannot be manipulated by the insured, it is less costly to administer than general insurance. However, index insurance suffers from "basis risk," which is the difference between the losses actually incurred and the losses insured. Since the indemnity provided by index insurance is based on an index rather than on verifiable losses, the insured can suffer a significant loss without an insurance contract payout. Others may receive payouts without suffering losses, at a financial loss to the program. Because of basis risk and the high cost of reinsurance, the results of many index insurance pilot programs have been disappointing. Significant uptake of index insurance has usually occurred only if subsidies are provided or if it is bundled with other benefits, such as low-interest loans (Miranda and Farrin 2012).

Many index insurance pilot projects are assessing different ways to overcome the problem of basis risk. Dercon et al. (2014) examine the potential of offering weather insurance contracts to groups. Groups could improve understanding of the product, could be better placed to enforce insurance contracts, and could be a means to manage basis risk, if basis risk is not perfectly correlated among its members. The empirical results of Dercon et al. (2014) show that demand for insurance was increased when groups were provided training that encouraged sharing of insurance within groups, and one mechanism for this higher demand may come from the ability of groups to mitigate some of the basis risk inherent in these products. Other approaches being tested to alleviate basis risk include gap insurance, which insures against basis risk. Recent research shows significant potential for reducing the cost of weather-based index insurance through picture-based insurance using smartphones (Ceballos, Kramer, and Robles 2019). Widespread adoption of index insurance requires a solution to basis risk (see Chapter 17 for further details). Long-term public goods investments, such as weather-reporting stations and basic data collection and analysis, can help to create the conditions and infrastructure for robust insurance markets. Complementary investments are also needed in basic methods of mitigating risk through low-cost irrigation, drought-resistant seed varieties, improved sanitation, and preventive healthcare (Meyer, Zander, and Fritsche 2011).

Natural Resource Management and Sustainable Productivity Growth

Land, Water, and Climate Change

LAND

Globally, the largest area of managed land is used for production of crops and livestock. Agricultural land covers 38 percent of the world's land surface. The largest amount of this area, nearly 70 percent, is for land use activities related to livestock rearing, including feed crop production and pasture. With growing demand, farmers and herders can intensify production either through technological improvements or by bringing new land into production. The potential for further expansion of land for food and feed production is limited by physical and economic constraints to conversion of currently unused land and is also limited by competition for land use from biofuels, urban expansion, forests, climate change mitigation, and biodiversity. Expansion of farmland into forests increases greenhouse gas (GHG) emissions, exacerbating climate change. Under these conditions, sustainable intensification of existing croplands is essential.

Further challenges arise from land degradation, which is the loss of goods and services from ecosystems and includes desertification, deforestation, salinization, and soil erosion. Farm-level land degradation, often driven by unsustainable management practices, has direct on-farm production costs in the form of lower crop yields or greater production costs as more fertilizers and other inputs are applied to offset yield losses. It also has off-site costs in the form of excessive runoff of fertilizers and pesticides, siltation of dams and irrigation systems, eutrophication of lakes and oceans, and damage to mangrove swamps (Rosegrant, Nkonya, and Valmonte-Santos 2009). Underlying causes of land degradation include high population density, poverty, insecure and unclear land rights, and lack of access to extension, infrastructure, and markets (Nkonya et al. 2011). Distorting trade policies, output price policies, and input subsidies, particularly for water and fertilizer, have also driven excessive use of these inputs, leading to land degradation.

A fundamental underlying condition for sustainable land management is effective land rights. Secure property rights to land provide incentives for producers to invest in land improvement, new crop varieties, improved crop management, and appropriate levels of input use (Otsuka and Place 2001). Place (2009) examines the results from economic studies of the relationships between land tenure security and agricultural productivity in Africa and how these results have been incorporated into policies. Property rights need to be clearly defined, understood, and secure. The need for secure tenure is particularly important for women, who currently face insecure tenure, which leads not only to inefficient use of resources but also to weak bargaining position within households (see Chapter 15).

A highly debated tenure policy for Africa and other regions is land registration and titling. Land registration is sometimes promoted as the only approach to guaranteeing secure land tenure, but in many cases traditional land tenure systems can provide adequate security without introducing costly, contentious, and complex land registration and titling systems (Place 2009). In other cases, traditional or communal systems lead to highly ineffective management, with centralized or disputed ownership that prevents individual farmers from investing in and managing their land efficiently. "In these cases, instead of costly titling systems, an alternative can be land banks that serve as land aggregators that can resolve conflicts related to ownership, separate property use rights from ownership rights, and reduce the transaction costs of transferring use rights," recommends Udry (2012). Land banks encourage participatory processes for delineating customary land, terms of leases, and processes for ensuring adequate compensation and can be a tool to enable customary groups to participate in the formal economy while retaining their group ownership and identity. The wide diversity of land tenure types and outcomes suggests that there is no "best practice" model for land tenure reform; rather, context is key, with land tenure regimes suited to particular locations and driven by community leadership as well as national concerns. The issue of sustainable use of farm land will be taken up in Chapter 18.

WATER

Rapidly increasing nonagricultural demands for water, changing food preferences, global climate change, and new demands for biofuel production put increasing pressure on scarce water resources. Challenges of growing water scarcity for agriculture are heightened by the transfer of water from irrigation to household and industrial uses, the increasing costs of developing new water sources, soil degradation, groundwater depletion, increasing water pollution, the degradation of water-related ecosystems, and wasteful use of already developed water supplies. If these challenges are to be met and if enough food is to be produced to meet growing demand, a strong irrigation sector remains vital. The rapid increases in agricultural yields and outputs over the past three decades could not have been achieved without expansion of irrigation. Although irrigation is sometimes associated with adverse environmental and sometimes also negative social impacts, it remains one of the most critical inputs into farming. Irrigation also contributes to poverty reduction, affordable food prices, and—through its significant multiplier effects—improvements in many other livelihood outcomes, such as health and nutrition (Rosegrant, Ringler, and Zhu 2009). A major challenge is how to make irrigation more effective for increased agricultural production while reducing socially wasteful use of irrigation water. Because new investments in irrigation and water supply are increasingly expensive and politically sensitive, large-scale water infrastructure investment has a reduced role globally compared with past decades (Rosegrant 2019). But small-scale systems, using both surface water and groundwater, have substantial potential. The World Bank's Africa Infrastructure Country Diagnostic study concluded that Africa has the hydrological and economic potential to add at least 16 million hectares of profitable, large-scale irrigation and 7 million hectares for farmer-led irrigation. The internal rate of return was substantially larger for individual systems and those managed by farm communities (You et al. 2011). An even larger potential was shown for profitable smallholder irrigation expansion in Africa south of the Sahara, up to 30 million hectares for motor pumps (Xie et al. 2014).

While cost-effective irrigation development remains important, past mistakes in development need to be avoided and alleviated. De Fraiture, Molden, and Wichelns (2010) point out that poorly conceived or poorly implemented water management interventions can have high social and environmental costs, including inequity in the allocation of benefits and undesirable impacts on natural resources. Irrigation (and domestic and industrial water use) has encroached on common pool resources such as rivers and wetlands. Dam building has displaced some communities, often without adequate compensation (de Fraiture, Molden, and Wichelns 2010).

A promising avenue for addressing water shortfalls and avoiding environmental damage from wasteful water use is to implement incentive policy reform to enhance the efficiency of existing water use, supported by infrastructure investment to modernize and upgrade existing irrigation and water delivery systems. Establishment of clear water rights for water users is an important tool, inducing users to consider the full opportunity cost of water, including its value in alternative uses, thus providing incentives to economize on the use of water. Improvements in the irrigation sector to increase water use efficiency must be made at the technical, managerial, and institutional levels. Establishing physical controls on water usage within the river basin, including rationing or quotas, through enforcement of water rights would increase the likelihood that local irrigation efficiencies from irrigation technology and management reform translate to basin-level efficiency and real water savings. Building on water rights, introducing market (or market-style) incentives into water use decision-making can also improve the efficiency of water use. Marginal cost pricing is not likely to be successful in the irrigation sector. Prices high enough to induce significant changes in water allocation or recover capital costs will severely reduce small farm income in developing countries (Perry 2001). In addition, measurements of water use by individual farmers are difficult to obtain, and without them price incentives to save water cannot be provided to individual users (Kajisa and Dong 2017). The water rights approach is therefore preferred to water pricing to support water efficiency in the irrigation sector (Rosegrant 2019). With appropriate policies and water management practices, much can be done in the irrigation sector to conserve water, and enhanced domestic sewage treatment and industrial recycling and effluent treatment will also be important in improving water availability in many developing countries (Rosegrant, Ringler, and Zhu 2009).

Under some conditions, community management of water resources is more effective than market-based approaches. Water can often be considered a common pool resource (CPR), as defined by Ostrom et al. (1999): "CPRs include natural and human constructed resources in which exclusion of beneficiaries through physical and institutional means is especially costly, and exploitation by one user reduces resource availability for others." In this situation, individuals following their own short-term interests can cause overexploitation of the commons and collapse of the resource, the so-called tragedy of the commons. But the tragedy of the commons is not inevitable. Ostrom et al. (1999) lay out the key principles for success in managing the commons. Cox, Arnold, and Tomás (2010) and Holden and Tilahun (2018) further assess these principles empirically. Although tragedies of the commons have occurred, for thousands of years people have also self-organized to manage CPRs, devising long-term, sustainable institutions for governing these resources (Ostrom et al. 1999). Community management of water resources with establishment of customary water rights and rules governing water allocation is one such approach to governing the commons that can lead to effective water management. How to deal with growing scarcity of water is further discussed in Chapter 20.

CLIMATE CHANGE

The Intergovernmental Panel on Climate Change (IPCC) has shown that global average temperatures have risen by roughly 0.13°C per decade since 1950, and they expect a faster pace of about 0.2°C per decade over the next two to three decades (Lobell, Schlenker, Costa-Roberts 2011). Modeling by Lobell et al. shows that "global maize and wheat production declined by 3.8 and 5.5 percent, respectively, relative to a counterfactual without climate trends." The estimated change in crop production excluding and including carbon dioxide (CO_2) fertilization translates into average commodity price increases of 18.9 and 6.4 percent. "Climate trends were large enough in some countries to offset a significant portion of the increases in average yields that arose from new improved technology, CO_2 fertilization and other factors" (Lobell et al. 2011). Thus, climate change is becoming a major challenge to agricultural development.

A recent systematic review by Knox et al. (2012) summarized by Wheeler and von Braun (2013) found that average crop yields may decline across Africa and Asia by 8 percent by the 2050s. "Across Africa, yields are predicted to change by −17 percent (wheat), −5 percent (maize), −15 percent (sorghum), and −10 percent (millet) and, across South Asia, by −16 percent (maize) and −11 percent (sorghum) under climate change. No mean change in yield was detected for rice" (Wheeler and von Braun 2013). A global review of crop model results by Challinor et al. (2014) found similar orders of magnitude for wheat, maize, and rice (with slightly negative impacts on rice), and greater negative yield effects in tropical regions than in temperate regions.

Nelson et al. (2014) presented climate change results from a model comparison exercise with 10 of the leading global economic models that have significant representation of agriculture. These models account for the initial biophysical impacts of climate change on crop yields estimated from crop models but also take into account the subsequent feedback effects from economic responses that will tend to dampen the initial crop yield impact. The results show a great deal of variation across models, but the average yield impacts of climate change across models and scenarios in 2050 after economic adjustments are −12.5 percent for coarse grains (mainly maize), −9.3 percent for rice, and −9.9 percent for wheat. Prices are projected to increase on average by 11.8 percent for maize, 13.9 percent for rice, and 15.9 percent for wheat (Nelson et al. 2014). An earlier review of climate change results from economic models (Parry et al. 2009) showed slightly higher increases in food prices, with a range of price increases of about 10 to 30 percent by 2050 due to climate change, and a median estimate of around 20 percent. The projected impact of climate change on the risk of hunger showed increases in hunger of about 10 to 20 percent by 2050. A detailed assessment of climate change impacts, and the possibility of mitigation of and adaptation to climate change in agriculture, is provided in Chapter 19.

In addition to the production and price effects, frequent occurrence of extreme weather events, such as drought and floods, is likely to increase variability in food supply, which in turn affects variability of household incomes and food security. Wheeler and von Braun (2013) conclude that the "strong evidence on the negative impacts of climate change on food security supports the need for considerable investment in adaptation and mitigation actions toward a 'climate-smart food system' that is more resilient to climate change influences on food security."

Sustainable Food System for the Future

Garnett et al. (2013) address key challenges for a sustainable food system. Food needs to be produced in ways that place far less pressure on the environment and that sustain its capacity to continue producing food in the future. Increased production must be met through higher yields because increasing the area of land in agriculture imposes major environmental costs. Expansion of agricultural land will in many places encroach on forests, wetlands, or grasslands, whose conversion would increase GHG emissions and the loss of biodiversity and important ecosystem services. But sustainably increasing production is not enough. Richer countries should reduce consumption of resource-intensive foods (such as meat and dairy products) and food loss and waste should be reduced in both developing and developed countries (Garnett et al. 2013).

Godfray et al. (2010) provide a broad overview of the key components of sustainable production growth. *First is reducing the yield gap* by moving farmers toward existing yield potential by reducing constraints to farmer access to seeds, water, nutrients, pest management, soils, biodiversity, finance, and knowledge. *Second is increasing production limits* through modern genetic techniques and a better understanding of crop physiology to allow for a more directed approach to selection across multiple traits. Gene sequencing and marker-assisted selection are speeding up the development of new, stress-tolerant and high-yielding varieties. Genetically modified (GM) crops also have potential to provide a broader combination of desirable traits, including the introduction of new traits such as drought tolerance. *Third, waste can be reduced.* Food losses can occur at any point in the value chain—from production (crop damage, spillage) to postharvest and processing stages (attacks from insect or microorganisms during storage), distribution (poor infrastructure, cold storage), and retail sale through home consumption (spoilage, table waste). In developing countries, losses occur mainly due to constraints in harvesting methods and techniques, lack of storage and/or cooling facilities, and poor marketing and transport systems. In developed countries, the biggest losses occur mainly after food reaches retail, restaurants, and home consumption. The potential scope for economically reducing food losses remains uncertain, but investments in reducing food loss and waste can be a complement to increasing productivity growth (Rosegrant et al. 2018). *Fourth, diets should change*, to include more fruits and vegetables and less meat in those regions that currently consume high-meat diets. *Fifth, aquaculture should be expanded on a sustainable basis* through advances in hatchery systems, feed production and feed-delivery systems, disease management, and better stock selection and potentially through larger-scale production technologies, aquaculture in open seas and larger inland water bodies, and the culture of a wider range of species (Godfray et al. 2010).

In the past few years, gene editing has emerged as a key new tool for crop improvement, with great potential for enhancing agricultural productivity, food security, and resilience to climate change. Gene editing differs from GMOs because it does not splice foreign genes into an organism. It may therefore be subject to an appropriately lower regulatory burden than GMOs and may avoid the negative public perception surrounding genetically altered crops, although debates are continuing. Gene editing technology such as CRISPR is more accessible to smaller companies and labs than genetic modification technology, allowing far broader participation and competition, particularly if the regulatory costs are indeed significantly lower, as argued in Chapter 21.

Thornton (2010) provides a comprehensive summary of trends in the livestock sector. Livestock systems have a global asset value of at least US$1.4 trillion, employ at least 1.3 billion people globally, directly support the livelihoods of 600 million poor smallholder farmers in the developing world, and account for about one-third of agricultural value-added. Livestock products contribute 17 percent to kilocalorie consumption and 33 percent to protein consumption globally, with far higher per capita consumption in rich countries. Rapid growth in livestock production is being driven by the rapidly increasing demand for livestock products, with this demand being driven by population growth, urbanization, and increasing incomes in developing countries (Thornton 2010).

The expansion of livestock production needs to take place in a way that allows poorer consumers to benefit from increased production and that reduces its impact on the environment, while balancing trade-offs between food security, poverty, equity, environmental sustainability, and economic development. Key innovations are needed in breeding and feeding programs that will focus not only on production and productivity, but also on product quality, animal welfare, disease resistance, and reduction in GHG emissions and other environmental impacts (Thornton 2010).

Pingali and Rosegrant (2000) review the ecological consequences of intensive rice production as typified by the Green Revolution. In the long run, if not managed properly, intensification can impose significant environmental costs. Environmental consequences of lowland rice intensification include "(i) the build-up of salinity and waterlogging; (ii) the depletion/pollution of (ground) water resources; (iii) the formation of a hardpan (subsoil compaction); (iv) changes in soil-nutrient status, nutrient deficiencies, and increased incidence of soil toxicities; and (v) increased pest build-up, pest-related yield losses, and associated consequences of increased and injudicious pesticide use" (Pingali and Rosegrant 2000). However, intensification per se is not the root cause of environmental and ecological damage. Environmental degradation in intensified agriculture occurs mainly when incentives are incorrect, due to bad policy, missing markets, or a lack of knowledge of the underlying processes of degradation.

Policies that support more sustainable and intensified agriculture while protecting the environment include "the removal of incentive-distorting subsidies and taxes; the establishment of secure property rights; increased investments in research, education, and training; improved public infrastructure; better integration of international commodity markets; and a greater inclusion of populations in developing countries into these markets. Appropriate policies will provide farmers with the incentives to invest in more sustainable land and crop-management practices" (Pingali and Rosegrant 2000).

Halberg et al. (2006) note that modern noncertified organic farming is a potentially sustainable approach in regions such as parts of Africa, which have low yields due to poor access to inputs or low yield potential, because it involves lower economic risk than technologies based on purchased inputs. In these regions, locally adapted agroecological methods can improve the yields and sustainability of smallholder farms and increase local food security. But in higher-yielding production regions, yields of organic farming are between 15 and 35 percent lower than present yields when comparing single crops (Halberg et al. 2006). Two recent metastudies showed that yields of organic agriculture average 20 to 25 percent less than conventional agriculture, but with large variations (de Ponti, Rijk, and van Ittersum 2012; Seufert, Ramankutty, and Foley 2012). Nitrogen leaching and emissions of nitrous oxide and ammonia are lower in organic agriculture than in conventional agriculture on an area basis, due to lower nitrogen inputs, but they are usually larger per unit of product because of conventional agriculture's higher crop yields per hectare (Pimentel et al. 2005; Balmford, Green, and Phalan 2012; Tuomisto et al. 2012).

As was pointed out in Chapter 1, how we build a new food system that is simultaneously nutrition and health driven, productive and efficient, environmentally sustainable and climate smart, inclusive, and business friendly is a challenge we must face. We will discuss key elements of such a food system and strategy to achieve it in Chapter 22.

References

Aker, J. C., I. Ghosh, and J. Burrell. 2016. "The Promise (and Pitfalls) of ICT for Agriculture Initiatives." *Agricultural Economics* 47 (S1): 35–48.

Aker, J. C., and I. M. Mbiti. 2010. "Mobile Phones and Economic Development in Africa." *Journal of Economic Perspectives* 24 (3): 207–232.

Alderman, H., J. Hoddinott, and B. Kinsey. 2006. "Long Term Consequences of Early Childhood Malnutrition." *Oxford Economic Papers* 58 (3): 450–474.

Alston, J. M., M. A. Andersen, J. S. James, and P. G. Pardey. 2010. *Persistance Pays: U.S. Agricultural Productivity Growth and the Benefits from Public R&D Spending*. Natural Resource Management and Policy series, vol. 34. New York: Springer-Verlag.

Alston, J. M., C. Chan-Kang, M. C. Marra, P. G. Pardey, and T. J. Wyatt. 2000. *A Meta-analysis of Rates of Return to Agricultural R&D: Ex pede Herculem?* IFPRI Research Report 113. Washington, DC: IFPRI.

Anderson, K., J. Cockburn, and W. Martin. 2011. "Would Freeing Up World Trade Reduce Poverty and Inequality? The Vexed Role of Agricultural Distortions." *The World Economy* 34 (4): 487–515.

Anderson, K., G. Rausser, and J. Swinnen. 2013. "Political Economy of Public Policies: Insights from Distortions to Agricultural and Food Markets." *Journal of Economic Literature* 51 (2): 423–477.

Arouna, A., J. D. Michler, and J. C. Lokossou. 2019. *Contract Farming and Rural Transformation: Evidence from a Field Experiment in Benin*. Working Paper 25665. Cambridge, MA: National Bureau of Economic Research.

Balmford, A., R. Green, and B. Phalan. 2012. "What Conservationists Need to Know about Farming." *Proceedings of the Royal Society B: Biological Sciences* 279: 1–12.

Banerjee, A., E. Duflo, R. Glennerster, and C. Kinnan. 2013. *The Miracle of Microfinance: Evidence from a Randomized Evaluation*. MIT Working Paper, MIT, Cambridge, MA.

Banerjee, A., D. Karlan, and J. Zinman. 2015. "Six Randomized Evaluations of Microcredit: Introduction and Further Steps." *American Economic Journal: Applied Economics* 7 (1): 1–21.

Barker, R., and M. W. Rosegrant. 2007. "Establishing Efficient Use of Water Resources in Asia." In *Reasserting the Rural Development Agenda: Lessons Learned and Emerging Challenges in Asia*, edited by A. Balisacan and F. Nobuhiko, 414. Laguna, Philippines: Institute of the South East Asian Studies and Southeast Asia Regional Center for Graduate Study and Research in Agriculture (SEARCA).

Barrett, C. B., M. E. Bachke, M. F. Bellemare, H. C. Michelson, S. Narayanan, and T. F. Walker. 2012. "Smallholder Participation in Contract Farming: Comparative Evidence from Five Countries." *World Development* 40 (4): 715–730.

Barrett, C. B., M. R. Carter, and C. P. Timmer. 2010. "A Century-Long Perspective on Agricultural Development." *American Journal of Agricultural Economics* 92 (2): 447–468.

Beamon, B. 1999. "Measuring Supply Chain Performance." *International Journal of Operations and Production Management* 19 (3): 275–292. Reprinted in the *Journal of Enterprise Resource Management* 7 (3): 46–63.

Bellemare, M. F., and J. R. Bloem. 2018. "Does Contract Farming Improve Welfare? A Review." *World Development* 112: 259–271.

Binswanger, H. 1986. "Agricultural Mechanization: A Comparative Historical Perspective." *World Bank Research Observer* 1 (1): 27–56.

Byerlee, D., T. S. Jayne, and R. J. Myers. 2006. "Managing Food Price Risks and Instability in a Liberalizing Market Environment: Overview and Policy Options." *Food Policy* 31 (4): 275–287.

Caballero, B. 2005. "A Nutrition Paradox, Underweight and Obesity in Developing Countries." *New England Journal of Medicine* 352: 1514–1516.

———. 2007. "The Global Epidemic of Obesity: An Overview." *Epidemiologic Reviews* 29 (1): 1–5.

Ceballos, F., B. Kramer, and M. Robles. 2019. "The Feasibility of Picture-Based Insurance (PBI): Smartphone Pictures for Affordable Crop Insurance." *Development Engineering* 4: 100042.

Challinor, A. J., J. Watson, D. B. Lobell, S. M. Howden, D. R. Smith, and N. Chhetri. 2014. "A Meta-Analysis of Crop Yield under Climate Change and Adaptation." *Nature Climate Change* 4: 287–291.

Christiaensen, L., and W. Martin. 2018. "Agriculture, Structural Transformation and Poverty Reduction: Eight New Insights." *World Development* 109: 413–416.

Colman, D. 2013. "Productivity Growth in Agriculture: An International Perspective," review of *Productivity Growth in Agriculture: An International Perspective*, edited by K. O. Fuglie, S. L. Wang, and V. E. Ball. *European Review of Agricultural Economics* 40 (3): 531–534.

Cox, M., G. Arnold, and S. V. Tomás. 2010. "A Review of Design Principles for Community-Based Natural Resource Management." *Ecology and Society* 15 (4): 38.

Craig, B. J., P. G. Pardey, and J. Roseboom. 1997. "International Productivity Patterns: Accounting for Input Quality, Infrastructure, and Research." *American Journal of Agricultural Economics* 79 (4): 1064–1076.

Davies, F. T., and B. N. Garrett. 2018. "Technology for Sustainable Urban Food Ecosystems in the Developing World: Strengthening the Nexus of Food–Water–Energy–Nutrition." *Frontiers in Sustainable Food Systems*. 2: 84.

de Fraiture, C., D. Molden, and D. Wichelns. 2010. "Investing in Water for Food, Ecosystems, and Livelihoods: An Overview of the Comprehensive Assessment of Water Management in Agriculture." *Agricultural Water Management* 97 (4): 495–501.

de Ponti, T., B. Rijk, and M. K. van Ittersum. 2012. "The Crop Yield Gap between Organic and Conventional Agriculture." *Agricultural Systems* 108: 1–9.

Dercon, S., and D. Gollin. 2014. "Agriculture in African Development: Theories and Strategies." *Annual Review of Resource Economics* 6 (1): 471–492.

Dercon, S., R. Vargas Hill, D. Clarke, I. Outes-Leon, and A. S. Taffesse. 2014. "Offering Rainfall Insurance to Informal Insurance Groups: Evidence from a Field Experiment in Ethiopia." *Journal of Development Economics* 106: 132–143.

Diao, X., J. Silver, and H. Takeshima. 2016. *Agricultural Mechanization and Agricultural Transformation*. IFPRI Discussion Paper 1527. Washington, DC: IFPRI.

Diao, X., J. Thurlow, S. Benin, and S. Fan, eds. 2014. *Strategies and Priorities for African Agriculture: Economywide Perspectives from Country Studies*. Washington, DC: IFPRI.

Dorosh, P., and J. Thurlow. 2018. "Beyond Agriculture versus Non-Agriculture: Decomposing Sectoral Growth–Poverty Linkages in Five African Countries." *World Development* 109: 440–451.

Duflo, E., M. Kremer, and J. Robinson. 2011. "Nudging Farmers to Use Fertilizer: Theory and Experimental Evidence from Kenya." *American Economic Review* 101 (6): 2350–2390.

Ecker, O. 2019. "Reshaping Agriculture to Reduce Obesity." In *Agriculture for Improved Nutrition: Seizing the Momentum*, edited by S. Fan, S. Yosef, and R. Pandya-Lorch, Chapter 8. Wallingford, UK: IFPRI and CABI.

Evenson, R. 2001. "Economic Impacts of Agricultural Research and Extension." In *Handbook of Agricultural Economics*, edited by B. Gardner and G. Rausser. Amsterdam: Elsevier Science.

Evenson, R. E., and D. Gollin. 2003a. "Assessing the Impact of the Green Revolution, 1960 to 2000." *Science* 300 (5620): 758–762.

———, eds. 2003b. *Crop Variety Improvement and Its Effect on Productivity: The Impact of International Agricultural Research*. Wallingford, UK: CABI.

Fan, S., P. Hazell, and S. Thorat. 2000. "Government Spending, Growth and Poverty in Rural India." *American Journal of Agricultural Economics* 82 (4): 1038–1051.

Fan, S., and R. Pandya-Lorch. 2012. *Reshaping Agriculture for Nutrition and Health*. Washington, DC: International Food Policy Research Institute.

Fan, S., S. Yosef, and R. Pandya-Lorch. 2019a. "Seizing the Momentum to Reshape Agriculture for Nutrition." In *Agriculture for Improved Nutrition: Seizing the Momentum*, edited by S. Fan, S. Yosef, and R. Pandya-Lorch, Chapter 1. Wallingford, UK: IFPRI and CABI.

———. 2019b. "The Way Forward for Nutrition-Driven Agriculture." In *Agriculture for Improved Nutrition: Seizing the Momentum*, edited by S. Fan, S. Yosef, and R. Pandya-Lorch, Chapter 20. Wallingford, UK: IFPRI and CABI.

Feed the Future. 2016. *Low-Cost Sensors for Agriculture. Key Findings Report.* USAID.

Fiedler, J., T. Sanghvi, and M. Saunders. 2008. "A Review of the Micronutrient Intervention Cost Literature: Program Design and Policy Lessons." *International Journal on Health Planning and Management* 23: 373–397.

Fuglie, K., and D. Schimmelpfennig. 2010. "Introduction to the Special Issue on Agricultural Productivity Growth: A Closer Look at Large, Developing Countries." *Journal of Productivity Analysis* 33: 169–172.

Garnett, T., M. C. Appleby, A. Blamford et al. 2013. "Sustainable Intensification in Agriculture: Premises and Policies." *Science* 341 (6141): 33–34.

Giller, K. E., P. Tittonell, M. C. Rufino et al. 2011. "Communicating Complexity: Integrated Assessment of Trade-Offs Concerning Soil Fertility Management within African Farming Systems to Support Innovation and Development." *Agricultural Systems* 104: 91–203.

Godfray, C. H., J. R. Beddington, I. R. Crute et al. 2010. "Food Security: The Challenge of Feeding 9 Billion People." *Science* 327 (5967): 812–818.

Gollin, D., C. W. Hansen, and A. Wingender. 2018. *Two Blades of Grass: The Impact of the Green Revolution.* Working Paper 24744. Cambridge, MA, US: National Bureau of Economic Research.

Gómez, M. I., C. B. Barrett, T. Raney et al. 2013. "Post-Green Revolution Food Systems and the Triple Burden of Malnutrition." *Food Policy* 42 (2013): 129–138.

Halberg, N., T. B. Sulser, H. Høgh-Jensen, M. W. Rosegrant, and M. Trydeman Knudsen. 2006. "The Impact of Organic Farming on Food Security in a Regional and Global Perspective." In *Global Development of Organic Agriculture: Challenges and Prospects*, edited by N. Halberg, H. F. Alrøe, M. T. Knudsen, and E. S. Kristensen, 277–322. New York: CABI.

Hawkes, C., J. Harris, and S. Gillespie. 2017. "Changing Diets: Urbanization and the Nutrition Transition." In *2017 Global Food Policy Report*, 34–41. Washington, DC: IFPRI.

Hayami, Y., and V. W. Ruttan. 1985. *Agricultural Development: An International Perspective.* Baltimore: Johns Hopkins University Press.

Hazell, P. B. R. 2009. "The Asian Green Revolution." IFPRI Discussion Paper 00911. In *Millions Fed: Proven Successes in Agricultural Development Compendium*, 1–31. Washington, DC: IFPRI.

Headey, D., S. Malaiyandi, and S. Fan. 2009. *Navigating the Perfect Storm: Reflections on the Food, Energy, and Financial Crisis.* IFPRI Discussion Paper 00889. Washington, DC: IFPRI.

Holden, S. T., and M. Tilahun. 2018. "The Importance of Ostrom's Design Principles: Youth Group Performance in Northern Ethiopia." *World Development* 104 (C): 10–30.

IFPRI (International Food Policy Research Institute). 2011. *Policies, Institutions, and Markets to Strengthen Food Security and Incomes for the Rural Poor.* A proposal submitted to the CGIAR Consortium Board. Washington, DC.

IFPRI, China Center for Agricultural Policy (CCAP), and Asian Development Bank (ADB). 2019. *Information and Communication Technology for Agriculture in the People's Republic of China.* Manila: Asian Development Bank.

InterAcademy Council. 2004. *Realizing the Promise and Potential of African Agriculture: Science and Technology Strategies for Improving Agricultural Productivity and Food Security in Africa.* Netherlands.

Jagger, P., and J. Pender. 2003. *Impacts of Programs and Organizations on the Adoption of Sustainable Land Management Technologies in Uganda.* EPTD Discussion Paper 101. Washington, DC: IFPRI.

Jin, S., J. Huang, and S. Rozelle. 2010. "Agricultural Productivity in China." In *The Shifting Patterns of Agricultural Production and Productivity Worldwide.* Iowa State University, Midwest Agribusiness Trade Research and Information Center (MATRIC).

Kajisa, K., and B. Dong. 2017. "The Effect of Volumetric Pricing Policy on Farmers' Water Management Institutions and Their Water Use: The Case of Water User Organization in an Irrigation System in Hubei, China." *World Bank Economic Review* 31 (1): 220–240.

Kearney, J. 2010. "Food Consumption Trends and Drivers." *Philosophical Transactions of the Royal Society B: Biological Sciences* 365 (1554): 2793–2807.

Knox, J., T. Hess, A. Daccache, and T. Wheeler. 2012. "Climate Change Impacts on Crop Productivity in Africa and South Asia." *Research Letters* 7 (034032): 1–8.

Lin, J. 1991. "The Household Responsibility System Reform and the Adoption of Hybrid Rice in China." *Journal of Development Economics* 36: 353–372.

Lobell, D., W. Schlenker, and J. Costa-Roberts. 2011. "Climate Trends and Global Crop Production since 1980." *Science* 333: 616–620.

Lohman, C., L. Fortuin, and M. Wouters. 2004. "Designing a Performance Measurement System: A Case Study." *European Journal of Operational Research* 15 (6): 67–286.

López, R., and G. I. Galinato. 2007. "Should Governments Stop Subsidies to Private Goods? Evidence from Rural Latin America." *Journal of Public Economics* 91 (5–6): 1071–1094.

Masters, W. A. 2005. "Paying for Prosperity: How and Why to Invest in Agricultural Research and Development in Africa." *Journal of International Affairs* 58 (2): 35–36.

Merrey, D. J., and N. Lefore. 2018. *Improving the Availability and Effectiveness of Rural and "Micro" Finance for Small-Scale Irrigation in Sub-Saharan Africa: A Review of Lessons Learned.* IWMI Working Paper 185. Colombo, Sri Lanka: International Water Management Institute.

Meyer, R. 2011. *Subsidies as an Instrument in Agriculture Finance: A Review.* Joint Discussion Paper. World Bank, BMZ, FAO, GIZ, IFAD, and UNCDF.

Michelson, H. C. 2013. "Small Farmers, NGOs, and a Walmart World: Welfare Effects of Supermarkets Operating in Nicaragua." *American Journal of Agricultural Economics* 95 (3): 628–649.

Minten, B., L. Randrianarison, and J. Swinnen. 2009. "Global Retail Chains and Poor Farmers: Evidence from Madagascar." *World Development* 37 (11): 1728–1741.

Miranda, M. J., and K. Farrin. 2012. "Index Insurance for Developing Countries." *Applied Economic Perspectives and Policy* 34 (3): 391–427.

Ncube, B., W. Mupangwa, and A. French. 2018. "Precision Agriculture and Food Security in Africa." In *Systems Analysis Approach for Complex Global Challenges*, edited by P. Mensah, D. Katerere, S. Hachigonta, and A. Roodt, 159–78. Springer International.

Nelson, G. C., D. van der Mensbrugghe, H. Ahammad et al. 2014. "Agriculture and Climate Change in Global Scenarios: Why Don't the Models Agree?" *Agricultural Economics* 45: 85–101.

Nin-Pratt, A. 2015. *Inputs, Productivity, and Agricultural Growth in Africa South of the Sahara.* IFPRI Discussion Paper 1432. Washington, DC: IFPRI.

Nkonya, E., N. Gerber, P. Baumgartner et al. *The Economics of Land Degradation: Toward an Integrated Global Assessment.* Bern, Switzerland: Peter Lang.

Ostrom, E., J. Burger, C. B. Field, R. B. Norgaard, and D. Policansky. 1999. "Revisiting the Commons: Local Lessons, Global Challenges." *Science* 284 (5412): 278–282.

Otsuka, K., and D. Larson, eds. 2013. *An African Green Revolution: Finding Ways to Boost Productivity on Small Farms.* Dordrecht, Netherlands: Springer.

———. 2016. *In Pursuit of an African Green Revolution: Views from Rice and Maize Farmers' Fields.* Dordrecht, Netherlands: Springer.

Otsuka, K., and R. Muraoka. 2017. "A Green Revolution for Sub-Saharan Africa: Past Failures and Future Prospects." *Journal of African Economies* 26 (S1): 73–98.

Otsuka, K., Y. Nakano, and K. Takahashi. 2016. "Contract Farming in Developed and Developing Countries." *Annual Review of Resource Economics* 8: 353–376.

Otsuka, K., and F. Place. 2001. *Land Tenure and Natural Resource Management: A Comparative Study of Agrarian Communities in Asia and Africa.* Baltimore: Johns Hopkins University Press.

Parry, M., A. Evans, M. W. Rosegrant, and T. Wheeler. 2009. *Climate Change and Hunger: Responding to the Challenge*. Rome; Washington, DC; New York; London; UK World Food Programme; IFPRI; New York University Center on International Cooperation; Grantham Institute at Imperial College London; Walker Institute at University of Reading.

Perry, C. 2001. "Water at Any Price? Issues and Options in Charging for Irrigation Water." *Irrigation and Drainage* 50: 1–7.

Pimentel, D., P. Hepperly, J. Hanson, D. Douds, and R. Seidel. 2005. "Environmental, Energetic, and Economic Comparisons of Organic and Conventional Farming Systems." *BioScience* 55 (7): 573–582.

Pingali, P. 2006. "Westernization of Asian Diets and the Transformation of Food Systems: Implications for Research and Policy." *Food Policy* 32 (2): 281–298.

Pingali, P., and M. W. Rosegrant. 2000. "Intensive Food Systems in Asia: Can the Degradation Problems Be Reversed?" In *Tradeoffs or Synergies? Agricultural Intensification, Economic Development and the Environment*, edited by D. R. Lee and C. B. Barrett, 383–397. New York: CABI.

Place, F. 2009. "Land Tenure and Agricultural Productivity in Africa: A Comparative Analysis of the Economics Literature and Recent Policy Strategies and Reforms." *World Development* 37 (8): 1326–1336.

Qamar, K. 2002. *Global Trends in Agricultural Extension: Challenges Facing Asia and the Pacific Region*. Paper presented at FAO regional expert consultation on agricultural extension, Bangkok.

Reardon, T., K. Z. Chen, B. Minten, L. Adriano, T. A. Dao, J. Wan, and S. D. Gupta. 2014. "The Quiet Revolution in Asia's Rice Value Chains." *Annals of the New York Academy of Sciences* 1331: 106–118.

Reardon, T., and C. P. Timmer. 2012. "The Economics of the Food System Revolution." *Annual Review of Resource Economics* 4 (1): 225–264.

Ricker-Gilbert, J., T. Jayne, and G. Shively. 2013. "Addressing the 'Wicked Problem' of Input Subsidy Programs in Africa." *Applied Economic Perspectives and Policy* 35 (2): 322–340.

Rivera, W. 2001. "Whither Agricultural Extension Worldwide? Reforms and Prospects." In *Knowledge Generation and Technical Change: Institutional Innovations in Agriculture*, edited by S. Wolf and D. Zilberman. Berkeley: University of California Berkeley.

Rosegrant, M. W. 2014. *Food Security*. Thousand Oaks, CA: SAGE.

———. 2019. *From Scarcity to Security: Managing Water for a Nutritious Food Future*. Chicago: Chicago Council on Global Affairs.

Rosegrant, M. W., and P. B. R. Hazell. 2000. *Transforming the Rural Asian Economy: The Unfinished Revolution*. Hong Kong: Oxford University Press.

Rosegrant, M. W., N. Leach, and R. V. Gerpacio. 1999. "Alternative Futures for World Cereal and Meat Consumption." *Proceedings of the Nutrition Society* 58 (2): 219–234.

Rosegrant, M. W., E. Magalhaes, R. A. Valmonte-Santos, and D. Mason-D'Croz. 2018. "Returns to Investment in Reducing Postharvest Food Losses and Increasing Agricultural Productivity Growth." In *Prioritizing Development: A Cost-Benefit Analysis of the United Nations' Sustainable Development Goals*, edited by B. Lomborg. Cambridge: Cambridge University Press.

Rosegrant, M. W., E. Nkonya, and R. A. Valmonte-Santos. 2009. "Food Security and Soil Water Management." *Encyclopedia of Soil Science* 1: 1–4.

Rosegrant, M. W., C. Ringler, and T. Zhu. 2009. "Water for Agriculture: Maintaining Food Security under Growing Scarcity." *Annual Review of Environmental Resources* 34: 205–222.

Rosegrant, M. W., S. Tokgoz, and P. Bhandary. 2013. "The Future of the Global Food Economy: Scenarios for Supply, Demand, and Prices." In *Food Security and Socio-Political Stability*, edited by C. B. Barrett, 35–63. Oxford: Oxford University Press.

Rozelle, S., and J. Swinnen. 2004. "Success and Failure of Reform: Insights from the Transition of Agriculture." *Journal of Economic Literature* 42 (2): 404–456.

Ruel, M. T., J. L. Garrett, and S. Yosef. 2017. "Food Security and Nutrition: Growing Cities, New Challenges." In *2017 Global Food Policy Report*, 24–33. Washington, DC: IFPRI.

Santos, P., and C. B. Barrett. 2011. "Persistent Poverty and Informal Credit." *Journal of Development Economics* 96 (2): 337–347.

Sen, A. 1981. "Ingredients of Famine Analysis: Availability and Entitlements." *Quarterly Journal of Economics* 96 (3): 433–464.

Seufert, V., N. Ramankutty, and J. A. Foley. 2012. "Comparing the Yields of Organic and Conventional Agriculture." *Nature* 485: 229–234.

Smith, V. H., and J. W. Glauber. 2012. "Agricultural Insurance in Developed Countries: Where Have We Been and Where Are We Going?" *Applied Economic Perspectives and Policy* 34 (3): 363–390.

Suri, T. 2017. "Mobile Money." *Annual Review of Economics* 9: 497–520.

Suri, T., and W. Jack. 2016. "The Long-Run Poverty and Gender Impacts of Mobile Money." *Science* 354 (6317): 1288–1292.

Takahashi, K., R. Muraoka, and K. Otsuka. 2020. "Technology Adoption, Impact, and Extension in Developing Country Agriculture: A Review of the Recent Literature." *Agricultural Economics* 51 (1): 31–45.

Thirtle, C., L. Lin, and J. Piesse. 2003. "The Impact of Research-Led Agricultural Productivity Growth on Poverty Reduction in Africa, Asia and Latin America." *World Development* 31 (12): 1959–1975.

Thornton, P. K. 2010. "Livestock Production: Recent Trends, Future Prospects." *Philosophical Transactions of the Royal Society B: Biological Sciences* 365 (1554): 2853–2867.

Ton, G., W. Vellema, S. Desiere, S. Weituschat, and M. D'Haese. 2018. "Contract Farming for Improving Smallholder Incomes: What Can We Learn from Effectiveness Studies?" *World Development* 104: 46–64.

Tuomisto, H. L., I. D. Hodge, P. Riordan, and D. W. Macdonald. 2012. "Does Organic Farming Reduce Environmental Impacts? A Meta-Analysis of European Research." *Journal of Environmental Management* 112: 309–320.

Udry, C. 2012. "Land Tenure." In *The Oxford Companion to the Economics of Africa*, edited by E. Aryeetey, S. Devarajan, R. Kanbur, and L. Kasekende. Oxford: Oxford University Press.

UN Standing Committee on Nutrition (UNSCN). 2010. "Sustainable Food and Nutrition Security (Summary)." In *UNSCN 6th Report on the World Nutrition Situation: Progress in Nutrition*, 77–97. Geneva: UNSCN Secretariat.

USAID (US Agency for International Development). 2017. *How Digital Tools Impact the Value Chain*. Washington, DC.

Valeeva, N. I., M. P. M. Meuwissen, and R. B. M. Huirne. 2004. "Economics of Food Safety in Chains: A Review of General Principles." *NJAS–Wageningen Journal of Life Sciences* 51 (4): 369–390.

Wheeler, T., and J. von Braun. 2013. "Climate Change Impacts on Global Food Security." *Science* 341 (6145): 508–513.

Winters, L. A., N. McCulloch, and A. McKay. 2004. "Trade Liberalization and Poverty: The Evidence So Far." *Journal of Economic Literature* 42: 72–115.

Wright, B. D. 2012. "International Grain Reserves and Other Instruments to Address Volatility in Grain Markets." *World Bank Research Observer* 27 (2): 222–260.

Xie, H., L. You, B. Wielgosz, and C. Ringler. 2014. "Estimating the Potential for Expanding Smallholder Irrigation in Sub-Saharan Africa." *Agricultural Water Management* 131: 183–193.

Yang, J., Z. Huang, X. Zhang, and T. Reardon. 2013. "The Rapid Rise of Cross-Regional Agricultural Mechanization Services in China." *American Journal of Agricultural Economics* 95 (5): 1245–1251.

You, L., C. Ringler, U. Wood-Sichra et al. 2011. "What Is the Irrigation Potential for Africa? A Combined Biophysical and Socioeconomic Approach." *Food Policy* 36 (6): 770–782.

PART II

Regional Issues in Agricultural Development

Chapter 3

CHANGING FARM SIZE AND AGRICULTURAL DEVELOPMENT IN EAST ASIA

Futoshi Yamauchi, Jikun Huang, and Keijiro Otsuka

When labor is abundant relative to land in the early stage of economic development, labor-intensive methods of cultivation are socially efficient. In such cultivation systems, no major indivisible inputs are used and, hence, there is no major source of scale economies. Roughly speaking, a farm of 1–2 hectares can be managed efficiently by family labor consisting of a few workers. Beyond that scale, hired labor must be employed. However, the monitoring cost of hired labor arises, which increases more than proportionally with the cultivation size (Feder 1985; Otsuka, Chuma, and Hayami 1992; Hayami and Otsuka 1993). This explains why the family farm dominates in agriculture in most countries in the world (Berry and Cline 1979; Eastwood, Lipton, and Newell 2010). Thus, the optimum farm size in low-wage economies is bound to be small because of the limited availability of family labor and the costly substitution of capital for labor. This situation was predominant in East Asia (that is, Northeast and Southeast Asia), which justifies the dominance of relatively small operational sizes.

In the process of the economic development in East Asia, which accompanies the continuous increases in the real wage rate, the comparative advantage of the economy in most Asian countries has been shifting from agriculture to nonagricultural sectors. A part of the reason could be the small farm size, predominant in this region, which requires labor-intensive cultivation (Estudillo and Otsuka 2016). Farm size expansion, however, is difficult to realize due to the imperfection of land markets (Otsuka 2007). As a result, high-income countries or economies in East Asia, for example, Japan and the Republic of Korea, have been increasing imports of grains.

The striking feature of East Asia is its historically unprecedented rapid and successful industrialization, realized outside European and North American continents, which transformed the economies, including agriculture. A large part of the rural population had a chance to migrate to urban sectors that are highly able to absorb surplus labor from rural origins. As a result, the real wage rate and therefore the opportunity cost of family labor in agriculture also increased continuously. Labor abundance disappeared in most of East Asian agriculture. To reduce the labor cost, farmers need to substitute machines for labor. To operate machines more efficiently, farm size must be expanded. Since large machines are indivisible, scale advantages arise. Thus, larger farms become more efficient than smaller farms, and so the land must be transferred from the smallholder farmers to the larger farmers.

In light of these emerging challenges, the second section of this chapter overviews historical paths taken by agriculture in the region following Hayami and Ruttan (1985, Chapter 5). We cover both Northeast Asia (China, Japan, and the Republic of Korea) and Southeast Asia (Cambodia, Indonesia, Lao PDR, Malaysia, Myanmar, the Philippines, Thailand, and Viet Nam). The third section summarizes what changes industrialization (and income growth) has brought to agriculture and food demand structure in East Asia. In other words, the section describes the emerging challenges in detail. The fourth section examines technological, institutional, and market innovations that have recently emerged in response to the challenges mentioned above. The final section is the conclusion of this chapter.

Productivity Growth in Agriculture

In this section, we follow Chapter 5 of Hayami and Ruttan (1985) to overview long-term changes experienced by East Asian countries. Asian paths in Hayami and Ruttan (1985) are characterized by continuous efforts to increase land productivity by intensifying labor and other input use, especially through biochemical technological innovations, given that the initial condition was that the majority of farmers were small family-based cultivators including owner and tenant farmers. Arable land per person, land-labor ratio, and the average farm size were generally small under high population density (though there are some differences in the initial factor endowment between Northeast Asia and Southeast Asia). As described in the next section, rapid and successful industrialization in this region has absorbed a large share of the labor force, which has resulted in a shortage of labor in agriculture. The rapidly rising real wage makes it necessary to substitute for labor. Divergence from historical paths observed in the past has been recently confirmed in Japan and is expected to happen soon in many other countries in the region.

Factor Endowment

First, we characterize factor endowments in the region by looking at changes in arable land per person residing in rural areas and the average farm size.

TABLE 3.1 **Agricultural land in square kilometers per person (rural population)**

Country	1961	1980	2000	2016
Japan	0.002093	0.002178	0.001941	0.005798
Rep. of Korea	0.001147	0.001362	0.002060	0.001907
China	0.006225	0.005404	0.006447	0.008856
Cambodia	0.006677	0.004395	0.004821	0.004378
Indonesia	0.005027	0.003308	0.003845	0.004794
Lao PDR	0.007764	0.005625	0.004343	0.005809
Malaysia	0.005091	0.006205	0.007964	0.011231
Myanmar	0.006049	0.004093	0.003212	0.003692
Philippines	0.004089	0.003586	0.002768	0.002161
Thailand	0.005147	0.005459	0.004592	0.006625
Viet Nam	0.002200	0.001562	0.001446	0.001958

Source: Data from World Bank (2019).

Table 3.1 shows agricultural land per rural population from 1961 to 2016. In 1961, Japan and the Republic of Korea in Northeast Asia had relatively small amounts of arable land per rural population. Countries in Southeast Asia, except Viet Nam, had relatively large amounts of arable land per rural population. China had a relatively large area of arable land per rural population. Interestingly, we observe a clear contrast between the above two groups. While the second group (Southeast Asia) decreased arable land over time, the first group marginally increased arable land or maintained the same level over time. In Japan, arable land per rural population was almost the same in 1961 and 2000, but it increased after 2000. China and the Republic of Korea experienced an increase in arable land over time, largely due to fast and large-scale urbanization. In contrast, the countries in Southeast Asia, except Malaysia, monotonically decreased arable land per rural population in 1961 to 2000, although the initial size differed across countries (the largest size was observed in Cambodia, Lao PDR, and Myanmar, the smallest in Viet Nam). Except for the Philippines, however, all of them experienced a slight increase after 2000. In East Asia, there was a turning point in agricultural land per rural population around 2000, and since then the emerging trend has been toward increase.

Figure 3.1 shows changes in the average farm size in the region. Though the initial size differs between Northeast and Southeast Asia and many countries showed decreasing trends until 2000, Japan clearly showed an increase in the average size recently. Farm size expansion has also been taking place in China since the early 2000s through land rental markets (Huang and Ding

FIGURE 3.1 **Changes in farm size (hectares) in East Asia, 1960 to 2014**

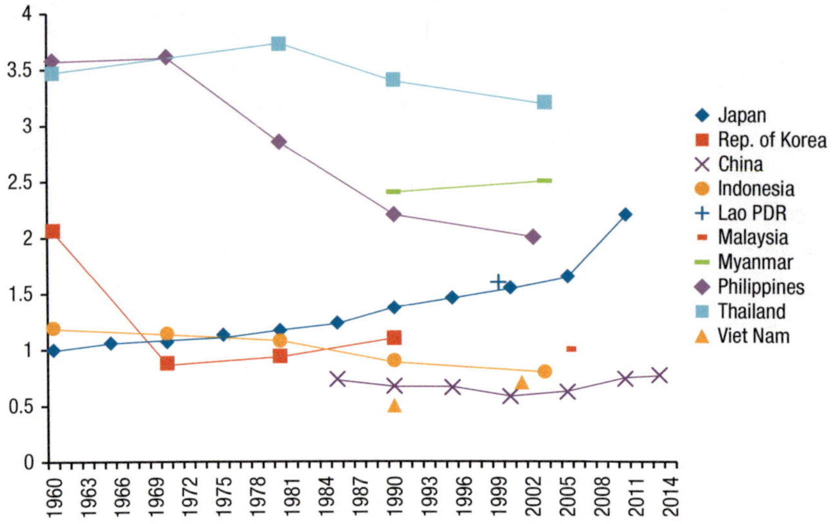

Source: FAO (2010); for China, Huang and Ding (2016); for Japan, Ministry of Agriculture, Forestry and Fisheries (various years), https://www.maff.go.jp/e/data/stat/index.html.

2016). The decreasing trend in the Philippines and Thailand was especially clear after 1980.

Obviously, whether farm size increases in Asia is a serious issue. The average operational farm size was already small in the 1970s, ranging from 1 hectare in Indonesia and Japan to 3 or 4 hectares in the Philippines and Thailand. It has declined in subsequent periods in all these countries, except in Japan and China, because of rapid population growth in rural areas. Farm size increased in Japan and China, but it is far below the average farm size in high-income economies in Europe and North America where farms of more than 100 hectares are common. If small farms continue to dominate and become a major constraint on large-scale mechanization in high-wage Asian economies, the continent could become a gigantic importer of food grains. This has already happened in high-income economies in Asia, such as Japan and the Republic of Korea (Otsuka 2013). In an attempt to address this issue, the Chinese government has started facilitating land consolidation through the rental market and introduced subsidies for mechanization. Therefore, the farm size in China is expected to increase much faster than generally assumed.[1]

1 It is noted that China food grains (rice and wheat) consumption has been declining and will fall more rapidly in the coming decades. The recent projections show that China may not need to

FIGURE 3.2 Output per worker versus output per hectare of agricultural land in East Asia, 1961 to 2014

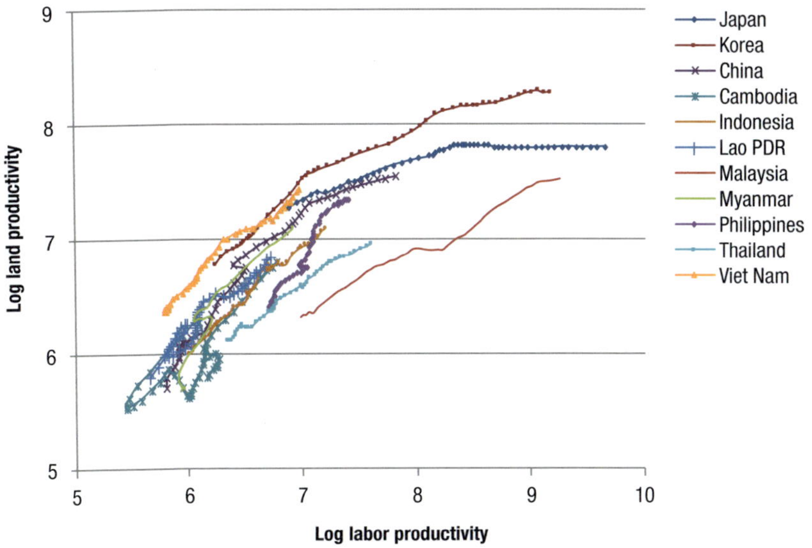

Source: Data from USDA (2019).

Note: Gross agricultural output in 2005 US dollars per the number of economically active adults in agriculture is on the *x*-axis; gross agricultural output in 2005 US dollars per total agricultural land in hectares of rainfed cropland equivalents is on the *y*-axis. They are both log transformed. The period covered for Japan is 1961–2000.

Productivity Growth

This subsection characterizes productivity growth, both partial and total, following Hayami and Ruttan (1985). Figure 3.2 shows the relationship between output per worker (horizontal axis) and output per agricultural land (vertical axis), both of which are log transformed. First, overall, countries in the region followed the so-called Asian path, described in Hayami and Ruttan (1985), in which land productivity increases faster than labor productivity in the early period followed by fairly rapid growth of labor productivity, even after the mid-1980s. The clearest case is the Republic of Korea. The two economies of the Republic of Korea and Japan started almost from an identical path (initial condition), whereas development stages were different across them. Second, divergence is clearly detected in Japan away from this path; labor productivity

import significant food grains, but imports of soybean, maize, edible oils, sugar, and dairy will rise in the future (Huang et al. 2017).

FIGURE 3.3 Land-labor ratio and machine use per worker, 1961 to 2014

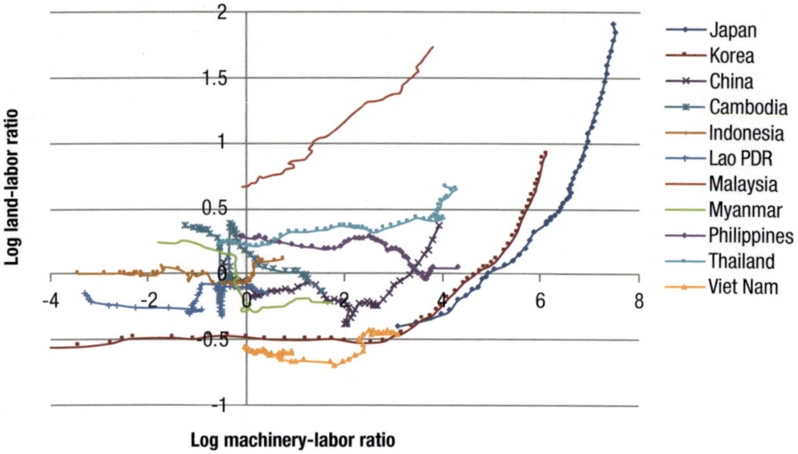

Source: Data from USDA (2019).

Note: Metric horsepower of machinery per the number of economically active adults in agriculture is on the *x*-axis; total agricultural land in hectares of rainfed cropland equivalents per the number of economically active adults in agriculture is on the *y*-axis. Both are log transformed. The period covered for Japan is 1961–2000.

keeps increasing while land productivity remains constant. That is, Japan is approaching the European path, described by Hayami and Ruttan (1985), which is closely related to an increase in farm size and mechanization. Third, Malaysia follows a unique path; labor productivity is clearly higher than other countries at the same level of land productivity. Although the analysis is descriptive and not rigorous, trends observed in Figure 3.2 seem largely consistent with the Hayami-Ruttan induced innovation hypothesis.

Figure 3.3 shows the relationship between land-labor ratio (vertical axis) and machine use per worker (horizontal axis), both of which are log transformed. First, Japan, the Republic of Korea, and Viet Nam seem to follow a similar path (starting from a nearly identical land-labor ratio), though machine use in Viet Nam clearly is still at the early stage relative to Japan and the Republic of Korea. Machine use per worker in Viet Nam is comparable to that of many countries in Southeast Asia. Second, Malaysia stands alone in starting from a very high land-labor ratio and achieving high machine use per worker (almost the same level of land-labor ratio as in Japan). Third, other countries, including China, fall between the above two cases. In China, an

FIGURE 3.4 Land productivity and fertilizer input per hectare, 1961 to 2014

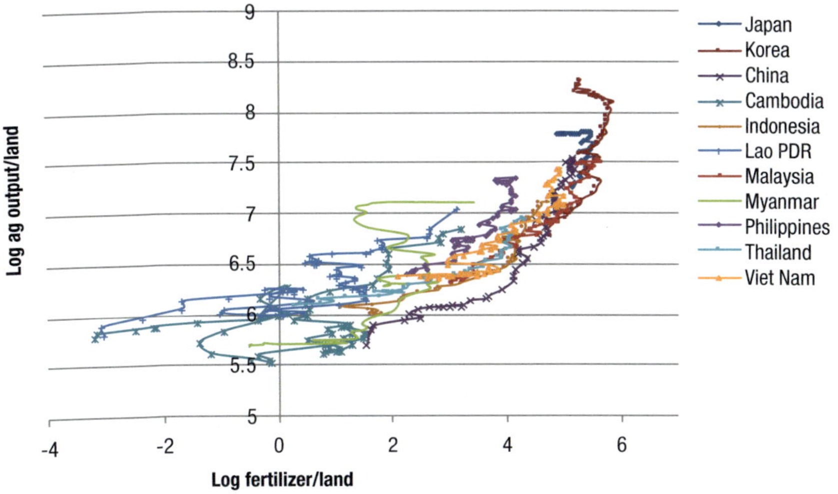

Source: Data from USDA (2019).

Note: Metric tons of fertilizer per total agricultural land in hectares of rainfed cropland equivalents is on the *x*-axis; gross agricultural output (2005 US dollars) per total agricultural land in hectares of rainfed cropland equivalents is on the *y*-axis. Both are log transformed. The period covered for Japan is 1961–2000.

increase in machine use per worker recently seems to happen with an increase in land-labor ratio. In the Philippines, Indonesia, and Thailand, machine inputs per worker increased while land per worker has been relatively constant. Nonmonotonicity observed in Cambodia and Myanmar seems to be related to data problems.

According to Figure 3.3, the land-labor ratio is significantly related to machine use per worker, which can be interpreted as a proxy for mechanical technology. Interesting observations are that land-labor ratio is higher in Malaysia compared with machine use per worker, and it is lower in Japan and the Republic of Korea compared with machine use per worker. These observations suggest that Japan and the Republic of Korea failed in expanding land area per worker because of the protective agricultural policies, which deter the exodus of the rural population. It seems that not only factor endowment, but also other factors affect the mechanization.

Figure 3.4 shows the relationship between fertilizer use per agricultural land area (horizontal axis) and output per agricultural land area (vertical axis), both of which are log transformed. According to this figure, land productivity is consistently and almost completely explained by fertilizer use per agricultural land area. Fertilizer use per agricultural land area can be interpreted as a proxy for biological-chemical technology. Strikingly similar paths across East Asian countries suggest that similar biological-chemical technologies, represented by fertilizer-using and high-yielding Green Revolution–type technologies of wheat and rice, have been developed and diffused in this region, beginning in Japan before World War II and transferred to tropical Asia primarily in the 1970s and 1980s (see Chapter 11). The elasticity of land productivity growth to fertilizer input is relatively low when fertilizer input level (per land area) is still low (below 4 log transformed), but land productivity growth appears to accelerate in response to fertilizer inputs once fertilizer use is intensified. This observation suggests that fertilizer-using and land-saving technology is developed in the process of economic development that accompanies growing scarcity of land. In other words, consistent with the Hayami-Ruttan induced innovation hypothesis, fertilizer-using technology is developed to save an increasingly scarce factor of production, that is, land. In this context, Japan, the Republic of Korea, China, Viet Nam, Indonesia, the Philippines, and Malaysia are all above the threshold. In Myanmar, Cambodia, and Lao PDR, land productivity increased without intensification of fertilizer use, which is somewhat puzzling.

Table 3.2 compares average annual total factor productivity (TFP) growth rate estimates in East Asia (Nin-Pratt 2018). First, China shows outstanding performance after 1980. In crop production, Viet Nam had the highest TFP growth after 1980, while Malaysia shows high TFP growth in livestock before and after 1980. Many countries in Southeast Asia had also experienced high TFP growth in 2001 to 2008. High TFP growth observed in Southeast Asia before 1980 by some studies (for example, Dias Avila and Evenson 2010) is not confirmed in Table 3.2, which implies that the Green Revolution was not mainly driven by TFP growth, but associated with intensive utilization of inputs such as fertilizers. See Pingali (2012) for more about the Green Revolution. Fan and Brzeska (2010, Sections 4 and 5) reviewed TFP in Northeast Asian countries, which points to the importance of institutional and policy reforms in determining TFP.

In sum, countries in the region followed the so-called Asian path, described in Hayami and Ruttan (1985), that is, land productivity increasing faster than labor productivity in the early period, but Japan seems to divert from the path by rapidly increasing labor productivity, approaching the European path, which is closely related to an increase in farm size

TABLE 3.2 **Average TFP growth rate for different periods**

Country	1961–1970	1971–1980	1981–1990	1991–2000	2001–2008	2009–2014
Japan	0.00090	0.00496	0.00680	0.00445	0.01168	0.00059
Rep. of Korea	−0.02684	−0.00528	0.00864	0.02373	0.00791	0.00709
China	0.00736	−0.00299	0.03125	0.03284	0.02181	0.01261
Cambodia	−0.02254	−0.03897	−0.00997	0.02180	0.01478	0.01118
Indonesia	0.00103	−0.00068	−0.00966	−0.00144	0.02579	0.00416
Lao PDR	−0.02425	−0.01032	0.00039	0.02613	−0.01723	0.02890
Malaysia	0.00492	0.00575	0.00692	0.01058	0.03615	−0.01082
Myanmar	−0.03009	0.00050	−0.00060	0.01963	0.05342	−0.04323
Philippines	−0.01365	0.00943	−0.00363	0.00159	0.02047	−0.00095
Thailand	−0.01397	0.00106	−0.02145	0.01070	0.01740	0.01304
Viet Nam	−0.01498	0.00882	0.01395	−0.00395	0.02036	0.01051

Source: Nin-Pratt (2018).
Note: TFP = total factor productivity.

and mechanization in recent years. Many countries in Southeast Asia had favorable land endowments in the initial conditions, but more recently both land-labor ratio and machine use per worker increase together in these countries following the cases of Japan and the Republic of Korea due to significant labor shortage common in the region.

Agricultural Growth, Urbanization, and Food Demand

In this section, we discuss details of recent economywide changes highly relevant to agriculture, including the development of nonagricultural sectors, intersectoral and interregional labor mobility, and changes in factor prices. The fastest and most successful industrialization experienced in this region inevitably accompanied rapid urbanization and created scarcity of labor in rural areas and therefore increased real rural wages. In rural areas, nonagricultural sectors also developed and absorbed labor from agriculture. As a result, the opportunity cost of farming has dramatically increased in the region. Income growth has contributed to large poverty reduction (though the head counts of the poor are large in this region due to its large population size) and transformation of food demands.

Industrialization, Urbanization, Employment, and Poverty Reduction

Declining shares of agriculture in gross domestic product (GDP) and employment occurred very rapidly in Northeast Asia. Rural labor was absorbed by nonagricultural sectors largely located in urban areas and newly created local cities and towns. Table 3.3 shows the share of the population that was urban from 1960 to 2018. Except in Japan where the share was already above 60 percent in 1960, the shares ranged between 8 and 30 percent in 1960. The Philippines had the highest share, 30 percent, and Lao PDR had the lowest share, 8 percent, in the region. By 2010, Japan reached above 90 percent, followed by the Republic of Korea, 82 percent. Since the share in the Republic of Korea was about 28 percent in 1960, the speed of that country's urbanization was very fast.

When the countries are grouped by percentage of urban population, there are three groups in the region. The first group consists of high-income countries, that is, Japan, the Republic of Korea, and Malaysia, in which more than 75 percent of the population was urban by 2018. The second group, China, Indonesia, the Philippines, and Thailand, had 45 to 60 percent of the population in urban areas in 2018. The last group consists of low-income countries, that is, Viet Nam, Myanmar, Lao PDR, and Cambodia, where the share of urban population was around or below 35 percent in 2018. Since the share of urban population is correlated with the share of employment in agriculture, the table likely indicates that these countries went through a declining share of population engaged in agriculture with the development of economies.

Rapid industrialization and urbanization in the region have increased stress on natural resources, particularly fresh water, which is essential to agricultural production. Table 3.4 shows freshwater withdrawal by sectors: agricultural, industrial, and municipal (domestic) use. The share of agriculture is relatively low in Japan (67 percent), the Republic of Korea (55 percent), China (66 percent), and Malaysia (22 percent). Indonesia and the Philippines remain at 81–82 percent, and the rest of Southeast Asian countries are at about 90 percent. Except in Malaysia, the share of agriculture is inversely correlated with industrial development and urbanization in the region. The issue of increasing water scarcity will be discussed in Chapter 20.

In Figure 3.5, we examine the relationship between shares of GDP and employment in agriculture in the period from 1961 to 2015. Both shares tended to simultaneously decrease over time, though there are variations across countries. The Republic of Korea showed the largest changes in both

TABLE 3.3 Share of urban population (%) in East Asia, 1960 to 2018

Country	1960	1970	1980	1990	2000	2010	2018
Japan	63.3	71.9	76.2	77.3	78.6	90.5	91.7
Rep. of Korea	27.7	40.7	56.7	73.8	79.6	81.9	81.5
China	19.5	17.4	19.4	26.4	36.2	50.0	59.2
Cambodia	10.3	16.0	9.9	15.5	18.6	19.8	23.4
Indonesia	14.6	17.1	22.1	30.6	42.0	49.9	55.3
Lao PDR	7.9	9.6	12.4	15.4	22.0	33.1	35.0
Malaysia	26.6	33.5	42.0	49.8	62.0	70.9	76.0
Myanmar	19.2	22.8	24.0	24.6	27.0	31.4	30.6
Philippines	30.3	33.0	37.5	48.6	48.0	45.3	46.9
Thailand	19.7	20.9	26.8	29.4	31.4	44.1	49.9
Viet Nam	14.7	18.3	19.2	20.3	24.4	30.4	35.9

Source: Data from World Bank (2019).

TABLE 3.4 Freshwater use by sectors in East Asia in the early 21st century

Country	Agricultural (100 m³/year)	Industrial (100 m³/year)	Municipal (100 m³/year)	Total (100 m³/year)
Japan	54.4 (2009)	11.6 (2009)	15.4 (2009)	81.5 (2009)
Rep. of Korea	16.0 (2003)	4.5 (2002)	6.9 (2005)	29.2 (2005)
China	385.2 (2015)	133.5 (2015)	79.4 (2015)	598.1 (2015)
Cambodia	2.0 (2006)	0.0 (2006)	0.1 (2006)	2.2 (2006)
Indonesia	92.8 (2000)	24.7 (2005)	14.0 (2005)	113.3 (2000)
Lao PDR	3.2 (2005)	0.2 (2003)	0.1 (2003)	3.5 (2005)
Malaysia	2.5 (2005)	4.8 (2005)	3.9 (2005)	11.2 (2005)
Myanmar	29.6 (2000)	0.5 (2005)	3.3 (2000)	33.2 (2000)
Philippines	67.1 (2009)	8.3 (2009)	6.2 (2009)	81.6 (2009)
Thailand	51.8 (2007)	2.8 (2007)	2.7 (2007)	57.3 (2007)
Viet Nam	77.8 (2005)	3.1 (2005)	1.2 (2005)	82.0 (2005)

Source: Data from FAO (2018).

GDP and employment shares. Starting from low levels of GDP and employment shares, the paths of Japan and Malaysia also overlap that of the Republic of Korea. A similar path is followed by the Philippines, Indonesia, China, and Viet Nam. In contrast, Thailand showed a unique pattern, in which the share of GDP decreased first while that of employment stayed relatively constant. This indicates that the share of nonfarm income of agricultural households increased. Subsequently the employment share started decreasing,

FIGURE 3.5 GDP share and employment share of agriculture (%) in East Asia, 1961 to 2015

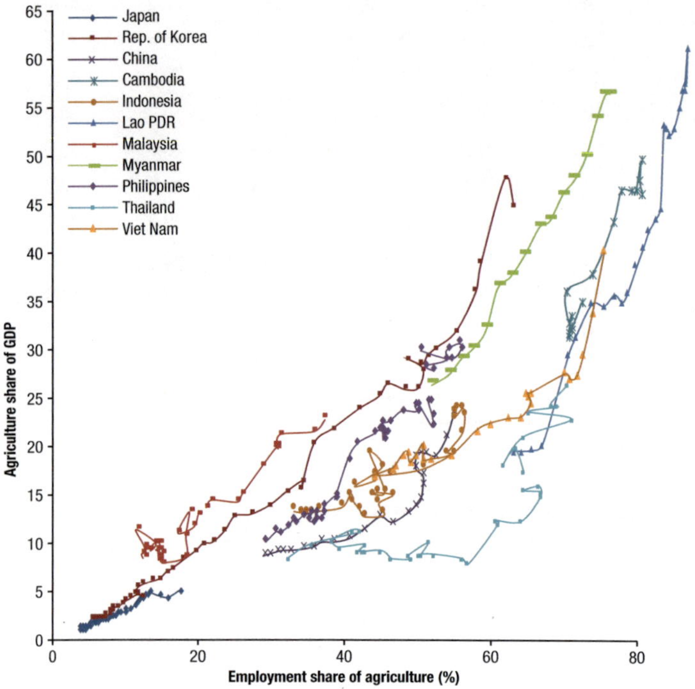

Source: Data from World Bank (2019).

Note: The database includes 2015 because the recent data include forestry and fishery together with agriculture for value-added percent of GDP. GDP = gross domestic product.

which brought Thailand back to a situation comparable to the Philippines and Indonesia.

Figures 3.6 and 3.7 show changes in employment and GDP shares, respectively, from 1961 to 2015, to confirm the decreasing trends in these shares. In employment share, Japan, the Republic of Korea, and Malaysia were the first to reach a very low level. The speed of the decrease in the Republic of Korea is remarkable, as its initial condition was comparable to that of many Southeast Asian countries. The second group, consisting of the Philippines, Thailand, Indonesia, China, and Viet Nam, still keep the share of employment in agriculture between 30 and 50 percent. In Indonesia, the Philippines, and Thailand the reduction in employment (share) accelerated after 1980. The employment share in the rest of the countries, namely, Cambodia, Lao PDR, and Mayanmar, seems to remain relatively high. In Figure 3.7, aside from Japan, we clearly observe two groups: the Republic of Korea, China, Indonesia,

FIGURE 3.6 Changes in employment share of agriculture (%) in East Asia, 1961 to 2015

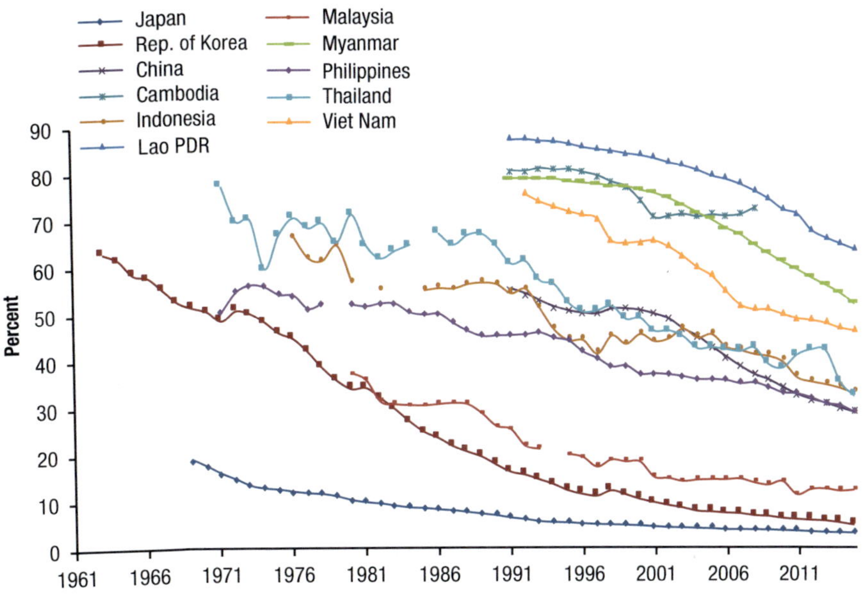

Source: Data from World Bank (2019).

the Philippines, Thailand, Malaysia, and Viet Nam as the first group, and Cambodia, Lao PDR, and Mayanmar as the second group. The first group started from 25–45 percent in 1961 and reached 10–15 percent in 2015. Note that the path of the Republic of Korea recently converged with the path of Japan, which started from an exceptionally low share of GDP in 1961. The second group still maintains about 30 percent of the GDP share in agriculture.

Along with successful industrialization, this region has experienced the above-mentioned transformation of the agricultural sector in the economy and unprecedented income growth and poverty reduction. Figures 3.8 and 3.9 show changes in the poverty gap and the head count ratio of poverty in rural areas, respectively. Both graphs clearly show a reduction of poverty level as well as its convergence. China, Indonesia, and Viet Nam showed remarkable reductions in their poverty gaps. Although international comparability in the head counts remains questionable, as they use national-level thresholds, overall all countries in which the statistics are available experienced a relatively fast reduction of rural poverty. Therefore, there seems to be no question that a decrease in the share of GDP and employment in agriculture was associated with a reduction of rural poverty.

FIGURE 3.7 Changes in GDP share of agriculture (%) in East Asia, 1961 to 2015

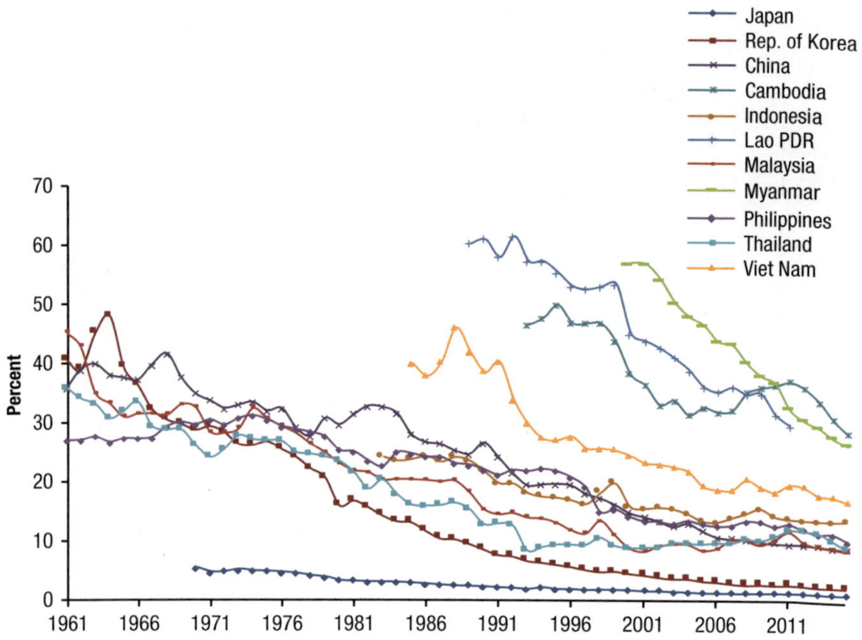

Source: Data from World Bank (2019).

Note: The database includes 2015 because the recent data include forestry and fishery together with agriculture for value-added percent of GDP.

Production Side—Labor Shortage

Successful industrialization not only led to the large-scale transformation of the agricultural sector through a reduction in its contribution to GDP and employment (which also resulted in a substantial reduction in the poverty level), but also created a serious labor shortage and thus a rapid increase in real wages. As reported in Wiggins and Keats (2014), rural wages in the region have been increasing fast recently.

In this context, whether or not the Lewis turning point has been passed is an interesting empirical issue. It is known that Japan and the Republic of Korea passed the Lewis turning point around 1960 and 1975, respectively (Minami 1968; Bai 1982). More recently, Thailand, China, and Indonesia appeared to have passed the turning point (Zhang, Yang, and Wang 2011). Viet Nam seems to be approaching the turning point relatively quickly.

Figure 3.10 shows changes in rural real wages in six provinces in China: Heben, Hubei, Jilin, Jiangxi, Shandong, and Sichuan (Wang, Yamauchi, and

FIGURE 3.8 Poverty gap (%) at $1.90 a day (2011 PPP) in East Asia, 1980 to 2017

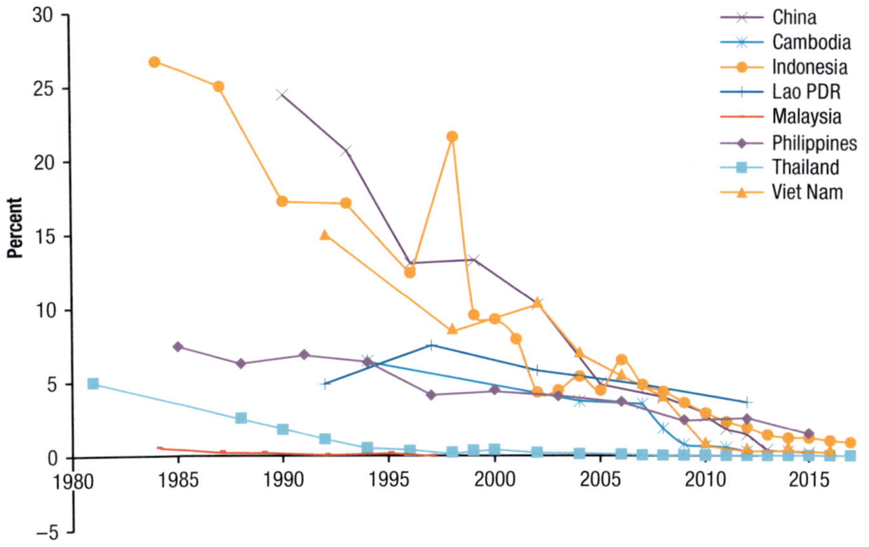

Source: Data from World Bank (2019).

Note: PPP = purchasing power parity.

FIGURE 3.9 Rural poverty head count (% in rural population) in East Asia, 1995 to 2015

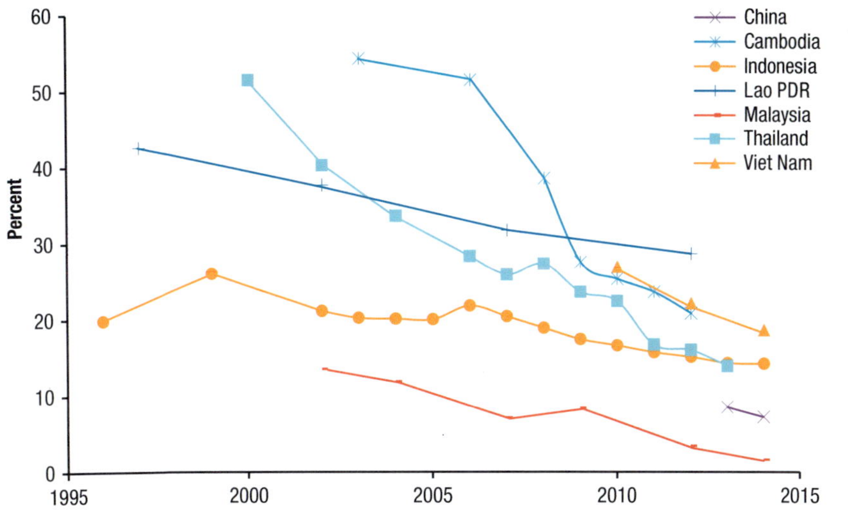

Source: Data from World Bank (2019).

FIGURE 3.10 Real agricultural wages in selected provinces in China: 1984 to 2014

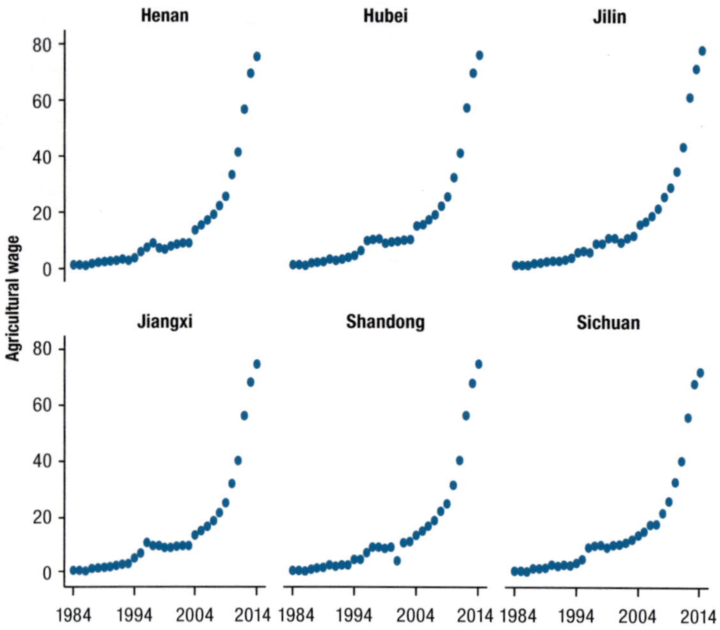

Source: Wang, Yamauchi, and Huang (2016).

Huang 2016). Strikingly, an acceleration of rising real wages after 2000 seems to be common in most provinces. How to tackle labor shortage (and rising real wages) is an emerging issue after 2000, which is consistent with the turning point reported for China (Zhang, Yang, and Wang 2011).

Outflow of labor from agriculture also involves intergenerational issues. In some countries, such as the Republic of Korea, China, and Japan, farmers cannot find successors to take over their farming, which exacerbates the labor shortage problem but promotes farm size expansion in the long run.

Demand Side—Food Demand and Nutrition Transition

This subsection discusses nutrition transitions (Popkin 2002a, 2002b, 2006) in the context of East Asia (Kelly 2016). Diets dominated by starchy, low-variety, low-fat, and high-fiber foods are being replaced by consumption of processed foods higher in fats, sugars, and salt and, in some cases, accompanied by rises in diet-related noncommunicable disease (NCD). The issue of nutrition transition is discussed in Chapter 10.

As described above, most of East Asia has experienced high income growth through rapid industrialization in recent decades, which is reflected in increased consumer purchasing power. This is a main driving force for the nutrition transition from complex carbohydrate consumption to more fat, salt, sugar, and processed-food consumption. This is apparent in higher-income countries, though we are beginning to observe such a change in lower-income countries (Drenowski and Popkin 1997; Popkin, Adair, Ng 2012). Another important factor relevant in the context of East Asia is, as emphasized previously, the rapid pace of urbanization. The opportunity cost of home cooking tends to be high in urban settings, especially when family size is much smaller than rural counterparts and many women work outside the home (Pingali 2007). Urbanization and nutrition will be discussed further in Chapters 9 and 10.

The existing studies seem to identify some common elements of the manifestation of the nutrition transition in East Asia. First, the most common change in the region is a dramatic increase in the proportion of dietary energy derived from oils and fats, mainly oilseed based, replacing complex carbohydrate sources (Kelly 2016). For example, in China, annual vegetable oil consumption per capita has increased from 1 kilogram in 1963 to 11 kilograms in 2003 (Kearney 2010). Second, the predominant rice-based diets are also transformed, with more wheat consumption. There seems to be a pattern: with rising incomes, people tend to increase rice consumption first when income increases, and then the dependence on rice decreases afterward. For example, in China, rice consumption increased rapidly in the 1960s and 1970s from 172 grams per capita per day in 1963 to 260 in 1983 before falling again to 213 in 2003. In the same period, wheat consumption increased from 71 grams per person per day to 167 (Kearney 2010). Third, consumption of meat has also been increasing as part of the nutrition transition. The largest change here has been seen in pork and poultry consumption. In China, for example, from 1990 to 2014 pork consumption more than doubled, from 15 to 31 kilograms per capita per year, and poultry consumption rose from 2 kilograms to 11.5 kilograms per capita per year during the same period.

Finally, we observed a dramatic increase in consumption of processed, packaged, and convenience foods, which overlaps the above-mentioned changes. In China, consumption of processed foods increased from 20 kilograms per capita annually in 1999 to over 80 kilograms per capita in 2012, and in Thailand it rose from 60 to 110 kilograms during the same period. In Thailand and the Philippines, soft drink consumption was particularly high and approached high-income-country level, and in Malaysia and Indonesia processed high-fat foods were most prevalent. In higher-income Asian countries such as Japan, consumption of processed foods has stopped increasing since the 1990s, but it is growing fastest in the upper-middle-income countries such as Thailand and China (Baker and Friel 2014).

While malnutrition has been receding in Asia overall, owing to rapid and successful economic growth, rates of overweight and obesity are rising rapidly, particularly in Southeast Asia (Chopra, Galbraith, and Darnton-Hill 2002; Popkin 2006). For example, from 1980 to 2008, prevalence of adult overweight and obesity combined increased from 12 to 39 percent in Thailand, from 15 to nearly 50 percent in Malaysia, and from 5 to around 25 percent in Indonesia (Stevens et al. 2012). In Southeast Asia, the burden of diet-related disease, including diabetes, cardiovascular disease, and diet-related cancers, is also quickly rising.

By focusing on the population of children under age 5, Figures 3.11 and 3.12 contrast prevalence of stunting to prevalence of overweight. Figure 3.11 shows that the prevalence of stunting has been decreasing over time in all countries, which reflects a substantial improvement of nutrition intake among young children as well as pregnant women in the region.

In contrast, the prevalence of overweight has had an upward trend in recent years. This is consistent with recent trends observed in adult obesity in Thailand, Malaysia, and Indonesia (reported above). Thailand and Indonesia reached an alarming level of child overweight in recent years. The prevalence has also increased in the Philippines and Viet Nam. Strikingly, Japan is an interesting case that shows the lowest prevalence of child overweight (which is consistent with the pattern of processed foods consumption). The gap between stunting and overweight is particularly large in Thailand and the Republic of Korea; the prevalence of overweight is much higher than that of stunting.

NCD Risk Factor Collaboration (2016) used 1,698 population-based data sources, with more than 19.2 million adult participants (9.9 million men and 9.3 million women) in 186 out of 200 countries for which estimates were made, to see changes in age-standardized mean body mass index (BMI). They found that global age-standardized mean BMI (in kilograms per square meter) increased from 21.7 (with the 95 percent credible interval being 21.3–22.1) in 1975 to 24.2 (24.0–24.4) in 2014 in men, and it increased from 22.1 (21.7–22.5) in 1975 to 24.4 (24.2–24.6) in 2014 in women. In the dynamics of adult BMI, East Asia (including Southeast Asia) is very interesting. Strikingly, China's obese populations in both men and women increased dramatically in

FIGURE 3.11 Prevalence of stunting among children under age 5 (%) in East Asia, 1972 to 2016

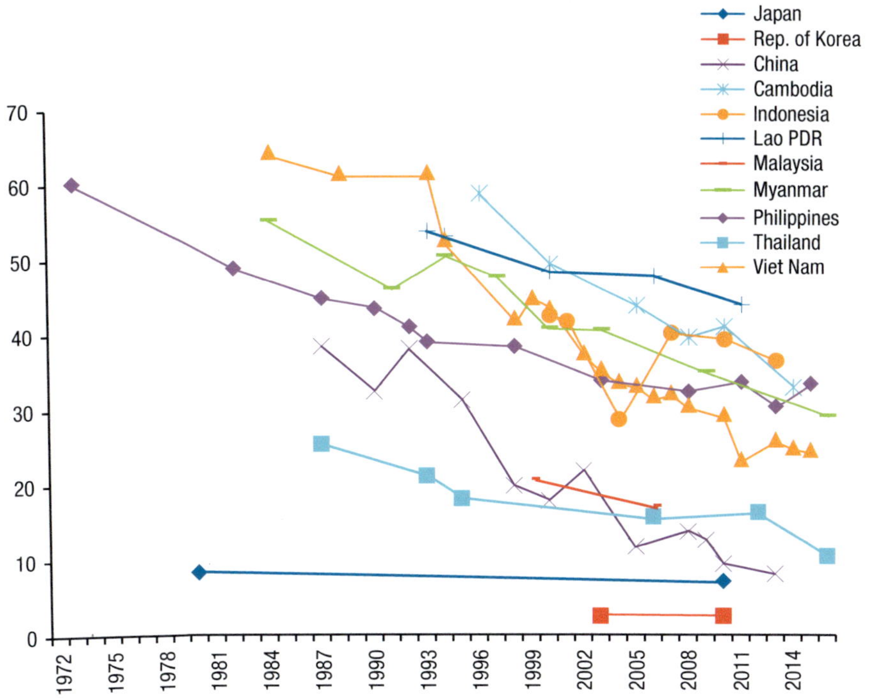

Source: Data from World Bank (2019).

the same period, from 0.7 to 43.2 million in men and 1.7 to 46.4 million in women, elevating China to the top of the obese population ranking in 2014 (NCD Risk Factor Collaboration 2016, Figure 7). In severe obesity, change in the global ranking of China is even more dramatic: from number 60 to number 2 in men and from 41 to 2 in women. Interestingly, some countries from this region, such as China, Indonesia, Viet Nam, the Philippines, Thailand, and Japan, remained among the top in the ranking of underweight population in 1975 and 2014.

There are only two countries in the world that have no hunger and no problem of rising overweight; they are Japan and the Republic of Korea (Laborde, Mamun, and Vos 2019). A variety of reasons unique to these countries explain this phenomenon. Further research on the agrifood system is needed to factor out the causes for achieving no hunger and no overweight.

FIGURE 3.12 Prevalence of overweight among children under age 5 (%) in East Asia, 1972 to 2016

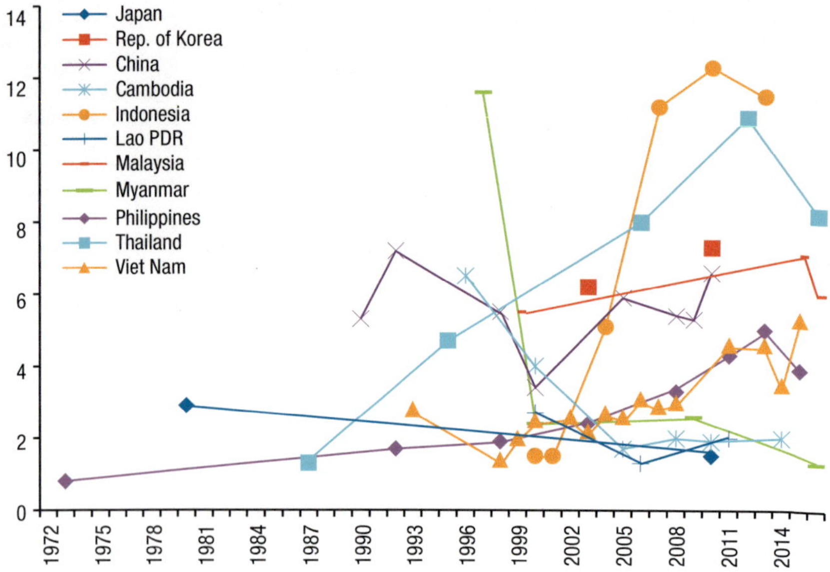

Source: Data from World Bank (2019).

Note: In China, 1995 and 2002 were omitted because they showed too high prevalence, outlying from the trend.

Innovations and Agricultural Transformation

Given rising real wages in rural areas, the conventional view that small farms are more productive is currently challenged. Labor-intensive production methods are not an optimal choice anymore in Northeast Asia. However, transition to more capital-intensive methods using machines requires a realization of scale economies, and a critical constraint is the relatively small farm size in the region (Figure 3.1). We discuss technological and institutional innovations under such circumstances in this region. They include activation of land rental markets, land consolidation, and mechanization in different forms (including the emergence of machine service providers), which together introduce more labor-saving production methods. The emergence of the advantage of large-scale farming is discussed here.

Technological and Institutional Innovations

Although small farms in Asia achieved higher land productivity than their larger counterparts in the past, they are facing more and more challenges as structural transformation has been occurring in most developing countries in Asia. In these countries, the rapid growth of the nonfarm sector has created more lucrative employment opportunities, which has resulted in a higher real wage in both farm and nonfarm sectors. Coupled with technological advances in manufacturing industries, the price ratio of labor to machine use is increasing, which renders the substitution of labor by machine profitable. The rapidly growing machine rental markets in China and Viet Nam represent a response to such trends in recent decades (see, for example, Liu et al. 2020; Zhang et al. 2011). As farming systems gradually change from labor-intensive to capital-intensive systems, the advantage of small farms relying on family labor is declining, while large farms' advantages are enhanced by the use of farm machinery.

Recently, we also observed an evolution of machine service providers as an institutional response to reduce the user cost of machine use. For example, in China (Yang et al. 2013), small farms can contract with the provider to use machine services, rather than renting or purchasing machines, in order to save on labor costs. By contracting with a large number of small farms, the provider can enjoy scale economies, provided that the transaction cost of machine service provision is sufficiently low. If it is high, small farms cannot save labor as much as large farms.

It is still also possible that farmland consolidation can be facilitated by market transactions as well as institutional arrangements. Historically, consolidation was largely achieved through market transactions in many of the Organisation for Economic Co-operation and Development (OECD) countries. Previously in China land transactions were seriously constrained by the insecurity in farmers' individual land rights, but recently farmland rental arrangements have been facilitated through both more secure land contract rights and online services provided by county and township governments in every province in China. In some areas, land banks are also established to facilitate borrowing and lending (renting in and renting out) of farmland. In this way, farm size expansion is taking place in China (Huang and Ding 2016).

In sum, when the real wage rate is low, the optimum farm size is small and the inverse correlation between farm size and productivity tends to emerge. When the wage rate increases, mechanization will take place to save labor. Since machines and land are complementary and machines are indivisible to some extent, the optimum farm size tends to increase. If farm size adjustments take place smoothly through market transactions and institutional arrangements, efficient large farms emerge. In practice, however, land markets may not function smoothly, so a positive relationship can arise between farm size and productivity in high-wage economies.

We examine the validity of our arguments by drawing on recent empirical evidence available from East Asian countries (Otsuka, Liu, Yamauchi 2016). They include case studies in Indonesia (Yamauchi 2016), Viet Nam (Liu et al. 2020), and China (Wang, Yamauchi, Huang 2016; Wang et al. 2016, 2018). These recent studies explicitly look at the impact of rising real wages on land and machine service transactions and the dynamically changing disadvantage of small farms (or the emerging advantage of large farms). This is a major departure from the earlier literature that assessed factors that were considered to explain the inverse relationship between farm size and productivity observed in cross-sectional data.[2]

Indonesia is an interesting case for the purpose of our study due to the coexistence of small farms in Java and relatively large farms in the outer islands. Yamauchi (2016) examines the dynamically changing patterns of land use, capital investments, and real wages by using farm household panel data from seven provinces collected in 2007 and 2010. His regression analyses show that an increase in real agricultural wages induced the substitution of labor by machines, either through machine rentals or machine service providers, notably among relatively large farmers. The total amount of payments for hired-in machines or services or both has increased significantly in response to rising real agricultural wages, especially among relatively large farms. They tend to increase the scale of operation by renting in more land when real agricultural wages increase. Furthermore, the effect of an increase in farm size on crop productivity per hectare becomes positive among relatively large farms. Thus, the Indonesian case study clearly supports our hypothesis that the efficiency of large farms increases with rising real agricultural wage rates.

The case of Viet Nam is similar to that of Indonesia. Liu et al. (2020) use data from 1992 and 1998 Vietnam Living Standards Surveys (VLSS) and from four rounds of Vietnam Household Living Standards Survey (VHLSS) data between 2002 and 2008. This makes it possible to investigate machine use and the farm size–productivity relationship from the 1990s to the 2000s. Descriptive analysis suggests that tractor rental has become more common: in 2008 more than 60 percent of farms rented in machines, whereas less than 20 percent did so in 1992. Rapid increase in tractor use would be associated

2 The existing empirical tests on the inverse relationship are grouped into several types. The first type investigates whether the advantage of small farms can be attributed to imperfect factor markets, particularly the labor market. The emerging reality that real wages are rising rapidly in East Asia is related to this. The second type inquires whether unobserved factors such as soil quality can explain the inverse relationship if such factors are unevenly distributed between small and large farms (Benjamin 1995). The third is concerned with the effect of measurement errors of farm size on the inverse correlation between farm size and productivity (Lamb 2003).

with an increase in the relative advantage of large farms. Consistent with such an expectation, large farmers are more likely to use agricultural machines, pointing to the scale economies arising from machine use. Interestingly, machine use was not responsive to the real agricultural wage in 1992 or 1998 but became significantly responsive in 2006 through 2008, suggesting the emergence of a clear substitution relationship between machine and labor in recent years when the wage rate has become high. Such differences may be attributed to the development of machine rental markets over time. The estimation results of the paddy yield regression demonstrate that the inverse relationship between farm size and land productivity has significantly lessened: when farm size doubled, the expected paddy yield is estimated to have decreased by 15.6 percent in the 1990s but only by 6.1 percent in the late 2000s. Thus, the inverse relationship is lessened but not reversed. Yet, another interesting finding is that the inverse relationship may be reversed in areas where farm size is larger and the wage rate is higher. This indicates that a positive relationship has emerged between farm size and productivity in advanced areas where the wage rate is higher. This result is also consistent with the observation of Estudillo and Otsuka (2016) that the average farm size among their sample households increased from 1.0 hectare in 1996 to 1.4 hectares in 2009 in the Mekong Delta region.

In China, the economy has been rapidly growing over the last three and one-half decades, and the wage rate has been rising sharply, particularly since 2003 (Zhang, Yang, and Wang 2011). Correspondingly, the use of riding tractors and combine harvesters is increasing (Yang et al. 2013). Yet the average farm size remained at 0.6 hectare in 2010, increasing only by 0.05 hectares per year since 2000, even though land rental markets have become increasingly active (Kimura et al. 2011; Huang, Wang, and Qui 2012). More recently, Huang and Ding (2016) pointed out that farm size in China is somewhat underestimated.

Using farm household panel data from China, collected in six provinces in 2000 and 2008, Wang et al. (2016) analyze the dynamic changes in land rental transactions, machine investments, and the use of machine services. Their study looks at the effects of nonagricultural and agricultural wage growth, changes in the migration rate, and the proportion of nonagricultural income, all of which are estimated at the village level, on changes in self-cultivated farm size, rented-in land areas, machine services used, and machine investments. The regression results show that increases in nonagricultural wages, the proportion of nonagricultural income, and the migration rate lead to the expansion of the operational farm size. Consistently, the demand for machine services also increased along with increases in agricultural wages and migration rates. This effect is larger for relatively large farms.

As in the case of the Indonesia study, the regression results of crop income equations support the hypothesis of complementarities between rented-in land and machine services demanded, especially among relatively large landholders in China. In other words, the possibility of renting in land and the availability of machine service providers led to expansion of farm size to take advantage of scale economies.

Wang, Yamauchi, and Huang (2016) used province-level crop-wise panel data in China to investigate the substitutability between agricultural labor and machine service. They support the conjecture that machines are increasingly used to substitute for labor under the circumstances where the real wage has been rapidly increasing. Figure 3.13 shows the relationship between machine use per hectare (expenses) and the relative wage of labor to machine cost in wheat and (japonica) rice production. In the production of both crops, an increase in agricultural wage relative to machine cost led to intensive machine use (measured by expenses on mechanical operations).

Wang et al. (2020) used the same household panel used in Wang et al. (2016) to investigate the role of land fragmentation in mechanization. It is clearly shown that fragmented farmlands discourage mechanization. All conditions being equal, the consolidation of fragmented lands could improve production efficiency by lowering transaction costs in mechanization.

Two recent movements in China deserve special attention. First, in northeast China, machine service providers directly attempt to consolidate farmland by renting in land from smallholders to realize scale economies. However, such land rental contracts with small farmers are mainly short-term and often subject to annual renewals, because small farmers feel insecure about renting out their land under long-term contracts, due to the lack of private land ownership. They also expect the land rent to increase over time, and thus they hesitate to sign long-term contracts. Second, recently, farmland rental arrangements have been facilitated through the Internet by local governments in nearly every province. In all likelihood, these new institutional arrangements are induced by the increasing optimum size of farm operation in China (Huang 2017).

In contrast to the conventional view of the small-farm advantage, the evidence shown above supports the dynamic shift of the relative advantage from small to large farms in Asia. That is, the inefficiency of small farms increases with rising real wages, whereas large farms increase their productivity by utilizing large-scale machines efficiently, saving increasingly costly labor, and

FIGURE 3.13 Expenses for mechanical operations in China, 1984 to 2014

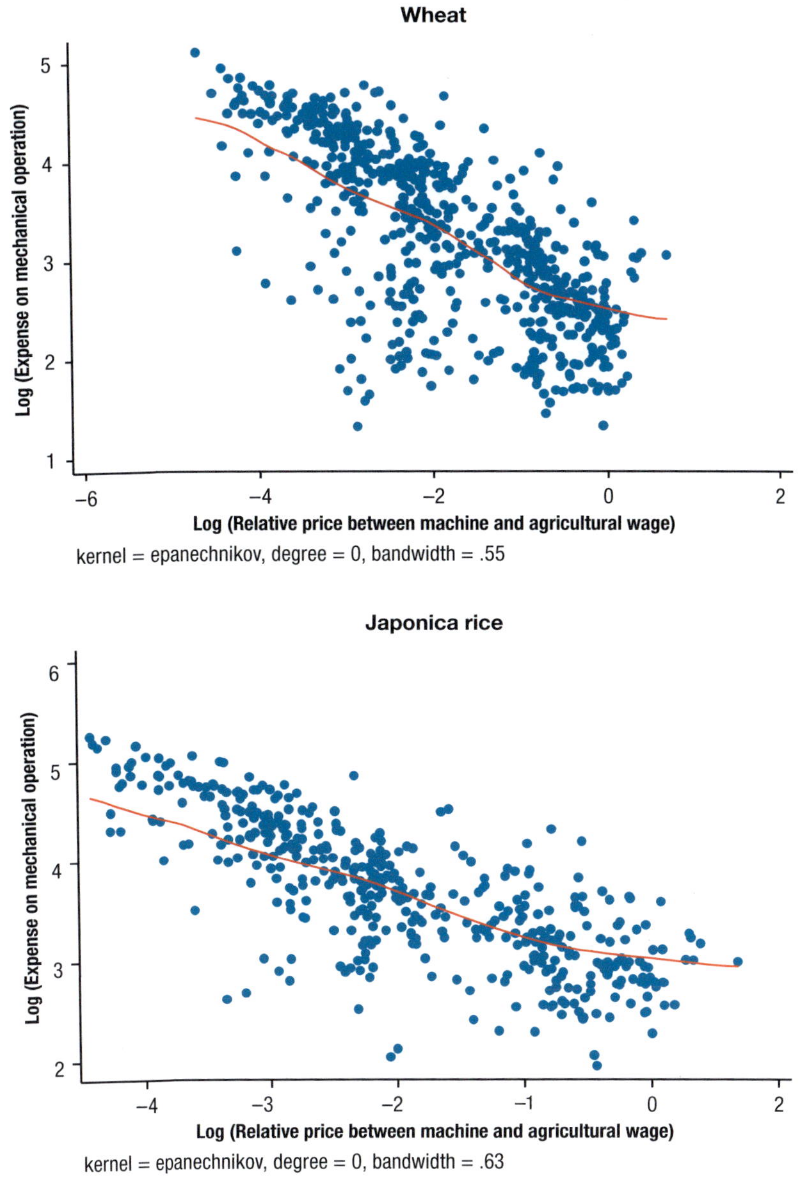

Source: Wang, Yamauchi, and Huang (2016).

thereby realizing scale economies. How far and how rapidly this trend continues will significantly affect the efficiency of farming in East Asia in coming decades. Smooth transfer of farmland from those who exit from farming to those who expand farming is critically important, and this would be a solution for the labor shortage problem, including the intergenerational transfer of farming, discussed in the previous section.

These analyses suggest that the relationship between the land-labor ratio and the use of machinery is not simple, because the development of land and land rental markets and that of machinery rental and service markets affect such a relationship. This is likely to explain why the relationship between land-labor ratio and machinery use per unit of land portrayed in Figure 3.3 varies from country to country even in East Asia.

Market Innovations—Diversification and Modernization of Food Value Chains

Successful economic growth and rapid urbanization have also changed the nature of food demands, which is intimately related to the nutrition transition. Consumers, especially in urban areas, demand high-value and high-quality foods (Reardon, Timmer, and Berdegue 2008; see Chapter 12 for further discussions on value chains). The rising demand for safe, high-value, and differentiated agricultural products has created large opportunities for farmers to participate in value chains to potentially improve productivity (for example, by contract farming, which improves access to technology, information, and capital) and increase incomes.

Modern value chains typically coexist with traditional value chains. For example, supermarkets extend networks to directly procure from producers near urban areas, whereas most rural communities depend on traditional value chains (Reardon et al. 2003; Wang et al. 2009). In East Asia, the duality of modern and traditional value chains is clear partly because of the very fast structural transformation the region went through and the initial dominance of relatively small farmers who, for various reasons, including high transaction costs, might have failed to take advantage of the modern version of value chains. The speed with which modern value chains penetrate in rural communities in the region remains a key empirical question.

In the same line, the emergence of contract farming is significant in East Asia (see Chapter 11 for further discussions on contract farming). Its impact on farm income tends to be positive, according to the literature reviews by Bellemare and Bloem (2018) and Ton et al. (2018). While the participation in contract farming makes it harder to select qualified empirical studies, Cahyadi and Waibel (2016) and Trifković (2014, 2016) show positive income effects from the participation in oil palm contract farming in Indonesia and in catfish contract farming in Viet Nam, respectively. Bellemare and Bloem, however, argue that there is considerable heterogeneity across contexts that limits external validity, and there is an identification issue of treatment effects due to the lack of appropriate instruments that limit internal validity. Furthermore, Ton et al. (2018) point out the survivor biases leading to the publication of mostly successful cases of contract farming. On the other hand, the impact of contract farming on productivity seems to be positive, especially if farmers are assisted to access better technologies and management practices (Otsuka, Nakano, and Takahashi 2016). Saenger et al. (2013) and Saenger, Torero, and Qaim (2014) point to the importance of actual contract designs in improving product quality from field experiments conducted in the dairy sector in Viet Nam.

There may be potential scale bias favoring large farmers in modern value chains as well as contract farming (Barrett et al. 2012). For example, the contractor may prefer to reduce the number of farmers in the contract to minimize transaction costs. To meet quality standards, farmers may be required to make investments prior to participation. If this is the case, large farmers have an advantage over smallholders. Given that East Asia holds a large number of small farmers whose financial and technical capacities are limited, such a setting may lead to low participation in modern value chains or contract farming (Simmons, Winters, and Patrick 2005; Stringer, Sang, and Croppenstedt 2009). This is a clear contrast to the dominance of supermarkets observed in Latin America (Reardon, Timmer, and Berdegue 2008). In the future, however, large farmers may dominate in the production of high-value products in East Asia. It must also be pointed out that Ton et al. (2018) found through their meta-analysis that smallholders can also benefit from contractual arrangement, even though the poorest farmers are rarely included.

Conclusions

This chapter has reviewed historical paths of agricultural development and described economywide changes that had undeniable impacts on agricultural transformation in East Asia. Though the region was heterogeneous in initial endowment, labor was relatively abundant (land is relatively scarce) and land productivity was augmented through intensifications such as labor-intensive production methods as well as applications of biochemical technologies. However, more recently, the region has encountered a dramatic change in relative factor prices due to successful industrialization by which labor was absorbed in nonagricultural sectors. In many countries, labor is not abundant anymore, and thus labor-saving methods had to be introduced, for example, replacing increasingly expensive labor by machines. Under such circumstances, the inverse productivity-size relationship is becoming less likely to hold and is being replaced by the positive relationship. Some countries are showing a divergence from the Asian path to the European path, both initially described by Hayami and Ruttan (1985).

Consistently, a reversal of the declining trend of average farm size has also been confirmed in some countries. For example, active land rental markets enable some farmers to increase operational size, whereas other farmers exit from agriculture by renting out their lands. Emerging labor shortage in the region starts creating the advantage of large-scale farming, in contrast to the conventional small-farm advantage.

High income growth and fast urbanization introduced some other fundamental transformations, such as nutrition transitions on the demand side and modernization of value chains and emergence of contract farming on the supply side, that respond to diverse and new food demands especially driven by rapid urbanization and economic growth. However, East Asia remains characterized by the duality of modern and traditional systems due to the sustained dominance of a large number of smallholders who may not meet the conditions required to enter the modern value chains. In order to sustain agricultural production in this region, large-scale institutional and technological innovations beyond the purview of Hayami and Ruttan (1985) are called for.

References

Bai, M. 1982. "The Turning Point in the Korean Economy." *Developing Economies* 20: 117–140.

Baker, P., and S. Friel. 2014. "Processed Foods and the Nutrition Transition: Evidence from Asia." *Obesity Reviews* 15: 564–577.

Barrett, C. B., M. E. Bachke, M. F. Bellemare, H. C. Michelson, S. Narayanan, and T. F. Walker. 2012. "Smallholder Participation in Contract Farming: Comparative Evidence from Five Countries." *World Development* 40: 715–730.

Bellemare, M. F., and J. R. Bloem. 2018. "Does Contract Farming Improve Welfare? A Review." *World Development* 112: 259–271.

Benjamin, D. 1995. "Can Observed Land Quality Explain the Inverse Productivity Relationship?" *Journal of Development Economics* 46: 51–84.

Berry, R. A., and W. R. Cline. 1979. *Agrarian Structure and Productivity in Developing Countries.* Baltimore: Johns Hopkins University Press.

Cahyadi, E. R., and H. Waibel. 2016. "Contract Farming and Vulnerability to Poverty among Oil Palm Smallholders in Indonesia." *Journal of Development Studies* 52: 681–695.

Chopra, M., S. Galbraith, and I. Darnton-Hill. 2002. "A Global Response to a Global Problem: The Epidemic of Overnutrition." *Bulletin of the World Health Organization* 80: 952–958.

Dias Avila, H. F., and R. E. Evenson. 2010. "Total Factor Productivity Growth in Agriculture: The Role of Technological Capital." In *Handbook of Agricultural Economics*, edited by R. E. Evenson and P. Pingali, 3769–3822. Amsterdam: Elsevier.

Drenowski, A., and B. M. Popkin. 1997. "The Nutrition Transition: New Trends in the Global Diet." *Nutrition Reviews* 55: 31–43.

Eastwood, R., M. Lipton, and A. Newell. 2010. "Farm Size." In *Handbook of Agricultural Economics*, edited by R. E. Evenson and P. Pingali, 3323–3397. Amsterdam: Elsevier.

Estudillo, J. P., and K. Otsuka. 2016. *Moving out of Poverty: An Inquiry into Inclusive Growth in Asia*. London: Routledge.

Fan, S., and J. Brzeska. 2010. "Production, Productivity, and Public Investment in East Asian Agriculture." In *Handbook of Agricultural Economics*, edited by R. E. Evenson and P. Pingali, 3401–3434. Amsterdam: Elsevier.

FAO (Food and Agriculture Organization of the United Nations). 2018. AQUASTAT. Accessed April 18, 2018. www.fao.org/nr/water/aquastat/data/query/index.html?lang=en.

———. 2010. World Census of Agriculture (various volumes up to 2000). www.fao.org/world-census-agriculture/en/.

Feder, G. 1985. "The Relation between Farm Size and Productivity." *Journal of Development Economics* 18: 297–313.

Hayami, Y., and K. Otsuka. 1993. *The Economics of Contract Choice: An Agrarian Perspective.* Oxford: Clarendon Press.

Hayami, Y., and V. Ruttan. 1985. *Agricultural Development: An International Perspective.* Baltimore: Johns Hopkins University Press.

Huang, J. 2017. "Land Transaction Service Centers in China: An Institutional Innovation to Facilitate Land Consolidation." Unpublished, China Center for Agricultural Policy, Peking University, Beijing.

Huang, J., and J. Ding. 2016. "Institutional Innovation and Policy Support to Facilitate Small-Scale Farming Transformation in China." *Agricultural Economics* 47: 227–237.

Huang, J., X. Wang, and H. Qui. 2012. *Small-Scale Farmers in China in the Face of Modernization and Globalization*. London: International Institute for Environment and Development.

Huang, J., W. Wei, Q. Cui, and W. Xie. 2017. "The Prospects for China's Food Security and Imports: Will China Starve the World via Imports?" *Journal of Integrative Agriculture* 16: 2933–2944.

Kearney, J. 2010. "Food Consumption Trends and Drivers." *Philosophical Transactions of the Royal Society B: Biological Sciences* 365: 2793–2807.

Kelly, M. 2016. "The Nutrition Transition in Developing Asia: Dietary Change, Drivers and Health." In *Eating, Drinking: Surviving*, edited by P. Jackson, W. Spiess, and F. Sultana, Chapter 9. Switzerland: Springer.

Kimura, S., K. Otsuka, T. Sonobe, and S. Rozelle. 2011. "Efficiency of Land Allocation through Tenancy Markets: Evidence from China." *Economic Development and Cultural Change* 59: 485–510.

Laborde, D., A. Mamun, and R. Vos. 2019. *Progress towards Ending Hunger and Malnutrition: A Cross-Country Cluster Analysis*. IFPRI technical paper prepared for the Food and Agriculture Organization of the United Nations (FAO).

Lamb, R. L. 2003. "Inverse Productivity: Land Quality, Labor Markets, and Measurement Error." *Journal of Development Economics* 71: 71–95.

Liu, Y., C. Barrett, T. Pham, and W. Violette. 2020. "The Intertemporal Evolution of Agriculture and Labor over a Rapid Structural Transformation: Lessons from Vietnam." *Food Policy* 94 (July): 101913.

Minami, R. 1968. "The Turning Point in the Japanese Economy." *Quarterly Journal of Economics* 82: 380–402.

Ministry of Agriculture, Forestry and Fisheries, Japan. (Various years). Statistics of Arable Land and Crop Production (for years in 1960–2005).

———. 2017. Survey of Agricultural Structural Changes (for years in 2010–2017). https://www.maff.go.jp/j/tokei/kouhyou/noukou/.

NCD Risk Factor Collaboration (NCD-RisC). 2016. "Trends in Adult Body-Mass Index in 200 Countries from 1975 to 2014: A Pooled Analysis of 1698 Population-Based Measurement Studies with 19.2 Million Participants." *Lancet* 387: 1377–1396.

Nin-Pratt, A. 2018. *TFP Estimates: A Global Update*. Washington, DC: IFPRI.

Otsuka, K. 2007. "Efficiency and Equity Effects of Land Markets." In *Handbook of Agricultural Economics*. Vol. 3, edited by P. Pingli and R. E. Evenson, 2671–2703. Amsterdam: Elsevier.

———. 2013. "Food Insecurity, Income Inequality, and the Changing Comparative Advantage in World Agriculture." *Agricultural Economics* 44: 7–18.

Otsuka, K., H. Chuma, and Y. Hayami. 1992. "Land and Labor Contracts in Agrarian Economies: Theories and Facts." *Journal of Economic Literature* 30: 1965–2018.

Otsuka, K., Y. Liu, and F. Yamauchi. 2016. "Growing Advantage of Large Farms in Asia and Its Implications for Global Food Security." *Global Food Security* 11: 5–10.

Otsuka, K., Y. Nakano, and K. Takahashi. 2016. "Contract Farming in Developed and Developing Countries." *Annual Review of Resource Economics* 8: 353–376.

Pingali, P. 2007. "Westernization of Asian Diets and the Transformation of Food Systems: Implications for Research and Policy." *Food Policy* 32: 281–298.

———. 2012. "Green Revolutions: Impacts, Limits and the Path Ahead." *Proceedings of the National Academy of Sciences* 109: 12302–12308.

Popkin, B. M. 2002a. "What Is Unique about the Experience in Lower and Middle Income Less-Industrialised Countries Compared with the Very-High Income Industrialised Countries: The Shift in Stages of the Nutrition Transition in the Developing World Differs from Past Experiences." *Public Health Nutrition* 5: 205–214.

———. 2002b. "An Overview of the Nutrition Transition and Its Health Implications: The Bellagio Meeting." *Public Health Nutrition* 5: 93–103.

———. 2006. "Global Nutrition Dynamics: The World Is Shifting Rapidly toward a Diet Linked with Noncommunicable Diseases." *American Journal of Clinical Nutrition* 84: 289–298.

Popkin, B. M., L. S. Adair, and S. W. Ng. 2012. "Global Nutrition Transition and the Pandemic of Obesity in Developing Countries." *Nutrition Reviews* 70: 3–21.

Reardon, T., and J. Berdegue. 2008. "The Rapid Rise of Supermarkets in Latin America: Challenges and Opportunities for Development." *Development Policy Review* 20: 371–388.

Reardon, T., C. P. Timmer, C. Barrett, and J. Berdegue. 2003. "The Rise of Supermarkets in Africa, Asia, and Latin America." *American Journal of Agricultural Economics* 85: 1140–1146.

Reardon, T., C. P. Timmer, and J. Berdegue. 2008. "The Rapid Rise of Supermarkets in Developing Countries: Induced Organizational, Institutional and Technological Change in Agri-Food Systems." In *The Transformation of Agri-Food Systems: Globalization, Supply Chains and Smallholder Farms*, edited by E. B. McCullough, P. L. Pingali, and K. G. Stamoulis, 47–65. Rome: FAO.

Saenger, C., M. Qaim, M. Torero, and A. Viceisza. 2013. "Contract Farming and Smallholder Incentives to Produce High Quality: Experimental Evidence from the Vietnamese Dairy Sector." *Agricultural Economics* 44: 297–308.

Saenger, C., M. Torero, and M. Qaim. 2014. "Impact of Third-Party Contract Enforcement in Agricultural Markets: A Field Experiment in Vietnam." *American Journal of Agricultural Economics* 96: 1220–1238.

Simmons, P., P. Winters, and I. Patrick. 2005. "An Analysis of Contract Farming in East Java, Bali, and Lombok, Indonesia." *Agricultural Economics* 33: 513–525.

Stevens, G. A., G. M. Singh, Y. Lu et al. 2012. "National, Regional, and Global Trends in Adult Overweight and Obesity Prevalences." *Population Health Metrics* 10: 22.

Stringer, R., N. Sang, and A. Croppenstedt. 2009. "Producers, Processors, and Procurement Decisions: The Case of Vegetable Supply Chains in China." *World Development* 37: 1773–1780.

Ton, G., W. Vellema, S. Desiere, S. Weitushat, and M. D'Haese. 2018. "Contract Farming for Improving Smallholder Incomes: What Can We Learn from Effectiveness Studies?" *World Development* 104: 46–64.

Trifković, N. 2014. "Governance Strategies and Welfare Effects: Vertical Integration and Contracts in the Catfish Sector in Vietnam." *Journal of Development Studies* 50: 949–961.

———. 2016. "Vertical Coordination and Farm Performance: Evidence from the Catfish Sector in Vietnam." *Agricultural Economics* 47: 547–557.

USDA (US Department of Agriculture). 2019. Economic Research Service. Accessed October 7, 2019. www.ers.usda.gov.

Wang, H., X. Dong, S. Roselle, J. Huang, and T. Reardon. 2009. "Producing and Procuring Horticultural Crops with Chinese Characteristics: The Case of Northern China." *World Development* 37: 1791–1801.

Wang, X., F. Yamauchi, and J. Huang. 2016. "Rising Wages, Mechanization and the Substitution between Capital and Labor: Evidence from Small Scale Farm System in China." *Agricultural Economics* 47: 309–317.

Wang, X., F. Yamauchi, J. Huang, and S. Rozelle. 2020. "What Constrains Mechanization in Chinese Agriculture? Roles of Farm Size and Fragmentations." *China Economic Review* 62 (August): 101221.

Wang, X., F. Yamauchi, K. Otsuka, and J. Huang. 2016. "Wage Growth, Landholding and Mechanization in Chinese Agriculture." *World Development* 86: 30–45.

Wiggins, S., and S. Keats. 2014. *Rural Wages in Asia*. ODI Report, Overseas Development Institute.

World Bank. 2019. World Development Indicators database. Accessed September 10, 2019. http://databank.worldbank.org.

Yamauchi, F. 2016. "Rising Real Wages, Mechanization and Growing Advantage of Large Farms: Evidence from Indonesia." *Food Policy* 58: 62–69.

Yang, J., Z. Huang, X. Zhang, and T. Reardon. 2013. "The Rapid Rise of Cross-Regional Agricultural Mechanization Services in China." *American Journal of Agricultural Economics* 95: 1245–1251.

Zhang, X., J. Yang, and S. Wang. 2011. "China Has Reached the Lewis Turning Point." *China Economic Review* 22: 542–554.

Chapter 4

AGRICULTURAL DEVELOPMENT AND MODERNIZATION IN SOUTH ASIA

Hiroyuki Takeshima, Anjani Kumar, Akhter U. Ahmed, and P. K. Joshi

Agriculture, natural resources, and the nutrition landscape in South Asia (SA) are unique. As in other Asian regions, SA's agriculture is largely smallholder based and continues to employ a large share of the workforce, compared with Latin America. Compared with Africa south of the Sahara (SSA), agricultural modernization and intensification in SA have progressed more substantially. However, compared with East Asia (that is, Northeast and Southeast Asia), the speed of agricultural transformation has remained slower in terms of the exit of labor from farming, despite comparable intensification levels. The region is still home to almost 300 million of the poor, a majority of whom live in rural areas, engaging in agriculture. SA is also one of the regions with the scarcest natural resource endowments per capita, including water resources.[1] Finally, the multiple burden of malnutrition persists, as the region remains one of the largest contributors to global undernutrition, while simultaneously overnutrition continues to emerge (Meenaksh 2016). Therefore, in SA, understanding the evolution and implications of agricultural development is important particularly in the context of agricultural transformation, natural resource management, poverty, and food/nutrition security improvement.

Research in the agricultural economics field has shed important light on various aspects of agricultural development patterns in the last few decades in the region, ranging from productivity growth, changes in resource endowments, intensification/modernization, urbanization and nutritional transition, and their drivers as well as inhibitors. The contribution of this chapter is to provide, through the review of productivity growth patterns and seminal papers in the literature, narratives of how these aspects may be interconnected in terms of their collective implications for the challenges and opportunities of future agricultural development in SA.

1 Renewable internal freshwater per capita in SA is greater than that of only the Middle East and North Africa and is a fraction of the resources of East Asia and the Pacific, SSA, or Latin America (World Bank 2018).

The second section of this chapter highlights that labor productivity (which is closely related to returns to labor) seems inherently low in SA and that agriculture seems to have contributed to income growth through land productivity growth. But the region may be approaching a point where further growth must come from greater labor productivity growth rather than land productivity growth. We explore this issue by examining the changes in labor and land productivity growth, as well as the land-labor ratio. The third section highlights that, in SA, because of low-income status and limited resources for food imports, domestic production remains an important source of food for consumers. Demand for diversification and improved nutritional intake might have been and may continue to be met by diversification in agricultural production in the region, while growing exportable high-value agricultural products may induce some specialization.

The fourth section then discusses key technological and institutional innovations, emerging rural employment as well as challenges, and growing inefficiency of small farms, with potential implications for the future labor productivity increase in SA. The section highlights innovations in technologies and institutions that have emerged, and related policies. Growing evidence of the inefficiency of small farms is reviewed. Emerging rural employment growth patterns, where the agricultural sector continues to be dominant but the nonfarm sector, such as the garment sector, is gradually expanding as a source of employment, are also discussed for their implications for labor productivity. It also touches on some inequality issues that are associated.

The last section summarizes the main messages and discusses some forward-looking emerging issues.

A Review of Productivity Growth in Agriculture

This section reviews the key patterns of agricultural productivity growth in SA, important aspects of the growth patterns of total factor productivity (TFP), and changes in natural resource endowments. The section aims to show that the past labor productivity growth in SA has been associated largely with the growth in land productivity, although TFP growth, particularly since the 1980s, has contributed to it as well. The section also shows that natural resource endowments in SA (water, soil, land) have been gradually eroded, potentially limiting substantial future land productivity growth in the region.

Factor Endowments and Partial Productivity Growth

This section extends analyses by Hayami and Ruttan (1985, Chapter 5) of factor endowments and partial productivity growth, focusing on key SA countries. Table 4.1 summarizes the changes in partial productivity of land and labor, as well as land-labor ratios, in seven SA countries from the 1960s to 2010s. This table corresponds to Table 5-1 in Hayami and Ruttan (1985). Figure 4.1 then plots the labor and land productivity for the seven countries presented in Table 4.1, corresponding to Figure 5-1 in Hayami and Ruttan (1985). To highlight the differences between SA and East Asia, which were analyzed in Chapter 3, Figure 4.1 also plots trajectories for East Asian countries (countries shown without labels). Table 4.1 and Figure 4.1 can be interpreted in the following way: many countries from the 1960s to the 2010s exhibited slopes that were steeper than the uni-*A*/*L* lines (dotted lines in Figure 4.1 along which agricultural output per agricultural land ([*A*]) and agricultural output per worker ([*L*]) change at the same proportions), indicating that increases in labor productivity were smaller than those in land productivity. Afghanistan and Bhutan even had absolute declines in labor productivity during these periods. While Bangladesh has been an exception, labor productivity growth has been only marginally higher than land productivity growth. Even in Pakistan, where land per labor has been historically high, the recent trend has exhibited greater land productivity growth than labor productivity growth. These patterns suggest that increases in land productivity also contributed to the increase in labor productivity between the 1960s and 2010s. Few of these countries have experienced growth paths that were more toward greater labor-productivity improvement than land-productivity improvement, which many higher-income countries have experienced. However, shifts from paths steeper than the uni-*A*/*L* line to paths that are parallel to it, which Bangladesh has experienced, suggest that those countries have managed to move out of the conditions prevalent in earlier periods, when high growth rates of the labor force in agriculture largely eroded the gains in labor productivity growth arising from land productivity growth.

Comparisons between SA countries and East Asian countries suggest that the trajectories of many SA countries are similar to those of East Asia, with SA lagging behind rather than following a different path. The differences between SA and East Asia might have been due to factors that have led to slower growth in human capital and land quality in SA, such as education and irrigation, as described in more detail in the later section.

TABLE 4.1 Estimates of land and labor productivity and endowments in agriculture, South Asian countries, 1960s–2010s

	1960s	1970s	1980s	1990s	2000s	2010s
Country			**Labor productivity (Y/L)**			
Afghanistan	756	782	818	755	624	626
Bangladesh	348	339	348	377	527	714
Bhutan	655	614	631	722	563	460
India	453	492	562	669	752	928
Nepal	335	345	392	460	468	515
Pakistan	868	972	1132	1521	1554	1554
Sri Lanka	561	584	650	617	644	758
			Land productivity (Y/A)			
Afghanistan	137	158	144	159	183	222
Bangladesh	698	681	724	751	949	1199
Bhutan	361	362	396	452	598	748
India	289	335	417	539	632	817
Nepal	646	600	583	716	901	1154
Pakistan	148	193	260	380	477	553
Sri Lanka	525	594	716	718	788	878
			Land per labor (A/L)			
Afghanistan	5.54	4.95	5.73	4.84	3.44	2.82
Bangladesh	0.50	0.50	0.48	0.50	0.55	0.60
Bhutan	1.82	1.70	1.60	1.60	0.96	0.61
India	1.57	1.47	1.35	1.24	1.19	1.14
Nepal	0.52	0.58	0.67	0.64	0.52	0.45
Pakistan	5.86	5.05	4.37	4.01	3.27	2.81
Sri Lanka	1.07	0.99	0.91	0.86	0.82	0.86

Source: Authors' calculations based on data from USDA (2020).

Table 4.2 summarizes the growth rates of labor productivity and land productivity, and the land-labor ratio of the 1960s, 1980s, and 2010s. This table corresponds to Table 5-2 of Hayami and Ruttan (1985). There is some intraregion diversity in productivity growth and the changes in factor endowments. Land productivity (*Y/L*) has exhibited some convergence, in which countries with low initial levels (India and Pakistan) have experienced faster growth compared with countries with higher initial levels (Bangladesh, Sri Lanka). On the other hand, signs of convergence are much weaker for labor productivity. Although Pakistan, with the highest labor productivity, experienced

FIGURE 4.1 **Labor and land productivity growth in agriculture in South Asia in comparison to East Asia**

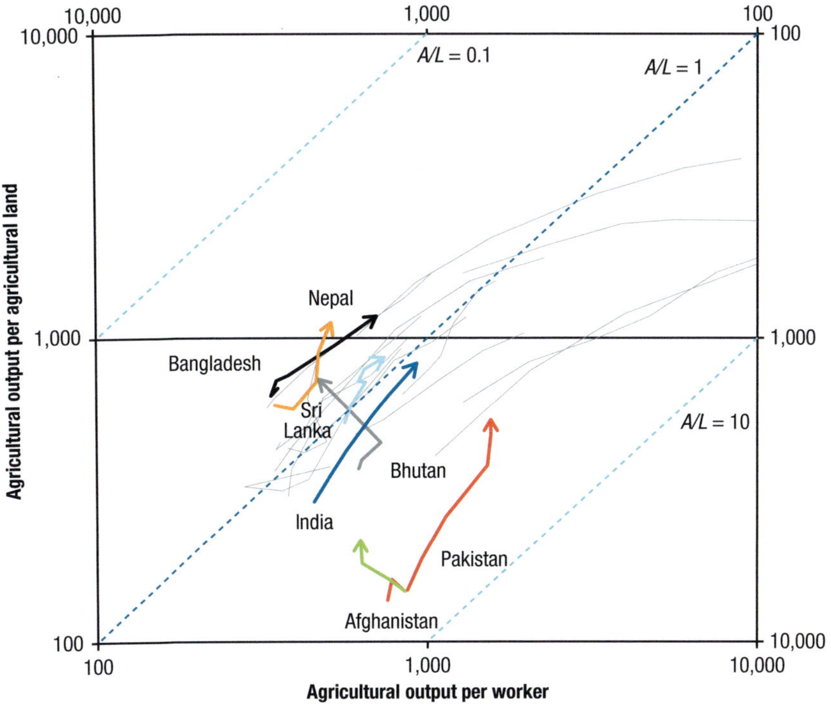

Source: Authors' calculations based on Table 4.1 for South Asian countries and Chapter 3 of this book for East Asian countries.
Note: Gray lines indicate the trajectories of countries in East Asia.

lower growth rates than other countries in the 2000s and 2010s, its growth rate had been high up to the 1990s. In contrast, Nepal and Sri Lanka, whose labor productivity was lower in the 1960s, have seen slower growth rates up to the 2010s.

While land-labor ratios have remained relatively constant in Bangladesh, Nepal, and Sri Lanka from the 1960s to the 2010s, those in India and Pakistan have declined more sharply. Because of this pattern, labor productivity growth in India and Pakistan has been much lower than the land productivity growth during this period (1.4 and 1.2 percent as opposed to 2.1 and 2.7 percent). The absorption of the growing rural labor force in nonfarm sectors in India and Pakistan has been much slower than in the other three countries in SA. Sri Lanka has experienced particularly slow labor and land productivity growth rates, with labor and land productivity being caught up

TABLE 4.2 Growth rate (%/year) of agricultural output per worker (Y/L), agricultural output per hectare (Y/A), and land-labor ratio (A/L)

Years		Afghanistan	Bangladesh	Bhutan	India	Nepal	Pakistan	Sri Lanka
1960s–2010s	Y/L	−0.4	1.4	−0.7	1.4	0.9	1.2	0.6
1960s–2010s	Y/A	1.0	1.1	1.5	2.1	1.2	2.7	1.0
1960s–2010s	A/L	−1.3	0.4	−2.2	−0.6	−0.3	−1.5	−0.4
1960s–1980s	Y/L	0.4	0.0	−0.2	1.1	0.8	1.3	0.7
1960s–1980s	Y/A	0.2	0.2	0.5	1.9	−0.5	2.9	1.6
1960s–1980s	A/L	0.2	−0.2	−0.6	−0.8	1.3	−1.5	−0.8
1980s–2010s	Y/L	−0.9	2.4	−1.0	1.7	0.9	1.1	0.5
1980s–2010s	Y/A	1.5	1.7	2.1	2.3	2.3	2.5	0.7
1980s–2010s	A/L	−2.3	0.7	−3.2	−0.6	−1.3	−1.5	−0.2

Source: Author's compilations based on Table 4.1.

by Bangladesh (labor productivity) and India or Nepal (land productivity) by the 2010s. The slow productivity growth in Sri Lanka in the last three decades has been partly due to the slow output growth rates of nonrice commodities that have accounted for about two-thirds of total outputs, including vegetables and fruits (plantains, etc.), tea, coconuts, and rubber (these are discussed more at the beginning of the third section).

Figure 4.2 and Figure 4.3 illustrate the changing relationship between the land-labor ratio and tractor horsepower (HP) per worker, and between land productivity and fertilizer input per hectare, corresponding to Figures 5-3 and 5-4 in Hayami and Ruttan (1985). These figures illustrate the agricultural growth process that involves the substitution of man-made inputs for labor and land. Fertilizer and tractor HP are used as proxies for the factors that substitute for land and labor, respectively.

Most SA countries have seen substantial increases in tractor HP per worker, without increases in land-labor ratios in the past five decades (Figure 4.2). In some countries, like India and Pakistan, the land-labor ratios have declined alongside the increase in tractor HP per worker. These patterns are, however, still consistent with Hayami and Ruttan (1985). In Hayami and Ruttan (1985, Figure 5-3), at a lower level of tractor HP per male worker, land-labor ratios are generally flat. Once it exceeds a threshold, the increase in tractor HP per male worker becomes increasingly more positively associated with the land-labor ratio. In the past five decades, while tractor HP per

FIGURE 4.2 Comparison of land-labor ratio and tractor horsepower per worker, with arrows connecting data points for decade averages from the 1960s to 2010s

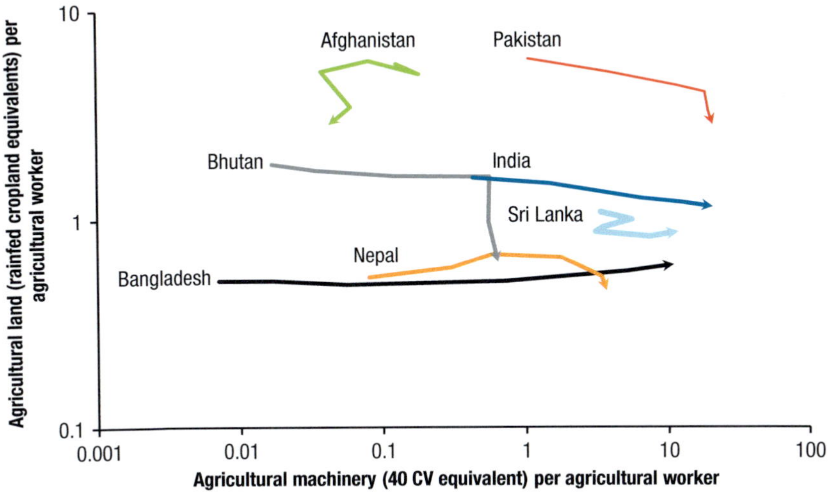

Source: Authors' calculations and FAO (2020a).

Note: CV (chevaux vapeur), used in FAO (2020a), is a unit of metric horsepower (HP), approximately equivalent to mechanical horsepower. CV is assumed to be 10 and 40 on average for two-wheel tractors (2wt) and four-wheel tractors (4wt), respectively, unless otherwise stated. Importantly, however, average tractor CV has generally increased over time, and ignoring that trend could underestimate the overall trend of tractor CV growth. For Bangladesh and India, interpolation used a constant growth rate (not linear). For Nepal, FAO figures seem to closely match with other sources and therefore FAOSTAT figures were used for 4wt. For 2wt, interpolation used a constant growth rate (not linear); there were also around 1,000 minitillers in 2010, but they were excluded. For Pakistan, several studies indicate CV to be slightly higher than 40. Also, information about 2wt in Pakistan has been very difficult to obtain. Therefore, the tractor CVs used for Pakistan are more conservative estimates. For Sri Lanka, information about 4wt has been relatively limited. Therefore, the growth rate between 1980 and 2002 for which information is available was applied to extrapolate the growth from 2002.

worker has increased considerably in SA countries, its level still has been too low to translate into significant increases in land-labor ratios. However, by the 2010s, many countries had reached the threshold of tractor HP per worker. Therefore, these countries are expected to see growth of the land-labor ratio as tractor HP per worker further increases in the future. As shown in the fourth section, tractor usage for land preparation has reached almost 100 percent in most SA countries (except Nepal), primarily through tractor rentals. Therefore, further increases in tractor HP per worker may accompany substantial reductions in the absolute number of agricultural workers.

Figure 4.3 illustrates the process where constraints in agricultural production due to land resource endowments have been mitigated through the increased use of fertilizer (and other yield-enhancing inputs) per unit of land. Figure 4.3 suggests that many SA countries have seen clearer trends in this aspect of land-constraint mitigation as indicated by strongly positive

FIGURE 4.3 Comparison of land productivity and fertilizer input per hectare, with arrows connecting data points for decade averages from the 1960s to 2010s

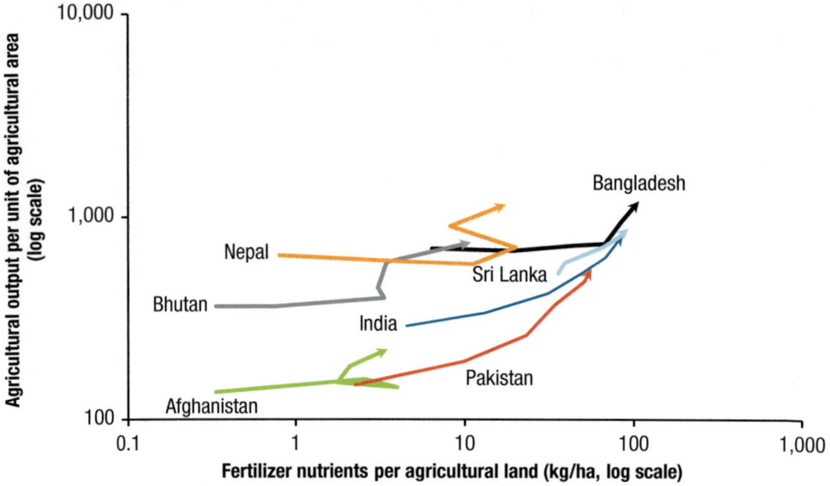

Source: Data from FAO (2020a).

Note: Fertilizer nutrients = the sum of the quantities of three major nutrients: nitrogen, phosphate, and potassium.

slopes. These patterns seem to have become reinforced since the 1980s. It is interesting to note that curves are convex rather than concave, which indicates that decreasing marginal productivity of fertilizer is more than offset by fertilizer-using technological changes. This is similar to the pattern observed in East Asia.

Altogether, the investigations of factor endowments, land and labor productivity growths, and the trends in the land- and labor-substituting input uses have revealed that SA still follows the agricultural growth trajectory hypothesized for land-scarce countries by Hayami and Ruttan (1985). Pre-1980s had been characterized by modest land productivity growth and fairly small labor productivity growth due to the growth of the number of agricultural workers relative to land. However, since the 1980s, the land-labor ratio has become relatively constant and the continuous growth of land productivity seems to have largely translated into the increase in labor productivity. The land productivity growth might have been primarily led by land-saving inputs like fertilizer and other yield-enhancing inputs. The use of power and machinery has continued growing, and SA seems to have reached the point where further increases in power and machinery use per worker may be associated with a significant increase in the land-labor ratio.

Total Factor Productivity Growth

Key evidence on agricultural TFP growth in SA countries is mostly provided in the literature as part of TFP analyses for a larger set of countries. For SA as a whole, TFP growth had been lower between the 1960s and mid-1980s and turned upward since the mid-1980s, although it has been generally lower than in most other regions throughout the periods since the 1960s, largely due to the slow efficiency growth (despite comparable technical improvements), including the post-1980s period (Nin-Pratt and Yu 2010). TFP growth between 1965 and 1996 was also lower than in East Asia, due to slower efficiency growth (Suhariyanto and Thirtle 2001). Both studies, however, also suggest that since the 1980s, technical changes in SA have been faster than in other parts of Asia, while slower growths of TFP in SA during the last several decades have been largely due to slower technological progress.

Coelli and Rao (2005) suggest substantial variations in TFP growth rates within SA between 1980 and 2000. During this period, Bangladesh and Pakistan observed the highest TFP growth, India and Nepal saw medium TFP growth, and Sri Lanka saw the lowest growth.

More detailed country-level investigations reveal spatial, regional, and temporal variations. In Indian Punjab, TFP growth would have been considerably higher, about 3 percent per year during the Green Revolution era (1966–1975) compared with 1 to 2 percent per year in the post–Green Revolution era (1986–1994) (Murgai 2001). TFP growth in the Indo-Gangetic Plain (IGP) region of India was 1.2 percent per year from 1981 to 1996, with more impressive TFP growth during the 1980s than in the 1990s (Kumar, Kumar, and Mittal 2004). TFP growth seems to have been particularly high in early periods. This is potentially because the shift from traditional varieties to modern high-yielding varieties (HYVs), compared with subsequent shifts from the original Green Revolution varieties to the second- and third-generation varieties, might have been much more drastic in terms of the technological progress. Lower TFP growth in the post–Green Revolution era might have been partly due to land quality, including depletion/pollution of groundwater resources, a buildup of soil salinity and waterlogging, nutrient mining, micronutrient deficiencies, deteriorating water quality, the formation of subsoil compaction, and increased pest buildup (Kumar, Kumar, and Mittal 2004). The development of roads and investment in irrigation infrastructure had positive causal effects on TFP growth in rural India between 1971 and 1994 (Zhang and Fan 2004).

In India, the TFP level has remained diverse across states (Mukherjee and Kuroda 2003). The TFP growth rates have also varied across regions and over time (Rada 2016), reflecting important dynamics. Since the 1980s, the northern, eastern, central, and northeastern regions have still grown the fastest, due to the remaining effects of the aforementioned Green Revolution technologies. The western and southern regions experienced more rapid growth since the 2000s—partly contributing to their catching up. Rada (2016) speculates that these patterns have been associated with shifts away from grains (commonly grown in the north) to higher-value crops including horticulture and livestock products. Meanwhile, the northern and eastern regions also saw recovery in TFP growth in the 2000s. This had been partly through reduced water use intensity enabled by investments in efficient irrigation methods, including drip irrigation (Rada 2016), which partly mitigated the overexploitation of groundwater in these regions, as described in a later subsection. In the northeastern region, deficiencies in transportation and market infrastructure, among others, relatively suppressed TFP growth even during this period (Rada 2016).

Similarly, in Bangladesh the TFP growth between 1948 and 2008 exhibited heterogeneity across regions (Rahman and Salim 2013). Growth had been the fastest in Chittagong (largely due to technical-efficiency gains and some scale-efficiency change). It had been slow in the Chittagong Hill Tracts, largely due to inefficiency in exploiting economies of scope in production achievable through diversifying into the production of crops suitable for hilly environments. According to Rahman and Salim (2013), regional variations in the efficiency in exploiting economies of scope explain significant variations in TFP growth across Bangladesh.

Changes in Natural Resources (Water, Soil)

The productivity gains in SA agriculture might have been achieved at the cost of key natural resource endowments, including soil (land degradation) and water. Such a decline in natural resource endowments suggests that future land productivity growth in SA faces considerable challenges.

FAO (1994) suggests that by the 1990s, land degradation in SA had intensified primarily through erosion of soils by water and wind, the decline of soil fertility, and deterioration of soil quality through waterlogging or salinization, among others. These conditions are likely to have remained since then. FAO (1994) estimates that by the early 1990s, 25 percent of areas under crop and pasture in SA had been affected by water erosion (Table 4.3).[2] More than

---

2 The figures were estimated based on the map created by the global assessment of soil degradation (GLASOD) project conducted by the International Soils Research and Information Centre (ISRIC), as well as secondary sources including government reports (FAO 1994).

TABLE 4.3 Share (%) of areas under crop and pasture affected by different types of land degradation, early 1990s

Country	Water erosion	Wind erosion	Soil fertility decline	Water logging	Salinization
Afghanistan	29	5	0	0	3
Bangladesh	15	0	65	0	0
Bhutan	10	0	0	0	0
India	18	6	2	2	2
Nepal	34	0	0	0	0
Pakistan	28	42	0	4	15
Sri Lanka	46	0	61	0	2
Region	25	18	3	1	13

Source: Data from FAO (1994).

half of the land in Bangladesh and Sri Lanka had suffered soil fertility decline. More than one-third of the areas were affected by different types of land degradation in different countries. Land degradation has become costly in SA. The estimated economic losses due to land degradation in SA by the 1990s were US$5.4 billion from water erosion, $1.8 billion from wind erosion, $0.6–1.2 billion from soil fertility decline, $0.5 billion from waterlogging, and $1.5 billion from salinization—the total being equivalent to 7 percent of agricultural gross domestic product (GDP) then (FAO 1994).

SA, which is the most water-scarce region in the world, has also suffered from a declining groundwater table lately. Since 1960, SA has used the most groundwater in the world, and this pattern has accelerated since the 1990s (Shah et al. 2009). Combined with the rising pumping costs due to rising diesel prices, the lowering of the groundwater table has further raised irrigation costs, particularly in the IGP. This has lately led farmers to shift from more water-intensive crops to less water-intensive crops or to rainfed farming systems (Shah et al. 2009).

Agricultural Growth, Urbanization, and Nutritional Transition

This section describes the key characteristics of agricultural growth, patterns of urbanization, and nutritional transitions in SA, with the aim of highlighting how they may be interlinked with each other, creating unique challenges and opportunities for further agricultural development and food and nutrition security improvements in the region.

Agricultural Growth

Unique agricultural growth patterns in SA can be highlighted by the analyses of the growth patterns of value-added, modernization, linkages with domestic and international markets, and diversifications.

GROWTH OF AGRICULTURAL VALUE-ADDED

Agricultural value-added in SA has increased steadily in the past several decades, and the analysis of its growth patterns in comparison with other regions like East Asia or SSA reveals important differences in underlying conditions. The growth rates of agricultural value-added in SA have been 3.2 percent in the past three decades starting in 1985, higher than the rate of 2.4 percent between 1961 and 1985 (Table 4.4). Such acceleration in growth is consistent with the TFP growth patterns described (Nin-Pratt and Yu 2010) in the previous section.

However, the growth rate may have been somewhat slower than in East Asia during both periods. This relatively low growth rate of value-added in SA is consistent with the slow labor productivity growth discussed earlier. This is related to persistently high shares of the workforce being employed in the agricultural sector in many SA countries, except in Sri Lanka. While the share has constantly declined in SA in the past three decades (from 62 percent in 1991 to 44 percent in 2017), the speed had been slower than the pace in East Asia, where the share declined from 55 percent in 1991 to 29 percent in 2017 (Table 4.4). While the agricultural sector in SA has undergone substantial modernization in the past few decades, as described in the next section, such modernization might have remained labor-intensive. In other words, SA seems to have been slow in transforming the economic structure away from agriculture.

MODERNIZATION

Modernization of agriculture in SA is pronounced in terms of increased adoptions of modern varieties and increased modern inputs use, including chemical fertilizer, machines, and agrochemicals, as well as more systematic uses of water through irrigation (Figure 4.4), although, as is discussed later, use of irrigation on major crops like rice in SA has lagged behind East Asia. Chemical fertilizer use intensity grew substantially up to the 1980s, and it kept growing afterward. By the 2000s, most countries exceeded the intensity of 100 kilograms of nutrients per hectare. The shares of cultivated areas prepared by tractors and irrigated areas have varied more widely across the countries than fertilizer use. Pakistan, Sri Lanka, and India led the growth of tractor uses,

TABLE 4.4 Growth rates of agricultural value-added, and shares of employment in the agricultural sector in South Asian countries

	Annual growth rates of agricultural value-added (%)								Share (%) of employment in the agricultural sector		
	Periods		Decades								
Country/Region	Pre-1985	1985	1961–1970	1971–1980	1981–1990	1991–2000	2001–2010	2011	1991	2005	2017
Afghanistan	—	3.0	—	—	—	—	3.9	1.8	62	65	62
Bangladesh	1.7	3.4	3.2	−0.1	2.4	3.0	4.2	3.4	61	48	41
Bhutan	—	—	—	—	5.6	1.7	2.1	3.0	62	63	57
India	2.4	3.2	2.5	1.8	3.5	2.8	3.3	2.9	63	56	44
Nepal	2.1	3.0	2.9	0.2	4.6	2.5	3.3	2.7	81	77	73
Pakistan	3.7	3.4	5.1	2.4	4.1	4.5	2.6	2.2	48	43	42
Sri Lanka	3.3	2.3	3.5	2.4	2.9	1.7	3.0	2.8	42	33	27
South Asia	2.4	3.2	2.7	1.7	3.5	3.0	3.2	2.8	62	54	44
East Asia & Pacific	4.3	3.6	4.9	2.7	5.2	3.3	3.8	3.6	55	43	29

Source: Authors' compilations from World Bank (2018).

Note: Figures for Afghanistan are only from 2003; figures for East Asia and Pacific exclude high-income countries; — = data not available.

FIGURE 4.4 Trends of modern input uses in agriculture in selected South Asian countries

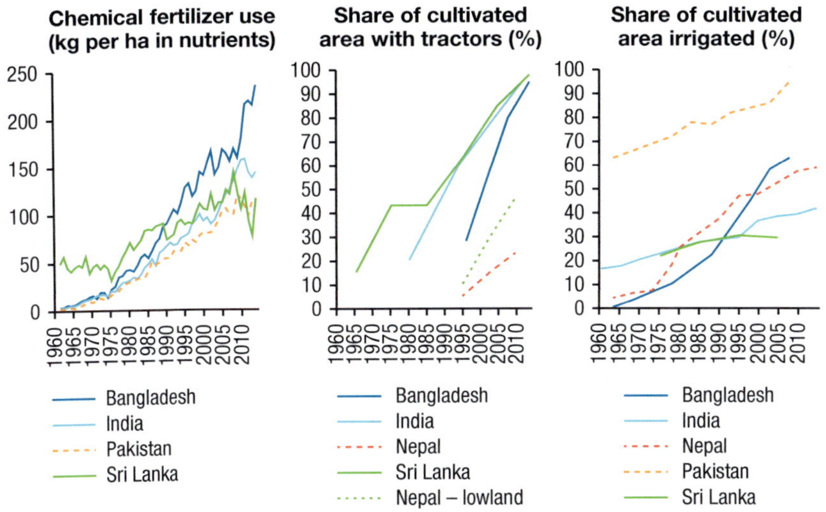

Source: FAO (2020a) for chemical fertilizer; various other sources for tractor uses and irrigated area shares.

which have risen continuously since the 1980s. In Bangladesh, tractor use is quickly catching up to other SA countries, mostly through increased use of power tillers. In all four countries, almost all farmland is now prepared by tractors. The tractor use growth in Nepal has been somewhat slower but has grown at comparable speed in the lowland (Terai) zone. Irrigation area coverages have also expanded steadily, in both early adopters, like Pakistan, and late adopters, like Bangladesh and Nepal. Last, the use of agrochemicals (pesticide, herbicide) has also gradually intensified. In India, herbicide use grew 2.5 times between 2005/2006 and 2015/2016, reaching the consumption level of 1 kilogram per hectare (Gupta et al. 2017), which is still considerably lower than in countries like Japan but comparable to use in the United States.

Contrary to the progress in production of staple crops, the modernization of the livestock sector has remained somewhat limited. The dairy sector in SA is still challenged by the limited supply of improved indigenous cattle breeds, limited artificial insemination to provide crossbred cattle, and the generally low quality of crop residue used for most feed resources (McDermott et al. 2010).

Modernization at the postharvest stage has been more heterogeneous, with areas closer to major urban markets experiencing faster modernization. The adoption of modern processing facilities has grown gradually around major urban areas. For example, Bangladesh has seen the emergence of large (automatic) rice mills displacing the traditional small huller mills extensively in parts of the Chittagong Division (Reardon et al. 2012). In the country as a whole, however, the majority of paddy is still processed by traditional mills. Recently, demand has shifted from cheap, coarse rice to higher-value (medium and fine) rice, and there are increasing price premiums for the latter in major urban centers like Dhaka, Bangladesh, accompanied by the growing sales by the modern retail sector (Minten, Murshid, and Reardon 2013). The shift for these modern agrifood systems, however, kept modernization to the postharvest stage, and there was relatively little effect on modernization at the farm level (Minten, Murshid, and Reardon 2013). Similarly, branding by the private sector is still often plagued with incompleteness or misinformation (Minten, Singh, and Sutradhar 2013).

GROWING BUT LIMITED CONTRIBUTIONS OF EXTERNAL TRADE

Domestic production and consumption seem to continue to be dominant sources of supply and demand for agricultural commodities in SA, despite growing external trade. Table 4.5 shows the shares of the net exported quantities as a percentage of the total quantities domestically produced for each group of food commodities (negative values indicate net imports). For example, 12 indicates that the net exports were equivalent to 12 percent of the quantities domestically produced.

Table 4.5 indicates that the quantities of cereals, starchy roots, vegetables, fruits, meat, eggs, milk, and fish products that have been exported or imported have remained fairly small compared with the domestic production of these commodities (generally accounting for 20 percent or less). Only pulses and tree nuts, and other food items that may be relatively minor nutrient sources, have seen substantial increases in the contributions of trade. Generally, domestic productions of key agricultural commodities have remained important sources of domestic food and nutrition, while domestic markets have remained the major drivers of agricultural growth, in SA. This pattern is in contrast to some developing countries in SSA, where the reliance on cereal imports has increased substantially in the past few decades.

DIVERSIFICATION

SA, as a whole, has seen a gradual decline of the share of cereals, and increasing shares for vegetables, and milk, in net production values (Table 4.6). The trends in India, given its dominant size, largely represent the overall patterns in SA. However, there have been considerable variations and divergence over time across countries outside India (Figure 4.5). In Bangladesh, diversification has been relatively limited, with the share of cereals remaining around 60 percent throughout the past five decades, while the shares of higher-value crops have not changed significantly.[3] Sri Lanka has actually experienced an increased share of cereals during the last five decades, accompanied by decreasing shares of oil crops, fruits, stimulants (tea in particular), and tobacco/rubber. In contrast, Nepal has seen a rapid shift from cereals and milk to vegetables and fruits and, to a lesser extent, spices. Pakistan, with historically low shares of cereals and higher shares of milk, meat, and fiber crops, has largely maintained the same level of agricultural diversification.

This diversification pattern might have been driven by both the changes in demand and deliberate efforts on the supply side. For example, Sri Lanka had been a major rice importer in the region in the 1960s but has gradually reduced import reliance, partly through the government's effort in achieving

3 It is, however, important to note that this indicator may not capture the important diversifications within each commodity group or crop. For example, Bangladesh has seen a gradual shift from cheap-quality (coarse) rice to more expensive, higher-value (medium and fine) rice over time (Minten, Murshid, and Reardon 2013).

TABLE 4.5 Share (%) of net export to domestic production of food commodities

	South Asia excluding India					
Food commodities	1961–1970	1971–1980	1981–1990	1991–2000	2001–2010	2011–2013
Cereals, excluding beer	−8	−12	−14	−15	−11	−11
Starchy roots	−3	−1	−1	−1	−1	0
Vegetables	−1	1	0	1	0	1
Fruits, excluding wine	5	4	4	4	2	1
Sugar crops	0	0	0	0	0	0
Pulses	−5	−2	−6	−13	−44	−62
Treenuts	22	18	8	15	5	−5
Oilcrops	10	5	3	−7	−22	−17
Meat	0	−4	−8	−2	−2	−2
Eggs	0	−4	−3	0	1	−2
Milk, excluding butter	−2	−6	−10	−3	−2	−3
Fish, seafood	−4	−10	−15	−11	1	2
Aquatic products, other	0	0	18	75	16	6
Offals	0	0	0	0	0	2
Alcoholic beverages	−11	−13	−13	−8	48	55
Animal fats	−26	−38	−50	−25	−16	−11
Spices	−10	1	−2	−6	−15	−20
Stimulants	72	51	33	20	22	−5
Sugar and sweeteners	−21	−18	−26	−28	−49	−41
Vegetable oils	−25	−94	−214	−268	−267	−362

rice self-sufficiency. Another contributing factor might have been the declining real price of rice and other grains, which is likely to have induced a shift of the production toward high-value crops.

Aquaculture production has grown in the last few decades in many SA countries, often 10-fold between the 1980s and the 2010s in gross production value. While the production value is still small compared with the entire agricultural production, it has grown to the equivalent of 20 percent of gross agricultural production in countries like Bangladesh (Table 4.6). The sector has contributed to poverty reduction through reduced price of fish products and their increased consumption by low-income consumers (Toufique and Belton 2014), through improved economic activities of women (Kawarazuka and Béné 2010), and through employment (Paul and Vogl 2011). Its growth has, however, also caused environmental damage, including mangrove

	India					
Food commodities	1961–1970	1971–1980	1981–1990	1991–2000	2001–2010	2011–2013
Cereals, excluding beer	−8	−3	0	1	3	8
Starchy roots	0	0	0	0	0	0
Vegetables	1	0	1	1	2	2
Fruits, excluding wine	−1	0	0	0	0	0
Sugar crops	0	0	0	0	0	0
Pulses	0	0	−4	−3	−16	−19
Treenuts	5	18	18	−20	−19	−42
Oilcrops	−2	0	0	1	1	2
Meat	0	1	2	4	9	20
Eggs	0	0	0	1	3	2
Milk, excluding butter	−2	−1	−1	0	1	1
Fish, seafood	3	4	4	6	7	12
Aquatic products, other	100	35	0	1	44	60
Offals	0	0	0	0	1	6
Alcoholic beverages	0	0	0	1	−2	3
Animal fats	−11	−10	−5	0	0	1
Spices	7	7	6	7	10	16
Stimulants	53	43	38	34	31	36
Sugar and sweeteners	3	4	−2	−1	4	7
Vegetable oils	0	−17	−27	−24	−70	−101

Source: Authors' calculations based on FAO (2020a).

degradation, saltwater intrusion, sedimentation, pollution, and disease outbreaks (Paul and Vogl 2011).

The diversification patterns have also been associated with cross-country variations in productivity growth, particularly the slower growth in Sri Lanka than in other SA countries described earlier. Table 4.7 summarizes the output growth between the 1980s and the 2010s by commodity groups in Sri Lanka and other SA countries. Output growth for vegetables and fruits has been particularly slow in Sri Lanka (0.3 percent per year) compared with SA as a whole (3.6 percent per year). In other SA countries like India, Nepal, and Pakistan, while rice output growth rates have been similar to that of Sri Lanka, nonrice commodities have grown by 3 percent or more (for example, milk in India and Pakistan; meat and cotton in Pakistan; meat, vegetables, and fruits in Nepal). In Bangladesh, rice output growth has been faster, at 2.8 percent per year.

FIGURE 4.5 Share (%) of total production of commodity groups, by year, in South Asia

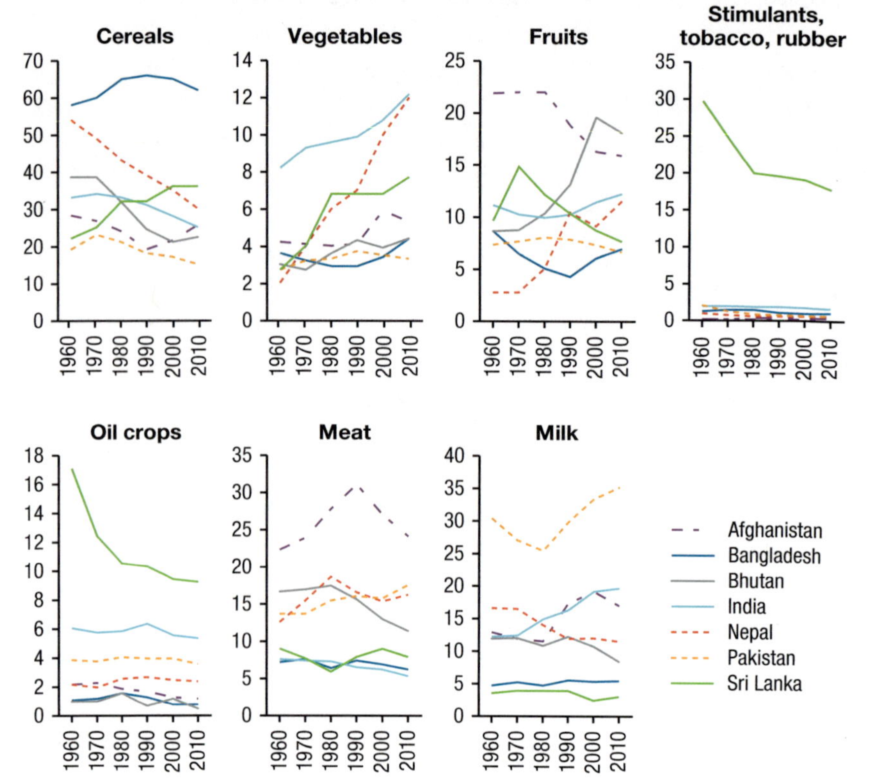

Source: Data from FAO (2020a).

Urbanization

In relative terms, the pace of urbanization in SA seems to have been slow compared with both growth in the East Asia and Pacific region and the historical experiences of developed countries. Urbanization levels are the lowest in Nepal (19 percent) and Sri Lanka (18 percent), while the Maldives (47 percent), Bhutan (39 percent), and Pakistan (39 percent) are the most urbanized countries in the region (Table 4.8). The urban population accounted for 35 percent in Bangladesh and 33 percent in India. The agricultural growth and modernizations in SA described in the previous section, therefore, seem to have largely taken place in the rural sector.

However, in absolute terms, SA's urban population growth has been and is likely to be significant. The urban population increased from about

TABLE 4.6 Agricultural diversification patterns in South Asia

Groups	1960s	1970s	1980s	1990s	2000s	2010s
Share (%) of production values of each commodity group						
Cereals/Country	33	33	32	30	28	25
Root crops	1	2	2	2	3	3
Pulses	6	5	4	3	3	3
Sugar crops	5	5	5	5	4	4
Oil crops	5	5	5	5	5	5
Vegetables	7	8	9	9	10	11
Fruits	11	10	11	11	12	12
Meat	9	9	9	9	9	8
Milk	13	13	15	16	19	19
Others	10	10	8	10	7	10
Aquaculture production value in South Asia (2010 US$ in thousands)						
Bangladesh	—	—	384,062	889,142	1,560,988	4,177,766
Bhutan	—	—	8	54	93	350
India	—	—	1,522,334	2,776,292	4,215,351	8,988,443
Nepal	—	—	10,219	16,084	33,740	94,732
Pakistan	—	—	20,645	16,849	138,071	209,918
Sri Lanka	—	—	4,806	37,348	37,805	51,103
Proportion of gross aquaculture production to gross agricultural production in values (agricultural production value = 100)						
Bangladesh	—	—	5	10	12	20
Bhutan	—	—	0	0	0	0
India	—	—	1	2	2	4
Nepal	—	—	1	1	1	2
Pakistan	—	—	0	0	1	1
Sri Lanka	—	—	0	2	2	2

Source: Authors' calculations based on FAO (2020a), FAO (2019) for aquaculture production values, and FAO (2020a) for commodity groups.

Note: Gross aquaculture production value does not include capture fishery. The "proportion" simply compares the sizes and does not mean the share of aquaculture production to agricultural production, since the latter figure does not include the former. — = data not available.

390 million in 2001 to 590 million in 2016—an addition of 200 million—and is expected to expand by almost 250 million more in the next 10 years (Table 4.8). Urbanization still has been one of the drivers of dietary diversifications in SA in recent years (Joshi, Gulati, and Cummings 2007).

Urbanization might have also posed certain challenges to food security. The 2007–2008 food crisis demonstrated the vulnerability of urban

TABLE 4.7 Growth of outputs by major commodity groups in South Asian countries, 1980s–2010s

	Annual growth rates, 1980s–2010s (%)				Shares to total outputs (average share, 1980s–2010s)	
Country	Land	Rice output	Nonrice output	Vegetables and fruits output	Nonrice (%)	Key nonrice commodities (%)
Bangladesh	1.2	2.8	3.2	4.2	36	Vegetables and fruits (8), meat (6), milk (5)
India	0.6	1.8	3.1	3.6	79	Vegetables and fruits (21), milk (17), meat (6), wheat (6), sugarcane (5)
Nepal	0.9	1.8	3.8	5.7	72	Vegetables and fruits (17), meat (16), milk (12), maize (5), wheat (5), potato (5)
Pakistan	0.4	1.0	3.1	2.6	93	Milk (31), meat (15), cotton lint/seed (13), wheat (10), vegetables and fruits (10)
Sri Lanka	0.6	1.8	1.0	0.3	66	Vegetables and fruits (17), tea (12), coconuts (9), meat (7), rubber (6)

Source: Authors' calculations based on FAO (2020a) and USDA (2020).

TABLE 4.8 Urbanization trend in South Asian countries

	Total population (millions)			Urban population (millions)			Urban population (%)		
Country	2001	2011	2016	2001	2011	2016	2001	2011	2016
Afghanistan	20.97	29.71	34.66	4.53	7.45	9.40	22	25	27
Bangladesh	134.11	153.91	162.95	32.31	48.06	57.09	24	31	35
Bhutan	0.59	0.74	0.80	0.16	0.26	0.31	26	36	39
India	1,071.48	1,247.24	1,324.17	299.14	390.09	438.78	28	31	33
Maldives	0.29	0.38	0.42	0.08	0.16	0.19	29	41	47
Nepal	24.16	27.33	28.98	3.37	4.69	5.51	14	17	19
Pakistan	141.60	174.18	193.20	47.36	64.46	75.78	33	37	39
Sri Lanka	18.80	20.27	21.20	3.46	3.71	3.90	18	18	18
South Asia	1,411.99	1,653.76	1,766.38	390.42	518.88	590.97	28	31	33

Source: Data from World Bank (2017).

populations, especially slum dwellers, to shocks in agricultural markets (OECD-FAO 2010). Large urban settlements in SA are marked by widespread slums. The share of the urban population living in slums in SA is high (except in Bhutan and Sri Lanka), ranging from 17 percent in India to 89 percent in Afghanistan (Ellis and Roberts 2016). At least 130 million people were living in informal urban settlements in SA in 2009 (ADB 2014). Slum populations often do not have access to water and sanitation facilities, making them more likely to suffer from disease and malnutrition. At the same time, where it has occurred, urbanization might have been associated with the dramatic increase in the proportion of dietary energy derived from oils and fats and considerable increases in the consumption of fruits, vegetables, and dairy and other animal products. Food consumption patterns have been changing across regions, with consumption of noncereal crops growing in both rural and urban areas. Rural and urban food consumption patterns are not uniform across SA, but some common trends emerge. Food accounts for a smaller share of consumption expenditure in urban areas than in rural areas, and urban households have more diverse diets than rural households (Joshi, Parappurathu, and Kumar 2016).

Nutrition Transition

SA seems to be going through a nutrition transition (Chapter 10)—defined as the process of changes in the food environment, physical activity, and lifestyle that result in declining levels of undernutrition and increasing levels of overnutrition over time (Popkin 1993). Reduced staple crop prices in countries like India and Bangladesh, realized through investments in public research and development, have led to significant reductions in poverty, and increases in disposable incomes have led to increased dietary diversification through consumption of higher-value nonstaple commodities (Pingali 2012). Considerable progress has also been made on improving nutritional outcomes. For instance, Nepal has recorded the fastest reduction in child stunting in the world recently, from 56 percent in 2001 to 36 percent in 2016 (Nepal, Ministry of Health 2016).

Despite such transitions and progress, challenges remain. SA remains the global hub of nutrition insecurity. In 2016, two out of every five of the world's stunted children lived in SA, and more than 15 percent of children under age 5 were wasted. In fact, stunting levels in SA (38 percent) were comparable to those in SSA (37 percent) and over three times higher than in East Asia and the Pacific (12 percent) or Latin America (11 percent).

Investments into staple crops are also said to have crowded out the required investments for traditional crops that were important sources of critical nutrients, such as legumes and pulses (Pingali 2012). Consequent increases in their prices in India are considered to have reduced the consumption of these commodities (Pingali 2012).

FIGURE 4.6 **Incidence of poverty, undernourishment, stunting, and overweight in South Asia (%)**

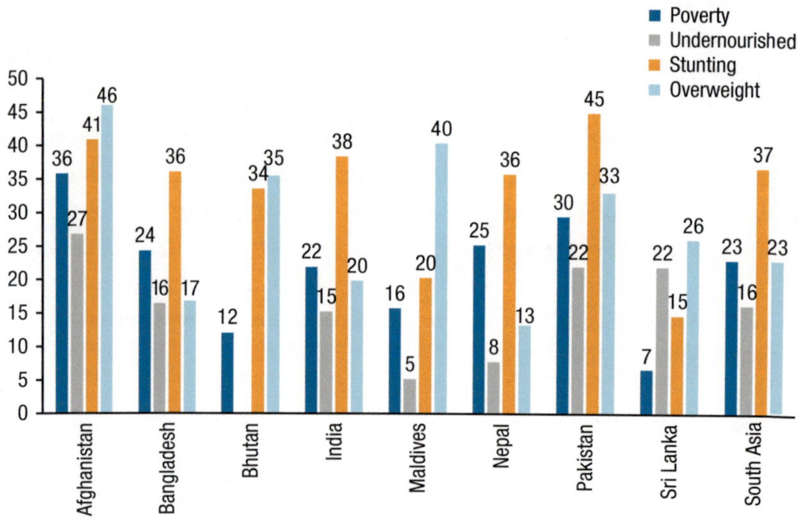

Source: Data from World Bank (2017) (poverty, undernourishment, stunting); IHME (2014) (overweight).

Note: The poverty head count data pertain to different years: Afghanistan, 2011; Bangladesh, 2016; Bhutan, 2012; India, 2011; Maldives, 2009; Nepal, 2010; Pakistan, 2013; and Sri Lanka, 2012. Undernourishment figures pertain to 2015. Stunting figures pertain to various years: Afghanistan, 2013; Bangladesh, 2014; Bhutan, 2010; India, 2015; Maldives, 2009; Nepal, 2016; Pakistan, 2012; Sri Lanka, 2012; and South Asia, 2015. Poverty head count ratio and undernourishment are given as percentage of population; stunting figures pertain to children under 5 years of age. The prevalence of overweight data are all from 2013.

South Asia has also seen a rapid emergence of overnutrition (Figure 4.6). Obesity reached 20 percent in India and 13 percent in Nepal by 2013. In India, overnutrition and associated noncommunicable diseases have also gradually spread from urban to rural areas lately (Meenakshi 2016). In SA, about one in four adults are currently overweight.

Innovations and Agricultural Transformation

This section highlights key technological and institutional innovations and key policies that might have partly enabled the agricultural growth in SA described in the previous sections. The section also discusses that these innovations have so far been associated with labor-intensive agricultural transformation in SA. Despite the gradual shift in the comparative advantages from smallholders to larger farms, such effects might not have been strong enough to overcome constraints for farm size expansions. Similarly, despite the growing wage gap between the agricultural sector and the nonagricultural sector, the former has remained one of the primary sources of rural employment.

Technological and Institutional Innovations

SA has seen considerable progress in technological and institutional innovations.

VARIETAL TECHNOLOGIES

For varietal technologies, public-sector-led conventional plant breeding has continued producing newer generations of improved varieties (Hossain et al. 2003). Genetic improvement efforts had been significant, not only for widely documented breeding for staple crops, but also for vegetables—including, for example, application of molecular-based methods for breeding virus-resistant tomatoes in southern India (Weinberger and Lumpkin 2007).

Over the years in the region, the public sector has accumulated plant-breeding technologies and knowledge, as well as a supply of superior germplasms, which have raised the potential of developing hybrid varieties. At the same time, there has been increased participation of the private sector in agricultural research and development over the past few decades. Recently, hybrid varieties have been developed for maize, rice, and wheat (Matuschke, Mishra, and Qaim 2007), as well as pearl millet (Matuschke and Qaim 2008). The adoption rates of hybrid maize and hybrid rice have grown considerably. Furthermore, the increased applicability of biotechnologies (including genetic engineering) has induced private-sector participation, which has resulted in the successful development of Bt cotton and vegetable seed developments (see, for example, Krishna and Qaim 2007). Strengthening intellectual property rights is also likely to have accelerated private-sector participation (Pray and Nagarajan 2014). In countries like India, this progress has coincided with the deregulation of the seed sector that started in the 1990s (Pray and Nagarajan 2014).

MECHANIZATION RELYING ON SMALL-SCALE MACHINES

Mechanical tools have spread widely in SA over the past few decades, but possibly in ways not displacing substantial labor force in agriculture, which is appropriate in raising labor productivity in the labor-abundant environment. This has been achieved by the spread of various types of small-scale machines, including mechanical water-lifting tools, tens of millions of which are in SA (Mandal, Biggs, and Justice 2017). Since the 1990s, the use of small-scale shallow tube wells and motorized pumps has increased the extraction of groundwater and surface water in Bangladesh, where importation of shallow tube wells was liberalized in the 1990s (Ahmed 1995), as well as Nepal and the IGP of India (Shah et al. 2009). Power tillers have spread as well for land preparation and local transportation. At the same time, four-wheeled, riding tractors that have been more common in India and Pakistan might have been more suitable in rice-nonrice crop rotation common in parts of SA like India, and mechanization might have led to significant area expansions, sustaining the overall demand for labor (Pingali 2007). The growth of the domestic manufacturing industry, such as in India, facilitated the development of small four-wheeled tractors that are more suitable on smallholder-dominated upland in the region (Pingali 2007). These mechanization patterns may explain why the land-labor ratio did not increase with the increases in tractor HP per worker shown in Figure 4.2.

INSTITUTIONAL INNOVATIONS

SA has seen various institutional innovations in the past few decades. The growth of market institutions has been observed in the growing participation of new actors in processing, trading, and wholesale for both staple crops like rice and potato (Reardon et al. 2012) and nonstaple crops (Fafchamps, Hill, and Minten 2008). Vertical integration has also been strengthened for crops like sugarcane in India (Patlolla, Goodhue, and Sexton 2015). However, constraints remain for food sanitation along the value chain for vegetables/fruits and dairy value chains in India (Fafchamps, Hill, and Minten 2008; Kumar, Wright, and Singh 2011). Markets for minor crops have also expanded. Increased market participation has had positive effects on on-farm varietal diversity for millet in India (Takeshima and Nagarajan 2012). Marketing institutions including contract farming and cooperatives have become increasingly important for production and marketing of high-value vegetables and fruits, poultry, and higher-quality milk, which has led to the adoption of improved food safety practices (Joshi, Gulati, and Cummings 2007; Kumar et al. 2018), as will be discussed in Chapter 11 in depth.

Market institutions have also developed regionally. The South Asian Free Trade Area (SAFTA) agreement in 2004 potentially increased market integration across SA countries. Trade liberalization in the early 1990s might have improved the resilience of countries like Bangladesh against food crises caused by natural disasters like floods (del Ninno and Dorosh 2001).

Institutional innovations have also been widespread for common resource management. Transfer of forest use rights to local communities reduced forestry resource extraction in Nepal (Edmonds 2002). In India, increased demand for wood products has led to increased forestry cover, reversing the trend of deforestation (Foster and Rosenzweig 2003). Community irrigation systems have also grown (Bardhan 2000). Environmental efficiency has also been raised by new production technologies, such as Bt cotton that has reduced chemical pesticide use in Pakistan (Kouser and Qaim 2015).

Key Policies That Partly Facilitated the Innovations

INFRASTRUCTURE DEVELOPMENT

The increased stocks of physical infrastructure (for roads, electricity, information and communications technology, etc.) as well as institutional infrastructure have been important drivers of agricultural diversification away from traditional low-value crops to higher-value crops (Joshi, Gulati, and Cummings 2007). Better road infrastructure has often reduced the transaction costs for marketing perishable commodities like vegetables and fruits (Joshi, Gulati, and Cummings 2007). Infrastructure for veterinary institutions and artificial insemination centers for livestock was likely to have encouraged diversification in the production of livestock products (Joshi, Gulati, and Cummings 2007).

Many SA countries, including Bangladesh, India, and Pakistan, have expanded key infrastructure during the past several decades (Table 4.9). Paved roads expanded by more than five times between the 1970s and 2000s in all three countries. The number of ground line telephones multiplied by almost 5 in Bangladesh and almost 10 in India and Pakistan during the same periods. The public expenditures on infrastructure during these periods have been substantial.

MACRO POLICIES ON LIBERALIZATION

The series of innovations are also likely to have been partly induced by the economic and macro policy reforms introduced in many SA countries in the past few decades. Many SA countries, including India, Pakistan, and Sri Lanka, have gradually reduced the distortions of their exchange rates, which had typically been fixed and overvalued (Sahoo, Nataraj, and Dash 2013). India since the 1980s and Pakistan throughout the 1980s saw significant devaluations that helped raise the competitiveness of some of the export crops. Unlike some other developing countries (particularly those in SSA), these SA countries have managed to stabilize the exchange rates even after the shift to the floating exchange rate system. For example, the real exchange rate in Pakistan remained largely unchanged from the late 1980s to the 2010s (Spielman et al. 2017).

TABLE 4.9 Improvement in infrastructure and communication in selected South Asian countries, 1970–2010

Infrastructure/communication	Bangladesh	India	Pakistan
Paved roads (1,000 km)			
1970s	3.7	334	19
1990s	16.4	1,258	104
2000s	21.2	1,818	167
Percent increase between 1970s and 2000s	573	544	954
Landline telephones (per 1,000 people)			
1970s	1.52	2.51	3.18
1980s	2.55	4.03	4.98
1990s	2.55	13.06	14.87
2000s	6.55	35.91	27.98
Percent increase between 1970s and 2000s	470	1,058	1,044
Cellular phones (per 1,000 people)			
1990s	0.3	0.9	0.7
2000s	141.4	154.6	226.8
2010–2012	552.0	674.0	618.0
Public expenditure on infrastructure (2005 billion US$)			
1980s	1.29	2.72	2.19
1990s	1.20	3.40	1.74
2000s	1.05	8.85	1.11

Source: Rashid et al. (2013). Reproduced by permission of the publisher;

Investment policies have also been reformed in many SA countries. For example, in 1997, Pakistan allowed foreign direct investment (FDI) in the agricultural sector, though it had restricted FDI to the manufacturing sector in the previous period (Sahoo, Nataraj, and Dash 2013). Similarly, in Nepal, the share of FDI flowing into the agricultural sector started increasingly, from almost zero in 2000 to 4 percent in 2010 (Sahoo, Nataraj, and Dash 2013).

However, the extent of the effectiveness of macro policies in the agricultural sector remains uncertain. While trade liberalization policies have significantly reduced the tariff rates for nonagricultural goods, average tariffs for agricultural goods have remained higher, around 40 percent in India and 30 percent in Sri Lanka (Sahoo, Nataraj, and Dash 2013). While SA countries have seen increased shares of tax revenues originating from income tax and value-added tax, rather than seigniorage and tariff (Aizenman and Jinjarak

2009), the taxation of agricultural incomes appears to have remained incomplete, especially in India and Pakistan.

SUBSIDIES ON CREDIT, INSURANCE, AND INPUTS

SA countries have also supported provisions of institutional credit, insurance, and agricultural inputs. For example, India has made efforts to allocate credit to the agricultural sector by increasing the overall volume of credit that goes to the sector, by waiving institutional debt for small farmers and allowing one-time settlement, and through interest-subvention schemes that reward prompt repayment by borrowers (Narayanan 2016). Consequently, in India the flow of formal-sector agricultural credit has been consistently increasing, with the ratio of agricultural credit to agricultural GDP rising from 10 percent in 1999–2000 to about 38 percent in 2012–2013 (Kumar et al. 2017), and the share of credit-financed portion out of the spending on agricultural inputs increasing from 21 percent in 1995/1996–2003/2004 to 69 percent in 2004/2005–2011/2012 (Narayanan 2016). Institutional credit has been positively associated with agricultural development in various ways in countries like India (see, for example, Narayanan 2016). Between the mid-1990s and early 2010s, greater credit flows had been associated with the increased use of chemical fertilizer, use of pesticides, or investments into agricultural capital like tractors. At the same time, the effects of credit on agricultural GDP, efficiency, and productivity have been relatively limited (Narayanan 2016). These patterns suggest that institutional credit might have been more associated with the increase in agricultural productivity where capital inputs (purchased by credit) offered higher productivity than noncapital inputs (see Chapter 16 for further discussion).

In Bangladesh, formal credit for machinery purchase has been increasingly provided by the importers that have comparative advantages in monitoring and loan payment collections, rather than directly through the bank. Khandker (2007) found that microfinance institutions in Bangladesh enhanced flood-affected households' access to finance and thereby played a central role in enhancing their coping ability after devastating floods.

The uptake of agricultural insurance has been rising, albeit slowly, in India. By 2016, 57 million farmers in India had been covered by some type of insurance (Gulati, Terway, and Hussain 2018). The cost-effectiveness of these subsidized insurance programs remains ambiguous with limited empirical evidence (see Chapter 17). In parts of India, indexed insurance increased investments into potentially higher-return cash crops that are more sensitive to rainfall (Cole et al. 2013). In parts of Bangladesh, indexed insurance significantly enhanced risk-management effects and expanded cultivated area and spending on agricultural inputs (Hill et al. 2019). However, further studies are needed on the effects of agricultural insurance on agricultural productivity, profitability, and farm incomes, for which evidence remains thin.

Among input subsidies, fertilizer subsidies often accounted for 2–3 percent of total public expenditures and often more than 30 percent of agricultural expenditures in Bangladesh, India, and Pakistan between 1980 and 2010 (Rashid et al. 2013). These fertilizer subsidies were effective in reducing fertilizer prices during this period—as much as 30 percent in India, 9 percent in Bangladesh, and 3 percent in Pakistan (Rashid et al. 2013). However, it is also important to note that these subsidies also led to significant distortions of fertilizer markets. Combined with exchange rate distortion, fertilizer prices often deviated substantially from international prices (Rashid et al. 2013). The subsidies have also been rather regressive; that is, larger farms generally benefited more than smallholders. Furthermore, fertilizer subsidies have also been considered partly responsible for various environmental damages caused by their overuse, including groundwater pollution in Bangladesh (Alauddin and Quiggin 2008) and India (Rashid et al. 2013), as well as air pollution in India (Aneja et al. 2012).

Farm Size and Production Efficiency

The average farm size in SA has been declining over the past several decades. This trend has been consistent across many SA countries, including Bangladesh and Nepal with very small farm sizes (less than 1 hectare on average), India with slightly larger farm sizes (around 1 hectare on average), and Pakistan with modestly larger farm sizes (around 3 hectares on average) (Table 4.10).

Declining farm sizes and the process of land fragmentation in SA countries have been partly influenced by government policies in these countries, including the law of inheritance of paternal property, lack of progressive tax on inherited land, heterogeneous land quality, and an underdeveloped land market (Niroula and Thapa 2005). Pakistan has imposed a progressive land tax, where larger farms are taxed at a higher rate than smaller farms (Adamopoulos and Restuccia 2014). In India and Nepal, a land reform program called the land-to-the-tiller program was implemented, albeit with a limited implementation capacity, whereby the land was transferred from large landholders to tenants who actually cultivate the land, and ceilings were set for the land ownership.

TABLE 4.10 Declining farm sizes in selected South Asian countries

Country	Year	Average holding size (ha)
Bangladesh	1983	0.80
	1996	0.47
	2008	0.37
India	1971	2.28
	1986	1.69
	1991	1.57
	1995–1996	1.41
	2001	1.33
	2005–2006	1.23
	2010–2011	1.15
Nepal	1980	1.12
	1990	0.95
	2000	0.79
	2011	0.66
Pakistan	1970	5.29
	1980	4.70
	1990	3.78
	2000	3.08
	2010	2.59
Sri Lanka	1960	1.61
	1970	1.24
	1980	1.08
	2002	0.47

Source: Data from FAO (2020c).

Evidence suggests potentially negative effects of declining farm size, as well as fragmentation, on productivity and efficiency. Adamopoulos and Restuccia (2014) estimated that the aforementioned tax policies in Pakistan reduced average farm size and agricultural productivity by 3 percent. The aforementioned policies on the land-to-the-tiller program in India and Nepal discouraged optimal land concentration that would achieve economies of scale in production (Otsuka, Liu, and Yamauchi 2016).

The negative productivity and efficiency effects of declining farm size have also been aggravated by accelerating land fragmentation. Land fragmentation was found to increase the cost of cultivation in India (Deininger 2017) and had negative effects on rice productivity and efficiency in Bangladesh (Rahman and Rahman 2009). Niroula and Thapa (2005) conclude, based on reviews of the literature in SA, that fragmentation of small landholdings and tiny land parcels have been detrimental to land conservation and economic gain and that they have discouraged farmers' adoption of agricultural innovations.

Meanwhile, relations between farm size and productivity have been changed by growing adoptions of mechanization in SA in the last few decades. Even though the machine rental markets providing hiring services have been relatively efficient and have managed to enable not only larger farmers but also smallholders to adopt mechanization, mechanization, particularly recent large-scale mechanization, has shifted the comparative advantage to larger farms. In Nepal, the adoptions of mechanical technologies (tractors, in particular) have directly raised the returns to scale in agriculture, shifting the comparative advantage from smallholders to larger farms (Takeshima 2017a). Aside from mechanization, Deininger et al. (2018) argue, the labor market imperfections in India have dissipated over time, and this too has weakened the inverse relationship between farm size and output per area. According to them, previously family labor was overused for farming because they were rationed out of off-farm labor markets, which were less functioning, and the extent of overuse was greater among smaller farms with greater family labor endowments. In the last few decades, however, an increasingly functioning labor market has started absorbing more family labor into the off-farm market, while reducing on-farm family labor use to a more optimal level.

With labor market transformation, increased infusion of capital into the agricultural sector in the form of mechanization, and the resulting shift in comparative advantages toward larger farms, the institutional failures in developing efficient land markets are now becoming increasingly important bottlenecks in SA.

Emerging Rural Employment Opportunities

In SA, the agricultural sector has remained the major source of rural employment in the past few decades, despite the gradual expansion of the nonfarm sector. The persistently high contribution of the agricultural sector as the source of rural employment is consistent with the presentations in earlier sections, where labor use per hectare has remained unchanged, even though labor productivity in agriculture has risen substantially in many SA countries. In India, the agricultural sector has long competed with the industrial and service sectors for workers. Where labor mobility has been low, the industrial sector has often moved to areas with lower agricultural productivity and thus lower wages (Foster and Rosenzweig 2004). However, rural labor mobility was not always low. In Nepal and Pakistan, the adoption of modern varieties in favorable areas benefited marginal areas through increased labor demand and migration (Upadhyaya, Otsuka, and David 1990; Renkow 1993).

Rural employment in the agricultural sector has remained high and, in some cases, has increased in absolute terms. The crop sector rather than the livestock sector has often been the major source of agricultural employment, as in, for example, Pakistan (Spielman et al. 2017). New agricultural technologies in SA, particularly in India, have been relatively more skill intensive, raising the returns to primary schooling (Foster and Rosenzweig 1996) and thereby raising the demand for moderately educated labor. Output growth has increased demand for harvesting labor, even for labor-saving varieties like Bt cotton in India (Subramanian and Qaim 2009). Such growth for harvesting labor is, however, also a reflection of the fact that wages are still low, preventing the spread of mechanical harvesters that has been widely observed in East Asia lately. Once the wage level rises to the critical threshold, rural employment in SA is likely to shift considerably to the nonfarm sector.

NONFARM SECTOR

In SA, the share of rural employment in the nonfarm sector has remained low and has just started rising gradually. Table 4.11 compares the shares of rural employment among different economic activities in the five SA countries in recent decades. Direct comparisons across countries or over time should be made with caution, due to the differences in the classifications of activities, measurements, and periods. However, these figures collectively provide useful insights.

The construction sector has been one of the fastest-growing sources of rural employment among the nonfarm sector, particularly in India and Nepal. To a lesser extent, rural employment in the manufacturing of farm products has grown in parallel with the overall farm output growth. The rice-milling sector, which is still dominated by small-scale mills in India and Bangladesh, and potato cold storage facilities in Bangladesh have provided significant sources of employment despite relatively high capital intensity (Reardon et al. 2012).[4]

4 For example, in Bangladesh, despite the rising rice yield and productivity, traditional small-scale mills still provide the majority of milled rice (Reardon et al. 2012).

TABLE 4.11 **Shares of rural employment in selected South Asian countries, by activities (%)**

Country	Period	Agriculture	Construction	Transport	Manufacture	Trade etc.	Other services
Bangladesh[a]	2000	58	3	17	10	—	12
India[b]	2004–2005	73	5	3	8	6	5
	2009–2010	68	10	3	7	6	6
Nepal[c]	1995	83	1	1	2	0	13
	2010	61	7	2	4	1	25
Pakistan[d]	2005/2006	43	—	—	—	Industry: 21, Services: 36	
	2013/2014	44	—	—	—	Industry: 23, Services: 34	
Sri Lanka[e]	2003	33	—	—	27	25	14

Source: Authors' compilations based on sources in the last column. [a] Haggblade, Hazell, Reardon (2010); [b] Chowdhury (2011, table 5); [c] Takeshima (2017b); [d] Spielman et al. (2017); [e] Deininger, Jin, Sur (2007).

Note: For Bangladesh, "transport" includes "commerce"; "other services" includes "personal, financial, and community services"; "construction" includes "construction and utilities mining." Transport includes "transport and commerce."

For India, "construction" includes "mining"; "trade" includes "trade, hotel, and restaurant." "Other services" includes "electricity, gas, and water supply; finance; real estate, etc.; and public administration and other services."

For Nepal, figures for the urban and rural sectors, as well as male and female, and across agroecological belts in Takeshima (2017b) are combined using the population weights. Figures are also based on the share of hours engaged in each activity.

For Pakistan, figures for 2005/2006 and 2013/2014 include urban sector, which account for about one-third of the country's population.

For Sri Lanka, agriculture data is from ADB (2007). Manufacture, trade, and other services includes nonfarm "enterprises" and not wage employment. Figures are based on Deininger, Jin, Sur (2007), multiplied by the share of nonfarm economic activities to total rural employment.

— = data not available.

Employment growth in the manufacturing of nonfarm products has been driven by textile industries. For example, the garment industry has grown considerably in Bangladesh, and by 2007–2008, 2.5 million workers had been employed in the sector (Mottaleb and Sonobe 2011). The growth of the Bangladesh garment industry has been enabled primarily by international transfer of production technologies and skills in management and international marketing, coupled with enhanced education levels (Mottaleb and Sonobe 2011). Furthermore, Bangladesh succeeded in creating an enabling environment for garment-sector growth by maintaining stability and predictability of garment-industry policies, rather than focusing on heavier interventions that may be more substantial in scope but can also bring significant unpredictability from the private sector's perspectives (Ahmed, Greenleaf, and Sacks 2014). Where these conditions exist, export-oriented manufacturing may grow substantially even in the rural areas in SA (see Chapter 11).

Other major sources of rural employment include government services. In the 1990s, government jobs provided about 25 percent of rural nonfarm earnings in rural Pakistan and nearly 20 percent of rural nonfarm employment in rural India (Haggblade, Hazell, and Reardon 2010).

The expansion of public works social protection policies has had certain effects as well on rural employment in SA. A notable example is India's Rural Employment Guarantee program, which has offered employment on local public works to any rural adult residents demanding work for up to 100 days. This program has been one of the largest public works schemes and had covered about 30 percent of rural households in India from 2009 to 2015, generating more than 2 billion person-days of work per year from 2008 to 2014 (Drèze and Khera 2017). Although there are still issues with these programs, including corruption and the lack of accountability in program implementation, the program has been found to significantly reduce lean-season poverty, to increase consumption particularly among scheduled castes (Bose 2017; Drèze and Khera 2017), and to increase wages (Merfeld 2019).

Potential Factors Limiting Structural Transformation in South Asia

Despite the agricultural and rural development described in SA in this chapter, the region has still lagged behind Northeast Asia (such as the Republic of Korea) and some countries in Southeast Asia in terms of structural transformation. This is despite the fact that these countries had started with similar factor endowments and level of economic development in the 1950s and had similar access to the Green Revolution technologies. While it is difficult to pinpoint exact factors that have led to such divergence, two potential factors are irrigation and education.

The difference in the use of irrigation for Green Revolution technologies between SA and East Asia might have been responsible for the differences in productivity growth between these regions. In Bangladesh and India, which account for the majority of the rice area in SA, the share of the total rice area that was irrigated remained lower than in many East Asian countries, and this has been consistent with differential rice yields in these regions (Figure 4.7). Often, irrigation had been the most important factor affecting the adoption of improved rice varieties in Asia during the Green Revolution (Estudillo and Otsuka 2012). This was because Green Revolution technologies had been developed more for irrigated areas, while the development of varieties for

FIGURE 4.7 Share of rice area irrigated (%) and rice yield (tons per hectare) in selected Asian countries and regions

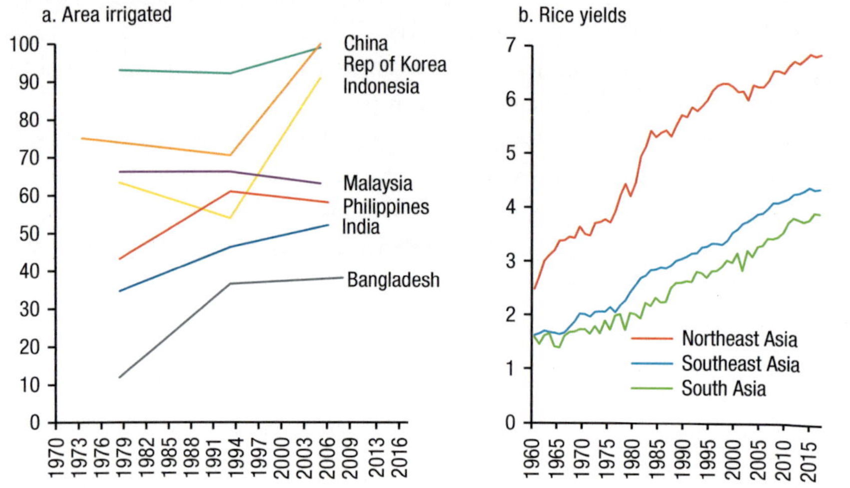

Source: FAO (2020b) for post-2000 shares of irrigated areas; FAO (2020a) for rice yield; Pingali and Hussain (1998) for shares of irrigated areas for 1978 and 1992–1994.

rainfed areas came much later. Importantly, however, SA countries have also been catching up with East Asia. In the first 30 years of the country's independence, Bangladesh invested US$5 billion (in real terms) in large-scale irrigation projects, in part through the financial loans from the development partner (Rashid et al. 2013). Similarly, India invested more than $9 billion in public irrigation projects between 1971 and 1995 (Rashid et al. 2013). These investments may help narrow the transformation gap between SA and East Asia.

Slower labor movement out of the agricultural sector in SA might have been partly associated with a historically lower human capital level in the region. Average education levels among adults have remained lower than those in East Asia over the last half century, despite the gradual increases during this period (Figure 4.8). These differences are likely to have affected the overall economic growth in these regions. Siddiqui and Rehman (2017) suggest that educational differences largely affected the growth difference between East Asia and SA. In India and China, education has had similar effects on growth, and thus the gap in education between these two countries has accounted for the gap in economic transformation (Bosworth and Collins 2008).

FIGURE 4.8 Average years of education among adults (15 years old and older) in selected South Asian countries contrasted with East Asian countries

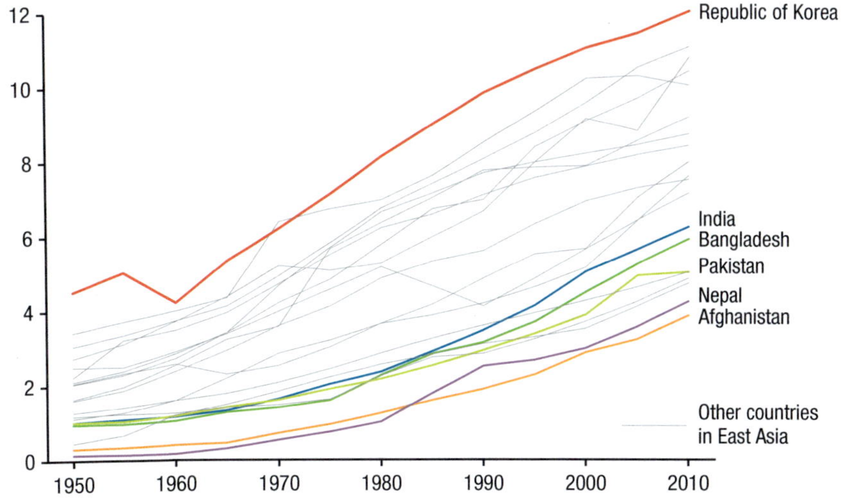

Source: Authors' calculations based on Barro and Lee (2013).

Concluding Remarks

SA has one of the highest labor-land ratios in the world. However, the region seems to have followed the Asian-type agricultural growth path characterized by Hayami and Ruttan (1985). Since the 1980s, even the most land-scarce countries, like Bangladesh and Nepal, seem to have started reversing the previous trend of declining labor productivity up to the 1970s. The region appears to be finally shifting toward the beginning of the transformation process with greater increases in labor productivity than land productivity, which many advanced countries have experienced in the past. The underlying technology landscape seems to have shifted as well; mechanization growth in the past three decades has been considerable and seems to have reached the stage where further increases in power and machinery use are associated with a significant increase in the land-labor ratio. This is consistent with the growing evidence in the region that smaller farms are becoming inefficient relative to the larger farms. Institutional failures in developing land markets can become increasingly important bottlenecks.

Land productivity growth in SA has followed diverse patterns, with some countries shifting from cereals to higher-value vegetables and fruits, some countries shifting from traditional export crops to cereals, and some countries shifting from traditional cereals to higher-value cereals. Yet, there has been a convergence of agricultural output per unit of land within SA.

Technological innovations, ranging from the adoption of machines and chemicals to improved varieties, have been led by both the public and the private sector in SA. Public support and institutional innovations seem to be associated with expanded provisions of agricultural finance and insurance, as well as collective actions in milk production, improved food safety practices, and management of common resources including irrigation facilities and forests. Liberalization and regional integration might have facilitated the trade growths of agricultural commodities within the region, and they have potentially improved national resilience against food crises.

SA, however, still faces various emerging challenges. Natural resources, including water resources, continue to degrade. Nutritional improvements still seem to lag behind agricultural growth. The varying agricultural growth might also have important implications for inequality, between urban or peri-urban areas experiencing fast modernization and rural areas where the low-paying agricultural sector remains the dominant source of employment. Raising the labor productivity in agriculture is likely to be critical in SA, and it will require, among other things, careful balancing of support for rural nonfarm employment, facilitating labor movement from farm to the nonfarm sector, and continued support for the farm sector in ways that lead to productivity enhancement without resource degradation.

References

Adamopoulos, T., and D. Restuccia. 2014. "The Size Distribution of Farms and International Productivity Differences." *American Economic Review* 104 (6): 1667–1697.

ADB (Asian Development Bank). 2007. *Sri Lanka Country Assistance Program Evaluation: Agriculture and Natural Resources Sector Assistance Evaluation. Supplementary Appendix A.* Operations Evaluation Department, ADB.

———. 2014. *Urban Poverty in Asia.* Mandaluyong, Philippines.

Ahmed, F., A. Greenleaf, and A. Sacks. 2014. "The Paradox of Export Growth in Areas of Weak Governance: The Case of the Ready Made Garment Sector in Bangladesh." *World Development* 56: 258–271.

Ahmed, R. 1995. "Liberalization of Agricultural Input Markets in Bangladesh: Process, Impact, and Lessons." *Agricultural Economics* 12 (2): 115–128.

Aizenman, J., and Y. Jinjarak. 2009. "Globalisation and Developing Countries: A Shrinking Tax Base?" *Journal of Development Studies* 45 (5): 653–671.

Alauddin, M., and J. Quiggin. 2008. "Agricultural Intensification, Irrigation and the Environment in South Asia: Issues and Policy Options." *Ecological Economics* 65 (1): 111–124.

Aneja, V. P., W. H. Schlesinger, J. W. Erisman, S. N. Behera, M. Sharma, and W. Battye. 2012. "Reactive Nitrogen Emissions from Crop and Livestock Farming in India." *Atmospheric Environment* 47: 92–103.

Bardhan, P. 2000. "Irrigation and Cooperation: An Empirical Analysis of 48 Irrigation Communities in South India." *Economic Development and Cultural Change* 48 (4): 847–865.

Barro, R, and J. W. Lee. 2013. "A New Data Set of Educational Attainment in the World, 1950–2010." *Journal of Development Economics* 104: 184–198.

Bose, N. 2017. "Raising Consumption through India's National Rural Employment Guarantee Scheme." *World Development* 96: 245–263.

Bosworth, B., and S. M. Collins. 2008. "Accounting for Growth: Comparing China and India." *Journal of Economic Perspectives* 22 (1): 45–66.

Chowdhury, S. 2011. "Employment in India: What Does the Latest Data Show?" *Economic and Political Weekly* 46 (32): 23–26.

Coelli, T., and D. Rao. 2005. "Total Factor Productivity Growth in Agriculture: A Malmquist Index Analysis of 93 Countries, 1980–2000." *Agricultural Economics* 32 (S1): 115–134.

Cole, S., X. Gine, J. Tobacman, P. Topalova, R. Townsend, and J. Vickery. 2013. "Barriers to Household Risk Management: Evidence from India." *American Economic Journal: Applied Economics* 5 (1): 104–135.

Deininger, K., S. Jin, Y. Liu, and S. K. Singh. 2018. "Can Labor Market Imperfections Explain Changes in the Inverse Farm Size-Productivity Relationship? Longitudinal Evidence from Rural India." *Land Economics* 94 (2): 239–258.

Deininger, K., S. Jin, and M. Sur. 2007. "Sri Lanka's Rural Non-Farm Economy: Removing Constraints to Pro-Poor Growth." *World Development* 35 (12): 2056–2078.

Deininger, K., D. Monchuk, H. Nagarajan, and S. Singh. 2017. "Does Land Fragmentation Increase the Cost of Cultivation? Evidence from India." *Journal of Development Studies* 53 (1): 82–98.

del Ninno, C., and P. A. Dorosh. 2001. "Averting a Food Crisis: Private Imports and Public Targeted Distribution in Bangladesh after the 1998 Flood." *Agricultural Economics* 25 (2–3): 337–346.

Drèze, J., and R. Khera. 2017. "Recent Social Security Initiatives in India." *World Development* 98: 555–572.

Edmonds, E. 2002. "Government-Initiated Community Resource Management and Local Resource Extraction from Nepal's Forests." *Journal of Development Economics* 68 (1): 89–115.

Ellis, P., and M. Roberts. 2016. *Leveraging Urbanization in South Asia: Managing Spatial Transformation for Prosperity and Livability.* Washington, DC: World Bank.

Estudillo, J. P., and K. Otsuka. 2012. "Lessons from the Asian Green Revolution in Rice." In *An African Green Revolution: Finding Ways to Boost Productivity on Small Farms*, edited by K. Otsuka and D. Larson, Chapter 2. Switzerland: Springer.

Fafchamps, M., R. V. Hill, and B. Minten. 2008. "Quality Control in Nonstaple Food Markets: Evidence from India." *Agricultural Economics* 38 (3): 251–266.

FAO (Food and Agriculture Organization of the United Nations). 1994. *Land Degradation in South Asia: Its Severity, Causes and Effects on People.* World Soil Resources Report 789. Rome.

———. 2019. "Fishery and Aquaculture Statistics. Global Aquaculture Production 1950–2017 (FishstatJ)." FishStatJ—Software for Fishery and Aquaculture Statistical Time Series. www.fao.org/fishery/statistics/software/fishstatj/en.

———. 2020a. FAOSTAT database. Accessed 2020. http://www.fao.org/faostat/en/#data.

———. 2020b. AQUASTAT: FAO's Global Information System on Water and Agriculture. Accessed June 8, 2020. http://www.fao.org/aquastat/en/databases/.

———. 2020c. "World Programme for the Census of Agriculture." Accessed June 8, 2020. http://www.fao.org/world-census-agriculture/en/.

Foster, A., and M. Rosenzweig. 1996. "Technical Change and Human Capital Returns and Investments: Evidence from the Green Revolution." *American Economic Review* 86 (4): 931–953.

———. 2003. "Economic Growth and the Rise of Forests." *Quarterly Journal of Economics* 118 (2): 601–637.

———. 2004. "Agricultural Productivity Growth, Rural Economic Diversity, and Economic Reform: India, 1970–2000." *Economic Development and Cultural Change* 52 (3): 509–542.

Gulati, A., P. Terway, and S. Hussain. 2018. *Crop Insurance in India: Key Issues and Way Forward.* Working Paper 352. Indian Council for Research on International Economic Relations. New Delhi.

Gupta, S., B. Minten, N. Rao, and T. Reardon. 2017. "The Rapid Diffusion of Herbicides in Farming in India: Patterns, Determinants, and Effects on Labor Productivity." *European Journal of Development Research* 29 (3): 596–613.

Haggblade, S., P. Hazell, and T. Reardon. 2010. "The Rural Non-Farm Economy: Prospects for Growth and Poverty Reduction." *World Development* 38 (10): 1429–1441.

Hayami, Y., and V. W. Ruttan. 1985. *Agricultural Development: An International Perspective.* Baltimore: Johns Hopkins University Press.

Hill, R., N. Kumar, N. Magnan, S. Makhija, F. de Nicola, D. Spielman, and P. Ward. 2019. "Ex Ante and Ex Post Effects of Hybrid Index Insurance in Bangladesh." *Journal of Development Economics* 136: 1–17.

Hossain, M., D. Gollin, V. Camanilla et al. 2003. "International Research and Genetic Improvement in Rice: Evidence from Asia and Latin America." In *Crop Variety Improvement and Its Effect on Productivity: The Impact of International Agricultural Research*, edited by R. Evenson and D. Gollin. Wallingford, UK: CABI.

IMHE (Institute for Health Metrics and Evaluation). 2014. *Global Burden of Disease Study 2013.* Seattle.

Joshi, P. K., A. Gulati, and R. Cummings. 2007. *Agricultural Diversification and Smallholders in South Asia.* Chicago: Academic Foundation.

Joshi, P. K., S. Parappurathu, and P. Kumar. 2016. "Dynamics of Food Consumption and Nutrient Insecurity in India." *Proceedings of Indian National Science Academy* 82 (5): 1587–1599.

Kawarazuka, N., and C. Béné. 2010. "Linking Small-Scale Fisheries and Aquaculture to Household Nutritional Security: An Overview." *Food Security* 2 (4): 343–357.

Khandker, S. 2007. "Coping with Flood: Role of Institutions in Bangladesh." *Agricultural Economics* 36 (2): 169–180.

Kouser, S., and M. Qaim. 2015. "Bt Cotton, Pesticide Use and Environmental Efficiency in Pakistan." *Journal of Agricultural Economics* 66 (1): 66–86.

Krishna, V., and M. Qaim. 2007. "Estimating the Adoption of Bt Eggplant in India: Who Benefits from Public–Private Partnership?" *Food Policy* 32 (5): 523–543.

Kumar, A., A. K. Mishra, S. Saroj, and P. K. Joshi. 2017. "Institutional vs Non-Institutional Credit to Agricultural Households in India: Evidence on Impact from a National Farmers' Survey." *Economic Systems* 41: 420–432.

Kumar, A., S. Saroj, P. K. Joshi, and H. Takeshima. 2018. "Does Cooperative Membership Improve Household Welfare? Evidence from a Panel Data Analysis of Smallholder Dairy Farmers in Bihar, India." *Food Policy* 75: 24–36.

Kumar, A., I. Wright, and D. Singh. 2011. "Adoption of Food Safety Practices in Milk Production: Implications for Dairy Farmers in India." *Journal of International Food & Argibusiness Marketing* 23 (4): 330–344.

Kumar, P., A. Kumar, and S. Mittal. 2004. "Total Factor Productivity of Crop Sector in the Indo-Gangetic Plain of India: Sustainability Issues Revisited." *Indian Economic Review* 39 (1): 169–201.

Mandal, S., S. Biggs, and S. Justice. 2017. *Rural Mechanization: A Driver in Agricultural Change and Rural Development.* Dhaka: Institute for Inclusive Finance and Development.

Matuschke, I., R. Mishra, and M. Qaim. 2007. "Adoption and Impact of Hybrid Wheat in India." *World Development* 35 (8): 1422–1435.

Matuschke, I., and M. Qaim. 2008. "Seed Market Privatisation and Farmers' Access to Crop Technologies: The Case of Hybrid Pearl Millet Adoption in India." *Journal of Agricultural Economics* 59 (3): 498–515.

McDermott, J., S. Staal, H. Freeman, M. Herrero, and J. Van de Steeg. 2010. "Sustaining Intensification of Smallholder Livestock Systems in the Tropics." *Livestock Science* 130 (1): 95–109.

Meenakshi, J. V. 2016. "Trends and Patterns in the Triple Burden of Malnutrition in India." *Agricultural Economics* 47 (S1): 115–134.

Merfeld, J. D. 2019. "Spatially Heterogeneous Effects of a Public Works Program." *Journal of Development Economics* 136: 151–167.

Minten, B., K. A. S. Murshid, and T. Reardon. 2013. "Food Quality Changes and Implications: Evidence from the Rice Value Chain of Bangladesh." *World Development* 42: 100–113.

Minten, B., K. M. Singh, and R. Sutradhar. 2013. "Branding and Agricultural Value Chains in Developing Countries: Insights from Bihar (India)." *Food Policy* 38: 23–34.

Mottaleb, K., and T. Sonobe. 2011. "An Inquiry into the Rapid Growth of the Garment Industry in Bangladesh." *Economic Development and Cultural Change* 60 (1): 67–89.

Mukherjee, A., and Y. Kuroda. 2003. "Productivity Growth in Indian Agriculture: Is There Evidence of Convergence across States?" *Agricultural Economics* 29 (1): 43–53.

Murgai, R. 2001. "The Green Revolution and the Productivity Paradox: Evidence from the Indian Punjab." *Agricultural Economics* 25 (2–3): 199–209.

Narayanan, S. 2016. "The Productivity of Agricultural Credit in India." *Agricultural Economics* 47 (4): 399–409.

Nepal Ministry of Health. 2016. *Nepal Demographic and Health Survey 2016.* Kathmandu, Nepal: MOH/Nepal, New ERA/Nepal, and ICF.

Nin-Pratt, A., and B. Yu. 2010. "Getting Implicit Shadow Prices Right for the Estimation of the Malmquist Index: The Case of Agricultural Total Factor Productivity in Developing Countries." *Agricultural Economics* 41 (3–4): 349–360.

Niroula, G., and G. Thapa. 2005. "Impacts and Causes of Land Fragmentation, and Lessons Learned from Land Consolidation in South Asia." *Land Use Policy* 22 (4): 358–372.

OECD-FAO (Organisation for Economic Co-operation and Development–Food and Agriculture Organization of the United Nations). 2010. *OECD-FAO Agricultural Outlook 2010.* Paris.

Otsuka, K., Y. Liu, and F. Yamauchi. 2016. "Growing Advantage of Large Farms in Asia and Its Implications for Global Food Security." *Global Food Security* 11: 5–10.

Patlolla, S., R. Goodhue, and R. Sexton. 2015. "Managing Quantity, Quality, and Timing in Indian Cane Sugar Production: Ex Post Marketing Permits or Ex Ante Production Contracts?" *World Bank Economic Review* 29 (3): 606–630.

Paul, B., and C. Vogl. 2011. "Impacts of Shrimp Farming in Bangladesh: Challenges and Alternatives." *Ocean & Coastal Management* 54 (3): 201–211.

Pingali, P., and M. Hossain, eds. 1998. *Impact of Rice Research.* Los Baños, Philippines: Thailand Development Research Institute and International Rice Research Institute.

———. 2007. "Agricultural Mechanization: Adoption Patterns and Economic Impact." In *Handbook of Agricultural Economics*, edited by R. Evenson and P. Pingali, 2779–2805. Amsterdam: Elsevier.

———. 2012. "Green Revolution: Impacts, Limits, and the Path Ahead." *Proceedings of the National Academy of Sciences* 109 (31): 12302–12308.

Popkin, B. 1993. "Nutritional Patterns and Transitions." *Population and Development Review* 19 (1): 138–157.

Pray, C., and L. Nagarajan. 2014. "The Transformation of the Indian Agricultural Input Industry: Has It Increased Agricultural R&D?" *Agricultural Economics* 45 (S1): 145–156.

Rada, N. 2016. "India's Post-Green-Revolution Agricultural Performance: What Is Driving Growth?" *Agricultural Economics* 47 (3): 341–350.

Rahman, S., and M. Rahman. 2009. "Impact of Land Fragmentation and Resource Ownership on Productivity and Efficiency: The Case of Rice Producers in Bangladesh." *Land Use Policy* 26 (1): 9–103.

Rahman, S., and R. Salim. 2013. "Six Decades of Total Factor Productivity Change and Sources of Growth in Bangladesh Agriculture (1948–2008)." *Journal of Agricultural Economics* 64 (2): 275–294.

Rashid, S., P. Dorosh, M. Malek, and S. Lenma. 2013. "Modern Input Promotion in Sub-Saharan Africa: Insights from Asian Green Revolution." *Agricultural Economics* 44 (6): 705–721.

Reardon, T., K. Chen, B. Minten, and L. Adriano. 2012. *The Quiet Revolution in Staple Food Value Chains in Asia: Enter the Dragon, the Elephant, and the Tiger.* Washington, DC: Asian Development Bank and International Food Policy Research Institute.

Renkow, M. 1993. "Differential Technology Adoption and Income Distribution in Pakistan: Implications for Research Resource Allocation." *American Journal of Agricultural Economics* 75: 33–43.

Sahoo, P., G. Nataraj, and R. Dash. 2013. *Foreign Direct Investment in South Asia: Policy, Impact, Determinants and Challenges.* Switzerland: Springer Science & Business Media.

Shah, T., M. Hassan, M. Khattak, P. Banerjee, O. Singh, and S. Rehman. 2009. "Is Irrigation Water Free? A Reality Check in the Indo-Gangetic Basin." *World Development* 37 (2): 422–434.

Siddiqui, A., and A. Rehman. 2017. "The Human Capital and Economic Growth Nexus: In East and South Asia." *Applied Economics* 49 (28): 2697–2710.

Spielman, D., S. Malik, P. Dorosh, and N. Ahmad. 2017. *Agriculture and the Rural Economy in Pakistan: Issues, Outlooks, and Policy Priorities.* Philadelphia: University of Pennsylvania Press.

Subramanian, A., and M. Qaim. 2009. "Village-Wide Effects of Agricultural Biotechnology: The Case of Bt Cotton in India." *World Development* 37 (1): 256–267.

Suhariyanto, K., and C. Thirtle. 2001. "Asian Agricultural Productivity and Convergence." *Journal of Agricultural Economics* 52 (3): 96–110.

Takeshima, H. 2017a. "Custom-Hired Tractor Services and Returns to Scale in Smallholder Agriculture: A Production Function Approach." *Agricultural Economics* 48 (3): 363–372.

———. 2017b. "Overview of the Evolution of Agricultural Mechanization in Nepal, with a Particular Focus on Tractors and Combine Harvesters," IFPRI Discussion Paper 01662, International Food Policy Research Institute, Washington, DC.

Takeshima, H., and L. Nagarajan. 2012. "Minor Millets in Tamil Nadu, India: Local Market Participation, On-Farm Diversity and Farmer Welfare." *Environment and Development Economics* 17 (5): 603–632.

Toufique, K., and B. Belton. 2014. "Is Aquaculture Pro-Poor? Empirical Evidence of Impacts on Fish Consumption in Bangladesh." *World Development* 64: 609–620.

Upadhyaya, H., K. Otsuka, and C. C. David. 1990. "Differential Adoption of Modern Rice Technology and Regional Wage Differential in Nepal." *Journal of Development Studies* 26 (3): 450–468.

USDA (United States Department of Agriculture). 2020. Economic Research Service Agricultural Productivity Project. Accessed June 8. https://www.ers.usda.gov/data-products/international-agricultural-productivity/.

Weinberger, K., and T. A. Lumpkin. 2007. "Diversification into Horticulture and Poverty Reduction: A Research Agenda." *World Development* 35 (8): 1464–1480.

World Bank. 2017. Data. Accessed November 15, 2017. http://data.worldbank.org/.

———. 2018. World Development Indicators database. Accessed 2018. http://wdi.org.

Zhang, X., and S. Fan. 2004. "How Productive Is Infrastructure? A New Approach and Evidence from Rural India." *American Journal of Agricultural Economics* 86 (2): 492–501.

Chapter 5

AFRICA'S UNFOLDING AGRICULTURAL TRANSFORMATION

Ousmane Badiane, Xinshen Diao, and Thomas Jayne[1]

For decades, observers of Africa have referred to the region's economic transformation in the future tense.[2] Today, most development scholars agree that Africa has experienced an extraordinary period of economic development in the 21st century, even while the underlying causes are still not fully understood.

This chapter on Africa documents the region's unfolding agricultural and broader economic transformations, explores the underlying drivers and implications of these transformations, and considers the new policy priorities dictated by these developments. It starts with an analysis of the shifting role of agriculture in the development of African economies, and the history and status of agricultural transformation and its implications for economic growth, poverty reduction, food systems, and nutrition. The second section reviews changes in agricultural productivity growth over time. The third section discusses patterns of poverty reduction and urbanization and related changes in diets, as well as their effects on agrifood systems and nutrition outcomes. The fourth section summarizes four broad trends that illustrate Africa's agricultural transformation to date and briefly discusses their possible future evolution.

Changing Roles of Agriculture[3]

Agricultural transformation is an important element in overall economic transformation processes (Mellor 1976; Timmer 1988). During the course of agricultural transformation, agrifood systems shift from traditional low-productivity subsistence production systems to commercially oriented, high-productivity systems with significant value addition occurring off the farm (Timmer 1988). Agricultural transformation contributes to overarching

1 This chapter has benefited from excellent research assistance by Julia Collins.

2 *Africa* refers to "Africa south of the Sahara," which is the regional focus of this chapter.

3 This section draws heavily on Jayne, Chamberlin, and Benfica (2018).

economic and rural transformation processes, which are associated with increased heterogeneity of livelihood opportunities, stronger links between rural and urban areas, and changing roles for agriculture (IFAD 2016).

As is discussed in Chapter 1, agriculture plays a central role in economic transformation as the main source of employment, income, and fiscal resources, all of which are essential in fueling demand for goods produced in the nascent industrial sector and financing public goods and services necessary to foster broader growth in the economy. As the nonagricultural sector expands, labor moves from agriculture to manufacturing and services, resulting in higher overall productivity and incomes as the economy matures.

Many of these transformation processes are now clearly visible in Africa. After decades of stagnation, much of Africa has enjoyed sustained agricultural growth—4.73 percent per year on average between 2000 and 2018.[4] Real per capita gross domestic product (GDP) grew by over one-third on average between 2000 and 2014, with faster growth of 100 percent or more in some countries (Barrett et al. 2017). Poverty rates have fallen significantly—the share of people in Africa living on less than US$1.90 a day declined from 55 percent in 1990 to 42 percent in 2015 (World Bank 2020b).[5] Africa's workforce is shifting, in some cases quite rapidly, from farming to off-farm sectors. The number of medium- and large-scale farms is increasing, and in some countries they account for a sizable and rising portion of total farmland (Jayne et al. 2016). Agribusiness and downstream food systems are responding dynamically to population growth and urbanization. Changing food diets associated with income growth and urbanization are raising the demand for processed food products and creating new employment opportunities in agrifood systems (Tschirley et al. 2015b). Governments that have invested robustly in their agricultural sectors, such as in Ethiopia, Rwanda, and Burkina Faso, are reaping the benefits—stronger economic growth, declining poverty rates, and better nutritional status (Badiane and Makombe 2015). The countries with the highest rates of agricultural productivity growth during this period also tended to experience the most rapid diversification of the labor force into off-farm and nonfarm employment (Yeboah and Jayne 2018).

---

4 Based on World Development Indicators annual growth of agricultural value-added, defined as constant local currency unit value of agriculture, forestry, and fishing outputs minus input costs (WDI code NV.AGR.TOTL.KD.ZG). Africa's impressive agricultural growth rate over the 2000–2018 period is weighted by population and is led by high rates of growth in Nigeria and Ethiopia, but many African countries also achieved annual growth rates well over the world average of 2.75 percent in this period.

5 All dollars are US dollars.

The economic landscapes in which small farmers have traditionally operated are therefore shifting rapidly. Urbanization and development of food systems to feed growing cities are reshaping African farmers' access to markets, starting with those closest to towns and moving outward into hinterland areas (Richards et al. 2016). The rise of secondary cities has expanded market access and extended value chains into previously hard-to-reach areas (Chamberlin and Jayne 2013). Access to output and input markets is increasing for many farmers, and the share of rural populations living in isolated areas—while still high—has fallen sharply (Masters et al. 2013). In many countries, average farm sizes are increasing as a growing group of urban-based investor farmers acquire medium- and large-scale farms. Markets for agricultural factors, including land, labor, and labor-saving inputs such as fertilizer, pesticides, herbicides, and mechanization, are seeing increased farmer participation in many areas (Deininger, Savastano, and Xia 2017; AGRA 2016; Jayne et al. 2019). Agricultural commercialization is increasingly attracting private investment. In areas with growing numbers of medium-scale farms, large-scale traders are also increasing investments and expanding operations (Sitko, Burke, and Jayne 2018). Africa's rapid urbanization has created sharply rising demand for food in urban markets, leading to growing opportunities for farmers, traders, and food processing along the value chain (Reardon 2015).

While these are unmistakable positive developments, there remain at least three major challenges that, if not proactively addressed, will impede the pace of Africa's economic transformations. First, unlike in the stylized Asian structural transformation process, a large majority of African countries have not achieved meaningful growth in manufacturing. In Asian countries such as Bangladesh, China, the Republic of Korea, and Viet Nam, labor-intensive manufacturing, often for export, offered a higher-productivity and more remunerative alternative to farming. People are leaving farming in Africa as well, but most of the former agricultural labor force enters informal sectors that are not able to provide sufficient numbers of well-paying jobs (Diao, Magalhaes, and McMillan 2018). The fastest-growing occupations are often nontradable, such as construction, the food trade, cooking goods, and personal care services. Private-sector employment, whether informal or formal, is growing more rapidly in most countries than public-sector jobs (Diao, Harttgen, and McMillan 2017; Yeboah and Jayne 2018).

Second, the region's transformation in recent years masks great heterogeneity across countries. Countries experiencing particularly severe challenges are those in which fast population growth and urbanization are taking place in the absence of significant productivity growth and dynamism in either agricultural or nonfarm sectors. This process, termed *urbanization without industrialization*, has occurred in countries such as Angola, Equatorial Guinea, Zambia, and others where natural resource exports (for example, oil and mining products) are a main driver of growth but do not show strong linkages with rural areas. In these countries, growth has resulted in little or no poverty reduction (McMillan, Rodrick, and Verduzco 2014; Gollin, Jedwab, and Vollrath 2016; IFAD 2016).

Third, agricultural growth in Africa has been driven mainly by area expansion rather than productivity increases (Fuglie and Rada 2013). In addition, the intensification that has taken place has resulted in widespread land degradation and soil nutrient depletion (Drechsel et al. 2001; Barbier and Hochard 2012; Tittonell and Giller 2013). Sustainable output growth will result not from exhausting natural resources but from increased productivity, producing more per unit of land and labor. The limits of production growth based on land expansion will be reached quickly, given continued population growth (Masters et al. 2013; Chamberlin, Jayne, and Headey 2014). Relatedly, Africa's recent agricultural growth has been supported by a period of unprecedentedly high world commodity prices. Unlike in the past, however, actors all along the commodity value chains were able to respond to the relatively high post-2007 food prices, thanks to sectoral commodity market reforms undertaken during the 1990s and early 2000s that cleared away major policy barriers to private investment in the agrifood systems of many African countries (Badiane, Benin, and Makombe 2016; Jayne, Chamberlin, and Benfica 2018).

How Important Is Agriculture Still?

A major emerging question about Africa's unfolding economic transformation is about the role of agriculture. The decades-long stagnation or even decline of the agricultural sector has delayed structural transformation and undermined livelihoods across the continent. In nearly all countries, agriculture as a share of GDP has declined so rapidly that African economies entered the new millennium with markedly smaller agricultural sectors than would be expected based on their level of development (Figure 5.1). And countries where this *stunting* of the agricultural sector was the strongest also historically experienced the highest rates of poverty (Figure 5.2).

As downstream agrifood systems and nonfarm employment grow rapidly in Africa, agriculture could be viewed as a declining sector and hence less crucial as a driver of continued economic transformation. As noted earlier, a growing proportion of young people are entering into off-farm and nonfarm

FIGURE 5.1 Actual versus expected agricultural sector GDP shares (2010–2018)

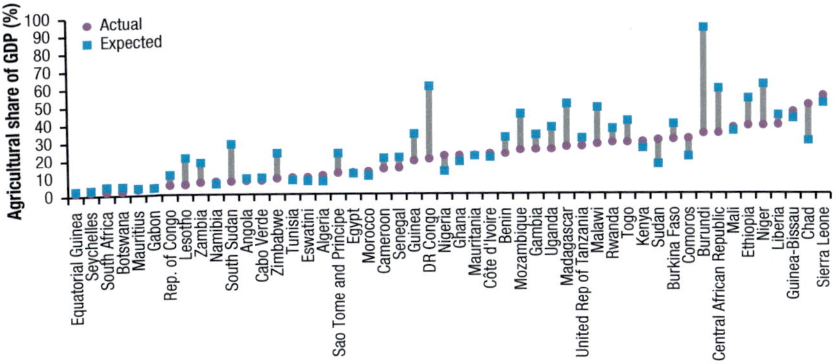

Source: Authors' calculations.

Note: Expected agricultural sector GDP shares are calculated based on a random effects model of the relationship between GDP per capita and agricultural sector GDP share for a sample of 197 countries over the 1980–2018 period. For further methodological details, see Badiane (2011).

FIGURE 5.2 Agricultural sector underperformance and poverty levels

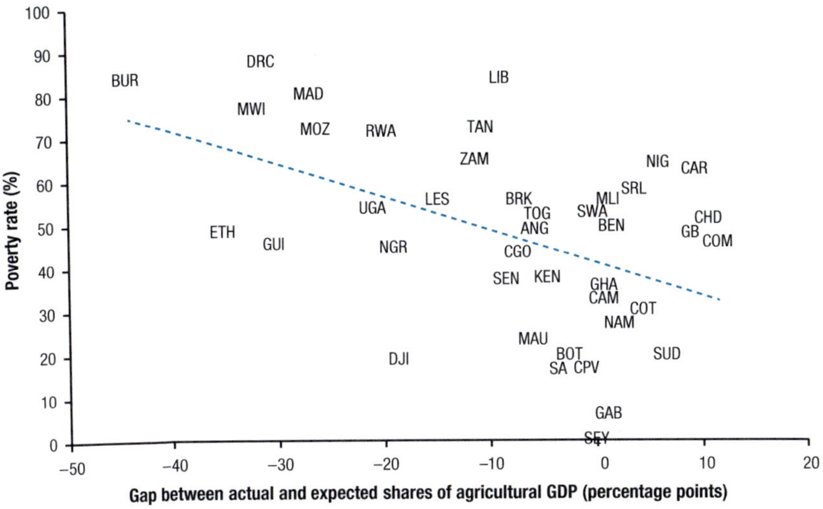

Source: Badiane and Collins (2016).

Note: ANG = Angola; BEN = Benin; BOT = Botswana; BRK = Burkina Faso; BUR = Burundi; CAM = Cameroon; CAR = Central African Republic; CGO = Republic of Congo; CHD = Chad; COM = Comoros; COT = Côte d'Ivoire; CPV = Cabo Verde; DJI = Djibouti; DRC = Democratic Republic of the Congo; ETH = Ethiopia; GAB = Gabon; GB = Guinea Bissau; GHA = Ghana; GUI = Guinea; KEN = Kenya; LES = Lesotho; LIB = Liberia; MAD = Madagascar; MAU = Mauritania; MLI = Mali; MOR = Morocco; MOZ = Mozambique; MWI = Malawi; NAM = Namibia; NGR = Niger; NIG = Nigeria; RWA = Rwanda; SA = South Africa; SEN = Senegal; SEY = Seychelles; SRL = Sierra Leone; SUD = Sudan; SWA = Eswatini; TAN = Tanzania; TOG = Togo; UGA = Uganda; ZAM = Zambia.

FIGURE 5.3 **Agricultural transformation and rural poverty, 1991–2018**

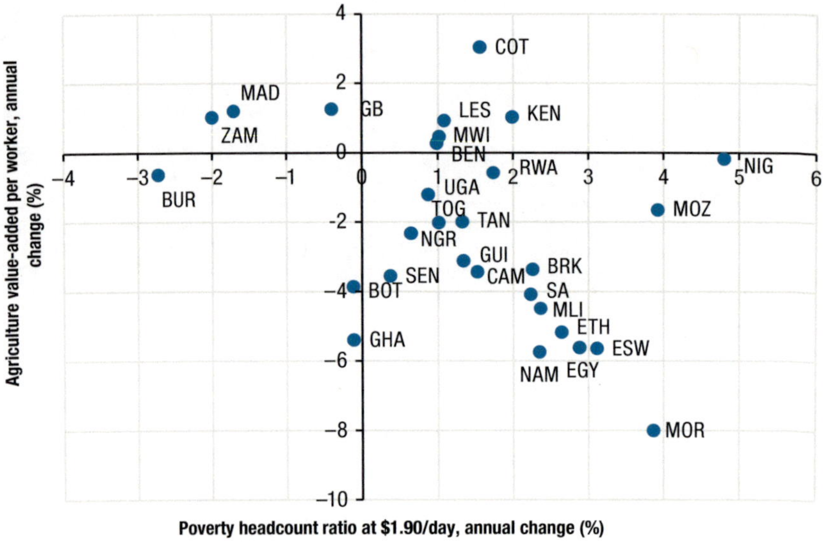

Source: Authors' calculations base on ReSAKSS (2020).
Note: For country names, see the note for Figure 5.2.

employment activities, and the share of the labor force primarily engaged in farming has declined rapidly. In many countries, farming constitutes less than 50 percent of the total labor force when computed in terms of full-time equivalents (Yeboah and Jayne 2018). Secondary cities and towns have produced the most rapid employment growth in some countries and are thus likely to accelerate structural transformation (Christiaensen and Todo 2014).

However, the causal relationships between growth in agricultural production, downstream value-chain segments, and nonfarm employment have yet to be clearly disentangled. The region's impressive agricultural productivity growth since 2000 may have had a big hand in driving the growth of nonfarm and downstream agrifood systems in recent years. Yeboah and Jayne (2018) find for a set of African countries that those registering the greatest growth in agricultural productivity per worker in farming have also tended to experience the most rapid shifts in the labor force out of farming as well as faster labor productivity growth in nonfarm sectors. Countries with faster agricultural labor productivity growth have been observed to experience greater poverty reduction than countries with lower productivity growth (Figure 5.3). And Imai et al. (2017) find that agricultural growth remains strongly associated with poverty reduction and that greater emphasis on promoting rural development (both agricultural and nonagricultural) may still provide the most effective means of reducing poverty in most African settings. In addition, some evidence suggests that sectoral productivity differences may be exaggerated: Hicks et al. (2017) found, using data from Kenya, that a large part of observed productivity differences between the farm and nonfarm sectors were explained by differences in the education and ability of the individuals who chose to migrate out of agriculture. When individual fixed effects were controlled for, the estimated labor productivity gap decreased by around 90 percent. Growth dynamics in Africa's agricultural sector and their implications for development and poverty reduction are examined in greater depth in the next section.

A Review of Productivity Growth in Africa's Agriculture

The 21st century is undeniably an unprecedented new era for Africa. Because of the disappointing growth performance in the 1980s and early 1990s, for many countries, the current levels of per capita GDP or per capita agricultural GDP/agricultural production are still below their historical peak in the 1960s or 1970s. Table 5.1 summarizes this situation. Among 44 African countries with GDP data available for the 1960s to 1980s, there are 29 countries for which the best performance of per capita GDP growth was in the early years, before growth faltered and per capita GDP fell in the late 1970s. A similar situation is also seen for agricultural growth performance, with more countries having a long growth-declining period in agriculture. Among these countries, some have fully recovered in the 21st century, with the new high in per capita GDP or agricultural GDP (AgGDP) per output in the 2000s above the peak level in the past. Column 7 of Table 5.1 lists the total number of such countries. In terms of gross production of agriculture, the number of countries that have yet to recover to their historical high in per capita agricultural output is still more than the countries that have recovered (Table 5.1).

Comparisons of Africa's Agricultural Productivity

Against the background of the longer-term growth performance in Africa highlighted in Table 5.1, we compare agricultural productivity between the period of 1986–1989 and the period of 2010–2014. Following Hayami and Ruttan (1985), agricultural labor productivity is measured as agricultural

TABLE 5.1 **Number of African countries with peak levels of per capita GDP or per capita AgGDP/output in the past**

Product/output (per capita)	Number of countries with maximum value in the past: 1960s	1970s	1980s	Peak year in 1960s–1980s	No data in 1960s–1980s	No decline following 1960s–1980s	Whether the new high created in the 2000s was above the historical peak: Yes	No
	(1)	(2)	(3)	(4)	(5)	(6)	(7)	(8)
GDP	5	12	12	29	4	15	14	15
AgGDP	8	11	7	26	11	11	7	19
Ag output	21	19	1	41	2	5	17	24

Source: Authors' calculation using GDP, AgGDP, and population data from World Bank (2017) and agricultural output data from FAO (2017).

Note: GDP = gross domestic product; AgGDP = agricultural GDP; Ag output = gross production of agriculture. The numbers reported in the first three columns are the numbers of countries with maximum levels of per capita GDP or agricultural GDP/agricultural output having occurred in the 1960s, 1970s, or 1980s respectively. The fourth column is the sum of the first three columns. The fifth column reports the number of countries with no data in the 1960s–1980s. The sixth column reports the number of countries that did not experience a decline in growth and in which the maximum levels of per capita GDP or agricultural GDP/agricultural output did not occur in the past.

output per worker,[6] while agricultural land productivity is output per hectare. There are significant differences in both labor and land productivity across African countries, as well as the difference between the two periods within a country. We want to assess whether such differences and change in land and labor productivity can be partially explained by the situation of agricultural endowment proposed in the induced technical innovation hypothesis in Hayami and Ruttan (1985).

COMPARISONS IN LABOR AND LAND PRODUCTIVITY ACROSS COUNTRIES

Figure 5.3 reports labor and land productivity for 46 African countries in 1985–1989 and 2010–2014 and the land-labor ratio in the two periods. Significant variation exists across countries for both labor and land

6 This section uses data from FAOSTAT on the number of agricultural workers to calculate agricultural labor productivity. The FAOSTAT data have the advantage of covering a large number of countries over time, but they also present two major disadvantages. First, FAOSTAT estimates of numbers of agricultural workers are often larger than those suggested by household survey data. Second, the data do not account for the fact, apparent from survey data, that rural people are devoting a declining share of their time to agriculture as engagement with rural nonfarm activities increases. Thus, even though the numbers of agricultural workers are rising, the actual labor time devoted to agriculture is not rising as quickly. The implication of the first issue is that labor productivity may be underestimated in this analysis; the implication of the second issue is that the growth of labor productivity over time is also likely to be underestimated.

productivity in the past and at present. In Figure 5.4, South Africa and Mauritius stand out as the two countries with the highest labor productivity in the past and at present, and in Figure 5.4 Mauritius is the country with the highest land productivity both in the past and at present. A few countries at the bottoms of both figures changed over time, including Malawi, Mozambique, and Angola in Figure 5.4 and Angola, Sao Tome and Principe, and Equatorial Guinea in Figure 5.4, while the agricultural land and labor ratio seems to be relatively stable over time for most countries (Figure 5.4). In the late 1980s, Ethiopia had the lowest labor productivity (Figure 5.4), and Niger had the lowest land productivity (Figure 5.4) in Africa, while in the 2010s, Ethiopia moved up the ladder slightly as shown in Figure 5.4, while Niger remained at the bottom in Figure 5.4. Moreover, the cross-country differences are larger in labor productivity than in land productivity, and such differences are further widened in recent years. We use standard deviations (SD) to measure such differences. In the late 1980s, the SD in labor productivity is 1.6 times higher than the SD in land productivity, while the difference in SD between labor and land productivity increases to 2.7 times in recent years.

The association between the level of per capita income (measured by GDP per capita) and labor and land productivity is observed internationally. Table 5.2 shows that the correlation between GDP per capita and labor productivity in Africa is stronger than the correlation between GDP per capita and land productivity in both periods, while the two correlation coefficients converge in the recent period. We also report the correlation between per capita GDP and the land-labor ratio, which seems to be stable over time.

COMPARISONS BETWEEN LABOR AND LAND PRODUCTIVITY WITHIN COUNTRIES

The difference between labor and land productivity within a country is associated with the country's land-labor ratio, which represents the pattern of agricultural endowment in the country. Given that $(Y/L) = (A/L) \times (Y/A)$, where Y is agricultural output, L agricultural workers, and A agricultural land, when $A/L \geq (<)1$, labor productivity is either higher than or equal to (lower than) land productivity. When the ratio of land to labor is greater than one, land is a relatively abundant endowment compared with agricultural labor, and hence, labor productivity is higher than land productivity. The ratio of land and labor (A/L) is shown in Figure 5.4.

However, the ratio of land and labor is not constant over time and is affected by both the expansion of agricultural areas through agricultural investment and increased employment opportunities outside agriculture.

FIGURE 5.4 Agricultural land and labor productivity and land-labor ratios, 1985–1989 and 2010–2014

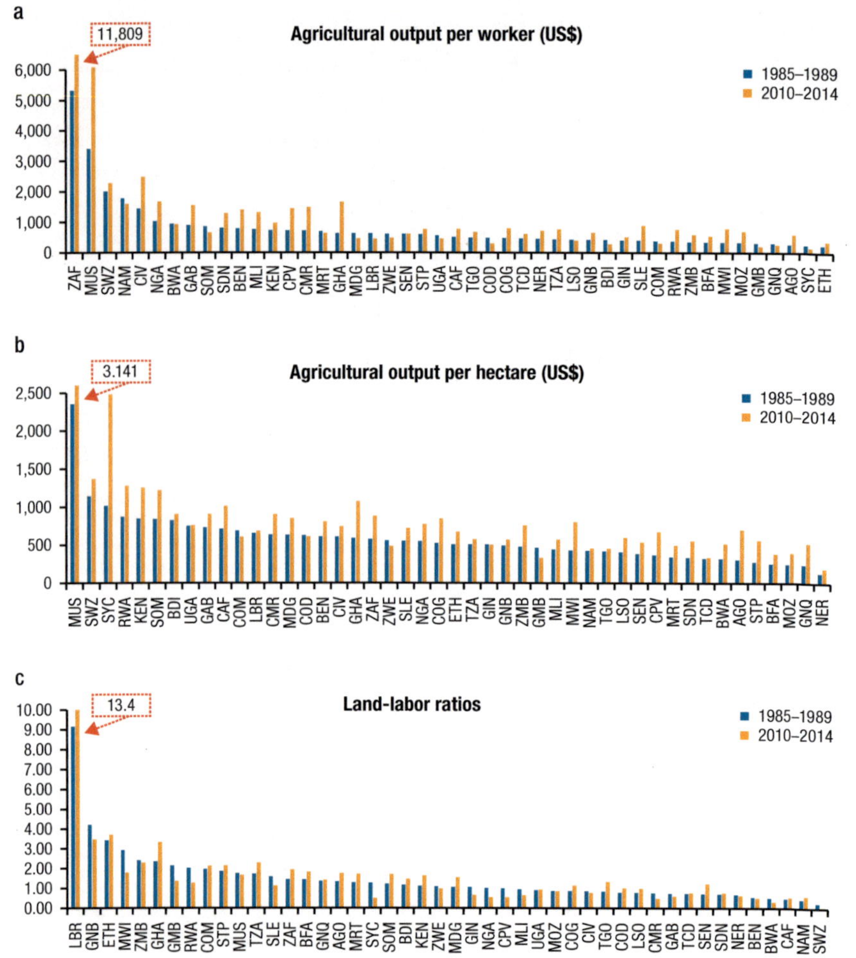

Source: Authors' calculations from agricultural output obtained from FAO (2017); Groningen Growth and Development Centre (GGDC) 10-sector database (Timmer, de Vries, and de Vries 2015); USDA-ERS (2017).

Note: AGO = Angola; BDI = Burundi; BEN = Benin; BFA = Burkina Faso; BWA = Botswana; CAF = Central African Republic; CIV = Côte d'Ivoire; CMR = Cameroon; COD = DR Congo; COG = Congo; COM = Comoros; CPV = Cabo Verde; ETH = Ethiopia; GAB = Gabon; GHA = Ghana; GIN = Guinea; GMB = Gambia; GNB = Guinea-Bissau; GNQ = Equatorial Guinea; KEN = Kenya; LBR = Liberia; LSO = Lesotho; MDG = Madagascar; MLI = Mali; MOZ = Mozambique; MRT = Mauritania; MUS = Mauritius; MWI = Malawi; NAM = Namibia; NER = Niger; NGA = Nigeria; RWA = Rwanda; SDN = Sudan; SEN = Senegal; SLE = Sierra Leone; SOM = Somalia; STP = Sao Tome and Principe; SWZ = Eswatini; SYC = Seychelles; TCD = Chad; TGO = Togo; TZA = United Rep of Tanzania; UGA = Uganda; ZAF = South Africa; ZMB = Zambia; ZWE = Zimbabwe. Values are the averages in the corresponding periods. The FAOSTAT agricultural output is measured as gross production of agriculture in constant 2004–2006 international dollars. Agricultural workers are economically active adults (15+) in agriculture reported by FAO, while for some countries with employment data available in the GGDC dataset (Timmer, de Vries, and de Vries 2015), they are the number of individuals employed in agriculture. Agricultural land is compiled by Fuglie (USDA-ERS 2017) and is in rainfed cropland equivalents.

TABLE 5.2 Correlation values of per capita GDP and labor and land productivity in agriculture

Indicator	1985–1989	2010–2014
Labor productivity	0.450	0.430
Land productivity	0.261	0.372
Land-labor ratio	0.364	0.369

Source: Authors' calculations.

Because of rapid urbanization that creates more jobs in urban areas outside agriculture, a decline in the share of agricultural employment can reverse the land-labor ratio from less than one to greater than one, as in Tanzania (Table 5.3).

Table 5.3 puts African countries into four groups. For the 22 countries in group one, labor productivity is consistently higher than land productivity during both periods (1985–1989 and 2010–2014). In group two, with 16 countries, land productivity is consistently higher than labor productivity in both periods. Countries in group three had higher labor productivity than land productivity in 1985–1989, but land productivity becomes higher in 2010–2014. The reverse is true for countries in group four. There are fewer countries in groups three and four, indicating that the endowment ratio is relatively stable for most African countries.

Table 5.3 also shows that there are more African countries with relatively abundant land as an agricultural endowment. However, apart from South Africa, which has land-labor ratios of 9.2 and 13.4 in the two periods, which is close to what the UK had in the 1960s, there are only 7 other countries with land-labor ratios greater than 2 in the two periods, a situation similar to the one in Asia.[7] On the other hand, out of 26 European countries, there are 24 with land-labor ratios higher than 2 and many countries with ratios similar to South Africa's, and out of 31 American countries, there are 23 with land-labor ratios higher than 2 and many with ratios much higher than South Africa's.

7 Land is measured as rainfed cropland equivalents, with higher weights for irrigated cropland than for permanent pastureland. Irrigated areas in Asia are three times those in Africa, while there is much more pastureland area in Africa than in Asia. This leads to higher measured land-labor ratios in many Asian countries and lower measured ratios in many African countries than actual land-labor ratios in these countries.

TABLE 5.3 Comparison between labor and land productivity, 1985–1989 and 2010–2014

Groups	Countries
Labor productivity > land productivity in both periods	Benin, Botswana, Burkina Faso, Cabo Verde, Cameroon, Chad, Côte d'Ivoire, Eswatini, Gabon, Ghana, Mali, Mauritania, Mauritius, Mozambique, Namibia, Niger, Nigeria, Sao Tome and Principe, Senegal, South Africa, Sudan, Togo
Labor productivity < land productivity in both periods	Angola, Burundi, Central African Republic, Comoros, Democratic Republic of Congo, Republic of Congo, Ethiopia, Gambia, Kenya, Liberia, Madagascar, Malawi, Rwanda, Seychelles, Uganda, Zambia
Labor productivity > land productivity in period 1 and reversed in period 2	Equatorial Guinea, Lesotho, Somalia, Zimbabwe
Labor productivity < land productivity in period 1 and reversed in period 2	Guinea, Guinea-Bissau, Sierra Leone, United Republic of Tanzania

Source: Authors' calculations.

Changes in Agricultural Productivity over Time

The trends of labor and land productivity between 1985–1989 and 2010–2014 across all African countries are presented in Figure 5.5. Both labor (along the *x*-axis) and land (*y*-axis) productivity are in log scale. The three diagonal dashed lines represent ratios of labor to land of 0.1, 1, and 100, with the middle line dividing countries into relatively land-abundant (above the middle line) and labor-abundant (below the middle line) groups. All 22 countries in group one of Table 5.3 locate in the area below the middle dashed line, and the 16 countries in group two locate in the area above the middle line. Eight countries that are in groups three and four have crossed the middle line and switched from land (labor) abundant to labor (land) abundant over time.

According to the induced technical innovation hypothesis (Hayami and Ruttan's hypothesis), countries with favorable land-labor ratios tend to have faster growth in labor productivity than land productivity as they adopt labor-saving technologies to take advantage of their relatively abundant land endowment. This implies that for the countries below the middle dashed line in Figure 5.5, trend lines are expected to be flatter than the diagonal dashed lines, while for the countries above the middle line, trend lines are expected to be steeper than the diagonal dashed lines. The productivity movements in Figure 5.5 are consistent with Hayami and Ruttan's hypothesis for many countries but with quite a few exceptions. For example, Rwanda is a land-scarce country locating above the middle dashed line, but the trend for its productivity movement is flatter than the diagonal dashed line. The same is also

FIGURE 5.5 Agricultural labor and land productivity between 1985–1989 and 2010–2014

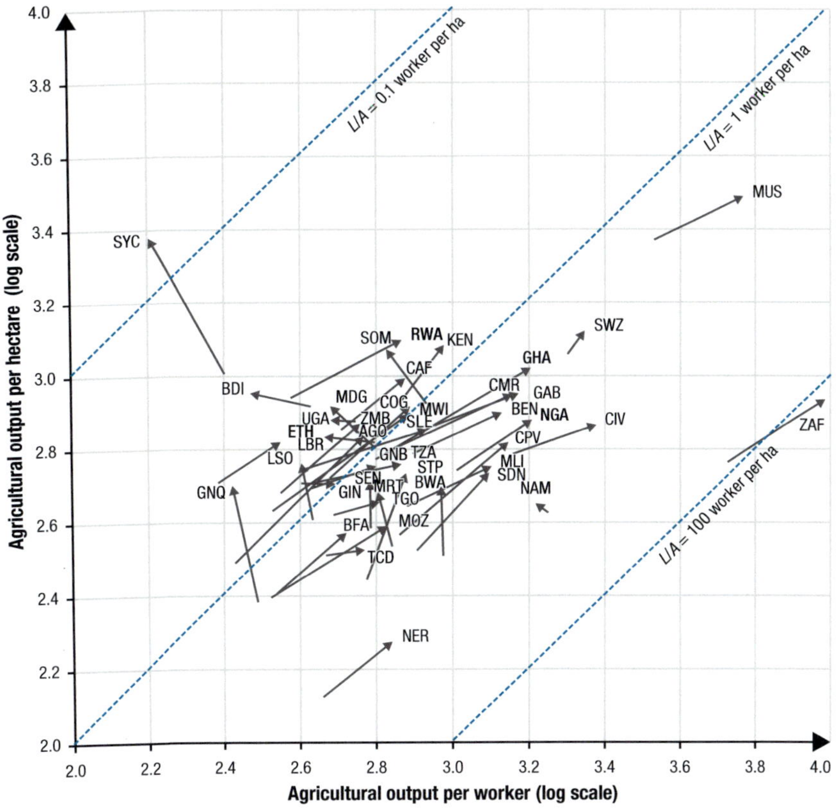

Source: Authors' calculation using data from USDA-ERS (2017).

Note: For country names, see the note for Figure 5.4. Both agricultural output per hectare and per worker are measured in purchasing power parity dollars ($PPP) (on a log scale). Four countries with negative changes in both land and labor productivity are excluded from the figure. Three are in group two and one in group three of Table 5.3.

true for Ethiopia and Malawi. The high growth in agricultural labor productivity in these countries is partially driven by rapid structural change (Diao, McMillan, and Rodrik 2017). When more labor moves from agriculture with low productivity to nonagricultural sectors with higher productivity, structural change has enhanced economywide labor productivity growth. In this process, not only does the land-labor ratio rise, but also labor productivity in agriculture grows more rapidly than land productivity.

Process of Substitution of Labor and Land

Figure 5.6 is used to assess whether increases in land productivity are associated with use of a modern input—fertilizer. The trends of output per hectare (*y*-axis) and fertilizer per hectare (*x*-axis) between 1985–1989 and 2010–2014 are presented in the figure. With Mauritius as an exception, the level of fertilizer use per hectare is extremely low among most African countries in the initial period. While for some countries the trend lines in Figure 5.6 are relatively flat, indicating that the pace of growth in fertilizer use is faster than land productivity growth, a steep trend is seen for many countries. In some countries, such as Zambia and South Africa, use of fertilizer per hectare even falls over time.

Measures of Total Factor Productivity

The productivity comparisons in the previous subsections are based on partial factor (land or labor) measures of productivity. Total factor productivity (TFP) is measured by combining the effects of all the resources used in agricultural production, which include not only labor and land, but also capital and other inputs. TFP measures critically depend on the calculation of capital stock and capital services, which are unobservable and arguably the most difficult production factors to measure. Moreover, when multiple factors are combined to derive the residue that is not explained by any such factors, it requires an assumption on the production function, and shares of labor, land, and capital are also unobservable and have to be assumed. Keeping such limitations in measuring TFP in mind, the TFP measures used in this subsection come from USDA-ERS (2017).

USDA-ERS (2017) considers inputs of land, labor, machinery power, livestock capital, synthetic nitrogen-phosphorus-potassium (N-P-K) fertilizers, and animal feed, which are weighted by their cost shares. The cost shares are assumed to be constant and the same across countries in Africa. TFP growth is the difference between growth in agricultural output and growth in weighted inputs. We report growth rates for two periods, 1981–1990 and 2001–2014, for which average TFP growth is available in USDA-ERS (2017). Figure 5.7 shows the annual growth rate of TFP in these two periods for the same countries we included in the previous analysis on partial productivity growth. Countries with land-labor ratios greater than one in 2010–2014 are represented by red dots, and the other countries are represented by blue triangles.

FIGURE 5.6 **Land productivity and fertilizer use per hectare of land between 1985–1989 and 2010–2014**

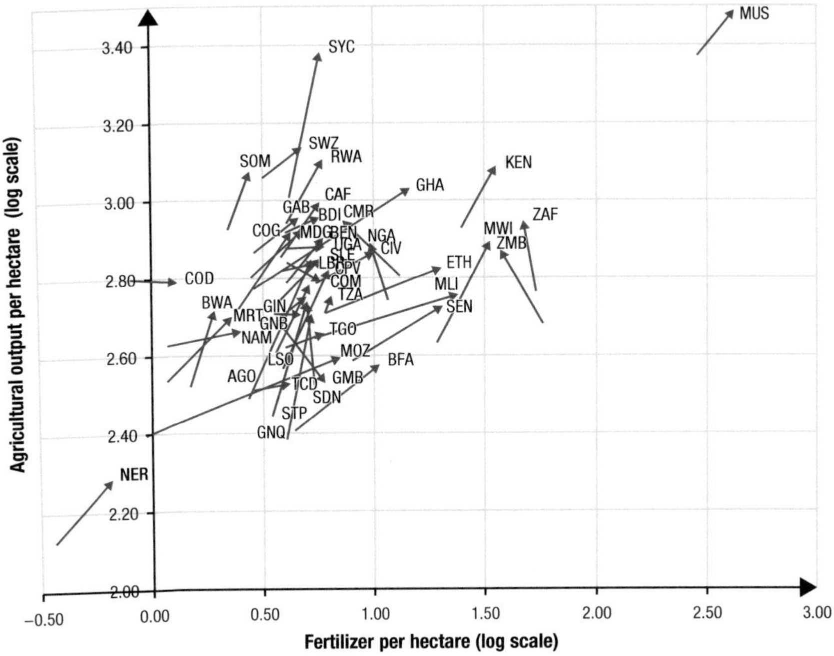

Source: Authors' calculation using data from USDA-ERS (2017).

Note: For country names, see the note for Figure 5.4. Fertilizer per hectare is measured in kilograms (in a log scale), and agricultural output per hectare is measured in purchasing power parity dollars (log scale). A negative log value of fertilizer use per hectare indicates that the absolute level of fertilizer use per hectare is less than 10 kilograms (in Niger).

The majority of countries locate in the upper right corner of Figure 5.7, implying that they have positive growth in TFP in both periods. However, for the 16 countries locating in the areas above the 45-degree line in this corner, TFP growth in the recent period (2001–2014) is higher than in the early period (1981–1990). There are 4 other countries that locate in the lower right corner of Figure 5.7, indicating that their TFP growth rate was positive in the past but turned negative in recent years. There are 5 countries with negative TFP growth in both periods, and they locate in the lower left corner of Figure 5.7.

FIGURE 5.7 TFP growth rate in 1981–1990 versus 2001–2014 (annual percentage)

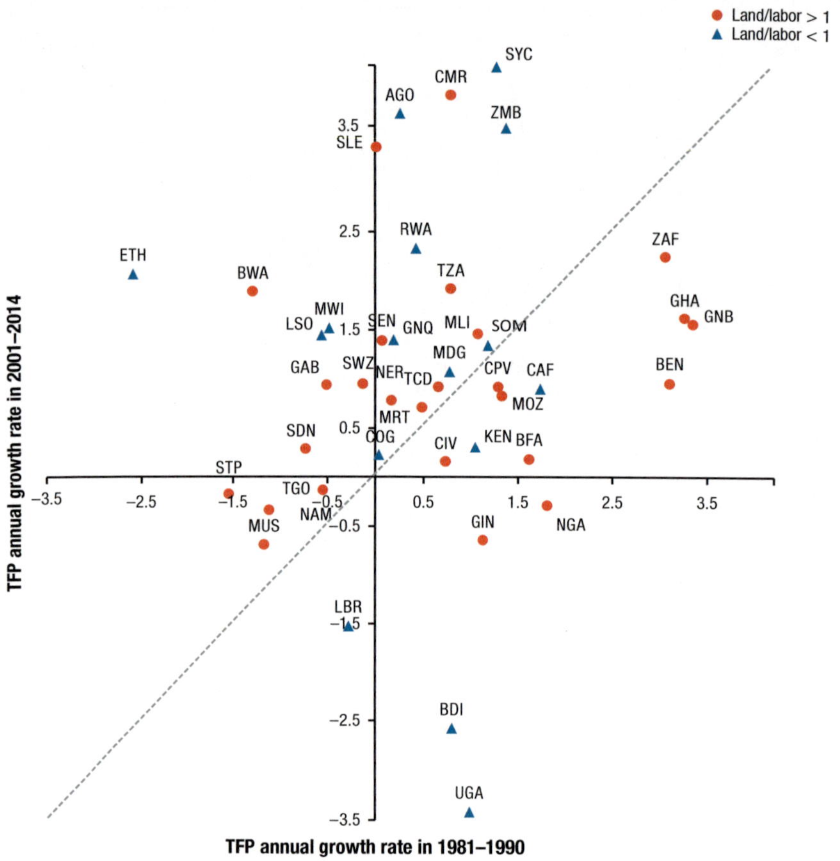

Source: Authors' calculation using data from USDA-ERS (2017).
Note: TFP = total factor productivity; for country names, see the note for Figure 5.4.

Summary

While it should not be expected that TFP and partial productivity growth will be highly consistent, because they measure different things and utilize different data, we do see a similar pattern of productivity growth across countries. In general, more countries have positive and higher growth in both partial and total factor productivity in recent years than in the past. More countries have positive land productivity growth than labor productivity growth. Land-labor ratios can partially explain the endowment situation of a country, and substitutions in land are generally consistent with the Hayami and Ruttan hypothesis. However, high labor productivity growth among some land-scarce countries is associated with growth-enhancing structural change through which labor moves out of agriculture into nonagricultural sectors.

Agricultural Growth, Urbanization, and Nutrition Transition

Income Growth and Poverty Trends

Poverty reduction in Africa also accelerated with growth acceleration, particularly in the 21st century. However, with poverty rates in Africa being the highest of any world region, recent improvements have not been sufficient to significantly reduce the absolute numbers of poor. In the meantime, Africa's middle class is growing. The share of the middle class jumped to 34 percent of the population in 2010, from 26–27 percent in the 2000s, and numbers rose to 327 million people, an increase of 122 million over the preceding decade (AfDB 2011). The definition for *middle class* here is rather broad, including those with income at \$2–\$4 per day. This "floating middle class" is still susceptible to falling back into poverty, and the purchasing power of this group of vulnerable middle class people is still limited (*Economist* 2015).

Africa is one of the fastest-urbanizing regions of the world. The urban share of the continent's population increased from 30.8 percent in 2000 to 37.9 percent in 2015, an increase of more than 160 million people in 15 years (UNDESA 2014). Both cities and rural towns are growing, with secondary cities and towns of less than 1 million inhabitants among the fastest-growing urban areas (UNDESA 2015). This pattern of less-concentrated urbanization may have implications for the inclusiveness of future growth; it has been argued that the expansion of the rural nonfarm economy and secondary cities has a greater effect on poverty reduction than growth concentrated in megacities (Christiaensen and Todo 2014). The rapid rise of secondary and tertiary towns in Africa is improving market access for many rural farmers by extending the reach of value chains into areas formerly considered remote (Richards et al. 2016).

Nutrition Transition

Dietary changes referred to as the nutrition transition have been widely observed in the developing world (Popkin, Adair, and Ng 2012). Africa is no exception. Numerous studies have documented different aspects of the nutrition transition, showing declining shares of staples in food expenditures and higher shares for higher-value food, including processed and perishable food and animal-sourced foods. For example, analysis of household consumption data in Ethiopia from 2000 to 2011 showed that households increased their relative consumption of animal products, fruits and vegetables, and oils and sugars at the expense of cereals; however, processed cereals increased as a share of food expenditures even as the overall share of cereals declined (Hassen et al. 2016).

Many of these changes are associated with urbanization, which raises the demand for convenient, easy-to-prepare foods and food away from home (Hollinger and Staatz 2015). However, transformation is occurring across a broad range of consumers in Africa, in both rural and urban areas and among different income groups. Studies of dietary change in eastern and southern African countries have found that purchased and processed foods account for a large share of diets even among the poor; demand for processed and perishable foods increases quickly as incomes rise in both urban and rural areas (Tschirley et al. 2015a).

Food balance sheets constructed by FAO provide estimates of food supply over time based on national data on production, trade, and other activities affecting food availability. The FAO data reveal that total food availability in Africa has increased significantly over time, from around 2,000 kilocalories per person per day in the mid-1980s to nearly 2,500 kilocalories in the early 2010s (Figure 5.8, left axis). The availability of fat and protein similarly increased over the same period (Figure 5.8, right axis).

Implications for Nutrition Outcomes

Increasing food consumption and wider dietary diversity are expected to lead to better nutrition outcomes in Africa. The Global Hunger Index (GHI), which combines data on child mortality and undernourishment and population undernourishment into a single indicator of hunger, improved markedly for Africa between 2000 and 2016, with several countries improving their GHI score by over 40 percent (Figure 5.9). However, levels of undernourishment and child malnutrition are still unacceptably high in Africa. Over a third of children under five were stunted in 2016, while around 8 percent

FIGURE 5.8 Supply of protein and fat (grams per capita per day) and total food supply (kilocalories per capita per day) in Africa

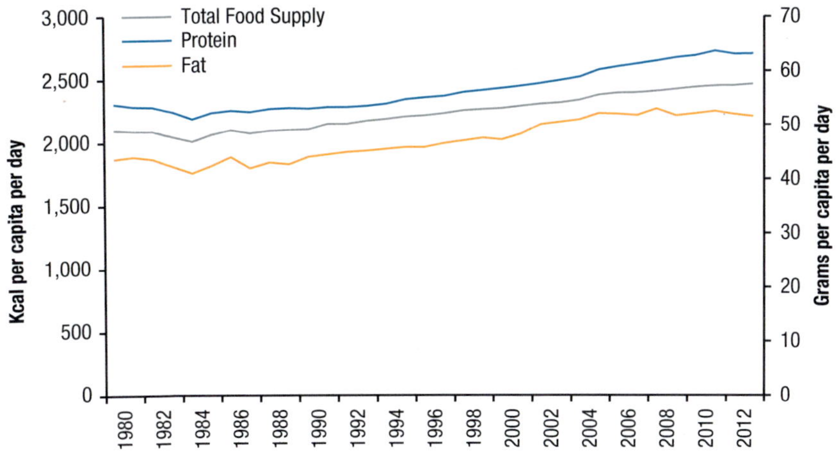

Source: Authors' calculations based on FAO (2017).

were wasted (Table 5.4). Of the 24 countries with sufficient time series data, only 5 are considered on track to meet the World Health Assembly (WHA) target of reducing child stunting by 40 percent between 2012 and 2025; another 13 countries had made at least some progress. Countries perform somewhat better on addressing child wasting, with 11 African countries (of 26 with sufficient data) on track to meet the WHA target of reducing wasting to under 5 percent (Development Initiatives 2017).

Despite increasing diversity in diets, micronutrient deficiencies remain widespread in Africa. An estimated 48 percent of preschool-age children suffered from vitamin A deficiencies in 2013 (Development Initiatives 2017). Rates of anemia are estimated at 40 percent and 60 percent in women of reproductive age and children under five, respectively (Table 5.4). No African countries are considered to be on track to meet the WHA target of reducing anemia in women by half by 2025, while 23 countries have made some progress (Development Initiatives 2017).

Much progress remains to be made in improving infant and young child feeding practices in Africa. The minimum acceptable diet (MAD) indicator for children under two years combines measures of dietary diversity and meal frequency; the 2017 Global Nutrition Report found that only 2 African countries (of 33 with available data) had shares of children benefiting from

FIGURE 5.9 Improvements in Global Hunger Index, 2000–2016

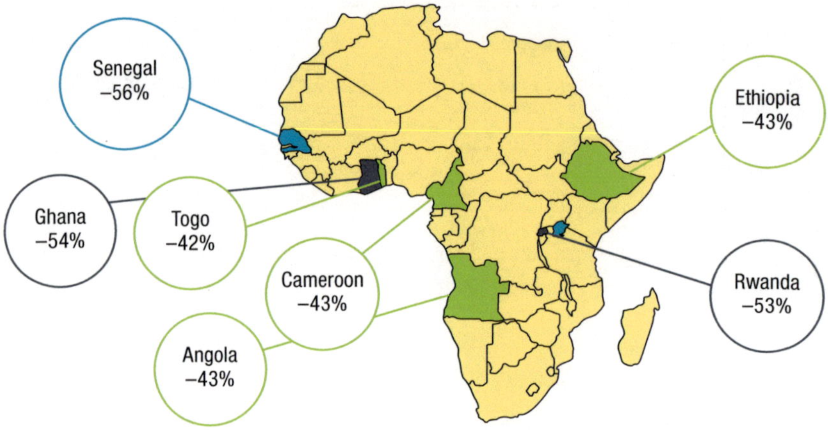

Source: von Grebmer et al. (2016).

TABLE 5.4 Selected nutrition indicators for Africa, 1990–2016

	1990	2000	2010	2016
Child stunting	49.2	43.2	37.5	34.1
Child wasting				7.8
Child underweight	30.3	25.3	20.8	18.5
Child overweight	4.8	4.4	4.1	3.9
Undernourishment		27.0	19.8	20.2*
Women's anemia	51.1	45.7	40.1	38.9
Child anemia	75.9	71.3	63.7	59.9

Source: Data from World Bank (2017).
Note: * Value is for 2015.

the MAD of over 20 percent. In 9 countries, the share was under 10 percent (Development Initiatives 2017).

In addition to still worryingly high rates of undernourishment and child undernutrition, Africa increasingly suffers from a double burden of malnutrition, with overnutrition coexisting with undernutrition. The World Health Organization estimates that adult overweight has increased over time in every African country. Rates of overweight are significantly higher for women, with the gap increasing over time (Figure 5.10). While the overall overweight rate of 28 percent in 2016 is lower than the world average of 39 percent, Africa's rate is increasing faster than the global average as well.

FIGURE 5.10 Prevalence of adult overweight, Africa south of the Sahara (%)

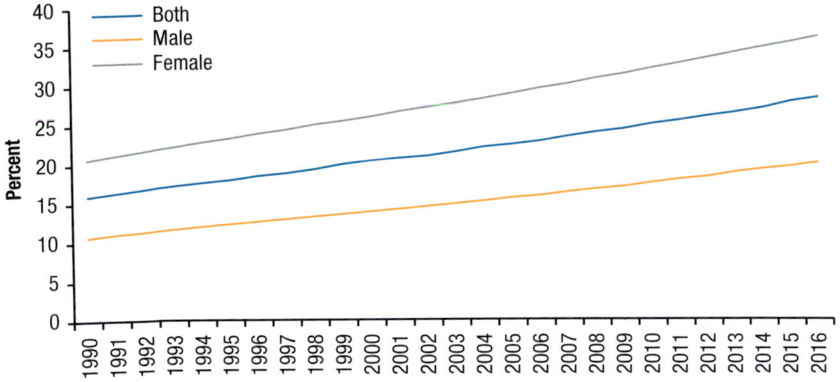

Source: Authors' calculations based on WHO (2017).

Higher rates of adult overweight and obesity are associated with increased risk for nutrition-related noncommunicable diseases, including heart disease and diabetes. In 2014, an estimated 6.5 percent of adults in Africa had diabetes or elevated blood glucose levels. The 2017 *Global Nutrition Report* classified every African country for which estimates were available as "off course" regarding progress in meeting the WHA 2025 targets of no increase in obesity and diabetes (Development Initiatives 2017).

Projected Growth and Diet Trends

The income and urbanization trends giving rise to the nutrition transition in Africa seem likely to continue into the medium term. Simulations based on current patterns of income and production growth suggest that the majority of countries in Africa will achieve middle-income status by 2030 (Sulser et al. 2015). Rapid urbanization is also expected to continue in Africa, with the urban population share projected to increase from 37.9 percent in 2015 to 54.8 percent by 2050 (UNDESA 2014).

Widespread dietary and nutrition transformation will also continue. Studies by Zhou and Staatz (2016) and Tschirley et al. (2015a) suggest that overall demand for higher-value foods will increase sharply. Projected increases in demand exceed increases in domestic supply for all products examined, particularly for meat and dairy products, followed by vegetable oil and fruits and vegetables (Zhou and Staatz 2016). Tschirley et al. (2015a) project rising shares for perishable foods and for processed foods, particularly for high-value-added processed foods.

Implications

The nutrition transition has broad implications for health. The increased availability of food is slowly reducing the most visible forms of undernutrition, but micronutrient deficiencies persist and problems of overnutrition are on the rise. Animal-sourced foods provide protein and micronutrients that can improve children's nutritional status, but consumption of larger quantities of meat is associated with an increased risk of chronic health issues (IFPRI 2015). As the nutrition transition continues, the burden of nutrition-related noncommunicable diseases is likely to increase, making more pronounced the multiple burdens of malnutrition faced by African countries.

African leaders have increasingly recognized the importance of concerted efforts to improve nutrition. The 2014 Malabo Declaration on Accelerated Agricultural Growth and Transformation for Shared Prosperity and Improved Livelihoods included a commitment to ending hunger by 2025; recent work by African governments to design new National Agricultural Investment Plans is taking nutrition into account to a much greater extent than previously. The Malabo Montpellier Panel's 2017 report on nutrition in Africa finds that countries that achieved impressive improvements in nutrition were able to make policy innovations in several areas. These include positioning nutrition as a top priority for public policy; building partnerships with the private sector, development partners, and national research institutions; mainstreaming nutrition into strategies and actions across sectors; mitigating threats associated with conflict and extreme weather and building resilience to crises; improving regulation and controls of food systems to promote production of safe and healthy foods; enhancing national agricultural and nutrition research systems; strengthening nutrition education to promote healthy food consumption; building women's leadership capacities; improving national data systems to better track nutrition outcomes; capitalizing on synergies between agriculture, water, sanitation, and health; and prioritizing the emerging issue of obesity. Scaling up these practices could bring significant progress in improving nutrition in Africa (Malabo Montpellier Panel 2017).

Drivers of Agricultural Transformation and Innovation

The changes in agricultural productivity and food demand described above are manifestations of the agricultural and structural transformation currently underway in Africa. The structural transformation process is unfolding at different paces across African countries. We see signs of rapid economic transformation in parts of Ghana, Ethiopia, and Rwanda, featuring a rise in the workforce engaged in nonfarm sectors, major self-investments by households in youth education and skill training, and a rapid reduction in poverty rates (McMillan and Harttgen 2014). Other countries have made variable progress. This section will examine the dynamics underlying the agricultural transformation in greater detail by highlighting four major trends: (1) greater vibrancy of agricultural factor markets; (2) the rise of commercialized African investor farmers; (3) major new investment in agricultural value chains by African entrepreneurs, leading to profound agrifood systems transformation; and (4) increasing linkages between science and policymaking to support the use of evidence in policy innovation.

Greater Participation and Vibrancy in Agricultural Factor Markets

Markets for land, labor, agricultural inputs (chemicals, fertilizers), and mechanization are rising rapidly in much of Africa. As land values rise, land rental markets are growing in importance (Holden, Otsuka, Place 2009). The research evidence generally finds that land markets are positive developments—they shift land from less productive to more productive users and support overall agricultural productivity growth (Holden, Otsuka, and Place 2009; Jin and Jayne 2013; Chamberlin and Ricker-Gilbert 2016; Deininger, Savastano, and Xia 2017).

There is also growing evidence of rising land rental values in areas of agricultural commercialization with favorable access to markets. Table 5.5 provides illustrative examples of a broader trend over the past decade in parts of Africa: that land prices have risen dramatically in areas of high agroecological potential within reasonable proximity of urban areas (Jayne et al. 2016; Wineman and Jayne 2018). These trends have created new stresses on the ability of customary tenure systems to protect small-scale farmers' land from encroachment or appropriation. The region has experienced rising demand for agricultural land by both international and national companies (Deininger and Byerlee 2011), as well as urban investor farmers (Sitko and Jayne 2014; Jayne et al. 2016).

Other agricultural factor markets are also increasing in importance in Africa. Despite conventional wisdom suggesting thin or absent agricultural labor markets, nationally representative data collected in 2008–2011 from Ethiopia, Niger, Malawi, Tanzania, and Uganda show that a significant share of farm households in each country hired labor to help with preharvest

TABLE 5.5 Land rental rates and purchase values

		Land rental rates			Land purchase values		
		Market access conditions			Market access conditions		
Country	Year	Remote	Near	Total	Remote	Near	Total
Tanzania	2008/2009	96	104	100	89	132	100
	2014/2015	137	155	146	148	227	179
Malawi	2010/2011	91	115	100	74	172	100
	2016/2017	162	199	190	137	522	334
Zambia	2011/2012	97	110	100	—	—	—
	2014/2015	127	113	123	—	—	—
Nigeria	2012/2013	62	112	100	64	136	100
	2018/2019	148	195	163	140	202	195

Source: Land rental and purchase prices for Tanzania, Malawi, and Nigeria come from household respondents in World Bank LSMS/ISA national survey data; in Zambia, the data are from the nationally representative Rural Agricultural Livelihoods Survey conducted by the Indaba Agricultural Policy Research Institute.

Notes: — = data not available. Reported prices are means within the 20th and 80th percentiles of the data. Prices are deflated by national consumer price indices published by the national statistical offices of each country and then reported as indices with 100 being the national mean in the first survey year. For example, land rental rates for the full sample of rural households in Tanzania were 46 percent higher in real terms in 2014/2015 than for the full sample in 2008/2009. Market access conditions were based on distance or travel time to urban areas of 100,000 or more people. Real land rental rates in rural Tanzania near urban areas were 55 percent higher than the national mean in 2008/2009. Rental rates and purchase values in Tanzania, Malawi, and Nigeria are the subjective views of plot operators and not necessarily based on actual transactions.

agricultural work, ranging from 30 percent in Ethiopia and Tanzania to 49 percent in Niger (Dillon and Barrett 2017). Increasingly active rural labor markets are also evident from the great rise of rural–rural migration, which has now become more important than rural-to-urban migration in many African countries (Mercandalli and Losch 2017).

The circumstances affecting the development of land and labor markets also impact markets for other factors of production. As land-labor ratios change due to the expansion of agricultural area or shifts of labor to other sectors, the relative costs of factors of production change and different land- or labor-saving technologies become profitable, giving rise to growing markets. In particular, labor-saving and capital-using technologies are becoming more important in areas affected by labor shortages due to low population density or the migration of workers to opportunities in other sectors (Jayne et al. 2019). Such technologies, including fertilizers, agrochemicals, and mechanization, are clearly becoming more prevalent in many areas. Sheahan and Barrett (2017) show that 35 percent of farming households in Ethiopia, Niger, Nigeria, Malawi, Tanzania, and Uganda apply inorganic fertilizer, with application rates per hectare significantly higher than previous estimates. In addition, 16 percent of households use other agrochemicals such as pesticides and herbicides during the main growing season, a significant increase over use rates before 2000 (FAO 2019).

The development of agricultural equipment markets remains impeded by constraints related to importing or manufacturing equipment. However, some countries have experienced more dynamic growth in mechanization, often by emphasizing equipment rental or service hiring markets, as in Ethiopia and Malawi, and by carrying out interventions to increase mechanization in the form of public-private partnerships (Malabo Montpellier Panel 2018). Mechanization appears to be rising especially rapidly in areas experiencing robust economic growth and out-migration of labor (Jayne et al. 2019).

The existence of factor markets and their functioning at higher levels than previously believed are positive developments. However, input use intensity in Africa still lags behind most other regions (Sheahan and Barrett 2017). Evidence suggests that factor markets often function poorly due to high transaction cost, weak information flows, and in some cases, government behavior. For example, Dillon and Barrett (2017) provide evidence of multiple factor market failures in their five countries of analysis. Other sources of continued slow increases in technical innovation are due to chronically underfunded and poorly performing national agricultural science and extension systems (World Bank 2011). Removing constraints to technology adoption and market performance is a priority in efforts to advance agricultural transformation.

The Rise of Commercialized African Investor Farmers[8]

The development of land rental markets described above has paved the way for major changes in farm size distributions. The most salient development is rapid growth of an entrepreneurial, educated, and relatively capitalized class of African investor farmers.

Medium-scale farms of 5 to 100 hectares account for around 20 percent of total farmland in Kenya, over 40 percent in Tanzania, over 50 percent in Ghana, and over 60 percent in Zambia (Table 5.6). Many owners of medium-scale farms are urban professionals or well-connected rural residents; around half purchased farmland later in life, using income from nonfarm sources. A greater share of savings in urban areas is being reinvested in farming and agribusiness.

---

8 This section draws on Jayne et al. 2016.

TABLE 5.6 Changes in farm structure in Ghana (1992–2005), Tanzania (2008–2012), Zambia (2008–2014), and Kenya (1994–2006) based on official national survey data

Farm size category (hectares)	Percentage of total operated land on farms between 0–100 hectares		Percentage growth in number of farms between initial and latest year
Ghana	**1992**	**2005**	
0–5	60.7	48.9	37.1
5–100	39.3	51.1	194.3
Total	**100.0**	**100.0**	**49.5**
Tanzania	**2008**	**2012**	
0–5	62.4	56.3	12.8
5–100	37.6	43.7	37.2
Total	**100.0**	**100.0**	**14.5**
Zambia	**2008**	**2014**	
0–5	54.1	38.8	15.9
5–100	45.9	61.2	148.7
Total	**100.0**	**100.0**	**227.2**
Kenya	**1994**	**2006**	
0–5	61.5	72.0	34.0
>5	38.5	28.0	–80.2
Total	**100.0**	**100.0**	**25.2**

Source: Jayne et al. (2016). Reproduced by permission of the publisher; this table is not covered by the CC By 4.0 license.
Note: Data for Zambia shown in the two middle columns are for land owned; Ghana, Kenya, and Tanzania are for operated farm size.

The change in farm size distributions is likely to have complicated effects that are difficult to generalize. Larger and more commercialized farms benefit from economies of scale related to skills, technology, and learning; finance and access to capital; and marketing, logistics, and storage (Collier and Dercon 2014). Rising numbers of medium-size farms may thus facilitate technology adoption, accelerate productivity growth, and ultimately catalyze agricultural transformation. However, their short- and medium-term effects on welfare in the rural sector are unknown. They may help to link smallholders to commercial opportunities, but they may also aggravate land scarcity in densely populated areas. Investor farmers often have outsize influence in farm lobbies and, in turn, on agricultural policy; while strong agricultural lobbies may have positive effects on macroeconomic management in resource-dependent economies (Gelb and Grasmann 2010), their effects on overall rural development are yet to be explored. Data from nationally representative Demographic and Health Surveys in six countries (Ghana, Kenya, Malawi,

Rwanda, Tanzania, and Zambia) suggest that between 10 and 35 percent of agricultural land is owned by urban households; in countries with multiple survey years, this share has risen over time (Jayne et al. 2016).[9]

Agrifood System Transformation: Inclusive and Driven by Local African Entrepreneurs

The changes in income, urbanization, and food consumption described earlier have significant implications for agrifood systems in Africa. Rising food expenditures have increased the volume of food being handled by African value chains, in particular for the midstream and downstream segments of food value chains (Reardon et al. 2015). Value chains are expanding to serve rapidly growing cities, and artisanal, micro, and small enterprises processing domestic crops have expanded significantly in the past several decades (Hollinger and Staatz 2015; AGRA 2019).

Higher food basket shares for processed food are reflected in increased imports of processed products, but also in the growing importance of various processed ready-to-cook or ready-to-eat traditional staples. Retail inventories carried out in Mali (Thériault et al. 2017), Ghana (Andam et al. 2015), and Tanzania (Snyder et al. 2015) demonstrate the enormous variety of domestically processed food products that have appeared in African countries. Around one-quarter of the processed products inventoried in Mali and Ghana were locally produced; in Tanzania, domestically produced products accounted for the majority of processed products inventoried. Locally produced products were strongly represented among grain products and traditional dairy products in Mali; products based on starchy root crops (for example, fufu flour, gari) in Ghana; and milled grain products in Tanzania, with over 60 Tanzanian brands of maize flour identified in Dar es Salaam alone. The retail inventories do not examine the spending or consumption shares of locally produced products, but they do provide initial evidence that dietary shifts are changing domestic value chains and giving rise to new products.

Agrifood system changes raise many questions about the implications for agriculture, nutrition, and overall economic growth in Africa. To meet rising demand for higher-value foods, productivity increases are urgently needed both on-farm and in downstream value-chain segments. The emerging processing sector provides great opportunities for job creation in processing,

9 Medium-scale farm holdings in the four countries studied by Jayne et al. (2016) were found to utilize a lower percentage of their landholdings than small-scale farms, and the share of medium-scale farms in national agricultural production was therefore somewhat lower than the share of medium-scale farms in total agricultural land.

distribution, packaging, and marketing, as well as increased incomes for farmers. However, the small and medium-enterprises (SMEs) undertaking these new activities face significant challenges. Rising numbers of firms combined with low levels of innovation will lead to a situation with an abundance of small firms with persistently low productivity and profitability and limited ability to drive agricultural transformation (Badiane and Ulimwengu 2017).

Strengthening the links between producers and processors is an important intervention to facilitate firm growth as well as benefit smallholders. One means of connecting producers with midstream processors is contract farming, an arrangement in which processing firms, most commonly medium or large enterprises, sign contracts with smallholders to buy farm outputs at specified prices. Contract farming has been welcomed as having the potential to alleviate farmers' constraints related to the lack of information and capital and the presence of risk (Minot and Sawyer 2016). Although many studies have identified positive effects on farmer incomes, nearly all research on contract farming is observational and suffers from selection bias: since farmers and processors self-select into contract farming, any differences between participants and nonparticipants may be driven by the same factors that determined their participation (Bellemare and Bloem 2018). However, experimental findings from Arouna, Michler, and Lokossou (2019) suggest that contracts produced positive effects on farmer welfare when participants were randomly selected into a contract farming arrangement in Benin.

SMEs have by far the largest role in African agrifood systems and will for many years. Most African value chains are benefiting from local SME investment by African entrepreneurs, resulting in a competitive and inclusive situation for most agricultural commodity value chains. Smallholder farmers are the main source of supply for SMEs. Therefore, strengthening the ability of SMEs to innovate in terms of product quality and business practices, increase their scale of operation and profitability, and compete in output markets is central to strengthening smallholder farmer links to agribusiness.

Evidence-Based Policymaking and Links between Science and Policy

Despite the importance of technological innovations, perhaps nothing affects African farmers more than the policy environment they operate within, which affects both the generation of technological innovations and the farmers' incentives to adopt them. Much of the agricultural stagnation and decline of the 1970s and 1980s can be attributed to policy bias against agriculture, which showed abrupt shifts in focus, at times inflicting genuine harm upon agriculture sectors in an effort to protect industry (Badiane and Makombe 2014). In many countries, agriculture was neglected, taxed heavily, or tightly regulated, or all of these, such that farmers were barred from selling their output to buyers and at prices of their own choosing. During much of this time, agricultural policymaking was completely delinked from scientific evidence, with policies dictated by still-forming development theory and uninformed by empirical knowledge of outcomes.

The broadness of the current recovery owes much to policy reforms that improved the agriculture sector and macroeconomic management and gave greater freedom to the private sector. The situation has changed markedly since the years of agricultural and economic decline in the 1970s and 1980s: the growth recovery has afforded greater fiscal space to governments, and political systems are more responsive. These are important and welcome developments, but coupled with a new generation of leaders and limited institutional memory, they pose the risk of policy reversal. The increasing use of costly input subsidies, export bans, price controls, and public agricultural agencies could all signal the beginnings of a return to the failed policies of the past.

The best guard against bad policymaking is the use of locally relevant evidence in policy formulation, as well as the establishment of robust systems for review, monitoring and evaluation, and consultation and dialogue. The development of mechanisms for evidence-based policy formulation and review, as well as their increasing use, is itself one of the most important innovations of the past two decades affecting the prospects of Africa's agriculture sector. In 2003 the African Union launched the Comprehensive Africa Agriculture Development Programme (CAADP) as the continent-wide framework for agriculture-led growth and development. African leaders committed to a set of key CAADP principles and values, including African ownership and leadership; inclusivity of different stakeholder groups; evidence-based policy planning, implementation, and review; and mutual accountability for actions and results. Key CAADP mutual accountability platforms include agricultural Joint Sector Reviews at the national and regional levels, which bring together state and nonstate actors for annual reviews of agricultural commitments and outcomes, and the Biennial Review (BR), a continent-wide review of progress against CAADP agricultural transformation goals. The first BR was carried out in January 2018, with 47 of 55 African Union member states participating and sharing data on national-level progress.

The emphasis of CAADP on evidence-based policymaking and careful review of outcomes has created stronger demand for data and analysis and

FIGURE 5.11 Selected performance and outcome indicators by CAADP implementation status

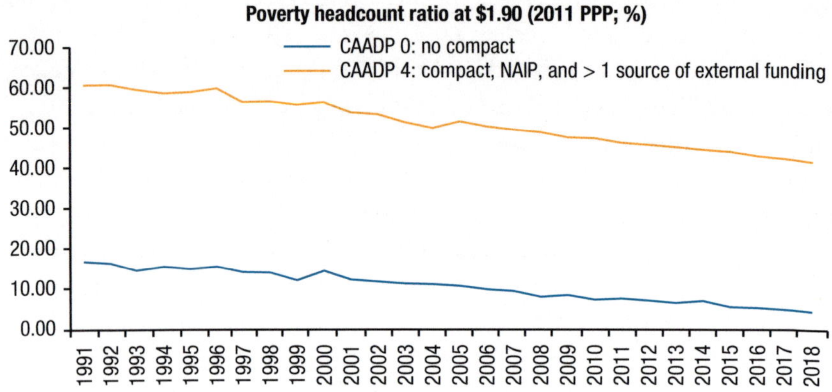

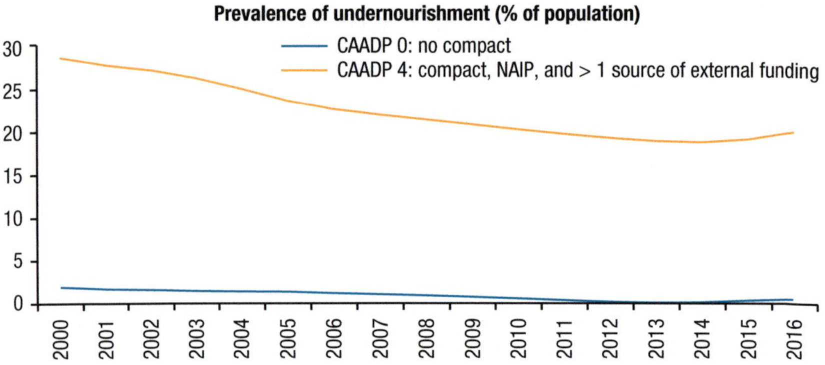

new incentives to better link science with policy. Institutions including the Regional Strategic Analysis and Knowledge Support System (ReSAKSS) are building connections between ministries of agriculture and local universities and think tanks capable of carrying out research to guide governments' actions. Greater use of evidence has enabled agricultural stakeholders to make a successful case for increased agricultural investment. Figure 5.11 shows poverty, hunger, agricultural expenditure, and agricultural growth outcomes in CAADP-0 countries—those that have not begun implementation of CAADP—and in CAADP-4 countries—those that are most advanced in the implementation process. The CAADP-4 group has had more success in maintaining agricultural expenditures and has seen faster agricultural growth and hunger reduction and slightly faster progress in reducing poverty. A detailed

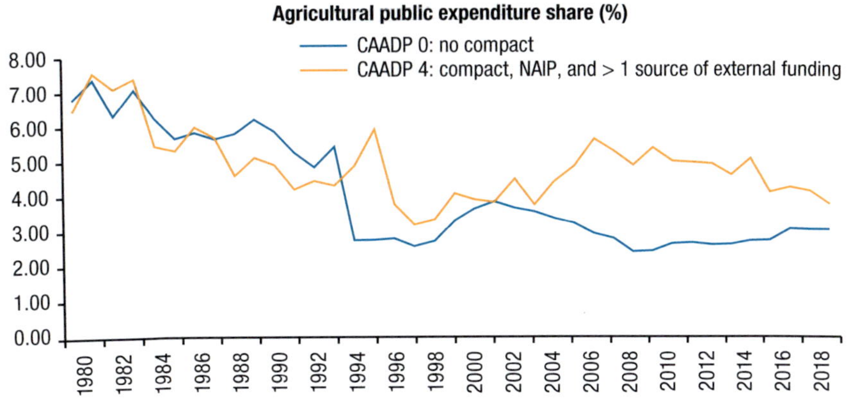

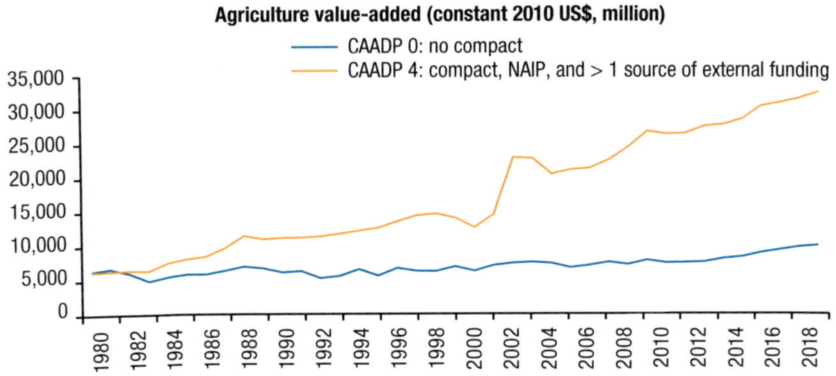

Source: ReSAKSS (2019).

Note: CAADP = Comprehensive Africa Agriculture Development Programme; NAIP = National Agriculture Investment Plan; PPP = purchasing power parity.

impact analysis of CAADP by Benin (2018) found that CAADP implementation had enabled countries to raise government agricultural expenditure, receive higher levels of agricultural development aid, and increase land and labor productivity compared with countries that were less advanced in the implementation process.

A Hidden "Spoiler"—What Could Arrest Progress if Not Adequately Addressed: Land Degradation

Since the 1960s, agricultural production growth in Africa south of the Sahara (SSA) has occurred primarily through area expansion. Yield growth contributed less than 20 percent of SSA's total agricultual production growth between 2000 and 2008 (Fuglie and Rada 2013). But rising population densities in many parts of Africa are making continued reliance on area expansion untenable for millions of African farmers. The land frontier has already been reached in many smallholder areas, causing farms to become subdivided, fragmented, and increasingly small. Smallholders have responded to shrinking farm sizes by more continuously cropping their fields every year, mainly growing their priority staple foods. Fallows have largely disappeared in densely populated areas, and for the overall SSA region, fallowed land as a proportion of total farmland has declined steadily from 40 percent in 1960 to 15 percent in 2011 (Fuglie and Rada 2013). It will be harder to sustain production growth on existing smallholder farms through area expansion, putting more pressure on African farming systems to raise yields and the value of farm output per hectare and per labor unit.

The challenge of achieving sustainable yield growth in SSA in the context of rising land scarcity is complicated by mounting evidence of yield-depressing soil degradation arising from unsustainable intensification in SSA's densely populated areas (Stoorvogel and Smaling 1990; Drechsel et al. 2001; Barbier and Hochard 2012; Tittonell and Giller 2013; Montpellier Panel 2014). Continuous cultivation of existing plots would not pose problems for sustainable intensification if farmers were able to maintain or improve soil quality over time through sufficient use of fertilizers, soil amendment practices, and other land-augmenting investments. However, there is growing evidence of a significant relationship between population pressure, reduced fallow periods, and land degradation, pointing to an unsustainable dynamic between population, agriculture, and the natural resource base (Drechsel et al. 2001; Lal 2011). Losses of soil organic matter and acidification pose special problems, because they cannot be ameliorated by the application of conventional fertilizers and because they tend to depress the efficiency of inorganic fertilizer in contributing to crop output. Smallholder farmers are often unable to benefit from the current yield gains offered by plant genetic improvement, due to their farming on depleted soils that do not respond well to fertilizer application (Tittonell and Giller 2013). Given UN projections that rural SSA will contain 52 percent more people in 2050 than it did in 2017, the challenge of helping millions of African smallholders to raise the productivity of their existing farmland in sustainable ways is an urgent priority.

Conclusions

Africa is now on the move. Agricultural productivity has shown a strong recovery from the stagnation of previous decades. Per capita incomes have nearly doubled since the 1990s. Poverty rates have declined significantly since 2000. Nutritional indicators are improving in most countries. The share of the labor force engaged in small-scale farming has declined substantially. Today, farming accounts for 40 to 65 percent of primary employment in Africa's working-age population, down from 70 to 80 percent just 10 years ago. The share of the workforce engaged in farming has declined most rapidly in countries enjoying the highest rates of agricultural productivity growth. These developments point to unmistakable economic transformation. SSA's agricultural system has been an important driver of the region's transformation, having experienced the most rapid agricultural growth of any region in the world.

Over the last decade, African governments have brought agriculture back to the top of their development agenda and are investing an increased proportion of their budgets from a growing national revenue base. The private sector is increasingly investing in agriculture, and the foundations have been laid for long-run dynamism in Africa's agrifood systems, powered by the enormous progress increasingly evident in farmers who have more options in the seeds they plant, the fertilizers they use, and the markets seeking to purchase their produce. So far, this is just a glimpse of success, and it is still largely a fragile success dependent on more decisive support from many African governments. But it offers an inspiring vision of a future Africa in which farming as a struggle to survive gives way to farming as a business that thrives.

However, despite the unprecedented decade of impressive growth across the continent, much more remains to be done to sustain these gains and truly drive the agricultural transformation needed for Africa's development and to ensure a better life for all its people as laid out in the Malabo Declaration and the Sustainable Development Goals. Africa is still facing tremendous challenges. The continent is the world's most food-insecure continent, with relatively low levels of agricultural productivity, low rural incomes, high rates of malnutrition, and a worsening food trade balance. It is a region challenged by climate change, the daunting prevalence of poverty, and an urgent need for jobs. In many countries, agriculture remains the predominant sector of the economy, accounting on average for 25 percent of the GDP in SSA and well above this level for many countries. The sector makes up close to half the GDP, on average, considering the broader agribusiness sector—including input supply, processing, and market access. Therefore, stronger agricultural growth can act as a powerful multiplier for economic growth.

The good news is that a vibrant agricultural sector, while not the solution to all of these problems, will clearly promote food security and economic opportunities for all Africans.

References

AfDB (African Development Bank). 2011. *The Middle of the Pyramid: Dynamics of the Middle Class in Africa*. Market Brief, April.

AGRA (Alliance for a Green Revolution in Africa). 2016. *Africa Agriculture Status Report 2016: Progress toward Agricultural Transformation in Africa*. Nairobi.

———. 2019. *The Hidden Middle: A Quiet Revolution in the Private Sector Driving Agricultural Transformation*. Nairobi.

Andam, K., R. M. Al-Hassan, S. B. Asante, and X. Diao. 2015. *Is Ghana Making Progress in Agro-Processing? Evidence from an Inventory of Processed Food Products in Retail Shops in Accra*. Ghana Strategy Support Program Working Paper 41. Washington, DC: IFPRI.

Arouna, A., J. D. Michler, and J. C. Lokossou. 2019. *Contract Farming and Rural Transformation: Evidence from a Field Experiment in Benin*. NBER Working Paper 25665. Cambridge, MA: National Bureau of Economic Research.

Badiane, O. 2011. *Agriculture and Structural Transformation in Africa*. Stanford Symposium Series on Global Food Policy and Food Security in the 21st Century. Stanford, CA: Stanford University.

Badiane, O., S. Benin, and T. Makombe. 2016. "Strengthening the Continental Agricultural Agenda and Accountability Framework: The Road from Maputo to Malabo." In *Africa Agriculture Status Report 2016: Progress Towards Agricultural Transformation*, 24–47. Nairobi: Alliance for a Green Revolution in Africa.

Badiane, O., and J. Collins. 2016. "Agricultural Growth and Productivity in Africa: Recent Trends and Future Outlook." In *Agricultural Research in Africa: Investing in Future Harvests*, edited by J. Lynam, N. Beintema, J. Roseboom, and O. Badiane, 3–30. Washington, DC: IFPRI.

Badiane, O., and T. Makombe. 2014. *The Theory and Practice of Agriculture, Growth, and Development in Africa*. WIDER Working Paper 2014/061. Helsinki: UN University World Institute for Development Economics Research.

———, eds. 2015. *Beyond a Middle Income Africa: Transforming African Economies for Sustained Growth with Rising Employment and Incomes*. ReSAKSS Annual Trends and Outlook Report 2014. Washington, DC: IFPRI.

Badiane, O., and J. Ulimwengu. 2017. "Business Pathways to the Future of Smallholder Farming in the Context of Transforming Value Chains." In *Africa Agriculture Status Report 2017: The Business of Smallholder Agriculture in Sub-Saharan Africa*, edited by D. Sumba, 25–44. Nairobi: Alliance for a Green Revolution in Africa.

Barbier, E., and J. Hochard. 2016. *Poverty and the Spatial Distribution of Rural Population*. Policy Research Working Paper, WPS 7101. Washington, DC: World Bank Group.

Barrett, C. B., L. Christiaensen, M. Sheahan, and B. Shiferaw. 2017. "On the Structural Transformation of Rural Africa." *Journal of African Economies* 26 (AERC S1): i11–i35.

Bellemare, M. F., and J. R. Bloem. 2018. "Does Contract Farming Improve Welfare? A Review." *World Development* 112: 259–271.

Benin, S. 2018. *From Maputo to Malabo: How Has CAADP Fared?* ReSAKSS Working Paper 40. Washington, DC and Dakar, Senegal: IFPRI.

Chamberlin, J., and T. S. Jayne. 2013. "Unpacking the Meaning of 'Market Access': Evidence from Rural Kenya." *World Development* 41: 245–264.

Chamberlin, J., T. S. Jayne, and D. Headey. 2014. "Scarcity Amidst Abundance? Reassessing the Potential for Cropland Expansion in Africa." *Food Policy* 48: 51–65.

Chamberlin, J., and R. Ricker-Gilbert. 2016. "Participation in Rural Land Rental Markets in Sub-Saharan Africa: Who Benefits and by How Much? Evidence from Malawi and Zambia." *American Journal of Agricultural Economics* 98 (5): 1507–1528.

Christiaensen, L., and Y. Todo. 2014. "Poverty Reduction during the Rural–Urban Transformation: The Role of the Missing Middle." *World Development* 63: 43–58.

Collier, P., and S. Dercon. 2014. "African Agriculture in 50 Years: Smallholders in a Rapidly Changing World?" *World Development* 63: 92–101.

Deininger, K., and D. Byerlee with J. Lindsay, A. Norton, H. Selod, and M. Stickler. 2011. *Rising Global Interest in Farmland: Can It Yield Sustainable and Equitable Benefits?* Washington, DC: World Bank.

Deininger, K., S. Savastano, and F. Xia. 2017. "Smallholders' Land Access in Sub-Saharan Africa: A New Landscape?" *Food Policy* 67: 78–92.

Development Initiatives. 2017. Global Nutrition Report 2017 Dataset. Accessed December 2017. www.globalnutritionreport.org/the-data/dataset-and-metadata/.

Diao, X., K. Harttgen, and M. McMillan. 2017. "The Changing Structure of Africa's Economies." *The World Bank Economic Review* 31 (2): 412–433.

Diao, X., E. Magalhaes, and M. McMillan. 2018. "Understanding the Role of Rural Nonfarm Enterprises in Africa's Economic Transformation: Evidence from Tanzania." *Journal of Development Studies* 54 (5): 833–855.

Diao, X., M. McMillan, and D. Rodrik. 2017. *The Recent Growth Boom in Developing Countries: A Structural Change Perspective.* NBER Working Paper 23132. Cambridge, MA: National Bureau of Economic Research.

Dillon, B., and C. B. Barrett. 2017. "Agricultural Factor Markets in Sub-Saharan Africa: An Updated View with Formal Tests for Market Failure." *Food Policy* 67: 64–77.

Drechsel, P., L. Gyiele, D. Kunze, and O. Cofie. 2001. “Population Density, Soil Nutrient Depletion, and Economic Growth in Sub-Saharan Africa.” *Ecological Economics* 38: 251–258.

Economist. 2015. “Few and Far Between: Africans Are Mainly Rich or Poor, but Not Middle Class. That Should Worry Democrats.” October 22.

FAO (Food and Agriculture Organization of the United Nations). 2017. Food Balance Sheets. Accessed December 2017. www.fao.org/faostat/en/#data/FBS.

———. 2019. FAOSTAT Database. Accessed September 2019. http://www.fao.org/faostat/en/#data.

Fuglie, K., and N. Rada. 2013. *Resources, Policies, and Agricultural Productivity in Sub-Saharan Africa*. Economic Research Report 145. Washington, DC: Economic Research Services, United States Department of Agriculture.

Gelb, A., and S. Grasmann. 2010. *How Should Oil Exporters Spend Their Rents?* Center for Global Development Working Paper 221. Washington, DC: Center for Global Development.

Gollin, D., R. Jedwab, and D. Vollrath. 2016. “Urbanization with and without Industrialization.” *Journal of Economic Growth* 21 (1): 35–70.

Hassen, I. W., M. Dereje, B. Minten, and K. Hirvonen. 2016. *Diet Transformation in Africa: The Case of Ethiopia*. Ethiopia Strategy Support Program Working Paper 87. Washington, DC: IFPRI.

Hayami, Y., and V. W. Ruttan. 1985. *Agricultural Development: An International Perspective*. Baltimore: Johns Hopkins University Press.

Hicks, J. H., M. Kleemans, N. Y. Li, and E. Miguel. 2017. *Reevaluating Agricultural Productivity Gaps with Longitudinal Microdata*. NBER Working Paper 23253. Cambridge, MA: National Bureau of Economic Research.

Holden, S., K. Otsuka, and F. Place, eds. 2009. *The Emergence of Land Markets in Africa*. Washington, DC: Resources for the Future.

Hollinger, F., and J. Staatz. 2015. *Agricultural Growth in West Africa: Market and Policy Drivers*. Rome: Food and Agriculture Organization of the United Nations, African Development Bank, Economic Community of West African States.

IAPRI (Indaba Agricultural Policy Research Institute). 2012. Rural Agricultural Livelihoods Survey 2012. Lusaka.

———. 2015. Rural Agricultural Livelihoods Survey 2015. Lusaka.

IFAD (International Fund for Agricultural Development). 2016. *Rural Development Report: Fostering Inclusive Rural Transformation*. Rome.

IFPRI (International Food Policy Research Institute). 2015. *Actions and Accountability to Advance Nutrition and Sustainable Development*. Global Nutrition Report 2015. Washington, DC.

Imai, K., R. Gaiha, and A. Garbero. 2017. "Poverty Reduction During the Rural–Urban Transformation: Rural Development Is Still More Important Than Urbanisation." *Journal of Policy Modeling* 39: 963–982.

Jayne, T. S., J. Chamberlin, and R. Benfica. 2018. "Africa's Unfolding Economic Transformation." *Journal of Development Studies* 54 (5): 777–787.

Jayne, T. S., J. Chamberlin, L. Traub et al. 2016. "Africa's Changing Farm Size Distribution Patterns: The Rise of Medium-Scale Farms." *Agricultural Economics* 47: 197–214.

Jayne, T. S., D. Mather, and E. Mghenyi. 2010. "Principal Challenges Confronting Smallholder Agriculture in Sub-Saharan Africa." *World Development* 38 (10): 1384–1398.

Jayne, T. S., S. Snapp, F. Place, and N. Sitko. 2019. "Sustainable Agricultural Intensification in an Era of Rural Transformation in Africa." *Global Food Security* 20: 105–113.

Jin, S., and T. S. Jayne. 2013. "Land Rental Markets in Kenya: Implications for Efficiency, Equity, Household Income, and Poverty." *Land Economics* 89 (2): 246–271.

Lal, R. 2011. "Managing the Soils of Sub-Saharan Africa." *Science* 236 (4805): 1069–1076.

Malabo Montpellier Panel. 2017. *Nourished: How Africa Can Build a Future Free from Hunger and Malnutrition*. Dakar.

———. 2018. *Mechanized: Transforming Africa's Agriculture Value Chains*. Dakar.

Masters, W. A., A. A. Djurfeldt, C. De Haan, P. Hazell, T. Jayne, M. Jirström, and T. Reardon. 2013. "Urbanization and Farm Size in Asia and Africa: Implications for Food Security and Agricultural Research." *Global Food Security* 2: 156–165.

McMillan, M., and K. Harttgen. 2014. *The Changing Structure of Africa's Economies*. NBER Working Paper 20077. Cambridge, MA: National Bureau of Economic Research.

McMillan, M., D. Rodrik, and I. Verduzco. 2014. "Globalization, Structural Change and Productivity Growth, with an Update on Africa." *World Development* 63: 11–32.

Mellor, J. 1976. *The New Economics of Growth: A Strategy for India and the Developing World*. Ithaca, NY: Cornell University Press.

Mercandalli, S., and B. Losch, eds. 2017. *Rural Africa in Motion: Dynamics and Drivers of Migration South of the Sahara*. Rome: Food and Agriculture Organization of the United Nations and Centre de coopération internationale en recherche agronomique pour le développement.

Minot, N., and B. Sawyer. 2016. "Contract Farming in Development Countries: Theory, Practice, and Policy Implications." In *Innovation for Inclusive Value-Chain Development: Successes and Challenges*, edited by A. Devaux, M. Torero, J. Donovan, and D. Horton. Washington, DC: IFPRI.

Montpellier Panel. 2014, December. *No Ordinary Matter: Conserving, Restoring and Enhancing Africa's Soils*. London: Imperial College.

Popkin, B. M., L. S. Adair, and S. W. Ng. 2012. "Global Nutrition Transition and the Pandemic of Obesity in Developing Countries." *Nutrition Reviews* 70 (1): 3–21.

Reardon, T. 2015. "The Hidden Middle: The Quiet Revolution in the Midstream of Agrifood Value Chains in Developing Countries." *Oxford Review of Economic Policy* 31: 45–63.

Reardon, T., D. Tschirley, B. Minten et al. 2015. "Transformation of African Agrifood Systems in the New Era of Rapid Urbanization and the Emergence of a Middle Class." In *Beyond a Middle Income Africa: Transforming African Economies for Sustained Growth with Rising Employment and Incomes*, edited by O. Badiane and T. Makombe, 62–74. ReSAKSS Annual Trends and Outlook Report 2014. Washington, DC: IFPRI.

ReSAKSS (Regional Strategic Analysis and Knowledge Support System). 2019. Tracking Indicators. Accessed July 2019. www.resakss.org/node/11.

Richards, P., T. Reardon, D. Tschirley, T. Jayne, J. Oehmke, and D. Atwood. 2016. "Cities and the Future of Agriculture and Food Security: A Policy and Programmatic Roundtable." *Food Security* 8 (4): 871–877.

Sheahan, M., and C. B. Barrett. 2017. "Ten Striking Facts about Agricultural Input Use in Sub-Saharan Africa." *Food Policy* 67: 12–25.

Sitko, N., W. Burke, and T. S. Jayne. 2018. "The Quiet Rise of Large-Scale Trading Firms in East and Southern Africa." *Journal of Development Studies* 54 (5): 895–914.

Sitko, N., and T. S. Jayne. 2014. "Structural Transformation or Elite Land Capture? The Growth of 'Emergent' Farmers in Zambia." *Food Policy* 48: 194–202.

Snyder, J., C. Ijumba, D. Tschirley, and T. Reardon. 2015. *Stages of Transformation in Food Processing and Marketing: Results of an Initial Inventory of Processed Food Products in Dar es Salaam, Arusha, and Mwanza*. Food Security Policy Innovation Lab Tanzania Policy Research Brief 3. Dar es Salaam.

Stoorvogel, J. J., and E. M. A. Smaling. 1990. *Assessment of Soil Nutrient Depletion in Sub-Saharan Africa: 1983–2000*. Report 28. Wageningen, Netherlands: Winand Staring Centre for Integrated Land, Soil and Water Research.

Sulser, T. B., D. Mason-D'Croz, S. Islam, S. Robinson, K. Wiebe, and M. W. Rosegrant. 2015. "Africa in the Global Agricultural Economy in 2030 and 2050." In *Beyond a Middle Income Africa: Transforming African Economies for Sustained Growth with Rising Employment and Incomes*, edited by O. Badiane and T. Makombe, 5–37. ReSAKSS Annual Trends and Outlook Report 2014. Washington, DC: IFPRI.

Thériault, V., A. Assima, R. Vroegindewey, D. Tschirley, and N. Keita. 2017. *A City-Retail Outlet Inventory of Processed Dairy and Grain Foods: Evidence from Mali*. Feed the Future Innovation Lab for Food Security Policy Research Paper 65. East Lansing: Michigan State University.

Timmer, C. P. 1988. "The Agricultural Transformation." In *Handbook of Development Economics, Volume 1*, edited by H. Chenery and T. N. Srinivasan, 275–331. Amsterdam: North Holland/ Elsevier.

Timmer, M. P., G. J. de Vries, and K. de Vries. 2015. "Patterns of Structural Change in Developing Countries." In *Routledge Handbook of Industry and Development*, edited by J. Weiss and M. Tribe, 65–83. Routledge.

Tittonell, P., and K. Giller. 2013. "When Yield Gaps Are Poverty Traps: The Paradigm of Ecological Intensification in African Smallholder Agriculture." *Field Crops Research* 143 (1): 76–90.

Tschirley, D., T. Reardon, M. Dolislager, and J. Snyder. 2015a. "The Rise of a Middle Class in East and Southern Africa: Implications for Food System Transformation." *Journal of International Development* 27 (5): 628–646.

Tschirley, D., J. Snyder, M. Dolislager et al. 2015b. "Africa's Unfolding Diet Transformation: Implications for Agrifood System Employment." *Journal of Agribusiness in Developing and Emerging Economies* 5 (2): 102–136.

UNDESA (UN Department of Economic and Social Affairs). 2014. World Urbanization Prospects: The 2014 Revision dataset. Accessed November 13, 2017. https://esa.un.org/ Unpd/Wup/DataQuery/.

———. 2015. *World Urbanization Prospects: The 2014 Revision*. New York.

USDA-ERS (US Department of Agriculture Economic Research Service). 2017. International Agricultural Productivity Data: Agricultural Total Factor Productivity Growth Indices for Individual Countries, 1961–2016. Accessed November 2017. https://www.ers.usda.gov/ data-products/international-agricultural-productivity/.

von Grebmer. K., J. Bernstein, D. Nabarro et al. 2016. *2016 Global Hunger Index: Getting to Zero Hunger*. Bonn, Washington, DC, and Dublin: Welthungerhilfe, IFPRI, and Concern Worldwide.

WHO (World Health Organization). 2017. Global Health Observatory Data Repository. Accessed December 2017. http://apps.who.int/gho/data/view.main.CTRY2430A?lang=en.

Wineman, A., and T. S. Jayne 2018. "Land Prices Heading Skyward? An Analysis of Farmland Values Across Tanzania." *Applied Economic Perspectives and Policy* 40 (2): 187–214.

World Bank. 2011. *Agricultural Innovation Systems: An Investment Sourcebook*. Washington, DC.

———. 2017. World Development Indicators Database. Accessed December 2017. http:// databank.worldbank.org/data/reports.aspx?source=world-development-indicators.

———. 2020a. Living Standards Measurement Study (LSMS) Database. Accessed April 2020. https://microdata.worldbank.org/index.php/catalog/lsms.

———. 2020b. PovcalNet Database. Accessed June 2020. http://iresearch.worldbank.org/PovcalNet/povDuplicateWB.aspx.

Yeboah, K., and T. S. Jayne. 2018. "Africa's Evolving Employment Trends." *Journal of Development Studies* 54 (5): 803–832.

Zhou, Y., and J. Staatz. 2016. "Projected Demand and Supply for Various Foods in West Africa: Implications for Investments and Food Policy." *Food Policy* 61: 198–212.

Chapter 6

DUALITY, URBANIZATION, AND MODERNIZATION OF AGRIFOOD SYSTEMS IN LATIN AMERICA AND THE CARIBBEAN

Eugenio Díaz-Bonilla and Ruben G. Echeverría

The agriculture sector in Latin America and the Caribbean (LAC) is certainly not homogeneous, covering a variety of very different agroecological and climate zones, along a south-north axis.[1] There are three large agricultural producers: Brazil (close to 48 percent of total agricultural production in the region on average during the 2010s), Argentina (almost 14 percent), and Mexico (about 12 percent), along with several intermediate and small producers, which, added together, have as much agricultural production as Argentina and Mexico combined.

Within that diversity, it is possible to identify three broad agricultural situations, a product of geography and climate, the historical occupation of space during the period of discovery and settlement of the Americas, and the different cycles of integration in global markets. A first group (exemplified by Brazil) was based on tropical agriculture for exports, which started with sugar produced in large plantations with slave labor in the 1600s and then expanded to other crops such as coffee, cocoa, tobacco, and bananas, eventually moving to salaried work in modern times.

A second group (which includes Mexico and Peru) was based on local staple crops (such as corn and potatoes) and livestock production oriented to local markets in economies dominated by mineral production for exports (gold and silver originally, but later other products such as copper, tin, and oil). In this second group, during colonial times land was also occupied in large productive units that used indentured labor from indigenous communities,

1 Latin America and the Caribbean comprises 33 nations according to the United Nations. Listed in alphabetical order they are Antigua and Barbuda, Argentina, Bahamas, Barbados, Belize, Bolivia, Brazil, Chile, Colombia, Costa Rica, Cuba, Dominica, Dominican Republic, Ecuador, El Salvador, Grenada, Guatemala, Guyana, Haiti, Honduras, Jamaica, Mexico, Nicaragua, Panama, Paraguay, Peru, Saint Kitts and Nevis, Saint Lucia, Saint Vincent and the Grenadines, Suriname, Trinidad and Tobago, Uruguay, and Venezuela. The region also has 15 territories that are dependencies of other countries, of which Puerto Rico is the largest. Here we refer to the 33 nations.

who were allowed to produce for self-consumption in smaller plots. Over time, these large units also moved toward more commercial operations based on salaried work.

The third group has been the temperate-climate agriculture on the southern plains of South America (basically Argentina and Uruguay). Lacking minerals and the possibility of producing highly priced tropical products, these regions attracted far less migration during colonial times. Agricultural activities started with extensive livestock production in large production units (given the low population density), which, following changes in world demand, eventually evolved to grain production and more specialized livestock activities toward the end of the 19th century and beginning of the 20th century. But by then, the land had already been occupied in large tracts, and landowners started to rent the land to sharecroppers for agriculture.

In the end, all three types of agriculture generated a dichotomy between a small number of large establishments ("latifundios") and a large number of small ones ("minifundios"), the traditional LAC dual agrarian structure.[2]

Another characteristic of agriculture in LAC is that since the region got integrated into world markets during colonial times, the evolution of the global economy determined to a large extent the performance of the agriculture sector. This has obviously been the case for countries with agriculture sectors oriented to export markets since colonial times, where the cycles of demand for and supply of their products in the world economy shaped the overall functioning of the countries' economies. But also in the case of those countries whose economies were based on exports of mineral products and oil, their agriculture sector was heavily influenced by global developments in those other primary products through their impacts on exchange rates and international trade, public revenues and expenditures, and the influence on the evolution of aggregate domestic demand.

With those two characteristics in mind (inequality in land structures and the strong influence of global markets), this chapter will look at the evolution of the agrifood system and related policies in LAC during the last decades.

2 The unequal agrarian structures in all three types of agricultural groups (very different from the more egalitarian occupation of land by family farms in the United States and Canada) have affected, although to different degrees depending on the countries, the development of both democracy and broad-based domestic markets (Engerman and Sokoloff 2002).

Changing Roles of Agriculture

The role of the agricultural sector and related policies must be considered in the context of the whole economy, with four levels of analysis: farmers (supply side), the performance of the whole economy and opportunities for trade (demand side), the links between supply and demand through value chains, and the geographical setting where those activities take place. From a geographical point of view, farmers are part of a rural and regional economy. From an economic point of view, they are part of agricultural value chains. These levels are embedded in the general economy, which determines the overall governance and policy setting, as well as the final demand for agricultural products in both domestic and external markets (Díaz-Bonilla 2015).

In LAC, as in other developing countries, a key policy issue has been the balance between the agriculture sector vis-à-vis other productive sectors in their economic strategies. A related aspect has been how LAC policies approached the traditional food policy dilemma of trying to maintain high prices for agricultural producers or keeping them low for consumers, with its impact also on the economic and social balance between rural areas and urban centers.

Another policy dilemma in LAC, resulting from its unequal agrarian structure, has been whether to pursue growth and production prioritizing larger agricultural units or to emphasize poverty reduction and food security with a focus on small farmers.[3] As briefly discussed immediately, there have been different policy approaches to the triple dichotomy (agriculture versus other sectors, rural versus urban, and large versus small producers).

After World War II many countries in LAC followed policies that considered the role of agriculture to be subordinate to the needs of industrialization and of the urban population, in what was called the "import substitution industrialization" (ISI) strategy. Economic arguments in favor of industrialization included the idea of declining terms of trade of countries exporting agricultural products (or primary products, in general) compared with those exporting industrial goods (Singer 1950; Prebisch 1950, 1968).

Within the context of the ISI strategy, agricultural policies were based mostly on the use of administered prices at different stages of the market chain, the existence of public and parastatal enterprises operating in product and input markets, and the establishment of public agricultural banks to

3 There are several studies arguing the positive impacts of an agrarian structure based on family farms on the emergence of democratic governance and on the formation of larger domestic markets that support the development of industry and other activities (Engerman and Sokoloff 2002).

supply subsidized credit. By and large those policies tried to support "modern," large producers. The policies followed included the technological component embedded in the Green Revolution, which led to the creation of the national agricultural research institutes and extension services starting in the late 1950s. The challenge of poverty in rural areas was mainly addressed through community development and land or agrarian reform.

The economic crises of the 1980s led to a change of strategy in LAC, in line with different studies that argued that ISI was economically inefficient and led to high inflation and macroeconomic crises (Little, Scitovsky, and Scott 1970; Balassa 1971). Also, ISI was criticized because it discriminated against agriculture and negatively affected employment, income distribution, and poverty. Agricultural sectoral policies were negatively evaluated as well for what was considered a maze of inefficient and at times contradictory interventions, subsidies, and public enterprises operating on input and output markets.

During the 1980s and 1990s, there was a change in macroeconomic, trade, and sectoral policies in LAC that led to the reduction or elimination of industrial protection and of the overvaluation of the exchange rate, along with the phasing out of export taxes on agriculture and a substantial revamping and scaling down of government's involvement in agricultural markets (World Bank 1986). The idea was that given that incentives would shift in favor of agriculture with the change in the general macroeconomic and trade framework, then all those other sectoral public interventions trying to compensate the previous negative bias were not necessary. The poor, particularly in rural areas, were supposed to benefit from more sustainable growth once the capital-intensive and antiagricultural development strategy was corrected.

However, while price distortions were reduced or eliminated during the 1980s and 1990s, other developments were moving against the agricultural sector in LAC.[4] The 1980s debt crisis led to a significant decline in growth in the region, dubbed "the lost decade," which hurt domestic demand for agricultural goods. Also, some of the reforms, while eliminating many of the inefficient and contradictory public-sector interventions, ended up, at the same

---

4 Anderson and Nelgen (2013) calculated the levels of support (or taxation) for the agriculture sector for several countries and regions at the world level. Those estimates calculate the support to producers that can come from the rest of the society as consumers (mainly through trade policies) or as taxpayers (through fiscal policies). By their estimates, LAC's overall and sectoral policies implied a *tax* on the agriculture sector of about 13 percent of the value of the production during the 1950s and until the 1980s (see also Chapter 13). After that, the authors calculate, the bias against agriculture changed into positive support for the sector in LAC, with a net *subsidy* equivalent to about 3.5 percent of the value of production on average during the period 1990–2010.

time, dismantling the institutional infrastructure that provided technical assistance and some key inputs to agricultural production, without ensuring the creation of private-sector institutions that could provide similar services and inputs.

Additionally, there were two important changes in external conditions. First, particularly since the 1980s, the extensive support and protection of agriculture in rich countries led to surpluses that were sold in world markets with subsidies, depressing world prices. Second, expanded capital flows led to a more volatile economic environment for developing countries, with the sequence of crises in Mexico in 1995, Asia in 1997, Russia in 1998, Brazil in 1999, and Argentina in 2001, which had negative implications for world growth and agricultural demand.

In terms of poverty alleviation, in the second half of the 1990s a new type of poverty reduction program began to be implemented in LAC, such as Mexico's "Progresa," Brazil's "Bolsa Scola," and similar programs. They basically consisted of income transfers given mostly to heads of households (many of them women), but with specific commitments related to attendance at school and health controls for their children. The programs appear to have had positive impacts on local economic activity and on accumulation of physical and human capital (Adato and Hoddinot 2010).

After the policy reforms in the 1980s and 1990s reduced, or even eliminated, the past bias in macro incentives against agriculture, the framework of incentives in LAC began to move, although with variations, in support of agriculture.

Table 6.1 shows nominal rates of protection (NPRs), which are the percentages by which the domestic prices paid by consumers and received by producers are above (or below) the equivalent world price.[5]

Table 6.1 shows that until the last year available (2015), Argentina was the only country in the region that placed domestic prices below those prevailing in world markets. A second group of countries maintained domestic prices broadly in line with world markets (between 0 and 10 percent). But other countries have tilted domestic prices in favor of agricultural producers and against domestic consumers (10 percent or more), with cases of NPRs of more than 30 percent (Jamaica, Haiti, and El Salvador), and even reaching 55.5 percent in the case of the Dominican Republic.

5 The measure reported in Table 6.1 represents mostly transfers from consumers, unlike the Anderson and Nelgen (2013) estimates, which also include transfers from taxpayers. The NPR numbers presented here are an aggregation of different products included in the database (details can be found at http://www.ag-incentives.org/).

TABLE 6.1 Nominal rates of protection

Country	Above 20 (%)	Country	10–20 (%)	Country	0–10 (%)	Country	Negative (%)
Dominican Republic	55.5	Colombia	18.7	Ecuador	6.2	Argentina	–26.5
Jamaica	37.1	Honduras	17.5	Peru	4.5		
Haiti	33.6	Suriname	13.8	Mexico	2.9		
El Salvador	32.1	Nicaragua	13.6	Brazil	2.2		
Guatemala	22.3	Belize	13.1	Uruguay	1.0		
		Bolivia	13.1	Chile	0.6		
		Guyana	12.5	Paraguay	0.0		
		Costa Rica	10.3				

Source: AgIncentives (2020).
Note: The years of the calculations vary by country, but they are within the period 2005–2015.

The policy changes were accompanied by a rebound of growth in LAC's agricultural production during the last two decades (see Production, in the second section), which also transformed LAC into the main *net* exporter of agricultural products, as can be seen in Figure 6.1, with the region surpassing the United States, Canada, Australia, and New Zealand combined.

The net trade surplus has been generated mainly by Brazil and Argentina, with some contribution from the rest of LAC (where some countries are net importers).

The structure of agricultural trade for LAC also changed significantly during recent decades, with increases in exports of oilseeds and related meal and oil products and, to a smaller degree, fruits and vegetables. Other historically important products, such as sugar and coffee, have declining shares both in LAC's and the world's share of exports. Also to be noted is that the direction of trade (historically more directed to Europe and the United States) has become more diversified, with an increased presence of Asian markets, particularly China (see also Chapter 13).[6]

Since the price shocks of 2008 and 2011, several LAC countries seem to have reverted in part to greater involvement of the public sector in agriculture related to concerns about food security. However, the main malnutrition

6 In general, China has become the top destination for South American exports and the second destination for all of LAC exports, after the United States. However, the composition of LAC exports to China is heavily tilted toward primary commodities, much more than the overall composition of LAC exports to the world, while LAC is buying mostly manufacturing goods and running an important deficit with China. This has led to debates about the benefits for LAC of such structure of trade.

FIGURE 6.1 Net agricultural trade (million current US dollars)

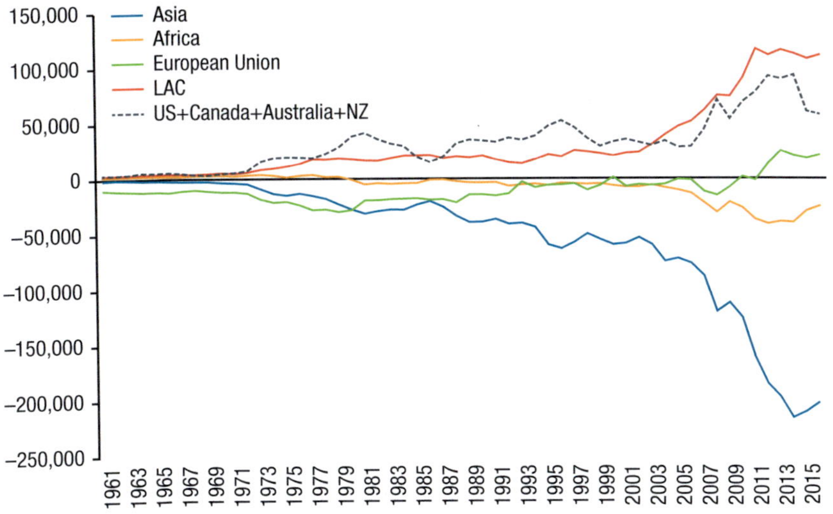

Source: Authors' calculation based on FAO (2018).

Note: LAC = Latin America and the Caribbean. NZ = New Zealand.

problems in LAC are becoming less those of undernourishment and food insecurity, and more those of the two other burdens of malnutrition—deficiencies in different micronutrients and vitamins, and overweight and obesity (see Nutritional Transition, in the third section).[7]

Part of these changes have been related to the emergence of LAC as the most urbanized developing region (Table 6.2): the percentage of urban population increased from 52 percent in the 1960s to almost 79 percent in the 2010s, way above the world's average of 53 percent (Díaz-Bonilla and Torero 2016).

It should be noted that although LAC has several megacities, almost 60 percent of the region's urban population is located in intermediate and small towns of less than 1 million inhabitants, as is shown in Table 6.2 (see also Chapter 9).

Therefore, the process of overall urbanization has been relatively more evident in the development of small and intermediate towns, which have stronger links to rural spaces and differential impacts on agricultural value chains and the structure of food production, employment, and consumption (Berdegué, Proctor, and Cazzuffi 2014). The large percentage of the population living

7 All three problems constitute the "triple burden of malnutrition" (Pinstrup-Andersen 2011).

TABLE 6.2 Changing shares of urban population in LAC and other regions

Regions	Variables	1960s	1970s	1980s	1990s	2000s	2010s
Latin America and Caribbean (excluding high income)	Urban population less than 1 million (% of total)	27.4	30.7	34.8	38.8	41.3	42.9
Low and middle income	Urban population less than 1 million (% of total)	15.1	16.8	19.9	22.8	25.3	28.0
World	Urban population less than 1 million (% of total)	20.5	21.6	23.9	26.0	28.1	30.5
Latin America and Caribbean (excluding high income)	Urban population (% of total)	52.0	59.6	66.5	72.2	76.4	79.0
Low and middle income	Urban population (% of total)	25.4	28.4	33.0	37.7	42.7	47.6
World	Urban population (% of total)	35.2	37.6	40.9	44.5	48.7	52.9

Source: Data from World Bank (2019).
Note: LAC = Latin America and the Caribbean.

in intermediate and small towns distinguishes LAC from other developing regions, with the exception of developing countries in Eastern Europe and Central Asia (Díaz-Bonilla and Torero 2016).[8]

All those changes present a variety of challenges for agricultural policies in LAC. Therefore, the overall policy framework for agriculture and food production is changing, and the policy objectives related to the sector have also become more complex. With differences depending on the levels of development, the agricultural sector is expected to contribute to the whole economy and society on several fronts. For instance, expanding on PIADAL (2013), it is possible to identify five objectives: (1) growth in agricultural production and productivity as a means to contribute to growth, employment, and economic development in general; (2) reduction of poverty, vulnerability, and food insecurity, considering gender issues; (3) food safety and improvements in nutrition conditions; (4) environmental sustainability and protection of biodiversity; and (5) regional and territorial development, helping to eliminate large income disparities between urban and rural areas.

8 Also, urban centers in LAC are among those most affected by violence and crime, apart from countries at war. In fact, measured by the number of homicides per 100,000 population, the 8 most dangerous cities in the world and 42 of the top 50 are in LAC. See www.worldatlas.com/articles/most-dangerous-cities-in-the-world.html.

In consequence, the challenges for policies related to the agriculture sector go beyond what ministries of agriculture can do. The focus has expanded from agricultural development, to broader views of rural and value-chain development, and more recently to food systems, encompassing the previous levels plus retail and consumption aspects (HLPE 2017).

Production and Productivity Growth in Agriculture

Production

Table 6.3 shows the growth rates of agricultural production (which is the sum of the two components of the FAOSTAT database labeled "crops" and "livestock") for the three main producing countries, for the LAC region, and for the world, in decades since the 1960s.[9]

LAC's total agricultural production performed better than the world average in all decades. The worst performance for LAC was during the 1980s, the "lost decade" of the debt crisis, when the macro, trade, and sectoral policy reforms that were supposed to help the agriculture sector were in part countered by slow domestic growth and the decline in world prices. In the 2000s and 2010s, growth levels returned to those seen in the 1960s and 1970s. This was the result of better performance in South America and particularly Brazil.

In general, crops have been growing faster in more recent decades, pushed mostly by Argentina and Brazil, with the expansion of oilseeds and grains. Livestock grew faster in the 1990s due to Brazil and Mexico, while in Argentina the 2010s were negative for aggregate livestock production, in part because of adverse domestic policies and relative prices that led to the expansion of soybeans to the detriment of pastures (Díaz-Bonilla et al. 2014).

Those differential growth rates changed the structure of agricultural production, as shown in Table 6.4 (divided into the two components of the FAOSTAT database—crops and livestock). The values for agriculture, crops, and livestock are presented for the three main producing countries, for the rest of LAC, for the LAC aggregate, and for the world, in decades since the 1960s.[10]

---

9 Value of production is measured in constant dollars at purchasing power parity (PPP) values, which FAOSTAT calls "international dollars" (which are different from constant common dollars). This measure avoids fluctuations in the total value of aggregates due to changes in market exchange rates, and it uses a form of adjustment for PPP by which each commodity has a single world price per relevant unit of volume, irrespective of the country where it was produced. This approach facilitates aggregations and comparisons across countries.

10 Data reflect the value of production in constant dollars at PPP values (which are different from constant common dollars).

TABLE 6.3 **Value of agricultural production (average annual growth rates in constant 2004–2006 PPP dollars)**

Region	1960s	1970s	1980s	1990s	2000s	2010s
Argentina	3.0	2.2	–0.4	4.1	1.4	4.8
Brazil	3.8	3.5	4.5	3.2	4.4	3.1
Mexico	4.5	3.8	1.9	3.4	1.9	2.8
LAC	3.2	3.0	2.5	3.0	3.1	3.2
World	2.8	2.4	2.2	2.1	2.4	2.3

Source: Authors' calculations based on FAO (2018).

Note: The average for the decades was affected by the strong decline of 2009 and sharp rebound of 2010. LAC = Latin America and the Caribbean; PPP = purchasing power parity.

TABLE 6.4 **Share (%) of value of world agricultural production (in constant 2004–2006 PPP dollars)**

Country or region	Variable	1960s	1970s	1980s	1990s	2000s	2010s
Argentina	Agriculture	2.0	1.8	1.7	1.7	1.8	1.8
Brazil	Agriculture	3.0	3.4	4.0	4.5	5.5	6.1
Mexico	Agriculture	1.3	1.4	1.6	1.6	1.6	1.6
Rest of LAC	Agriculture	3.2	3.3	3.2	3.2	3.3	3.2
LAC	Agriculture	9.5	9.9	10.5	11.0	12.2	12.7
World	Agriculture	100	100	100	100	100	100
Argentina	Crops	1.3	1.4	1.4	1.5	1.8	1.8
Brazil	Crops	3.4	3.6	4.3	4.3	5.0	5.6
Mexico	Crops	1.3	1.4	1.5	1.4	1.4	1.3
Rest of LAC	Crops	3.4	3.4	3.2	3.2	3.1	3.1
LAC	Crops	9.4	9.8	10.4	10.4	11.3	11.8
World	Crops	100	100	100	100	100	100
Argentina	Livestock	3.0	2.6	2.2	2.0	1.8	1.7
Brazil	Livestock	2.5	2.9	3.7	4.8	6.4	7.0
Mexico	Livestock	1.2	1.5	1.8	1.8	2.0	2.0
Rest of LAC	Livestock	3.0	3.0	3.0	3.4	3.7	3.8
LAC	Livestock	9.7	10.0	10.7	12.0	13.9	14.5
World	Livestock	100	100	100	100	100	100

Source: Authors' calculations based on FAO (2018).

Note: LAC = Latin America and the Caribbean; PPP = purchasing power parity.

Overall, the relatively stronger growth of LAC's agriculture (Table 6.4) led to the increase in the region's share in world agriculture: it moved from 9.5 percent of world agricultural production in the 1960s (a combination of a share of 9.4 percent in crops and 9.7 percent in livestock) to 12.7 percent in the 2010s (combining 11.8 percent in crops and 14.5 percent in livestock). LAC's increased share of more than 3 percentage points has been due mostly to Brazil's performance, considering that Argentina lost some share, while Mexico did not gain much, and the rest of LAC stayed about the same. In terms of composition of agricultural production, the increase in LAC's share at the world level is related more to the increase of livestock production, as opposed to crops.

A point worth noting is that by the 2010s, LAC's agricultural production (measured in international dollars) had grown somewhat bigger in size than both that of the European Union, on the one hand, and that of the United States and Canada, on the other (both with about 11 percent of global agricultural production each). As a comparison, all Asia represents about 50 percent, and Africa close to 9 percent (Díaz-Bonilla, Saini, Henry, Creamer, and Trigo 2014).

Land, Deforestation, and Greenhouse Gas Emissions

Agricultural land (including crops and pastures) at the global level increased by about 380 million hectares between the 1960s and the 2010s.[11] To place LAC in the global context, Table 6.5 shows how much of that change occurred in different producing regions. The LAC region is presented as a whole and also disaggregated into Argentina, Brazil, Mexico, and the rest of LAC. Table 6.5 includes the same calculations as Table 6.3, related to increases in the global share for the value of agricultural production in billions of constant dollars of equivalent purchasing power (or purchasing power parity, PPP).[12]

LAC represented 44.1 percent of the world increase in agricultural land during the decades between the 1960s and 2010s, while the share of

11 FAOSTAT uses "agricultural area" as the general category, which has different components such as "arable land," "permanent crops," and "permanent meadows and pastures." In the table we utilize the general category of "agricultural area" because it seems the most comprehensive estimation of land use, and it is particularly relevant for LAC, given the large share of livestock production in the region.

12 Value of production is measured in constant dollars at purchasing power parity (PPP) values, which FAOSTAT calls "international dollars" (which are different from constant common dollars). This measure avoids fluctuations in the total value of aggregates due to changes in market exchange rates, and it uses a form of adjustment for PPP by which each commodity has a single world price per relevant unit of volume, irrespective of the country where it was produced. This approach facilitates aggregations and comparisons across countries.

TABLE 6.5 Increases in production and area

Country or region	Net production value (constant 2004–2006 billion PPP dollars)				Agricultural area (million hectares)			
	1960s	2010s	Percentage of world change	Change from 1960s to 2010s (%)	1960s	2010s	Percentage of world change	Change from 1960s to 2010s (%)
Asia	248.9	1,247.7	62.0	401.4	1,094.6	1,648.4	145.4	50.6
China	81.1	569.4	30.3	601.8	355.1	514.6	41.9	44.9
India	63.3	251.6	11.7	297.4	177.0	179.6	0.7	1.5
LAC	79.7	311.8	14.4	291.2	582.5	750.4	44.1	28.8
Argentina	16.5	43.0	1.6	160.8	132.6	148.6	4.2	12.1
Brazil	25.7	148.6	7.6	478.8	170.9	277.2	27.9	62.2
Mexico	10.8	38.1	1.7	253.9	98.0	106.7	2.3	8.9
Rest of LAC	26.8	82.1	3.4	206.6	181.1	217.9	9.7	20.3
Africa	58.0	210.5	9.5	262.9	1,048.1	1,130.2	21.6	7.8
European Union	188.4	269.4	5.0	43.0	209.9	186.6	−6.1	−11.1
USA and Canada	131.9	268.9	8.5	103.9	508.9	471.6	−9.8	−7.3
Australia and New Zealand	16.9	37.7	1.3	123.7	486.8	414.6	−19.0	−14.8
World	841.3	2,452.5	100	191.5	4,504.3	4,885.1	100	8.5

Source: Authors' calculations based on FAO (2018).
Note: LAC = Latin America and the Caribbean; PPP = purchasing power parity.

the increase in world agricultural production during that same period was 14.4 percent. Therefore, LAC's increase in agricultural and food production and exports (see below), although benefiting from improvements in productivity, was also associated with an important expansion of agricultural area based on land-use changes that will be difficult to repeat in the future without compromising sustainability.

The increase in agricultural land happened mostly in Brazil and, to a smaller degree, in the rest of LAC, while Mexico and Argentina experienced relatively minor expansions.

The expansion in agricultural land has been closely linked to deforestation (Table 6.6). LAC lost about 10 percent of its forest between 1990 and 2015, while the world lost about 3 percent. Overall, the region represented more than 80 percent of all the forest lost at the world level during that period.

TABLE 6.6 Forest area (million hectares)

Region	1990	2015	Lost forest	Lost forest as percentage of total forest in 1990
World	4,128.3	4,007.4	−120.9	−2.9
LAC	1,007.0	906.3	−100.7	−10.0
LAC as percentage of world	24.4	22.6	83.3	

Source: Authors' calculations based on FAO (2018).
Note: LAC = Latin America and the Caribbean.

Therefore, even though LAC represents only 10 percent of all greenhouse gas (GHG) emissions at the world level, the region has comparatively higher levels of GHG emissions related to agriculture and land-use change and forestry (LUCF). According to the database of emissions from FAOSTAT, in the decade of the 2010s LAC represented about 22 percent of all world emissions, combining agriculture and LUCF emissions, with Brazil accounting for some 44 percent of LAC's total for that aggregate.

Therefore, the loss of forest cover should be monitored to ensure long-term sustainability (as is done with the satellite data collected by Terra-I, a project of CIAT).[13] Rapid land-use change is putting pressure on LAC's role as a major provider of global environmental public goods, including biodiversity, oxygen, and carbon sinks.

Labor

Table 6.7 shows that the region has far smaller shares of employment in agriculture than both developing countries and the world, but with great variations: along with Brazil and Mexico, which are the largest countries in the region, two other countries are included, as representatives of lower (Venezuela) and higher (Haiti) levels of agricultural employment in the region.[14]

The fact that LAC has expanded the land area utilized but reduced employment in agriculture is reflected in the differential productivity levels for land and labor, as discussed below.

13 See www.terra-i.org/terra-i.html for information about Terra-I.

14 Some small island countries in the Caribbean show lower levels than Venezuela. Data for Argentina is not included because of the limited coverage of household surveys.

TABLE 6.7 Employment in agriculture (% of total employment)

Countries and regions	1990s	2000s	2010s
LAC	22.1	19.0	14.9
Brazil	21.2	19.2	11.6
Mexico	23.4	15.5	13.5
Venezuela	12.0	9.8	7.5
Haiti	50.6	50.4	50.0
Low and middle income	50.3	44.8	36.0
World	41.7	37.3	30.2

Source: Data from World Bank (2019).
Note: LAC = Latin America and the Caribbean.

Productivity

The worldwide efficiency in land and labor use increased on average between 1961 and 2010 (Pardey 2012). In particular, LAC increased output per worker by about 270 percent during that period, and output per hectare some 205 percent (measured in 2004–2006 international dollars). But there are clear differences between regions and countries: Latin America shows less labor and land productivity than in all developed regions (except for Australia and New Zealand, which have less land productivity than LAC) while being above the world average and that of all developing regions (except Eastern Europe) in labor productivity, but only exceeding Africa south of the Sahara and countries of the former Soviet Union when considering land productivity (Pardey 2012; see also Chapter 1).

Figures 6.2, 6.3, and 6.4, following Hayami and Ruttan (1985), present several indicators of partial productivity.[15] The indicators are presented for the region only, comparing the averages for three decades—1960s, 1990s, and 2010s—and separating Argentina, Brazil, Mexico, and the rest of LAC (which is the simple average of the remaining countries in the database).[16]

Argentina, due to its large land area (Figure 6.3), has the lowest ratio of product per unit of land (Figure 6.2), but a low labor-land ratio (the inverse

15 Data come from the IFPRI (2019) database maintained by Alejandro Nin-Pratt, which also uses data from the USDA Economic Research Service Agricultural Productivity project and FAOSTAT (see also Fuglie 2012 and 2015). Figures 6.2–6.4 follow Hayami and Ruttan (1985; see also Kawagoe, Otsuka, and Hayami 1986). The variables are defined in the figure notes. The data is presented in normal units (not logarithms) to facilitate interpretation.

16 Those countries are Bolivia, Chile, Colombia, Costa Rica, Dominican Republic, Ecuador, El Salvador, Guatemala, Guyana, Haiti, Honduras, Jamaica, Nicaragua, Panama, Paraguay, Peru, Suriname, Trinidad and Tobago, Uruguay, and Venezuela.

FIGURE 6.2 Product per unit of land (Q/A) and product per unit of labor (Q/L)

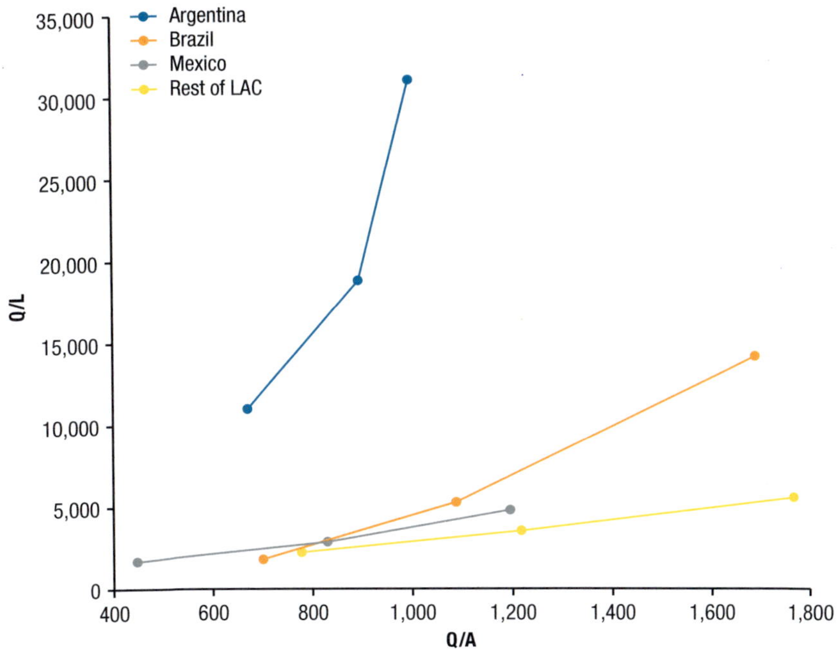

Source: Authors' calculations using IFPRI (2019).

Note: A = land (total agricultural land in hectares of "rainfed cropland equivalents"); L = labor (persons economically active in agriculture); Q = output (FAO gross agricultural output, measured in international 2005 dollars); LAC = Latin America and the Caribbean.

of the indicator in Figure 6.3) combined with larger mechanization ratios (Figure 6.3) and natural fertility leads to the largest ratio of product per unit of labor (Figure 6.2), even with low levels of use of fertilizers (Figure 6.4). Mexico, far more constrained in land per worker (Figure 6.3) while still having an important agricultural population with lower levels of capital use (Figure 6.3) and fertilizer application (Figure 6.4), shows the lowest ratios of product per land and labor. Brazil and the rest of LAC are in between those two countries in terms of partial productivity of land (Figure 6.2), but Brazil is clearly above the rest of LAC in the ratio of product per unit of labor (Figure 6.2), resulting from larger indices of capital per unit labor (Figure 6.3) and fertilizer use (Figure 6.4). Also noticeable in Brazil is the strong growth of the level of production per unit of labor (almost 680 percent since the 1960s), supported by large increases in land and capital per unit of labor (Figure 6.3)

FIGURE 6.3 Land per unit of labor (A/L) and capital per unit of labor (K/L)

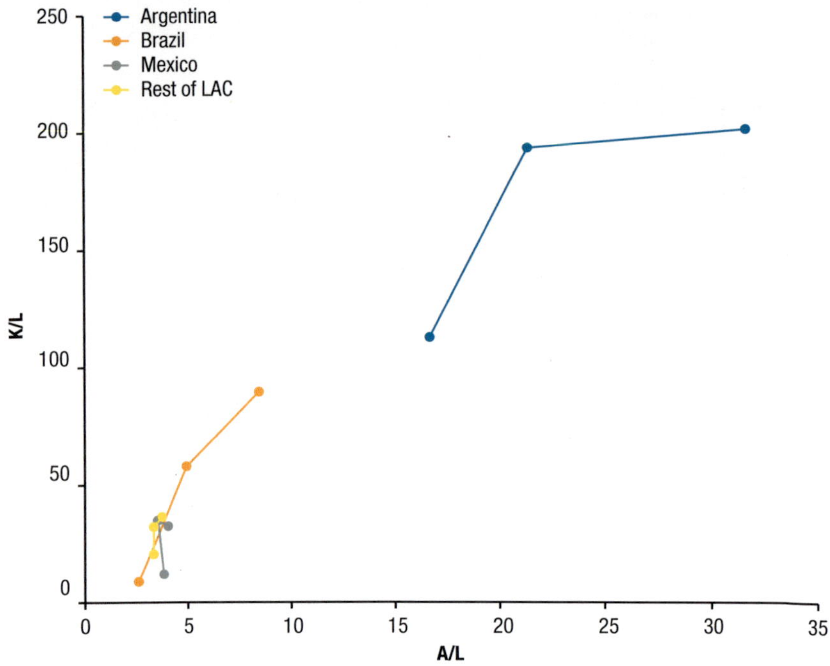

Source: Authors' calculations using IFPRI (2019).

Note: A = land (total agricultural land in hectares of "rainfed cropland equivalents"); K = capital (total stock of farm machinery in "40-CV tractor equivalents" (CV is a measure of horsepower). LAC = Latin America and Caribbean.

and in the use of fertilizers (Figure 6.4).[17] To sum up, the observation that high (or low) land-labor ratio is associated with high (or low) capital-labor ratio and high (or low) labor productivity is consistent with the Hayami-Ruttan (1985) induced innovation hypothesis.

A more comprehensive approach to the evolution of productivity in LAC is to analyze the growth rate of total factor productivity (TFP).[18] Nin-Pratt et al. (2015), looking at TFP growth in LAC agriculture between 1980 and 2012, show that regional agricultural output per worker and TFP increased 82 and

17 For instance, Bustos, Caprettini, and Ponticelli (2016) show significant technical change in soybean production in Brazil that has been strongly laborsaving and led to structural transformation.

18 It should be noted that the calculation of total agricultural production is a composite of products measured in constant prices of a certain base year. Therefore, that "quantity" can increase also because of composition effects (for example, if a country reduces the share of lower-price products and increases the share of higher-value ones).

FIGURE 6.4 Productivity per unit of land (Q/A) and fertilizer use per unit of land (F/A)

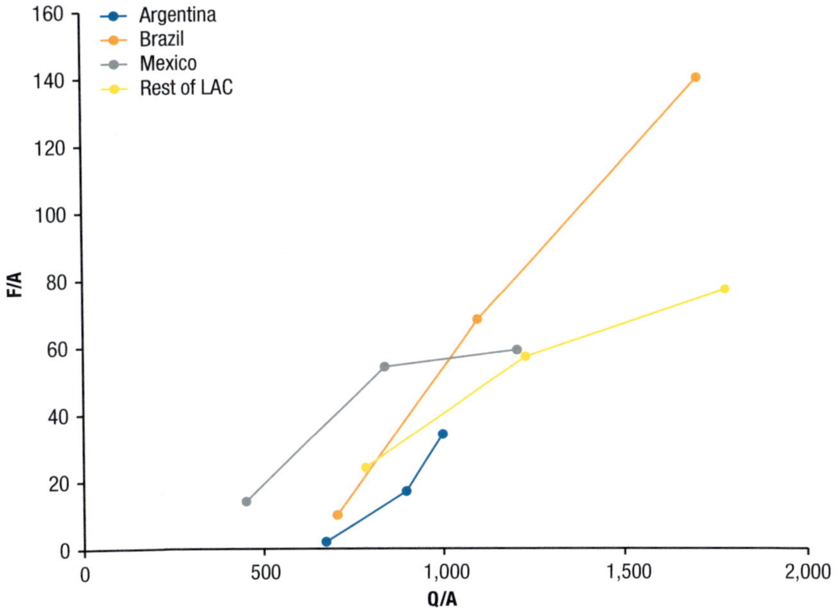

Source: Authors' calculations using IFPRI (2019).

Note: A = land (total agricultural land in hectares of "rainfed cropland equivalents"); F = fertilizer (metric tons of N–P_2O_5–K_2O fertilizer consumption measured in "N-fertilizer equivalents"); LAC = Latin America and the Caribbean; Q = output (FAO gross agricultural output, measured in international 2005 dollars).

45 percent, respectively. The authors attribute the improved performance to the use of fertilizer, increases in land productivity, and growth in the use of capital that expanded cultivated area per worker, as well as to higher productivity of the animal stock.

Table 6.8 shows the latest estimates from IFPRI (2019) for LAC compared with other regions and differentiating Argentina, Brazil, Mexico, and the rest of LAC.[19]

LAC was behind all developing regions in TFP growth except the Middle East and North Africa during 1991–2000 (affected by Argentina's low growth) but outperformed all regions during the next two periods shown in Table 6.8. Brazil and Mexico maintained a reasonably strong TFP growth during the years considered.

19 The countries are the same as for Figures 6.2 to 6.4, except Venezuela.

TABLE 6.8 Growth of TFP (% annual)

Region/country	1991–2000	2001–2010	2011–2015
Africa south of the Sahara	1.5	1.0	0.4
Asia	1.7	1.7	1.5
Middle East and North Africa	1.3	1.5	1.2
Latin America and Caribbean	1.3	2.3	1.9
Argentina	0.2	0.9	3.4
Brazil	1.6	3.3	1.6
Mexico	2.2	2.0	2.2
Rest of LAC (average)	0.8	1.7	1.9

Source: IFPRI (2019).
Note: LAC = Latin America and the Caribbean; TFP = total factor productivity.

The positive trends in LAC's partial and total factor productivity in the recent past took place within a relatively favorable global macroeconomic environment and higher commodity prices. Sustaining future agricultural productivity trends in the region under conditions that will likely be less favorable will require important investments in agricultural research (Pardey and Beintema 2001; Fan 2008).

Food Security and Nutritional Transition

Food Availability

Food availability per capita has increased between the 1960s and the current decade in LAC and worldwide (Table 6.9). While in the 1960s the average number of daily calories per capita was between somewhat more than 2,100 and 2,300 depending on the subregions of LAC, the averages for the decade of the 2010s were between about 2,700 and more than 3,000 calories per day per person. Daily proteins per capita increased from 51–64 grams (1960s) to about 67–85 (2010s), while fats moved from 47–52 grams per capita (1960s) to 70–100 (2010s). Mexico and Central America, and South America (the largest LAC subregions considered in the data) have maintained absolute values of food availability above world averages; the Caribbean region, where Haiti has a large influence in the aggregates due to its large population, is below the world average, but availability per capita has still grown about 27 percent in calories, 31 percent in proteins, and almost 50 percent in fats.

TABLE 6.9 Food availability

Regions	Variable	1960s	2010s	Change (%)
Caribbean	Fat supply quantity (grams/capita/day)	47.0	70.4	49.6
Mexico and Central America	Fat supply quantity (grams/capita/day)	48.9	85.0	73.7
South America	Fat supply quantity (grams/capita/day)	52.2	99.6	90.7
World	Fat supply quantity (grams/capita/day)	50.9	82.4	62.0
Caribbean	Food supply (kcal/capita/day)	2,127.6	2,699.0	26.9
Mexico and Central America	Food supply (kcal/capita/day)	2,271.6	2,920.0	28.5
South America	Food supply (kcal/capita/day)	2,393.8	3,014.0	25.9
World	Food supply (kcal/capita/day)	2,291.6	2,869.3	25.2
Caribbean	Protein supply quantity (grams/capita/day)	50.9	66.6	30.9
Mexico and Central America	Protein supply quantity (grams/capita/day)	61.1	81.7	33.6
South America	Protein supply quantity (grams/capita/day)	64.0	85.2	33.2
World	Protein supply quantity (grams/capita/day)	63.2	80.6	27.5

Source: Authors' calculations based on FAO (2018).

Income Growth, Poverty, and Inequality

During recent decades LAC countries also showed other important socioeconomic changes. Since the 1960s, gross domestic product (GDP) per capita (in constant $2010 dollars) increased by 130 percent to almost $9,500 dollars per capita (average of the 2010s), clearly above the less than $4,200 dollars for all developing countries, but still far below the more than $41,000 dollars per capita of developed countries (World Bank 2018).

That growth of income was accompanied by the traditional process of structural transformation in which agriculture declines as percentage of total value-added: though the sector had represented about 14.6 percent of total GDP for LAC during the 1960s, it declined to 5.5 percent by the 2010s. There is still an important variation across countries in LAC, with Haiti, Paraguay, Nicaragua, Belize, Honduras, Bolivia, El Salvador, Guatemala, and Suriname showing percentages of agricultural GDP above 10 percent of total GDP, while in Chile, Mexico, and Trinidad and Tobago those shares are below 5 percent. Rural population was also falling, moving from 48 percent of total population in the 1960s to 21 percent in the 2010s.

The percentage of population suffering from poverty declined from almost 17 percent on average in the 1980s to 5.8 percent for the 2010s (using a poverty line of US$1.9 per day in PPP terms), levels that are less than half the average for developing countries (Table 6.10).

In fact, the region has achieved the Millennium Development Goal (MDG) to cut the incidence of poverty by half by 2015.[20]

Another point to be noted is that in LAC, poverty has become mostly urban since the 1990s, when the number of urban poor started to exceed the rural poor; early in the 2010s about two-thirds of the poor lived in urban areas. Still, the incidence of poverty is higher in rural areas. Also, the incidence of poverty is higher among the indigenous population and among households that depend on agricultural income or on government transfers (CEPAL, FAO, and IICA 2012).

With regard to food security, LAC also shows better indicators than other developing regions (von Grebmer et al. 2016). The region has achieved as well two other 2015 MDGs, by cutting in half the *percentage* of both underweight children under five years of age and undernourishment in the total population. Latin America (without the Caribbean) has also reached the goal set by the 1996 World Food Summit of cutting the total *number* of undernourished people in half (Díaz-Bonilla and Torero 2016).

However, while the region shows significant improvements as a whole, food security conditions are still worrisome in some countries, such as Haiti, Honduras, and Guatemala, and in disadvantaged regions in bigger countries.

Several factors appear to have supported the achievement in poverty reduction and food security, including the relatively strong performance of the agriculture sector in recent decades (as discussed above), the decline in poverty resulting from economic growth, and the expansion of safety nets based on cash transfers for the poor and vulnerable, since the second half of the 1990s. Compared with other regions, LAC shows the highest coverage by social safety nets of the poorest 20 percent of the population: in LAC more than 50 percent of the population in that quintile are covered, while in Eastern Europe and Central Asia and East Asia and the Pacific less than 50 percent are covered, and in Africa, South Asia, and the Middle East and Northern Africa less than 30 percent are covered (World Bank 2014; Díaz-Bonilla and Torero 2016). Other factors that have arguably contributed to the

20 With a poverty line of US$1.25 (PPP measured in 2005 prices).

TABLE 6.10 Poverty indicators (poverty head count ratio, % of population)

Regions	Poverty line (in 2011 PPP)	1980s	2010s
Latin America and Caribbean	At $1.90/day	16.6	5.8
Low and middle income	At $1.90/day	47.1	15.4
World	At $1.90/day	38.5	13.0
Latin America and Caribbean	At $3.10/day	30.8	12.5
Low and middle income	At $3.10/day	67.6	35.8
World	At $3.10/day	—	—

Source: Data from World Bank (2019).
Note: — = not available; PPP = purchasing power parity.

region's achievements include advances in the education and status of women; improvements in water, sanitation, and health infrastructure; and the spread of democracy in the region since the 1980s and 1990s (Díaz-Bonilla and Torero 2016).

At the same time, it should be noted that for all its advances in income and declines in poverty and food insecurity, LAC remains the most unequal region in the world (an average Gini of about 0.51 for the countries with data in the World Bank database), closely followed by Africa south of the Sahara (Gini of 0.47).

Nutritional Transition

Helped by improvements in income, food security, and health care, the average inhabitant of LAC countries reached about 76 years of life expectancy at birth in the latest year with data in the World Development Indicators (World Bank 2018) (adding 20 years since 1960), compared with 71 years of life expectancy for developing countries as a whole, but some 81 years for developed countries.

However, the decline in poverty and food insecurity in the region has been accompanied by problems of overweight and obesity, part of the triple burden of malnutrition mentioned above (Pinstrup-Andersen 2011). Table 6.11 shows that LAC suffers from a high incidence of overweight and obesity compared with the world average.

Popkin and Reardon (2018) have clearly shown that the region faces a major diet-related health challenge with great economic and social costs. The significant increase—among all ages—of overweight and obese levels can be explained by a shift toward consumption of less-healthful low-nutrient-density foods and sugary beverages as well as changes in away-from-home eating and

TABLE 6.11 Overweight and obesity, 2016 (%)

	Average in LAC		Average in world	
Gender and age	Overweight	Obese	Overweight	Obese
Adolescents, male	30.0	10.4	17.5	5.6
Adolescents, female	31.2	13.5	19.2	7.8
Adults, male	58.5	20.2	24.8	11.6
Adults, female	60.1	28.0	28.4	15.7

Source: Authors' calculations from Development Initiatives (2018).
Note: LAC = Latin America and the Caribbean.

snacking. As a result, governments in the region have started to adopt stricter labeling regulations (such as in Chile, Peru, and Ecuador) and to impose taxes on unhealthy food (Mexico). Moreover, deficiencies in critical nutrients, such as iron and vitamin A, also constitute a problem (see the 2018 Global Nutrition Report by Development Initiatives).

Innovations and Agricultural Transformation

Global and regional agrifood systems are undergoing remarkable changes, while facing a series of major challenges, including sustainable intensification, contribution to global and regional food security, adaptation to and mitigation of climate change, territorial development and the role of "family agriculture," and many more. Facing such challenges requires new approaches and innovative institutional arrangements as well as strengthened agrifood research and innovation.

Investments in Research and Development

In 2011, only 5 percent of total global investments in public and private research and development (R&D) were directed to food and agriculture (Pardey 2012). That small percentage represented US$69.3 billion (in 2009 purchasing-power-adjusted currency) with a decreasing trend (for a total of 55 percent) in high-income countries and significant growth (representing 43 percent) in middle-income countries, compared with 1980.

Stads et al. (2016) show that agricultural R&D spending in LAC (excluding the private for-profit sector), after following a period of volatility and declines until the early 2000s, recovered after 2004. By 2013, the region spent US$5.1 billion on agricultural R&D (2011 PPP prices), representing a 75 percent increase over levels recorded in the early 1980s; total researcher

FIGURE 6.5 Long-term trends in agricultural researchers and research spending in LAC, 1981–2013

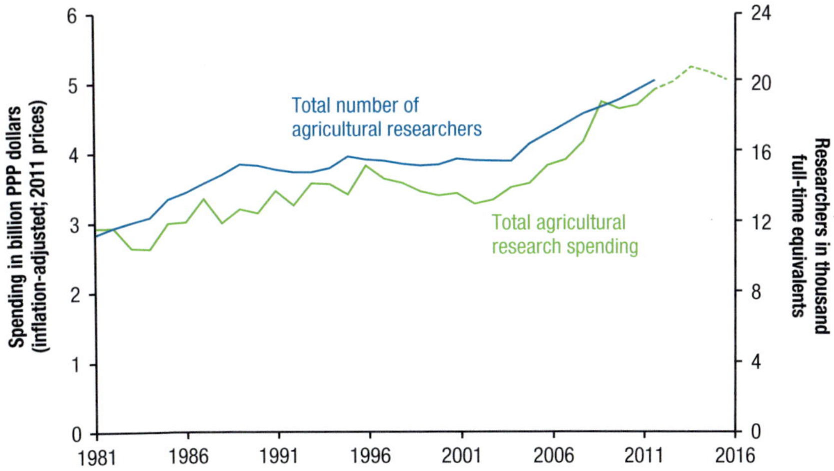

Source: Underlying datasets from Beintema, Nin-Pratt, and Stads (2020).

Note: LAC = Latin America and the Caribbean. The 2013–2016 spending trend is an estimate using various secondary data sources. PPP = purchasing power parity.

numbers (measured in full-time equivalents) also recovered after a period of stagnation during 1990–2004: in 2013, there were about 20,600 agricultural researchers (in full-time equivalents, FTEs), about double the number in the early 1980s. Figure 6.5 shows the evolution of expenditures and researchers in agricultural R&D.

These trends have been driven mostly by Brazil (53 percent of the total in LAC), and to a smaller degree by Argentina (13.5 percent) and Mexico (14 percent), which together represented about 80 percent of LAC's spending and some 75 percent of agricultural researchers (Stads et al. 2016). The Agricultural Science and Technology Indicators (ASTI) database, on which Stads et al. (2016) is also based, shows that by 2013, LAC expenditures in agricultural R&D represented almost 26 percent of the total by developing countries. As a comparison, in the same year China represented 48.2 percent and India 16.5 percent. Stads et al. (2016) summarize the source of LAC agricultural research funding and its use in 2012/2013. Three features portray important challenges for the future of the region: governments are still the main source of funding for agricultural research; most of that funding is being spent to cover researchers' salaries; and most researchers did not hold a PhD in the mid-2010s. Approximately 55 percent of agricultural researchers in

2012/2013 were public-sector employees; universities accounted for 40 percent and nonprofit agencies for 5 percent.[21]

Another useful indicator is the intensity ratio of R&D (that is, expenditures in R&D over the value of agricultural production).[22] Calculations by Pardey (2012) show that the research intensity in LAC for the public sector marginally improved in the 1990s and mid-2000s, from 0.8 percent to somewhat more than 1 percent, but that is far less than in high-income countries, particularly if public and private expenditures are considered together. The level of patenting and publications in LAC is also lower than that of developed countries and even some developing countries, such as China (Pardey 2012). On the other hand, LAC shows better ratios of R&D intensity than the average for developing countries (Pardey 2012). In that regard, Latin America has been an institutional laboratory on funding mechanisms for agricultural research (Echeverría, Trigo, and Byerlee 1996). Some research agencies conduct research on a contract basis for the private sector and through the commercial sale of improved seed. In addition, several countries have established levies on the value of agricultural production.[23]

However, estimates of the intensity ratio (Table 6.12) show an important dispersion in the region (it is the average of the decade of the 2010s). Also, those intensity ratios, even in the countries with the highest value, do not reach 2 percent of agricultural GDP, while in developed countries they usually go above 4–5 percent.

Furthermore, a significant number of countries are below the 1.0 percent intensity (the minimum recommended by the UN). This remains a big challenge for these countries where government support for agricultural R&D is too low to sustain viable agricultural research programs capable of addressing current and future priorities. Traditionally, such a research capacity gap has been to a certain extent filled by multicountry initiatives such as subregional cooperative research programs (generically referred to in Spanish as PROCIs, for *programas cooperativos de investigación*) complementing weak national

---

21 It is worth noting that universities have gained prominence in agricultural research, especially in Argentina, Bolivia, Costa Rica, Paraguay, and Uruguay. Producer organizations in Colombia and Honduras explain 40 percent of the total number of researchers.

22 Like any indicator, the intensity of R&D has its problems of interpretation, starting with the fact that it may increase not because R&D has gone up, but because agricultural GDP has gone down (Beintema et al. 2012).

23 For example, the Colombian coffee, sugarcane, and oil palm producer organizations fund most of their research using a levy on production. Uruguay's National Agricultural Research Institute (INIA) is partly funded by a commodity tax on the total value of the sale of the country's agricultural commodities, and the government matches the funds generated by the tax.

TABLE 6.12 Investments in public research and development as percentage of agricultural gross domestic product (average 2010s)

1% or more		0.5%–0.9%		Less than 0.5%	
Brazil	1.9	Bolivia	0.9	Nicaragua	0.4
Chile	1.7	Jamaica	0.9	Peru	0.4
Uruguay	1.4	Panama	0.8	Dominican Republic	0.3
Argentina	1.2	Belize	0.7	Paraguay	0.3
Mexico	1.1	Colombia	0.7	Venezuela	0.3
Costa Rica	1.0			Ecuador	0.2
				Honduras	0.2
				Guatemala	0.1
All countries in ASTI database					
Average	0.9	Median	0.5		

Source: ASTI (2017).

capacities, as well as regional and subregional funding mechanisms (for example, FONTAGRO).

It should be noted that the region is an important user of biotechnology products. Globally, it has been estimated that 26 countries planted almost 190 million hectares in 2017 (ISAAA 2017). After the United States (75 million hectares, 40 percent of the world total), the next two countries were Brazil (50.2 million hectares, 26 percent of the world total) and Argentina (23.6 million hectares, 12 percent). In total, there are 10 countries in LAC with more than 79 million hectares planted with biotech products (ISAAA 2017).[24]

Technological and Institutional Innovations

The institutional structure and policies for funding and executing public agricultural research have undergone important changes in Latin America and the Caribbean since the early 1960s, when national agricultural research institutes (NARIs) were conceived and when international centers based in the region were set up.[25] In addition, the region has been prolific in the

24 In addition to Brazil and Argentina, in 2017 the next countries in order of importance were Paraguay (3 million hectares); Bolivia (1.3 million hectares); Uruguay (1.1 million hectares); Mexico and Colombia with about 0.1 million hectares each; and Honduras, Chile, and Costa Rica, with less than 0.1 million hectares each.

25 Also, starting earlier than other developing regions, LAC has hosted since the 1960s three of the international centers of the CGIAR system: the International Center for Tropical Agriculture

establishment of various regional and subregional cooperative research mechanisms to execute and also to fund research.

As already pointed out, in the 1980s and 1990s macroeconomic crises led to important changes in agricultural policies in the LAC region, with the dismantling of many agricultural public institutions. Some NARIs started to focus more on small farmers and poverty issues, as part of a broader strategy for rural development.

The regional institutional framework for the development and diffusion of agricultural technology also expanded over time to include other regional institutions.[26] There was also increased participation of multinational companies, producer associations, universities, and civil society (Byerlee and Echeverría 2002).

As the agrifood systems modernized, they exerted new demands on all these traditional research structures.

The relatively recently substantial transformation of global food systems has several key implications for agricultural research strategies of international and national research systems, particularly in LAC countries well connected to international markets (Reardon et al. 2019). A key requirement is to consider the whole food system and its transformation in order to adopt strategies and designs of innovations in technologies and products related to overall food systems transformations. For instance, emerging requirements in terms of product types, quality and safety attributes, shelf life, cost, consistency, and seasonality should influence research priorities. Farmers themselves are far more involved in markets and commercialization than just a few decades ago, having significantly intensified and diversified farming; therefore, public research strategies need to keep up with that change. In addition, and because innovations are part of a supply chain, it is important to understand the strategies and needs of private-sector actors other than farmers, who are essential in the entire food system. Hence, public research organizations face a growing demand to transform themselves to the new food systems' realities, considering the presence of more actors, which requires new forms of collaboration (Byerlee and Echeverría 2002).

All these changes in the context where agricultural research takes place, new demands on food systems as well as new global challenges such as climate

(CIAT), the International Maize and Wheat Improvement Center (CIMMYT), and the International Potato Center (CIP).

26 Cooperation programs, such as the PROCISUR, PROCIANDINO, PROCITROPICOS, SICTA, PROMECAFE, PROCICARIBE, and PROCINORTE; FORAGRO; regional centers such CATIE and CARDI; and FONTAGRO (Regional Fund for Agricultural Technology).

change, and new funding opportunities and alternative providers of research solutions imply that, among other challenges, the public research community needs to take into consideration the importance of policy contexts as well as research on the off-farm components of the food system. Reardon et al. (2018) suggest that "research on and productivity of processing, packaging, logistics, and commerce technologies have equal weight in the performance of the food system relative to the farm sector, and investment in research and development value chains for these technologies and value chains for the inputs to these segments need a much higher profile in the context of the transformed food system where post farm segments occupy 40–70% of value added." In addition, returns to investments at the farm level are very well linked to related innovations in the whole supply chain, from inputs to marketing of final products.

Other challenges include the convergence of life sciences (including those related to agriculture) with physics, chemistry, computer sciences, mathematics, and engineering, leading to the emergence of new interdisciplinary research areas (Committee on a New Biology for the 21st Century 2009; MIT 2011). This challenges NARIs' institutional structures, organized into separate compartments of traditional disciplines related to agricultural R&D. Not only does such convergence require collaboration between disciplines, but more fundamentally, it calls for true disciplinary integration as well as new forms of funding (Committee on a New Biology for the 21st Century 2009; MIT 2011).

Further, there is the multiplicity of demands now placed on agriculture, in addition to increasing supply and alleviating poverty, which include health and nutrition, equity and gender integration, environmental sustainability, and a broader view of the bioeconomy. The implication of these multiple demands is that agricultural R&D needs to go beyond an exclusive focus on primary production to include the forward and backward linkages of the value chain, considering the views of a variety of social actors. All of this requires moving from more limited R&D strategies to a broader innovation approach and implies new organizational approaches, particularly to coordinate across multiple actors and networks, and, also, innovative funding arrangements (Trigo 2012).

Irrigation

Irrigation is another important source of productivity. Table 6.13 shows the area equipped for irrigation in LAC compared with the world. Although the irrigated area in LAC has increased more than the world average since the

TABLE 6.13 Irrigation

Regions and countries	Area with irrigation (1,000 hectares) 1960s	2010s	Increase from 1960s to 2010s (1,000 hectares)	Increase (%)
LAC	9,059.9	24,240.3	15,180.4	167.6
Argentina	1,112.2	2,337.4	1,225.2	110.2
Brazil	612.2	5,400.0	4,787.8	782.0
Mexico	3,216.7	6,496.0	3,279.3	101.9
Rest of LAC	4,118.8	10,006.9	5,888.1	143.0
World	170,668.3	327,245.8	135,898.0	91.7
LAC/world	5.3%	7.4%		
	Irrigated land as share of total land (%)			
LAC	1.6	3.2		
World	3.8	6.7		

Source: Authors' calculations based on FAO (2018).
Note: LAC = Latin America and the Caribbean.

1960s, it is still only 3.2 percent of the total agricultural area in that region, while 6.7 percent of the world's agricultural area is irrigated.

Farm Size and Land Tenure Issues

As discussed above, the unequal land structure in LAC, with the traditional dichotomy of "latifundio-minifundio," has been a trait of many LAC countries since the colonial period. Notwithstanding earlier land reforms (many linked to peasants' revolts and revolutions in the first part of the 20th century), the agrarian reforms that took place during the Alliance for Progress in the 1960s and 1970s, and a more recent phase since the return of democracy, inequality of land ownership remains high. At the same time, it should be noted that more recently the greater integration of rural and urban markets, along with the expansion of value chains, and the growth of exports have led to a more diversified land structure and the emergence of a segment of middle-sized farms.

Besides the issue of inequality, LAC continues to be the developing region with the largest average landholdings. Table 6.14 shows the average holding size, based on the most recent data compiled by the FAO and analyzed in Lowder, Skoet, and Singh (2014).[27] The table includes all LAC countries

27 Data correspond to the "2000" column in Table 2 of Lowder, Skoet, and Singh (2014), except for Mexico, Paraguay, Peru, and Honduras; that group corresponds to the "1990" column.

TABLE 6.14 Average size of agricultural holdings in LAC in comparison with other countries

Latin American countries		Non–Latin American countries	
Country or region	Area (hectares)	Country or region	Area (hectares)
Argentina	582.5	Australia	3,243.2
Uruguay	287.4	Canada	273.4
Chile	83.7	New Zealand	223.4
Paraguay	77.5	United States	178.4
Brazil	72.8	United Kingdom	70.9
Venezuela	60.0	France	45.0
Mexico	41.4	Germany	40.5
Nicaragua	31.3	Spain	23.9
Colombia	25.1	Saudi Arabia	16.7
Peru	20.1	Tunisia	10.5
Ecuador	14.7	Senegal	4.3
Panama	11.7	Thailand	3.2
Honduras	11.2	Myanmar	2.5
Guatemala	4.5	India	1.3
Trinidad and Tobago	4.4	Japan	1.2
Jamaica	2.2	Viet Nam	0.7
LAC average	83.2	World average	76.7
LAC median	28.2	World median	4.7

Source: Based on data from Lowder, Skoet, and Singh (2014).
Note: LAC = Latin America and the Caribbean.

covered in Lowder, Skoet, and Singh (2014) and, as a comparison, a sample of other countries in the same study.

As noted, another characteristic of landholdings in LAC is the inequality: in the late 1990s and early 2000s, the concentration measured by the Gini coefficient for landholdings in LAC was about 0.82, against 0.53 in Africa, 0.57 in Asia (developing), 0.59 in the European Union, and 0.64 in Canada (Diao et al. 2005). In some countries, such as Brazil, the inequality is large: the data reported in Lowder, Skoet, and Singh (2014) indicate that farms larger than 1,000 hectares represented 1 percent of all units but included 45 percent of the land (with an average size of some 3,200 hectares), while those with less than 2 hectares represented 20 percent of the units but occupied only 0.25 percent of the land (average 0.9 hectares).[28] This heterogeneity explains

28 It was estimated that 8 percent of farms produced 85 percent of the value of agricultural production (Ribeiro Vieira Filho, Garcia Gasques, and Gervásio de Sousa 2011).

the relatively intermediate average value for the whole country. In fact, the disparities are so large that a previous government created two separate ministries, one for commercial farms and another for small farmers (IFAD 2016).

During the last decade, the process of concentration may have increased further in several countries, but it has declined in others, such as Mexico (CEPAL, FAO, and IICA 2012). A clear picture will only be available when the latest round of censuses is processed.

Other trends and facts related to land issues include an expansion of land buying in the region, mainly by regional firms and local groups expanding into neighboring countries; the concentration of production (but not land ownership) through schemes to achieve economies of scale, such as "agricultural planting pools" in the Southern Cone; and the expansion of contract farming in most LAC countries (Dirven 2011; CEPAL, FAO, and IICA 2012; FAO 2012).[29]

All in all, data from Lowder, Skoet, and Singh (2014) suggest that there were about 7 million farmers with less than 5 hectares of land, on about 11 million hectares.[30] In the next bracket there was a variety of intermediate family farms in the range of 5 to 200 hectares, with different levels of assets and market access, amounting to some 5–6 million units with about 210 million hectares.[31] Finally there were about 500,000 commercial farms with more than 200 hectares per unit, occupying somewhat more than 500 million hectares (Lowder, Skoet, and Singh 2014, annex 6).

Policymaking in LAC has always been complicated by this heterogeneity. Public policies will have to acknowledge this fact and consider differentiated policies by types of producers, possibly defining and implementing adequate public programs in support of small farmers and family farms and monitoring concentration of land and ensuring that "land grabbing" does not take place. While for commercial farms the most important need is to develop an adequate business policy environment (Díaz-Bonilla, Orden, and Kwieciński 2014), for small and family farms public policies should not only eliminate biases against small farmers in land, labor, inputs, and credit markets, but also make sure that value chains operate in ways that allow an adequate integration

29 These agricultural planting pools are financial and operational vehicles managed by agricultural professionals, where different people contribute capital to finance a productive cycle (which may be an annual crop or longer crop and livestock operations). When the cycle finishes, the net profits are distributed among the investors.

30 A more adequate classification than hectares of land would be based not only on size but also on other variables, such as the levels and origins of their incomes (Berdegué and Fuentealba 2011).

31 Berdegué and Fuentealba (2011) estimated a group of family farms to be around 5 million with some 300 million hectares.

of those producers. Also, more funds for public R&D in support of family and small farms, diversified crops and livestock activities, and mitigation and adaptation research linked to climate change will be needed.

Transformation of Rural Economies, Rural-Urban Linkages, and Value Chains

The process of urbanization (discussed in more detail in the first section and Table 6.2; see also Chapter 9), along with the expansion of infrastructure and intermediate cities, and the greater integration of rural and urban markets have led to important changes in rural labor markets and nonagricultural rural activities, as well as the expansion of and greater complexity in value chains, domestically and for exports.

Regarding rural labor markets, Reardon, Berdegué, and Escobar (2001) summarize the evidence of 11 country studies in LAC by the late 1990s and early 2000s, noting that rural nonfarm employment (RNFE) and rural nonfarm incomes (RNFI) have grown significantly during the past three decades, with RNFI averaging 40 percent of total rural incomes in the countries analyzed. Reardon, Berdegué, and Escobar (2001) also observe, in terms of shares of rural incomes, "(1) nonfarm wage incomes exceed self-employment incomes; (2) RNFI far exceeds farm wage incomes; (3) local RNFI far exceeds migration incomes; (4) service-sector RNFI far exceeds manufactures RNFI." Their conclusion is "the need for more development program attention to wage employment in the service sector, versus the traditional focus on small enterprise manufactures." They mention that although there are other new development engines in rural areas (such as tourism, mining, and forestry), agricultural production maintains its crucial importance as an economic anchor in rural areas.

Related work by CEPAL, FAO, and IICA (2012) and IFAD (2016) confirmed and expanded evidence regarding several of those trends, including the increase in nonagricultural rural activities, with a larger presence of women's employment and growth in urban residence among agricultural workers (CEPAL, FAO, and IICA 2012). The latter trend has been helped by better rural infrastructure: for instance, the level of rural electrification in LAC has reached 88 percent of the population (against 71 percent for all developing countries and 73 percent for the world as a whole), while the number of mobile phone subscriptions in LAC is above the average for developing countries and the world (World Bank 2018).

Value chains have also been significantly modified by different waves of changes in the processing and retail segments (see Chapter 12). The larger

South American countries, which were more advanced in their development and urbanization, saw a surge of foreign direct investment (FDI) in the processing sector in the mid-1980s to early 1990s (after the process of liberalization and privatization of those years), while retail transformation started in the early 1990s. Mexico and Central America followed later, with retail transformation starting only in the middle to late 1990s (Reardon and Timmer 2012).

In the processing sector, in addition to the traditional milling and meat-packing industries, there has been a more recent advance of large conglomerates, particularly in beef, poultry, and pork production. The expansion of FDI related to processed and packaged products has been transforming several value chains (Bolling and Gehlhar 2005). On the input side, international seed companies and other providers have expanded in the region, providing technology mainly for cereals and oilseeds. Machinery and irrigation companies have also extended their operations in the region.

On the retail side, the most important change has been the supermarkets restructuring whole food chains, including both processed and fresh products (Reardon and Timmer 2012). In LAC the expansion of supermarkets started earlier and has proceeded further than in other developing regions: in the 1990s, retail was dominated by domestic firms, with supermarkets covering about 10–20 percent of national food retail sales, but by 2000 supermarkets had increased their share to 50–60 percent and, in many countries, have lately been penetrating national food retail at rates that took several decades to achieve in the United States. Brazil has the highest share of supermarkets, followed by Argentina, Chile, Costa Rica, Colombia, and Mexico. The advance of supermarkets was faster in processed, dry, and packaged foods but has also been increasing in fresh products, including vegetables, fruits, and different types of meats. Still, the share of supermarkets in fresh foods was estimated to be about half their share in packaged foods (Reardon and Berdegué 2002). But supermarkets in Latin America buy about 2.5 times more of some fresh products, such as fruits and vegetables, from local producers than are exported to world markets (Reardon and Berdegué 2002). It has also been noted that the expansion of supermarkets has been driven by FDI, and according to some estimates in LAC, multinational chains constitute about 70–80 percent of the top five chains in several countries (Reardon and Berdegué 2002).

A related debate has been about whether small and family farms may benefit from the advance of supermarkets and from integrating in value chains such as those for fruits and vegetables. Several studies (see Chapter 11) suggest that small farmers (and not just large farmers) sell to supermarket chains because the chains use a combination of specialized wholesalers and traditional wholesale markets; those small farmers with the assets or investments needed for consistent and quality supply (irrigation, road access, and education) tend to benefit more (see also Michelson 2013 for Nicaragua); and the farmers selling to supermarkets report moderate to substantial gains in incomes compared with nonparticipants (although at times the gain may not be higher prices but risk reduction).[32]

Concluding Remarks

Although, as noted, LAC is a heterogeneous region, it still has enough internal similarities in its geography and history to present a clear contrast with other developing regions considered in this book.

First, the pattern of occupation during colonial times and the integration with the world economy has defined the unequal structure of land ownership and the evolution of the agriculture sector in LAC. World markets and global developments, instead of domestic markets, have a stronger impact on agricultural developments. Changes in production took place within agrarian structures showing large inequalities in land tenure, with small farms fragmenting further and large landholdings expanding, which at times put pressure on land owned by family farms and local communities. There have also been important developments in rural labor markets, with more salaried employment, the expansion of nonagricultural rural activities, and a differentiated role for women's labor. The combination of high inequality and strong outward orientation is peculiar to LAC.

Second, LAC has seen greater experimentation with, and changes of, economic and social policies. The policies of supporting import substitution industrialization after WWII were later reversed, and now the region is following macro, trade, and agricultural policies that in many countries may have shifted toward greater support for the agriculture sector. Regarding social policies, LAC pioneered conditional cash transfers that helped reduce poverty and inequality (but see below).

32 And additional debate relates to whether increased globalization of value chains may raise prices of specialty food products and have negative welfare effects at the household level in some countries, as it has been argued in the case of quinoa in Peru. However, regarding this particular product, Bellemare, Fajardo-Gonzalez, and Gitter (2018) do not find evidence of harmful effects on local households, and rather the opposite seems to be true.

Third, with regard to socioeconomic conditions, LAC experienced improvements in income per capita and education and health indicators, along with declines in poverty rates, which places LAC at clearly better levels than the average for Asian and African developing countries. The incidence of poverty in rural areas is larger than in the cities, but because LAC countries have become the most urbanized in the world (another difference from other regions), the greatest concentration of poverty occurs in urban centers in the region. Indigenous populations and Afro-descendants are still among the poorest groups in the society. At the same time, LAC shows some of the highest levels of inequality in the world, which starts with large inequalities in land tenure and wealth in general.

Fourth, LAC has moved further along than other developing regions in the nutrition transition. Undernourishment has declined significantly, but in some cases the lack of micronutrients is still important. Furthermore, the large economic and social costs associated with the diet-related health problem of growing overweight and obesity rates constitute some of the most significant food policy challenges ahead for the region.

Fifth, value chains have also evolved significantly, with increased urbanization and a larger presence of intermediate cities and with agricultural and agro-industrial activities increasingly controlled by large agricultural operators, input companies, agro-industrial processors, and supermarket chains. In general, agricultural markets are more modernized and integrated in LAC than in other developing regions.

Sixth, LAC's agriculture sector has had relatively strong performance during the last five to six decades, outpacing global growth in food availability, while agricultural production increased its share of global output from about 10 percent in the 1960s to about 13 percent in the 2010s. This increase resulted to a large extent from agricultural expansion in Brazil. During the 2000s, LAC also became the world's main net food-exporting region, supported mainly by the net trade surpluses generated by Brazil and Argentina. Policy changes related to macroeconomic stabilization and liberalization led to a significant diversification of agricultural production, which has been reflected in important changes in the structure of exports: the share of traditional products (such as coffee, cocoa, sugar, and textiles) in total exports declined, while that of fruits and vegetables, oilseeds, and meat products increased.

Seventh, those gains in LAC's agriculture have been driven in part by productivity improvement but also resulted from a significant expansion of agricultural area over the last half century. This is another differential characteristic of LAC: the region has contributed to a third of the global increase in agricultural land (crops and pastures) since the 1960s and also accounted for some 80 percent of global deforestation from 1990 to 2015, which has resulted in land-use change contributing more to LAC's greenhouse gas (GHG) emissions than any other source. Another worrisome consequence of rapid land-use change is the pressure on LAC's globally important reservoirs of biodiversity. A similarly fast-paced expansion of LAC's agricultural area will be difficult to repeat in the future without affecting climate change and biodiversity.

The combination of the previous two points is another crucial difference between LAC and other regions, developing or developed: the region's performance has important global implications because of the dual role it plays by contributing both to the world's food security (as the main net food exporter region) and to environmental sustainability at the national and global levels (as the main provider of a variety of environmental public goods such as carbon sinks, oxygen, and biodiversity). Over the long term, sustaining LAC's dual role will require substantial investment in agricultural R&D, infrastructure, and governance of natural resources (Díaz-Bonilla et al. 2014). Failure to take adequate measures regarding efficiency, productivity, and sustainability will have far-reaching implications for the world if LAC cannot continue performing its double global functions in support of food security and environmental public goods.

Public investment in agricultural R&D has increased somewhat, particularly over the last decade. But LAC's average ratios are well below the levels of developed nations, and a few countries, notably Brazil, account for much of the improvements, as investment has declined in the smaller and poorer countries that are most in need of agricultural R&D.

This review tried to highlight the complex challenges for LAC countries regarding their agricultural and rural development. Powerful socioeconomic drivers could keep this region and the world on a business-as-usual path that appears unsustainable. Reshaping those trends requires multiple interventions that include but go beyond the agricultural sector. Governments need to take a broader agrifood systems approach to attain the desired objectives related to growth, employment, poverty and inequality alleviation, health and nutrition, and environmental sustainability, acknowledging trade-offs and strengthening R&D, innovation, and knowledge activities.

If LAC countries fail to address those challenges, not only will the region suffer, but the negative consequences will be felt globally.

References

Adato, M., and J. Hoddinott, eds. 2010. *Conditional Cash Transfers in Latin America*. Baltimore: Johns Hopkins University Press.

AgIncentives. 2020. AgIncentives database facilitated by IFPRI. http://www.ag-incentives.org/.

Anderson, K., and S. Nelgen. 2013. *Updated National and Global Estimates of Distortions to Agricultural Incentives, 1955 to 2011*. Washington, DC: World Bank.

ASTI. 2017. ASTI database facilitated by IFPRI. https://www.asti.cgiar.org/.

Balassa, B., and Associates. 1971. *The Structure of Protection in Developing Countries*. Baltimore: Johns Hopkins University Press.

Beintema, N., A. Nin-Pratt, and G.-J. Stads. 2020. "Global Agricultural Research Investments: Key Trends.". ASTI Program Note. Washington, DC: IFPRI.

Beintema, N., G. Stads, K. Fuglie, and P. Heisey. 2012. *ASTI Global Assessment of Agricultural R&D Spending*. Washington, DC: IFPRI, ASTI, and Global Forum on Agricultural Research.

Bellemare, M. F., J. Fajardo-Gonzalez, and S. R. Gitter. 2018. "Foods and Fads: The Welfare Impacts of Rising Quinoa Prices in Peru." *World Development* 112: 163–179.

Berdegué, J. A., and R. Fuentealba. 2011. "Latin America: The State of Smallholders in Agriculture." In *Conference on New Directions for Smallholder Agriculture, 24–25 January 2011, Rome*. Rome: International Fund for Agricultural Development.

Berdegué, J. A., and F. J. Proctor with C. Cazzuffi. 2014. *Inclusive Rural–Urban Linkages*. Working Paper Series 123. Working Group: Development with Territorial Cohesion. Territorial Cohesion for Development Program. Santiago, Chile: Latin American Center for Rural Development (Rimisp).

Bolling, C., and M. Gehlhar. 2005. *Global Food Manufacturing Reorients to Meet New Demands*. Agriculture Information Bulletin Number 794. Washington, DC: USDA/Economic Research Service.

Bustos, P., B. Caprettini, and J. Ponticelli. 2016. "Agricultural Productivity and Structural Transformation: Evidence from Brazil." *American Economic Review* 106 (6): 1320–1365.

Byerlee, D., and R. Echeverría, eds. 2002. *Agricultural Research Policy in an Era of Privatization*. Wallingford, UK: CABI.

CEPAL (United Nations Economic Commission for Latin America and the Caribbean), FAO (Food and Agriculture Organization of the United Nations), and IICA (Inter-American Institute for Cooperation on Agriculture). 2012. *Perspectivas de la agricultura y del desarrollo rural en las Américas: una mirada hacia América Latina y el Caribe*. Santiago, Chile: FAO.

Committee on a New Biology for the 21st Century. 2009. *A New Biology for the 21st Century. Ensuring the United States Leads the Coming Biology Revolution*. National Research Council. Washington, DC: National Academies Press.

Development Initiatives. 2018. *2018 Global Nutrition Report: Shining a Light to Spur Action on Nutrition*. Bristol, UK: Development Initiatives.

Diao, X., E. Diaz-Bonilla, S. Robinson, and D. Orden. 2005. *Tell Me Where It Hurts, An' I'll Tell You Who to Call: Industrialized Countries' Agricultural Policies and Developing Countries*. MTID Discussion Paper 84. Washington, DC: IFPRI.

Díaz-Bonilla, E. 2015. *Macroeconomics, Agriculture, and Food Security: A Guide to Policy Analysis in Developing Countries*. Washington, DC: IFPRI.

Díaz-Bonilla, E., D. Orden, and A. Kwieciński. 2014. *Enabling Environment for Agricultural Growth and Competitiveness: Evaluation, Indicators and Indices*. OECD Food, Agriculture and Fisheries Paper 67. Organisation for Economic Co-operation and Development.

Díaz-Bonilla, E., E. Saini, G. Henry, B. Creamer, and E. Trigo. 2014. *Global Strategic Trends and Agricultural Research and Development in Latin America and the Caribbean: A Framework for Analysis*. CIAT Publication 400. Cali, Colombia: Centro Internacional de Agricultura Tropical.

Díaz-Bonilla, E., and M. Torero. 2016. "Regional Developments: Latin America and Caribbean." In *2016 Global Food Policy Report*, 104–107. Washington, DC: IFPRI.

Dirven, M. 2011. *Dinámicas del mercado de tierras en los países del Mercosur y Chile: una mirada analítica-crítica*. 2011. Document prepared at the request of the FAO Regional Office for Latin America and the Caribbean for presentation at the workshop "Dinámicas en el mercado de la tierra en América Latina." November 14–15. Santiago, Chile: FAO.

Echeverría, R., E. Trigo, and D. Byerlee. 1996. *Institutional Change and Effective Financing of Agricultural Research in Latin America: Findings of a Workshop Organized by the Inter-American Development Bank and the World Bank in August 1995*. World Bank Technical Paper 330. Washington, DC: World Bank.

Engerman, S., and K. Sokoloff. 2002. *Factor Endowments, Inequality, and Paths of Development among New World Economies*. NBER Working Paper 9259. Cambridge, MA: National Bureau of Economic Research.

Fan, S., ed. 2008. *Public Expenditures, Growth, and Poverty: Lessons from Developing Countries*. Washington, DC: IFPRI.

FAO (Food and Agriculture Organization of the United Nations). 2012. *Dinámicas del mercado de la tierra en América Latina y el Caribe: concentración y extranjerización*. Santiago, Chile: Editores Fernando Soto Baquero y Sergio Gómez.

———. 2018. FAOSTAT database. Accessed in 2017 and 2018. http://faostat.fao.org.

Fuglie, K. O. 2012. "Productivity Growth and Technology Capital in the Global Agricultural Economy." In *Productivity Growth in Agriculture: An International Perspective*, edited by K. Fuglie, S. L. Wang, and V. E. Ball, 335–368. Wallingford, UK: CABI.

———. 2015. "Accounting for Growth in Global Agriculture." *Bio-based and Applied Economics* 4 (3): 221–254.

Hayami, Y., and V. W. Ruttan. 1985. *Agricultural Development: An International Perspective* (revised and expanded). Baltimore: Johns Hopkins University Press.

HLPE (High Level Panel of Experts). 2017. *Nutrition and Food Systems. A Report by the High Level Panel of Experts on Food Security and Nutrition*. Rome.

IFAD (International Fund for Agricultural Development). 2016. *Rural Development Report 2016: Fostering Inclusive Rural Transformation*. Rome.

IFPRI (International Food Policy Research Institute). 2019. "Agricultural Total Factor Productivity (TFP), 1991–2015: 2019 Global Food Policy Report Annex Table 4." https://dataverse.harvard.edu/dataset.xhtml?persistentId=doi:10.7910/DVN/9IOAKR.

ISAAA (International Service for the Acquisition of Agri-biotech Applications). 2017. *Global Status of Commercialized Biotech/GM Crops: 2017.* ISAAA Brief 53. Ithaca, NY.

Kawagoe, T., K. Otsuka, and Y. Hayami. 1986. "Induced Bias of Technical Change in Agriculture: The United States and Japan, 1880–1980." *Journal of Political Economy* 94 (3 Pt. 1): 523–544.

Little, I., T. Scitovsky, and M. Scott. 1970. *Industry and Trade in Some Developing Countries*. Paris: Oxford University Press.

Lowder, S. K., J. Skoet, and S. Singh. 2014. "What Do We Really Know about the Number and Distribution of Farms and Family Farms Worldwide?" Background paper for *The State of Food and Agriculture 2014*. ESA Working Paper 14-02. Rome: FAO.

Michelson, H. 2013. "Small Farmers, NGOs and a Walmart World: Welfare Effects of Supermarkets Operating in Nicaragua." *American Journal of Agricultural Economics* 95 (3): 628–649.

MIT (Massachusetts Institute of Technology). 2011. *The Third Revolution: The Convergence of the Life Sciences, Physical Sciences and Engineering.* Washington, DC: MIT.

Nin-Pratt, A., C. Falconi, C. E. Ludena, and P. Martel. 2015. *Productivity and the Performance of Agriculture in Latin America and the Caribbean: From the Lost Decade to the Commodity Boom*. IDB Working Paper Series N.608. Washington, DC: Inter-American Development Bank.

Pardey, P. G. 2012. "Foresight in LAC Agriculture: Agricultural Technology & Intellectual Property." Paper presented at the Workshop on Future Challenges and Opportunities for Latin America and the Caribbean, IADB–CIAT, March 2012.

Pardey, P., and N. Beintema. 2001. *Slow Magic: Agricultural R&D a Century after Mendel*. Food Policy Report. October 26, 2001. Washington, DC: IFPRI.

PIADAL (Panel Independiente sobre la Agricultura para el Desarrollo de América Latina). 2013. *Agricultura y Desarrollo en América Latina: Gobernanza y Políticas Públicas*. Buenos Aires: Editorial Teseo.

Pinstrup-Andersen, P. 2011. *The Food System and Its Interaction with Human Health and Nutrition*. IFPRI 2020 Conference Brief 13. Washington, DC: IFPRI.

Popkin, B. M., and T. Reardon. 2018. *Obesity and the Food System Transformation in Latin America. Obesity Reviews*. Hoboken, NJ, US: John Wiley & Sons on behalf of World Obesity Federation.

Prebisch, R. 1950. *The Economic Development of Latin America and Its Principal Problems*. New York: United Nations.

———. 1968. "Development Problems of the Peripheral Countries and the Terms of Trade." In *Economics of Trade and Development*, edited by J. D. Theberge. New York: John Wiley and Sons.

Reardon, T., and J. A. Berdegué. 2002. "The Rapid Rise of Supermarkets in Latin America: Challenges and Opportunities for Development." *Development Policy Review* 20 (4): 371–388.

Reardon, T., J. A. Berdegué, and G. Escobar. 2001. "Rural Nonfarm Employment and Incomes in Latin America: Overview and Policy Implications." *World Development* 29 (3): 395–409.

Reardon, T., R. Echeverría, J. Berdegué, B. Minten, S. Liverpool-Tasie, D. Tschirley, and D. Zilberman. 2019. "Rapid Transformation of Food Systems in Developing Regions: Implications for Agricultural Research Strategies." In "Agricultural Research for Rural Prosperity: Rethinking the Pathways," edited by T. Tomich, P. Lidder, and P. Carberry. Special issue, *Agricultural Systems* 172 (Feb.): 47–59.

Reardon, T., and P. Timmer. 2012. "The Economics of the Food System Revolution." *Annual Review of Resource Economics* 4: 14.1–14.40.

Ribeiro Vieira Filho, J., J. Garcia Gasques, and A. Gervásio de Sousa. 2011. *Agricultura e crescimento: cenários e projeções*. IPEA. 1642. Texto para Discussao. Brasilia: Institute of Applied Economic Research.

Singer, H. 1950. "The Distribution of Gains between Investing and Borrowing Countries." *American Economic Review* 40: 473–485.

Stads, G. J., N. Beintema, S. Perez, K. Flaherty, and C. Falconi. 2016. *Agricultural Research in Latin America and the Caribbean: A Cross-Country Analysis of Institutions, Investment, and Capacities.* ASTI-IDB Document. Washington, DC: IFPRI.

Trigo, E. 2012. "Los Nuevos Escenarios para la Institucionalidad de la Investigación Agroalimentaria en América Latina y el Caribe. Borrador." Draft prepared for the Inter-American Development Bank, Buenos Aires.

von Grebmer, K., J. Bernstein, D. Nabarro et al. 2016. *2016 Global Hunger Index: Getting to Zero Hunger.* Bonn, Washington, DC, and Dublin: Welthungerhilfe, IFPRI, and Concern Worldwide.

World Bank. 1986. *World Development Report 1986.* New York: Oxford University Press.

———. 2014. *The State of Social Safety Nets 2014.* Washington, DC: World Bank.

———. 2018. World Development Indicators database. Accessed at various times in 2018. https://databank.worldbank.org/reports.aspx?source=world-development-indicators#.

Chapter 7

AGRICULTURAL DEVELOPMENT AND FOOD SECURITY IN EASTERN EUROPE AND CENTRAL ASIA

Saule Burkitbayeva, William Liefert, and Johan Swinnen

The transition countries of Eastern Europe and Central Asia (ECA) are often jointly grouped (as they are in this chapter) because of their common institutional history. However, this is a very heterogeneous set of countries, in terms of their geography, culture, political systems, general economic development, and agricultural development and potential, as shown in Table 7.1. This table groups the countries by (sub)region, and we will use these groups to examine some of the main changes during transition and future potential developments.

These countries' economic and institutional transformation in the 1990s and 2000s had a dramatic impact on their land use, food production, poverty, and food security. However, the pace and current stage of transition from socialist-planned to market economy differ strongly between countries. Some are now full market economies (such as the Eastern European countries within the European Union [EU]). However in others (such as Belarus and Uzbekistan), state regulation and control are still widespread. Yet, for the vast majority of countries in the region, the economic liberalization and transition caused dramatic changes in agricultural productivity, output, and consumption, as well as in the food system.

In many ECA countries, food security worsened and poverty increased during the transition process in the 1990s. However, since 2000 the ECA region has experienced significant growth and rising incomes, which have substantially reduced poverty and undernourishment and improved food security. Yet, undernourishment remains a problem in some of the poorer Caucasus and Central Asian states. Diets are also of low quality in many of these countries, resulting in micronutrient deficiencies. On the other hand, as ECA

This study was supported by the Methusalem program of KU Leuven and by the Economic Research Service of the US Department of Agriculture (USDA). The findings and conclusions in this chapter are those of the authors and should not be construed to represent any official USDA or US Government determination or policy.

TABLE 7.1 Characteristics of Eastern European and Central Asian countries by regional groups

	GDP/capita in 2018 (constant 2010 prices)	Share of agriculture in employment in 2018 (%)	Agricultural land use in 2016 (million hectares)	Agricultural production in 2016 (million $)	Agricultural land in individual farms (%)	Labor/land (persons/hectare)	Wheat production in 2017 (million tons)
Central Asia							
Kazakhstan	11,165.5	15.0	217.0	10,116	39	0.010	14.8
Kyrgyzstan	1,087.2	26.5	10.6	2,100	95	0.074	0.6
Tajikistan	1,073.0	51.1	4.9	1,963	83	0.306	0.9
Turkmenistan	7,647.9	22.8	33.8	2,861	93	0.028	1.0
Uzbekistan	2,026.5	33.4	26.8	15,034	98	0.139	6.1
Caucasus							
Armenia	4,406.7	33.3	1.7	1,301	99	0.278	0.2
Azerbaijan	5,769.0	36.1	4.8	3,038	94	0.356	1.8
Georgia	4,469.2	42.9	2.6	738	72	0.423	0.1
European CIS							
Belarus	6,744.5	9.6	8.6	7,863	10	0.046	2.6
Republic of Moldova	2,684.1	32.2	2.5	1,866	50	0.135	1.3
Russian Federation	11,729.1	5.8	216.8	63,994	31	0.022	85.9
Ukraine	3,110.2	15.3	41.3	34,151	45	0.077	26.2
Baltics							
Estonia	19,948.9	3.5	1.0	656	44	0.026	0.7
Latvia	16,405.7	6.8	1.9	1,124	87	0.036	2.1

Lithuania	17,669.7	7.7	2.9	2,365	87	0.042	3.9
Central Europe							
Czech Republic	23,344.2	2.8	4.2	4,283	31	0.032	4.4
Hungary	16,503.5	5.0	5.3	6,013	53	0.034	5.2
Poland	16,639.7	10.1	14.4	22,469	91	0.127	11.7
Slovakia	20,669.9	2.7	1.9	1,772	19	0.043	1.8
Balkans							
Albania	5,075.4	38.0	1.2	1,482	90	0.402	0.3
Bulgaria	8,651.1	6.9	5.0	3,615	38	0.041	6.1
Romania	11,534.9	22.6	13.9	10,133	56	0.162	10.0
Slovenia	26,758.9	5.5	0.5	715	95	0.148	0.1

Source: FAO (2019), World Bank (2019), Eurostat (2018), IAMO (2018).

Note: CIS = Commonwealth of Independent States; GDP = gross domestic product. Agricultural production in 2016 is in constant 2004–2006 dollars. Agricultural land in individual farms refers to "single holder farms" for Central and Eastern European (CEE) countries in Eurostat statistics and "households" and "individual farms" combined for CIS statistics, latest years available. According to FAO and World Bank, the amounts of agricultural land used in the Russian Federation and Kazakhstan are almost the same. Agricultural land use includes arable land, under permanent crops and under permanent meadows and pastures.

countries become richer and work becomes more sedentary, they are experiencing the challenge of overweight and obesity.

Eight Eastern European countries joined the EU in 2004, and three more have joined since. The agricultural development issues in these countries are now intertwined with the agricultural policies and general development challenges in the EU. These are quite different from the challenges and opportunities facing the transition countries farther east. For the purposes of this chapter, we therefore focus mostly on the non-EU transition countries and refer to these with the term *ECA*. However, within these, in more eastern ECA countries there is still much heterogeneity. Some of these countries have relatively higher incomes, have a comparative advantage in grain production, and are dominated (especially in land use) by large farms. Examples are the Russian Federation, Ukraine, and Kazakhstan. Another group have much lower incomes, face more severe food security problems, and are dominated by small-scale farms. Examples are Albania, Tajikistan, Kyrgyzstan, and Armenia. As Figure 7.1 shows, these poorer countries have many people working in agriculture and a large share of gross domestic product (GDP) from agriculture.

The ECA region is important both for its production potential and food security challenges on the consumption side. The region is a major food producer, in particular of grain and dairy goods. For example, it produces 12 percent of the world's milk, the same as the United States. The region's share in global agricultural trade is just below 10 percent (8 percent of world imports and 9 percent of exports). The region's status as a major producer and exporter of grain, and especially wheat, has attracted particular attention during the global food security debate over the past decade. The region accounts for 9 percent of world grain production and 18 percent of world wheat production (twice the US production of 9 percent). It is a major player in international grain markets, as it supplies 15 percent and 22 percent of global grain and wheat exports. The Russian Federation, Ukraine, and Kazakhstan (RUK) account for almost all of those exports.

Some specialists have argued that these countries could become a "breadbasket" for the world because of their potential to increase their already large volumes of grain production and exports by exploiting their "immense land and yield reserves" (Glauben et al. 2014). The decline in agricultural land use during the transition years of the 1990s was huge: between about 50 and 60 million hectares of land were abandoned, equal to around 50 percent of current land use within the RUK. However, there is considerable uncertainty about the economic and environmental feasibility of returning all this land to

FIGURE 7.1 Share of agriculture in GDP and employment in 2016

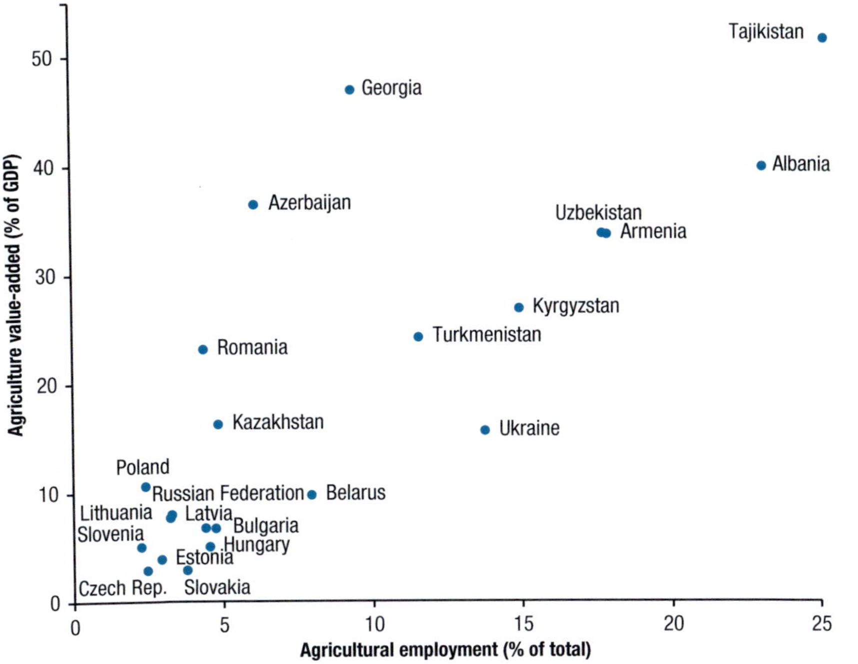

Source: World Bank (2019); ILO (2014).
Note: Data for Turkmenistan is for 2010.

production, as well as actual yield potential (Kraemer et al. 2015; Liefert and Liefert 2015).

Analyzing the region's potential to increase world food production, and how its agricultural development can contribute to world poverty reduction and food security, requires an understanding of its recent economic and agricultural transformation. Therefore we begin our chapter with a conceptual framework of the transition process and its implications, followed by an empirical examination of how agricultural production and productivity have evolved during transition and how these evolutions are related to changing farm structures and value chains. We then draw on these insights to discuss the future production and export potential. Then, we discuss the improvements, persisting problems, and new challenges in food security and nutrition of the region. Next, we discuss key agricultural, food, and nutrition security policies. The last section draws conclusions.

The Transition Process and Its Implications for Agricultural Development

Conceptual Framework

The ECA countries have a common systemic and institutional history, and the move from a centrally planned to a more market-oriented economy that began around 1990 strongly affected all these states' agricultural sectors. To illustrate and examine key aspects of the agricultural transition process important for this chapter, we use a simplified version of the conceptual model of Swinnen and Rozelle (2006).

Consider a socialist farming system in which farms produce output q with input x. Socialist farms were technically inefficient because (1) the government was typically less efficient than the market in how it allocated inputs, and (2) production was organized in state-owned collective farms, where managers and workers had less incentive to give optimal effort (Johnson and Brooks 1983). This second source of technical inefficiency—poor incentives—is illustrated in Figure 7.2a where the production functions are $q = f(x)$ for a farm operating privately in a market economy and $q_o = f_o(x)$ for a socialist farm. A second form of inefficiency (allocative) was that prices set by the government for inputs and output deviate from market prices. In ECA, planners typically set output-input price ratios higher than what a market-based system would generate, thereby subsidizing producers. The output-input combination of a farm facing the distorted prices represented by the relative (output-input) price line tangent to the production frontier is given in Figure 7.2a by point *A*.[1] Price and market liberalization removed subsidies, and the farm moved from point *A* to point *B*, with the (steeper) price line now representing market prices. Output and input use both fell. With property rights reform, the production function shifted from $f_o(.)$ to $f(.)$, that is, from point *B* to point *C*.

However, the shift from *A* to *B* and *C* did not go smoothly. Price liberalization and property rights reforms occurred in an environment characterized by imperfect information and weak institutional support. Transition, especially in the early reform years, was characterized by the breakdown of institutions of exchange and the rise of transaction costs. The disruptions can be modeled as a downward shift in the production function, to $q = f_d(x)$. A "perfect" property rights reform would induce a shift from point *B* to *C*. However, with disruptions, the initial reform impact induced a move from point *B* to *D*; instead

1 Even if planners did not set output prices high relative to input prices, price ratios within the closed economy would differ from those of goods traded on world markets.

FIGURE 7.2 Transition process and its implications

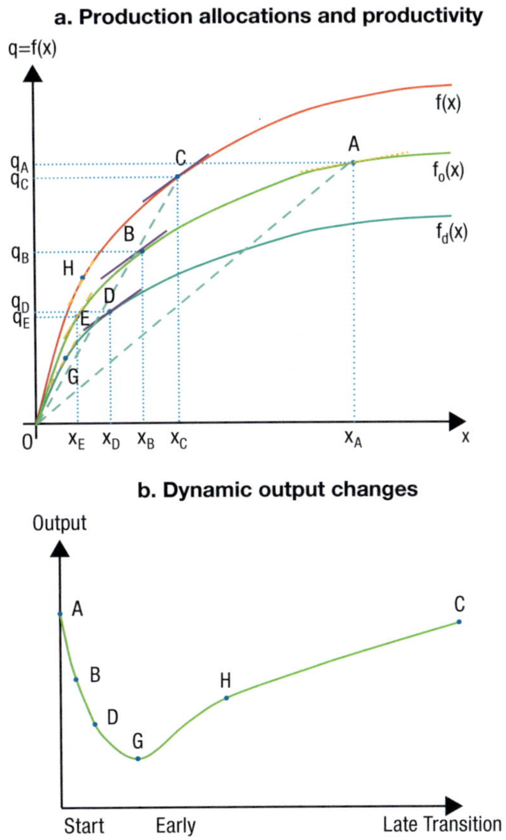

of increasing inputs and output, property rights reforms caused a decline. These disruption effects were particularly important in countries with capital-intensive agricultural systems that were part of a complex network of input supply chains and output procurement or processing channels.

Price (and market) liberalization was also accompanied by disruptions due to the elimination of planning and the disorganization of institutions of exchange. Hence, when liberalization occurred, relative price shifts were driven not only by the move toward world market prices but also by institutional disruption caused by increased transaction costs. Rising transaction costs depressed output prices and raised farm input prices, thereby leading to deteriorating terms of trade (between output and input prices) at the farm level. This is reflected in a steeper relative price line and a move to point *E* or *G*. While the mechanism is different from that caused by the disruptions associated with property rights reform, the effect is the same: a fall in input use and output.

There are also important dynamic effects, as the disruptions from transition were likely to have a different time sequence than the efficiency gains. In the words of Kornai (2000): "Transition calls for creative destruction. Because destruction is rapid, whereas creation proceeds more slowly, the two processes led to deep recession." Figure 7.5b illustrates this dynamic (with the letters consistent with those in Figure 7.2a). Prior to reform, farms are at point *A*. With price reform, output declines and farms move toward the allocatively efficient point *B*. Property rights reform initially mainly creates disruptions, such that in the early phase of transition, farms actually become more inefficient (point *D*). Market liberalization also creates initial disorganization, and the rising transaction costs exacerbate the falling terms of trade, moving farms to point *G*. As the reforms proceed and succeed, property rights reforms strengthen incentives, and the reorganization of input supply chains, output procurement, and processing channels improves the provision of inputs that originally were supplied by the state, allowing farms to increase their technical efficiency. Also, emerging and more effective institutions of exchange reduce transaction costs, improving farms' real terms of trade and allocative efficiency, inducing a shift from *G* to *H* and *C*.

After this dynamic process, technical and allocative efficiency increased at the end of the process, as average productivity. In Figure 7.2a, the average productivity, q/x, is represented by the slopes of the *OA* and *OC* lines. Both price and property rights reforms increase average productivity from *OA* to *OC*, with the two reforms reinforcing each other in terms of productivity growth.

However, while average productivity increases, the input and output effects are ambiguous. Price and property rights reform work in opposite directions on input use and output. In Figure 7.2a, the net impact of the combined reforms on output is negative ($q^C < q^A$), and on input use strongly negative ($x^C << x^A$). This net effect depends on the shape of the production function and on the relative size of the technical inefficiencies and pre-reform subsidies. Output and input use decline due to the price reform, but they increase with technical efficiency gains stemming from property rights reforms. This implies that efficient output and input use (including land) may well be (substantially) below the levels under the distorted socialist system.

In the rest of this section we present data that show an evolution consistent with this model, albeit the dynamics and levels of decline and recovery differed among countries and commodities.

Decline during Transition

Actual agricultural production followed the **J**-curve of Figure 7.2b but with considerable variation among countries. In the first years of transition, gross agricultural output strongly decreased throughout ECA (Figure 7.3), varying from −25 percent in the Balkan countries to −45 to −55 percent in the Baltics and the European Commonwealth of Independent States (CIS, which includes the Russian Federation and Ukraine).

The large-scale reduction of subsidies resulted in a huge deterioration in agricultural producers' terms of trade: input prices rose relative to output prices, and output prices fell to world prices. For example, from 1991 to 1997, the terms of trade of Russian farms fell by about 75 percent, and input use (for example, fertilizer) declined by 70 percent (OECD 1999). Fertilizer use declined dramatically in all ECA countries in the early transition years, in response to price changes and supply chain problems. The decline in fertilizer use and crop prices generated a strong drop in crop yields. In most ECA countries, grain yields decreased between 20 and 30 percent (Figure 7.4).

Especially hard hit was the highly subsidized livestock sector. The downsizing of the livestock sector substantially reduced countries' demand for animal feed, which in turn lowered local grain production as well as imports of grain and soybeans needed for feed (Liefert and Swinnen 2002). As a result, countries such as the Russian Federation became meat importers and grain exporters (discussed more below). These countries thereby restructured their livestock and grain sectors consistent with their agricultural comparative advantage in the world (Liefert 2002).

Recovery

Countries also differed as to when recovery started. In some countries (for example, the Czech Republic, Poland, Albania), recovery began around 1993, while others (for example, the Baltics and European CIS) experienced more than a decade of falling output, with the rebound beginning as late as 2001. The production recovery was also uneven. As Figure 7.3 illustrates, current agricultural output in ECA varies from −30 to +30 percent of the pre-reform level, with the Caucasus and Central Asian regions doing best by this measure.

The magnitude of the output rebound has also varied by commodity. For example, wheat production in the region fell from 101 to 86 million tons from the beginning to end of the 1990s but rose strongly in the following years.[2]

---

2 Tons are metric tons throughout.

FIGURE 7.3 Evolution of gross agricultural output (GAO) (index, 1990 = 0)

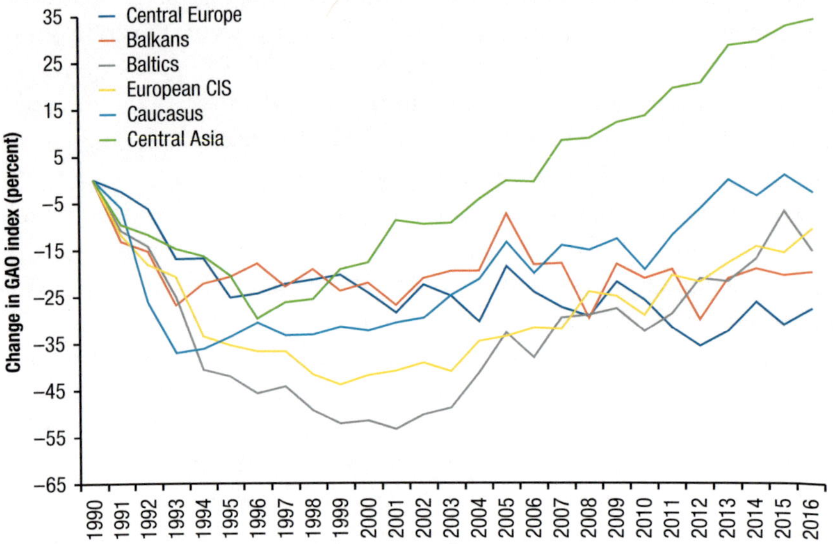

Source: Data from FAO (2019).

FIGURE 7.4 Evolution of grain yields (index, 1990 = 0)

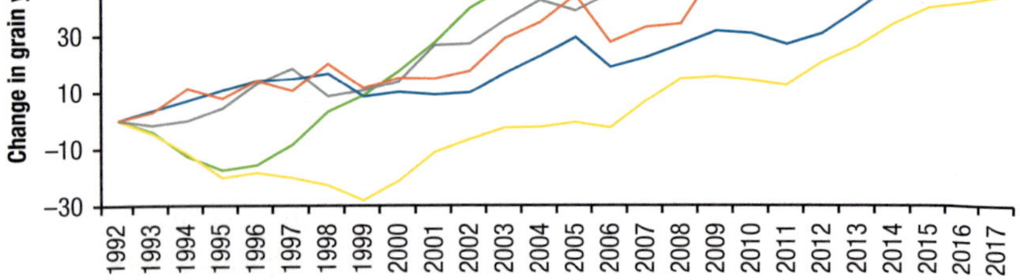

Source: Data from FAO (2019).

Note: The grain yield index is based on a three-year moving average of grain yields.

TABLE 7.2 Agricultural production in Eastern Europe and Central Asia (ECA), 1989–2015 (three-year average in million tons)

	1989–1991	1995–1997	2001–2003	2007–2009	2013–2015
Grain production					
Kazakhstan	19.76	10.97	15.55	18.86	17.89
Russian Federation	103.54	73.77	79.63	95.62	99.70
Ukraine	46.96	31.33	32.91	42.83	63.03
Wheat production					
ECA	—	94.10	107.86	135.43	—
Kazakhstan	11.29	7.71	12.31	15.35	14.17
Russian Federation	44.17	36.43	43.90	58.29	55.90
Ukraine	26.31	16.07	15.17	20.24	24.30
Meat production					
ECA	—	16.67	15.31	17.66	—
Kazakhstan	1.55	0.85	0.67	0.87	1.57
Russian Federation	9.86	5.33	4.7	6.25	9.07
Ukraine	4.27	2.09	1.63	1.91	2.37
Poland	—	2.84	3.09	3.38	4.07
Milk production					
ECA	—	91.56	87.63	91.02	—
Kazakhstan	5.57	3.86	4.12	5.19	5.00
Uzbekistan	—	3.48	3.80	5.43	—
Russian Federation	54.48	36.42	33.26	32.37	30.80
Ukraine	23.76	15.62	13.74	11.88	11.07
Poland	—	11.82	11.90	12.34	12.97

Source: FAO (2015a); Kazstat (2016); Rosstat (2016); Ukrstat (2016).
Note: — = data not available.

The increase was most spectacular in the main producing countries (the Russian Federation, Ukraine, and Kazakhstan; Table 7.2).

The recovery in grain production has been driven largely by increased yields and fertilizer use since 2000 (though use per hectare is still considerably below the pre-reform levels). Depending on the country and region, grain yields are currently between −10 and +35 percent of the pre-reform level. Figure 7.5b and Figure 7.5c show the correlation between the growth in fertilizer use and grain yields in RUK.

Livestock production has not recovered as well in the larger ECA countries, consistent with the argument that these countries have a comparative

FIGURE 7.5 Land use, fertilizer use, and grain yields in the Russian Federation, Ukraine, and Kazakhstan

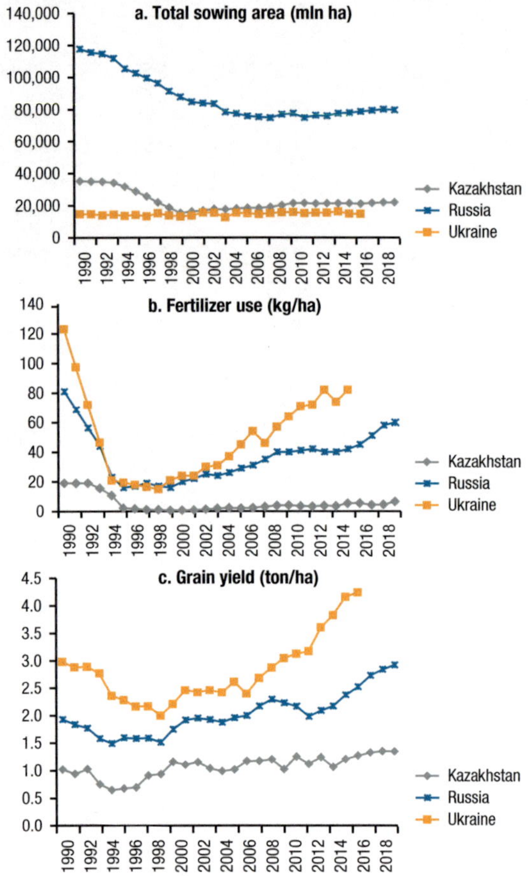

Source: Rosstat (2019); Kazstat (2019); Ukrstat (2019).
Note: Yields are expressed as three-year moving averages.

disadvantage in producing livestock products. In many ECA countries, meat and dairy output in 2013–2015 was still below the levels of 1989–2001. After falling from 20.8 million tons in 1992–1994 to 15.0 million tons, ECA meat production rose to 19.8 million tons in 2010–2012. However, that figure is still well below the late socialist level. Total ECA milk output had the same trajectory, dropping from 107 million tons in 1992–1994 to 85 million tons in 1998–2000, and then rising to 91 million tons by 2012–2014. Yet, that volume is also still below the pre-reform level.

FIGURE 7.6 Land and labor productivity in Eastern Europe and Central Asia

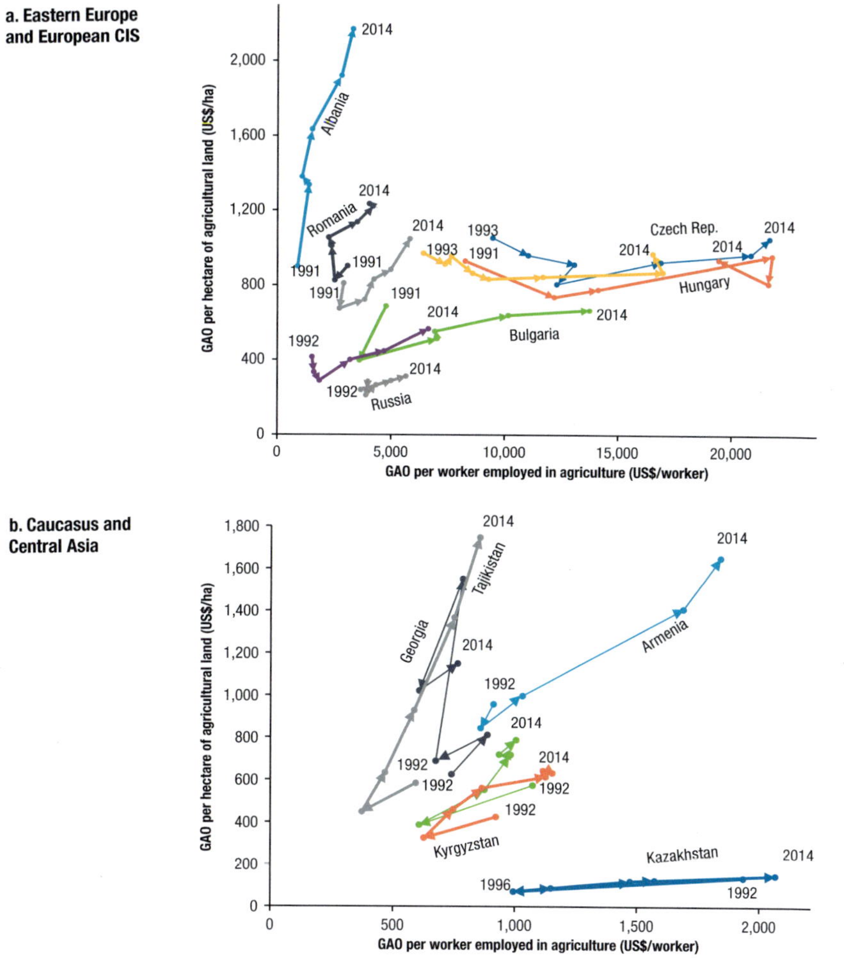

Source: FAO (2019); ILO (2014).

Note: GAO = gross agricultural output. Agricultural land refers to arable land and permanent crops. Gross production value is measured in constant 2004–2006 million US dollars.

Diverging Growth Patterns

As already emphasized, the development of agriculture in recent years differs quite strongly among countries. The heterogeneity in recovery and growth patterns is clearly illustrated with the use of labor-land productivity dynamics, following Hayami and Ruttan (1971). Figure 7.6 summarizes the labor and land productivity development.

One growth pattern is that of an agriculture sector where large farms play a significant role and growth has come mostly from (very strong) labor productivity growth. This pattern has been followed by Slovakia, the Czech Republic, and Hungary, while recently Bulgaria and Ukraine are joining it.

A very different pattern is followed by poorer countries dominated by small farms (such as Albania, Armenia, Tajikistan, Georgia, and Kyrgyzstan). However, in these countries output per hectare has risen much more than in the first pattern, due to intensive use of inputs on smaller farms.

Countries such as Poland (and to a lesser extent Romania) are a mix, as growth has come from both higher land and labor productivity, though productivity growth has been limited in both directions. These countries are characterized by a majority of smaller farms, a sizable share of large farms, and relatively rich economies (compared with the ECA average).

To sum up, in the ECA region, a high (or low) land-labor ratio is more likely to be associated with a high (or low) capital-labor ratio and growth in labor productivity (or land productivity) as a result of labor shedding (absorption).

Growth Patterns and Farm Structures

As is clear from the discussion above, changes in countries' agricultural productivity are associated with their farm structures. The ECA region is now characterized by strong heterogeneity in farm structures: in some countries smallholders and family farms dominate, while in others large farms play an important role. There is no simple East-West divide in this. In Central Europe, large farms hold most of the land in Slovakia and the Czech Republic, while family farms dominate in Poland and Slovenia. In Central Asia, large farms are important in the northern parts of Kazakhstan while small farms are significant in southern Kazakhstan, Kyrgyzstan, and Tajikistan.

The emergence of different farm structures is to an important extent "endogenous" to the transition process. The nature of the commodity and factor market imperfections appears to have played a significant role in the evolution of the farm structures. Large farms dominate in extensive crop cultivation (including grains) in the Russian Federation, Ukraine, and Kazakhstan. Many of these farms are part of large-scale vertically integrated agroholdings. These farms and agribusiness structures have emerged as a consequence of the specific privatization programs in these countries, the simultaneous financial constraints in agriculture, and the inflow of capital from other sectors of the economy (Serova 2007). Vertical integration resulted from the need to overcome financial constraints at the farm level and the better access to capital in commodity trading (World Bank 2005; Swinnen 2009). The shift to small-scale farming has been lowest in labor-extensive production systems and strongest in labor-intensive production systems. Small farms have also served as a labor-absorbing institution during transition, thereby contributing to a divergence of farm structures (Dries and Swinnen 2002; Swinnen, Dries, and Macours 2005). This was most important in the poorer countries, so there is some correlation between farm sizes and GDP per capita (Figure 7.7).[3]

In Central Europe, a rapid restructuring of farms combined with labor shedding has been strongly associated with labor productivity growth (Figure 7.8). In other ECA countries, agricultural labor productivity (ALP) was substantially lower and agriculture provided a buffer during transition, in terms of both labor allocation and food security. In some of the poorest countries, such as Tajikistan, agricultural employment remains very high (Figure 7.1) with high labor use on small farms. In fact, Figure 7.9 illustrates the negative correlation between the share of smallholder farms and labor productivity in ECA agriculture.

Today, most countries have a mix of large- and small-scale farms. Some of these differences reflect disparities in natural resources or commodity characteristics.[4] Table 7.3 documents the share of smallholders in land use and agricultural production. For most CIS countries, statistics distinguish between "agricultural enterprises," "individual farms," and "households." For the Central and Eastern European (CEE) countries, statistics distinguish "family farms" (as a subcategory), "farms less than 2 hectares," and larger farms. If we identify "smallholders" as "households" in CIS and "farms less than 2 hectares" in the CEE, these farms use less than 3 percent of land in CEE and have produced around 6 percent of output in recent years. The main exception

3 Slovenia is an outlier in Figure 7.7, with high incomes and only small farms. This is because in Slovenia, as part of the former Yugoslavia, farms were never collectivized into large-scale communist farms as in the other ECA countries (the other exception being part of Poland, where small-scale farms also survived Communist rule).

4 There is also remarkable variation in how farms have adjusted labor use during transition. In some countries, farms absorbed labor, while in other countries farms massively shed labor, even in the early years of transition. The differences are large: from a strong increase in some of the Central Asian countries to a dramatic decline of more than 50 percent of official employment in Central Europe. We observe such large differences not only between very different regions, but also within regions and even within countries. For example, on average the outflow of labor in Poland was much lower than in neighboring Slovakia and the Czech Republic, but within Poland there were huge differences between regions (Dries and Swinnen 2004; Swinnen, Dries, and Macours 2005). See also the discussion in the next section.

FIGURE 7.7 GDP per capita versus share of large farms in land use in Eastern Europe and Central Asia

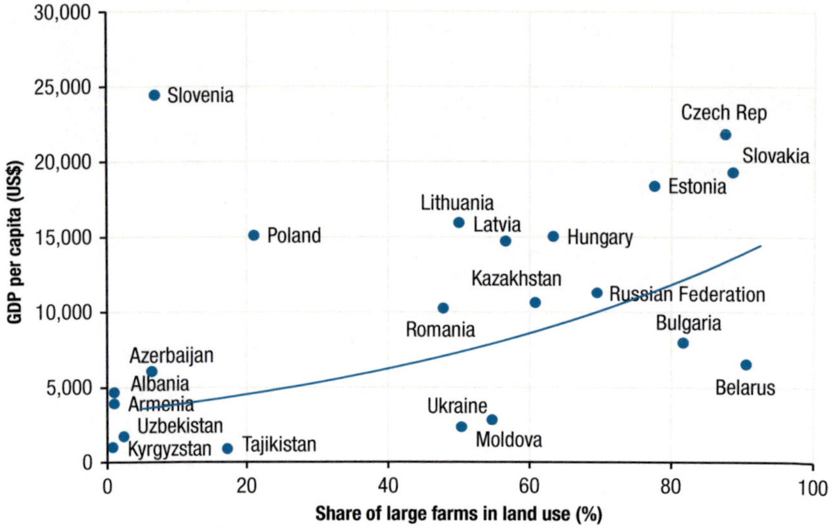

Source: World Bank (2019); Eurostat (2018).

Note: Data for Central and Eastern Europe (CEE) are for 2016; data for the Commonwealth of Independent States (CIS) are for 2015. GDP = gross domestic product.

FIGURE 7.8 Agricultural labor productivity (ALP) (index, 1990 = 0)

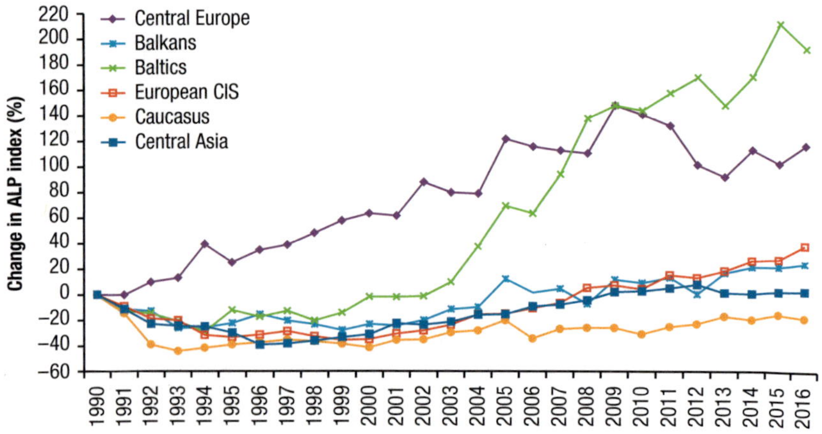

Source: ILO (2014); World Bank (2019); FAO (2019).

FIGURE 7.9 Agricultural labor productivity and smallholders in Eastern Europe and Central Asia

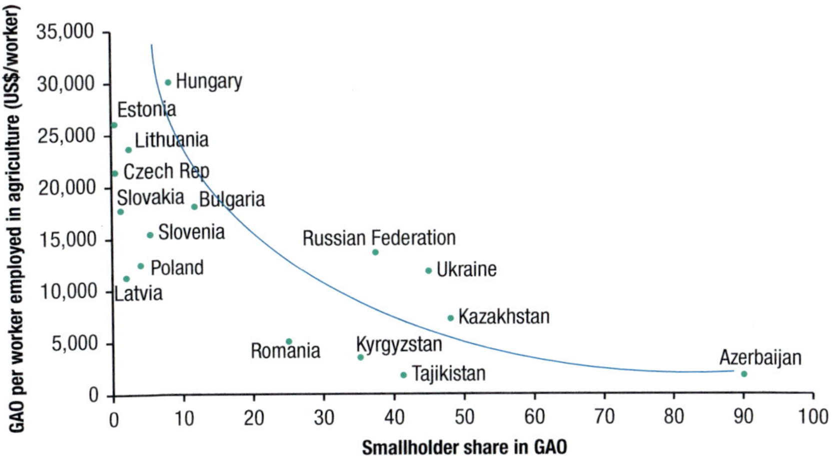

Source: Eurostat (2018); Burkitbayeva and Swinnen (2018); FAO (2019).

Note: GAO = gross agricultural output. "Smallholders" refers to "farms with less than 2 ha of land" for CEE and "households" in CIS statistics.

is Romania, where they use 12 percent of land and produce 24 percent of output.[5]

Household producers are more important in CIS. On average, they use 13 percent of land and produce 50 percent of output. As the output-land indicator shows, smallholders typically produce around twice as much output per unit of land in both CIS and CEE.

These transition processes thus contributed to large differences in labor productivity in agriculture. Figure 7.9 illustrates how today's agricultural labor productivity, on average, is much higher in countries with fewer smallholders.

Almost everywhere, the importance of smallholders in agriculture has fallen and often strongly so, as shown in Figure 7.10 for CIS and CEE on average. Between 2003 and 2016, on average, the share of smallholders in both land use and output fell by more than 50 percent in Eastern Europe (Figure 7.10). With one exception (Romania), their share in land use is now less than 5 percent in all Eastern European countries. In the CIS, their share

5 Family farms as a whole are much more important in CEE, using the majority of the land in several countries and producing between 17 and 91 percent of output. There are large differences between countries, with family farms more important in Poland and Slovenia and less important in Slovakia and the Czech Republic (Table 7.1).

TABLE 7.3 Area used and agricultural production by smallholders and family farms

	Smallholders		Family farms		
	Utilized agricultural area (% of total)	Agricultural output (% of total)	Utilized agricultural area (% of total)	Agricultural output (% of total)	Smallholder output/land ratio
Bulgaria	1.6	7.6	38.0	51.8	4.7
Czech Republic	0.1	0.5	30.5	22.8	5.0
Estonia	0.2	0.3	44.3	29.3	1.5
Latvia	0.6	1.6	87.4	80.1	2.7
Lithuania	1.1	2.3	86.8	77.8	2.1
Hungary	2.0	5.6	53.0	48.7	2.8
Poland	2.9	3.5	90.9	91.1	1.2
Romania	12.3	23.9	55.7	70.5	1.9
Slovenia	4.0	6.2	94.6	89.9	1.6
Slovakia	0.4	0.8	19.4	17.4	2.0
Average CEE	**2.5**	**5.2**	**60.1**	**57.9**	**2.1**
Russian Federation	4.3	37.4	30.5	48.5	8.7
Ukraine	30.5	44.9	45.3	—	1.5
Belarus	7.6	19.8	9.5	21.7	2.6
Republic of Moldova	—	—	49.6	58.0	—
Azerbaijan	—	—	93.6	92.7	—
Kazakhstan	1	52.1	39.2	79.4	52.1
Kyrgyzstan	10.2	36.3	95.2	98.1	3.6
Tajikistan	25.0	41.3	82.7	94.2	1.7
Uzbekistan	12.9	65.0	97.6	98.0	5.0
Average CIS	**13.1**	**49.7**	**60.4**	**73.8**	**3.8**

Source: Burkitbayeva and Swinnen (2018). Reprinted by permission of the publisher; this table is not covered by the CC BY 4.0 license.

Note: — = data not available; CEE = Central and Eastern European countries; CIS = Commonwealth of Independent States. Numbers for smallholders refer to "farms with less than 2 ha of land" for CEE in Eurostat statistics and "households" in CIS statistics. Numbers for family farms refer to "single holder farms" for CEE in Eurostat statistics (latest available year 2013) and "households" and "individual farms" combined for CIS statistics.

in output declined by 17 percent between 2000 and 2015, and the decline was most rapid in recent years (–10 percent between 2010 and 2015).

The more extensive reduction of smallholders in the past decade in CEE (compared with CIS) is most likely a combination of several effects. First, stronger growth in CEE was associated with stronger demand for labor (or, in other words, the supply of better-paying jobs) in the rest of the economy. Second, while CEE agriculture as a whole benefited from increased subsidies through the EU's Common Agricultural Policy, these subsidies were biased

FIGURE 7.10 **Share of smallholders in production in Eastern Europe and Central Asia**

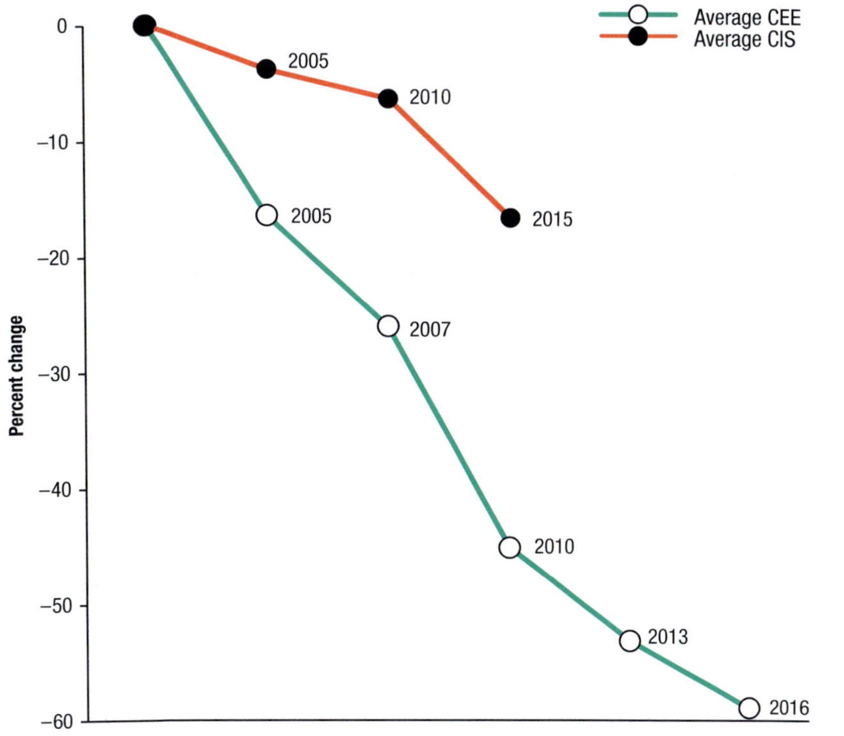

Source: Burkitbayeva and Swinnen (2018); Eurostat (2018). Reprinted by permission of the publisher; this table is not covered by the CC BY 4.0 license.

Note: CEE = Central and Eastern Europe; CIS = Commonwealth of Independent States. For CIS, percent change starts with 2000 = 0; for CEE, percent change starts with 2000 = 0. "Smallholders" refer to "farms with less than 2 ha of land" for CEE and "households" in CIS statistics.

toward larger farms and stimulated higher input costs (including land prices), both of which may have hurt smallholders' competitive position. Third, EU accession required the introduction of strict safety and quality standards throughout the value chains, both through public (EU) regulations. Fourth, (the expectation of) EU accession stimulated large inflows of foreign direct investment (FDI) into the CEE economies and the food system. This was associated with the introduction of stringent private standards that were sometimes difficult for smallholders to implement. Fifth, extensive FDI in the food system also involved the arrival of hard discounters in CEEs, which pushed down the price of basic products, including food, thereby reducing the attractiveness of "make versus buy," that is, household subsistence food production. All these factors have contributed to the decline of smallholders.

Market Imperfections, Value Chains, and Investments in Agribusiness and Food Companies

Agricultural productivity growth has been strongly influenced by value-chain investments, up- and downstream from the farms. These investments have been motivated by a combination of market imperfections (especially in capital, inputs, and technology markets) and the introduction of quality standards by modern food processing and retail companies. Access to output and input markets was an important constraint for farms in ECA and still is in some countries. The evolution of the food value chain is discussed in Chapter 12, whereas the role of contract farming between food processors/agribusiness/supermarkets and farmers is discussed in Chapter 11.

Private investments in agribusiness, food processing, and retail companies led to important restructuring in production, trading, and marketing systems and in the integration of farms in value chains in ECA. Contracting and vertical coordination in these supply chains played an important role in this process (Gow and Swinnen 1998). Dries et al. (2009) document the rapid growth of vertical coordination in Eastern European value chains to overcome input and capital constraints of the farms and ensure high-quality raw material for the dairy processors. One reason for the emergence of large vertically integrated agroholdings in the Russian Federation, Ukraine, and Kazakhstan was to counter the high transaction costs and supply uncertainty that existed because of weak infrastructure (physical and institutional) and commercial support services for market agriculture.

Vertical coordination in the ECA region included a variety of institutional innovations, such as provision of credit, inputs, and quality control. These programs are believed to have triggered major technology transfer, productivity growth, and improvements in product quality at the farm level (Gow, Streeter, and Swinnen 2000; Dries and Swinnen 2004, 2010; Noev, Dries, and Swinnen 2009).

Investments in the value chains often came from foreign companies, especially in Eastern Europe where Western European agribusiness, food processing, and retail companies were eager to invest. They were attracted by the closeness of the markets, the relatively high incomes of the population, and the (expected) integration into the EU. In the Russian Federation, some of the most important investments in the value chains came not from foreign companies but from domestic financiers who got access to capital from profits in other sectors, such as trading, banking, or oil and gas.[6] These investments reduced capital constraints and thereby made access to inputs and technology easier and stimulated productivity growth in these chains.

Last, studies on the dynamics of smallholder integration in modern value chains in ECA find that (1) smallholders integrated into supply chains were upgrading in terms of technology, productivity, and size; and (2) the ones that were not growing were older households, keeping animals for household consumption and without product standards required by the market (Van Herck and Swinnen 2015). This process of small though upgrading farms led by younger and more entrepreneurial types and the decline of small farms run by the elderly, either because they retire or can no longer (physically) farm, seems to be a common development across the more developed countries in ECA. In contrast, small poor farmers remain very important in the poorest ECA countries, such as Tajikistan, where investments in value chains are lagging.

Constraints on Land and Water Use

Transition led to substantial changes in land use. Land abandonment was caused by (1) the reduced profitability of farming from the drop in agricultural subsidies and move to market prices and (2) uncertainty on land property rights. Both developments affected how much land was used and how intensely it was cultivated (that is, how many other inputs were used along with it in production). Vranken et al. (2011) showed how land plots with uncertain ownership were more likely to be left abandoned or used less intensively. Land abandonment was widespread in some countries.[7] According to Kraemer et al. (2015), 45 percent of cropland in the former Soviet Union was converted to grassland between 1990 and 2000.[8]

6 In the RUK countries, much of the investment financing (and management) of the agroholdings came from sectors outside of agriculture.

7 Official data do not always present an accurate picture of the changes because they often do not distinguish between cropland and pastures.

8 For Russia alone, estimates of cropland abandonment vary from 20 million hectares to more than 40 million hectares: the estimate is 43.5 million hectares by Lambin et al. (2013), 32 million hectares by Alcantara et al. (2013), and 20 million hectares by Ioffe, Nefedova, and Zaslavsky (2004). The disparities arise for several reasons: differences in the time periods and regions of measurement, the definitions of land abandonment used, and the quality of the data and heterogeneity of abandonment patterns in the regional estimates (Alcantara et al. 2013).

Estimates of abandoned cropland in RUK during transition vary between 50 and 60 million hectares (Schierhorn et al. 2013; Meyfroidt et al. 2016; and RUK official statistics).

In addition, the process of transition led to a significant deterioration of infrastructure, most notably involving irrigation. Irrigation is a major component of agricultural production, particularly in certain arid regions of Central Asia and the Caucasus subregion.

Irrigation problems were reinforced in some of these regions in Central Asia and the Caucasus by the shift to household farms. This process led to land fragmentation, which brought further challenges related to irrigation, as the existing infrastructure had been constructed for much larger units of production, such as kolkhozes and sovkhozes (Lerman 2010; Akramov and Shreedhar 2012). On-farm irrigation was left to be managed by smaller plots, creating problems in access to and management of irrigation infrastructure. This led to a complex set of formal and informal arrangements and problems, such as rent seeking, information asymmetry, and distrust between community members (Akramov and Omuraliev 2009).

In order to deal with the changing landscape of farm structures and irrigation needs, water users associations (WUAs) were established in some countries of Central Asia and the Caucasus. These were intended to regulate water allocation at the community level and ensure operation and maintenance of the on-farm irrigation and drainage infrastructure via in-kind and cash contributions from the members. However, due to weak governance structures, financial mismanagement, and dwindling membership numbers, these proved to be largely unsuccessful (Wegerich 2008; Akramov and Shreedhar 2012; Zinzani 2015). In some countries, the introduction of irrigation water fees led to "elite capture" allowing certain groups to gain better access to irrigation (Akramov and Omuraliev 2009). Some argue that the failure of WUAs, particularly in Central Asia, is associated with the way they were established–via top-down approaches that precluded grassroot initiatives, experience, and incentive among the members (Abdullaev et al. 2010; Abdullaev and Atabaeva 2012). Most recent experimental evidence from Uzbekistan and Kazakhstan by Amirova, Petrick, and Djanibekov (2019) calls into question the top-down approach in current water policies of the region and suggests the superiority of endogenous cooperation and hence encourages the implementation of self-governed water management systems.

Problems with irrigation infrastructure also compound challenges associated with soil quality, salinization, and waterlogging. With the increasing frequency of weather extremes and aridity, improving irrigation water management and efficiency of irrigation infrastructure remains important, not only for raising agricultural productivity, but also for improving the climate resilience of the sector.

Potential for and Constraints to Feeding the Region (and the World)

Agricultural output can grow through the use of more inputs (extensive growth) or through increased productivity (intensive growth) (Babcock 2015). As explained above, agricultural productivity covers many inputs, including labor and knowledge (human capital). However, the vast bulk of the literature on agricultural productivity in the ECA region focuses on land use (especially for grain) and yields. Many studies have pointed out that the widespread cropland abandonment in the ECA countries during the transition years represents large untapped agricultural production potential (EBRD and FAO 2008; Lambin and Meyfroidt 2011; Glauben et al. 2014; Meyfroidt et al. 2016).

Land Use Potential

There has been some recultivation of abandoned cropland in the past decade, primarily in areas with good agronomic conditions. However, land use is still much lower than during the pre-reform period. Compared with 1990, land use in 2015 was still 40 percent and 30 percent lower in Kazakhstan and the Russian Federation, though about the same in Ukraine (Figure 7.5a).

The recultivation of all idled cropland would dramatically increase total crop area. However, much of the abandoned land was farmed during the socialist period only because of the decisions of state planners, backed by heavy subsidies. Hence, there is no strong economic rationale for returning all this land to production. Liefert and Liefert (2015) find that most of the Russian Federation's abandoned grain area was in high-cost regions of the country, especially in the north and east. Uzun et al. (2014) estimate that returning 19 million hectares of abandoned cropland to production in the Russian Federation would require grain export prices to rise to $400 per ton (with the average world market price for wheat from 2012–2015 being $200 per ton). The authors also point out that the degree of grain recultivation on abandoned croplands in northern European Russia, that is, outside the fertile black soil areas, is low mainly because of biophysical constraints. Liefert and Liefert (2015) also argue that cropland recultivation in the Russian Federation might not necessarily lead to much additional grain production because of competing land demand from other crops, particularly oilseeds.

Moreover, crop area that was abandoned in the 1980s or 1990s is often currently infested with deep-rooting vegetation that renders recultivation expensive (Larsson and Nilsson 2005). Such costs should be kept in mind because by 2012, approximately 3.5 million hectares of agricultural land from the Soviet period was covered by forest in European Russia alone (Potapov et al. 2015). The secondary vegetation is important for biodiversity and ecosystem services (Kamp et al. 2011), and it stores substantial amounts of carbon in soil and vegetation, much of which would be emitted to the atmosphere with recultivation (Schierhorn et al. 2013; Kurganova, Lopes de Gerenyu, and Kuzyakov 2015).

Hence, it appears that only a fraction of the abandoned ECA (and mainly RUK) cropland can be put back into production without significant costs or major environmental trade-offs. Meyfroidt et al. (2016) identified that from 60 million hectares of RUK crop area that was initially abandoned, about 12 million hectares had been returned to production by 2009 and that only 8.5 million hectares of the remaining idled land qualifies economically for being returned to the plow, in that the land has high soil quality and recultivation would involve low environmental trade-offs and few accessibility and socioeconomic constraints (5.3 million hectares in the Russian Federation, 2.4 million in Kazakhstan, and 0.9 million in Ukraine).

Agroenvironmental and socioeconomic constraints limit reusing abandoned cropland in RUK. The vast bulk of high-quality land is already back in cultivation. Other more marginal lands may be more suitable for other uses, such as livestock grazing, development of a livestock fodder base, or ecosystem services.

Yield Potential

Since 2000, there have been significant improvements in farm access to yield-increasing inputs used in grain production. Fertilizer use has recovered significantly from its low point in the late 1990s, especially in the Russian Federation and Ukraine. Important structural and institutional changes over the past decade helped to overcome systemic constraints in the major grain-producing regions and contributed to the emergence of large-scale and vertically integrated farming operations, such as the agroholdings (Gataulina et al. 2005; Serova 2007; Swinnen 2009; Nefedova 2016). Since 2005, the Russian government has also increased subsidies to agriculture (Liefert and Liefert 2012). Increased government support, strong depreciation of the ruble in 2014/2015 (largely because of falling world oil prices, which cut the Russian Federation's export earnings), and high agricultural world market prices have all contributed to increased investments in booming export-oriented Russian grain production (Kingwell et al. 2016).

Increasing investments and higher export returns have helped to generate significant growth in fertilizer use and yields. Grain yields started recovering in the late 1990s and have since risen in the RUK by 50 to 70 percent. However, current fertilizer use per hectare is still considerably below the pre-transition levels, reflecting, among other things, the cut in the large fertilizer subsidies and possibly inefficient use of fertilizer in the planned socialist economy. Figure 7.5b also shows how current fertilizer use per hectare is much higher in Ukraine than in the Russian Federation, and substantially higher in the Russian Federation than in Kazakhstan.

Several studies have identified significant potential for yield growth in RUK grain production due to low fertilizer use, low-quality seeds, poor extension services, volatile weather resulting in frequent crop failures, and inadequate insurance schemes (FAO 2009; Schierhorn et al. 2014; Uzun et al. 2014; Kingwell et al. 2016).

Climate Change

The impact of climate change on RUK grain production potential varies across the different parts of this vast region (Dronin and Kirilenko 2011; Müller et al. 2016; Fehér et al. 2017).[9] The northern parts may benefit from warmer weather and longer growing seasons, but the soil quality there limits growth potential. Production in the southern regions, which has most of the good soils, is likely to become more vulnerable with climate change, especially from a drop in precipitation.

Production Potential

Swinnen et al. (2017) review a series of studies on RUK grain production potential under various scenarios. Table 7.4 summarizes their conclusions. Under the recultivation scenario, grain area would expand by 8.5 million hectares, mostly in the Russian Federation, and this would result in an extra 12.5 million tons of RUK output. Closing the yield gap (the intensification scenario) to 60 percent of potential yield on existing croplands would generate additional grain production of 23.9 million tons (of which 12.2 million is in the Russian Federation, 7.7 million in Ukraine, and 4.0 million in

9 See Akramov, Park, and Ilyasov (2017) for possible climate change effects on Central Asian agriculture.

TABLE 7.4 Potential wheat production in the Russian Federation, Ukraine, and Kazakhstan (RUK) under different scenarios

	Area (million tons)	Yield (tons/hectare)	Production (million tons)
Baseline	77.6	2.1	161.5
Recultivation	94.7	2.0	174.0
Intensification at 60% of the yield potential	77.6	2.4	185.4
Intensification at 60% of the yield potential + recultivation + climate change	87.6	2.4	203.8
Growth in production			**42.3%**
Intensification at 80% of the yield potential	77.6	3.2	246.9
Intensification at 80% of the yield potential + recultivation + climate change	87.6	3.2	271.5
Growth in production			**110.0%**

Source: Authors' calculations based on Swinnen et al. (2017).

Kazakhstan).[10] This potential seems realistic, partly because relatively small increases in input use could result in substantial yield growth. This output rise equals 34 percent of the average annual RUK grain exports during 2013–2015, which totaled 71 million tons. Closing the RUK yield gap to a more challenging 80 percent of the yield potential would increase grain production on existing cropland by 85.4 million tons, compared with the baseline, an increase of more than 50 percent.

Adding the impact of climate change to the intensification scenario generates a further 4–5 million tons of grain output, depending on the yield gap assumptions. This production rise comes from the area added to grain production in northern regions, despite low yields on the new land.

The combination of the three effects (recultivation of land, climate change, and closing of yield gaps) generates a total production of 203.8 million tons. The 60 percent intensification scenario yields an increase compared with baseline output of 42.3 million tons (26 percent). The (perhaps unobtainable) 80 percent intensification scenario generates total production of 271.5 million

10 In this scenario, the additional grain production in Ukraine is 63 percent of the additional output of the Russian Federation, while total area under grain cultivation in Ukraine is only 35 percent of the cultivated area in the Russian Federation. Yield gap closure in Ukraine generates more additional production relative to the Russian Federation because of both higher yield potential and higher yield gaps in Ukraine (see Figure 7.5b). Despite the higher share of grain cultivation in total sown area in Kazakhstan (16 percent higher than in the Russian Federation and Ukraine), the production potential in Kazakhstan on existing cropland is relatively small because of the low current yields and small yield gaps.

tons of wheat, 110 million tons above current production (and a rise of 68 percent).[11] In all scenarios, most of the gains come from yield growth.

In summary, output potential for grain in the RUK could be somewhere between 200 and 270 million tons per year (of which more than 120 to 160 million tons would be in the Russian Federation alone). This would be an increase of approximately 40 to 100 million tons above "current" (baseline) production, which implies that even under more pessimistic scenarios, the RUK could satisfy a substantial share of the projected increase in global wheat and grain demand.

Will the RUK Be a Reliable Grain Supplier to ECA and the World?

A key question is whether this production and export potential is also a reliable source of food for other countries. Grain is a major component in the diet of all ECA countries, and several of the poorer Central Asian and Caucasus countries rely heavily on imports from the Russian Federation, Ukraine, and Kazakhstan for their supply. For example, more than 50 percent of the grain consumed in Georgia and Armenia during the period 2010–2017 was imported, almost exclusively from the RUK (Figures 7.11 and 7.12).

Figure 7.13 shows that since 2000 RUK grain exports (net as well as gross) have increased dramatically, especially of wheat.[12] Average annual RUK gross grain exports increased from 10.0 million tons in 1998–2000 to 71.3 million tons in 2013–2015. However, Figure 7.13 also shows that these countries' grain exports are quite volatile, especially compared with other major suppliers such as the United States and the EU. This unreliability of supply is important in assessing the RUK countries' potential to become a "world breadbasket," especially to food-insecure countries that rely on the imports.

One factor that reduces the RUK countries' grain export reliability is volatile weather, with bad conditions (such as high temperature or low precipitation) reducing the domestic surpluses available for export. Figure 7.13 shows how closely the annual volumes of RUK grain exports track the weather-driven fluctuation in production. RUK trade policies (in particular, export restrictions) also contribute to their grain export volatility. A common catalyst for the export controls is the reduced domestic harvests from poor weather, which motivate governments to keep grain within the country. The controls

---

11 For comparison, EBRD and FAO (2008) project maximum grain production potential for the RUK countries of 230 million tons, a 42 percent increase over the same period.

12 Given that the RUK countries import little grain, their net grain export volumes are only slightly below their gross exports (for the RUK collectively, only 2 percent lower in 2013–2015).

FIGURE 7.11 Import dependency for cereals

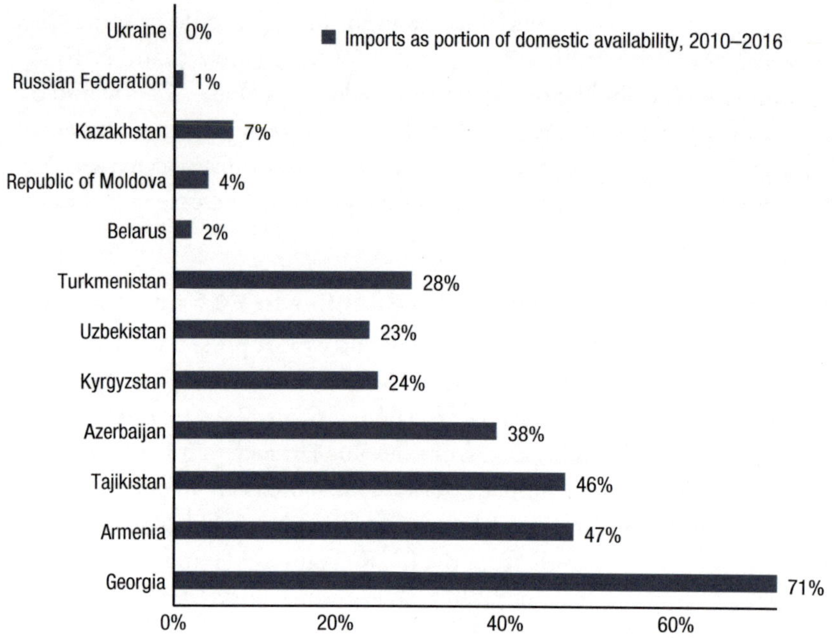

Source: FAO (2019).

FIGURE 7.12 Wheat imports by source country (% of total wheat imports, ave. 2010–2017)

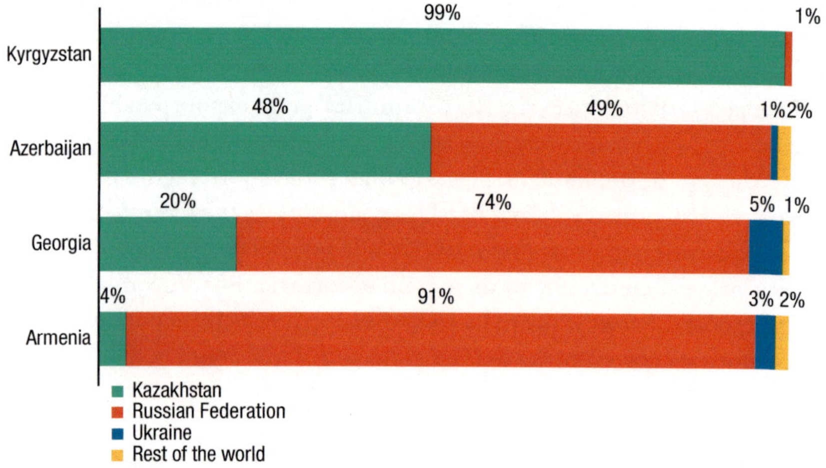

Source: Data from FAO (2019).

FIGURE 7.13 Grain production and exports in the Russian Federation, Ukraine, and Kazakhstan

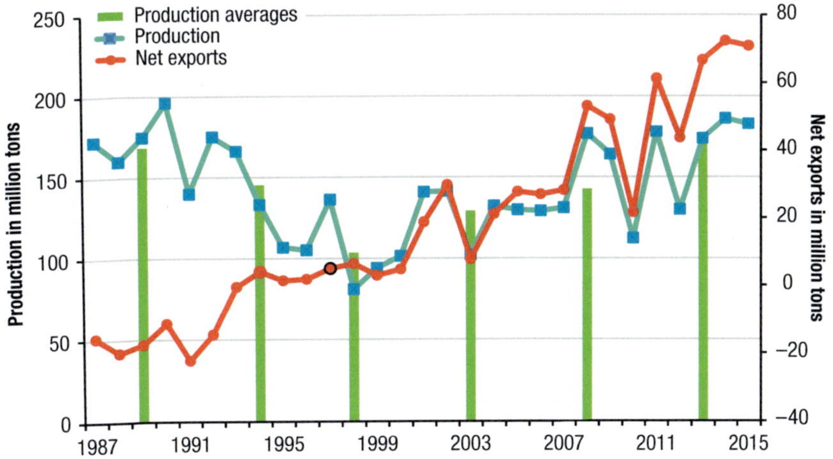

Source: USDA, Foreign Agricultural Service (2020).

Note: The bars give average annual grain production over the periods 1987–1991, 1992–1995, 1996–2000, 2001–2005, 2006–2010, and 2011–2015. Negative net grain exports are net imports.

are intended to mitigate the rise in domestic grain prices, and thereby help domestic food consumers and the feed-consuming livestock sector. For example, prompted by a poor harvest in 2010, the Russian government banned all grain exports from August 2010 through June 2011.

During the surge in world agricultural and food prices in 2007–2008 and again in 2011–2012, all the ECA grain exporters (not just the RUK but also Tajikistan), as well as grain net-importers such as Belarus and Kyrgyzstan, imposed export restrictions to secure domestic supply and help domestic consumers, including the livestock sector (von Cramon and Raiser 2006; Dollive 2008; Jones and Kwiecinski 2010; World Bank 2011). An FAO study (Sedik 2013) found that a third of the surveyed ECA countries imposed agricultural export restrictions in some form, while a third reduced import taxes.[13]

The Russian Federation's economic crisis of 2015–2016, generated by a large drop in world oil prices (the Russian Federation's main export), subsequent depreciation of the ruble, geopolitical conflict with Ukraine and

13 Sedik (2011, 2013) showed that the harmful effect of RUK grain export restrictions on ECA importing countries was mitigated by their rapid shift toward greater imports of flour and other grain products whose exports were not restricted (or less so).

the West, and an agricultural import ban imposed on the major Western countries, led to serious food price inflation. The Russian policy package to counter the price jump also included restricting grain exports (Liefert et al. 2019).

Food and Nutrition Security in ECA

Food prices spiked in 2007–2011, there was an economic crisis in 2009–2010, and real GDP and remittances declined as a result.[14] Despite these occurrences, incomes in the ECA region, and thereby also food security, continue to improve steadily (although the income growth rates differ substantially between subregions). The income growth over the past 15 years has greatly reduced poverty, undernourishment, and micronutrient deficiency (Table 7.5 presents several food security indicators for selected ECA countries that experienced the strongest poverty decline). The ECA region has made major progress, and currently the majority of the ECA countries are below 2.5 percent prevalence of undernourishment. Nevertheless, undernourishment remains a challenge in a few of the Caucasus and Central Asian countries, in particular Tajikistan (where the prevalence is 30 percent).

In addition, poor diets and micronutrient deficiencies are a problem for a larger part of the ECA region. Poor diets can lead to an insufficient intake of nutrients, particularly iron, vitamin A, and zinc (FAO 2017a). Micronutrient deficiencies occur in both rich and poor ECA countries, but they are again more prevalent in the poorer countries of the Caucasus and Central Asia, as well as in the Republic of Moldova. This can be seen from high rates of stunting and anemia among children under five (Table 7.5).

As in other regions, economic transition and growing incomes generate a so-called nutritional transition where changes in diets and a more sedentary lifestyle lead to changes in nutritional and anthropomorphic outcomes (see Chapter 10). Now that food is becoming more readily available in the ECA region and undernutrition is less prevalent, a new challenge facing the ECA countries is growing obesity.

Figure 7.14 shows the relationship between income and nutritional problems in ECA. As per capita income grows, so too does overnutrition, in the form of overweight and obesity. According to FAO (2017a), overnutrition tends to increase alongside per capita incomes until the latter reach a level

14 In some ECA countries, remittances remain an important source of income. In Kyrgyzstan and Tajikistan they equal up to 30 percent of GDP (World Bank 2019).

TABLE 7.5 Food security indicators for selected Eastern Europe and Central Asia (ECA) countries

	GDP per capita (PPP, in 2011 international $)		Poverty ratio at $1.90 a day in 2011 PPP (%)		Prevalence of undernourishment (%)		Prevalence of anemia in children < 5 years old (%)	Stunting in children < 5 years old (%)
	In 2018	Change from 2001 (%)	In 2015	Change from 2001	2014–2016	Change from 1999–2001	2016	Latest years available
Armenia	9,178	185	2	−17	24	−19	32	21
Georgia	10,152	170	8	−13	7	−7	24	11
Azerbaijan	16,011	219	< 1	−2	< 2.5	−21	24	18
Republic of Moldova	6,490	117	0	−28	9	−11	26	6
Kyrgyzstan	3,447	60	3	−34	6	−10	38	13
Tajikistan	3,061	140	5	−26	30	−12	31	27
Uzbekistan	6,240	141	27	−40	6	−10	37	19
Turkmenistan	17,129	211	—	—	6	−3	28	20

Source: FAO (2017); FAO (2019); World Bank (2019).

Note: — = data not available; GDP = gross domestic product; PPP = purchasing power parity. Latest years available: Armenia 2010, Azerbaijan 2013, Georgia 2009, Kyrgyzstan 2014, Tajikistan 2012, Uzbekistan and Turkmenistan 2006, Albania 2009, Moldova 2004. Poverty data for Uzbekistan is for 2012 and 2002; Tajikistan data is for 2015 and 2003.

FIGURE 7.14 GDP per capita versus prevalence of obesity among males (+18)

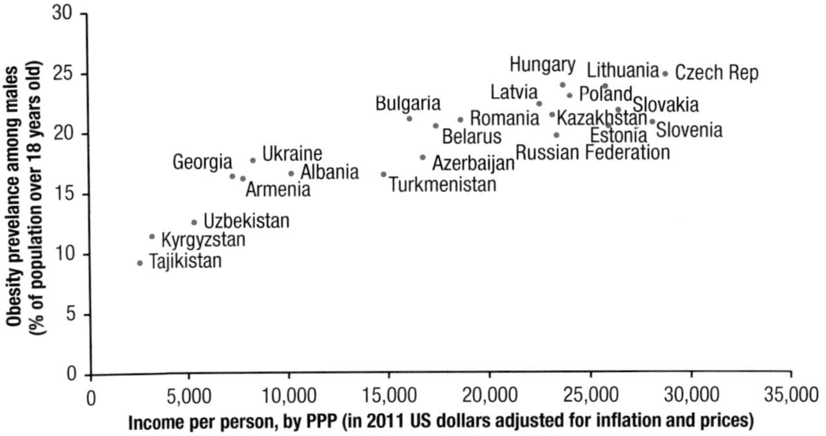

Source: WHO (2014); Gapminder (2019).

Note: "Obese" refers to individuals with body mass index greater than or equal to 30. PPP = purchasing power parity.

of between $30,000–$40,000 (in 2010 US dollars), after which the trend is reversed. Due to the nutritional transition currently underway in the ECA countries, combatting food insecurity in the future will no longer be a question of simply providing access to food, but also of providing access to healthy food in order to reduce micronutrient deficiencies and health risks caused by an overweight and obese population.

Numerous programs, policies, and strategies exist toward achieving food security, especially given that countries interpret food security in different ways. On this question, the ECA countries can be divided roughly into two groups: those that aim for food security through independence and food self-sufficiency, and those that seek to improve food availability without the use of food self-sufficiency targets (FAO 2015b).

Countries belonging to the first group include Belarus, Kazakhstan, Kyrgyzstan, the Russian Federation, Tajikistan, Turkmenistan, and Uzbekistan. In these countries, the concept of food security is primarily viewed as "food independence" or "food self-sufficiency," which is also reflected in the CIS Concept of Food Security, adopted by the CIS member countries in 2010. It states that "food security is the state of the economy of the countries in which their own production has to ensure food independence for not less than 80 percent of the annual demand in food production in accordance with physiological nutrition standards."[15] The main policy directions to achieve self-sufficiency in these countries include (1) producer support for agricultural inputs; (2) trade interventions to favor domestic production and protect the country from imports, or discourage exports to protect domestic consumers from soaring prices; (3) price controls for basic food items; and (4) market interventions and management of commodity stocks (FAO 2015b). There are differences among countries in the mix of policy instruments and degree to which they are used to attain food self-sufficiency.

For example, we discussed earlier that large grain-exporting ECA countries such as the Russian Federation and Kazakhstan have used export restrictions to keep grain within their countries during periods of high world prices or bad weather that reduced the domestic harvest. Many ECA countries have also used agricultural import controls to protect domestic producers and promote self-sufficiency. For example, the Russian government regarded the large-scale contraction of the livestock sector during the 1990s as a disaster that

15 Decision of Heads of Governments of CIS, "About the complex of joint efforts on increase in food security of the State Parties of the CIS," November 19, 2010 (available at https://zakon.rada.gov.ua/laws/show/997_m96#Text).

state policy should help reverse. In 2003, the government imposed a system of restrictive tariff rate quotas on imports of meat (beef, pork, and poultry). The government also began enacting an array of sanitary restrictions, including complete bans, on imports of meat and other livestock products (Liefert and Liefert 2012). These policies were consistent with the Russian Federation's Food Security Doctrine of 2010, which advocated agricultural import substitution and self-sufficiency.

Another policy is the use of state-run production and procurement of certain strategic commodities to fulfill food self-sufficiency goals. Elements of such a centrally planned system remain in Uzbekistan, Turkmenistan, and to some extent, Belarus, where state support is extended in the form of subsidized inputs to producers who fulfill the state order.

Although the approaches implemented toward achieving food self-sufficiency might be successful in strategic terms, they may not always be economically rational. This can be demonstrated in the case of Uzbekistan, where in pursuit of food self-sufficiency, wheat became another state-ordered crop. As a result, the area planted to wheat more than doubled between 1991 and 2017, most notably by shifting production to irrigated areas. Although this policy shift did increase wheat production, it came at a cost, as irrigation water was diverted away from high-value fruits and vegetables to wheat, a relatively low-value crop (FAO 2015b). Despite these efforts, the bulk of wheat consumed in Uzbekistan continues to be imported from Kazakhstan.

The second group, which includes Armenia, Azerbaijan,[16] Georgia, the Republic of Moldova, and Ukraine, more closely follows the FAO approach to food security. This is outlined in the World Food Summit Plan of Action, which defines food security as being achieved "when all people, at all times, have physical and economic access to sufficient, safe and nutritious food to meet their dietary needs and food preferences for an active and healthy life." This interpretation of food security is much more inclusive and focuses more on improving food availability and access, food safety, and the nutritional state of the population.[17] Policies adopted by these countries mainly focus on macroeconomic stability, investments in infrastructure, ensuring a reliable and stable food supply by both producing food domestically and recognizing that imports are necessary for food security and social development, supporting the agricultural sector within a liberal trade environment, finding new markets for agricultural products, and promoting exports (FAO 2015b).

---

16 With the exception of Azerbaijan's self-sufficiency in grain production.

17 For more detailed information on food-security-related policies in ECA, see FAO (2017a).

Another way of influencing food and nutrition security is through agricultural subsidies that affect production and value chains. There have been attempts to assess the importance of these policies by either measuring the benefits and costs for those who are paying for the policies or looking at price regulations affecting consumers and producers. The Producer Support Estimate (PSE) and Consumer Support Estimate (CSE) are well-known indicators calculated by the Organisation for Economic Co-operation and Development (OECD). The nominal rate of assistance to agriculture (NRA) and the relative rate of assistance to agriculture (RRA) are well-known indicators, calculated by the World Bank, of the amount of support (taxation) to agriculture as a consequence of government regulations and subsidies (Anderson 2009; OECD 1999).[18] These indicators are widely used in policy discussions and political economy analyses (see, for example, Anderson, Rausser, and Swinnen 2013). Unfortunately, for the Eurasian countries, updated PSE and CSE indicators are only available for the Russian Federation, Ukraine, and Kazakhstan. The average support given to agriculture in those countries (and thus the average tax on the budget and on food consumers) has fluctuated around 10 percent for the past 20 years and has stayed mostly between 0 and 20 percent. This means that, on average, farmers in those countries have benefited from government policies, but not extensively: on average, farms received around 10 percent more income than they would have without government interventions.

However, the average numbers hide important differences among commodities within the countries. For example, in Kazakhstan, livestock producers were subsidized in recent years (around 15 percent PSE), while maize, rice, and sunflower producers were taxed significantly (the PSE was between −20 and −80 percent). In the Russian Federation, the livestock sector was supported subtantially (around 25 percent PSE), while most grains and sunflowers had negative PSEs (between −5 and −30 percent) and were thus taxed.

For other countries, NRA and RRA indicators have been calculated for some commodities in Central Asia but have not been updated after 2004 (Anderson and Swinnen 2008; see the chapters by Christensen and Pomfret [2008] on Kyrgyzstan; by Pomfret [2008a] on Kazakhstan; and by Pomfret [2008b] on Tajikistan, Uzbekistan, and Turkmenistan). The (more limited) NRA studies on Central Asia point out the important differences

18 Negative PSE or CSE values imply taxation—see OECD (1999) for more details. NRA and RRA are indicators of the amount of support (taxation) to agriculture there is as a consequence of government regulations and subsidies—see Anderson (2009) for more details.

FIGURE 7.15 Share of agriculture in total government expenditures (average of 2012–2017)

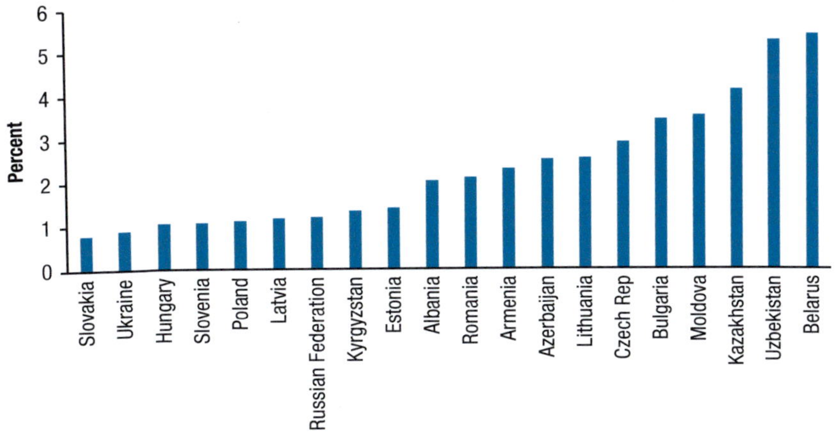

Source: Data from FAO (2019).

among countries in their policies vis-à-vis the cotton sector. Cotton—a very important (export) commodity especially in Uzbekistan, Turkmenistan, and Tajikistan—was heavily taxed by the governments in these countries in the 1990s and 2000s, as it was an important source of government tax revenue. In contrast, it was much less taxed in other countries in the region where it was less important as an export commodity—such as in Kazakhstan and Kyrgyzstan (Pomfret 2008b).

Budget allocations are also indicators of government support, though they are more limited (because they do not provide information about transfers between consumers and producers through price and trade regulations). Figure 7.15 shows the average share of agriculture in total government expenditures between 2012 and 2017 in ECA. Some countries, like Belarus and Uzbekistan, spent over 5 percent of their state budgets on agricultural support. Others, like Slovakia, spent less than 1 percent.

Conclusion

Although the ECA countries are agriculturally heterogeneous in many ways, they share a common institutional history and, in certain respects, a common reform experience. The economic and institutional transformation that took place in the 1990s and 2000s had a dramatic effect on all these countries. However, the pace of transition from planned to market economy has differed greatly among them. Using a conceptual model, we illustrate and examine the key aspects of the agricultural transition process, some of which are still playing out in various ECA countries. In almost all countries, agricultural output fell substantially during the transition decade of the 1990s, especially in the livestock sector. Key reasons for the decline included the institutional and allocative disruptions and high transaction costs stemming from the move from planned to market economies, as well as the severe reduction in the large subsidies to agriculture under the planned system that maintained high-cost production. A dramatic decline in fertilizer use and crop prices in all ECA countries led to a strong drop in crop yields.

Agricultural output began to rebound in many ECA countries in the mid to late 1990s, and by the early 2000s production was increasing almost everywhere. Current agricultural output in ECA varies by country, from −30 to +30 percent of the pre-reform level. The output rebound has also varied by commodity. While wheat production declined during the 1990s and recovered strongly in the 2000s, especially in the RUK region, livestock production in many ECA countries still remains below the pre-reform level.

The rise in output was driven largely by input productivity growth. There were two general patterns involving farm structure and the rise in productivity and output. The first pattern included countries (such as Slovakia, the Czech Republic, and Hungary) where large farms dominated and most of the production increase came from growth in labor productivity. The second pattern occurred in poorer countries (such as Albania, Armenia, and Tajikistan) dominated by small farms, where the productivity of land (output per hectare) rose due to the intensive use of labor and other inputs. There are further countries (such as Poland) that have experienced some growth from an increase in both land and labor productivity.

Agricultural growth patterns are strongly related to farm structures. The ECA region is now characterized by strong heterogeneity of farm structures, and most countries have a mix of large- and small-scale farms. The emergence of different farm structures was largely "endogenous" to the transition process, due to differences in natural resources, commodity characteristics, and factor market imperfections.

Value-chain investments have played an important role in the growth of agricultural productivity in the ECA countries. Investments in value chains, often from foreign companies, led to major restructuring in production, trading, and marketing systems. Contracting and vertical coordination in value chains helped farmers to overcome production constraints through the provision of credit and inputs and resulted in major technology transfer, productivity growth, and improvements in production quality.

The transition process led to substantial changes in land use and a significant deterioration of infrastructure, most notably involving irrigation, which is of particular importance for agriculture in Central Asia and the Caucasus.

One of the main questions for agriculture in ECA is whether the RUK region has the potential to become a "breadbasket" to improve world food security. Some observers argue that by increasing yields or reusing 50–60 million hectares of abandoned land or both, the RUK region could boost the production and export of grain, and especially wheat. However, only a fraction of the abandoned land could actually be recultivated in a way that is economically profitable and environmentally sustainable, as most of it is in high-cost marginal regions where grain production is not commercially viable. On the other hand, we also project that the RUK countries will act on their more feasible potential to further increase grain yields, mainly by continuing to adopt better management practices that reduce waste, improve efficiency, and utilize Western technology.

Yet, due to climate change and weather volatility, uncertainties exist concerning the reliability of the RUK countries as grain export suppliers. Volatile RUK weather (both temperature and precipitation) creates major fluctuation in annual grain production, with bad weather reducing domestic surpluses for export. Low weather-driven harvests in turn often motivate the RUK governments to impose export controls to keep grain within the country in order to mitigate domestic price increases and thereby help food consumers and the feed-consuming livestock sector. For example, prompted by a poor harvest in 2010, the Russian government banned all grain exports from August 2010 through June 2011. During the surges in world agricultural and food prices in 2007–2008 and 2011–2012, all three RUK countries imposed various controls on grain exports (taxes, quotas, complete bans).

Food and nutrition security has improved substantially within the ECA region. Since around 2000, rising incomes have reduced poverty and undernourishment. However, in some of the poorer countries of the Caucasus and Central Asia, undernourishment remains a challenge, and even in countries with growing incomes, poor diets can cause micronutrient deficiencies. Also, as countries get richer, they experience the new problem, common to developed countries, of overnutrition, generating overweight and obesity.

Various programs, policies, and strategies exist for achieving food security in the ECA countries, depending on how they interpret food security. In countries where food security is primarily viewed as "food self-sufficiency," market interventionist agricultural policies have been employed to stimulate the domestic supply of staple foods, grains in particular. These policies were especially visible during the 2007–2012 period when food prices spiked on international markets. Another way of influencing food and nutrition security is through agricultural subsidies. However, not many indicators of agricultural support are calculated across the ECA countries that allow comparison, nor do they fully capture the important differences between commodities in different regions.

References

Abdullaev, I., and S. Atabaeva. 2012. "Water Sector in Central Asia: Slow Transformation and Potential for Cooperation." *International Journal of Sustainable Society* 4 (1–2): 103–112.

Abdullaev, I., J. Kazbekov, H. Manthritilake, and K. Jumaboev. 2010. "Water User Groups in Central Asia: Emerging Form of Collective Action in Irrigation Water Management." *Water Resources Management* 24 (5): 1029–1043.

Akramov, K., and N. Omuraliev. 2009. *Institutional Change, Rural Services, and Agricultural Performance in Kyrgyzstan*. IFPRI Discussion Paper 904. Washington, DC: IFRPI.

Akramov, K., A. Park, and J. Ilyasov. 2017. *Review of Agricultural Development in Central Asia: Trends, Challenges, and Opportunities*. Washington, DC: IFPRI.

Akramov, K., and G. Shreedhar. 2012. *Economic Development, External Shocks, and Food Security in Tajikistan*. IFPRI Discussion Paper 1163. Washington, DC: IFPRI.

Alcantara, C., T. Kuemmerle, M. Baumann et al. 2013. "Mapping the Extent of Abandoned Farmland in Central and Eastern Europe Using MODIS Time Series Satellite Data." *Environmental Research Letters* 8: 035035.

Amirova, I., M. Petrick, and N. Djanibekov. 2019. "Long- and Short-Term Determinants of Water User Cooperation: Experimental Evidence from Central Asia." *World Development* 113: 10–25.

Anderson, K. 2009. *Distortions to Agricultural Incentives, A Global Perspective, 1955–2007*. Washington DC: World Bank and Palgrave Macmillan.

Anderson, K., G. C. Rausser, and J. Swinnen. 2013. "Political Economy of Public Policies: Insights from Distortions to Agricultural and Food Markets." *Journal of Economic Literature* 51: 423–477.

Anderson, K., and J. Swinnen, eds. 2008. *Distortions to Agricultural Incentives in Europe's Transition Economies*. Washington, DC: World Bank.

Babcock, B. A. 2015. "Extensive and Intensive Agricultural Supply Response." *Annual Review of Resource Economics* 7: 333–348.

Burkitbayeva, S., and J. Swinnen. 2018. "Smallholder Agriculture in Transition Economies." *Journal of Agrarian Change* 18 (4): 882–892.

Christensen, G., and R. Pomfret. 2008. "The Kyrgyz Republic." In *Distortions to Agricultural Incentives in Europe's Transition Economies*, edited by K. Anderson and J. Swinnen. Washington DC: World Bank.

Dollive, K. 2008. *The Impact of Export Restraints on Rising Grain Prices*. US International Trade Commission, Office of Economics Working Paper 2008-09-A. Washington, DC.

Dronin, N., and A. Kirilenko. 2011. "Climate Change, Food Stress, and Security in Russia." *Regional Environmental Change* 11: 167–178.

Dries, L., E. Germenji, N. Noev, and J. Swinnen. 2009. "Farmers, Vertical Coordination, and the Restructuring of Dairy Supply Chains in Central and Eastern Europe." *World Development* 37 (11): 1742–1758.

Dries, L., and J. Swinnen. 2002. "Institutional Reform and Labor Reallocation during Transition: Theory Evidence from Polish Agriculture." *World Development* 3 (3): 457–474.

———. 2004. "Foreign Direct Investment, Vertical Integration and Local Suppliers: Evidence from the Polish Dairy Sector." *World Development* 32 (9): 1525–1544.

———. 2010. "The Impact of Interfirm Relationships on Investment: Evidence from the Polish Dairy Sector." *Food Policy* 35 (2): 121–129.

EBRD (European Bank for Reconstruction and Development) and FAO (Food and Agriculture Organization of the United Nations). 2008. *Fighting Food Inflation through Sustainable Investment*. Agricultural Outlook. Investment Centre/EBRD Cooperation Programme. Report Series March 10. London.

EU Parliament. 2014. *Family Farming in Europe, Challenges and Prospects: In-Depth Analysis*. European Union.

Eurostat. 2018. Database. Accessed 2018. https://ec.europa.eu/eurostat/data/database.

FAO (Food and Agriculture Organization of the United Nations). 2009. *The State of Agricultural Commodity Markets: High Food Prices and the Food Crisis—Experiences and Lessons Learned*. Rome.

———. 2015a. FAOSTAT database. Accessed 2015. http://faostat.fao.org.

———. 2015b. *Regional Overview of Food Insecurity: Europe and Central Asia*. Budapest.

———. 2017a. *Europe and Central Asia: Regional Overview of Food Insecurity*. Budapest.

———. 2017b. *Food Security Indicators*. Rome. www.fao.org/economic/ess/ess-fs/ess-fadata/en/#.We-gWWVCL-Y.

———. 2019. FAOSTAT database. Accessed 2019. http://faostat.fao.org.

Fehér, I., J. Lehota, Z. Lakner, Z. Kende, C. Bálint, S. Vinogradov, and A. Fieldsend. 2017. "Kazakhstan's Wheat Production Potential." In *The Eurasian Wheat Belt and Food Security: Global and Regional Aspects*, edited by S. Gomez y Paloma, S. Mary, S. Langrell, and P. Ciaian, 177–194. Switzerland: Springer.

Gapminder. 2019. Data. Accessed 2019. https://www.gapminder.org/data/.

Gataulina, E. A., V. Y. Uzun, A. V. Petrikov, and R. G. Yanbykh. 2005. "Vertical Integration in an Agroindustrial Complex: Agrofirms and Agroholdings in Russia." In *The Dynamics of Vertical Coordination in Agrifood Chains in Eastern Europe and Central Asia, Case Studies*, edited by J. F. M. Swinnen, 45–71. Washington, DC: World Bank.

Glauben, T., M. Belyaeva, I. Bobojonov et al. 2014. *Eastern Breadbasket Obstructs Its Market and Growth Opportunities*. IAMO Policy Brief 16. Halle, Germany: Leibniz Institute of Agricultural Development in Transition Economies.

Glowny Urzad Statystyczny. 2016. Online Statistical Database of the Central Statistical Office of the Republic of Poland. Accessed 2017. http://stat.gov.pl/en/.

Gow, H. R., D. H. Streeter, and J. Swinnen. 2000. "How Private Contract Enforcement Mechanisms Can Succeed Where Public Institutions Fail: The Case of Juhockur a.s." *Agricultural Economics* 23: 253–265.

Gow, H., and J. Swinnen. 1998. "Agribusiness Restructuring, Foreign Direct Investment, and Hold-Up Problems in Agricultural Transition." *European Review of Agricultural Economics* 25 (4): 331–350.

Hayami, Y., and V. W. Ruttan. 1971. *Agricultural Development: An International Perspective*. Baltimore: Johns Hopkins University Press.

IAMO (Leibniz Institute of Agricultural Development in Transition Economies). 2018. AGRICISTRADE. Accessed January 12, 2018. https://www.iamo.de/en/research/projects/details/agricistrade/.

ILO (International Labour Organization). 2014. Global Employment Trends. www.ilo.org/global/research/global-reports/global-employment-trends/2014/WCMS_234879/lang--en/index.htm.

Ioffe, G., T. Nefedova, and I. Zaslavsky. 2004. "From Spatial Continuity to Fragmentation: The Case of Russian Farming." *Annals of the Association of American Geographers* 94: 913–943.

Johnson, D. G., and K. Brooks, eds. 1983. *Prospects for Soviet Agriculture in the 1980s*. Bloomington, IN: Indiana University Press.

Jones, D., and A. Kwiecinski. 2010. *Policy Responses in Emerging Economies to International Agricultural Commodity Price Surges*. OECD and Food, Agriculture and Fisheries Working Paper 34. Paris: Organisation for Economic Co-operation and Development.

Kamp, J., R. Urazaliev, P. F. Donald, and N. Hölzel. 2011. "Post-Soviet Agricultural Change Predicts Future Declines after Recent Recovery in Eurasian Steppe Bird Populations." *Biological Conservation* 144: 2607–2614.

Kazstat. 2016. *Statistical Yearbook*. Almaty, Kazakhstan: Agency of Statistics of Kazakhstan.

———. 2019. Statistics of Agriculture, Forestry, Hunting and Fisheries. Main Indicators. Accessed 2019. https://stat.gov.kz/official/industry/14/statistic/7.

Kingwell, R., C. Carter, M. P. Elliott, and P. White. 2016. *Russia's Wheat Industry: Implications for Australia*. Perth: Australian Export Grains Innovation Centre.

Kornai, J. 2000. "Ten Years After 'The Road to a Free Economy': The Author's Self Evaluation." Working paper. Cambridge, MA, US: Harvard University.

Kraemer, R., A. Prishchepov, D. Müller et al. 2015. "Long-Term Agricultural Land-Cover Change and Potential for Cropland Expansion in the Former Virgin Lands Area of Kazakhstan." *Environmental Research Letters* 10: 054012.

Kurganova, I., V. Lopes de Gerenyu, and Y. Kuzyakov. 2015. "Large-Scale Carbon Sequestration in Post-Agrogenic Ecosystems in Russia and Kazakhstan." *Catena* 133: 461–466.

Lambin, E. F., H. Gibbs, L. Ferreira et al. 2013. "Estimating the World's Potentially Available Cropland Using a Bottom-Up Approach." *Global Environmental Change* 23: 892–901.

Lambin, E. F., and P. Meyfroidt. 2011. "Global Land Use Change, Economic Globalization, and the Looming Land Scarcity." *Proceedings of the National Academy of Sciences* 108: 3465–3472.

Larsson, S., and C. Nilsson. 2005. "A Remote Sensing Methodology to Assess the Costs of Preparing Abandoned Farmland for Energy Crop Cultivation in Northern Sweden." *Biomass and Bioenergy* 28 (1): 1–6.

Lerman, Z. 2010. "Agricultural Recovery and Individual Land Tenure: Evidence from Central Asia." In *Changing Landscape of European Agriculture. Essays in Honour of Professor Csaba Csaki*, edited by I. Ferto, C. Forgacs, and A. Jambor, 95–113. Budapest: Agroinform.

Liefert, W. M. 2002. "Comparative (Dis?)Advantage in Russian Agriculture." *American Journal of Agricultural Economics* 84 (3): 762–767.

Liefert, W. M., and O. Liefert. 2012. "Russian Agriculture during Transition: Performance, Global Impact, and Outlook." *Applied Economic Perspectives and Policy* 34 (1): 37–75.

———. 2015. "Russia's Potential to Increase Grain Production by Expanding Area." *Eurasian Geography and Economics* 56 (5): 505–523.

Liefert, W. M., O. Liefert, R. Seeley, and T. Lee. 2019. "The Effect of Russia's Economic Crisis and Import Ban on Its Agricultural and Food Economy." *Journal of Eurasian Studies* 10 (2): 119–135.

Liefert, W. M., and J. Swinnen. 2002. *Changes in Agricultural Markets in Transition Economies*. Agricultural Economics Report AER806. Economic Research Service. Washington, DC: US Department of Agriculture.

Meyfroidt, P., F. Schierhorn, A. V. Prishchepov, D. Müller, and T. Kuemmerle. 2016. "Drivers, Constraints and Trade-Offs Associated with Recultivating Abandoned Cropland in Russia, Ukraine and Kazakhstan." *Global Environmental Change* 37: 1–5.

Müller, D., A. Jungadreas, F. Koch, and F. Schierhorn. 2016. *Impact of Climate Change on Wheat Production in Ukraine*. Agricultural Policy Report. Kiev: German-Ukrainian Agricultural Policy Dialogue.

Nefedova, T. G. 2016. "Russian Agricultural Resources and the Geography of Their Use in Import-Substitution Conditions." *Regional Research Russia* 6: 292–303.

Noev, N., L. Dries, and J. Swinnen. 2009. "Institutional Change, Contracts, and Quality in Transition Agriculture: Evidence from the Bulgarian Dairy Sector." *Eastern European Economics* 47 (4): 62–85.

OECD (Organisation for Economic Co-operation and Development). 1999. *Agricultural Policies in Emerging and Transition Economies*. Paris.

Pomfret, R. 2008a. "Kazakhstan." In *Distortions to Agricultural Incentives in Europe's Transition Economies*, edited by K. Anderson and J. Swinnen. Washington, DC: World Bank.

———. 2008b. "Tajikistan, Turkmenistan and Uzbekistan." In *Distortions to Agricultural Incentives in Europe's Transition Economies*, edited by K. Anderson and J. Swinnen. Washington, DC: World Bank.

Potapov, P. V., S. A. Turubanova, A. Tyukavina, A. M. Krylov, J. L. McCarty, V. C. Radeloff, and M. C. Hansen. 2015. "Eastern Europe's Forest Cover Dynamics from 1985 to 2012 Quantified from the Full Landsat Archive." *Remote Sensing of Environment* 159: 28–43.

Rosstat. 2016. Statistics. Moscow: Federal Service of State Statistics of the Russian Federation. Accessed 2016. https://gks.ru/statistic.

———. 2019. Official statistics. Agriculture, Hunting and Forestry. Accessed 2019. https://gks.ru/enterprise_economy.

Schierhorn, F., D. Müller, T. Beringer, A. V. Prishchepov, T. Kuemmerle, and A. Balmann. 2013. "Post-Soviet Cropland Abandonment and Carbon Sequestration in European Russia, Ukraine, and Belarus." *Global Biogeochemical Cycles* 27: 1175–1185.

Schierhorn, F., D. Müller, A. V. Prishchepov, M. Faramarzi, and A. Balmann. 2014. "The Potential of Russia to Increase Its Wheat Production through Cropland Expansion and Intensification." *Global Food Security* 3: 133–141.

Sedik, D. 2011. "Food Security and Trade in the ECA Region." Paper presented at the Course on Food and Agricultural Trade for ECA, Vienna, February 7–10.

———. 2013. "The New Wheat Exporters of Eurasia and Volatility." In *The Eurasian Wheat Belt and Food Security: Global and Regional Aspects*, edited by S. Gomez y Paloma, S. Mary, S. Langrell, and P. Ciaian, 119–138. Switzerland: Springer.

Serova, E. 2007. "Agro-Holdings: Vertical Integration in Agro-Food Supply Chains in Russia." In *Global Supply Chains, Standards and the Poor: How the Globalization of Food Systems and Standards Affects Rural Development and Poverty*, edited by J. Swinnen, 188–206. Wallingford, UK: CABI.

Swinnen, J. 2009. "Reforms, Globalization, and Endogenous Agricultural Structures." *Agricultural Economics* 40 (6): 719–732.

Swinnen, J., S. Burkitbayeva, F. Schierhorn, A. Prishchepov, and D. Muller. 2017. "Production Potential in the 'Bread Basket' of Eastern Europe and Central Asia." *Global Food Security* 14: 38–53.

Swinnen J., L. Dries, and K. Macours. 2005. "Transition and Agricultural Labor." *Agricultural Economics* 32 (1): 15–34.

Swinnen, J., and S. Rozelle. 2006. *From Marx and Mao to the Market. The Economics and Politics of Agricultural Transition*. Oxford: Oxford University Press.

Ukrstat. 2016. *Statistical Yearbook*. State Committee of Statistics. Kiev, Ukraine.

———. 2019. Statistical observations. Economic statistics. Agriculture, forestry and fishery. Accessed 2019. http://www.ukrstat.gov.ua.

USDA (US Department of Agriculture), Foreign Agricultural Service. 2020. FAS Production, Supply, and Distribution Online (USDA PS&D). https://apps.fas.usda.gov/psdonline/app/index.html#/app/home.

Uzun, V., V. Saraikin, E. Gataulina, N. Shagayda, R. Yanbykh, S. Mary, and S. Gomez y Paloma. 2014. *Prospects of the Farming Sector and Rural Development in View of Food Security: The Case of the Russian Federation*. Scientific and Technical Research Reports. European Commission.

Van Herck, K., and J. Swinnen. 2015. "Small Farmers, Standards, Value Chains, and Structural Change: Panel Evidence from Bulgaria." *British Food Journal* 117 (10): 2435–2464.

von Cramon, S., and M. Raiser. 2006. *The Quotas on Grain Exports in Ukraine: Ineffective, Inefficient, and Non-Transparent*. Kiev, Ukraine: Institute for Economic Research and Policy Consulting in Ukraine, German Advisory Group on Economic Reform.

Vranken, L., K. Macours, N. Noev, and J. Swinnen. 2011. "Property Rights Imperfections and Asset Allocation: Co-ownership in Bulgaria." *Journal of Comparative Economics* 39: 159–175.

Wegerich, K. 2008. "Blueprints for Water User Associations' Accountability versus Local Reality: Evidence from South Kazakhstan." *Water International* 33 (1): 43–54.

World Bank. 2005. *The Dynamics of Vertical Coordination in Agrifood Chains in Eastern Europe and Central Asia*. Washington, DC.

———. 2011. *Rising Food and Energy Prices in Europe and Central Asia*. Washington, DC.

———. 2019. World Development Indicators database. Accessed February 12, 2019. https://data.worldbank.org/data-catalog/world-development-indicators.

World Health Organization (WHO). 2014. Global Health Observatory (GHO) data. www.who.int/gho/ncd/risk_factors/overweight/en/.

Zinzani, A. 2015. "Irrigation Management Transfer and WUAs' Dynamics: Evidence from the South-Kazakhstan Province." *Environmental Earth Sciences* 73: 765–777.

Chapter 8

REGIONAL EXPERIENCES: WHAT HAVE WE LEARNED?

Keijiro Otsuka and Shenggen Fan

The five chapters on regional issues in agricultural development provide an overview of the various regional experiences and the transformation of agri-food systems. In these chapters, as well as in Chapter 1, the validity of the Hayami-Ruttan induced innovation hypothesis (Hayami and Ruttan 1985) was graphically examined, particularly by looking at growth paths of land and labor productivities in various countries and across major regions. Also examined were emerging trends of agricultural transformation in these regions. This chapter summarizes regional experiences in productivity growth and agricultural transformation.

Regional Experiences in Productivity Growth

As can be seen from Figure 1.1 in Chapter 1, there are enormous differences and changes in labor productivity among the seven major regions over the last several decades. Such large differences in labor productivity can be partly attributed to the difference in factor endowment represented by the land-labor ratio. Roughly speaking, there is a positive correlation between labor productivity and the land-labor ratio, with the possible exception of Latin America and the former Soviet Union countries. If land is abundant relative to labor, farm size is large and mechanized operation is economical, whereas if land is scarce, farm size is small and labor-intensive operation is advantageous. The close relationship between the land-labor ratio and labor productivity is clearly reported in Chapters 3 to 7.

The five regional chapters also show the close relationship between fertilizer use per unit of land and land productivity, which indicates the importance of fertilizer-using technology to boost land productivity. Land productivity has been growing rapidly and is second highest in South America, even though it is endowed with rich land resources. Land productivity in East Asia, where land is particularly scarce, was growing fast in the 1970s and

1980s due, in part, to the yield-enhancing Green Revolution technologies, but its level was not highest in this early period.

Figure 1.2 in Chapter 1 shows the changing relationship between labor and land productivities, with the logarithm of labor productivity on the horizontal axis and that of land productivity on the vertical axis. It is clear that East Asia took the high-land-productivity development path. In Asia where land is scarce, land-saving and yield-enhancing technologies tend to be developed. Subsequently, labor cost increases due to rapid development of the nonfarm sectors, which leads to the development of labor-saving and labor-productivity-enhancing technology more than that of land-productivity-enhancing technology. An interesting finding is that not only South Asia but also Africa south of the Sahara (SSA) seem to have been following the East Asian path. Indeed, it seems that South Asia is attempting to catch up with East Asia (see Chapter 4) and that SSA follows the South Asian path so long as smallholder-based agriculture is maintained (Chapter 5).

A close examination reveals that Eastern Europe improved primarily its labor productivity, whereas both South America and Mexico and Central America improved both labor and land productivity simultaneously. Although Latin American countries are characterized by land abundance, they have successfully introduced land-saving and land-productivity-enhancing technologies while maintaining the impetus to improve labor productivity growth. This seems inconsistent with the Hayami-Ruttan hypothesis. Endowed with extremely rich land resources, the former Soviet Union countries follow a unique path of low land productivity and high labor productivity.

The above analysis points out that the induced innovation hypothesis is broadly consistent with the recent data in many regions, even though there are also deviations from the expected patterns of land and labor productivity growth, particularly in Latin America and the former Soviet Union countries. Regional experiences also point to limitations of Hayami and Ruttan's narrow focus on grain productivity and growth. A major departure from the Hayami-Ruttan thesis can be found in the modern food system transformation at both regional and global levels, where the demands on agriculture now go beyond food to encompass health, nutrition, and environmental sustainability. For instance, the Hayami and Ruttan (1985) measures of productivity are at the farm level. But the food system transformation involves postfarm value addition. In addition, the essence of the modern value chains lies in the demand-driven activities throughout the whole chains, which in turn lead to innovations in technologies and institutions. Technologies used or adopted are increasingly coming from nonagriculture or the private food sector and are not induced by land-labor ratio changes: for example, information and communication technologies (ICTs). Finally, environmental externalities and climate change were not considered in the Hayami and Ruttan model. Clearly we must have broader perspectives than Hayami and Ruttan in order to understand the rapid transformation of agriculture over the last few decades and its implication for global issues.

Transformation of Agriculture

In East Asia, industrialization has led to the large-scale transformation of the agricultural sector on the production side through a reduction in its contribution to gross domestic product and employment, and it also has created a serious labor shortage and thus a rapid increase in real wages. At the same time, the rising demand for safe, high-value, and differentiated agricultural products driven by the nutrition transition has created large opportunities for farmers to participate in emerging value chains to potentially improve productivity and increase incomes. As farming systems gradually change from labor-intensive to capital-intensive systems, technological and institutional innovations have occurred, such as in farm machinery and machine service providers and emerging e-commerce. There has been a notable modernization of value chains and emergence of contract farming on the supply side, which respond to diverse and new food demands especially driven by rapid urbanization and the rise of ICTs.

South Asia has been shifting toward the beginning of agricultural transformation, with greater increases in labor productivity than in land productivity, as experienced by many developed countries in the past. There has been significant mechanization growth in the past three decades, coupled with technological innovations, ranging from the adoption of machines and chemicals to improved varieties, increasingly led by the private sector in the region. Institutional innovations have raised the efficiency of agricultural finance, insurance provision on a broader scale, and collective action in milk production, improved food safety practices, and management of common pool resources, including irrigation facilities and community forests. Liberalization and regional integration have also facilitated the trade of agricultural commodities within the region and improved national resilience against food crises.

Africa, after decades of stagnation, has enjoyed sustained agricultural growth since 2000, while poverty rates have also fallen significantly. Africa's workforce is shifting, in some cases quite rapidly, with a rising fraction of labor moving from farming to off-farm sectors, even if the rural population continues to grow and average farm sizes continue to fall across most of the continent. Africa's delayed demographic transition ensures a continuously rising absolute number of rural people despite urbanization, leading to falling land-labor ratios. Thus, there is need for more labor-using, land-saving innovation in most of Africa, even though the rest of the world proceeds to mechanize (Chapter 5; Masters et al. 2018). Agribusiness and downstream food systems are responding dynamically to population growth and urbanization. Changing food diets associated with income growth and urbanization are creating new employment opportunities in agrifood systems. Urbanization and development of food systems to feed growing cities are also reshaping African farmers' access to markets, as secondary cities are emerging and extending the reach of value chains into areas formerly considered remote. There has been greater vibrancy of agricultural land markets as well as a rise of commercialized African investor farmers, especially in products with scale economies, such as poultry and horticulture. At the same time, African governments have brought agriculture back to the top of their development agendas, while the private sector is also increasingly investing in agriculture.

The agriculture sector in Latin America and the Caribbean has had relatively strong performance during the last several decades. Policy changes related to macroeconomic stabilization and liberalization have led to a significant diversification of agricultural production. Gains in the region's agriculture have been driven by productivity improvement as well as significant expansion of agricultural area over the last half century. These changes have been accompanied by an increase in nonagricultural rural activities, with a larger presence of women's employment and growth in urban residence, helped by better rural infrastructure. Food value chains have also evolved significantly, with increased urbanization and a larger presence of intermediate cities, and with agricultural and agro-industrial activities increasingly controlled by large agricultural operators, input companies, agro-industrial processors, and supermarket chains. Waves of changes in the processing and retail segments, such as the expansion of supermarkets, have occurred with a surge of foreign direct investments (FDI).

Eastern Europe and Central Asia (ECA) has shown several patterns of agricultural growth, including strong labor productivity growth from large farms as well as land productivity growth in poorer countries dominated by small farms. Agricultural productivity growth in the region has been strongly influenced by value-chain investments, up- and downstream from the farms. Private investments in agribusiness, food processing, and retailing have also led to important restructuring in production and trading and marketing systems and in the integration of farms in value chains. Contract farming and vertical coordination in these supply chains have played important roles in the process of restructuring, together with a variety of institutional innovations such as provision of credit, inputs, transportation, and quality control. As in some other regions, investments in the value chains often came from FDI, which has reduced capital constraints, making access to inputs and technology easier and stimulating productivity growth.

Issues Addressed in Subsequent Chapters

It is clear that significant and rapid transformation of agriculture and food value chains, as well as nutrition transition, have been taking place in various regions associated with urbanization and income growth. Also observed are rapid expansion of agricultural trade and reduction in trade distortions associated with trade liberalization. These issues will be discussed in Chapters 9 to 14. Other major changes include women's status, development of credit and insurance institutions, and natural resource management including land, forest, water, and climate change. This book will address these issues in Chapters 15 to 20, before presenting the analysis of the future of agricultural research in Chapter 21 and the reshaping of agrifood systems to achieve multiple development goals in Chapter 22.

References

Hayami, Y., and V. W. Ruttan. 1985. *Agricultural Development: An International Perspective.* Baltimore: Johns Hopkins University Press.

Masters, W. A., N. Z. Rosenblum, and R. G. Alemu. 2018. "Agricultural Transformation, Nutrition Transition and Food Policy in Africa: Preston Curves Reveal New Stylised Facts." *Journal of Development Studies* 54 (5): 788–802.

PART III

Context for Agricultural Development

Chapter 9

AGRICULTURAL GROWTH, URBANIZATION, AND POVERTY REDUCTION

Paul Dorosh and James Thurlow

There are opposing views on the importance of agriculture and urban development in reducing national and global poverty. At the extreme, these perspectives can be separated into two schools of thought. The first is the "pro-agriculture" school, whose thinking is underpinned by long-standing theory and a reading of certain global trends. For instance, despite rapid urbanization, rural populations continue to grow in most late-transforming economies (Thurlow, Dorosh, and Davies 2018), placing tremendous pressure on agricultural land and rural economies to create sufficient jobs and income opportunities (Chamberlin, Headey, and Jayne 2014; Thurlow 2015). Fast population growth and slow economic growth largely explain why global poverty is concentrating in Africa's rural areas (Thurlow, Dorosh, and Davies 2019). Traditional development models afford agriculture an important role in stimulating national economic growth—via backward and forward linkages between sectors (Johnston and Mellor 1961). Linkages to the *rural* nonfarm economy are often even stronger, because farm incomes are more often used to purchase locally produced goods and services (see Haggblade, Hazell, and Dorosh 2007). Agriculture is therefore expected to continue playing a major role in promoting economywide growth in today's late-transforming economies. Finally, agriculture-led growth is said to have stronger linkages to poverty reduction than nonagricultural growth (Christiaensen, Demery, and Kuhl 2011; Diao, Hazell, and Thurlow 2010; Dorosh and Thurlow 2016). The evidence supporting this is varied, ranging from macroeconometric and structural modeling of intersectoral growth linkages, to microeconomic and value-chain analyses of growth opportunities (see Dercon and Gollin 2014). Given these trends and evidence, the pro-agriculture school usually recommends that governments in developing countries should maintain (or increase) investments and policy support for agriculture.

The second school of thought offers a "pro-urban" perspective, again by emphasizing certain global trends. The developing world is urbanizing rapidly and economic growth is fastest in urban centers. Although poverty is lower in urban areas, faster urban growth can create market opportunities for farmers and provides an indirect means of reducing rural poverty (Dorosh and Thurlow 2014). Urbanization also coincides with structural transformation, in which workers leave agriculture for more productive, and likely more remunerative, sectors. History shows that this process is strongly associated with long-term development (McMillan, Rodrik, and Verduzco-Gallo 2014). As agriculture's share of the economy falls, so too does its absolute contribution to national growth and poverty reduction. Simply put, if agriculture is not growing, then it does not matter how strong its growth-poverty linkages are. Fast-growing urban centers, on the other hand, could lift more people out of poverty, even though they have weaker linkages to poverty reduction than agriculture (see Dorosh and Thurlow 2016; Thurlow, Dorosh, and Davies 2019). Finally, at least some urban economic growth may be due to positive "agglomeration effects." These arise when an expansion of economic activity is concentrated within certain locations, giving rise to positive externalities between firms as product and labor markets deepen (see Fujita, Krugman, and Venables 1999). Alongside the above claims about urban-centered industrialization, the pro-urban school also questions the evidence supporting agriculture-led development, arguing, for example, that the ability to import food makes raising agricultural productivity less critical for sustaining economic development (see Dercon and Gollin 2014). The conclusion often reached by the pro-urban school is that current development strategies overemphasize agriculture and exhibit a "rural bias" (Collier and Dercon 2014) and that future investments should rather be directed toward cities as engines of growth and poverty reduction.

The two perspectives are perhaps best reflected in the World Bank's back-to-back publication of the 2008 and the 2009 *World Development Report*, which in turn advocated for agricultural and urban-oriented development strategies (World Bank 2008, 2009). The reports encapsulated an ongoing debate about the role of smallholder farmers in agricultural transformation (Dercon 2009; Hazell et al. 2010; Collier and Dercon 2014) and, more broadly, the role of agriculture in national development (Diao, Hazell, and Thurlow 2010; Dercon and Gollin 2014; Thurlow, Dorosh, and Davies 2019). There is also debate over the role of small towns vis-à-vis big cities in driving development and reducing poverty (Christiaensen and Todo 2014; Dorosh and Thurlow 2014; Ingelaere et al. 2018). Finally, agriculturalists have increasingly emphasized value-chain development as a means of creating jobs and incomes for the rural poor. Value chains and food system approaches are essentially a more business- or consumer-oriented reformulation of the production and demand linkages that underpin traditional development models (see Chapter 12).

Emerging from these debates is a more nuanced view—one that drops sharp dualistic distinctions between rural and urban areas and agricultural and nonagricultural sectors. Urban areas are viewed as a hierarchy in which major cities have different economic structures and rural linkages than secondary cities and towns. Investing in smaller urban centers rather than large cities could stimulate faster growth in the rural economy for reasons that are familiar to the proponents of rural farm-nonfarm linkages. Moreover, many downstream activities forming part of agricultural value chains are located within or close to towns. Investing in smaller towns and peri-urban areas is often equivalent to investing in agricultural value-chain development. Conversely, industrial activity can cluster in rural areas, as in China and Viet Nam (see Chapter 11), possibly leading to *rural* agglomeration effects. There is also recognition that economies do not neatly separate into the "modern" and "backward" sectors of traditional dualistic models. Agriculture comprises both traditional subsistence farming and higher-value commercial agriculture. Nonagriculture also consists of a wide range of activities, including labor-intensive informal services (for example, street vending), capital-intensive industries (for example, mining), and skill-intensive services (for example, banking). Informal food traders in urban areas may generate stronger backward linkages to the rural poor than even export-oriented commercial agriculture. A more nuanced perspective argues that economies do not fit standard dualistic structures. It emphasizes the overlaps and synergies between developing small towns and agricultural value chains and recommends greater alignment between agricultural and urban-oriented investments and policies. The Food and Agriculture Organization's 2017 *State of Food and Agriculture* report is a recent example of this more nuanced view (FAO 2017).

This chapter argues in favor of the more nuanced perspective. The second section reviews global development and demographic trends and discusses what is meant by "urbanization" in different parts of the developing world. The third section presents a conceptual framework for understanding rural-urban linkages across cities and towns. The fourth section presents a Malawian case study that operationalizes the conceptual framework. The study uses an economywide model to capture various farm-nonfarm linkages, as well as urban agglomeration economies and rural-urban migration. The model is used to simulate the effects of redirecting investments between rural areas, small towns, and major cities. The analysis reveals synergies (and trade-offs) between urbanization, agriculture, and poverty reduction. The last section summarizes the chapter and identifies where future research is needed.

Understanding Urbanization

Speed of Urbanization

It is generally held that the developing world is urbanizing rapidly and that this process is particularly fast in Africa south of the Sahara (SSA). However, there are two ways of gauging the speed of urbanization (Figure 9.1). From the point of view of urban areas or urban planners, the most relevant indicator is urban population growth (vertical axis). Municipal and local governments need to provide infrastructure and services to urban residents, and the scale of this challenge is determined by the number of people living in urban areas. Almost all developing countries have seen urban populations expand over the last decade. From this perspective, SSA is the fastest urbanizing region, with an average urban population growth rate of 4.2 percent per year. Urban populations are even expanding in Latin America and the Caribbean (LAC), even though this region has the slowest growth rate (1.6 percent). Urban population growth is fastest in low-income countries (LIC), most of which are in SSA, and is slowest in upper-middle-income countries (UMIC), which includes China.[1] Global projections indicate that the urban population in developing countries will almost double between 2015 and 2050 (from 2.9 to 5.1 billion) (UN DESA 2015) and that a third of this expansion will occur in SSA, whose urban population will triple in size (from 0.3 to 1.1 billion people). It is not surprising then that many studies about global urbanization emphasize Africa's demographic trajectory (Masters et al. 2013).

National planners and policymakers, however, are also concerned with how fast populations are growing outside of urban areas, since scarce public resources are also needed for rural inhabitants. A more relevant urbanization indicator for national planners is therefore the *share* of the total population living in urban areas. This can be derived from Figure 9.1, which also shows rural population growth rates (horizontal axis). Countries and regions above the diagonal line have urban populations growing faster than rural populations, and so the share of their population living in urban areas is rising.

---

1 LIC and UMIC, as well as lower-middle-income countries (LMIC), are World Bank country income categories, as of 2016, based on per capita gross domestic product measured using the Atlas method. Upper thresholds are $1,005 for LIC, $3,955 for LMIC, and $12,235 for UMIC.

FIGURE 9.1 Urban and rural population growth rates, 2005–2015

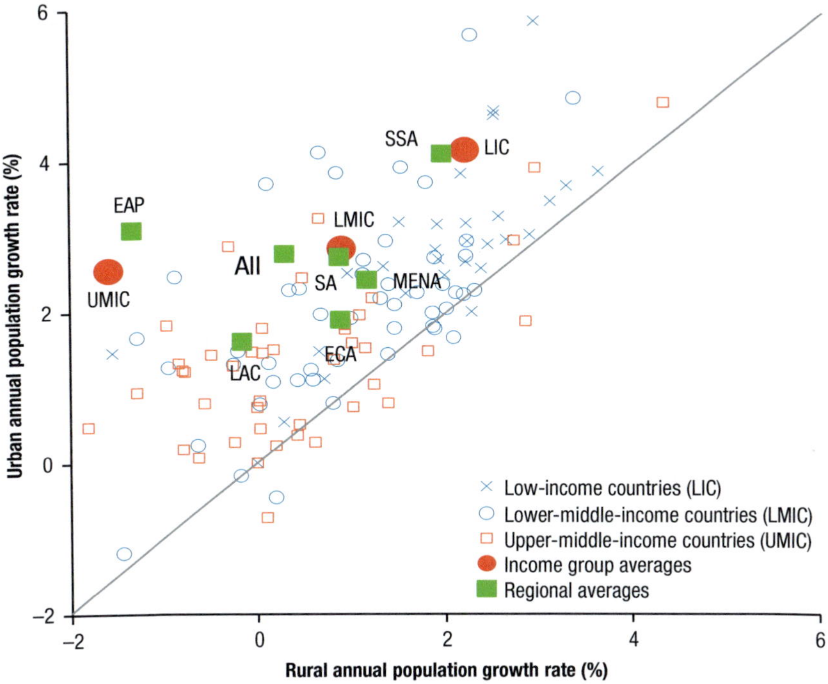

Source: Authors' calculations using population data from World Bank (2018).

Note: Sample includes 137 countries. Regions are East Asia and the Pacific (EAP), Europe and Central Asia (ECA), Latin America and the Caribbean (LAC), Middle East and North Africa (MENA), South Asia (SA), and Africa south of the Sahara (SSA).

Countries and regions that are farther from the diagonal line have larger gaps between urban and rural population growth rates, and hence faster increases in urban population shares. The figure shows that, even from this perspective, urbanization is occurring in almost all developing countries. However, rural populations continue to grow quite quickly in many LICs, causing slower increases in their urban population shares compared with most UMICs. Based on this indicator, SSA is no longer the fastest urbanizing region, which is instead East Asia and the Pacific (EAP), where the rural population is contracting. About half of developing countries' populations live in urban areas today, and projections indicate that this share will increase to about two-thirds by 2050 (UN DESA 2015). Even in SSA, where urban population shares are lowest, more than half of the population will live in urban areas by 2040.

In summary, while the developing world is urbanizing, there are differences in how the speed of urbanization is perceived, with implications for policy. For urban planners, particularly in SSA, absorbing the large influx of migrants into cities and towns is a major policy concern. National planners, on the other hand, are also tasked with raising incomes and living standards of people in rural areas, where much of the developing world still lives. National development plans inevitably involve difficult choices about how to allocate scarce resources between fast-growing urban centers and large rural populations.

Urbanization and Structural Change

Urbanization is closely associated with economic development, particularly with structural transformation. When workers move from rural to urban areas, this often involves an exit out of agriculture into other sectors of employment, leading to higher labor productivity (and incomes). Figure 9.2 provides information on urban population and agricultural employment patterns for a sample of 52 developing countries.[2] These countries were selected because they conducted at least one nationally representative household/labor force survey or population census since 2000, with information on employment by sector and rural/urban location. Normally the most recent survey or census is used, but preference is given to labor force surveys because they are better at capturing employment statistics. Since data are from different years, the income groups are simple cross-country averages (not weighted by population). In total, the countries represent 80 percent of the developing world's total population in 2015, with a slightly lower coverage rate for LICs.

The figure shows that more-developed countries tend to be more urbanized and have fewer of their workers engaged in agriculture. Although the strength of this relationship varies across countries, there is a consistent tradeoff between agricultural work and urban residence. In a typical LIC, around 60 percent of employed workers are in agriculture and 20 percent of the population live in urban areas. This is reversed for UMICs, that is, 20 percent

---

2 Country and survey/census years include LICs: Afghanistan 2008, DR Congo 2005, Ethiopia 2013, Haiti 2003, Liberia 2008, Malawi 2013, Mali 2009, Mozambique 2007, Nepal 2008, Rwanda 2016, Sierra Leone 2004, South Sudan 2008, Tanzania 2014, and Uganda 2012; LMICs: Armenia 2011, Bangladesh 2013, Bolivia 2001, Cambodia 2008, Cameroon 2005, Egypt 2006, El Salvador 2007, Ghana 2015, India 2009, Indonesia 2010, Jordan 2004, Kenya 2005, Kyrgyzstan 2009, Nicaragua 2005, Nigeria 2010, Pakistan 2015, Palestine 2007, Sudan 2008, Viet Nam 2009, and Zambia 2012; UMICs: Belarus 2009, Brazil 2010, China 2000, Colombia 2005, Costa Rica 2011, Dominican Republic 2010, Ecuador 2010, Fiji 2007, Iran 2011, Jamaica 2001, Malaysia 2000, Mexico 2015, Panama 2010, Paraguay 2002, Peru 2007, Romania 2011, Thailand 2000, and Venezuela 2001.

FIGURE 9.2 **Urban population and agricultural employment shares**

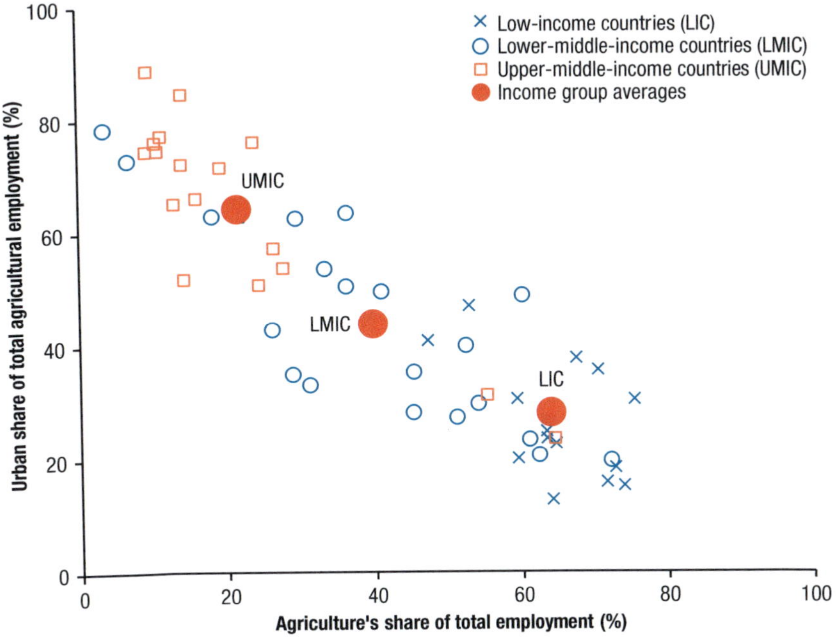

Source: Labor force and household surveys, and population censuses (all post-2000); censuses from IPUMS (2019).
Note: Sample includes 52 countries. The 2015 population coverage is 70% LIC, 84% LMIC, and 81% UMIC. The calculations use official urban definitions and simple unweighted averages for country income groups.

of workers are in agriculture and 60 percent of people are urbanized. Lower-middle-income countries (LMICs) fall in between, with 40 percent of the workforce and population in agriculture as well as in urban areas.[3]

The symmetry in Figure 9.2 can give the impression that urban areas and nonagricultural employment are interchangeable, as are rural areas and agricultural employment. This is not the case. Using the same dataset, Figure 9.3 shows that many agricultural workers live in urban areas (horizontal axis) and many nonagricultural workers live in rural areas (vertical axis). Almost half of all nonagricultural workers in LICs reside in rural areas, reflecting the important role of the rural nonfarm economy in these countries. Employment in the rural nonfarm sector declines with economic development and

3 The figure is based on primary occupation as stated by survey respondents and so does not capture secondary work or actual time spent (see discussion in the chapter's third section, Framing Rural–Urban Linkages).

FIGURE 9.3 Spatial distribution of agricultural and nonagricultural employment

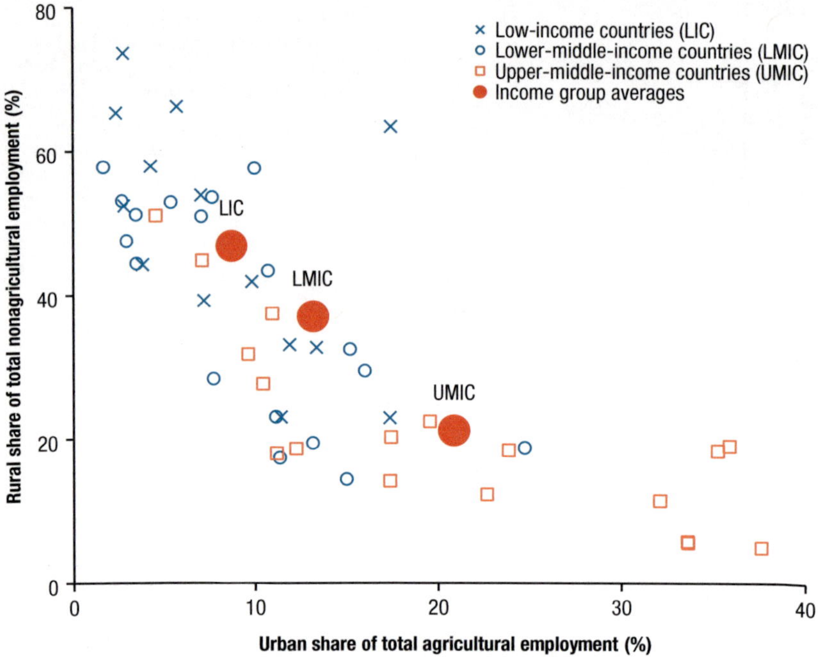

Source: Labor force and household surveys, and population censuses (all post-2000); censuses from IPUMS (2019).
Note: Sample includes 53 countries. The 2015 population coverage is 70% LIC, 84% LMIC, and 81% UMIC. Calculations use official urban definitions and simple unweighted averages for country income groups.

urbanization. More surprising, however, is the greater importance of urban agricultural employment in more-developed countries. About 20 percent of all agricultural jobs in UMICs are in urban areas, compared with just 9 percent in LICs. Urban-based farming seems to become a more important part of the agricultural sector as countries develop.

Lying behind the trends in Figure 9.3 is the declining importance of agriculture as countries develop. The top panel in Figure 9.4 reports the contribution of urban agriculture to national employment (vertical axis) and to total urban employment (horizontal axis). Urban agriculture is a small share of national employment in most countries, although in a few countries it does account for around 10 percent of all jobs. As at the national level, urban agriculture becomes less important as countries develop, falling from 5.5 percent in LICs to 3.5 percent in UMICs. Agricultural employment also becomes less important *within* urban economies. The share of agriculture in urban employment is 21.8 percent in LICS and 5.5 percent in UMICs. Although Figure 9.3

FIGURE 9.4 Sectoral distribution of urban and rural employment

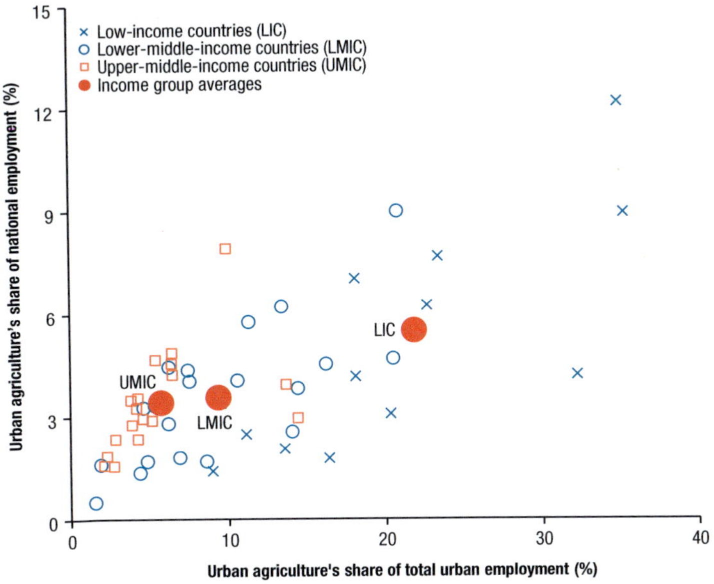

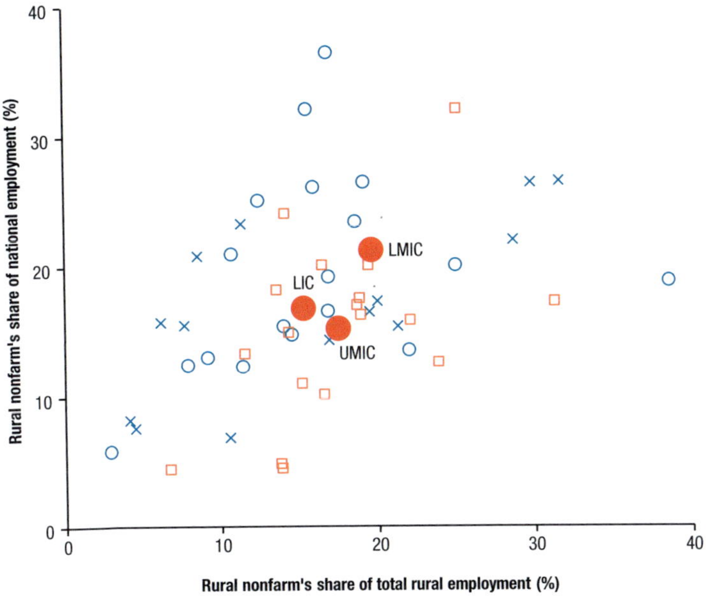

Source: Labor force and household surveys, and population censuses (all post-2000); censuses from IPUMS (2019).

Note: Sample includes 53 countries. The 2015 population coverage is 70% LIC, 84% LMIC, and 81% UMIC. Calculations use official urban definitions and simple unweighted averages for country income groups.

shows the share of agricultural employment in urban areas rising in more developed countries, this is caused by reductions in *rural* farming. Simply put, as countries develop, workers leave rural agriculture faster than they leave urban agriculture.

The lower panel in Figure 9.4 provides similar information for rural nonfarm employment. The link between economic development and the rural nonfarm economy is less clear. The share of rural nonfarm employment in total *rural* employment (horizontal axis) is slightly higher in middle-income countries, but there is no consistent trend between LMICs and UMICs (that is, rural nonfarm employment shares rise and then fall). Rural nonfarm employment appears to remain an important part of the economy, even in more-developed countries.

In summary, while *urban* and *nonagriculture* are not interchangeable terms, cross-country evidence suggests that *urbanization* is strongly associated with a movement out of rural agriculture into urban nonagriculture. In fact, the contributions of urban agriculture and the rural nonfarm economy do not change dramatically across the development spectrum—together they consistently account for around one-fifth of total employment. Of course, employment is not equivalent to labor productivity or value addition, and so the contributions of rural nonfarm and urban agricultural sectors to gross domestic product (GDP) may be quite different in low- and middle-income countries. Unfortunately, the data needed to test this only exists for a few countries (see, for example, Dorosh and Thurlow 2013). Moreover, the cross-country data presented here focus only on primary jobs. The early stages of rural transformation are often associated with workers moving from full- to part-time farming or with nonfarm income diversification within rural households (see Timmer 1988). Finally, as discussed next, the definition of *urban areas* may change over time, with unclear delineations between where rural areas end and urban areas begin.

Defining Urban and Rural Areas

The figures above are all based on countries' own official definitions of what constitutes an urban area. However, these definitions vary considerably. The supplementary material to the *United Nations Demographic Yearbook* (UN DESA 2016a) provides information on how countries define their urban centers. Table 9.1 summarizes the various criteria used in these definitions. Of the 215 countries listed, about half include population size thresholds. The typical or median population threshold is 2,500 people, but this ranges from 200 (Sweden) to 50,000 (Japan). A larger share of LICs include population

TABLE 9.1 Official definitions of urban areas across countries

Indicator	All countries	Income group			
		LIC	LMIC	UMIC	HIC
Number of countries (N)	215	31	52	56	76
Share of countries using these criteria (%)					
Population level	52.1	64.5	63.5	39.3	48.7
Population density	6.5	3.2	7.7	3.6	9.2
Nonfarm employment	16.3	9.7	32.7	17.9	6.6
Infrastructure and services	7.0	3.2	15.4	8.9	1.3
Administrative center	51.2	51.6	48.1	55.4	50.0
More than one criteria	27.0	25.8	48.1	23.2	15.8
Minimum population level thresholds (people)					
Lowest	200	1,500	500	400	200
Median	2,500	2,500	4,000	2,500	2,500
Highest	50,000	10,000	20,000	10,000	50,000

Source: Authors' calculations using urban definitions from UN DESA (2016a).

Note: LIC = low-income country; LMIC = lower-middle-income country; UMIC = upper-middle-income country; HIC = high-income country. Official criteria are usually expressed as minimum population levels and densities; nonfarm employment or activities as a share of total; existence of certain infrastructure or services (for example, roads, piped water, or health facilities); and administrative function.

thresholds in their urban definitions, and the minimum thresholds in LICs tend to be higher (that is, the smallest threshold is 1,500 people in LICs compared with 400 people in UMICs). A small share of countries (6.5 percent) use population density as a criterion—sometimes, but not often, in combination with population size.

The relationship between urbanization and structural change was discussed above. The positive association could result from how urban areas are defined. About one in six countries define urban areas based on how dependent a location is on agricultural activities. Some countries are quite specific, requiring that a minimum share of employment or economic activity be in nonfarm sectors. Middle-income countries are more likely to include nonfarm employment criteria in their definitions, although only a third of LMICs and a fifth of UMICs use such criteria. The unevenness in how this criterion is applied may also explain variations in urban agriculture across countries. It does not necessarily explain why urban agriculture becomes more important for the overall agricultural sector, since the nonfarm criterion is more often applied in middle-income countries.

About half of countries in each income group use a location's administrative function in their *urban* definition. For example, a country may decide that all provincial or district capitals should be classified as urban centers, irrespective of their population size or nonfarm employment share. Finally, a few countries require locations to have certain infrastructure and services, such as paved-road networks, piped-water systems, and health facilities.

A country's choice of criteria can affect the number of locations classified as urban. For example, countries that use administrative function, rather than minimum population size, may include smaller urban centers. Figure 9.5 reports the simple or unweighted average urban population shares across countries within each income group. There is no strong directional relationship between which criteria are used and the average population shares of the countries that use them.

Finally, we consider how using the same definition across countries affects relative urbanization rates. Uchida and Nelson (2010) used high-resolution spatial information on population size and location to apply a standard definition of an "urban area" to all countries. The resulting urban agglomeration index reported in Figure 9.6 requires that a location have a minimum population density of 150 people per square kilometer and be within one-hour travel time to an urban center of either 50,000 or 100,000 people. Travel time is calculated based on the location of roads, including their surface type and slope. The figure compares urban population shares for different definitions of urban areas, including countries' official definitions.

Applying a uniform definition of *urban* affects urban population shares differently across developing regions. In South Asia (SA), for example, the official definition puts the region's urban population share at 27 percent, which is the lowest of all developing regions. However, when the authors' urban agglomeration criteria are applied, the region's urban population share rises to 42 and 51 percent, depending on whether the core urban area needs to contain 100,000 or 500,000 people, respectively. Urbanization rates in the Middle East and North Africa (MENA) also increase, although the change is not as pronounced as in SA. In contrast, LAC, which by official definitions is the world's most urbanized region, sees its urban population fall considerably (from 75 percent to as low as 56 percent). There are smaller and inconsistent changes for the other three regions.

In summary, there are major differences in how countries define *urban areas*. That said, the types of criteria used do not greatly influence urban population shares. Rather, it is the threshold values of these criteria that matter. For instance, SA and LAC may both use population levels as criteria but

FIGURE 9.5 Urban criteria and population shares, 2015

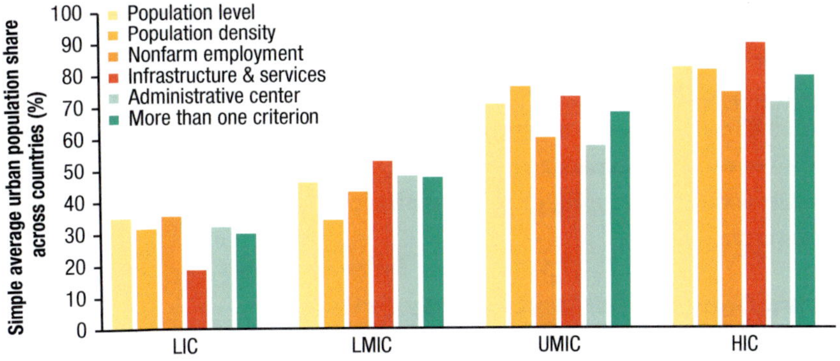

Source: Authors' calculations using urban definitions from UN DESA (2015) and populations from World Bank (2018).

Note: Countries are grouped into high-income countries (HIC), upper-middle-income countries (UMIC), lower-middle-income countries (LMIC), and low-income countries (LIC). Official criteria are usually expressed as minimum population levels and densities; nonfarm employment or activities as a share of total; existence of certain infrastructure or services (for example, roads, piped water, or health facilities); and administrative function.

FIGURE 9.6 Developing country populations within urban agglomerations, 2000

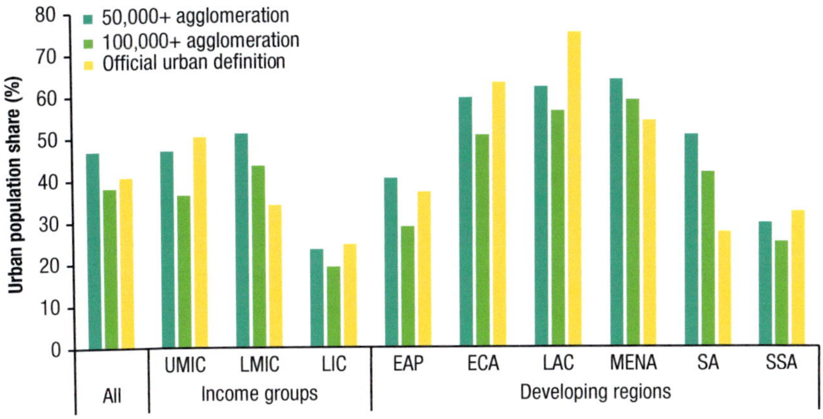

Source: Authors' calculations using data from Uchida and Nelson (2010).

Note: Urban agglomerations vary by population threshold; population density is constant at 150 people or more per square kilometer, with a maximum one-hour travel time to an urban center (given road network surface and slope). Sample includes 139 countries. Income groups are upper-middle-income countries (UMIC), lower-middle-income countries (LMIC), and low-income countries (LIC). Regions are East Asia and the Pacific (EAP), Europe and Central Asia (ECA), Latin America and the Caribbean (LAC), Middle East and North Africa (MENA), South Asia (SA), and Africa south of the Sahara (SSA).

apply very different threshold values. Applying common thresholds substantially reduces the gap in urban population shares between these two extreme regions. This sensitivity to how *urban area* is defined reveals the existence of an urban-rural "continuum," in which locations transition from (1) large densely populated areas (cities and towns) to (2) lower-density areas with substantial access to urban markets (peri-urban areas) to (3) remote rural areas in the hinterlands. We expect economic linkages and household livelihoods to vary across this continuum, implying different roles for agriculture in reducing poverty.

Spatial Consumption and Poverty Patterns

Poverty is typically measured in developing countries using information on household consumption spending, including the imputed value of own-produced goods and services. The Global Consumption Database (World Bank 2017) provides consistent estimates of household spending in 2010 for 77 developing countries. Using this information, Table 9.2 reports differences in consumption patterns across rural and urban areas. As expected, the table shows that per capita consumption is higher in middle-income countries and that consumption is consistently higher in urban areas, irrespective of the level of development. As a result, the urban consumption share is much higher than the urban population share. In LICs, for example, urban areas accounted for 25 percent of the population in 2010, but 36 percent of total consumption spending. Urban areas therefore account for a disproportionate share of total consumer demand.

Urban consumers generate backward demand linkages to rural agriculture. Even in LICs, where most of the population still live in rural areas, urban consumers account for about a third of total food consumption, most of which is produced in rural areas. This is despite urban households allocating a smaller share of their consumption basket to food products (mainly due to their higher incomes). For example, urban households in developing countries account for 68 percent of total spending on all goods and services, but only 59 percent of spending on food and beverages. This gap is present across developing regions and country income groups, as well as between lower- and higher-income households within developing countries (see the final two columns in Table 9.2).

Urban households consume different kinds of food products. This is evident from the low urban consumption share for cereals, fats, and oils but high shares for meat, dairy, and fish. Urban consumers are also more likely to

TABLE 9.2 Differences in consumption patterns between rural and urban areas, 2010

		Country income group			Household group	
Indicator	All countries	LIC	LMIC	UMIC	Lower income	Upper income
Per capita spending ($PPP)	1,645	738	983	2,703	639	3,197
Rural households	916	636	719	1,378	603	2,203
Urban households	2,628	1,044	1,444	3,800	751	3,593
Urban share of population (%)	42.6	25.1	36.4	54.7	23.9	71.5
Urban share of spending (%)	68.0	35.5	53.4	76.9	28.0	80.4
Nonfood goods and services	73.4	43.5	60.5	79.3	30.9	82.5
Food and beverages	59.4	29.6	45.8	72.1	25.6	76.1
Cereals and roots	40.8	21.7	35.9	62.5	23.1	63.6
Fats and oils	52.7	36.7	44.2	66.5	26.5	72.5
Fruits	64.5	36.1	51.7	72.1	24.3	77.0
Vegetables	57.1	26.4	43.6	72.6	26.3	76.4
Meat and eggs	66.0	36.0	53.9	71.3	26.2	77.4
Milk and dairy	58.4	26.0	45.6	77.4	25.7	77.5
Fish and seafood	59.3	44.6	46.8	69.1	26.7	75.2
Sugar and confectionary	49.8	36.3	40.6	64.3	24.4	69.1
Bread and baked goods	72.6	65.3	59.5	78.1	41.6	80.6
Other foods	56.9	26.5	44.6	69.2	24.4	74.7
Beverages	67.1	36.4	58.4	72.8	28.2	78.0
Restaurants and vendors	74.0	40.8	65.4	77.0	28.0	82.7

Source: Authors' calculations using the Global Consumption Database from World Bank (2017).

Note: LIC = low-income countries; LMIC = lower-middle-income countries; UMIC = upper-middle-income countries; PPP = purchasing power parity. Sample is 77 countries; 2015 population coverage is 85% LIC, 91% LMIC, and 80% UMIC. Table uses official urban definitions. Households are grouped (across countries) based on average level of per capita consumption spending (including consumption of own-produced products). Lower-income household group spends less than $2.97 per person per day after adjusting for 2005 purchasing power parity (PPP); upper-income household group spends more than $2.97.

consume processed products, such as bread, rather than less-processed products, such as cereals. Finally, urban households are the most important consumers of meals prepared outside of the home, such as at restaurants and from street vendors. Urbanization and urban income growth not only generate demand for agricultural products, but also alter the composition of the national agriculture-food system (Tschirley et al. 2015), with uncertain implications for poor households and smallholder farmers (Thurlow, Dorosh, and Davies 2019).

TABLE 9.3 Living conditions in rural and urban areas

Indicator	All countries	Country income group		
		LIC	LMIC	UMIC
Poverty headcount ratio (under $1.25) (%)	16.8	40.9	22.8	4.3
Urban areas	9.8	29.6	18.5	1.3
Rural areas	22.8	45.3	25.4	8.8
Urban share of poor population (%)	26.4	20.0	29.7	18.2
Approximate change, 2000s (percentage point)	8.4	0.9	7.8	5.2
Urban population in slums (%)	29.9	64.9	32.2	23.7
Urban population without access to facilities (%)				
Electricity	5.1	42.1	5.4	0.3
Improved sanitation	23.8	59.7	33.1	12.1
Improved water	4.9	13.5	6.6	2.6
Rural population without access to facilities (%)				
Electricity	29.7	83.2	30.4	1.8
Improved sanitation	54.0	76.7	58.8	32.5
Improved water	17.3	44.6	14.9	8.6

Source: Authors' calculations using data from IFAD (2016) and World Bank (2018).

Note: LIC = low-income countries; LMIC = lower-middle-income countries; UMIC = upper-middle-income countries. Sample is 94 countries. Poverty line is $1.25 per day adjusted for 2005 purchasing power parity (PPP). Population-weighted poverty ratios are estimated using surveys closest to 2012 (and to 2000 for change in urban share of poor population).

Finally, Table 9.3 provides a rough assessment of household poverty and living conditions in urban areas across developing countries. The table reports poverty head count rates based on the World Bank's $1.25 per day poverty line (adjusted for purchasing power parity). The rural and urban poverty ratios were computed by the World Bank for IFAD (2016) and provided to us for this study. The poverty ratios are for different years. Table 9.3 uses the most recent poverty estimate for each country—dropping those countries where recent surveys are unavailable.

As mentioned, urban households have higher consumption levels, and so urban poverty is almost always lower than rural poverty. National poverty ratios are also lower in more-developed countries. Urbanization, however, has, in recent years, caused the share of poor people living in urban areas to increase. This is driven either by an inflow of poor migrants from rural areas, or by overcrowding in low-income urban neighborhoods. Even when migrants are better educated and wealthier than the average rural household, they may still be poorer than the average urban household. Unfortunately, evidence on the characteristics of new urban migrants is limited, particularly in SSA. That said, the growth of informal trade services in urban Africa over the last decade suggests that new migrants' education and expertise may lend themselves more to some of the less-productive and lower-paid occupations within urban areas. Inadequate housing and public services also help explain urban poverty in Africa (Lall, Henderson, and Venables 2017).

The "urbanization of poverty" is most pronounced in middle-income countries, where rural populations are growing slowly or even contracting (see above). In LMICs, for example, urban areas are home to about 30 percent of poor people, and this share has risen over the last decade or more. In contrast, LICs have seen a much smaller expansion in urban poverty, despite more rapid urban population growth. This is because the gap between urban and rural poverty is much larger in these countries and, more importantly, because rural populations continue to grow quite rapidly. Long-run projections suggest that faster urban economic growth will outweigh urbanization, leading to an eventual concentration of the world's poor population in rural areas in SSA (Thurlow, Dorosh, and Davies 2019).

Urban households in LICs also lack access to basic services. Two-thirds of urban residents live in slums, and many do not have access to electricity and improved sanitation facilities and water sources. Even "middle-class" Africans, most of whom live in cities and towns, often lack the assets and income security typically associated with middle-class lifestyles in developed countries (for example, properly roofed dwellings, flush toilets, piped water, and formal-sector jobs) (Thurlow, Resnick, and Ubogo 2015). This underscores the huge infrastructure and service gaps that already exist in urban areas, and the pressure on local governments to provide for new migrants. Of course, household poverty and asset gaps are much larger in rural areas.

Given the pace of urbanization and existing living conditions in urban areas, it is perhaps not surprising that more than half of developing country governments (78 of 148) have implemented policies specifically aimed at reducing rural-to-urban migration (UN DESA 2016b). To avoid congestion in major cities, many governments (47 of 148) are also attempting to encourage populations in major cities to move to smaller towns and peri-urban areas. Almost every government has policies aimed at improving income opportunities and living conditions in rural areas, and some governments even aim to narrow the rural-urban divide. Urban and rural policies need to be closely aligned with broader national development strategies. This is especially true in LICs and SSA, where urban and rural populations are expanding fastest and where global poverty is concentrating.

Framing Rural–Urban Linkages

The previous section provided five key facts about urbanization, agriculture, and poverty. First, while urbanization is associated with workers leaving agriculture, the rural economy is not solely composed of farming. The nonfarm sector is a large part of the rural economy (Davis et al. 2010), and urban agriculture is also important for urban economies and the broader food system (Zezza and Tasciotti 2010). Davis et al. (2010), for example, find that off-farm activities account for almost half of all rural incomes in their sample of developing countries. Self-employment in informal trade services is the main rural nonfarm activity and is much larger than rural manufacturing, particularly in SSA. Second, at the national level, the movement of workers out of agriculture into nonagricultural sectors is a major driver of economic growth and structural change in developing countries. Third, many people live in peri-urban areas close to urban centers, especially in SA and LAC. Migrants from more-remote rural areas may choose peri-urban areas over urban centers in order to avoid living in slums. Similarly, rural agriculture and urban industries may choose to locate within peri-urban areas, where firms face lower congestion costs and farms have better access to urban consumers, particularly markets for high-value foods. Fourth, urban unemployment and poverty are major concerns, even though most of the poor live in rural areas. Finally, many governments are actively trying to reduce the speed of urbanization (and an urbanization of poverty) by investing in rural areas.

To be more effective, national development strategies need to have alignment between urban, industrial, and agricultural policies, as well as alignment between public and private sector incentives and objectives. The facts listed above indicate the importance of considering both spatial and sectoral transformation when evaluating policy options. This section provides a conceptual framework that is useful for linking these sectoral and spatial dimensions of economic transformation. The framework emphasizes the rural-urban continuum that extends across cities, towns, and villages and considers how strategies for agricultural transformation may vary depending on where they are implemented in relation to urban centers. The framework also recognizes that not all agricultural activities are low-productivity activities (and not all nonagricultural activities are high in productivity). Instead, the returns to employment in different sectors will vary depending on where the job is located and whether rural or urban markets are being supplied. The framework encourages agricultural economists to look beyond the farm and consider how consumption patterns and market potential vary across space. Similarly, it encourages urban economists to consider the broader role of smaller towns in promoting rural farm and nonfarm development.

The framework is presented in Figures 9.7 and 9.8. As agriculture-food systems transform, there is a gradual reorientation of farmers and their practices from subsistence to commercialization and finally to modernization (Figure 9.7). This involves a shift from production of diverse foods for subsistence consumption—first to selling largely unprocessed agricultural products in local markets, and then to more-intensive formal processing and retailing of food products. Agricultural transformation therefore involves an increasing number of value-chain actors (farmers, aggregators, processors, wholesalers, retailers, and consumers), as is stressed in Chapter 12.

This process closely follows economic transformation, in which workers move from low-productivity sectors, like agriculture, into more-productive nonagricultural sectors (Figure 9.8). Within the food system, this usually involves a shift out of farming into trading and processing. It is now recognized that, while average labor productivity in agriculture is lower than elsewhere in the food system, these averages hide considerable variation within these sectors. In other words, some agricultural workers or activities generate higher value-added per worker than some downstream activities or sectors, such as agro-processing. Economic development therefore also involves a move out of low- to higher-productivity activities within agriculture.

Higher-productivity agricultural activities may be unevenly distributed within a country, and so agricultural transformation may promote certain locations over others. As countries develop, they typically experience a spatial transformation along the rural-urban continuum. Transport networks improve and populations concentrate, creating urban "catchment areas" around major cities, large towns, and to some extent, even small towns. Fewer areas remain remote, and they account for a smaller share of the national population. Only in very remote rural areas does agriculture almost completely dominate the local economy, with the nonagricultural sector almost entirely absent. Most agricultural output in countries is produced in relatively well-connected rural areas, such as the "peri-urban" areas lying within the catchments of cities and towns. These areas also support substantial nonagricultural employment and production, including manufacturing, trade and transport, and various services. And, as discussed above, although agriculture is only a small part of the economies of cities and larger towns, there is still some agricultural employment in urban areas that may account for a major share of national agricultural employment.

FIGURE 9.7 **Agriculture's transformation process**

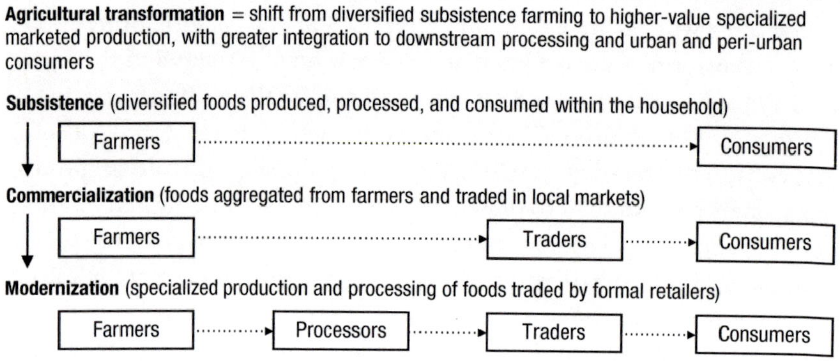

Source: Authors.

FIGURE 9.8 **Linking sectoral and spatial transformations**

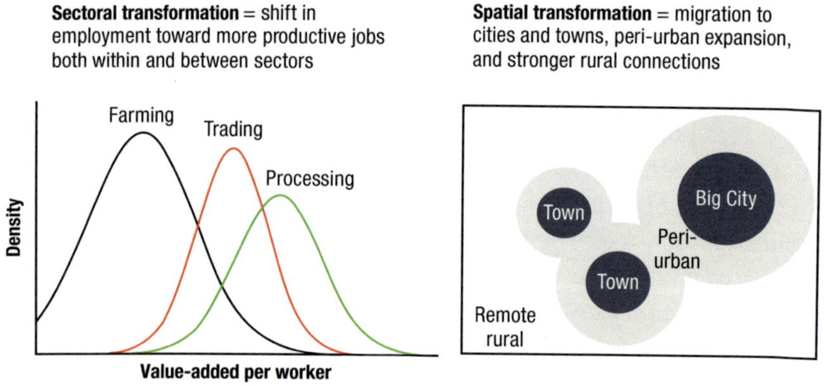

Source: Authors.

Policy priorities will vary across space and by the stage of agricultural transformation. In remote rural areas, where agricultural transformation is still in its earlier stages, raising farm productivity and reducing poverty and hunger among smallholder subsistence farmers is more likely to be the main policy objective. In peri-urban areas and small towns, farmers are generally more market oriented, and there is greater scope to promote nonagricultural activities, making agricultural commercialization and off-farm income diversification more important policy priorities. This may include investing in and

FIGURE 9.9 Developing country populations across the rural and urban hierarchy, 2000

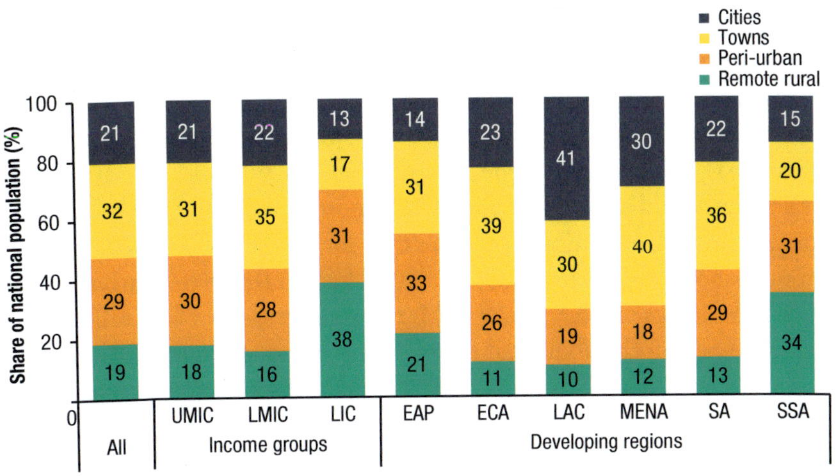

Source: Authors' calculations using data from FAO (2017).

Note: Cities (towns) are locations within an hour travel time of an urban agglomeration of 500,000 (50,000) people; peri-urban (remote rural) areas are one to three hours (more than three hours) from a 50,000-person agglomeration. Sample includes 125 countries. Income groups are upper-middle-income countries (UMIC), lower-middle-income countries (LMIC), and low-income countries (LIC). Regions are East Asia and the Pacific (EAP), Europe and Central Asia (ECA), Latin America and the Caribbean (LAC), Middle East and North Africa (MENA), South Asia (SA), and Africa south of the Sahara (SSA).

supporting food processors and traders. Finally, within the boundaries of cities and large towns, key policy issues are more likely to include food safety, regulation of urban traders, promoting more formal private sector operators, and managing import competition.

Figure 9.8 divided the population into four broad areas: cities, towns, peri-urban areas, and remote rural areas. Figure 9.9 provides some indication of the share of the total population living in these areas, although the data are from two decades ago. The figure draws on analysis conducted for FAO (2017) using the same spatial population data that underpinned the agglomeration indices discussed in the second section (see Uchida and Nelson 2010). Cities are defined as locations within one hour of an urban agglomeration of 500,000 people or more, whereas towns include locations within one hour of urban agglomerations with 50,000–500,000 people. Peri-urban areas are within three hours of cities and towns, and remote rural areas are more than three hours away.

About a fifth of the developing world's population live in cities, and another fifth live in remote rural areas. The remaining population is evenly divided across towns and peri-urban areas. Both UMICs and LMICs share this broad population distribution, partly because most large developing countries fall into one of these two groups. The spatial distribution is very different in LICs, where two-fifths of the population live in remote rural areas and only a third live in either cities or towns. Differences are even more pronounced across developing regions. About 70 percent of the populations in LAC and MENA live within one hour of a city or town, and only 10 percent live in remote rural areas. This pattern is almost reversed in SSA, where 65 percent of the population lives more than an hour away from a city or town. That said, about half of the rural population in SSA live in peri-urban areas within three hours of a city or town. Many urban centers in SSA fall below the 50,000-person threshold and so are reclassified as peri-urban areas in the figure (see Dorosh and Thurlow 2013). Nevertheless, the large peri-urban population suggests that there is scope for farmers to raise their incomes by supplying foods (even perishable foods) to urban markets or by working as seasonal or circular migrants within the urban economy.

Figure 9.8 also suggests that while average labor productivity may be higher in nonfarm activities (Gollin, Lagakos, and Waugh 2014), this hides the wide range of jobs within the agricultural and nonagricultural sectors. A large share of economic growth in Africa over the last decade was generated by wholesale and retail trade, and much of this was in the low-productivity informal sector (McMillan, Rodrik, and Verduzco-Gallo 2014; Thurlow, Dorosh, and Davies 2019). Parts of the agricultural sector, especially high-value crops and livestock, may be more productive than informal trade, manufacturing, and business services. There is evidence to support this. McCullough (2017) used time-use data from household surveys to estimate hourly labor returns for individual workers in four African countries. Findings suggest that, in some countries, hourly labor returns are higher in agriculture than in industry. Nagler and Naudé (2017) use the same household data to estimate how labor productivity in nonfarm enterprises changes, the further these enterprises are located from urban markets. Figure 9.10 reports their findings for Uganda. While average labor productivity is higher in urban areas, this hides the large overlap in labor productivity in rural and urban areas. Just as it is incorrect to equate rural areas with agriculture, it is also incorrect to equate rural areas with low-productivity jobs (at least for households' nonfarm enterprises).

The above framework captures some of the main arguments from both the pro-agriculture and pro-urban perspectives. The next section implements this framework using a model of Malawi's economy to evaluate the impacts of urbanization and investments along the rural-urban continuum.

FIGURE 9.10 Labor productivity of nonfarm enterprises in Uganda (by rural/urban area or distance to nearest urban population center)

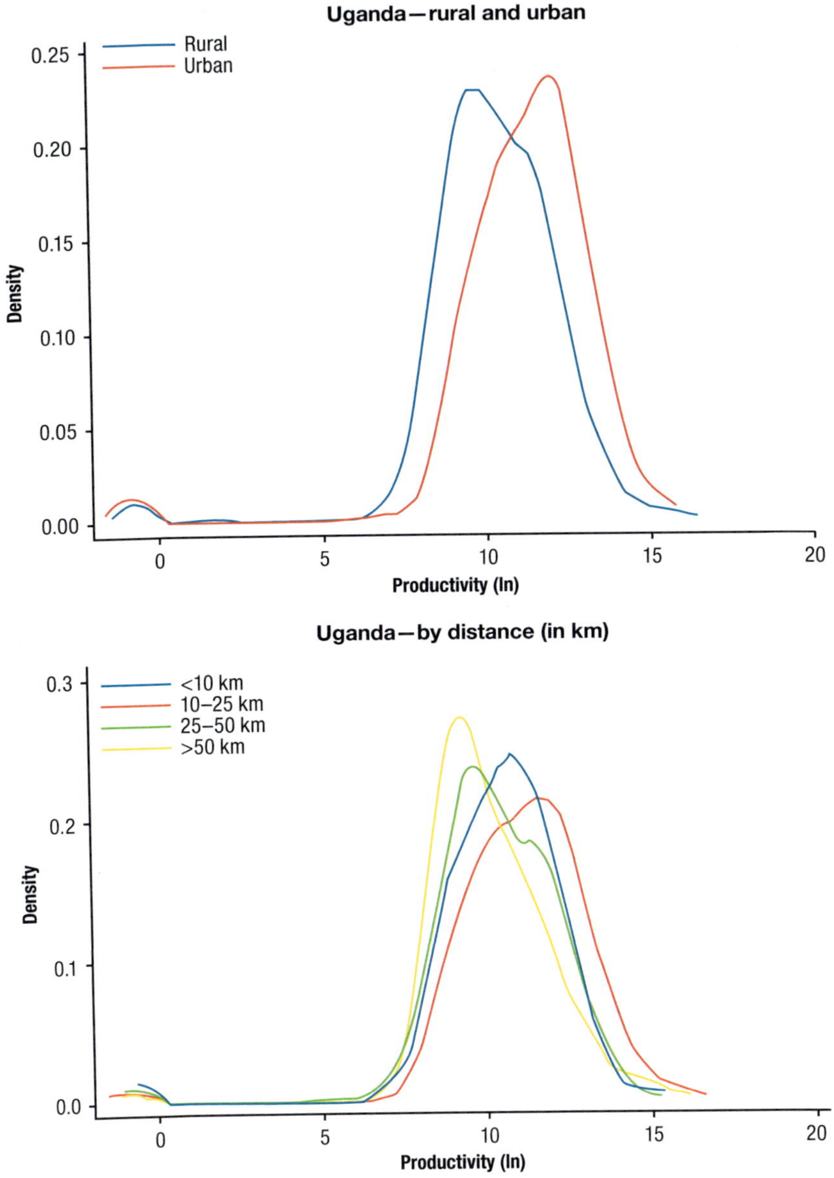

Source: Nagler and Naudé (2017).

Note: Distribution of labor productivity is measured as average monthly enterprise sales divided by number of workers. Urban and rural designations are based on official national definitions. Distance is measured in kilometers (km) to the nearest population center.

Investing in Agricultural Growth or Urban Development

An open-economy computable general equilibrium (CGE) model used in this section separates the Malawian economy into detailed sectors and household groups within cities, towns, and rural areas. Previous sections emphasized cross-country variation in urban definitions and the importance of peri-urban areas, but due to data constraints, we retain Malawi's official definition of *urban centers* and cannot separate peri-urban and more-remote rural areas. Nevertheless, the model still captures key aspects of the debate over the roles of rural agriculture and urban areas in development, including agriculture's production and demand linkages, internal rural-to-urban migration, and urban agglomeration economies and congestion effects (see Dorosh and Thurlow 2011, 2013). The model's parameters are calibrated to a 2010 social accounting matrix and to employment and migration data from national labor force surveys and population censuses (see Pauw, Schuenemann, and Thurlow 2015). Urban agglomeration effects are calibrated to cross-country estimates from Henderson and Wang (2005).

Workers in cities, towns, and rural areas are fully employed, but their labor markets are segmented by education levels. Full employment for lower-skilled workers in rural areas reflects seasonal labor shortages and distance barriers to circular migration. Full employment in urban areas reflects the greater skill intensity of urban employment and a scarcity of those skills within the urban labor force. Barriers to migration lead to trade-offs between farm and nonfarm work, which may explain why so few farm households in Malawi also earn incomes from nonfarm enterprises and off-farm work, despite the country's high population density (Benson, Erman, and Baulch 2019).

Even without full employment, migrants from rural areas in the model would still need to compete with urban workers and would have limited job opportunities in more skill-intensive sectors. Workers migrate in response to wage differentials, with initial migration replicating observed flows at prevailing wage gaps. Workers migrate with their families. Cities and towns benefit from migrant workers, but their labor dependency ratios rise. Finally, migrants adopt the consumption patterns of urban households at similar income levels, including the import intensity of their consumption baskets.

The model is dynamic and is used to evaluate future growth, urbanization, and investment trajectories until 2030. We run three policy scenarios. The "faster urban migration" scenario simulates faster rural-to-urban migration without reallocating public resources. The "higher urban investment" scenario again simulates faster migration but increases public investment in urban areas. Government resources are fixed in this scenario, and so higher urban investment reduces rural investment. Finally, the "maintaining rural investment" scenario simulates faster urbanization and higher urban investment (as above) but maintains rural investment levels by increasing urban taxation.

The model used in this analysis incorporates neoclassical microeconomic assumptions regarding consumer demand, producer profit maximization, and market-clearing prices, as well as the special features related to positive agglomeration effects on total factor productivity in urban areas. The empirical results are heavily dependent on the structure of the Malawian economy as captured in the detailed social accounting matrix, which does capture some of the country's macroeconomic rigidities and resource misallocations. The model also draws on econometric analysis of consumer behavior, such as by imposing income elasticities. However, the model abstracts from various important aspects of household decision-making that is often the focus of the more recent microeconomics literature. This includes incorporating risk and risk aversion, quasi-hyperbolic discounting and demand for commitment devices, status quo biases, and gender dynamics and related labor market frictions or allocation inefficiencies (see discussion in Dercon and Gollin 2014).

There is considerable scope to develop more elaborate structural models that incorporate the recent findings from household-level behavioral studies, especially on the factors influencing a worker's decision to migrate (urbanize) and the economic relationships that migrants maintain with their sending households (in rural areas). In fact, a major gap in the rural-urban linkages literature is a lack of sophisticated macro-micro economic models, especially for developing countries. Macroeconomic modeling would benefit from stronger microempirical foundations, whereas microeconomic studies could benefit from demonstrating the economywide importance of their findings about household and worker behavior.

Malawi Case Study

Before presenting the results of these scenarios, we introduce the Malawian case study and the country's baseline (or business-as-usual) growth path. Malawi has much in common with other low-income agrarian economies. Economic growth over the last 15 years was modest but was enough to raise average per capita incomes. Underlying the growth process were three major trends. First, while agriculture is the dominant sector, most recent growth came from urban areas and the rural nonfarm economy. Second, urban population growth exceeds rural population growth, but the urban population share is only increasing slowly. Third, urban poverty ratios have risen, even though national poverty ratios have declined.

Given the above trends, it is possible to reach different conclusions about where Malawi's government should focus. Investing in agriculture could raise incomes for smallholder farmers in rural areas, who make up most of the population. Alternatively, investing in cities or towns could generate even faster urban economic growth, with potential benefits for migrants and rural areas. Malawi's Growth and Development Strategy has traditionally favored rural agriculture and small villages. In fact, this series of strategies has explicitly aimed to reduce the pace of urbanization, which is considered a constraint to economic development. While the existing strategy is definitive, there is some debate in Malawi over whether agriculture, as opposed to industry and urban areas, should remain the focus of future policies (see World Bank 2016).

Malawi's rural and urban economies are compared using a social accounting matrix (SAM) built for this study. Sectors and households are disaggregated across rural and urban areas using household and economic survey data. Table 9.4 summarizes the main characteristics of the city, town, and rural economies. Rural areas contain 85 percent of the population but generate only 62 percent of GDP. Household consumption in rural areas is below average, which explains why most poor Malawians live in rural areas. Agriculture is concentrated in rural areas, but there is also a large rural nonfarm economy. Rural households spend two-thirds of their income on food and agricultural products.

Towns have 3 percent of the population and contribute 6 percent of the GDP. While there is some farming within town boundaries, the main sectors are industry and services, and these tend to employ slightly better-educated workers than rural agriculture. Compared with rural households, town households consume more processed foods and fewer unprocessed agricultural products. Households in cities have a similar preference for processed foods. Cities are the core of Malawi's economy—they contain a tenth of the population but generate a third of the GDP. Consumption levels are highest and poverty is lowest in cities. Services are concentrated in cities, and this is reflected in consumption patterns.

Table 9.4 also provides information on the size of rural-urban demand linkages. City households, for example, account for 18 percent of consumption spending on agricultural products, but they produce only 5 percent of agricultural output. Cities are therefore "net importers" of agricultural products from rural areas (even after accounting for urban food imports). Overall,

TABLE 9.4 Malawi's city, town, and rural economies, 2010

Indicator	Rural areas	Towns	Cities	All
Consumption per capita ($)	341	940	1,136	458
Poverty headcount rate (%)	45.3	18.4	9.4	40.0
Population share (%)	84.6	2.7	12.7	100
Poor population share (%)	95.8	1.2	3.0	100
Sector GDP shares (%)	100	100	100	100
Agriculture	49.2	9.5	4.5	32.3
Industry	15.5	20.0	17.7	16.5
Services	35.3	70.5	77.8	51.2
Regional GDP shares (%)	61.6	5.9	32.5	100
Agriculture	93.8	1.7	4.5	100
Industry	57.9	7.2	34.9	100
Services	42.5	8.1	49.4	100
Employment shares by education (%)	100	100	100	100
Secondary schooling completed	6.0	12.4	16.1	7.3
Primary schooling completed	30.8	37.9	40.9	32.0
Primary schooling incomplete	63.2	49.7	43.0	60.7
Total consumption shares (%)	100	100	100	100
Agriculture	49.3	28.7	22.1	39.6
Processed foods	14.3	13.7	12.8	13.8
Industrial goods	9.2	18.8	12.9	10.9
Services	27.3	38.8	52.1	35.7
Product consumption shares (%)	63.0	5.5	31.5	100
Agriculture	78.4	4.0	17.6	100
Processed foods	65.2	5.5	29.3	100
Industrial goods	53.1	9.5	37.4	100
Services	48.1	6.0	45.9	100

Source: Malawi social accounting matrix and CGE model.

Note: GDP = gross domestic product. Poverty line is upper threshold of second per capita consumption quintile.

urban areas import most of their agricultural products from rural areas, although they are almost self-sufficient for processed foods. Urban centers in turn export services to rural areas, including transport, finance, and business. Both rural and urban areas rely on foreign imports for industrial products (for example, machinery and chemicals). These economic structures and linkages will determine the impacts of urbanization and public investments in our model scenarios.

Baseline Scenario

The baseline assumes a continuation of recent trends (1998–2013). The population grows at 2.8 percent per year, with faster growth in cities and towns (Table 9.5). Labor supply grows more slowly at 1.4 percent per year, and land supply grows even more slowly at 0.5 percent (reflecting major land constraints). Private investment and capital accumulation rates are endogenous; public investment is exogenous and increases faster than the population.

Total factor productivity (TFP) growth has exogenous and endogenous components. The latter include urban agglomeration effects (and offsetting congestion effects) proxied by population density. Urban nonagricultural productivity grows faster than rural agricultural productivity in the baseline. Overall, national GDP grows at 4.0 percent per year (or 1.2 percent in per capita terms). There is only modest structural change in the baseline. The urban population share rises from 15 to 16 percent during 2010–2030, and agriculture's share of GDP falls. Poor households in the bottom two consumption quintiles enjoy modest welfare gains. Equivalent variation (EV), a consumption-based welfare measure that adjusts for changing prices, grows at 0.1 percent per year. Nonpoor households' welfare grows much faster (1.2 percent), implying a widening gap between poor and nonpoor households in the baseline.

Faster Urban Migration

The faster urban migration scenario increases rural-to-urban migration rates above historical values. The urban population share now rises to 21.1 percent in 2030, as compared with 16.2 percent in the baseline. Faster migration increases urban labor supply and narrows the gap between urban and rural wages. Urbanization reduces rural labor supply, but it also generates backward linkages to agriculture, mainly because higher urban incomes generate demand for food (Table 9.6). Overall, faster urbanization leads to a 0.7 percentage point increase in the national GDP growth rate such that by 2030 the national economy is 14.1 percent larger than it would have been without faster urbanization.

All additional economic growth from faster urbanization occurs in cities and towns. Despite increased agricultural production, the rural economy contracts (relative to the baseline) due to slower growth of rural nonfarm activities. This change in the composition of rural employment is mainly caused by higher demand from urban households for imported goods, a real exchange rate depreciation, and greater incentives to produce exports (and

TABLE 9.5 Baseline scenario, 2010–2030

Indicator	Rural areas	Towns	Cities	All
Annual GDP growth (%)	3.04	4.91	5.22	3.96
Labor	1.10	2.90	2.90	1.38
Crop land/livestock	0.50	0.00	0.00	0.48
Private capital	4.09	6.17	6.37	6.22
Public capital	3.67	3.93	2.95	4.00
TFP	1.28	1.43	1.64	1.30
Annual migrant flow (1,000s)	−16.98	3.35	13.63	0.00
Inflow	0.00	3.80	13.63	17.43
Outflow	−16.98	−0.45	0.00	−17.43
Change in share of workforce (%)	−0.31	1.62	1.71	0.00
Population growth rate (%)	2.70	3.27	2.98	2.75

Source: Malawi CGE model results.
Note: GDP = gross domestic product; TFP = total factor productivity.

TABLE 9.6 Economic growth and household welfare results, 2010–2030

		Urbanization scenarios (percentage point deviation from baseline)		
Indicator	Baseline annual growth rate (%)	Faster urban migration	Higher urban investment	Maintaining rural investment
Annual GDP growth	3.96	0.66	0.68	0.82
Agriculture	2.81	0.17	−0.03	0.14
Industry	4.69	0.83	0.95	1.10
Services	4.35	0.83	0.90	1.03
Rural areas	3.04	−0.29	−0.52	−0.33
Towns	4.91	1.23	1.22	1.54
Cities	5.22	1.60	1.83	1.91
Annual welfare change	1.01	1.52	1.48	1.36
Poor	0.13	0.25	0.14	0.25
Urban	0.95	−0.07	−0.16	0.00
Rural	0.06	0.23	0.12	0.22
Nonpoor	1.16	1.73	1.70	1.56
Urban	1.90	1.88	1.92	1.63
Rural	0.45	0.82	0.71	0.70

Source: Malawi CGE model results.
Note: GDP = gross domestic product.

other tradable goods) within the rural economy. Because of this change, labor allocations within agriculture shift toward export crops, such as tobacco and cotton, and toward food crops that are difficult to substitute with imports, such as livestock and fish. Urbanization in Malawi encourages a shift toward higher-value activities in agriculture, and this is consistent with recent survey-based studies suggesting that urbanization in SSA transforms national food systems (Tschirley et al. 2015).

Faster urbanization accelerates structural change in Malawi. Migration to urban centers causes agricultural employment to fall and nonagricultural employment to increase, particularly in trade services. Again, this is consistent with recent growth and employment trends in SSA (Thurlow, Dorosh, and Davies 2019). Migration also results in more manufacturing and construction jobs, in part because urban migrants demand more manufactured goods.

Finally, the increase in urban population exceeds the increase in urban GDP, leading to lower urban per capita GDP and welfare relative to the baseline. Conversely, the decline in the rural population is larger than the decline in rural GDP, and so rural welfare improves. Faster urbanization leads to higher national welfare and falling poverty, but most of the benefits accrue to nonpoor households. Thus, while urbanization raises national GDP and welfare and stimulates agricultural growth, it also leads to an "urbanization of poverty" as urban centers struggle to absorb migrant workers and their families. This might justify reallocating public resources from rural to urban areas.

Higher Urban Investment

The decline in urban welfare in the faster urban migration scenario is partly due to declining public capital (infrastructure) per capita (relative to the baseline) as urban populations increase. To avoid these adverse congestion effects, the higher urban investment scenario increases public investment in cities and towns so that urban public capital per capita remains unchanged from baseline levels. The overall level of government spending is fixed, however, and so increasing urban investment in this scenario reduces investment in rural areas, causing agricultural TFP to fall below baseline levels. In other words, we capture both the benefits and opportunity costs of raising urban investment.

As shown in Table 9.6, GDP growth in the industrial and service sectors is faster in the higher urban investment scenario than in the faster urban migration scenario. Agricultural growth decelerates, however, and is even slower than in the baseline. Faster urban GDP growth widens the gap between urban and rural wages and encourages more migration to urban centers. Not surprisingly, urban welfare is higher and rural welfare is lower when investments are reallocated toward urban centers.

Reallocating investment away from rural agriculture does not achieve the goal of preventing urban poverty from rising with urbanization. Instead, the model simulations show that reducing rural investment leads to even worse outcomes for poor urban households, because slower agricultural growth leads to higher real food prices. This finding is consistent with traditional development models that argue that raising agricultural productivity benefits the urban poor by reducing food prices (see Diao, Hazell, and Thurlow 2010). The results suggest that increasing urban investment is necessary to prevent an urbanization of poverty, but this should not come at the expense of rural investments (at least over the near term).

Maintaining Rural Investment

The final scenario increases government spending in urban areas (as in the previous scenario) but finances these investments by raising direct tax rates on urban enterprises and households rather than reducing rural investments. Raising domestic revenues is a challenge for countries like Malawi, where the tax base is small and the government has few tax instruments to apply. Higher tax rates could also lead to tax evasion. Nevertheless, this scenario allows us to assess the potential economywide benefits of a more balanced rural and urban investment strategy, while still accounting for both benefits and costs of investment decisions.

Maintaining rural investment, while also increasing urban investment, leads to faster agricultural growth relative to the baseline (that is, by 0.14 percentage points per year) and higher overall GDP growth. As shown in Table 9.6, secondary towns are the main beneficiaries from maintaining rural investment levels, since the economy of towns is more closely linked to rural agriculture, and town households tend to be poorer than city households with higher shares of food in their consumption baskets. Preventing the fall in agricultural productivity actually leads to faster rural-to-urban migration and most of the increase in the urban migrant population occurring in secondary towns.

Poor urban households are better off in the maintaining rural investment scenario than in either the faster urban migration or the higher urban investment scenarios. By design, urban welfare now remains unchanged, compared with the 1.3 and 3.1 percent declines in the previous scenarios. Poor rural households are also better off when rural investment is maintained at baseline levels. Only nonpoor urban households are worse off, because taxes are raised to finance urban investments. Even with higher taxes, however, nonpoor urban households are still the main beneficiaries from faster urbanization—their per capita welfare is still 38.1 percent higher in 2030 than it would have been without faster urbanization, and this is only slightly below the 46.4 percent welfare improvement they would have enjoyed had they not had to finance urban investments.

The Malawi case study provides some insights into the relationship between agriculture, urbanization, and poverty. As in many developing countries, Malawi's development plan views urbanization as a constraint to development. To reduce migration, the strategy focuses on raising agricultural productivity and value addition and creating markets and job opportunities in rural villages. The above analysis suggests that faster urbanization is not a constraint but rather a potential catalyst for long-term economic development in Malawi. Urbanization accelerates national economic growth and structural change and can help spur rural transformation and poverty reduction. However, reducing investments in rural agriculture to finance urban development can lead to an "urbanization of poverty" that underpins many of the concerns about urban migration in Malawi. Instead, findings suggest that urbanization can finance itself (via urban taxes) without preventing higher-income urban households from benefiting from urban development.

Conclusions

Rapid urbanization is a defining feature of low-income countries, even as rural populations continue to grow. Greater attention should therefore be paid to the spatial dimensions and urban drivers of agricultural transformation. Unfortunately, agricultural and urban policies are usually developed in isolation from each other and are often viewed as being in competition. This perceived competition between sectors has led many countries to adopt policies aimed at curbing internal migration or redirecting scarce public resources toward agriculture. However, urban populations are growing rapidly in low-income countries and a large share of urban residents are poor and live in slums. Even maintaining urban investments at current levels will be insufficient and could become a constraint to national development, including efforts to transform agriculture-food systems.

This chapter reviewed cross-country evidence on the relationship between urbanization, structural change, and poverty reduction. Urbanization is associated with workers leaving agriculture. That said, the nonfarm sector is an important part of the rural economy, as is agricultural employment within the urban economy. Many people, including farmers and other workers in the food system, live in peri-urban areas close to urban centers. Finally, urban poverty is a justified concern, even though most of the world's poor still live in rural areas.

The Malawi case study considered many of the concerns and arguments raised by the pro-agriculture and pro-urban schools. Findings suggest that, contrary to national policies in many developing countries, faster urbanization is not a constraint on, but instead a potential catalyst for, economic development. Urbanization can accelerate national economic growth and structural change and spur rural transformation and poverty reduction. The analysis, however, cautions against reducing investments in rural agriculture to finance urban development. This leads to an "urbanization of poverty" that underpins many concerns about urban migration in countries like Malawi. Instead, findings indicate that urbanization could finance itself without preventing higher-income urban households from benefiting from faster urbanization and urban development. A more balanced national strategy for developing countries like Malawi should limit the policy constraints placed on migration and instead ensure that urban policy includes investments that support urban enterprises and job creation, particularly in downstream components of the agriculture-food system.

Further research is needed to better understand the economies of small towns and their rural/agricultural linkages. More systematic research on agriculture-food systems is needed to gauge the urban extent of agricultural value chains (for example, processors and traders) and the interactions between these value chains (for example, competition for market space, barriers to entry). This research agenda would require use of both household and business surveys and the study of the incentives and needs of both firms and farms. Similarly, agriculture-food systems stretch across multiple sectors and locations, and so there needs to be better alignment between agricultural, industrial, and urban policies and investments. Adopting a more nuanced conceptual framework, like the one presented in this chapter, requires a broadening of food policy research beyond farming and value chains. It also needs better coordination of ministries and agencies tasked with promoting private sector development within food systems. While agriculture may indeed be in competition with urban development, it may also be that rapid urbanization provides the impetus needed to broaden existing food policies and research.

References

Benson, T., A. Erman, and B. Baulch. 2019. "Change and Rigidity in Youth Employment Patterns in Malawi." In *Youth and Jobs in Rural Africa: Beyond Stylized Facts*, edited by V. Mueller and J. Thurlow. Oxford: Oxford University Press.

Chamberlin, J., D. D. Headey, and T. S. Jayne. 2014. "Land Pressures, the Evolution of Farming Systems, and Development Strategies in Africa: A Synthesis." *Food Policy* 48 (C): 1–17.

Christiaensen, L., L. Demery, and J. Kuhl. 2011. "The (Evolving) Role of Agriculture in Poverty Reduction: An Empirical Perspective." *Journal of Development Economics* 96 (2): 239–254.

Christiaensen, L., and Y. Todo. 2014. "Poverty Reduction during the Rural–Urban Transformation: The Role of the Missing Middle." *World Development* 63 (1): 43–58.

Collier, P., and S. Dercon. 2014. "African Agriculture in 50 Years: Smallholders in a Rapidly Changing World?" *World Development* 63 (11): 92–101.

Davis, B., P. Winters, G. Carletto et al. 2010. "A Cross-Country Comparison of Rural Income Generating Activities." *World Development* 38 (1): 48–63.

Dercon, S. 2009. "Rural Poverty: Old Challenges in New Contexts." *World Bank Research Observer* 24 (1): 1–28.

Dercon, S., and D. Gollin. 2014. "Agriculture in African Development: Theories and Strategies." *Annual Review of Resource Economics* 6: 471–492.

Diao, X., P. Hazell, and J. Thurlow. 2010. "The Role of Agriculture in African Development." *World Development* 38 (1): 1375–1383.

Dorosh, P., and J. Thurlow. 2011. "Agglomeration, Growth and Regional Equity: An Analysis of Agriculture versus Urban-Led Development in Uganda." *Journal of African Economies* 21 (1): 94–123.

———. 2013. "Agriculture and Small Towns in Africa." *Agricultural Economics* 44: 449–459.

———. 2014. "Can Cities or Towns Drive African Development? Economywide Analysis for Ethiopia and Uganda." *World Development* 63: 113–123.

———. 2016. "Beyond Agriculture versus Non-Agriculture: Decomposing Sectoral Growth–Poverty Linkages in Five African Countries." *World Development* 109: 440–451.

FAO (Food and Agriculture Organization of the United Nations). 2017. *The State of Food and Agriculture: Leveraging Food Systems for Inclusive Rural Transformation*. Rome.

Fujita, M., P. Krugman, and A. J. Venables. 1999. *The Spatial Economy: Cities, Regions, and International Trade.* Cambridge, MA: MIT Press.

Gollin, D., D. Lagakos, and M. E. Waugh. 2014. "The Agricultural Productivity Gap." *Quarterly Journal of Economics* 129 (2): 939–993.

Haggblade, S., P. Hazell, and P. Dorosh. 2007. "Sectoral Growth Linkages between Agriculture and the Rural Nonfarm Economy." In *Transforming the Rural Nonfarm Economy: Opportunities and Threats in the Developing World*, edited by S. Haggblade, P. Hazell, and T. Reardon. Baltimore: Johns Hopkins University Press.

Hazell, P., C. Poulton, S. Wiggins, and A. Dorward. 2010. "The Future of Small Farms: Trajectories and Policy Priorities." *World Development* 38 (10): 1349–1361.

Henderson, J. V., and H. G. Wang. 2005. "Aspects of the Rural–Urban Transformation of Countries." *Journal of Economic Geography* 5: 23–42.

IFAD (International Fund for Agricultural Development). 2016. *Rural Development Report 2016: Fostering Inclusive Rural Transformation*. Rome.

Ingelaere, B., L. Christiaensen, J. De Weerdta, and R. Kanbur. 2018. *Why Secondary Towns Can Be Important for Poverty Reduction: A Migrant Perspective.* Jobs Working Paper 12. Washington, DC: World Bank.

IPUMS (Integrated Public Use Microdata Series). 2019. International: Version 7.2. Minnesota Population Center, Minneapolis, MN. https://doi.org/10.18128/D020.V7.2.

Johnston, D. G., and J. W. Mellor. 1961. "The Role of Agriculture in Economic Development." *American Economic Review* 51 (4): 566–593.

Lall, S. V., J. V. Henderson, and A. J. Venables. 2017. *Africa's Cities: Opening Doors to the World.* Washington, DC: World Bank.

Masters, W. A., A. A. Djurfeldt, C. De Haan, P. Hazell, T. Jayne, M. Jirström, and T. Reardon. 2013. "Urbanization and Farm Size in Asia and Africa: Implications for Food Security and Agricultural Research." *Global Food Security* 2 (3): 156–165.

McCullough, E. B. 2017. "Labor Productivity and Employment Gaps in Sub-Saharan Africa." *Food Policy* 67: 133–152.

McMillan, M., D. Rodrik, and I. Verduzco-Gallo. 2014. "Globalization, Structural Change, and Productivity Growth with an Update on Africa." *World Development* 63 (11): 11–32.

Nagler, P., and W. Naudé. 2017. "Non-Farm Entrepreneurship in Rural Sub-Saharan Africa: New Empirical Evidence." *Food Policy* 67: 175–191.

Pauw, K., F. Schuenemann, and J. Thurlow. 2015. *A 2010 Social Accounting Matrix for Malawi.* Washington, DC: International Food Policy Research Institute.

Thurlow, J. 2015. "Youth Employment Prospects in Africa." In *African Youth and the Persistence of Marginalization: Employment, Politics, and Prospects for Change*, edited by D. Resnick and J. Thurlow. New York: Routledge.

Thurlow, J., P. Dorosh, and B. Davies. 2019. "Demographic Change, Agriculture and Rural Poverty." In *Sustainable Food and Agriculture: An Integrated Approach*, edited by C. Campanhola and S. Pandey, Chapter 3. London: Elsevier and FAO.

Thurlow, J., D. Resnick, and D. Ubogo. 2015. "Matching Concepts with Measurement: Who Belongs to Africa's Middle Class?" *Journal of International Development* 27 (5): 588–608.

Timmer, C. P. 1988. "The Agricultural Transformation." In *Handbook of Development Economics*, Vol. I, edited by H. Chenery and T. N. Srinivasan. New York: Elsevier.

Tschirley, D., T. Reardon, M. Dolislager, and J. Snyder. 2015. "The Rise of a Middle Class in East and Southern Africa: Implications for Food System Transformation." *Journal of International Development* 27 (5): 628–646.

Uchida, H., and A. Nelson. 2010. *Agglomeration Index towards a New Measure of Urban Concentration*. Working Paper 2010/29. Helsinki: UNU-WIDER.

UN DESA (UN Department for Economic and Social Affairs). 2015. *World Urbanization Prospects: 2014 Revision*. New York.

———. 2016a. *United Nations Demographic Yearbook, 2015*. New York.

———. 2016b. *World Population Policies Database: The 2015 Revision*. New York.

———. 2017. *World Population Prospects: 2017 Revision*. New York.

World Bank. 2008. *World Development Report 2008: Agriculture for Development*. Washington, DC.

———. 2009. *World Development Report 2009: Reshaping Economic Geography*. Washington, DC.

———. 2016. *Malawi Urbanization Review: Leveraging Urbanization for National Growth and Development*. Washington, DC.

———. 2017. Global Consumption Database. Accessed January 25, 2018. http://datatopics.worldbank.org/consumption.

———. 2018. World Development Indicators database. Accessed January 25, 2018. http://databank.worldbank.org.

Zezza, A., and L. Tasciotti. 2010. "Urban Agriculture, Poverty, and Food Security: Empirical Evidence from a Sample of Developing Countries." *Food Policy* 35 (4): 265–273.

Chapter 10

AGRICULTURE AND UNDERNUTRITION

Derek Headey and William A. Masters

For most of modern history, agricultural development has focused on reducing poverty and food insecurity, typically by attempting to provide sufficient dietary energy for survival and work (see Chapters 1 and 9). For these basic needs, rapid population growth in the late 20th century, especially, required much greater productivity of staple cereals such as rice, wheat, and maize, as well as potatoes, cassava, and other root crops, to meet human energy requirements, especially in the lowest-income regions where people remain at the margin of subsistence. These few "starchy" staples have evolved and been selected to produce the most dietary energy per unit of cost in terms of labor, land, water, and other inputs.

Once a person's energy needs are met, health and longevity depend on other aspects of human nutrition, which is a complex biological process. Human foods contain several dozen essential nutrients needed for health that cannot be synthesized within the body. "Macronutrients" include essential amino acids (protein) and fatty acids (fats) that serve as building blocks for cellular activity and can also be burned for dietary energy, in contrast to vitamin and mineral "micronutrients" that are essential for health but do not directly provide energy for physical activity. A wide range of dietary components other than essential nutrients also contribute to health, notably fiber and water as well as phytochemicals (produced by plants), and the form of a nutrient may be important for bioavailability, such as the vitamin B-12 and heme iron in some animal-sourced foods. Each nutrient or other component often serves multiple biological functions affecting physical growth, cognitive development, immune system strengthening, and prevention of specific diseases. Many nutrients can also be harmful at high doses, sometimes in ratio terms relative to other dietary components. Foods may contain harmful toxins or antinutrients that block absorption, and nutrient utilization is heavily influenced by infection and other aspects of a person's health.

Improving nutrition goes far beyond the quantity of food. Moreover, while poverty and food security are often monitored on the basis of consumption per person, nutritional needs vary greatly by age, and undernutrition tends to be most costly and most pronounced among children less than two years of age (Victora et al. 2009). Undernutrition also has intergenerational dimensions, with poor maternal nutrition—especially during pregnancy—resulting in children who are malnourished even before birth.

The inherent complexity of human nutrition, combined with the logistical and ethical difficulty of conducting research on how diets affect health, makes it extremely challenging to predict the impacts of agricultural policies, structures, systems, and practices on nutrition outcomes such as height, weight, anemia, and other markers. Agricultural change can influence what people eat through income and price effects, but also influence nutrition through the disease environment via crops (for example, molds that spawn mycotoxins), livestock (for example, zoonotic diseases, fecal contamination of the environment), and exposure to hazardous chemicals (for example, pesticides). Moreover, even the dietary impacts of agricultural interventions are likely to be complex, especially since the main burden of undernutrition occurs among infants and very young children (for example, 6–24 months of age) rather than the broader family.

Obtaining rigorous empirical evidence on agriculture-nutrition linkages is made more challenging by institutional separation and methodological differences between agricultural and nutritional sciences. Nutrition research is a very new field, with most sources of data and analytical methods having been developed in the late 20th century. Agricultural research is much older, with large subfields such as plant genetics that were founded in the late 19th century. Economics spans both topics, studying farm household behavior, markets, and food choice since the late 18th century, but its techniques are typically not available to either agricultural or nutrition researchers and rarely include much biological detail about crops, livestock, or people. Hence it is little surprise that the scientific literature on agriculture-nutrition linkages has, for most of recent history, remained extremely limited (see von Braun [1995] for an early exception).

After 2005, sharp increases in world food prices, combined with gradually increasing concerns and new discoveries about diet-related diseases around the world, led to a burst of research about how agricultural projects could improve health. These studies were largely confined to evaluations of geographically focused interventions designed to help target households diversify their own production and nutritional knowledge to increase the household's own dietary diversity, diet quality, and nutritional status (Leroy and Frongillo 2007; Girard et al. 2012; Masset et al. 2012; Ruel and

Alderman 2013; Haselow, Stormer, and Pries 2016; Ruel, Quisumbing, and Balagamwala 2018). Reviews of that literature focus on how each household's nutritional outcomes relate to their own farm production, in contrast to systems-level research that tries to understand the nutritional impacts of larger-scale agricultural growth and transformation processes (Pinstrup-Andersen 2013). The economic development and agricultural economics literature generally focuses on these larger-scale processes, including the impacts of investments that drive the creation and adoption of new technologies, the development of new value chains, the evolution of retail outlets (including supermarkets), and changes in both domestic and international trade, but typically does not measure nutrition outcomes in particular (Pinstrup-Andersen 2013). This chapter aims to bridge these two literatures on agriculture and nutrition, including studies of geographically focused projects and programs, as well as the nutritional impacts of agricultural growth and transformation processes.

In light of these knowledge gaps, the objectives of this chapter are to

1. conceptualize the variety of linkages between agriculture and nutrition;
2. describe methodological challenges to evaluating these agriculture-nutrition linkages;
3. summarize existing evidence on these linkages, focusing on lessons from the program evaluation literature, microeconomic research (largely based on household surveys), and meso- or macroeconomic research on markets, agricultural growth, and transformation; and
4. identify important research and policy implications from the existing evidence.

At the outset we define the parameters of our review as follows.

First, we typically use the term *agriculture* to mean the farm production of crop and livestock commodities; *agribusiness* to mean commercial suppliers of farm inputs and wholesale purchasers of farm commodities; *food markets* to mean storage, shipment, and transformation of food into final products including packaged foods and meals away from home; and *policy* as a shorthand for any public sector or philanthropic intervention or program that aims to alter these activities. These are all interrelated influences on what we will call *livelihoods*, meaning all of a household's activities, including both agriculture and nonfarm work. Putting these pieces together, the chapter will focus on agricultural and food *systems*, meaning the aggregate outcome of interactions between these components at the scale of an entire village, region, country, or the world as a whole.

Second, this chapter focuses on undernutrition—defined as insufficient intake of nutrients and healthy foods relative to biological needs, and the resulting deficiencies such as anemia and failure to achieve a child's genetic potential such as linear growth—rather than problems linked to excessive intake of total energy, salt, and harmful foods that lead to obesity and diet-related diseases such as hypertension and diabetes. The two types of malnutrition are connected, as many people who suffer from dietary insufficiencies in some dimensions also have excesses in others (for example, stunting in childhood followed by obesity in later life, or ongoing micronutrient deficiencies during weight gain), but excessive intake of harmful foods is less likely than dietary insufficiency to be caused by agricultural production and supply constraints. As this chapter will show, agricultural change can raise food intake where deficiencies are most widespread, while excesses that lead to obesity and diet-related disease are more closely tied to nonfarm activity, market conditions, and food demand. Excessive intake clearly links back to agriculture through demand for ingredients in unhealthy foods such as sugar and other carbohydrates, but we would hypothesize a fundamental asymmetry in agriculture-nutrition linkages: improving agricultural production can help poor people fill deficiencies, especially for children in poor rural households, but after intake has risen, attempting to use agriculture against obesity can be like pushing on a string.

Third, our review of the linkages between agriculture and undernutrition is not intended to be universally comprehensive, but focused on results of policy-relevant research about systemic interventions or processes that alter nutrition outcomes at the scale of an entire population. As noted above, there are many existing reviews of agriculture-nutrition projects designed to help individual farm families grow nutrient-dense foods via homestead gardening or livestock transfers. These interventions often have some efficacy, especially in remote areas for families with limited access to purchased food, but they are likely very expensive per household reached, are rarely implemented on a large scale, especially by governments, and may well lack sustainability given their frequent dependence on highly intensive interactions between implementers and subjects. To improve nutrition among all people in a given location, systemic change in agriculture almost always involves specialization and trade through local, regional, and global food markets. Even in poor, isolated places like Ethiopia, rural households obtain most of their nonstaple food from markets (Sibhatu and Qaim 2017), since each farm's output is inevitably much less diverse and unstable than the requirements of a healthy diet. The project/program literature on agriculture and nutrition also says little about agriculture's role in driving income changes, food price changes, or livelihood-related changes in health and nutrition. The project/program literature is not a sufficient basis for designing nutrition-sensitive agricultural policies. This review focuses on the broader question of how agriculture—livelihoods, farming practices, food systems, and policy interventions—influences diets and nutrition outcomes.

A Conceptual Framework for Linking Agriculture to Nutrition

Several previous studies have sought to conceptualize the complex relationship between agriculture and nutrition (Hoddinott 2011; Headey, Chiu, and Kadiyada 2012; Kadiyala et al. 2014). Here we focus on three basic linkages that encompass a number of specific interconnections, as illustrated in Table 10.1.

The first basic linkage is the role of agriculture in influencing household income. In low-income countries there are far more workers than there are nonfarm employment opportunities, and those at greatest risk of undernutrition are the poorest people who have no choice but to do at least some farming to meet daily energy needs. Their total incomes determine affordability and utilization of a wide range of nutritionally relevant goods and services, including food, health care, and education as well as water, sanitation, and hygiene (WASH). Economywide analyses of productivity growth in different agricultural and nonagricultural sectors estimate the direct and indirect benefits to different socioeconomic groups (see Diao, Hazell, and Thurlow 2010 and Chapter 9 for a review). These studies tend to find that investments in staple cereals often have the largest benefits for poor groups. Productivity growth in cereals has direct benefits to adopting farm households, many of whom are very poor. But productivity growth in cereals clearly also affects the level and stability of local food prices if cereals are not perfectly tradable, which has real income implications for rural and urban populations. Increases in farmer income and production also raise demand for nonagricultural goods and services, including labor, leading to higher wages and higher incomes for nonadopting households, particularly in the rural nonfarm economy. Hence, investments in staple cereals tend to have highly progressive impacts on the income distribution, although their benefits to catalyzing broader economic growth are still much debated (Dercon and Gollin 2014).

TABLE 10.1 **Conceptualizing basic and specific linkages between agriculture and nutrition**

Basic linkage	Specific linkage	Comments
1. Agriculture as a source of income (growth, equality, poverty)	1a. Farm profits	• Majority of the poor still work in agriculture
	1b. Off-farm income (off-farm and nonfarm wage incomes and profits)	• Substantial numbers of poor people work in off-farm or nonfarm occupations
2. Agriculture as a source of (relative) food prices	2a. Changes in production affect relative prices of different foods	• Many foods are imperfectly tradable, implying prices are heavily influenced by local supply and demand
	2b. Changes in trade policies affect relative prices of different foods	• Many food sectors are still highly protected
	2c. Changes in transport and storage affect relative prices of different foods	• Many foods are highly perishable and not traded long distances
3. Agriculture as a livelihood	3a. Farming practices affect time use of parents and children	• Women's time use may be associated with less care toward young children
	3b. Farming practices can directly affect health through diseases, exposure to chemicals, and level of physical activity	• Zoonotic diseases, including enteric and pulmonary infections • Exposure to pesticides, herbicides • Physically arduous tasks during pregnancy
	3c. Agricultural livelihoods affect empowerment of women and children	• Women's control of agricultural assets is highly variable
	3d. Agricultural livelihoods influence access to nutrition-relevant goods and services	• Remoteness and low population density reduce access to health, education, family planning, and WASH services

Source: Authors' construction.
Note: WASH = water, sanitation, and hygiene.

A second linkage between agriculture and nutrition is through food prices. In coastal cities that trade freely with the rest of the world, easily shipped food commodities can originate anywhere, so prices are set by demand and supply among all trading partners. In the poorest rural areas, however, poor infrastructure and poor governance often limit access to distant markets, so local agriculture plays a larger role. Spatial and seasonal variation in prices tends to be greatest in the places where malnutrition is most widespread, especially for perishable products such as eggs, fresh dairy, and many fruits and vegetables.

For these items, transport and storage cost is so high that prices, even in coastal cities, will be determined by national agricultural systems (Headey and Alderman 2019). In these situations, production is "non-separable" from consumption (Singh, Squire, and Strauss 1986; de Janvry, Fafchamps, and Sadoulet 1991), in the sense that farm output depends on local demand, so agriculture-nutrition linkages work through prices as well as income. For example, fresh milk is highly perishable and expensive to transport, so farm families with a lactating cow have much lower-cost access to milk each day and are more likely than others to feed milk to their children (Hoddinott, Headey, and Dereje 2015). More generally, the imperfect tradability of many fresh foods means that we should expect to see significant associations between local production patterns and local consumption patterns, and even between household production and household consumption (Jones 2017a; Sibhatu and Qaim 2018).

A third basic linkage involves the broader non-income dimensions of agricultural livelihoods and encompasses farming practices and their effects on time use and exposure to health hazards, cultural institutions influenced by agriculture (such as women's control over assets), and the remoteness dimension of agricultural livelihoods as it pertains to access to nutrition-relevant goods and services. These specific linkages are complex and, unlike the income and price pathways, often characterized by negative connotations: agriculture as a risk factor for poor health, for poor care practices, for diseases stemming from livestock ownership. In a dynamic sense, however, the transformation of agriculture involves adoption of modern farming practices in place of traditional practices. Sometimes modern practices may be beneficial to health and nutrition (for example, more-modernized livestock management, labor-saving technologies, improved storage facilities), and sometimes they may not be, especially if improperly used (for example, pesticide use). Examples are provided below.

Measurement and Analysis of Agriculture-Nutrition Linkages

The conceptual framework presented above points to how empirically challenging it can be to evaluate the complex linkages between agriculture and nutrition. Here we emphasize some of the main indicators regularly used to analyze agriculture-nutrition linkages, as well as some of the analytical tools that can be used.

Anthropometric Markers of Nutrition

Many surveys collect height, weight, and age of children 0–59 months of age, for comparison with the sizes of healthy children in standardized anthropometric measures of malnutrition (WHO 2006). The most widely used metric of whether a population is achieving its genetic growth potential is height at each age, measured in standard deviation units known as z-scores. In a healthy population, the resulting height-for-age measure (HAZ) has a normal distribution around a mean of 0, with a negative or positive score for children who happen to be short or tall for genetic or other reasons. A common threshold for extreme shortness is HAZ < –2. By definition, in healthy populations about 2.5 percent of children would have such low HAZ scores, but in malnourished populations the fraction is often more than 40 percent. Not all short children have had their growth stunted by nutritional deprivation, but for practical purposes all children with HAZ < –2 are classified as such. HAZ is thought of as the best all-encompassing marker for chronic malnutrition, reflecting as it does the cumulative impact of repeated nutritional insults. In contrast, a low weight-for-height z-score (WHZ)—with WHZ < –2 defining wasting—is thought of as a marker of acute malnutrition, reflecting recent nutritional insults, such as inadequate food, or recent morbidities such as diarrhea. Many data sources such as the USAID-funded Demographic and Health Surveys (DHS) or UNICEF's Multiple Indicator Cluster Surveys (MICS) also record the height, weight, and age of adult women and sometimes adult men, and low body mass is often used as a marker of maternal undernutrition. Beyond anthropometry, the DHS also records hemoglobin levels and anemia status for women and children. Like stunting and wasting, anemia has many nonfood causes—related to infections, in particular.

Figure 10.1 illustrates how stunting, wasting, and anemia vary across the (young) life cycle of children 0–59 months of age in recent DHS rounds. The figure illustrates the critical importance of the so-called 1,000-day window of opportunity, which refers to the period from conception to a child's second birthday (Victora et al. 2009). The importance of undernutrition in utero is demonstrated by the fact that in this sample almost 20 percent of children are stunted at birth, and almost 15 percent are wasted at this age. From 0–6 months there is little systematic increase in stunting, as many children are protected from infections and inadequate nutrient intake by exclusive or at least predominant breastfeeding. However, from approximately 6–20 months, stunting rates increase precipitously from around 20 percent to 40 percent, marking this period as one of progressive growth faltering. Thereafter,

FIGURE 10.1 Stunting, wasting, and anemia prevalence in recent DHS rounds, by child age

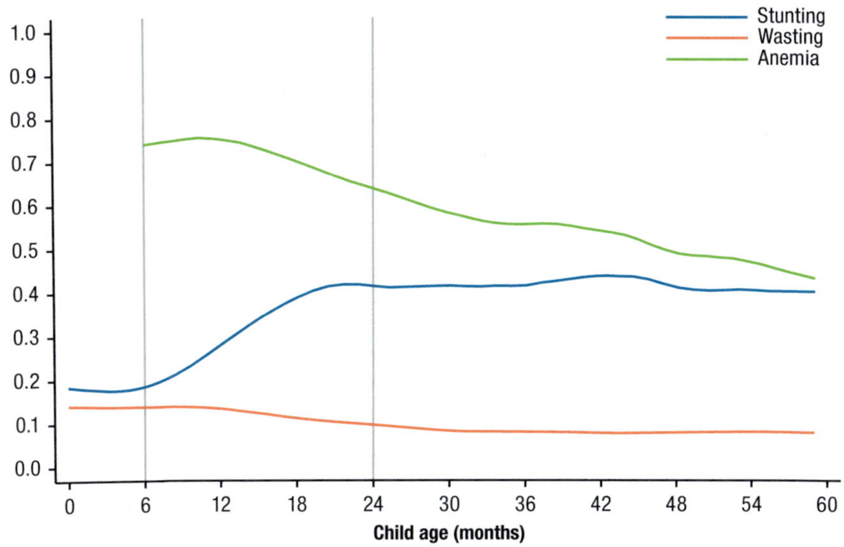

Source: Authors' estimates from DHS data from Headey, Stifel et al. (2017).

Note: DHS = Demographic and Health Surveys. Data are for 44 countries, with samples sizes of 296,370 (stunting), 284,784 (wasting), and 139,356 (anemia). Estimates are based on local polynomial smoothing using the LPOLY command in STATA 14.

stunting rates stabilize at a population level, even though individual children may experience catch-up growth after the first two years of life (or, conversely, become stunted). Wasting generally peaks at about 10–12 months of age, although wasting patterns vary substantially across countries. Anemia, which is only recorded from 6 months onward, also peaks at 10–12 months of age and gradually declines thereafter.

These dynamics of undernutrition in young children reveal important clues as to the determinants of undernutrition. First, a significant amount of growth faltering takes place in utero, suggesting maternal undernutrition is a major problem in developing countries. Second, most growth faltering, and much of the incidence of anemia and wasting, takes places in the 6-to-20-month window. In this window, children have weak immune systems, are increasingly exposed to pathogens and repeated infections (diarrhea, environmental enteric disorder, pulmonary infections, anemia), and are introduced to nutritionally deficient diets and poor feeding practices.

These dynamics also have important implications for nutritional research. Wasting is likely to be more sensitive to, or better explained by, recent shocks, and some studies have indeed linked wasting to agroecological shocks (for example, in Nepal). Stunting or HAZ, however, substantially reflects past shocks, and this measure should be more sensitive to shocks that occurred in the first 1,000 days of life. However, as Alderman and Headey (2017) point out, regression analyses of various underlying determinants should typically exclude children less than 24 months of age, since these children have not yet been fully exposed to the various nutritional insults or protective factors that determine their growth trajectory.

Survey Data on Individual and Household Diets

Diets are a key linkage between agriculture and nutrition, but measuring diets is extremely challenging. Economists have typically focused on household diets derived from food consumption or expenditure modules, and historically focused on per capita calorie intake rather than other nutrients. Nutritionists, however, have been more interested in measuring individual diets, particularly those of mothers and young children, and more concerned with measuring dietary quality, typically proxied by dietary diversity metrics (Ruel, Harris, and Cunningham 2013).

Both approaches face several significant problems. First, it is costly and time-consuming to measure actual quantities of foods consumed, particularly at the individual level. Second, in most cases researchers want a latent measure of "usual diets," but concerns over recall errors typically limit recall to very recent periods. This approach minimizes between-person errors at the expense of within-person errors, because diets vary seasonally and because many food items are simply not consumed at high frequency (Ruel, Harris, and Cunningham 2013; Thorne-Lyman, Spiegelman, and Fawzi 2014).

As a quick, inexpensive, and easily measured indicator of dietary quality, nutritionists have proposed dietary diversity metrics, usually based on 24-hour recall of consumption of major food groups (Ruel, Harris, and Cunningham 2013). An example of child dietary diversity data from the DHS is given in Table 10.2. The first column focuses on 7 aggregated food groups, while the second column refers to 14 more-disaggregated groups that respondents listed. The use of the 7 food groups has been extensively investigated and validated, and a 4-group threshold has been proposed as an indicator of minimum dietary diversity, as it is relatively successful at predicting mean micronutrient adequacy (FANTA 2006). Researchers have also advanced maternal dietary diversity scores and household dietary diversity, though a recent review finds a lack of standardization in their use in the agriculture-nutrition literature

TABLE 10.2 Food groups listed in phases 5 and 6 of the Demographic and Health Surveys (DHS)

Aggregated food groups in dietary diversity score (7 groups)	Disaggregated food groups (14 groups)
(1) Starchy staples	(1) Grains
	(2) Roots/tubers
(2) Legumes/nuts	(3) Legumes/nuts
(3) Vitamin A–rich fruits/vegetables	(4) Vitamin A–rich fruits
	(5) Vitamin A–rich vegetables
(4) Other fruits/vegetables	(6) Other fruits
	(7) Dark green leafy vegetables
	(8) Other vegetables
(5) Dairy	(9) Fresh/powdered cow's milk
	(10) Infant formula
(6) Eggs	(11) Eggs
(7) Flesh foods	(12) Meat/organs
	(13) Fish
	(14) Fortified infant cereals

Source: Authors' construction.

(Verger et al. 2019). Nevertheless an important area for future research is to advance more cost-effective ways of measuring usual diets.

Survey-Based Analyses of Agricultural Production and Agricultural Livelihoods

Unfortunately, nutrition surveys such as the DHS and MICS have very limited information on agriculture, including only occupational status of both parents and ownership of agricultural land and livestock. As the previous section illustrated, however, agricultural production and livelihoods have many different dimensions that may influence nutrition outcomes, including time use, farming practices as they relate to livestock, use of chemical inputs and levels of physical activity, women's empowerment, and access to nutrition-relevant goods and services. Information on these types of indicators is generally only contained in household surveys, particularly those with an agricultural focus such as the World Bank's Living Standards Measurement Study–Integrated Surveys on Agriculture (LSMS-ISA). Agriculture and nutrition data are also collected together in some USAID-funded Feed the Future surveys conducted by IFPRI and other international and national organizations. LSMS-ISA, for example, now include modules devoted to parental time use. IFPRI also developed the Women's Empowerment in Agriculture Index (WEAI) to increase understanding of women's roles in five key domains: (1) decisions about agricultural production, (2) access to and decision-making power about productive resources, (3) control of use of income, (4) leadership in the community, and (5) time allocation (Alkire et al. 2012). Other surveys have also incorporated expanded WASH modules, including a focus on livestock-related WASH (Headey and Hirvonen 2016; Headey, Hoddinott, and Park 2017).

Despite improvements in the availability of agricultural production and livelihood data, there are still many knowledge gaps. For example, relatively little research focuses on agricultural markets, or nutrition-related agricultural extension services, particularly as they relate to diversification out of staple crops into more nutrient-rich crops and livestock products. And many agricultural/economic surveys still do not include sufficient nutritional indicators. In light of this, Pingali and Ricketts (2014) propose that agricultural household surveys incorporate a minimum set of nutrition indicators, including anthropometric and biochemical indicators for children and mothers, household and individual dietary diversity data, and market-level food supply data on availability and price.

Several papers have also tried to explore the drivers of dietary diversity in national per capita terms by utilizing food supply data from the FAO Food Balance Sheets, which report estimates of calorie, fat, and protein consumption from different foods. Choudhury and Headey (2017) use these data to construct measures of the diversification of food supplies, particularly the share of calories sourced from nonstaple foods. The Global Dietary Database project has derived alternative estimates of food intake based on Bayesian modeling of individual dietary recall survey data and has found significant discrepancies between the resulting estimates of many countries' population means and the FAO's estimates of national average consumption, particularly for vegetables (Micha et al. 2015).

Food Prices and Markets

Price data are an underutilized resource for understanding the affordability of nutritious foods. All countries collect food price data through consumer price surveys (usually through national bureaus of statistics), but these data are rarely publicly available. The FAO also collates producer price data as well as market level data, but chiefly only for staple foods, and agricultural ministries also often collect agricultural market surveys. Globally, the International Comparison Program (ICP) of the World Bank also collects prices of standardized food products approximately every five years, as discussed by Headey and Alderman (2019). And at the most granular level, household survey data also collect price data from households and often from community survey modules. Our perception is that these data are underused, however.

Price data can be nutritionally informative in a number of different ways. Consumer price indices weight the prices of different foods according to prevailing consumption patterns, whereas Masters et al. (2018) construct novel price indices where weights are defined by nutritional criteria, such as achieving sufficient dietary diversity or adequate nutrient intake. Headey and Alderman (2019) use the ICP price data to measure the caloric costs of different foods relative to an index of the cost of a composite staple food calorie in a country. This measure captures how costly it is to diversify diets away from staples, but it also has the virtue of being a currency-free ratio, thereby circumventing problems of exchange rates and purchasing power parity conversion.

Price data are potentially very powerful tools for understanding the (mis)-functioning of markets and consumer behavior. As noted above, price data can be used to estimate the affordability of individual foods as well as whole diets, but they can also be used to explore consumer demand for different foods and the impacts of price shocks (Tiberti and Tiberti 2018), seasonality issues (Gilbert, Christiaensen, and Kaminhsk 2017), secular trends in food affordability (Wiggins and Keats 2015; Bachewe and Headey 2016), and integration across markets (Van Campenhout 2007). However, many previous studies using price data to examine these issues have focused on prices of staple foods or consumer prices as a whole, rather than prices of nutrient-dense foods or costs of nutritious diets. In light of the utility of price data for addressing important research questions on agriculture-nutrition linkages, it is surprising that there is not a larger body of research using these data.

Methods for Analyzing Agriculture-Nutrition Linkages

However, beyond econometric techniques, economywide simulation models are also now being applied to understand how different agricultural policies can influence both household income and dietary measures such as household calorie supply (Pauw and Thurlow 2010), dietary diversity (Arndt et al. 2017), or disease burdens associated with inadequate diets (Springmann et al. 2016). A key challenge here is reaching consensus on the most appropriate nutritional or dietary measures to incorporate into the simulation models, as well as to refine consumer demand systems to more realistically estimate the impacts of income and price changes on diets.

Methodological Limitations of the Literature on Agriculture-Nutrition Linkages

There are several significant methodological challenges to conducting research on nutrition and agriculture.

First, the experimental trials of various nutrition-sensitive agricultural interventions have been widely critiqued in the literature for their limitations, including small projects and sample sizes lacking statistical power, as well as the fact that many interventions combine agricultural components with nutritional education components, making it difficult to disentangle the mechanisms of impact (Masset et al. 2012; Ruel, Quisumbing, and Balagamwala 2018). Sustainability and scalability of these programs are major concerns. It is also notable that very few of these interventions have been implemented by governments (Hirvonen and Headey 2018). In addition, randomized control trials in the agricultural sector—and economic sectors in general—are fraught with complications, including high degrees of variance induced by idiosyncratic and covariate shocks, as well as seasonality (Barrett and Carter 2010; Rosenzweig and Udry 2016).

Second, the growing body of research linking farm production indicators to household consumption or individual nutrition outcomes is observational and essentially describes associations that may or may not be causal (Jones 2017a; Sibhatu and Qaim 2018). Some authors attempt to engage in quasi-experimental approaches, using agroecological variables as instruments, for example (Dillon, McGee, and Oseni 2015; Hirvonen and Hoddinott 2017), or conducting placebo tests to check for evidence of confounding (Hoddinott, Headey, and Dereje 2015; Choudhury and Headey 2018), but these approaches by no means rule out biases in the derived estimates. More ecological studies of markets, prices, or community characteristics face similar concerns, being largely descriptive (albeit usefully descriptive).

Finally, economywide simulation modeling faces both the aforementioned challenges pertaining to incorporating meaningful nutritional measures into their models, and also significant problems in modeling realistic food demand behaviors. Simulation models have the advantage of offering structural explanations of the linkages between agriculture and nutrition, but their deterministic structure means that outcomes can be sensitive to assumption and modeling procedures that are not always fully transparent. Even so, we feel that more research using these tools is strongly warranted given the potential importance of the economywide impacts of agricultural interventions on incomes, prices, and diets.

Agricultural Growth as a Driver of Incomes

A substantial body of literature explores linkages between income/wealth and child nutrition at both the household and economywide level, and most of it establishes at least moderately strong linkages (Bershteyn et al. 2014). Unsurprisingly, however, the associations between economic growth—typically, changes in gross domestic product (GDP) per capita—and changes in stunting are not as strong as they are for economic growth and poverty reduction, because poverty is definitionally related to the income distribution (Headey 2013a). In contrast, incomes have multiple indirect linkages to nutrition through microeconomic impacts on nutritionally relevant food and nonfood expenditures and through macroeconomic impacts on government expenditures on nutrition-specific and nutrition-sensitive investments (for example, WASH, education, agriculture).

At the household level, analyses of the demand for different foods confirm that people diversify out of staple foods as their income levels rise (Subramanian and Deaton 1996). However, meta-analyses of food demand estimates tend to suggest that households diversify most rapidly into animal-sourced foods (ASFs) and unhealthy processed foods, while diversification into fruits, vegetables, and pulses is typically less pronounced (Colen et al. 2017). These stylized results also appear to be true of national food systems, which often diversify only slowly, especially into fruits and vegetables (Choudhury and Headey 2017). We also know that governments in different countries allocate their investments across sectors in different ways. India has experienced rapid economic growth in recent decades but has only recently invested in WASH and community health/nutrition initiatives targeted at young children, which may provide an explanation of the often-weak linkages between growth and nutritional improvement in India. At the same time, the connections between economic growth at the macroeconomic level and improved economic status at the household level may simply be quite weak. Indeed, household incomes and wealth tend to be some of the strongest predictors of nutritional improvement (Haddad et al. 2003; Headey, Hirvonen, and Hoddinott 2017).

The literature cited above therefore establishes at least a moderately strong association between income/wealth and nutrition measures but does not specifically inform us about the role of agriculture. Only a handful of studies have specifically addressed the role of agricultural growth in influencing child nutrition. Headey (2013a) disaggregates economic growth into weighted growth in agricultural and nonagricultural value-added and finds no systematic difference in the strength of their associations with stunting, although growth in food production in food-insecure countries is significantly associated with stunting reduction. Another study by Headey and Hoddinott (2016) explores the impact of Bangladesh's late Green Revolution in rice production on linear growth (HAZ) and weight gain (WHZ) among young children, as well as infant and young child feeding (IYCF) indicators, using a panel of districts over 1996–2011. They find that growth in rice yields is significantly associated with weight gain and the earlier introduction of solid foods, but not with stunting or dietary diversity indicators. They suggest that growth in staple food production by itself may have limited impacts on stunting because of weak linkages to the diversification of children's diets. An implication is that diversification of food systems is important for improving nutrition, not just increased productivity of staple foods. Moreover, while improving productivity of staples may itself contribute to diversification through both demand and supply-side effects (for example, grains are an input into livestock production), governments themselves may have a role to play in accelerating diversification of production and consumption.

Although the evidence on the nutritional impacts of agricultural growth is limited, it suggests that agricultural growth may have important nutritional impacts in localities characterized by

1. high levels of basic food insecurity in which many households do not consume sufficient calories, and
2. early stages of structural transformation where most of the poor depend on agriculture for a living, particularly production of staple foods.

The first criterion ensures that increases in calorie availability from growth in staple food production could translate into nutritional improvements before hitting diminishing returns, while the second criterion ensures that agricultural growth has large impacts on the incomes of the poor.

In Figure 10.2 we used multicountry DHS data to illustrate these differences by examining stunting prevalence in Africa south of the Sahara (SSA) and South Asia disaggregated by locality (rural/urban), father's primary occupation (farmers, manual workers, in services, professional), and a simple measure of household wealth (the average number of six assets owned). A fourth dimension is population size as approximated by the share of each DHS

FIGURE 10.2 Stunting prevalence among children 24–59 months, by location, father's occupation, and asset ownership

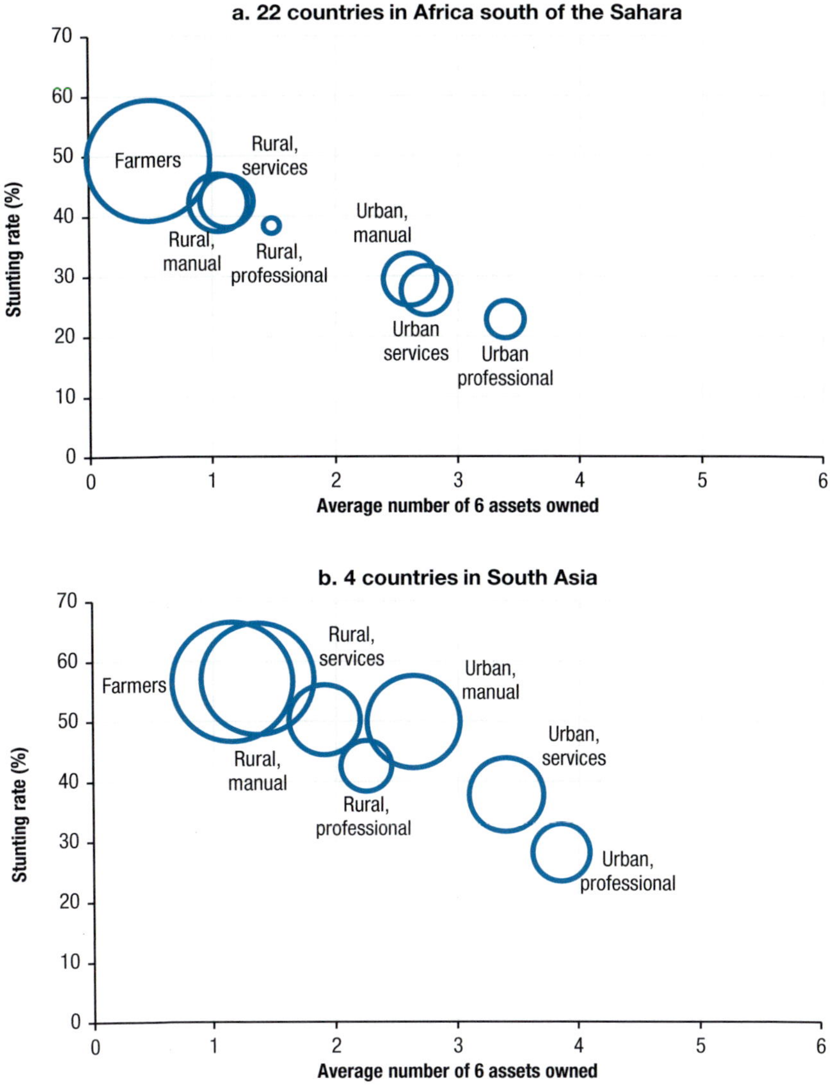

Source: Authors' estimates from Demographic and Health Survey data from Headey, Stifel et al. (2017).

Note: Rural/urban classifications follow the Demographic and Health Survey (DHS) (national) definitions, except that all farming households are classified as rural. Circle sizes reflect the sizes of the samples of children in each locality-occupation class. Ideally this measure should use sample weights, because the DHS oversamples urban areas, but many surveys are several years old, so we do not apply weights, in order to allow for recent urbanization trends. The six assets in question are a TV, motorbike, car, refrigerator, electrification, and an improved floor material. The four countries in South Asia are India, Pakistan, Bangladesh, and Nepal.

sample classified into a rural-occupation bin. The figure illustrates several striking findings.

First, SSA is still at a relatively early stage of structural transformation, whereas South Asia is at a significantly more advanced stage. In SSA, easily the largest share of children live in households where the father primarily relies on farming for a living, while the rural and urban manual and services occupation clusters all have broadly similar population sizes. In South Asia, more children live in rural nonfarm households (manual work plus services) than in farm households, although the latter group is still sizable. More of the population is also urbanized.

Second, there are very strong associations between locality-occupation categories and household wealth. In SSA farming households are the poorest, but all rural households are substantially poorer than urban households. In South Asia, where landlessness is more common, the wealth difference between farm and nonfarm occupation is not marked, and rural-urban differences in wealth are evident but not as pronounced as they are in SSA.

Third, it is notable that in SSA children of farming households have by far the highest rate of undernutrition, with almost 50 percent stunted in the 24-to-59-month range. Stunting rates are lower for rural nonfarm occupations and substantially lower for more urbanized livelihoods. Overall, however, the association between stunting and wealth is quite pronounced in SSA. In contrast, in South Asia the relationship between stunting rates, occupations, and wealth is more nuanced. Stunting is very high for all rural occupations, but also for urban manual worker households. Indeed, in absolute terms stunting is still very high in urban services and urban professional households.

Overall these findings suggest that agriculture growth is likely to have more impact on stunting in SSA, where stunting is closely associated with poverty, with farming as a livelihood, and with rural living in general. In the more structurally advanced economies of South Asia, the income role of agricultural growth is somewhat more limited.

Food Prices and Nutrition

Food Price Shocks and Nutrition

Food affordability has long been viewed as an important linkage between agricultural productivity and poverty, as well as basic food security. Studies of the 2007/2008 global food crisis primarily focused on the impacts of rising prices of staple foods on poverty (Ivanic and Martin 2008) or food security measures

(Headey 2013b). However, at least one earlier study of the 1998 Indonesian financial crisis showed that rising rice prices can also reduce dietary diversity and increase micronutrient deficiencies, as the real income shock associated with higher rice prices in Indonesia compelled consumers to cut back on more expensive sources of calories, such as eggs (Block et al. 2004).

To assess impacts of food price changes on household expenditure or poverty, economists have typically tried to assess whether consumers are net food consumers or net food producers, following Deaton (1989). Studies conducted after the 2007/2008 crisis often found that many of the poor were net food consumers, even in rural areas (Aksoy and Isik-Dikmelik 2008), suggesting that the poor suffer from higher prices. However, a recent review of these issues highlights two important caveats (Headey and Martin 2016). First, accurate measurement of net food consumption positions is very difficult because of the significant differences in recall periods for consumption and production in household surveys, potentially leading to systematic errors. Second, high food prices typically stimulate increased demand for labor, leading to higher wages and offsetting at least some of the damage of higher prices to nonfarming households.

While the research cited above is relevant to nutrition because of the associations between income/food security, diets, and nutrition outcomes, hardly any rigorous research examines the impacts of price volatility on nutrition outcomes or dietary diversity. This would seem to be an important area for future research.

International Variation in Food Prices and Dietary Affordability

While food prices are volatile over time, there are also strong theoretical reasons to expect prices to vary across countries, since many foods are not perfectly tradable or are very costly to trade. This is particularly true of highly perishable fruits, vegetables, and ASFs, but much less true for cereals, which are highly traded all over the world. Moreover, since cereals and other staples constitute the largest source of calories for poor populations, it makes sense to ask how costly it is for different populations to diversify away from staples.

Previous research only provides indirect answers to this question. Bennett (1941) provides a classic analysis of the diversification of Western diets away from wheat-based products, while Subramanian and Deaton (1996) show that consumers opt for more-expensive sources of calories as incomes rise. In the United States and Europe obesity researchers have also focused on measuring the costs of different sources of calories, typically noting that unhealthy processed foods (such as soft drinks) are much cheaper than healthy fresh foods

FIGURE 10.3 Relative caloric prices for various vegetable-based foods

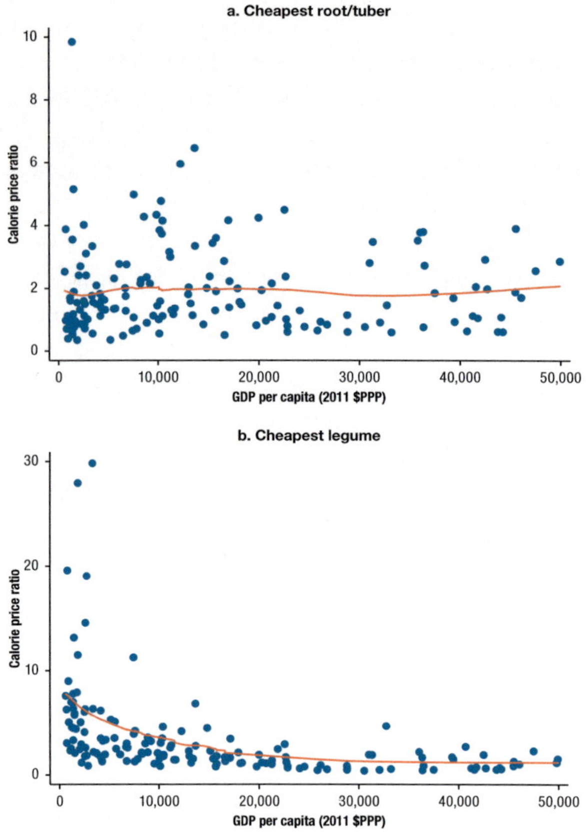

in caloric terms (Darmon and Drewnowski 2015). A recent study provides a more extensive quantification of the caloric costs of 657 different foods in 176 countries but measures each nonstaple food's caloric cost relative to that of a basket of country-specific staple foods (Headey and Alderman 2019). These relative caloric prices are therefore currency-free ratios that measure the costliness of diversifying away from the cheapest staple in any given country.

The results of that study reveal that prices of different foods are highly variable across countries and across different income levels, particularly for less tradable products such as eggs, fresh milk, and some (but not all) fresh fruits and vegetables. Figures 10.3 and 10.4 show scatter plots of these calorie price ratios against GDP per capita for the seven food groups listed in Table 10.2 and fortified infant cereals (which are a potential alternative means

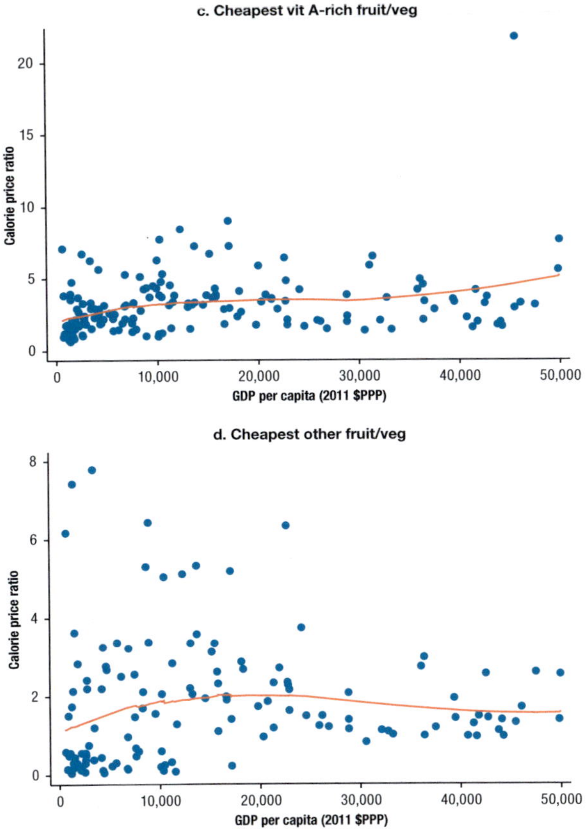

Source: Data are from Headey and Alderman (2019).

Note: GDP = gross domestic product; $PPP = purchasing power parity dollars. Red lines show LOWESS curves to allow for nonlinear functional forms.

of diversifying the diet with a complete range of essential nutrients). In this large sample of countries, per capita income varies from just $617 per capita in the Democratic Republic of Congo to approximately $50,000 per capita in the United States.

The results reveal several important findings.

First, among vegetal foods (Figure 10.3), legumes/nuts are a cheap source of calories (and typically the cheapest source of protein), but dark green leafy vegetables, vitamin A–rich fruits and vegetables, and other fruits/vegetables are relatively expensive. Interestingly, though, some vegetal foods tend to be more expensive in higher-income countries (especially dark green leafy vegetables), although in poor countries there is also substantial variation in relative caloric prices.

FIGURE 10.4 Relative caloric prices for various animal-sourced foods and fortified infant cereals

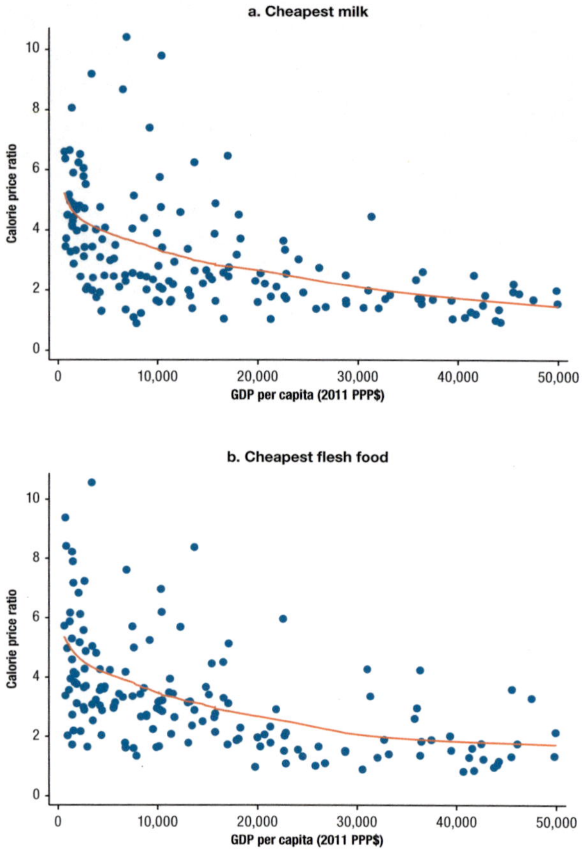

Second, dairy, eggs, and fortified infant cereals show a very different pattern (Figure 10.4), with very high prices in poorer countries (on average) that decline as incomes rise. Indeed, in many of the poorest countries, calories from milk, eggs, and fortified cereals are 10 times as expensive as calories from staple foods. Headey and Alderman (2019) also show that dairy prices, in particular, are robustly and positively associated with stunting rates across countries, meaning that the high cost of dairy might offer an important explanation of high stunting rates in Africa and Asia especially. In contrast to these other ASFs, meat (red or white) tends to be cheaper, with a much weaker income gradient, while fish (not shown in the table) is relatively cheap in Southeast Asia and western and central Africa where it is widely consumed.

c. Cheapest eggs

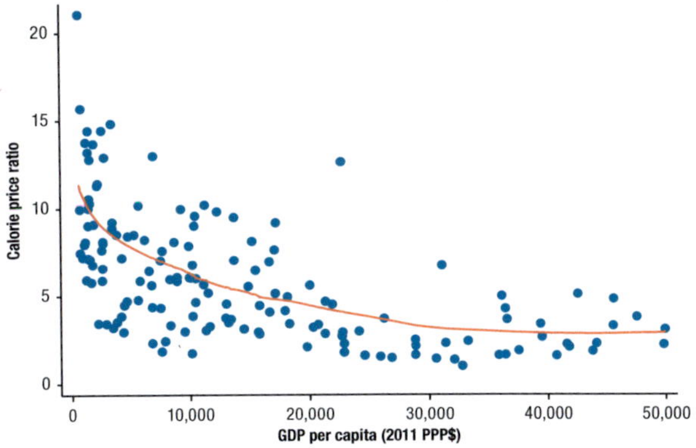

d. Fortified infant cereal

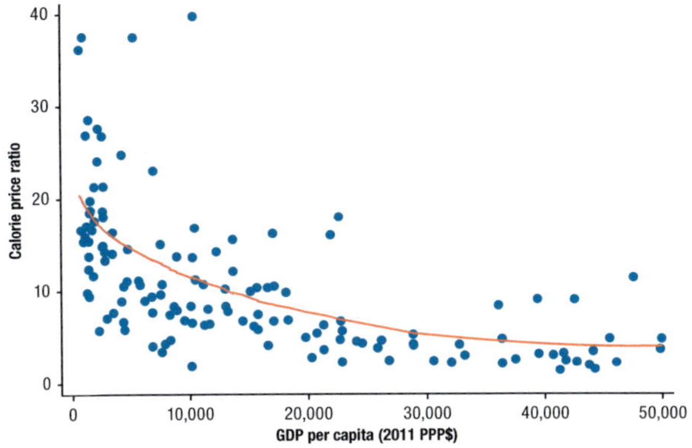

Source: Data are from Headey and Alderman (2019).

Note: GDP = gross domestic product; PPP$ = purchasing power parity dollars. Red lines show LOWESS curves to allow for nonlinear functional forms.

The variation in relative prices across countries is stark, but in several respects not surprising. First, the relative caloric prices across countries partly reflect variation in the denominator (the cost of staple foods, which rise as incomes rise because consumers choose more expensive staple foods when their incomes rise). Second, perishability and tradability clearly explain some of these patterns. Fresh milk and eggs are extreme cases of food products that are difficult to trade long distances (especially in countries with underdeveloped value chains), so the prices of these products are largely determined by productivity levels in the domestic dairy and poultry sectors, which are generally low in poorer countries. Finally, fortified infant cereals are highly tradable products but still extremely expensive in lower-income countries because they are informational goods: parents have little trust in locally produced infant cereals and instead opt for very expensive multinational brands if they can afford them (Masters, Nene, and Bell 2017).

These findings suggest that an important objective of nutrition-sensitive agricultural investments and reforms should be making nutrient-dense foods more affordable. This could stem from productivity investments in "non-tradable" sectors (for example, poultry or fresh milk), but also by addressing trade distortions that limit imports of nutrient-dense foods (for example, dairy powder).

Shadow Prices and Missing Markets for Nutritious Foods

The section above describes cross-country variation in prices, but there is also an extensive literature on the variation of implicit "shadow" prices within countries. Shadow prices are the implicit values or opportunity costs that households face when trying to obtain additional quantities of something for which an explicit price does not exist because of missing or incomplete markets. Markets might fail to provide an adequate supply of all foods throughout the year because of perishability—particularly of eggs, fresh milk, and many fruits and vegetables—and because local demand for these products is quite limited in low-income and low-density rural populations. Faced with imperfect supply from markets, households make production decisions that are designed to satisfy their consumption demands. For example, in Ethiopia over 90 percent of the milk produced by rural households is consumed by those same households (Hoddinott, Headey, and Dereje 2015). In this situation there are, unsurprisingly, strong associations between cattle ownership and children's milk consumption, but also between cattle ownership and child growth outcomes. Many other studies have linked livestock ownership to increased consumption of ASFs (Kabunga, Ghosh, and Webb 2017; Choudhury and Headey 2018) and local crop diversity to dietary diversity (Dillon, McGee, and Oseni 2015; Jones 2017b; Sibhatu and Qaim 2018).

In Figure 10.5 we report summary results from a multivariate regression analysis of an extensive multicountry synthetic dataset combining DHS surveys with geographic information system (GIS) data on agroecological characteristics and a "market access" indicator of travel times to markets (Heady, Nguyen et al. 2017). The figure shows the sign (+/–) and strength of each association with the goal of uncovering broad patterns of associations. For household-level factors from the DHS, we observe the expected strong and positive relationship between wealth and most dietary indicators, with vitamin A–rich fruits and vegetables and legumes/nuts being the sole exceptions,

FIGURE 10.5 Associations between children's food consumption patterns and various household (DHS) and community characteristics (GIS)

	Consumption indicators for children 6–23 months of age, over the past 24 hours							
	Minimum dietary diversity	Dairy	Egg	Red or white meat	Fish	Vitamin A-rich fruit/veg	Other fruit/veg	Legumes/ nuts
Wealth (DHS)	+++	+++	+++	++	+++	0	+	0
Education (DHS)	+++	+++	+++	+++	+++	0	+++	++
Healthcare (DHS)	+++	– –	++	+	+++	+++	+++	+++
Livestock proximity (DHS/GIS)	–	+++	++	0	+++	+/–	+/–	– – –
Market access (GIS)	+	++	0	– –	0	– – –	0	0
Crop potential (GIS)	+++	– –	++	– –	– –	+++	+++	+++
Tree cover (GIS)	+	– – –	+	+	++	+++	+++	– –
Elevation (GIS)	0	++	0	– –	– – –	++	– –	+++

Legend

Positive associations: + Weak ++ Moderate +++ Strong

Negative associations: – Weak – – Moderate – – – Strong

0 = No association

Source: Authors' estimates from a multicountry dataset linking DHS surveys to GIS indicators (Headey, Stifel et al. 2017).
Note: DHS = Demographic and Health Surveys; GIS = geographic information system.

suggesting these may be inferior goods. Similar patterns hold for parental education and access to healthcare. Of greater interest is the importance of community characteristics. Livestock proximity (ownership of livestock or proximity to fisheries at major lakes and rivers) is strongly associated with dairy, poultry, and fish consumption, but not with red or white meat consumption. A market access indicator—estimated travel time to 25,000-person cities—is associated with aggregate dietary diversity and with dairy consumption but has insignificant or negative associations with consumption of other foods. Crop potential is strongly and positively associated with minimum dietary diversity, egg consumption, and consumption of crop-based foods, while tree cover is generally also associated with consumption of most foods (see Johnson, Brown, and Jacob 2013 for a detailed study on this relationship). Elevation is associated with increased dairy consumption, which is unsurprising given the reduction in livestock diseases at higher altitudes, but has mixed associations with other variables. Another recent study explores these kinds of associations in multivariate regression models and shows that child dietary patterns are significantly related to environmental characteristics, even after controlling for individual and household characteristics. However, the magnitude of the associations is often moderate, consistent with a recent meta-analysis of farm-level studies finding generally modest links between production diversity and dietary diversity.

Another recent study links data on farm production, community production, and the availability and cost of different foods in local markets in poor rural areas of Ethiopia (Headey et al. 2019). As was the case with more aggregate price data, these authors show that ASFs and fruits and vegetables are generally very expensive in rural markets, while dairy is not sold in approximately half of the markets surveyed but is instead own-consumed or sourced from neighbors. Their results paint a somewhat complex picture: own-consumption is not overwhelmingly important except in the case of dairy, but markets were also incapable of providing nutrient-dense foods at an affordable price. A large analysis of food sources in Ethiopia shows that households predominantly rely on markets to source nonstaple foods (except dairy), suggesting that markets are important. But while these studies are interesting, there are remarkably few studies examining food markets in developing countries through a nutritional lens, and much more work is needed on this front.

Agriculture, Rural Livelihoods, and Nutrition

Rural Poverty and Limited Access to Infrastructure and Services

As we saw in Agricultural Growth as a Driver of Incomes, stunting rates and absolute numbers of stunted children are significantly higher in rural areas of SSA than in urban areas. Figure 10.2 also provided at least one explanation of this phenomenon: rural people are simply poorer than urban people, as measured by household assets. However, wealth differences provide only one likely explanation of the nutritional disadvantages of rural populations, since there are many other dimensions of rural living that might also contribute to undernutrition. Earlier research concluded that this rural-urban inequality in nutritional status stemmed from differences in "endowments"—such as wealth, education, and access to services—rather than from the nutritional returns to these endowments (Smith, Ruel, and Ndiaye 2004; Zanello, Srinivasen, and Shankar 2016). A recent paper extended this research for a broader range of SSA countries to examine both rural-urban differences and differences across different degrees of remoteness based on proximity to 25,000-person cities (Headey, Nguyen et al. 2017). Strikingly, the results imply that proximity to cities is not important independent of its associations with wealth, education, and access to services. Instead, remote villages—and rural villages in general—are simply characterized by poverty, low levels of human capital, and poor access to services. Indeed, these authors extend the rural-urban decompositions of earlier work to find that differences in endowments explain 77 percent of the observed difference in stunting rates across rural and urban areas, with differences in socioeconomic status (wealth and nonfarm employment) accounting for almost 40 percent of this difference, followed by parental education (19 percent) and health/infrastructural services (11 percent). They also find similar results for rural-urban differences in child dietary diversity. An implication of these results is that physical remoteness seems to primarily influence nutrition through its harmful impacts on multidimensional poverty.

A related strand of literature looks at "nutrition smoothing" to identify how access to markets, local infrastructure, and public services helps households buffer the impacts of nutritional shocks (Mulmi et al. 2016; Darrouzet-Nardi and Masters 2017; Shively 2017; Thapa and Shively 2018). The specific kind of shock most often investigated is seasonal fluctuations in rainfall and temperature around the time of a child's birth, which is much more closely linked to attained height for children in poorer, more remote places than in towns and cities. Some of the apparent link between month of birth and height for age is due to random errors in recorded birth months among children without birth registrations (Larsen, Headey, and Masters 2019), but even after adjusting for those errors, there is significant variation in attained height by season of birth (Finaret and Masters 2019). Research to date has shown how sanitation, food markets, and local infrastructure can help households protect their children from seasonal climate fluctuations (Mulmi et al. 2016; Shively 2017; Thapa and Shively 2018), leading to new work focusing on specific kinds of smoothing such as year-round access to a nutritious diet (Bai, Naumova, and Masters 2019).

Agricultural Production Hazards

Beyond problems with remoteness and poor access to services, most rural populations still primarily work in agriculture, which presents its own potential health and nutritional hazards. A mostly qualitative literature, particularly in South Asia, has explored associations between women's employment in agriculture and maternal nutrition, including substantial workloads during and soon after pregnancy (Balagamwala et al. 2015). One important concern is that physical exertion—especially in conditions of high temperatures—may be harmful for women's nutrition, including during pregnancy or lactation.

This literature also expresses concerns about the impacts of women's workload and time constraints on their ability to care for children and attain diverse diets (Johnston et al. 2018; Stevano et al. 2019). There is some evidence, for example, that agricultural interventions can increase women's workloads and time constraints, although there are few clear and consistent relationships to nutrition outcomes, perhaps because different household members respond in different ways or because income gains from agricultural intensification offset time constraints. Overall, though, there is little solid empirical evidence on whether agricultural employment is harmful to maternal or child nutrition beyond the obvious associations between employment in agriculture and general socioeconomic poverty.

Other research has examined specific biological mechanisms linking agricultural livelihoods to health. For example, Brainerd and Menon (2014) look at the long-standing concern that excess and inappropriate use of chemical inputs, particularly pesticides and herbicides, has harmful effects on health and nutrition. They find significant evidence of adverse impacts on maternal and child health, including birthweight. More recently, Sheahan, Barrett, and Goldvale (2017) examine pesticide use in five SSA countries and find that pesticide use is associated with greater health expenditures and more sick days, although they do not test associations with child nutrition outcomes.

Another strand of research has looked at health problems associated with livestock ownership. Formative research in a rural Zimbabwean village monitored children for 12-hour periods (Ngure et al. 2013). They found that a large proportion of children directly consumed or mouthed chicken feces and dirt that may have been contaminated with chicken feces. They also show that the bacterial loads of chicken feces are some 10,000 times higher than dirt and other surfaces in the household. They argue that ingestion of such high loads of bacteria may contribute to environmental enteropathy, a chronic but latent infection of the gut that has been strongly linked to child stunting. Subsequent research in Ethiopia (Headey and Hirvonen 2016) and Bangladesh (George et al. 2015) found that children living under the same roof as poultry were more likely to be stunted, and a three-country study by Headey, Hoddinott, and Park (2017) found that homesteads in Bangladesh and Ethiopia with observable animal feces in the compound had shorter children. A systematic review of diarrheal studies also found significant associations with livestock ownership in a majority of studies (Zambrano et al. 2014). Hence, despite the importance of ASF intake for child nutrition, and the clear associations between livestock ownership and ASF consumption, there are also negative pathways between livestock ownership and child growth. These pathways presumably operate mainly through fecal contamination and enteric infections, though pulmonary infections associated with poultry ownership are a potential concern (Gomaa et al. 2015), as is the connection between cattle ownership and malaria (Donnelly et al. 2015).

Implications for Policy and Research

Historical and comparative evidence suggests there are three main levers by which agricultural change and food policy influence diets and nutrition outcomes.

The first lever of agriculture's influence is through growth in real incomes and poverty reduction. Agricultural growth—including growth in staple food production—has been shown to be a historically important driver of poverty reduction because so many poor people directly and indirectly depend on agriculture for a living (Diao, Hazell, and Thurlow 2010). Evidence reported above suggests that agricultural growth is still likely to be an important driver of poverty reduction in South Asia and SSA, though much more so in the latter. Of course, whether agricultural investments offer the most-effective means of reducing poverty is another question (Dercon and Gollin 2014). Moreover, admittedly scant empirical evidence from Bangladesh (Headey and Hoddinott 2016) suggests that the nutritional benefits of cereals-oriented agricultural development strategies may be limited, because rapid increases in consumption of more nutrient-dense foods is not guaranteed by productivity growth in cereals. More research is needed on this issue, however.

The second lever is through influencing the relative price of nutritious foods and managing food price variation. Evidence from different settings suggests that urbanization, infrastructure, or market access or all of those may suppress various nutritional insults. And though not well documented, it is likely that market development plays an important role in reducing both average prices and price fluctuations. Recent research on relative price differences across countries emphasizes the imperfect tradability of nutrient-dense foods, including ASFs that may be critical for healthy growth and cognitive development in early childhood. This suggests that in many localities certain nutrient-rich foods are either not affordable or not accessible in the market, forcing farmers into potentially inefficient self-reliance. The costliness of highly nutritious foods in lower-income countries perhaps provides the strongest mandate for nutrition-focused agricultural development; other economic sectors may well drive income growth, but only food policies can influence the affordability of nutritious foods.

The third agricultural lever for nutrition is through the transformation of agricultural livelihoods. In developing countries rural populations pervasively have nutrition outcomes that are significantly worse than urban populations, and there is little mystery as to why this is so: rural populations are much poorer and have substantially less access to essential goods and services. Yet they also face specific hazards associated with agricultural living, including exposure to potentially harmful chemical inputs, physically arduous work (including mothers), and highly unhygienic environments due to poor livestock management practices.

What are the policy implications of these findings? More specifically, what should agricultural development actors do differently, if they care about nutrition?

First, agricultural policy instruments should still be used to improve incomes of rural populations, but they should also be used to reduce the prices of nutritionally important foods. These include ASFs of importance to young children (dairy, eggs, fish), as well as fresh fruits and vegetables. This might entail alterations in the portfolio of agricultural investments toward nonstaple foods, both in international organizations (such as CGIAR) and national research and implementation agencies. We estimate that at least 80 percent of CGIAR expenditures are allocated to crops—and many of these are starchy staples—and just 20 percent are allocated to livestock and fisheries, and we suspect that in many developing countries these ratios are even more skewed in favor of staple crops. Arguably this is a lopsided composition. However, it is also crucial to better understand the important role of cereals and soybeans as feed inputs into the livestock and aquaculture sectors. High feed costs in Africa, in particular, may offer one explanation as to why many ASFs remain prohibitively expensive there (Andam, Arndt, and Hartley 2017).

Beyond the composition of public investment, however, the development of higher-value sectors also entails a greater role for the private sector and public-private partnerships, including working with larger and more commercialized farms and firms. Nutritionally important foods are highly perishable (unlike staple cereals), meaning that value-chain bottlenecks are often binding constraints, including poor infrastructure and storage. Lengthier value chains involve many more private sector actors such as wholesalers, traders, and retailers. There are also important economies of scale for *some* high-value sectors, meaning that larger firms are a more efficient structure for achieving rapid productivity growth and larger reductions in prices for consumers (particularly poultry). There are also opportunities to use trade as an instrument for reducing the prices of nonperishable nutrient-rich foods. Many meat products are highly tradable when frozen or dried, while dairy powder is also highly tradable and relatively affordable but underconsumed in many low-income countries. Hence, efforts to improve consumption of nutrient-rich foods do not need to solely rely on production, especially for products in which the country has little or no comparative advantage. This will likely be the case for dairy production in most tropical countries, and East Asian experience has been instructive in this regard through its combination of trade and investment in the domestic dairy sector (FAO 2008).

Second, policies need to focus more holistically on the health implications of farming as a livelihood. One of the most persistent problems for rural populations has been access to nutritionally important goods and services, including education, basic maternal and child health services, and WASH. This is likely a combination of the high cost of extending these services to remote and dispersed rural populations (especially in Africa), urban-biased political systems, and various governance problems such as corruption and insufficient decentralization (Headey, Bezemer, and Hazell 2010). In recent times some countries have made impressive strides in extending basic services to remote rural communities (Global Health Workforce Alliance 2013), with Nepal being an exemplar in this regard. But for agricultural households, specifically, there are farming-related health hazards that have long been neglected, including inappropriate use of chemical inputs (Sheahan, Barrett, and Goldvale 2017), scavenging livestock systems combined with poor household hygiene (Headey and Hirvonen 2016), and (one suspects) excessively high workloads for women in general and mothers in particular.

In truth, we still know relatively little about the importance of these farm-level hazards for human nutrition, or how policies can most effectively combat them. Nevertheless, policymakers cannot always wait on rigorous evidence, and there is a strong case to be made for multisectoral policies designed to reduce the health hazards associated with agricultural livelihoods. This includes regulations and agricultural extension practices designed to reduce the health hazards associated with inappropriate chemical input use, integrated agriculture-WASH programs to reduce fecal contamination by livestock (for example, community hygiene, improved poultry housing, increased awareness), and integrated programs that combine social protection, nutrition, and agriculture to reduce women's workloads, especially during and soon after pregnancy.

In recent years our understanding of agriculture-nutrition linkages has vastly improved, but many knowledge and policy gaps remain. Much more emphasis is needed on how policies influence diets through income and price changes, what trade-offs may exist between poverty and nutrition targets, and on how multisectoral rural development efforts can improve access to basic services and reduce farming's hidden health risks.

References

Aksoy, M. A., and A. Isik-Dikmelik. 2008. *Are Low Food Prices Pro-Poor? Net Food Buyers and Sellers in Low-Income Countries*. Policy Research Working Paper 4642. Washington, DC: World Bank.

Alderman, H., and D. Headey. 2017. "The Timing of Growth Faltering Has Important Implications for Observational Analyses of the Underlying Determinants of Nutrition Outcomes." In *Stunting: Past, Present and Future*. London: London School of Economics and Political Science.

Alkire, S., R. Meinzen-Dick, A. Peterman, A. Quisumbing, G. Seymour, and A. Vaz. 2012. *The Women's Empowerment in Agriculture Index*. IFPRI Discussion Paper 1240. Washington, DC: IFPRI.

Andam, K. S., C. Arndt, and F. Hartley. 2017. *Eggs Before Chickens? Assessing Africa's Livestock Revolution with an Example from Ghana*. IFPRI Discussion Paper 1687. Washington, DC: IFPRI.

Arndt, C., et al. 2017. "Measuring Household Diet Quality: Revisiting the 'Diet Problem.'" Unpublished. IFPRI, Washington, DC.

Bachewe, F., and D. Headey. 2016. "Urban Wage Behaviour and Food Price Inflation in Ethiopia." *Journal of Development Studies* 53 (8): 1–16.

Bai, Y., E. Naumova, and W. Masters. 2019. *Seasonality in the Cost of Nutritious Diets in Malawi, Tanzania and Ethiopia*. CANDASA project working paper. Medford, MA: Tufts University.

Balagamwala, M., H. Gazdar, and H. B. Mallah. 2015. *Women's Agricultural Work and Nutrition in Pakistan: Findings from Qualitative Research*. Working Paper 2. Leveraging Agriculture for Nutrition in South Asia (LANSA).

Barrett, C. B., and M. R. Carter. 2010. "The Power and Pitfalls of Experiments in Development Economics: Some Non-Random Reflections." *Applied Economic Perspectives and Policy* 32: 515–548.

Bennett, M. 1941. "Wheat in National Diets." *Wheat Studies* 18: 37–76.

Bershteyn, A., H. Lyons, D. Sivam, and N. Myhrvold. 2014. "Association between Economic Growth and Early Childhood Nutrition." *Lancet Global Health* 3: e79–e80.

Block, S., L. Kiess, P. Webb, S. Kosen, R. Moench-Pfanner, M. Bloem, and C. P. Timmer. 2004. "Macro Shocks and Micro Outcomes: Child Nutrition during Indonesia's Crisis." *Economics and Human Biology* 2: 21–44.

Brainerd, E., and N. Menon. 2014. "Seasonal Effects of Water Quality: The Hidden Costs of the Green Revolution to Infant and Child Health in India." *Journal of Development Economics* 107: 49–64.

———. 2017. "What Drives Diversification of National Food Supplies? A Cross-Country Analysis." *Global Food Security* 15: 85–93.

———. 2018. "Household Dairy Production and Child Growth: Evidence from Bangladesh." *Economics & Human Biology* 30: 150–161.

Darmon, N., and A. Drewnowski. 2015. "Contribution of Food Prices and Diet Cost to Socioeconomic Disparities in Diet Quality and Health: A Systematic Review and Analysis." *Nutrition Reviews* 73: 643–660.

Darrouzet-Nardi, A. F., and W. A. Masters. 2017. "Nutrition Smoothing: Can Proximity to Towns and Cities Protect Rural Children against Seasonal Variation in Agroclimatic Conditions at Birth?" *PLOS ONE* 12: e0168759.

Deaton, A. 1989. "Rice Prices and Income Distribution in Thailand: A Non-Parametric Analysis." *Economic Journal* 99: 1–37.

de Janvry, A., M. Fafchamps, and E. Sadoulet. 1991. "Peasant Household Behavior with Missing Markets: Some Paradoxes Explained." *Economic Journal* 101: 1400–1417.

Dercon, S., and D. Gollin. 2014. "Agriculture in African Development: Theories and Strategies." *Annual Review of Resource Economics* 6: 471–492.

Diao, X., P. Hazell, and J. Thurlow. 2010. "The Role of Agriculture in African Development." *World Development* 38: 1375–1383.

Dillon, A., K. McGee, and G. Oseni. 2015. "Agricultural Production, Dietary Diversity and Climate Variability." *Journal of Development Studies* 51: 976–995.

Donnelly, B., L. Berrang-Ford, N. Ross, and P. Michel. 2015. "A Systematic, Realist Review of Zooprophylaxis for Malaria Control." *Malaria Journal* 14: 313.

FANTA (Food and Nutrition Technical Assistance). 2006. *Developing and Validating Simple Indicators of Dietary Quality and Energy Intake of Infants and Young Children in Developing Countries: Summary of Findings from Analysis of 10 Data Sets*. Washington, DC: FANTA and FHI.

FAO (Food and Agriculture Organization of the United Nations). 2008. *Improved Market Access and Smallholder Dairy Farmer Participation for Sustainable Dairy Development: Asia Smallholder Dairy Development Strategy and Outline Investment Plan*. Rome.

Finaret, A. B., and W. A. Masters. 2019. "Correcting for Artifactual Correlation between Misreported Month of Birth and Attained Height-for-Age Reduces but Does Not Eliminate Measured Vulnerability to Season of Birth in Poorer Countries." *American Journal of Clinical Nutrition* 110: 485–497.

George, C. M., L. Olja, S. Biswas et al. 2015. "Fecal Markers of Environmental Enteropathy Are Associated with Animal Exposure and Caregiver Hygiene in Bangladesh." *American Society of Tropical Medicine and Hygiene* 93: 269–275.

Gilbert, C. L., L. Christiaensen, and J. Kaminhski. 2017. "Food Price Seasonality in Africa: Measurement and Extent." *Food Policy* 67: 119–132.

Girard, A. W., J. Self, C. McAuliffe, and O. Olude. 2012. "The Effects of Household Food Production Strategies on the Health and Nutrition Outcomes of Women and Young Children: A Systematic Review." *Paediatric and Perinatal Epidemiology* 26: 205–222.

Global Health Workforce Alliance. 2013. *Global Experience of Community Health Workers for Delivery of Health Related Millennium Development Goals: A Systematic Review, Country Case Studies, and Recommendations for Integration into National Health Systems*. World Health Organization.

Gomaa, M. R., A. Kayed, M. Elabd et al. 2015. "Avian Influenza A(H5N1) and A(H9N2) Seroprevalence and Risk Factors for Infection among Egyptians: A Prospective, Controlled Seroepidemiological Study." *Journal of Infectious Diseases* 211: 1399–1407.

Haddad, L., H. Alderman, S. Appleton, L. Song, and Y. Yohannes. 2003. "Reducing Child Malnutrition: How Far Does Income Growth Take Us?" *World Bank Economic Review* 17: 107–131.

Haselow, N. J., A. Stormer, and A. Pries. 2016. "Evidence-Based Evolution of an Integrated Nutrition-Focused Agriculture Approach to Address the Underlying Determinants of Stunting." *Maternal & Child Nutrition* 12 (Suppl. 1): 155–168.

Headey, D. 2013a. "Developmental Drivers of Nutritional Change: A Cross-Country Analysis." *World Development* 42: 76–88.

———. 2013b. "The Impact of the Global Food Crisis on Self-Assessed Food Security." *World Bank Economic Review* 27: 1–27.

Headey, D. , and H. Alderman. 2019. "The Relative Caloric Prices of Healthy and Unhealthy Foods Differ Systematically across Income Levels and Continents." *Journal of Nutrition* 149 (11): 2020–2033.

Headey, D., D. Bezemer, and P. Hazell. 2010. "Agricultural Employment Trends in Asia and Africa: Too Fast or Too Slow?" *World Bank Research Observer* 25: 57–89.

Headey, D., A. Chiu, and S. Kadiyala. 2012. "Agriculture's Role in the Indian Enigma: Help or Hindrance to the Crisis of Undernutrition?" *Food Security* 4: 87–102.

Headey, D., and K. Hirvonen. 2016. "Is Exposure to Poultry Harmful to Child Nutrition? An Observational Analysis for Rural Ethiopia." *PLOS ONE* 11: e0160590.

Headey, D., K. Hirvonen, and J. Hoddinott. 2017. *Animal Sourced Foods and Child Stunting.* IFPRI Discussion Paper 1695. Washington, DC: IFPRI.

Headey, D., K. Hirvonen, J. Hoddinott, and D. Stifel. 2019. "Rural Food Markets and Child Nutrition." *American Journal of Agricultural Economics* 101 (5): 1311–1327.

Headey, D., and J. Hoddinott. 2016. "Agriculture, Nutrition and the Green Revolution in Bangladesh." *Agricultural Systems* 149: 122–131.

Headey, D., J. Hoddinott, and S. Park. 2017. "Accounting for Nutritional Changes in Six Success Stories: A Regression-Decomposition Approach." *Global Food Security* 13: 12–20.

Headey, D., and W. J. Martin. 2016. "Food Prices, Poverty, and Food Security." *Annual Review of Resource Economics* 8: 329–351.

Headey, D., P. Nguyen, S. Kim, R. Rawat, M. Ruel, and P. Menon. 2017. "Is Exposure to Animal Feces Harmful to Child Nutrition and Health Outcomes? A Multicountry Observational Analysis." *American Journal of Tropical Medicine and Hygiene* 96: 961–969.

Headey, D., D. Stifel, L. You, and Z. Guo. 2017. *Remoteness, Urbanization and Child Nutrition in Sub-Saharan Africa.* IFPRI Discussion Paper 1694. Washington, DC: IFPRI.

Hirvonen, K., and D. Headey. 2018. "Can Governments Promote Homestead Gardening at Scale? Evidence from Ethiopia." *Global Food Security* 19: 40–47.

Hirvonen, K., and J. Hoddinott. 2017. "Agricultural Production and Children's Diets: Evidence from Rural Ethiopia." *Agricultural Economics* 48: 469–480.

Hoddinott, J. 2011. *Agriculture, Health and Nutrition: Conceptualizing the Linkages, Leveraging Agriculture for Improving Nutrition and Health.* New Delhi: IFPRI.

Hoddinott, J., D. Headey, and M. Dereje. 2015. "Cows, Missing Milk Markets, and Nutrition in Rural Ethiopia." *Journal of Development Studies* 51: 958–975.

Ivanic, M., and W. Martin. 2008. "Implications of Higher Global Food Prices for Poverty in Low-Income Countries." *Agricultural Economics* 39 (S1): 405–416.

Johnson, K. B., M. E. Brown, and A. Jacob. 2013. "Forest Cover Associated with Improved Child Health and Nutrition: Evidence from the Malawi Demographic and Health Survey and Satellite Data." *Global Health: Science and Practice* 1 (2): 237.

Johnston, D., S. Stevano, H. Malapit, E. Hull, and S. Kadiyala. 2018. "Review: Time Use as an Explanation for the Agri-Nutrition Disconnect: Evidence from Rural Areas in Low and Middle-Income Countries." *Food Policy* 76: 8–18.

Jones, A. D. 2017a. "Critical Review of the Emerging Research Evidence on Agricultural Biodiversity, Diet Diversity, and Nutritional Status in Low- And Middle-Income Countries." *Nutrition Reviews* 75: 769–782.

———. 2017b. "On-Farm Crop Species Richness Is Associated with Household Diet Diversity and Quality in Subsistence- and Market-Oriented Farming Households in Malawi." *Journal of Nutrition* 147: 86–96.

Kabunga, N. S., S. Ghosh, and P. Webb. 2017. "Does Ownership of Improved Dairy Cow Breeds Improve Child Nutrition? A Pathway Analysis for Uganda." *PLOS ONE* 12: e0187816.

Kadiyala, S., J. Harris, D. Headey, S. Yosef, and S. Gillespie. 2014. "Agriculture and Nutrition in India: Mapping Evidence to Pathways." *Annals of the New York Academy of Sciences* 1331: 43–56.

Larsen, A. F., D. Headey, and W. Masters. 2019. "Misreporting Month of Birth: Diagnosis and Implications for Research on Nutrition and Early Childhood in Developing Countries." *Demography* 56: 707–728.

Leroy, J. L., and E. A. Frongillo. 2007. "Can Interventions to Promote Animal Production Ameliorate Undernutrition?" *Journal of Nutrition* 137: 2311–2316.

Masset, E., L. Haddad, A. Cornelius, and J. Isaza-Castro. 2012. "Effectiveness of Agricultural Interventions That Aim to Improve Nutritional Status of Children: Systematic Review." *BMJ* 344: d8222.

Masters, W. A., Y. Bai, A. Herforth, D. Sarpong, F. Mishili, J. Kinabo, and J. Coatesl. 2018. "Measuring the Affordability of Nutritious Diets in Africa: Price Indexes for Diet Diversity and the Cost of Nutrient Adequacy." *American Journal of Agricultural Economics* 100: 1285–1301.

Masters, W. A., M. Nene, and W. Bell. 2017. "Nutrient Composition of Premixed and Packaged Complementary Foods for Sale in Low- and Middle-Income Countries: Lack of Standards Threatens Infant Growth." *Maternal & Child Nutrition* 13: e12421-n/a.

Melo, P. C., Y. Abdul-Salam, D. Roberts et al. 2015. *Income Elasticities of Food Demand in Africa: A Meta-Analysis*. Publications Office of the European Union.

Micha, R., S. Khatibzadeh, P. Shi, K. Andrews, R. Engell, D. Mozaffarian, and Nutrition and Chronic Diseases Expert Group. 2015. "Global, Regional and National Consumption of Major Food Groups in 1990 and 2010: A Systematic Analysis Including 266 Country-Specific Nutrition Surveys Worldwide." *BMJ Open* 5: e008705.

Mulmi, P., S. Block, G. Shively, and W. Masters. 2016. "Climatic Conditions and Child Height: Sex-Specific Vulnerability and the Protective Effects of Sanitation and Food Markets in Nepal." *Economics & Human Biology* 23: 63–75.

Ngure, F. M., J. Humphrey, M. Mbuya et al. 2013. "Formative Research on Hygiene Behaviors and Geophagy among Infants and Young Children and Implications of Exposure to Fecal Bacteria." *American Journal of Tropical Hygiene & Medicine* 89: 709–716.

Pauw, K., and J. Thurlow. 2010. *Agricultural Growth, Poverty, and Nutrition in Tanzania, 2010 AAAE Third Conference/AEASA 48th Conference*. Cape Town: African Association of Agricultural Economists (AAAE); Agricultural Economics Association of South Africa (AEASA).

Pingali, P. L., and K. D. Ricketts. 2014. "Mainstreaming Nutrition Metrics in Household Surveys—Toward a Multidisciplinary Convergence of Data Systems." *Annals of the New York Academy of Sciences* 1331: 249–257.

Pinstrup-Andersen, P. 2013. "Nutrition-Sensitive Food Systems: From Rhetoric to Action." *Lancet* 382: 375–376.

Rosenzweig, M., and C. Udry. 2016. *External Validity in a Stochastic World*. Working Paper 22449. Cambridge, MA: National Bureau of Economic Research.

Ruel, M. T., and H. Alderman. 2013. "Nutrition-Sensitive Interventions and Programmes: How Can They Help to Accelerate Progress in Improving Maternal and Child Nutrition?" *Lancet* 382: 536–551.

Ruel, M. T., J. Harris, and K. Cunningham. 2013. "Diet Quality in Developing Countries." In *Diet Quality: An Evidence-Based Approach*, Vol. 2, edited by V. R. Preedy. New York: Springer.

Ruel, M. T., A. Quisumbing, and M. Balagamwala. 2018. "Nutrition-Sensitive Agriculture: What Have We Learned So Far?" *Global Food Security* 17 (June): 128–153.

Sheahan, M., C. B. Barrett, and C. Goldvale. 2017. "Human Health and Pesticide Use in Sub-Saharan Africa." *Agricultural Economics*. 48 (S1): 27–41.

Shively, G. E. 2017. "Infrastructure Mitigates the Sensitivity of Child Growth to Local Agriculture and Rainfall in Nepal and Uganda." *Proceedings of the National Academy of Sciences* 114: 903–908.

Sibhatu, K. T., and M. Qaim. 2017. "Rural Food Security, Subsistence Agriculture, and Seasonality." *PLOS ONE* 12: e0186406.

———. 2018. "Review: Meta-Analysis of the Association between Production Diversity, Diets, and Nutrition in Smallholder Farm Households." *Food Policy* 77: 1–18.

Singh, I., L. Squire, and J. Strauss. 1986. *Agricultural Household Models: Extension, Application and Policy*. Baltimore: Johns Hopkins University Press.

Smith, L. C., M. T. Ruel, and A. Ndiaye. 2004. *Why Is Child Malnutrition Lower in Urban Than Rural Areas?* IFPRI Discussion Paper 176. Washington, DC: IFPRI.

Springmann, M., D. Mason-D'Croz, S. Robinson et al. 2016. "Global and Regional Health Effects of Future Food Production under Climate Change: A Modelling Study." *Lancet* 387: 1937–1946.

Stevano, S., S. Kadiyala, D. Johnston, H. Malapit, E. Hull, and S. Kalamatianou. 2019. "Time-Use Analytics: An Improved Way of Understanding Gendered Agriculture-Nutrition Pathways." *Feminist Economics* 25: 1–22.

Subramanian, S., and A. Deaton 1996. "The Demand for Food and Calories." *Journal of Political Economy* 104: 133–162.

Thapa, G., and G. Shively. 2018. "A Dose-Response Model of Road Development and Child Nutrition in Nepal." *Research in Transportation Economics* 70: 112–124.

Thorne-Lyman, A., D. Spiegelman, and W. Fawzi. 2014. "Is the Strength of Association between Indicators of Dietary Quality and the Nutritional Status of Children Being Underestimated?" *Maternal & Child Nutrition* 10: 159–160.

Tiberti, L., and M. Tiberti. 2018. "Food Price Changes and Household Welfare: What Do We Learn from Two Different Approaches?" *Journal of Development Studies* 54: 72–92.

Van Campenhout, B. 2007. "Modelling Trends in Food Market Integration: Method and an Application to Tanzanian Maize Markets." *Food Policy* 32: 112–127.

Verger, E. O., T. Ballard, M. C. Dop, and Y. Martin-Prevel. 2019. "Systematic Review of Use and Interpretation of Dietary Diversity Indicators in Nutrition-Sensitive Agriculture Literature." *Global Food Security* 20: 156–169.

Victora, C. G., M. de Onis, P. C. Hallal, M. Blossner, and R. Shrimpton. 2010. "Worldwide Timing of Growth Faltering: Revisiting Implications for Interventions." *Pediatrics* 125 (3): e 473–480.

von Braun, J. 1995. "Agricultural Commercialization: Impacts on Income and Nutrition and Implications for Policy." *Food Policy* 20: 187–202.

WHO (World Health Organization). 2006. "WHO Child Growth Standards Based on Length/Height, Weight and Age." *Acta Paediatrica Supplement* 450: 76–85.

Wiggins, S., and S. Keats. 2015. *The Rising Cost of a Healthy Diet: Changing Relative Prices of Foods in High-Income and Emerging Economies*. Research Reports and Studies. London: Overseas Development Institute (ODI).

Zambrano, L. D., K. Levy, N. Menezes, and M. Freeman. 2014. "Human Diarrhea Infections Associated with Domestic Animal Husbandry: A Systematic Review and Meta-Analysis." *Transactions of the Royal Society of Tropical Medicine & Hygiene* 108: 313–325.

Zanello, G., C. S. Srinivasen, and B. Shankar. 2016. "What Explains Cambodia's Success in Reducing Child Stunting—2000–2014?" *PLOS ONE* 11: e0162668.

Chapter 11

TRANSFORMATION OF THE RURAL ECONOMY

Keijiro Otsuka and Xiaobo Zhang

The transformation of the rural economy is often accompanied by (1) increasing productivity of staple crop farming, (2) the introduction of more profitable high-value agricultural products, (3) more lucrative employment opportunities in rural nonfarm sectors (World Bank 2007; de Janvry, Sadoulet, and Murgai 2002), and (4) migration. Developing rural industries is a major way to increase employment opportunities in rural nonfarm sectors (Hayami 1998; Haggblade, Hazell, and Reardon 2007; Otsuka 1998, 2007). Although relationships among technological innovations in agriculture and the development of rural nonfarm sectors are highly complex, we attempt to sketch the farm-nonfarm sectoral linkages in the next section.

In low-income economies where land is relatively scarce and labor is abundant, notably in tropical Asia and increasingly in Africa south of the Sahara (SSA), the Green Revolution (the development and diffusion of high-yielding varieties of cereal crops) ought to be a major way to increase the productivity of rice, wheat, maize, and other staple crop farming (Evenson and Gollin 2003).[1] However, the Green Revolution has not taken place in most areas of SSA, where farm size is now almost as small as that in tropical Asia in the 1960s. It is also unclear to what extent the Green Revolution stimulated the development of nonfarm sectors through growth linkages. Next, we explore why SSA has failed to realize the Green Revolution and the extent to which the Green Revolution stimulated the development of nonfarm sectors in rural areas in tropical Asia.

In developing countries, markets of modern inputs—such as improved seeds, effective chemical fertilizer, and safe pesticides as well as credit markets—do not function well, nor does the market for high-value agricultural products, whose quality and safety cannot be easily identified by visual inspections of consumers. Also the market of production information services

1 In Africa, root crops like cassava and yams have experienced large productivity gains (Haggblade and Hazell 2010).

ubiquitously malfunctions, not to mention the malfunctioning of public-sector extension systems. Thus contract farming, which provides high-quality inputs on credit and production instruction, is considered to be an effective institutional device to overcome market failures.[2] Fears, however, are widely expressed that contract farming may primarily benefit large farmers because transaction costs can be saved by making a small number of contracts with large farmers rather than a large number of contracts with smallholder farmers (Otsuka, Nakano, and Takahashi 2016; Ton et al. 2018).

The chapter examines whether contract farming confers benefits primarily to large farmers in practice and how we may be able to make smallholders significantly better off by introducing new profitable crops and livestock products. More often than not, agriculture does not provide ample employment opportunities, largely due to land constraints on production expansion. One solution is to develop nonfarm sectors so as to provide more lucrative employment opportunities, in which working members of farm households increasingly find jobs. The chapter demonstrates the critical importance of increasing nonfarm income to improve the income of rural households and examines the roles of infrastructure and human capital in raising nonfarm income.

In general, the service sector is growing faster than the industrial sector in developing countries. One key driver is remittance sent home by migrants. It has been widely documented that most remittances are spent on housing, land purchase, and local service sectors (Ballard 2001). Broadly speaking, the service sector consists of traditional and modern service jobs. Jobs in traditional service sectors (for example, vending, rickshaw driving, and garbage picking) are low-paying, whereas jobs in modern service sectors (such as ICT-related service providers and finance) are high-paying but are offered only to highly educated workers. In order to provide ample and lucrative employment opportunities to a large, relatively uneducated and unskilled rural labor force, the development of labor-intensive industries in rural areas is indispensable. The chapter identifies the major characteristics of successfully developed rural industries and explores how we can stimulate the development of such industries. We conclude with a discussion of the role of rural-to-urban migration in the transformation of the rural economy and policy implications.

---

2 This is called the production contract, under which autonomy of farmers is restricted. However, farmers maintain greater autonomy under the market contract, which primarily specifies prices, quantity, and quality of products delivered as well as delivery dates.

A Conceptual Framework of Farm-Nonfarm Linkages

Following de Janvry and Sadoulet (2002) and Otsuka, Estudillo, and Sawada (2009), Figure 11.1 illustrates how technological innovations in agriculture, such as the Green Revolution and the introduction of high-value agricultural products, and investment in rural infrastructure affect rural income directly by increasing the efficiency and profitability of farming and indirectly through the growth linkage effects on nonfarm sectors, the inducement effect of improved infrastructure on the development of nonfarm sectors, and the reduction in prices of agricultural products.[3]

Although the growth linkage effects on the development of nonfarm sectors are expected to be positive, a reduction in agricultural prices will have negative effects on farm income. Another major factor that affects the income of rural households and the development of nonfarm sectors is human capital. Schooling, which is a major component of human capital along with health, is known to affect the efficiency of farming (for example, Foster and Rosenzweig 1996) and to have a positive effect on the choice of nonfarm jobs, rural-to-urban migration, and nonfarm income (for example, Matsumoto, Kijima, and Yamano 2006; Jolliffe 2004). Furthermore, the supply of an educated labor force to the nonfarm sectors will contribute to their development (Otsuka, Estudillo, and Sawada 2009; Estudillo and Otsuka 2016). The income of a rural household is a major determinant of its investment in human capital (Otsuka, Estudillo, and Sawada 2009). If the Green Revolution or a "high-value product revolution" takes place, thereby improving farm income and stimulating the development of the nonfarm sectors, there can be a virtuous circle of income growth, increased investment in human capital, and subsequent development of the farm and nonfarm sectors.

So far we have treated the Green Revolution, the introduction of high-value products, and investment in infrastructure as if they are exogenous, and responses of agricultural product markets, the development of nonfarm sectors, and human capital as partly endogenous and partly exogenous. The success or failure of the Green Revolution critically hinges on government policies, because new improved seeds, particularly if they can be reproduced, are public goods. New knowledge of improved production practices, which can be conveyed by the extension system, is also a local public good. Given

3 For simplicity, Figure 11.1 does not take into account the consumption linkage effect of agricultural innovations on nonfarm sectors that arises from an increase in farm income. Also the possible effect of human capital on the introduction of new agricultural technologies and products is ignored.

FIGURE 11.1 An illustration of farm-nonfarm sectoral linkages

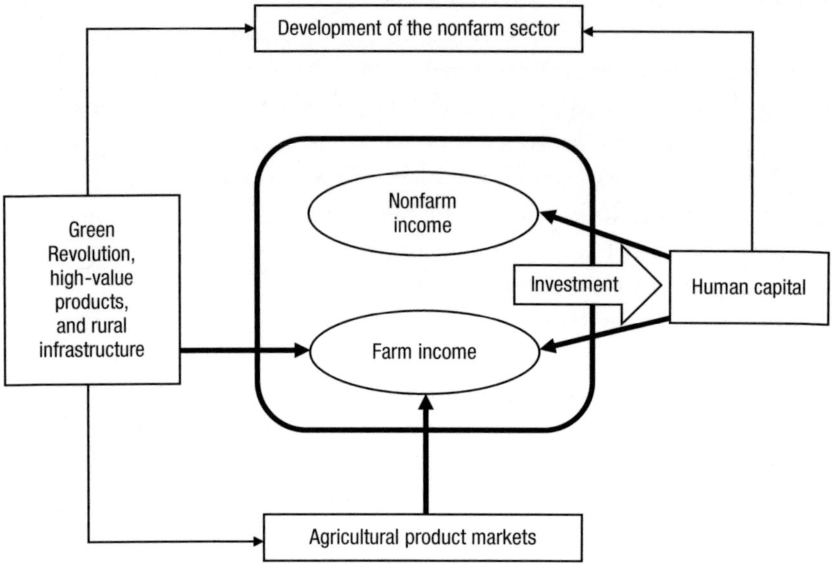

Source: Based on Estudillo and Otsuka (2016, Figure 1.6, p. 18).

inefficiency of public-sector extension systems, the role of social networks or farmer-to-farmer extension has received increasing attention (Takahashi, Muraoka, and Otsuka 2020).

While improved infrastructure may stimulate the development of both the farm and nonfarm sectors, the latter may increase the demand for infrastructure investments (Zhang and Fan 2004; Gibson and Olivia 2010). The introduction of high-value products in developing countries is often driven by rising income, led by the spread of supermarkets in both developing and developed countries, which offer contract farming, and sometimes supported by producer cooperatives (Reardon and Timmer 2012; Chapter 12, this volume). So far as contract farming is designed to overcome market failures, there is room for the government to assist in the introduction of high-value products. If agricultural products are freely traded internationally, the impacts of agricultural innovations on domestic prices of agricultural products will be small or negligible, as agricultural prices are largely determined by global supply and demand forces. However, if agricultural markets are mainly local, the price effect of supply changes could be substantial. The development of rural nonfarm sectors is affected not only by the development of agriculture but also by the linkage with urban/industrial sectors and even foreign companies. This chapter considers the causes and consequences of the Green Revolution, the introduction of high-value products, the improvement of rural infrastructure, the development of rural nonfarm sectors, and the role of migration in the transformation of the rural economy.

The Green Revolution and Rural Development

The Green Revolution

Although the best indicator of technological change or production efficiency is total factor productivity (Gollin and Udry 2019), an easily available proxy for production efficiency in agriculture in developing countries is physical yield per hectare of cultivated land.[4] As shown in Figure 11.2, yields of various crops increased substantially in tropical Asia (Southeast and South Asia) over the past half century, whereas yields were largely stagnant in Africa south of the Sahara, except for wheat, which is produced on large farms in temperate zones. Particularly noteworthy are the dramatic yield growth of rice since the mid-1960s, maize since the mid-1990s, and wheat since the mid-1970s. Such yield growth can be attributed to the Green Revolution, which is characterized by the continuous development of improved varieties and their dissemination (for example, Hayami and Otsuka 1994; Chapter 3, this volume). According to Gollin, Hansen, and Wingender (2018), a 10-percentage point increase in adoption of high-yielding varieties (HYVs) increases GDP per capita by about 15 percent, by the direct effect on crop yields, factor adjustment, and structural transformation. Their study also finds that HYV adoption reduced both fertility and mortality.

Major beneficiaries of the Green Revolution are consumers, because an increased supply of grains results in a large decrease in grain prices given price-inelastic demand for grains (Quizon and Binswanger 1986; Evenson and Gollin 2003).[5] Not only are farmers producers, they are consumers as well. On the one hand, they lose from decrease in grain prices; on the other hand, they gain as consumers. By comparison, urban consumers unambiguously benefited from the decline in grain prices. The gain for consumers can be most clearly seen in the declining real price of rice, primarily grown in Asia,

4 Alternatively, a change in profit is also a good indicator of the impact of new technology (Foster and Rosenzweig 2010). It is, however, practically difficult to measure the cost of unpaid family labor, which plagues the use of profit as the efficiency indicator.

5 Theoretically, aggregate benefits for consumers and producers depend critically on relative elasticities of supply and demand.

FIGURE 11.2 Average crop yield in tropical Asia and Africa south of the Sahara (tons per hectare), 1961–2018

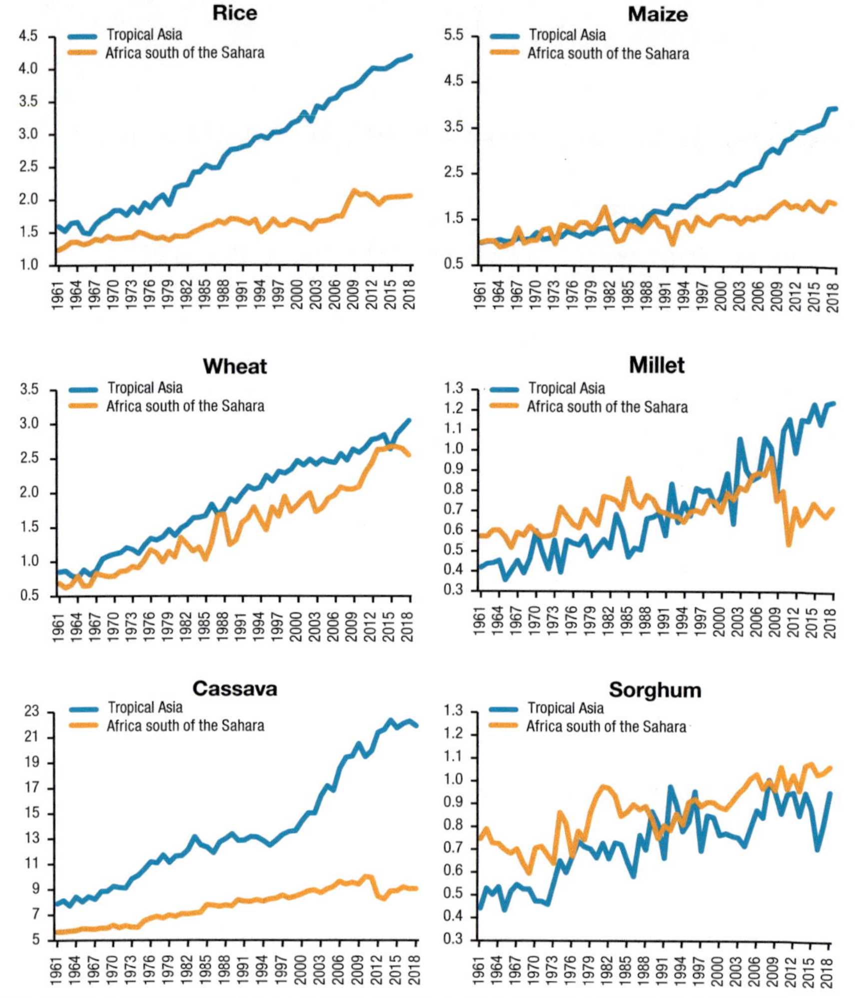

Source: Data from FAO, FAOSTAT various years.

where the Green Revolution has taken place widely (Figure 11.3). Aside from a chaotic hike in rice prices in 1973/1974, the real rice price declined from the late 1960s to the early 2000s, with the lowest price well below one-half of the level around 1970. Even though the productivity of rice farming doubled, rice farmers would not have been made better-off because of the huge price reduction. Since new varieties are high-yielding only in irrigated and favorable

FIGURE 11.3 World production and real prices of rice, 1960–2015

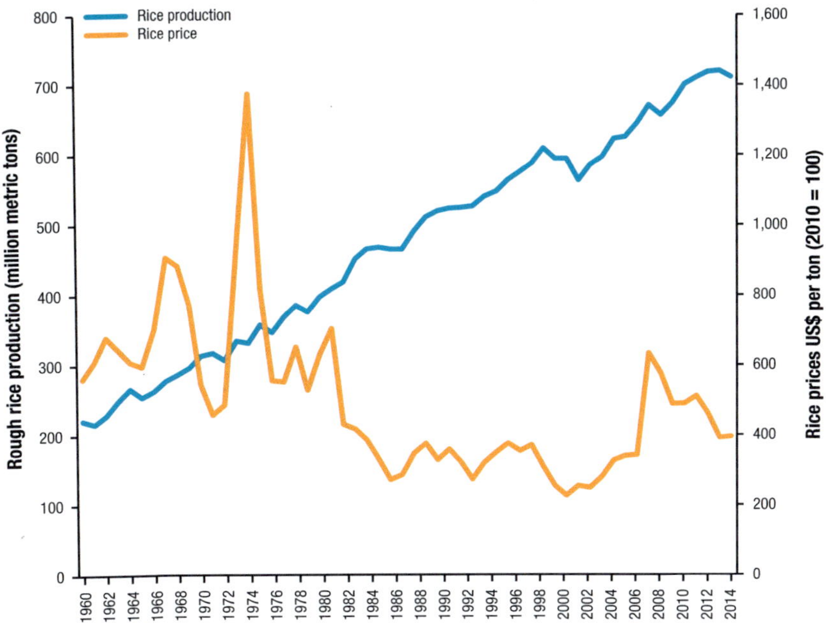

Source: USDA database.

rainfed areas (David and Otsuka 1994), they would have been adopted in 70 percent to 80 percent of the rice area in tropical Asia.

The clear losers are rice farmers who did not have access to improved rice varieties, including those in unfavorable areas subject to flood and drought in Asia and in almost the whole of Africa south of the Sahara. Due to price reductions, these farmers lost not only income but also incentives to grow these crops, whose production increased dramatically in other places. Although not shown in Figure 11.3, real prices of wheat and maize also declined as much as that of rice, due at least in part to the Green Revolution of these crops in Asia. Thus it is possible that one of the reasons for the failure of the Green Revolution in Africa south of the Sahara was the low grain prices that resulted from the successful Green Revolution in Asia.

Growth Linkage Effect

The agricultural growth linkage hypothesis postulates that modern agricultural technology propels the development of the nonfarm economy through several production and consumption linkages (Haggblade, Hazell, and

Dorosh 2007). On the production side, improved agricultural technology may spur the birth and development of industries engaged in the production of agricultural inputs (for example, implements) and service-related support to the agriculture sector (for example, repair shops for agricultural machinery and delivery of fertilizer). In addition, an increase in the supply of agricultural products breeds food processing and agro-based manufacturing industries. On the consumption side, increases in rural household income brought about by increased agricultural productivity stimulates consumer demand for locally produced nonfarm goods and services.

A critical assumption of the growth linkage hypothesis is that the increased *local* demand for nonfarm goods and services leads to the increased *local* supply of those goods and services. In fact, it is often assumed that supply curves of locally produced goods and services are elastic. Yet in the case of tradable goods, increased local demand may induce increased supply from other locations, such as industrial clusters elsewhere, unless rural areas have capacity to produce what Ranis and Stewart (1993) have called "modern Z-goods." Thus the growth linkage effects are likely to be pronounced in the case of nontradable services, such as retailing of agricultural inputs, milling services, and machine repair services (Haggblade, Hazell, and Dorosh 2007; Renkow 2007), as well as processing in the case of perishable products that must be processed immediately after harvesting.

There is lack of strong econometric evidence that supports the significant effect of agricultural growth on nonfarm sector development (for example, Ramos et al. 2012). Increased *local* demand for nonfarm goods and services may stimulate an increased supply of these goods and services in the entire economy but not necessarily in the locality where agricultural technological change takes place (Christiaensen, Demery, and Kuhl 2011). For example, the success of rural reform in the 1980s in China stimulated the demand for manufacturing goods, which boosted the subsequent rapid growth of township- and village-owned enterprises and private enterprises nationwide. But the growth is uneven across regions. Most manufacturing activities have been increasingly concentrated in some coastal areas, which are not necessarily the major cropping regions.

Recently Bustos, Caprettini, and Ponticelli (2016) found that the introduction of genetically engineered soybean seeds, a labor-saving agricultural technology, reduced labor demand in agriculture, inducing workers to reallocate toward the local manufacturing sector in Brazil. By contrast, the introduction of maize as a second-season crop increased agricultural labor use. Thus, if regional labor mobility is low, labor-saving technological change in agriculture may stimulate industrial development in rural areas, while land-saving technologies may not. In fact, the Green Revolution is largely labor-using, thereby likely not inducing workers to relocate to the local nonfarm sector.

Failure of the Green Revolution in Africa South of the Sahara

The Asian Green Revolution in rice and wheat began more than a half century ago in the 1960s, when the International Rice Research Institute (IRRI) and the International Maize and Wheat Improvement Center (CIMMYT) released the semi-dwarf, high-yielding modern rice and wheat varieties (MV), respectively (Dalrymple 1978; Hayami and Otsuka 1994; Byerlee and Moya 1993). According to Hayami and Godo (2005: 96–97), "As agriculture is strongly constrained by environmental conditions, it is difficult to transfer advanced technologies developed in the temperate zone to the tropical zone. . . . IR8 was modeled after the high-yielding Japanese varieties."[6] Although it is understandable that transfer of agricultural technology from the temperate zone to the tropical zone was difficult, it is puzzling why, after 50 years, agricultural technology developed in tropical Asia has not been successfully transferred to SSA, which is also located in the tropical zone.

According to Boserup (1965), population pressure on limited land areas stimulates the adoption of land-saving and labor-using production methods spontaneously by farmers, which leads to an increase in crop yield. Hayami and Ruttan (1985) extend the view of Boserup by arguing that population pressure induces not only technological innovations but also institutional innovations that support the development and diffusion of innovations. Following Schultz (1964), we assume that farmers in poor economies are rational and willing to adopt new technologies if they are profitable. If this is true, the failure of the new agricultural technologies to take hold in SSA could simply have been because they were not yet profitable in the conditions of SSA. Because of rapid population growth in SSA, arable land per capita in rural areas now is nearly comparable with that in Asia in the early 1960s (Figure 11.4). If Boserup's and Hayami-Ruttan's hypotheses are correct, there ought to be the sign of agricultural intensification and yield growth in this continent. A careful look at Figure 11.2 indicates that yield of grains,

6 IR8 was the first MV released by IRRI in 1966, whose parent came from Taiwan. Hsieh and Ruttan (1967) argued that new rice varieties being developed in tropical Asia in the 1960s resembled, in terms of fertilizer responsiveness and yield potential, the ponlai varieties introduced to Taiwan by the Japanese colonial government in the mid-1920s. Also see Hayami and Ruttan (1970).

FIGURE 11.4 Arable land per rural population in tropical Asia and Africa south of the Sahara, 1961–2016

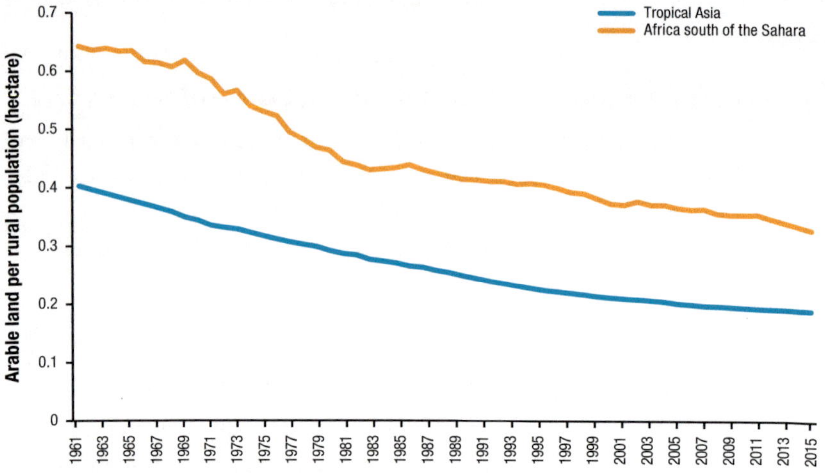

Source: FAO, FAOSTAT.

particularly rice and maize, began growing around the beginning of this century. Whether this is the initial phase of the Green Revolution in SSA is an extremely important open research question.

The first possible reason for the failure of the Green Revolution in Africa south of the Sahara is that agroclimates are so different between tropical Asia and SSA (Evenson and Gollin 2003) and vastly heterogeneous within SSA (Suri 2011), making Asian technology nontransferable. That is, without major changes in the plant, plant performance would be so poor as to be unprofitable in many regions of SSA. The hypothesis of different agroclimates, however, is not necessarily plausible because crop yields before the Green Revolution were comparable between SSA and tropical Asia (see Figure 11.2). Also, as far as rice is concerned, many Asian varieties and their offspring are grown well in SSA by means of adaptive research (Yamano et al. 2016).[7]

Yet heterogeneity of agroclimate can be a serious constraint on new technology adoption, particularly in Africa south of the Sahara, where crop mix is diverse and vastly different in different locations (Otsuka and Larson 2013). Because of the heterogeneity of agroclimate as well as farmers' abilities, the same technology may have different returns for different farmers on different

7 Otsuka and Larson (2013) point out that lowland rice technology is highly transferable from Asia to SSA.

plots (Foster and Rosenzweig 2010; Suri 2011).[8] In a study on Kenyan maize farming, Suri (2011) finds that farmers with high net returns adopted hybrid maize seeds, whereas farmers with low net returns did not. She suggests that removing supply and infrastructure constraints would be a cost-effective policy to raise hybrid seed adoption rates. Gollin and Udry (2019) demonstrate that the dispersion of total factor productivity is very high in Uganda and Tanzania even among plots cultivated by the same farmer, indicating large heterogeneity in production environments across plots. In contrast, however, Michler et al. (2019) find that adoption of improved cowpea seeds reduced production cost of a large number of farmers in Ethiopia.

Another possibility is that while scientific research was successful and hence suitable and profitable technologies were available in SSA, the technologies have not been disseminated widely due to lack of an effective extension system. If there were little effort to disseminate such technology, farmers simply would not have been aware that it was available. A failure to recognize that the high-yielding Asian farming practices are management-intensive, at least compared with the existing cultivation methods in SSA, seems to deter the establishment of effective public-sector extension systems. Important crop management activities, or agronomic practices, include using improved seed, seed selection, leveling fields and construction of bunds for water and weed management, row transplanting, and fertilizer application in lowland rice farming, and the application of manure and compost, intercropping with leguminous crops, which have capacity to fix nitrogen from the atmosphere, and mulching, in upland farming. African farmers seldom adopt these practices. Importantly Otsuka and Larson (2016) argue that the failure of the Green Revolution in SSA can be attributed to the lack of recognition that Green Revolution technology is management-intensive and to the absence of an effective public-sector extension system.[9]

It is well-known that new technological information is transmitted from farmers to farmers (see Conley and Udry [2010] for a recent study). Although

8 Because of the heterogeneity, we cannot assume that because a technology is profitable in one time or place, it is also profitable in another. Foster and Rosenzweig (2010) provide an excellent review of the literature on the assessment of the profitability of new technology.

9 This conflicts with a conventional view that the Green Revolution can be alternatively called the "seed-fertilizer revolution" (Johnston and Cownie 1969), which indicates that if improved seeds and fertilizer use are widely adopted, a Green Revolution automatically takes place. However, Murgai, Ali, and Byerlee (2001) and Ali and Byerlee (2002) find that total factor productivity in Indian and Pakistan Punjab grew faster from the mid-1970s to the mid-1980s, when the adoption of modern wheat and rice varieties was already high, than from the mid-1960s to the mid-1970s, when the adoption increased fast.

efficiency of a public-sector extension system may be improved by eliciting feedback from farmers about the quality and content of agricultural extension services (Jones and Kondylis 2018), a great deal of attention has been paid to the effectiveness of farmer-to-farmer extension and a number of randomized controlled trials (RCTs) have been conducted. Although Kondylis, Mueller, and Zhu (2017) find that training provided by extension agents to selected contact farmers for technology diffusion to other farmers does not work, BenYishay and Mobarak (2018) find that providing reward to contact farmers, who are ordinary farmers, for their successful technology dissemination leads to wider adoption of new technology. Shikuku et al. (2019), however, find that providing private material reward to contact farmers did not have any impacts, even though reputational gain for them from knowledge diffusion facilitated knowledge dissemination. Takahashi, Mano, and Otsuka (2019) report that a package of improved rice cultivation methods has been diffused from randomly selected training participants to other farmers without any material reward in Côte d'Ivoire. How to select contact farmers and under what conditions new knowledge can be diffused from them to other farmers are critically important yet unexplored issues in technology diffusion.[10]

There are several other possible constraints on new technology adoption, such as inadequate education of farmers for acquiring and decoding new technology information, and the incompleteness of insurance and credit availability for dealing with the production risk associated with the adoption of new technology (Foster and Rosenzweig 2010; Chapters 16 and 17, this volume). It is not clear, however, to what extent such constraints are particularly serious in SSA. According to Chapter 5 in this book, agricultural wage rates tend to be high in SSA, particularly during peak seasons, when labor shortage is acute in predominantly rainfed farming.[11] As a result, labor-intensive and management-intensive Green Revolution technology may not be highly profitable for farmers even if it boosts higher yield. If high wages are the root cause, mechanization has to be introduced, together with yield-enhancing technology, in order to realize the Green Revolution in Africa south of the Sahara. As a matter of fact, agricultural mechanization has begun to take off in SSA (Diao and Kolavalli 2014). Another possible reason for the failure of the Green Revolution was low grain prices, as was discussed earlier in this chapter.

10 See Takahashi, Muraoka, and Otsuka (2020) for a review of the recent literature on technology adoption and extension.

11 Since the timing of farming depends on the timing of rainfall under rainfed farming, labor demand of all the farmers changes simultaneously. As a result, peak-season labor demand is high relative to supply of labor.

Development of High-Value Product Sectors

To introduce high-value agricultural products such as fresh vegetables and fruits and livestock products, a variety of market failures, including the malfunctioning of markets of improved seeds, safe pesticides, and credit, as well as the absence of management service providers, must be dealt with (see Chapter 12 for a discussion of the transformation of food value chains). Furthermore, compared with staple crops, it is more difficult for consumers to identify the quality and safety of high-value products due to larger quality variations (Bai 2016). Thus consumers place more premium on truly high-value agricultural products if their quality and safety are verifiable and traceable or if production standards meet international standards (Ortega et al. 2011; Ifft, Roland-Holst, and Zilberman 2012).

Another problem is the riskiness of high-value products, whose prices fluctuate widely and unexpectedly. Thus sellers who can guarantee the quality and safety of such products to consumers and who can shoulder the price risk by offering a fixed product price (or similar pricing formulae) in advance are indispensable for product markets to work and for farmers to undertake new businesses.[12] Contract farming offered by large supermarkets and agro-processors is designed to overcome these market failures (Barrett et al. 2012; Otsuka, Nakano, and Takahashi 2016).[13] Supermarkets and agro-processors provide inputs on credit and production instruction to farmers and undertake marketing. Once new improved inputs and superior production technologies are introduced, the efficiency of farming of high-value products in developing countries is improved. Thus the World Bank (2007) strongly recommends the introduction of high-value products by contract farming in developing countries.

The transaction cost associated with contract farming may be substantial. If input prices charged by the contractor are lower than market prices, contract farmers have incentive to divert the use of inputs for other purposes. If product prices paid by the contractor are lower than market prices, contract farmers have incentive to sell products at the market rather than deliver them to the contractor as promised (Kunte, Wollni, and Keser 2017). Farmers

12 Michelson, Reardon, and Perez (2012) found that contract farming with Walmart supermarket significantly reduced price risk, even though mean prices paid to farmers are lower than the traditional market in Nicaragua.

13 As Casaburi and Macchiavello (2019) suggest, another advantage of contract farming over ordinary market exchange with traders may be that a farmer prefers to receive payments infrequently (for example, monthly rather than receiving payments every time a transaction is made), because the farmer can commit to saving under infrequent payment. This argument applies to milk and tea production rather than crops harvested infrequently.

also may not follow the production instructions specified by the contract. Therefore, the contractor will prefer to make a small number of contracts with large farmers, rather than a large number of contracts with smallholders. In this way, smallholder farmers may be excluded from contract farming.

A large number of studies, including recent studies by Mishra et al. (2016), Michelson (2013), and Ragasa, Lambrecht, and Kufoalor (2018) as well as a review article by Ton et al. (2018), found that farmers who are engaged in contract farming are larger landholders or wealthier farmers. According to the literature review by Otsuka, Nakano, and Takahashi (2016), although contract farmers are larger than noncontract farmers in terms of cultivation area, the difference between them is very small, indicating that farm size may not be the decisive factor affecting the choice of contract farmers. In order to make contracts with large buyers, smallholder farmers often establish producer cooperatives, which make contracts with buyers on behalf of small producers. Furthermore, producer cooperatives sometimes undertake market research to find out which products are highly demanded at the markets, introduce new profitable crops and technology for member farmers, negotiate better prices for input purchase, and guarantee the quality and safety of high-value products produced by member farmers to the buyer (Hashino and Otsuka 2016). To the extent that transaction costs among member farmers of producer cooperatives are sufficiently low, the emergence of producer cooperatives will help smallholder farmers participate in contract farming. In a nutshell, smallholder farmers are not always excluded from contract farming in practice. Furthermore, by receiving relevant market and technological information and services from cooperatives, small subsistence farmers may be transformed into profit-oriented farmers.

Another issue is the extent to which farmers are made better off by entering contract farming. The identification of the impact of contract farming on farmers' welfare is not easy, because contracts are endogenously chosen by the firm and farmers (Barrett et al. 2012; Bellemare and Bloem 2018). Because of the difficulty in finding a truly exogenous instrumental variable that is strongly correlated with participation in contract farming, the internal validity, or the ability of a study to identify treatment effects, may be limited (Bellemare and Bloem 2018). Even though Bellemare and Lim (2018) argue that the causality of the welfare impacts of contract farming has thus far been elusive, the available evidence suggests that contract farming increases the welfare of farmer households, as reflected in increased household or crop income (Maertens and Vande Velde 2017), profit (Mishra et al. 2016), community wealth (Gatto et al. 2017), productive assets (Michelson 2013), and subjective well-being (Väth, Gobien, and Kirk 2019). While the existing studies rely on observational data, a recent study by Arouna, Michler, and Lokossou (2019) conducted an RCT by offering fixed-price contracts to reduce price uncertainty, fixed-price with management training contracts, and fixed-price with management training and input provision contracts. They find significant and positive impacts of these contracts on rice area, rice yield, proportion of rice sold, and rice income. One difficulty of this study is the focus on rice income, as a measure of welfare effect, because additional family resources may be allocated to rice framing at the sacrifice of nonrice cultivation.

In contrast, Ragasa, Lambrecht, and Kufoalor (2018) and Soullier and Moustier (2018) did not find significant impacts of contract farming on profitability of rice farming.[14] Ton et al. (2018) point out through meta-analysis that assessment of the income effect of contract farming is upwardly biased due to mechanically excluding contract farming arrangements that failed and due to the tendency that primarily successful cases of contract farming are reported in published articles. Otsuka, Nakano, and Takahashi (2016) also question whether contract farmers, who passively accept inputs and follow the production instructions given by the contractor without making risky decisions under uncertainty, receive significantly higher income. As Schultz has argued (1974), high income accrues to the ability to make proper decisions under a dynamically changing setting. If this is the case, it is likely that contract farmers receive only marginally higher income than alternative income sources. To conclude, it is fair to argue that the effect of contract farming on the welfare of smallholder farmers needs to be explored further in a variety of settings, while carefully dealing with selection bias arising from endogenous choice of contractual arrangements.

By conducting an RCT in collaboration with an NGO, Ashraf, Gin, and Karlan (2009) offered contract farming with provision of services and credit and found certain effects on the production of export crops and income compared with the control group.[15] Bellemare and Bloem (2018) point out, however, that since nonprofit organizations are involved, the external validity of their study, or the ability to extrapolate their research findings across time and space, will be limited. Bernard et al. (2018) conducted an RCT in which a micronutrient-rich fortified yogurt is additionally provided to contract

14 Among staple crops, the quality variations are larger for rice than others, and rice is sometimes a commercial crop in Africa south of the Sahara.

15 See de Janvry, Sadoulet, and Suri (2017) for an excellent review of the literature for field experiments in agriculture.

farmers. They found that the provision of nutrient-based incentives had large and significant impacts on the frequency and amount of milk delivered during the dry season, particularly when suppliers are women. Casaburi and Willis (2018) conducted another RCT in which the contractor offers insurance to contract farmers and deducts the premium from farmers' revenues at harvest time. According to this study, delaying the premium payment until harvest results in a large increase in insurance demand. These studies indicate that there is room for further improving the nature of contract farming for the benefit of smallholders.

Otsuka, Nakano, and Takahashi (2016) argue that to make farmers truly better off, farmers must be transformed from ordinary farmers engaged primarily in production to "entrepreneurial" farmers engaged also in market research, production designs, technology and crop choices as well as marketing. If and only if farmers are engaged in such pre-production and post-production activities will the introduction of high-value products in developing countries lead to significant improvements in both efficiency and equity.

The Role of Nonfarm Sectors in Income Generation

There is no doubt that increasing nonfarm income is a major source of increasing rural household income. According to Haggblade, Hazell, and Reardon (2007), the share of nonfarm income in total income of rural households is 51 percent in Asia, 47 percent in Latin America, and 37 percent in Africa in the early 2000s. These numbers are surprisingly consistent with findings of many case studies. For example, de Janvry and Sadoulet (2001) found that its share was 55 percent in Mexico; Estudillo and Otsuka (1999) found a share of 51 percent in the Philippines; and Matsumoto, Kijima, and Yamano (2006) reported shares of 45 percent in Kenya, 30 percent in Uganda, and 13 percent in Ethiopia. It appears that the nonfarm income share is higher in more developed countries. Indeed, the nonfarm income share has been rising with income growth of the economy. For example, the nonfarm income share increased from 27 percent in 1966 to 51 percent in 1994 in the Philippines (Estudillo and Otsuka 1999), 34 percent in 1988 to 46 percent in 2004 in Bangladesh, and 6 percent in 1987 to 37 percent in Thailand's Central Plain, and 30 percent to 81 percent during the same period in the country's northeast region (Otsuka, Estudillo, and Sawada 2009).

There is consensus in the literature that the increasing nonfarm income share is pro-poor, thereby contributing to the equality of income distribution (de Janvry and Sadoulet 2010; de Janvry, Sadoulet, and Zhu 2005; Estudillo and Otsuka 1999, 2016; Hayami and Kikuchi 2000; Lanjouw and Lanjouw 2001; Matsumoto, Kijima, and Yamano 2006; Lanjouw and Murgai 2009; Otsuka, Estudillo, and Sawada 2009). A major reason why nonfarm jobs are pro-poor is that land-poor households—that is, landless and near-landless households that belong to the poorest segment of poor rural society—more actively seek nonfarm jobs than landed households. They rely more on labor income than on other income sources, such as land rental revenue and capital income. In Myanmar the shares of nonfarm income were 9 percent and 32 percent for farmers and landless households, respectively, in 1996, whereas the corresponding shares were 18 percent and 62 percent in 2012 (Estudillo and Otsuka 2016). In the Philippines the income of the landless was roughly one-half that of farmer households in 1985, but the incomes of their children were roughly equalized in 2008, owing to the more active participation of children of the landless in nonfarm jobs (Estudillo and Otsuka 2016).

Studies have also indicated that farm households engaging in nonfarm employment tend to earn higher household incomes and produce agricultural products more efficiently, which likely increase and stabilize food consumption. Nonfarm employment also allows farmers to reduce their vulnerability. Nonfarm activities serve as diversification of household earnings to sustain household income and stabilize consumption over time (Seng 2015). In fact, off-farm income is a risk management strategy, which is used for managing the production and income risk to ensure the stability of the farmer's income (Velandia et al. 2009; Ping et al. 2016). Akhtar et al. (2019) investigate the factors affecting the concurrent use of off-farm income diversification strategy and agricultural credit, which the farmers use to manage the risk in maize production in Pakistan. The authors find that education, livestock number, and the risk-averse nature of the farmers are among factors that encourage income diversification. Needless to say, the impact of nonfarm activities depends on the returns from the nonfarm income generation activities. Scharf and Rahut (2014) found that low-return nonfarm work is associated with lower income inequality, while high-return nonfarm activities seem to be linked with less equal household income distribution.

The accurate evaluation of the economic effects of nonfarm activities on farm households is analytically difficult, especially when using household-level cross-sectional data, due to self-selection bias and endogeneity of job choice. Some studies attempt to reduce these issues using econometric techniques, such as propensity score matching used by Owusu, Awudu, and Seini (2011) and Olugbire et al. (2011), to find positive impact of nonfarm employment on household income and food security. The development of the nonfarm sector also affects income distribution through the wage channel. Over time, as the nonfarm sector absorbs and eventually exhausts surplus rural labor, real wages tend to increase. This is the so-called Lewis turning point. The arrival of the Lewis turning point has implications for poverty reduction and agricultural production. As most of the poor rely on labor income, rising real wages mean higher family income for the poor even if they work for the same amount of time as before. After a country reaches the Lewis turning point, for example China, poverty incidence tends to decline and income distribution improves (Zhang et al. 2014; Wang et al. 2016). The pressure of labor shortage after the Lewis turning point also induces farmers to substitute labor for machinery in agricultural production. The rapid spread of agricultural mechanization in China after the arrival of its Lewis turning point is a noted example (Wang et al. 2016; Zhang, Yang, and Reardon 2017).

One of the most important factors affecting the choice of nonfarm jobs and the amount of nonfarm income is schooling (Ackah 2013; Atamanov and Van den Berg 2012; Bezu and Barrett 2012; Cunguara, Langyintuo, and Darnhofer 2011; Do, Nguyen, and Grote 2019; Essers 2016; Marchetta 2013; Nakajima, Otsuka, and Yamano 2017). Although Foster and Rosenzweig (1996) found that schooling affected the profitability of technological change in agriculture in the early phase of the Green Revolution in India, a large number of studies found that schooling affected nonfarm income positively but not farm income (de Janvry and Sadoulet 2010; Estudillo and Otsuka 1999, 2016; Jolliffe 2004; Matsumoto, Kijima, and Yamano 2006; Lanjouw and Murgai 2009; Otsuka, Estudillo, and Sawada 2009). Jolliffe (2004) also demonstrates that schooling positively affects the allocation of time to nonfarm jobs. There is a tendency that the magnitudes of the coefficient of schooling in the nonfarm income regression increase over time (Otsuka, Estudillo, and Sawada 2009; Estudillo and Otsuka 2016). The ownership of land or farm size has a generally negative effect on nonfarm income.

At the same time, increasing rural household income stimulates investment in the schooling of children, who can later find better nonfarm jobs (Otsuka, Estudillo, and Sawada 2009). As illustrated in Figure 11.1, there is therefore a virtuous circle of the development of nonfarm sectors, the increasing income of rural households, the increased investment in the schooling of children, and the supply of an educated young labor force to the nonfarm sectors, which, in turn, contribute to the development of the nonfarm sectors.

The effects of improved infrastructure, such as improved access to paved roads and a stable supply of electricity, on the development of local nonfarm sectors is not straightforward (see Figure 11.1). For example, road improvement has direct effects on agricultural production and consumption by reducing prices of purchased inputs and consumer goods and increasing agricultural product prices, and indirect effects on the development of farm and nonfarm sectors by inducing investments and institutional changes (Khandker, Bakht, and Koolwal 2009). As pointed out by Renkow (2007), however, infrastructure is a double-edged sword: While adequate roads, improved communication facilities, and a stable supply of electricity will stimulate development of the nonfarm sectors, connecting rural areas to urban areas via infrastructure improvement and expansion will strengthen competition with urban firms. The agglomeration effect in cities may attract more nonfarm jobs to cities. Although investment in rural infrastructure fosters overall economic growth, whether it promotes development of local rural nonfarm sectors is an empirical question.

As would be expected, increased road density and road access have positive impacts on total factor productivity in agriculture in India (Zhang and Fan 2004) and on greater specialization of crops and larger application of fertilizer in China (Qin and Zhang 2016). More important, there are several studies that report the positive effects of improved quality or quantity of rural infrastructure on nonfarm production and share of nonfarm income (Fan and Zhang 2004; Gibson and Olivia 2010; Khandker, Bakht, and Koolwal 2009; Qin and Zhang 2016; Ramos et al. 2012). As pointed out by Ramos et al. (2012), however, the impacts of rural infrastructure on the development of manufacturing sectors in remote rural areas are rarely found.

In contrast, Khandker and Koolwal (2010) find in rural Bangladesh that the expansion of irrigation, paved roads, electricity, and access to formal and informal credit are positively associated with farm and rural nonfarm incomes probably because of high population density. The authors also found that the income gains did not translate to significant poverty reduction for the poorest households. Senadza (2012) noted in a study of Ghana that gender composition of households, age, education, and access to credit, electricity, and markets are among important determinants of nonfarm activities and income. In addition, Demissie and Legesse (2013) found that livelihood assets (livestock holding, size of cultivated land) and livelihood diversifying strategy (crop-based diversification through number of crops grown and harvested) are important factors affecting successful income diversification.

Strategy for Rural Industrialization

More often than not, manufacturing industries are clustered, meaning that enterprises producing the same and related products (for example, part-suppliers and assemblers) are located in a small area (Sonobe and Otsuka 2006). Sonobe and Otsuka (2011) find numerous industrial clusters not only in Asia but also in Africa south of the Sahara, some of which are located outside urban areas. Otsuka (1998) reports that many garment and weaving clusters are located in rural areas in Asia. Renkow (2007) points out that small and medium-sized manufacturing enterprises tend to cluster together in remote rural locations. Long and Zhang (2012) find a large number of industrial clusters in China and conclude that "China's industrialization process is largely cluster-based." Ali, Peerlings, and Zhang (2014) report a large number of industrial clusters in Ethiopia. Galvez-Nogales (2010) reports that a variety of agro-based clusters exist in rural areas in Latin America, Asia, and Africa. There seems to be little exaggeration in arguing that manufacturing and agro-processing industries in developing countries are clustered in general and clustered in rural areas in particular. These clusters are often located in rural towns, which have been remarkably growing (see Chapter 9).

Why are manufacturing industries so often clustered? According to Marshall (1920), there are three advantages of industrial clusters, which may also be termed agglomeration economies or localization economies: (1) ease of division and specialization of labor between enterprises thanks to their proximity, (2) labor pooling or the development of labor markets, and (3) information spillovers. Ruan and Zhang (2009) and Long and Zhang (2011) point out and statistically confirm that fine division of labor in industrial clusters lowers the cost of initial investment, thereby lowering the entry barriers. In this way, capital constraints in developing countries can be lessened. In addition, thanks to the strong social capital embedded in local communities, trade credit is widely used in clusters, greatly reducing the reliance on working capital.

Sonobe and Otsuka (2006) argue that a variety of skilled workers, such as experienced engineers, designers, and traders, are available in industrial clusters, and are conducive to innovations, alternatively called "new combinations." There has also been a growing number of industrial clusters in China, or "Specialized Towns" with a high concentration of firms producing one specific item. Such industrial development has been facilitated to increase agglomeration economies and visibility for increased specialization as well as an expansion of industrial output, innovation, and regional economic growth

(Barbieri, Tommaso, and Bonnini 2012). To sum up, there are positive externalities among enterprises in industrial clusters, which attract the entry of new enterprises and contribute to the formation of industrial clusters. This implies that the social benefits of the entry of new enterprises and innovations exceed the private benefits. Rural industrial clusters are often agro-based, processing high-value agricultural products (Galvez-Nogales 2010). Agro-processors include not only food processors, such as producers of potato chips, fruit juice, and wine, but also "pack houses," which undertake washing, fumigation, sorting out, grading, packing, branding, and marketing of fresh vegetables and fruits. While geographical proximity between farmers and processors and among processors does not matter much in cluster development, access to markets and transportation infrastructure play key roles (Cavatassi et al. 2011; Chatterjee and Ganesh-Kumar 2016; Geldes et al. 2015). In order to develop such agro-based industrial clusters, linking the clusters with high-value product sectors through contract farming is indispensable.

According to Sonobe and Otsuka (2006, 2011, 2014), broadly speaking, there are dynamic and stagnant clusters. Although innovations take place in dynamic clusters that lead to the expansion of production and employment, an improvement of product quality, and the enlargement of innovative enterprises with sizable profits, no such innovations take place in stagnant clusters, in which many small enterprises employ similar production methods, produce similar low-quality products, sell such similar products at the same domestic markets, and earn hardly any profit. Producers in stagnant clusters face challenges of carrying out innovations that often require collective action. Similarly, Galvez-Nogales (2010) confirms that there are both dynamic and stagnant agro-processing clusters.

It is extremely important to recognize that the social return to innovation is higher than the private return due to imitation. This strongly suggests that industrial clusters that successfully internalize externalities by collective action can innovate and grow. Indeed, Hashino and Otsuka (2016) and Galvez-Nogales (2010) show that producer cooperatives play a critical role in introducing new technologies, acquiring new market information, and controlling the quality of products produced by member companies within clusters, not only historically in Japan and France but also in postwar Europe and the developing world. Similarly, Ruan and Zhang (2009, 2016) find that the reputation of low-quality producers in industrial clusters in Zhejiang province in China was reversed by the collective action of producer cooperatives, which introduced improved production methods, disseminated them to cooperative members, and forced those producers who produced low-quality products to exit. It is interesting to note that the activities of producer cooperatives in industrial clusters are so similar to those of producer cooperatives in contract farming. This is not surprising because the issues facing producers of high-value agricultural products and those of manufactured goods are qualitatively similar (that is, internalization of externalities).

Producer cooperatives as well as local governments played an important role in organizing collective actions and stimulating innovations (Hashino and Otsuka 2016). A spectacular example is provided by Zhang and Hu (2014), who find that the local government was instrumental in overcoming a series of bottlenecks facing Anding potato cluster, such as developing local seeds, expanding markets, and attracting starch processing plants. The central question of developing rural industrial clusters is how to stimulate the initiation of new businesses and growth of existing ones. Unfortunately, the existing literature does not provide definitive answers to this question. Otsuka (1998, 2007) points out that subcontracting between rural and urban enterprises is one of the ways to introduce new production knowledge to rural areas from outside. Sonobe and Otsuka (2006) report that township-village enterprises in China learned technological and managerial knowledge from state-owned enterprises (SOEs) by recruiting workers of SOEs or employing them as advisers. In China, local governments have embedded incentives to develop the local economy, because local officials' promotion hinges upon local economic growth relative to peers. In addition, local governments and the central government share revenues according to a fixed formula. Therefore, local officials are keen in helping figure out solutions, often indigenous, to overcoming the obstacles facing cluster development. However, local governments in many developing countries do not have such incentives. A research question is how to incentivize local governments so that they have embedded interest in fostering local economic development.

Considering that entrepreneurial human capital is missing in developing countries (Bloom et al. 2012; McKenzie and Woodruff 2014), Sonobe and Otsuka (2014) argue that management training of entrepreneurs (owners and managers) is likely to be an appropriate strategy to stimulate the entry of new enterprises and innovations. Several RCTs demonstrated that management training improves the enterprise performance significantly (Bloom et al. 2013; Mano et al. 2012; Higuchi, Mhede, and Sonobe 2019). Sanyang et al. (2016) support such a view in the context of agro-industries in Africa. In areas without any organically formed industrial clusters, governments may use industrial parks to attract new businesses, in addition to training of potential entrepreneurs.[16] While it is often a great challenge to improve the business environment and infrastructure for a country as a whole, it is viable for governments in developing countries to build industrial parks on a smaller scale to take advantage of agglomeration economies. In industrial parks the government can offer potential first movers various kinds of benefits, such as tax breaks and ready-to-move factory buildings (Zhang 2016). It is very important for the government to take exit strategies into account when designing the investment promotion policy as the government cannot afford to subsidize the firms forever. One should also be aware of the enormous challenges of establishing industrial parks from scratch.

Rural-to-Urban Migration

In many rural areas, nonfarm job opportunities are limited. Under such circumstances, seeking jobs elsewhere becomes an appealing option. Migration can have a profound effect on the local economy in two ways. First, migration can help increase the level of household consumption and reduce its volatility. After a family member migrates out, the number of people eating at home naturally drops, thereby increasing the food available for the remaining family members (Xing et al. 2009; de Brauw and Ambler 2018). There are also improvements in food security and nutritional status associated with remittance from migrants. Remittances can be spent on household consumption and children's education, serving as an important instrument for poverty reduction (Yang 2011; Kikkawa, Matsumoto, and Otsuka 2019).

In Guatemala and El Salvador, stunting prevalence among children under five appears to be lower among migrant-source households (most of which receive remittances) than nonsource households. In Tajikistan migrant-source households have been found to consume more calories, on average, than nonsource households (de Brauw and Ambler 2018). In northwest Bangladesh an NGO program provided migrant laborers money for bus tickets during the hungry seasons, allowing many to become seasonal migrants. This support led to a long-term increase in seasonal migration and an increase in per capita consumption among migrant households (de Brauw and Ambler 2018). However, there is also evidence that remittance has fueled the conspicuous consumption, such as gorgeous housing and gift-giving (Ballard 2001). Thus negative impacts of migration on the well-being of migrants and their households must be noted (Chen, Kosec, and Mueller 2019).

16 In such a case, what industries the government should support is a major question.

The second way migration can have a profound effect on the local economy is that remittance can also be used for productive investment. For example, Woodruff and Zenteno (2007) find that more than 20 percent of the capital invested in microenterprises in urban Mexico is financed by remittances from the United States. Yang (2008) shows that in the Philippines remittances are spent on children's education and small enterprises. In Guatemala households receiving either internal or international remittances spend more at the margin on two investment goods—education and housing—compared to what they would have spent on these goods without remittances. These findings support the growing view that remittances can help increase the level of investment in human and physical capital in remittance-receiving countries (Adams and Cuecuecha 2010; Kikkawa, Matsumoto, and Otsuka 2019). Similar findings were made in Ghana, where receiving remittances was linked to more investment in education, housing, and health as well as lower likelihood of poverty (Adams and Cuecuecha 2013).

In contrast, some other studies reveal that remittance has little to do with productive investment (de Brauw and Rozelle 2008). In communities heavily dependent on remittances, people tend to reduce their work effort. As a large proportion of remittances is channeled to local service sectors, such as construction, local manufacturing jobs become less competitive compared with service-sector jobs (Acosta, Lartey, and Mandelman 2009). In the Philippines, overseas migration cuts into the domestic labor force and the receipts of remittances further reduces labor supply, resulting in higher wages. While the impact of the higher wages on competitiveness can be mitigated by the increase in labor productivity, potential long-term negative effects, such as limited diversification of the economy, tend to persist (Bayangos and Jansen 2011).

Third, migration facilitates land consolidation and mechanization in rural areas. As a farmer migrates out, he or she either rents out or sells his or her land. Thus farm size tends to go up with the outflow of migrants (Deininger and Jin 2005). Over time, rural-to-urban migration bids up local wages, inducing farmers to use more machinery to substitute for labor, as in East Asia (Chapter 3; Yang et al. 2013; Wang et al. 2016; Zhang, Yang, and Reardon 2017). In the case of China, rural-to-urban migration saw land use transition accompanied by increased nonfarm income. These nonfarm income sources contribute to stimulation of technology adoption for farming to save agricultural labor and to allow earning of more incomes. Together with improved transportation and marketization, agriculture is gradually transformed from subsistence agriculture to commercial agriculture in China (Chen et al. 2014).

Overall, migration plays a positive role in expanding the scale of agricultural production and facilitating the adoption of agricultural mechanization by smallholder farmers.

Conclusion

The first step toward the transformation of the rural economy is to improve the productivity of staple crop production by technological change. This is particularly the case in low-income economies, where population pressure on limited land resources leads to an imbalance between food supply and food demand. The Green Revolution is a solution to this food problem (Otsuka 2013). Since the Green Revolution requires research on crop breeding and improved cultivation practices, as well as extension services, both of which are public goods, the role of government is critical in achieving revolutionary improvements in the productivity of staple crops. Investment in the human capital of farmers through agricultural extension is often neglected but must be an integral part of the strategy in realizing the Green Revolution. Since the Green Revolution pushes the production possibility frontier of the entire economy outward, there will be income gains for the economy as a whole, particularly through reallocation of resources, even though the Green Revolution may not directly stimulate local rural industrialization.

Shift of resources from staple food production to production of high-value products is another important strategy to transform the rural economy. The transformation crucially hinges upon increasing demand for high-value products. Cutting trade barriers and improving infrastructure can help lower the transaction cost from farmgate to consumers and increase market size, thereby improving the profit margin of high-value crop producers.

Contract farming is an effective institutional arrangement to organize the production of high-value crops. Yet, to bring about large income gains to smallholder farmers from the production of such products, it is essential for the government or private sector to provide training to transform ordinary farmers producing staple crops to entrepreneurial farmers who carry out not only production of high-value products but also pre-production and post-production activities. Otherwise, the income of farmers from the production of high-value products may not increase significantly, even if the production efficiency of farming improves dramatically. Another option is to encourage producer cooperatives. By joining cooperatives, smallholder farmers gain greater bargaining power when negotiating with buyers and obtain better access to inputs and services. However, organizing cooperatives involves coordination cost. Only when the benefits overtake the cost, do the producer cooperatives sustain.

The importance of nonfarm income has been increasing in the developing world. To provide lucrative nonfarm employment opportunities, governments should recognize the importance of indigenous rural industrial clusters and facilitate their growth through various local policies, such as improving access to markets, organizing training, and inspecting product quality. The clusters capitalize the existing strengths embedded in rural communities—that is, strong social trust and abundant labor. The social rate of return to investment in the training of workers and entrepreneurs exceeds the private rate of return, partly because of the turnover of trained workers and partly because of information spillovers from competent entrepreneurs to others. Thus, in many cases the government must support the investment in training on top of the training offered by the private sector, including producer cooperatives. In areas with an absence of any industrial clusters, the government may try to construct industrial parks, offering some incentives to attract promising entrepreneurs in addition to providing training to potential entrepreneurs. However, the odds of success are often much lower than promoting indigenous clusters.

In areas where local job opportunities are limited, migration is another choice. When migrants find better-paid jobs elsewhere, they send home remittances, which boost the consumption level of remaining family members and reduce poverty. As more rural people migrate out, agricultural land rental markets and mechanization service markets often emerge to cope with the resultant labor shortage. A side effect of remittance is that it often results in conspicuous consumption. In short, whether the rural economy is successfully transformed from subsistence farming areas producing primarily staple crops to a rural economy producing high-value agricultural products and diverse industrial goods depends critically on the role of local government as well as the collective action of producers and agro-processors. Policies that help expand markets and address externality problems should be greatly encouraged.

References

Ackah, C. 2013. "Nonfarm Employment and Incomes in Rural Ghana." *Journal of International Development* 25 (3): 325–339.

Acosta, P. A., E. K. K. Lartey, and F. S. Mandelman. 2009. "Remittances and the Dutch Disease." *Journal of International Economics* 79 (1): 102–116.

Adams, R. H., Jr., and A. Cuecuecha. 2010. "Remittances, Household Expenditure and Investment in Guatemala." *World Development* 38 (11): 1626–1641.

———. 2013. "The Impact of Remittances on Investment and Poverty in Ghana." *World Development* 50: 24–40.

Akhtar, S., G. Li, A. Nazir, A. Razzaq et al. 2019. "Maize Production under Risk: The Simultaneous Adoption of Off-Farm Income Diversification and Agricultural Credit to Manage Risk." *Journal of Integrative Agriculture* 18 (2): 460–470.

Ali, M., and D. Byerlee. 2002. "Productivity Growth and Resource Degradation in Pakistan's Punjab: A Decomposition Analysis." *Economic Development and Cultural Change* 50 (4): 839–863.

Ali, M., J. Peerlings, and X. Zhang. 2014. "Clustering as an Organizational Response to Capital Market Inefficiency: Evidence from Microenterprises in Ethiopia." *Small Business Economics* 43 (3): 697–709.

Arouna, A., J. D. Michler, and J. C. Lokossou. 2019. *Contract Farming and Rural Transformation: Evidence from a Field Experiment in Benin*. NBER Working Paper 25665. Cambridge, MA, US: National Bureau of Economic Research (NBER).

Ashraf, N., X. Gin, and D. Karlan. 2009. "Finding Missing Markets (and a Disturbing Epilogue): Evidence from an Export Crop Adoption and Marketing Intervention in Kenya." *American Journal of Agricultural Economics* 91 (4): 973–990.

Atamanov, A., and M. Van den Berg. 2012. "Participation and Returns in Rural Nonfarm Activities: Evidence from the Kyrgyz Republic." *Agricultural Economics* 43 (4): 459–471.

Bai, J. 2016. "Melons as Lemons: Asymmetric Information, Consumer Learning and Seller Reputation." Job Market Paper, Department of Economics, Massachusetts Institute of Technology.

Ballard, R. 2001. "The Impact of Kinship on the Economic Dynamics of Transnational Networks: Reflections on Some South Asian Developments." Centre for Applied South Asian Studies, University of Manchester, UK.

Barbieri, E., M. R. Di Tommaso, and S. Bonnini. 2012. "Industrial Development Policies and Performances in Southern China: Beyond the Specialised Industrial Cluster Program." *China Economic Review* 23 (3): 613–625.

Barrett, C. B., M. E. Bachke, M. F. Bellemare, H. C. Michelson, S. Narayanan, and T. F. Walker. 2012. "Smallholder Participation in Contract Farming: Comparative Evidence from Five Countries." *World Development* 40 (4): 715–730.

Bayangos, V., and K. Jansen. 2011. "Remittances and Competitiveness: The Case of the Philippines." *World Development* 39 (10): 1834–1846.

Bellemare, M. F., and J. R. Bloem. 2018. "Does Contract Farming Improve Welfare? A Review." *World Development* 112: 259–271.

Bellemare, M. F., and S. Lim. 2018. "In All Shapes and Colors: Varieties of Contract Farming." *Applied Economic Perspectives and Policy* 40 (3): 379–401.

BenYishay, A., and M. A. Mobarak. 2018. "Social Learning and Incentives for Experimentation and Communication." *Review of Economic Studies* 86: 976–1009.

Bernard, T., M. Hidrobo, A. Le Port, and R. Rawat. 2018. "Nutrition-Based Incentives in Dairy Contract Farming in Northern Senegal." *American Journal of Agricultural Economics* 101 (2): 404–435.

Bezu, S., and C. Barrett. 2012. "Employment Dynamics in the Rural Nonfarm Sector in Ethiopia: Do the Poor Have Time on Their Side?" *Journal of Development Studies* 48 (9): 1223–1240.

Bloom, N., B. Eifert, D. McKenzie, A. Mahajan, and J. Roberts. 2013. "Does Management Matter? Evidence from India." *Quarterly Journal of Economics* 128 (1): 1–51.

Bloom, N., C. Genakos, R. Sadun, and J. Van Reenen. 2012. "Management Practices across Firms and Countries." *Academy of Management Perspectives* 26 (1): 12–33.

Boserup, E. 1965. *The Conditions of Agricultural Growth: The Economics of Agrarian Change under Population Pressure*. London: George Allen and Unwin.

Bustos, P., B. Caprettini, and J. Ponticelli. 2016. "Agricultural Productivity and Structural Transformation: Evidence from Brazil." *American Economic Review* 106 (6): 1320–1365.

Byerlee, D., and P. Moya. 1993. *Impacts of International Wheat Breeding Research in the Developing World, 1966–1990.* Mexico, DF: International Maize and Wheat Improvement Center.

Casaburi, L., and R. Macchiavello. 2019. "Demand and Supply of Infrequent Payment as a Commitment Device: Evidence from Kenya." *American Economic Review* 109 (2): 523–555.

Casaburi, L., and J. Willis. 2018. "Time versus State in Insurance: Experimental Evidence from Contract Farming in Kenya." *American Economic Review* 108 (2): 3778–3813.

Cavatassi, R., M. Gonzales-Flores, P. Winters, J. Andrade-Piedra, P. Espinosa, and G. Thiele. 2011. "Linking Smallholders to the New Agricultural Economy: The Case of the Plataformas de Concertacion in Ecuador." *Journal of Development Studies* 4710: 1545–1573.

Chatterjee, T., and A. Ganesh-Kumar. 2016. "Geographical Neighborhood and Cluster Formation: Evidence from Indian Agriculture." *Journal of Development Studies* 52 (11): 1577–1592.

Chen, J., K. Kosec, and V. Mueller. 2019. "Moving to Despair? Migration and Well-Being in Pakistan." *World Development* 113: 186–203.

Chen, R., C. Ye, Y. Cai, X. Xing, and Q. Chen. 2014. "The Impact of Rural Out-Migration on Land Use Transition in China: Past, Present and Trend." *Land Use Policy* 40: 101–110.

Christiaensen, L., L. Demery, and J. Kuhl. 2011. "The (Evolving) Role of Agriculture in Poverty Reduction: An Empirical Perspective." *Journal of Development Economics* 96 (2): 239–254.

Conley, T. G., and C. R. Udry. 2010. "Learning about a New Technology: Pineapple in Ghana." *American Economic Review* 100 (1): 35–69.

Cunguara, B., A. Langyintuo, and I. Darnhofer. 2011. "The Role of Nonfarm Income in Coping with the Effects of Drought in Southern Mozambique." *Agricultural Economics* 42 (6): 701–713.

Dalrymple, D. G. 1978. *Development and Spread of High Yielding Varieties of Wheat and Rice in the Less Developed Nations*. USDA Foreign Agricultural Economics Report 95. Washington, DC: United States Department of Agriculture.

David, C. C., and K. Otsuka, eds. 1994. *Modern Rice Technology and Income Distribution in Asia*. Boulder, CO: Lynne Rienner.

de Brauw, A., and K. Ambler. 2018. "Tightening Borders and Threats to Food Security." In *2018 Global Food Policy Report*, 38–45. Washington, DC: International Food Policy Research Institute (IFPRI).

de Brauw, A., and S. Rozelle. 2008. "Migration and Household Investment in Rural China." *China Economic Review* 19 (2): 320–335.

Deininger, K., and S. Jin. 2005. "The Potential of Land Rental Markets in the Process of Economic Development: Evidence from China." *Journal of Development Economics* 78 (1): 241–270.

de Janvry, A., and E. Sadoulet. 2001. "Income Strategies among Rural Households in Mexico." *World Development* 29 (3): 467–480.

———. 2002. "World Poverty and Role of Agricultural Technology: Direct and Indirect Effects." *Journal of Development Studies* 38 (4): 1–26.

———. 2010. "Agricultural Growth and Poverty Reduction: Additional Evidence." *World Bank Research Observer* 25(1): 1–20.

de Janvry, A., E. Sadoulet, and R. Murgai. 2002. "Rural Development and Rural Policy." In *Handbook of Agricultural Economics: Volume 2A*, edited by B. L. Gardner and G. C. Rausser, 1593–1658. Amsterdam: North-Holland.

de Janvry, A., E. Sadoulet, and T. Suri. 2017. "Field Experiments in Developing Country Agriculture." In *Economic Field Experiments: Volume 2*, edited by A. V. Banerjee and E. Duflo, 427–466. Amsterdam: Elsevier.

de Janvry, A., E. Sadoulet, and N. Zhu. 2005. *The Role of Non-Farm Incomes in Reducing Rural Poverty and Inequality in China*. CUDARE Working Paper 1001. Berkeley, US: UC Berkeley Department of Agricultural and Resource Economics.

Demissie, A., and B. Legesse. 2013. "Determinants of Income Diversification among Rural Households: The Case of Smallholder Farmers in Fedis District, Eastern Hararghe Zone, Ethiopia." *Journal of Development and Agricultural Economics* 5 (3): 120–128.

Diao, X., and S. Kolavalli. 2014. "Mechanization in Ghana: Emerging Demand, and the Search for Alternative Supply Models." *Food Policy* 48: 168–181.

Do, T. L., T. T. Nguyen, and U. Grote. 2019. "Nonfarm Employment and Household Food Security: Evidence from Panel Data for Rural Cambodia." *Food Security*: 1–16.

Essers, D. 2016. "South African Labour Market Transitions since the Global Financial and Economic Crisis: Evidence from Two Longitudinal Datasets." *Journal of African Economies* 26 (2): 192–222.

Estudillo, J. P., and K. Otsuka. 1999. "Green Revolution, Human Capital, and Off-Farm Employment: Changing Sources of Income among Farm Households in Central Luzon, 1966–94." *Economic Development and Cultural Change* 47 (3): 497–523.

———. 2016. *Moving out of Poverty: An Inquiry into Inclusive Growth in Asia*. London: Routledge.

Evenson, R. E., and D. Gollin, eds. 2003. *Crop Variety Improvement and Its Effect on Productivity: The Impact of International Agricultural Research*. Wallingford, UK: CABI Publishing.

Fan, S., and X. Zhang. 2004. "Infrastructure and Regional Economic Development in Rural China." *China Economic Review* 15 (2): 203–214.

FAO (Food and Agriculture Organization of the United Nations). FAOSTAT, various years. http://www.fao.org/faostat/en/

Foster, A. D., and M. R. Rosenzweig. 1996. "Technical Change and Human Capital Returns and Investment: Evidence from the Green Revolution." *American Economic Review* 86 (4): 931–953.

———. 2010. "Microeconomics of Technology Adoption." *Annual Review of Economics* 2 (1): 395–424.

Galvez-Nogales, E. 2010. *Agro-Based Clusters in Developing Countries: Staying Competitive in a Globalized Economy*. Agricultural Management, Marketing and Finance Occasional Paper 25. Rome: FAO.

Gatto, M., M. Wollni, R. Asnawi, and M. Qaim. 2017. "Oil Palm Boom, Contract Farming, and Rural Economic Development: Village-Level Evidence from Indonesia." *World Development* 95: 127–140.

Geldes, C., C. Felzensztein, E. Turkina, and A. Durand. 2015. "How Does Proximity Affect Interfirm Marketing Cooperation? A Study of Agribusiness Cluster." *Journal of Business Research* 68 (2): 263–272.

Gibson, J., and S. Olivia. 2010. "The Effects of Infrastructure Access and Quality on Nonfarm Enterprises in Rural Indonesia." *World Development* 68 (5): 717–726.

Gollin, D., C. W. Hansen, and A. Wingender. 2018. *Two Blades of Grass: The Impact of the Green Revolution*. NBER Working Paper 24744. Cambridge, MA: NBER.

Gollin, D., and C. R. Udry. 2019. *Heterogeneity, Measurement Error and Misallocation: Evidence from African Agriculture*. NBER Working Paper 25440. Cambridge, MA: NBER.

Haggblade, S., and P. Hazell. 2010. *Successes in African Agriculture: Lessons for the Future.* Baltimore: Johns Hopkins University Press.

Haggblade, S., P. Hazell, and P. Dorosh. 2007. "Sectoral Growth Linkages between Agriculture and the Rural Nonfarm Economy." In *Transforming the Rural Nonfarm Economy,* edited by S. Haggblade, P. Hazell, and T. Reardon, 141–182. Baltimore: Johns Hopkins University Press.

Haggblade, S., P. Hazell, and T. Reardon, eds. 2007. *Transforming the Rural Nonfarm Economy.* Baltimore: Johns Hopkins University Press.

Hashino, T., and K. Otsuka, eds. 2016. *Industrial Districts in History and the Developing World.* Dordrecht, Netherlands: Springer.

Hayami, Y., ed. 1998. *Towards the Rural-Based Development of Commerce and Industry.* Washington, DC: World Bank.

Hayami, Y., and Y. Godo. 2005. *Development Economics: From the Poverty to the Wealth of Nations.* New York: Oxford University Press.

Hayami, Y., and M. Kikuchi. 2000. *A Rice Village Saga: Three Decades of Green Revolution in the Philippines.* London: Macmillan Press.

Hayami, Y., and K. Otsuka. 1994. "Beyond the Green Revolution: Agricultural Development Strategy into the New Century." In *Agricultural Technology: Policy Issues for the International Community,* edited by J. Anderson, 15–42. Wallingford, UK: CAB International.

Hayami, Y., and V. W. Ruttan. 1970. "Korean Rice, Taiwan Rice, and Japanese Agricultural Stagnation: An Economic Consequence of Colonialism." *Quarterly Journal of Economics* 84 (4): 562–589.

———. 1985. *Agricultural Development: An International Perspective.* Baltimore: Johns Hopkins University Press.

Higuchi, Y., E. P. Mhede, and T. Sonobe. 2019. "Short- and Medium-Run Impacts of Management Training: An Experiment in Tanzania." *World Development* 114: 220–236.

Hsieh, S.-C., and V. W. Ruttan. 1967. "Environmental, Technological, and Institutional Factors in the Growth of Rice Production: Philippines, Thailand, and Taiwan." *Food Research Institute Studies* 7 (3): 307–341.

Ifft, J., D. Roland-Holst, and D. Zilberman. 2012. "Consumer Valuation of Safety-Labeled Free-Range Chicken: Results of a Field Experiment in Hanoi." *Agricultural Economics* 43 (6): 607–620.

Johnston, B. F., and J. Cownie. 1969. "The Seed-Fertilizer Revolution and Labor Force Absorption." *American Economic Review* 59 (4): 569–582.

Jolliffe, D. 2004. "The Impact of Education in Rural Ghana: Examining Household Labor Allocation and Returns on and off Farm." *Journal of Development Economics* 73 (1): 287–314.

Jones, M., and F. Kondylis. 2018. "Does Feedback Matter? Evidence from Agricultural Services." *Journal of Development Economics* 131: 289–241.

Khandker, S. R., Z. Bakht, and G. B. Koolwal. 2009. "The Poverty Impact of Rural Roads." *Economic Development and Cultural Change* 57 (4): 685–722.

Khandker, S. R., and G. B. Koolwal. 2010. "How Infrastructure and Financial Institutions Affect Rural Income and Poverty: Evidence from Bangladesh." *Journal of Development Studies* 46 (6): 1109–1137.

Kikkawa, A., T. Matsumoto, and K. Otsuka. 2019. "An Inquiry into the Heterogeneous Outcomes of International Migration: Evidence from Rural Households in Bangladesh." *Journal of Development Studies* 55 (10): 2106–2128.

Kondylis, F., V. Mueller, and J. Zhu. 2017. "Seeing Is Believing? Evidence from an Extension Network Experiment." *Journal of Development Economics* 125: 148–169.

Kunte, S., M. Wollni, and C. Keser. 2017. "Making It Personal: Breach and Private Ordering in a Contract Farming Experiment." *European Review of Agricultural Economics* 44 (1): 121–148.

Lanjouw, J. O., and P. Lanjouw. 2001. "The Rural Nonfarm Sector: Issues and Evidence from Developing Countries." *Agricultural Economics* 26 (1): 1–23.

Lanjouw, P., and R. Murgai. 2009. "Poverty Decline, Agricultural Wages, and Nonfarm Employment in Rural India: 1983–2004." *Agricultural Economics* 40 (2): 243–263.

Long, C., and X. Zhang. 2011. "Cluster-Based Industrialization in China: Finance and Performance." *Journal of International Economics* 84 (1): 112–123.

———. 2012. "Patterns of China's Industrialization: Concentration, Specialization, and Clustering." *China Economic Review* 23 (3): 593–612.

Maertens, M., and K. Vande Velde. 2017. "Contract-Farming in Staple Food Chains: The Case of Rice in Benin." *World Development* 65: 73–87.

Mano, Y., A. Iddrisu, Y. Yoshino, and T. Sonobe. 2012. "How Can Micro and Small Enterprises in Sub-Saharan Africa Become More Productive? The Impacts of Experimental Basic Managerial Training." *World Development* 40 (3): 458–468.

Marchetta, F. 2013. "Migration and Nonfarm Activities as Income Diversification Strategies: The Case of Northern Ghana." *Canadian Journal of Development Studies* 34 (1): 1–21.

Marshall, A. 1920. *Principles of Economics*. London: Macmillan.

Matsumoto, T., Y. Kijima, and T. Yamano. 2006. "The Role of Local Nonfarm Activities and Migration in Reducing Poverty: Evidence from Ethiopia, Kenya, and Uganda." *Agricultural Economics* 35 (S3): 449–458.

McKenzie, D., and C. Woodruff. 2014. "What Are We Learning from Business Training Evaluations around the Developing World?" *World Bank Research Observer* 29 (1): 48–82.

Michelson, H. 2013. "Small Farmers, NGOs, and a Walmart World: Welfare Effects of Supermarkets Operating in Nicaragua." *American Journal of Agricultural Economics* 95 (3): 628–649.

Michelson, H., T. Reardon, and F. Perez. 2012. "Small Farmers and Big Retail: Trade-offs of Supplying Supermarkets in Nicaragua." *World Development* 40 (2): 342–354.

Michler, J. D., E. Jiernström, S. Verkaart, and K. Mausch. 2019. "Money Matters: The Role of Yields and Profits in Agricultural Technology Adoption." *American Journal of Agricultural Economics* 101 (3): 710–731.

Mishra, A. K., A. Kumar, P. K. Joshi, and A. D'Souza. 2016. "Impact of Contracts in High Yielding Varieties Seed Production on Profits and Yield: The Case of Nepal." *Food Policy* 62: 110–121.

Murgai, R., M. Ali, and D. Byerlee. 2001. "Productivity Growth and Sustainability in Post-Green Revolution Agriculture: The Case of the Indian and Pakistan Punjab." *World Bank Research Observer* 16 (2): 199–218.

Nakajima, M., K. Otsuka, and T. Yamano. 2017. "Jobs off the Farm: Wealth, Human Capital, and Social Group in Rural Eastern India." *Journal of Development Studies* 54 (1): 111–132.

Olugbire, O. O., A. O. Falusi, A. I. Adeoti, A. S. Oyekale, and O. A. Adeniran. 2011. "Non-Farm Income Diversification and Poverty Reduction in Nigeria: A Propensity-Score Matching Analysis." *Continental Journal of Agricultural Science* 5 (3): 21–28.

Ortega, D. L., H. Wang, L. Wu, and N. J. Olynk. 2011. "Modeling Heterogeneity in Consumer Preferences for Select Food Safety Attributes in China." *Food Policy* 36 (2): 318–324.

Otsuka, K. 1998. "Rural Industrialization in East Asia." In *The Institutional Foundations of East Asian Economic Development*, edited by Y. Hayami and M. Aoki, 447–475. London: Macmillan.

———. 2007. "Rural Industrialization in East Asia: Its Influences, Nature and Implications." In *Transforming the Rural Nonfarm Economy,* edited by S. Haggblade, P. Hazell, and T. Reardon, 216–233 Baltimore: Johns Hopkins University Press.

———. 2013. "Food Insecurity, Income Inequality, and the Changing Comparative Advantage in World Agriculture." *Agricultural Economics* 44 (S1): 7–18.

Otsuka, K., J. P. Estudillo, and Y. Sawada, eds. 2009. *Rural Poverty and Income Dynamics in Asia and Africa*. London: Routledge.

Otsuka, K., and D. Larson, eds. 2013. *An African Green Revolution: Finding Ways to Boost Productivity on Small Farms*. Dordrecht, Netherlands: Springer.

———. 2016. *In Pursuit of an African Green Revolution: Views from Rice and Maize Farmers' Fields*. Dordrecht, Netherlands: Springer.

Otsuka, K., Y. Nakano, and K. Takahashi. 2016. "Contract Farming in Developed and Developing Countries." *Annual Review of Resource Economics* 8: 353–376.

Owusu, V., A. Awudu, and A.-R. Seini. 2011. "Non-Farm Work and Food Security among Farm Households in Northern Ghana." *Food Policy* 36 (2): 108–118.

Ping, Q., M. A. Iqbal, M. Abid, U. I. Ahmed, A. Nazir, and A. Rehman. 2016. "Adoption of Off-Farm Diversification Income Sources in Managing Agricultural Risks among Cotton Farmers in Punjab Pakistan." *Journal of Applied Environment and Biological Sciences* 6 (8): 47–53.

Qin, Y., and X. Zhang. 2016. "The Road to Specialization in Agricultural Production: Evidence from Rural China." *World Development* 77: 1–16.

Quizon, J., and H. P. Binswanger. 1986. "Modeling the Impact of Agricultural Growth and Government Policy on Income Distribution in India." *World Bank Economic Review* 1 (1): 103–148.

Ragasa, C., I. Lambrecht, and D. Kufoalor. 2018. "Limitations of Contract Farming as a Pro-Poor Strategy: The Case of Maize Outgrower Schemes in Upper West Ghana." *World Development* 102: 30–56.

Ramos, C., J. P. Estudillo, Y. Sawada, and K. Otsuka. 2012. "Transformation of the Rural Economy in the Philippines, 1988–2006." *Journal of Development Studies* 48 (11): 1629–1648.

Ranis, G., and F. Stewart. 1993. "Rural Nonagricultural Activities in Development: Theory and Application." *Journal of Development Economics* 40 (1): 75–101.

Reardon, T., and P. Timmer. 2012. "The Economics of Food System Revolution." *Annual Review of Resource Economics* 4: 225–264.

Renkow, M. 2007. "Cities, Towns, and Rural Nonfarm Economy." In *Transforming the Rural Nonfarm Economy,* edited by S. Haggblade, P. Hazell, and T. Reardon, 183–198. Baltimore: Johns Hopkins University Press.

Ruan, J., and X. Zhang. 2009. "Finance and Cluster-Based Industrial Development in China." *Economic Development and Cultural Change* 58 (1): 143–164.

———. 2016. "Low-Quality Crisis and Quality Improvement: The Case of Industrial Clusters in Zhejiang Province." In *Industrial Districts in History and the Developing World*, edited by T. Hashino and K. Otsuka, 169–189. Dordrecht, Netherlands: Springer.

Sanyang, S., S. J. B. Taonda, J. Kuiseu, N. T. Coulibaly, and L. Konate. 2016. "A Paradigm Shift in African Agricultural Research for Development: The Role of Innovation Platforms." *International Journal of Agricultural Sustainability* 14 (2): 187–213.

Scharf, M. M., and D. B. Rahut. 2014. "Nonfarm Employment and Rural Welfare: Evidence from Himalaya." *American Journal of Agricultural Economics* 96 (4): 1183–1197.

Schultz, T. W. 1964. *Transforming Traditional Agriculture*. New Haven, CT: Yale University Press.

———. 1974. "The Value of the Ability to Deal with Disequilibria." *Journal of Economic Literature* 13 (3): 827–846.

Senadza, B. 2012. "Non-Farm Income Diversification in Rural Ghana: Patterns and Determinants." *African Development Review* 24: 233–244.

Seng, K. 2015. "The Effects of Nonfarm Activities on Farm Households' Food Consumption in Rural Cambodia." *Development Studies Research* 2 (1): 77–89.

Shikuku, K. M., J. Pieters, E. Bulte, and P. Läderach. 2019. "Incentives and the Diffusion of Agricultural Knowledge: Experimental Evidence from Northern Uganda." *American Journal of Agricultural Economics* 101 (4): 1164–1180.

Sonobe, T., and K. Otsuka. 2006. *Cluster-Based Industrial Development: An East Asian Model.* Hampshire, UK: Palgrave Macmillan.

———. 2011. *Cluster-Based Industrial Development: A Comparative Study of Asia and Africa.* Hampshire, UK: Palgrave Macmillan.

———. 2014. *Cluster-Based Industrial Developments: Kaizen Management for MSE Growth in Developing Countries.* Hampshire, UK: Palgrave Macmillan.

Soullier, G., and P. Moustier. 2018. "Impacts of Contract Farming in Domestic Grain Chains on Farmer Income and Food Security: Contrasted Evidence from Senegal." *Food Policy* 79: 179–198.

Suri, T. 2011. "Selection and Comparative Advantage in Technology Adoption." *Econometrica* 79 (1): 159–209.

Takahashi, K., Y. Mano, and K. Otsuka. 2019. "Learning from Experts and Peer Farmers about Rice Production: Experimental Evidence from Cote d'Ivoire." *World Development* 122: 157–169.

Takahashi, K., R. Muraoka, and K. Otsuka. 2020. "Technology Adoption, Impact, and Extension in Developing Countries' Agriculture: A Review of the Recent Literature." *Agricultural Economics* 51 (1): 31–45.

Ton, G., W. Vellema, S. Desiere, S. Weitushat, and M. D'Haese. 2018. "Contract Farming for Improving Smallholder Incomes: What Can We Learn from Effectiveness Studies?" *World Development* 104: 46–64.

USDA (United States Department of Agriculture). National Agricultural Statistics Service. Database. www.nass.usda.gov/.

Väth, S., S. Gobien, and M. Kirk. 2019. "Socio-Economic Well-Being, Contract Farming and Property Rights: Evidence from Ghana." *Land Use Policy* 81: 878–888.

Velandia, M., R. M. Rejesus, T. O. Knight, and B. J. Sherrick. 2009. "Factors Affecting Farmers' Utilization of Agricultural Risk Management Tools: The Case of Crop Insurance, Forward Contracting, and Spreading Sales." *Journal of Agricultural and Applied Economics* 41: 107–123.

Wang, X., Y. Futoshi, K. Otsuka, and J. Huang. 2016. "Wage Growth, Landholding, and Mechanization in Chinese Agriculture." *World Development* 86: 30–45.

Woodruff, C., and R. Zenteno. 2007. "Migrant Networks and Microenterprises in Mexico." *Journal of Development Economics* 82 (2): 509–528.

World Bank. 2007. *World Development Report 2008: Agriculture for Development*. Washington, DC: World Bank.

Xing, L., S. Fan, X. Luo, and X. Zhang. 2009. "Community Poverty and Inequality in Western China: A Tale of Three Villages in Guizhou Province." *China Economic Review* 20 (2): 338–349.

Yamano, T., A. Arouna, R. A. Labarta, Z. M. Huelgas, and S. Mohanty. 2016. "Adoption and Impacts of International Rice Research Technologies." *Global Food Security* 8: 1–8.

Yang, D. 2008. "International Migration, Remittances, and Household Investment: Evidence from Philippine Migrants' Exchange Rate Shocks." *Economic Journal* 118 (528): 591–630.

———. 2011. "Migrant Remittances." *Journal of Economic Perspectives* 25 (3): 129–152.

Yang, J., Z. Huang, X. Zhang, and T. Reardon. 2013. "The Rapid Rise of Cross-Regional Agricultural Mechanization Services in China." *American Journal of Agricultural Economics* 95 (5): 1245–1251.

Zhang, X. 2016. *Building Effective Clusters and Industrial Parks*. IFPRI Discussion Paper 1590. Washington, DC: IFPRI.

Zhang, X., and S. Fan. 2004. "How Productive Is Infrastructure? A New Approach and Evidence from Rural India." *American Journal of Agricultural Economics* 86 (2): 492–501.

Zhang, X., and D. Hu. 2014. "Overcoming Successive Bottlenecks: The Evolution of a Potato Cluster in China." *World Development* 63: 102–112.

Zhang, X., S. Rashid, K. Ahmad, and A. Ahmed. 2014. "Escalation of Real Wages in Bangladesh: Is It the Beginning of Structural Transformation?" *World Development* 64: 273–285.

Zhang, X., J. Yang, and T. Reardon. 2017. "Mechanization Outsourcing Clusters and Division of Labor in Chinese Agriculture." *China Economic Review* 43: 184–195.

Chapter 12

FOOD VALUE CHAIN TRANSFORMATION IN DEVELOPING REGIONS

Thomas Reardon and Bart Minten

The "food system" is defined by Reardon et al. (2019) as a "dendritic cluster" of value chains linking the following:

1. Input suppliers to farmers ("farm input value chains").
2. Farmers upstream to wholesalers and processors midstream, to retailers then consumers downstream ("farm output value chains").
3. "Lateral service value chains" to all segments of the above two value chains (such as the transport supply chain as input to the wholesale segment of the output value chain).
4. R&D&E (research, development, and extension) suppliers to all the segments of the above value chains (such as the generation of new crop varieties by breeders in research institutes, to extension agents to farmers).

In this chapter, for brevity, we focus on the output value chains, but the conceptual framework and most trends are also relevant to the input value chains, the lateral service value chains, and R&D&E suppliers. Output value chains in developing countries have transformed over the past 50 years but particularly quickly only in the past 25 years. In many countries the transformation of value chains has been abrupt, not gradual. Reardon and Timmer (2014), illustrating with Asian evidence, explain the drivers of this rapid change as a confluence of the following three sets of interlinked transformations:

1. Downstream demand side change (urbanization and diet change) "pulling" system transformation.
2. Midstream/downstream change (in the structure and conduct of retail, wholesale, logistics, and processing) "intermediating" system transformation.
3. Upstream change (intensification, diversification, and commercialization of farming) "feeding" system transformation.

This chapter focuses on the downstream and midstream transformations (leaving to other chapters the treatment of upstream farm sector change). The chapter proceeds as follows. The next section lays out a heuristic conceptual framework of the product cycle and stages of transformation of the food value chains (FVCs). Then the main lines of the evidence concerning rapid transformation of FVCs in the developing region are traced. The three waves of research on the transformation of FVCs mainly over the past 70 years are discussed. The chapter concludes with an assessment of the literature and its gaps.

A Heuristic Conceptual Framework Regarding the Stages of Transformation of Food Value Chains and the Product Cycle

We provide a simple heuristic framework to classify and discuss the transformation of FVCs. We start with the traditional "structure-conduct-performance" model to facilitate a "stylized facts" description of the stages of transformation of FVCs (de Figueirêdo, Meuwissen, and Oude Lansink 2014). That framework is neutral with respect to the product, but one encounters products in different phases of their evolution in the market. Thus we also discuss the concept of the "product cycle," which classifies products and thus FVCs. Finally, we combine these first two concepts by presenting a taxonomy of FVC transformation in three stages: traditional, transitional, and modern. The taxonomy is necessarily, for our purposes, heuristic, broad brushstroke, and simplifying, setting aside substantial nuance and heterogeneity across countries and products in the empirical reality.

Structure-Conduct-Performance

An "industry" can be equated here with a given food value chain, such as the rice value chain from rice farm to rice retailer. The framework's application to an FVC has the three sets of analytical questions: the structure, the conduct, and the performance of the FVC.

Structure of the Food Value Chain

The structure (in an industrial organization sense) of the FVC can be seen from several angles. First, it can be viewed "chain-wide":

1. What are the number of segments of the FVC and thus what we call its "intermediation length"? For example, the elimination of a segment,

such as wholesalers between retailers and growers, is "disintermediation" and reduces the intermediation length.

2. What is the FVC's "spatial length" (and geographical orientation)? For example, is it a long chain stretching from distant rural areas to cities, or a short local chain? Is it rural to rural, or urban to rural, or rural to urban? Is it domestic or international?

3. What is the distribution across the segments of the chain of the formation of costs and value-added? This can be thought of as "cross-segment concentration." Viewed another way, what is the share of the post-farmgate segments (and its complement, the share of farmers) in the total value of output of the FVC?

Second, structure can be viewed per segment:

1. How concentrated (over firms) is a given segment (such as measured by the Gini coefficient of, for example, processing firms in the FVC)? Another way to look at this is the share of the largest scale stratum of the actors in a segment in the total output of the segment. For example, what share do large mango farms have in the total output of mangoes in a province in Indonesia (compared with medium and small farms)? The share of output can differ from the share of a stratum in the number of firms or farms in that segment. For example, Qanti et al. (2017) showed that although marginal or "backyard" mango farms constitute 80 percent of mango farms on Java, they control only 20 percent of the output; the 10 percent of farms that are medium-sized control 60 percent of the output.

2. How specific is a given segment's actors to the product of that FVC versus supplying products or services "horizontally" across a number of product FVCs? For example, the vegetable trucks of the logistic segment of the tomato FVC may also carry cucumbers for the cucumber FVC. So vegetable trucks are both a (vertical) segment in one vegetable FVC but also a shared segment across a number of vegetable FVCs.

   But there can also be "asset specificity" for a segment. This term is usually used in the new institutional economics (NIE) literature to indicate, say, a piece of equipment that a firm has to buy to participate in a specialized supply chain, such as an ultra-high temperature (UHT) milk processor to supply milk to the Tetrapak packaged sterile milk market in Brazil (Farina et al. 2005). That machine is not of use for other purposes, so buying it means locking part of the firm's capital into that particular FVC. That can also go for products. A modern meat-cutting machine can be calibrated to process a number of different ruminant carcasses but cannot be used for processing fish into fish fillets. So the machine is specific to the ruminant FVCs, perhaps used multifunctionally across those meat chains, but not used for the fish FVC.

Similarly, a "trader" (in the commerce segment) might trade not just in maize but also in soy, such as in Nigeria (Liverpool-Tasie et al. 2017). But a milk trader might only trade in milk due to asset specificity of his or her vehicle (such as a tanker truck) or equipment (cold tank). Agglomerations of actors can also serve across FVCs. A good case is a wholesale market.

3. What is the spatial distribution of the actors in each segment? Are the firms or farms clustered or dispersed? Is this correlated with firm scale and thus with concentration in the industrial organization sense? An example is whether small dairy farms tend to be concentrated in hilly areas and larger farms in valley bottomland, or whether large feed mills are peri-urban and small feed mills are scattered in rural areas.

4. What is the capital ownership pattern, such as what share of the capital is foreign owned, or collectively owned?

Conduct of the Food Value Chain

The conduct or behavior of the FVC can be seen from several angles. First, what is the technology used by producers? Producers are the actors in all segments of the FVC—from input producers and sellers, to farmers, to wholesalers, retailers, and logistics agents who produce services, and processors who produce (food) manufactures. Technology is assessed via three measures: (1) the ratio of productive factors used, such as the intensity of labor use and capital per unit of output—for example, a supermarket may be more capital intensive than a traditional retailer; (2) the intensity and types of intermediate inputs (such as types of milk used to make cheese); and (3) the scale of production of the actor.

Second, what organization is used by farms and firms to produce, to buy inputs, and to market outputs? To what degree do firms and farms in a given segment vertically integrate (such as a processor also having its own farms or its own retail shops) versus acquire inputs or marketing services from others? To what degree do firms and farms horizontally cluster in cooperatives or other groupings? Is that for production or for input procurement or for marketing?

Third, what institutions do farms and firms in the segment (and between segments of the FVC) use to organize their purchase of inputs and sale of outputs? In general, institutions are used to coordinate intersegment relations and impose requirements on suppliers. To what degree do firms and farms mediate their intersegment relations with contracts? For example, how prevalent is contract farming by processors? Do public standards for product quality and safety regulate suppliers in the food value chain? Do midstream and downstream firms such as processors and retailers impose private standards on suppliers?

Fourth, what is the supply and demand of facilitating lateral supply chain services such as finance and information to and from the segments of the FVC? For example, FVC financial conduct analysis examines the sources of and destinations of finance to and from the actors in each segment and between the segments of the FVC. The types of finance by FVC actors include: value-chain finance, and in particular, buyers' credit to suppliers and suppliers' credit to clients; self-finance by the FVC actor; and credit that is not predicated on value-chain relationships, such as a bank loan or microcredit.

Performance of the Food Value Chain

As with structure and conduct, the performance of the FVC can be seen in terms of the FVC as a whole or of the specific segments (such as the performance of the farm segment per se). But both of these can be analyzed along similar axes, including but not limited to (1) the efficiency of the whole chain, which can include the marketing margin and price integration analysis; (2) the equity of the chain or its segments, measured for example in concentration of income, inclusion of the poor and disadvantaged (as producers or consumers or zones), and adequacy of producer prices and affordability of consumer prices; and (3) externality effects of the FVC, such as on nutrition and the environment.

Product Cycle

The product cycle is a concept from classical economists such as Ricardo (1817) and then formalized as part of a link to trade theory by Vernon (1966; 1979). Translated into the context of the agrifood economy, the essential idea is that a product has a life cycle that starts as a new product or niche, then that niche product is commoditized into a bulk widely distributed product, perhaps growing and diffusing over national markets, until it reaches "maturity." At the latter point its profitability has been competed down, and it either declines or "dies" or transitions into differentiated products that themselves are again niche products and grow into commodities.

We can think of the product cycle evolution as driven by the "Cochrane's treadmill" (Cochrane 1979), where suppliers enter at each stage and compete down the profit, with the innovator driving the product to the next stage of the product cycle. The first step is the product's introduction into a local market as a niche product. Then, as its production is increased and it is marketed more broadly (beyond the local market), its price is driven down, and it becomes a bulk commodity. Competition drives innovation to create differentiated products from the commodity base. These innovations can be, for example, in variety, color, taste, size, type or degree of processing, packaging, and so on. Each differentiated product becomes again a niche and the cycle continues by the drive to commoditize it.

An example is the product that used to be called a Chinese gooseberry, a niche fruit grown locally in China's mountains. In the second half of the 20th century it was renamed "kiwi fruit" by New Zealanders and the green variety grown extensively throughout the world as a bulk commodity. It was then bred by New Zealanders into sweeter varieties, such as the golden kiwi—a differentiated product that is sweeter and with edible skin. The golden kiwi is now being "commoditized" such as with a recent shift to mass production of it in Italy and California.

Three Stages of FVC Transformation

We conceive of three stages of transformation of the food value chain (Reardon et al. 2019). These stages are functions of characteristics of the structure and conduct of the FVC as well as the phase of the product cycle.

The Traditional Stage

In the traditional stage, the following patterns are common. The reader could imagine in current time cereals or dairy in rural village markets in Ethiopia and Nepal, or food markets in most developing regions in the 1970s, or such markets in the United States in the 1920s and 1930s. First, in terms of structure, traditional FVCs are spatially short because the urban share of the population in the food market is low. They are also intermediationally short because much of the market is in the rural area and even the same village. The share of grains and other staples in the food economy is very high: there is relatively little production of nongrain products, hence few supply chains for them, except in pockets of traditional cash crops. The share of value-added in postharvest segments of the FVCs is small: home processing reigns, and the wholesale and logistic sectors are small because food is not moved far. Most of the segments are fragmented. Concentration is mainly observed where governments assured grain supply to emerging urban populations at subsidized prices via parastatals.

Second, in terms of conduct, as the bulk of the food system is in niche phase for nongrains and niche-to-commodity phase for grains, there is little quality differentiation, standards, or economies of scope. Technologies are labor intensive per unit of output. Enterprise scale is tiny. Spot market relations dominate food value chains; contracts are not used.

The Transitional Stage

In the transitional stage, the following patterns are common. The reader could imagine in current time aquaculture fish, teff, and maize into urban markets in Bangladesh, Ethiopia, and Tanzania respectively, or food markets in the United States in the 1940s through the 1960s. First, in terms of structure, transitional FVCs are spatially long because the urban share of the population in the food market is moderately high. The exception, where short supply chains still dominate, is in highly perishable products such as leafy greens and farmed fish and dairy. Food value chains are intermediationally long as many small and medium-scale actors in the midstream of the FVCs emerge to add value and move food from rural areas to urban areas. In the transitional phase the share of value-added in postharvest segments of the FVCs is moderately large as wholesale, processing, and logistics sectors have blossomed.

Governments have by the transitional stage largely dismantled output and input parastatals and private firms, especially small and medium-sized enterprises (SMEs), have flooded in, as part of what Reardon (2015) calls a **J**-curve of concentration: in the traditional stage, there is moderate concentration (parastatals in the public sector alongside a fragmented "parallel market" private sector). Then, with liberalization and privatization of parastatals (in the 1980s and 1990s in developing regions), there is deconcentration with the proliferation of SMEs. By the transitional stage the share of grains and other staples in the food economy is now a minority share. Bennett's Law (1941)—stating that when incomes increase, people consume relatively fewer starchy staples and relatively more nutrient-dense (and more expensive) foods—in consumption and diversification in farming beyond grains has radically increased production of and FVCs in animal and horticultural products.

Second, in terms of conduct, the bulk of the food system is by the transitional stage in a major shift from the niche to the commodity phase of the product cycle. There is still little quality differentiation, but public standards have started to emerge for grades and quality, and in some cases for product safety. The exception is mostly on the side of value-added, as purchased-processed foods and differentiated products for grains (convenience foods and increases in quality) develop rapidly in this stage.

Technologies are still labor intensive, but machine use has emerged in farming and in the rest of the food value chain. For example, home processing such as hand pounding of grain found in the traditional phase has now given way to the proliferation of SMEs, milling companies, alongside a few emerging large-scale companies. SMEs also start playing a bigger role in logistics and wholesale. Spot market relations still dominate, but in a few cases contracts are beginning to emerge.

The Modern Stage

In the modern stage, the following patterns are common. The reader could imagine in current time food value chains for Senegal mangoes into export markets, rice into the main urban markets in China, and dairy products in Brazil, as well as US food markets in the 1970s, 1980s, and beyond. First, in terms of structure, modern FVCs are spatially long as the urban share is high. Even perishable products such as poultry and dairy and vegetables are by this stage produced far from cities and shipped frozen, chilled, packed, and so on. (But in the most advanced modern stages, there is a renaissance of peri-urban production of some of these products, such as in vertical horticulture.)

By the modern stage, food value chains have become intermediationally short with a trend toward "disintermediation"—as supermarkets and large processors transact directly between themselves and in some cases buy direct from farms. The right-hand side of the **J**-curve noted earlier has occurred, with substantial concentration in all segments at least upstream and downstream from farms, as well as in the farm sector in some countries. The SMEs that remain stayed competitive through product differentiation or a shield of high transaction costs (such as those in the hinterlands). Foreign direct investment (FDI) liberalization started in the transition stage has by the modern stage caused widespread multinationalization of a number of FVC segments.

The wholesale sector has evolved to be off-market (outside wholesale markets) with the rise of large logistics and wholesale firms. In the transitional phase the share of value-added in postharvest segments of the FVCs is large (in the United States around 90 percent) in long and complex supply chains. The "food service sector" such as fast-food chains has increased to a substantial share of the food economy (from a modest share in the transitional stage and a tiny share in the traditional stages). By the modern stage, the share of grains and other staples in the food economy is now a small share of overall food, about a quarter or less. Nongrain supply chains and processed food FVCs dominate the total food sector.

Second, in terms of conduct, the bulk of the food system is by the modern stage in a major shift from the commodity to the differentiated products phase of the product cycle. There is now substantial quality differentiation, and private standards for quality and in some cases safety have emerged to begin to eclipse public standards (Henson and Reardon 2005), such as occurred in the dairy sector in Brazil in the 1990s (Farina et al. 2005). Technologies are now largely capital intensive—that is, mechanized all along the FVC. Information-based systems (such as smart chips in packaging and logistics and drones in agriculture) have emerged. Spot market relations are relegated to some sectors like fruits and vegetables, but in meats, grains and dairy contracts have come to dominate.

Synthesis of Stages of Transformation

A synthesis of the features of the food economy in these three different stages of food value chain transformation is shown in Table 12.1. We see significant differences over the three stages in the urban share in food markets, the share of grain and staples, seasonality, the importance of the food service sector, and the stage in the product cycle as well as the reach of the FVC, which moves overall from local niche products in the traditional FVC to globalized differentiated products in the modern FVC.

The FVC lengthens in spatial and intermediation terms with transformation. Significant change is also seen in value-addition (from low to high) and concentration (from moderate, to low, to high) over the three different stages of FVC transformation (Table 12.2). Conduct changes from low-quality demands, spot market trade, and labor-intensive technologies in traditional FVCs to high-quality differentiation, the wide use of contracts, and capital-intensive technologies in modern FVCs.

TABLE 12.1 **Synthesis of features of the food economy for different stages of food value chains (FVC)**

Characteristics of the food economy	Traditional FVC	Transitional FVC	Modern FVC
Urban share in food market	Low	Medium	High
Share of grains and staples	High	Medium	Low
Seasonality	High	Medium	Low
Food service sector	Small	Modest	Large
Reach FVC	Local	National	Global
Product cycle	Local niche	National commodity	Differentiated product

Source: Authors.

TABLE 12.2 **Synthesis of features of structure and conduct for different stages of food value chain (FVC) transformation**

Characteristics of structure and conduct	Traditional FVC	Transitional FVC	Modern FVC
Structure			
Spatial length FVC	Short	Long	Long
Intermediation length FVC	Short	Long	Short
Value-addition	Low	Medium	Large
Concentration	Moderate	Low (parastatals; large number of small and medium-sized enterprises)	High (private large-scale food industry firms)
Conduct			
Quality differentiation	Low	Low	High
Quality and safety standards	Few	Public	Private
Technologies	Labor-intensive	Labor-intensive	Capital-intensive
Contracts	Spot markets only	Spot markets dominate; emergence of contracts	Spot markets small; contracts dominate

Source: Authors.

Table 12.3 focuses on the midstream and downstream segments. It shows that traditional FVCs are dominated by domestic processing and fragmented wholesale and retail markets. In transitional FVCs, the public sector takes a more important role for the organization of midstream and downstream segments, but we also note a fast proliferation of small and medium-sized enterprises in this stage. Modern FVCs see consolidation and concentration with large private supermarkets becoming dominant players in this stage.

TABLE 12.3 Synthesis of features of midstream and downstream segments for different stages of food value chain (FVC) transformation

Activities	Traditional FVC	Transitional FVC	Modern FVC
Midstream processing	Mostly domestic	Parastatals and proliferation of small and medium-sized enterprises	Consolidation of private sector
Midstream wholesale/logistics	Fragmented wholesale markets	Publicly run wholesale markets (with private wholesalers)	Concentration; dedicated wholesalers; large-scale logistics firms; direct linkages of retail firms
Downstream retail	Fragmented	Small retail shops dominant; government parastatal retailing; emergence of private supermarkets	Private supermarkets dominant; modernization of procurement systems of supermarkets

Source: Authors.

Evidence Concerning the Rapid Transformation of FVCs in Developing Regions

While the transformation from traditional to transitional to modern FVCs took about 100 years in the United States and western Europe, today's developing regions have in the main made the step to transitional in 20 years, and the frontrunners the step to emerging modern FVCs in 20 to 30 years. The two surprises are that the transformation is in its essence similar to what happened in the United States, just that it has been far faster, nearly abrupt and sudden, in the developing regions. A fitting image is a tidal wave, which is a small ripple in the ocean for a thousand miles then wells up as it comes to the continental shelf. That is the shape of the change in the developing world's food economy. The main reason FVCs changed so fast in developing regions was that there was a confluence of changes that were mutually reinforcing, magnifying, and accelerating—a coevolution of the different pieces. We explore that confluence from two angles.

On the one hand, there were systemic changes, which we classify as (1) "meta conditioners" (change in policy and transport/communications, urbanization, and change in diets) that spurred and facilitated the changes in the FVC segments; and (2) downstream and midstream segment transformations (in retail, processing, wholesale, logistics). We do not treat the upstream change in farms and input systems; those are covered in other chapters. On the other hand, there are a number of cases of local confluences of transformation within or over segments. Below we discuss two such cases: in potato cold storages in India and teff wholesale, processing, and transport in Ethiopia.

These episodes provide a deep understanding of why and how FVCs are changing so very fast.

Meta Conditioners and Downstream Pull Forces

Policy and Infrastructure

Policy liberalization and privatization occurred during the 1980s through the 2000s as part of transitions away from administered food economies in "transition" countries such as China and Viet Nam as well as "structural adjustment programs" in most of the developing countries in Africa, Asia, and Latin America. Food procurement and distribution parastatals were largely privatized, and internal and external trade in food at least partially liberalized. This spurred the spread both of small and medium-sized enterprises as well as the entry of modern firms via both foreign direct investment and domestic investments. The latter led to the spread of large processors, supermarkets, and fast food chains. FDI liberalization was particularly important as a key inducement to this process.

Infrastructural investment has encouraged lengthening of supply chains and transformation in midstream and downstream segments. Hard infrastructure encourages the development of FVCs. Combined with rising urban demand, infrastructural investment has encouraged private investments by SMEs in the midstream of value chains, such as by teff millers, transporters, and retailers in Ethiopia. Small farmer access to markets is conditioned by infrastructure and distance to market. Barrett (2008) found the latter much more influential than macroeconomic and trade policies on small farmer participation in markets. Stifel, Minten, and Koru (2016) show a doubling of commercial surplus for farmers connected to a market compared with more remote farmers.

Communication access to information has increased significantly with the widespread availability of mobile phones (Nakasone, Torero, and Minten 2014). A large share of farmers in the commercial areas, ranging from a high of 97 percent in China to a low of 27 percent in Ethiopia, owned mobile phones at the time of the survey (around 2010). Taking a simple average of crops and countries in Asia, the data show that almost a quarter of farmers in commercial zones reached a price agreement by phone in their last transaction (Nakasone, Torero, and Minten 2014). For rice and potato supply chains in Dhaka, rice chains in Beijing, and potato chains in Delhi, almost all farmers who used phones contacted multiple traders before engaging in a transaction. Overall, 40 percent of staple suppliers in these rural-urban supply chains had contacted multiple buyers by phone before their last transaction (Reardon et al. 2012).

Urbanization

Urbanization has advanced to the point where rural-urban food supply chains dominate food markets in Asia and Africa. The impacts of urban markets have transmitted deeper and deeper into rural areas, and value chains have grown longer, spurred by urbanization and aided by the spread of rural wholesale markets, rural roads, and rural electrical grids. The rapid urbanization under way in developing regions, especially in Africa and Asia, is discussed in detail in Chapter 9 of this volume. Here we simply add the point that urban population shares underestimate the share of urban areas in total food consumption and total food purchased. This is because urban incomes sufficiently exceed rural incomes, which compensates for the higher-income urban consumers (per Engel's Law [1857]) having lower shares of food in their total budgets.[1] In Eastern and Southern Africa, 25 percent of the population is urban, but cities consume 48 percent of food produced and sold in the countries. In Asia, Reardon, Hu et al. (2015) show that while 38 percent of the population is urban, 53 percent of the (purchased) food market is urban.

Diet Change

Whereas Chapter 10 in this volume focuses on nutrition, agriculture, and health linkages, this chapter emphasizes changes in diets in developing regions that have important effects on the transformation of value chains per se. Diets have gone from (traditionally) mainly home-produced to increasingly purchased. Even the rural poor are heavily engaged in the food market as buyers. In the developing countries of Eastern and Southern Africa, for example, Dolislager, Tschirley, and Reardon (2015) show that rural households bought 44 percent (in value terms) of food they consume. Reardon, Hu et al. (2015) show that in Nepal, Indonesia, and Viet Nam, rural households bought 73 percent of their food (in value terms).

---

1 Engel's Law states that as incomes increase, the share of those incomes that is spent on food decreases.

There has been substantial diet diversification into processed food with penetration first in urban but also in rural areas. In Eastern and Southern Africa (Tschirley et al. 2015) urban households dedicate 56 percent of food expenditures to processed foods, and rural households, 29 percent. In Asia (Reardon, Hu et al. 2015) urban households dedicate 73 percent of food expenditures to processed foods, and rural areas, 60 percent. In addition, there has been much diet diversification beyond grains, with only modest differences between urban and rural. In the Eastern and Southern Africa study countries, the share of nongrains in urban food expenditure was 66 percent, and rural, 61 percent. In Asia the figures were 74 percent for urban and 63 percent for rural.

The middle class at least in Asia and Latin America has increasing demand for food quality and safety, in particular for semiprocessed foods such as dairy and to a certain extent for perishable foods (Ortega et al. 2012; Pingali 2006). Diets remain basically local—with only a small share imported. For example, 80 percent to 90 percent of national food consumption is supplied by domestic supply chains. Based on our calculations from FAOSTAT food balance sheets and COMTRADE, imports are only 10 percent to 20 percent (in 2010 for Eastern and Southern Africa, 15 percent; West Africa, 11 percent; South Asia, 10 percent; and Southeast Asia, 21 percent).

There have been several drivers of diet change in developing regions. Diet change is driven by factors on the demand side. Income increase drives a relative shift toward nonstaples (per Bennett's Law). But that income increase is not only between poor and middle-class status: Dolislager, Tschirley, and Reardon (2015) show for Eastern and Southern Africa that sharp changes in diet occur over the segments of the poor, with the rate of change steeper than between the poor and nonpoor segments. With urbanization, women are increasingly working outside the home, and thus with rising opportunity costs of time to shop for, process, and prepare food. Men are increasingly working far from home. These trends spur purchase of "convenience cereals" like rice and wheat (Senauer, Sahn, and Alderman 1986) as well processed foods and restaurant-prepared foods.

Diet change has also been driven by factors on the supply side. The food processing sector has grown fast over the past several decades. Agriculture has rapidly diversified beyond grains into horticulture, dairy, livestock, fish, and pulses. Rural-urban food supply chains have developed enormously to move these products to urban as well as rural consumers; Reardon, Tschirley et al. (2015) calculated that African food supply chain volumes increased six- to eight-fold over 1970 to 2010, with most of the increase in the past 20 years.

Trends in Transformation of the Downstream and Midstream of Food Value Chains

Changes in the structure and conduct of FVCs have occurred over the whole length of the chain as a system, as well as at the level of each segment—from downstream, to midstream, to upstream.

Overall Changes in Food Value Chains

Urbanization and better transport infrastructure have induced spatial lengthening and deseasonalization of FVCs, to draw from an increasingly broad market-catchment area to feed cities. There has been a proliferation of traditional intermediaries and then a reduction of their numbers and a rise of modern intermediaries. Traditionally there was a short FVC (from farms to the local villages and towns). With urbanization, however, the FVCs grew longer, and there was a proliferation of rural brokers and wholesalers, urban wholesalers, urban semiwholesalers, transporters, warehouse firms, and retailers, all as small-scale firms. As consolidation in processing and retail occurred, there has been a shift, fastest in processed and semiprocessed foods, slowest in perishables, toward the exit (or absorption) of small rural brokers and small processors (Reardon 2015). With the rise of supermarkets and processors, there is also a "re-intermediation" with the rise of dedicated/specialized wholesalers (Reardon and Berdegué 2002).

In links between FVC segments, there is organizational and institutional change, albeit at very different paces depending on the product, the scale of the firm buying the product, and the country. There is a start of vertical coordination through de facto semicontractual relations, like supplier lists (Berdegué et al. 2005), and some formal (even if just verbal) contracts. The latter are still limited but the former appears to be spreading especially among large companies. There is a rise of private standards (Reardon et al. 1999) specified in the contracts. Moreover, a traditional method of intersegment linkage, tied output-credit markets (Bardhan 1980), where a trader advances funds to a farmer and then expects his harvest at the end of the season, have declined substantially. This has been shown in Asia for rice and potato sectors (Reardon et al. 2014) and Africa generally (Adjognon, Liverpool-Tasie, and Reardon 2017).

Waves of Diffusion of Downstream and Midstream Transformation

OVERVIEW OF WAVES OF DIFFUSION

Despite heterogeneous conditions, there is some regularity in waves of diffusion, over countries and within countries, over income classes, and over products. The first wave was in countries that started their post–World War II growth spurt, urbanized, and started industrializing earlier—in particular, South American countries, East Asia outside of China, and South Africa. The start of processing transformation occurred with FDI liberalization and the start of privatization in the mid-1980s to early 1990s. Retail transformation took off from the early 1990s.

The second wave was in countries that had their growth and urbanization spurts later and/or had prolonged internal sociopolitical pressure to limit FDI. In Central America, Mexico, and Southeast Asia, processing transformation took off in the 1980s, but retail transformation did not start until the mid to late 1990s. The third wave was in countries, such as in China, India, and Viet Nam, that had their growth and urbanization spurts mainly in the 1990s and 2000s and/or had lagged liberalization into the 1990s. Processing transformation occurred somewhat before retail, with the latter mainly in the late 1990s and the 2000s. There was also a late part of the third wave, or a fourth wave, an incipience of processing and retail transformation in East and Southern Africa.

THE DOWNSTREAM SEGMENT TRANSFORMATION: RETAIL

The retail segment has changed as the result of direct government action as well as by the relinquishing of government involvement and the rapid diffusion of private-sector supermarkets. The modern retailers themselves had several phases of change in their conduct—in particular the shift from traditional to modern procurement systems. There were several changes, as follows.

First, governments themselves induced directly a first stage of retail transformation from traditional, fragmented retail to state-run chain stores—prior to liberalization and privatization in the 1990s and 2000s, when most of the state chains were dismantled. Examples are the Fair Price Shops in India. Second, after the liberalization of retail FDI and privatization of state retail, there was a huge surge in the 1990s and 2000s in private investment in supermarket chains in developing countries (Reardon et al. 2003). The waves of diffusion emerged in the spatial pattern discussed earlier. The share of modern retail in overall food differs over the wave of diffusion, with the deepest penetration to date in the first-wave countries, where the share is near half by the late 1990s and 50 percent to 60 percent in the 2000s; in the second-wave countries, about 30 percent to 50 percent by the 2000s; and in the third-wave countries, some 10 percent to 30 percent. The fastest spread is in the third-wave countries in Asia, where the supermarket sector is growing at three to five times the rate of GDP per capita growth (Reardon et al. 2012).

Third, inside a country, diffusion has rolled out from large cities to small cities and finally into rural towns in adapted formats, from upper to middle to poorer classes, as well as from processed foods to semiprocessed foods to fresh produce. These paths are essentially the same as occurred in the United States and western Europe. Fourth, to become cost-competitive with traditional retail, supermarket chains have increasingly modernized their procurement systems. They have started to buy direct from processors including under contracts; (in some cases) specify private standards; use centralized procurement and logistics via distribution centers; and use specialized-dedicated wholesalers who distribute to their stores and also organize procurement from suppliers according to volume and quality and timing specifications (Reardon and Berdegué 2002). This has gone by far the furthest in processed foods but has started to be applied to fresh produce as well (see, for example, Berdegué et al. 2005 for Central America).

THE MIDSTREAM SEGMENT TRANSFORMATION: PROCESSING

Similar and in parallel to the retail sector, the processing sector has transformed in structure and conduct. The processed food sector has grown quickly over the past several decades. Packaged food sales are growing at only 2 percent to 3 percent annually in developed countries, versus 13 percent, 28 percent, and 7 percent in low, lower-middle, and upper-middle income developing countries (Wilkinson and Rocha 2009). As in retail and wholesale, the first stage of transformation of food processing was driven mainly by governments setting up parastatals, especially in grains (and in export crops like rubber). However, the actual effect on food systems was limited, as the parastatal processors were mainly confined to grain sold to urban markets while there were large "parallel markets" (not via parastatals). In addition, there was rapid privatization in the late 1980s or 1990s. Only a few countries still have substantial government food processing operations into the 2000s.

Privatization and liberalization combined with urbanization, and income increase led to two phenomena that we referred to as the **J**-curve. On the one hand, especially in the 1990s through the present, there was a proliferation of small and medium-sized enterprises (SMEs) in the processing of grain, dairy, meat, fish, and produce both to fill the gap left by the demise of public-sector operations and to meet growing urban demand. This is the transitional stage of FVC transformation. Examples include teff in Ethiopia (discussed below); dairy, wheat, and horticultural product processing SMEs in Brazil (Farina 1997; Farina et al. 2005); and maize, vegetable, and fruit processing in Africa (Broutin and Bricas 2006; Jaffee and Morton 1995; Rubey 1995; Tschirley et al. 2015).

On the other hand, privatization and FDI liberalization led to an avalanche of FDI from western Europe and the United States, then from Japan. The consequence was that foreign firms formed a major share of the processing sector in a number of first- and second-wave countries by the end of the 1990s, and the trend has started in third- and even fourth-wave countries in the 2000s. Regional multinationals like CP (Thailand) and Bimbo (Mexico) are also buying domestic processors in their regions in the 2000s (Wilkinson and Rocha 2009). This is starting in Africa, such as the 2015 purchase of Blue Ribbon (a large maize mill in Zimbabwe) by Bakhresa (a large wheat and maize mill in Tanzania) or the acquisitions by Olam International (a large agribusiness group started in Nigeria) of several mills in West Africa. Large regional multinationals have acquired large US and European processors; an example is the 2014 acquisition by Shuanghui (China) of Smithfield Foods (United States), which had been the largest pork processor in the world (Zhang, Rao, and Wang 2019).

This latter part of the **J**-curve is the modernization stage where consolidation is occurring. For example, by the early 2000s, Nestlé had a 61 percent market share in Latin America for packaged foods (confections, soups, pet food, baby food, dairy, and baked goods). This has been driven by large processing firms' having advantages in processing compared to small and medium-sized enterprises. Larger processors often have economies of scale; economies of scope; bargaining power, monitoring capacity, and "resource provision contract" capacity; access to cheaper credit; as well as more efficient marketing systems, such as via the use of distribution centers and logistics fleets. This has created a symbiosis between large-scale processors and supermarket chains.

In the modern stage, SMEs have found it hard, especially in the medium term, to compete with large processors. Examples are large tortilla firms displacing traditional women's tortilla firms in Mexico (Rello and Saavedra 2007). The emerging penetration of rural towns by modern retail, selling branded processed foods at a discount, may accelerate this competition (Reardon, Henson, and Berdegué 2007). With health crises, consumers have also moved away from small processors and wet markets as a result of food safety concerns (for Thailand, see Posri and Chadbunchachai 2006).

THE MIDSTREAM SEGMENT TRANSFORMATION: WHOLESALE AND LOGISTICS

While governments played a major role in the development and transformation of wholesale markets, the overall segment of wholesale and logistics underwent changes similar to that of processing, with an expansion, fragmentation following liberalization and privatization, and then concentration. We note these changes as follows. Governments induced directly a first stage of wholesale transformation from traditional, fragmented wholesale to government-run wholesale markets (of private wholesalers). This shift created economies of agglomeration and sometimes economies of scale relative to the traditional fragmented wholesale sector, such as in Africa (Tollens 1997). The large markets created by this investment are huge, such as that in Mexico City, the world's largest wholesale market. China's wholesale market volume increased 11,000 percent from 1990 to 2000 (Ahmadi-Esfahani and Locke 1998; Huang et al. 2007).

The "traditional" wholesale sector appears to be restructuring in several ways. First, the public-sector wholesale market segment is consolidating in some countries, over wholesale markets, such as South Africa (Louw et al. 2007), and over wholesalers within wholesale markets, such as in Mexico (Echánove and Reardon 2006) and Peru (Escobal and Agreda 1997). There is also in some countries evidence of a decline in the share of rural brokers upstream in the FVC, with the exit of village traders in Bangladesh, India, and China in rice and potato (Reardon et al. 2012). But in the segment from rural towns to cities, it appears that SMEs in wholesale and logistics are proliferating. Reardon et al. (2012) call this proliferation a "quiet revolution in food supply chains" that is being observed in Asia and increasingly in Africa (Reardon, Tschirley et al. 2015; Minten, Tamru et al. 2016).

Beyond the traditional wholesale sector, a modern wholesale sector is emerging, with the emergence of the specialized/dedicated modern wholesalers noted earlier, as well as large-scale foreign and domestic logistics firms. In some cases large processors and retailers are buying direct from suppliers, most commonly with respect to procurement from processors (such as Carrefour buying from Nestlé).

Local Confluences of Transformation of Food Value Chains: Two Examples

Cold Storages in Agra, India

The technology shift in the midstream can be rapid and dramatic, in particular when it is linked to increasing urban demand, improving infrastructure, and a policy of encouragement and support. The case of the rapid rise of potato cold storages in South Asia illustrates this—for example, in Bangladesh (Reardon et al. 2012), in Bihar (Minten, Reardon et al. 2014), and in western Uttar Pradesh near Delhi (Das Gupta et al. 2010). We focus on the last.

A survey of cold storages in Agra found a combination of the following factors to have caused the diffusion of cold storages: (1) the rapid development of vegetable demand in Delhi; (2) the improvement of the road link from Agra to Delhi; (3) the introduction of a disease-resistant and long-shelf-life potato variety; (4) the introduction of an electricity grid; (5) the partial subsidizing of irrigation pumps and cold storage equipment; and (6) the economy's generating investable funds among the intermediate city business sector. These changes drove rapid diffusion of modern cold storages and, in turn, a reduction in the seasonality and cost of potatoes in Delhi as well as shifts in the intermediation patterns in the rural areas.

In the early 1990s relatively few farmers grew potatoes in Agra and there were almost no modern cold storages. By the late 1990s, however, cold storages had risen to store 40 percent of the vastly larger potato output, and by 2009, 80 percent. Traditional on-farm storage went from ubiquitous to just 1 percent of the potato harvest. Delhi went from sharply seasonal potato consumption (from fresh harvest) to multiseason availability and 65 percent of consumption from cold-storage potatoes, mainly from Agra. Rural brokers were sidelined by the cold storages themselves becoming the main locus of intermediation, with urban wholesalers coming (despite formal regulations barring this) to buy potatoes from farmers at the storages.

Teff Supply Chain to Addis Ababa, Ethiopia

Teff is the leading cereal in Ethiopia. The marketed surplus of teff to domestic markets in 2013/2014 was $750 million, higher than that of coffee ($560 million), the most important Ethiopian export product. An explosion of growth in the teff value chain to Addis Ababa has occurred over the past decade, based on field surveys of farmers, rural and urban wholesalers, and truckers midstream, as well as field surveys of cereal retail shops, mills, and co-op retailers downstream (Minten, Tamru et al. 2016).

Addis has experienced a proliferation of SME mills-cum-retailers and rapid transformation all along the supply chain. The recent development of the teff value chain was found to be driven overall by the following: (1) significant growth in Addis and increase in incomes (with a doubling of income and a doubling of teff expenditure since the beginning of the 2000s); (2) increased opportunity cost of women's time (saving time cleaning and milling teff and making enjera [teff pancake]); (3) the diffusion of cell phones; (4) improvements in roads and reduction of transport costs; and (5) the provision of teff government extension services.

The development of the teff value chain was in turn correlated with the following: (1) increasing adoption of modern inputs (chemical fertilizers, improved varieties of seed, and herbicides) by farmers, especially by those living close to urban centers; (2) rising quality demands and important shifts from the cheap red varieties to the more expensive white teff varieties, with concomitant increases in productivity due to the uptake of improved varieties; (3) increasing consumer willingness to pay for convenience in urban areas, with the rapid emergence of one-stop retail shops that provide sales, cleaning, milling, and transport services as well as a sizable food service industry; and (4) declining share of the margins of rural-urban marketing, urban distribution, and milling in the final retail prices of teff, indicating improved marketing efficiency over time.

Traditionally, and still in rural areas and small cities and towns outside of Addis, consumers buy teff as a grain, clean it at home, have it custom-milled, and then prepare enjera at home. These practices have changed in Addis since the beginning of the 2000s, with a decline in custom milling and in cleaning grain at home. Instead, consumers are buying teff flour or enjera, driving a sharp increase (nearly 50 percent) of teff mills, enjera-making enterprises, and retail outlets in the neighborhoods.

Moreover, the wholesale marketing of teff has surged. This segment is seldom studied, as attention is usually paid only to the farm segment. Minten, Tamiru, and Stifel (2014) show that cereal wholesale market activity—including teff and other cereals, such as maize and sorghum—has recently been developing quickly. Focus group participants in a wholesale market survey in Ethiopia were asked about levels and trends concerning the numbers of traders and brokers in the markets as well as cereal trucks arriving in these markets. The reported numbers confirm that the marketed surplus of teff had increased rapidly over the past decade, and significantly more trade was reported on average in these markets over time. The reported number of trucks increased over the decade by almost 70 percent and 80 percent in the peak and lean periods, respectively.

These teff value chain growth rates are faster than the urban population growth rates in Ethiopia, indicating higher consumption levels in the cities over time, more trade between rural areas that might pass through these urban wholesale markets, and shifts from other means of transportation to trucks. There was a rapid shift from transport of teff by foot (head loads) to animal transport (donkey/horses, carts), to motorized transport, and then from small trucks of 4 to 5 tons to truck-trailers of 20 tons—a rapid transport change in a decade.

The Waves of the Research Literature on FVCs

The literature on value chains has developed in three waves since 1950, largely reflecting the real-world development of FVCs and the stages of the product cycle discussed earlier in this chapter.

First-Wave Literature: Food Supply Chains in the Commoditization Phase and Transitional Stage

TRENDS LEADING TO THE EMERGENCE OF THE FIRST WAVE

Developing regions in the 1980s through the 2010s were undergoing the following rapid changes: (1) urbanization; (2) shifting diets from grain-dominated to diversified diets with substantial consumption of nongrains (meat, fish, dairy, horticulture products, edible oils); (3) lengthening supply chains; (4) shifting from local niche to commodity products (and in some cases beyond that in the product cycle, into differentiated products adding variety and quality); and (5) shifting from home-grown and purchased-unprocessed products to relying on purchased food, including a substantial amount of processed products (as women took jobs outside the home).

Yet these very same trends and processes occurred in the United States (and western Europe) from the 1920s through the 1980s (Barkema, Drabenstott, and Welch 1991). The United States went from a traditional food system—with a low share of urbanization, grain-focused diets, bare emergence of purchased-processed food—in the 1920s to a food system that fits closely our description of a "transitional" food system by the 1950s (with the main part of the shift to modern food systems after 1950 through the 1980s). The urban share of the population was dominant by the 1950s, as diets had diversified into processed foods and meat as well as some commodity vegetables. Supply chains were increasing in length, becoming national instead of local. Large processors and supermarket chains had made their appearance, albeit still with a minority share of the food system.

The genesis of the first strand of FVC literature reflected researchers wrestling with economic and societal issues emerging with what seemed then as a rapid shift to the transitional stage of FVCs and the commoditization phase of the product cycle by the 1950s and the 1960s. But this first wave of literature continued and expanded with application in developing regions starting mainly in the 1980s and beyond, as those regions entered a similar stage.

First-Wave Literature: Meso Analysis of Supply Chains in the Commoditization Phase

The first wave of food value chain literature (which we call the "food supply chain strand") emerged in the 1950s and 1960s along with the transformation of the FVC in the United States and Europe from the traditional to the transitional stage and the emergence of the modern stage. The founder of this strand can be said to be Ray Goldberg of Harvard Business School, who coined the term "agribusiness systems" for food chains from seed to farm to consumer. Goldberg studied the formation of marketing margins and the use of coordination mechanisms such as contracts and vertical integration along the chain (Davis and Goldberg 1957; Goldberg 1965). Many agricultural economics studies ensued from this approach. Terms used by this strand of studies, with some modest differences of emphasis and method among them, include agribusiness commodity systems; food or commodity chains; *filières*; food supply chains; subsectors; and netchains. See, for example, Holtzman (1986); Lazzarini, Chaddad, and Cook (2001); Loader (1997); Raikes, Friis-Jensen, and Ponte (2000); and, for Africa, Jaffee et al. (2003).

The essence of nearly all the applied studies of this first strand were laid out in the ideas in the work of Goldberg in the 1950s and 1960s, which we summarize here. Goldberg took the basic "vertical chain" idea from earlier literature (such as the input-output literature of the 1930s), married it with the industrial organization literature's "structure-conduct-performance paradigm" discussed earlier, and applied the resulting framework to the rapidly emerging and commoditizing FVCs of that era. Goldberg and others in this first wave incorporated concepts from the growing work in economics (for example, Williamson 1979, 1981) on what were later termed by the NIE (new institutional economics) school as "institutional arrangements"—use of quality and safety standards, use of contracts (such as contract farming, discussed in Chapter 11), use of spot markets versus use of vertical coordination and integration, and other "chain coordination" arrangements. As supply chains developed, these institutional arrangements to coordinate them arose, and that reality penetrated the research literature. While Gereffi (1999) and others formalized the ideas in later work (in the value chain strand), Goldberg and others had researched the roles of different segments and lead companies in setting and enforcing coordination arrangements, and studied the role of regulations and policies.

The first wave focused on the burning issues of debate of the day, in the 1960s through the 1980s in the United States and in the 1980s on in developing regions. Questions naturally arise when supply chains shift from traditional to transitional system, in particular:

1. What are the paths and determinants of the proliferation of long FVCs operated by SMEs (in the transitional phase) and emerging large enterprises (in the initial steps of the modern phase)?
2. What are the constraints or bottlenecks faced in the development of commodity FVCs? A key part of this work was to measure marketing costs between segments—later termed "transaction costs" and built on by the NIE school (arising in the 1970s and 1980s, discussed more below).
3. Where are there "excessive" margins indicating noncompetitive behavior such as collusion in price-setting?
4. What is the path and the impact on FVCs of the emerging coordination mechanisms, the new private "institutions"?

Work on these issues brought to bear analysis of structure, conduct, and performance; the impacts of urbanization, developing road infrastructure, and changing diets; and the impacts of domestic and international policies and regulations (on competition, on taxes and subsidies, on marketing orders). At the risk of oversimplifying and ignoring exceptions, the analysis in this wave tended (and tends) to be at the "meso" level, using aggregated data at the segment and zone level, such as the PAM (Policy Analysis Matrix) work (Monke and Pearson 1989), and other supply chain "mapping" analysis. This can be compared with work in the second strand that tended to make more use of business management case studies.

Several recent strands of literature in the first wave have emerged in developing regions. Some descriptive studies have focused on the characteristics of the transitional stage of FVCs in developing countries, emphasizing the proliferation of a multiplicity of types of SMEs in the midstream of supply chains as supply chains grew longer. They point to the emergence of large wholesalers based in secondary cities, of third-party logistics enterprises linking urban markets to rural traders and farmers, and to clusters of SME processors. These studies emphasize that there is dynamism in transitional supply chains that is not associated either with large firms or formal contracts or global markets, but rather small firms, informal relations, and domestic urban markets. Examples include Minten, Tamru et al. (2016) for Ethiopia, Verhofstadt and Maertens (2013) for Rwanda, and Reardon et al. (2012) for Bangladesh, China, and India.

Other analytical studies examine the competitiveness of FVCs as agrifood markets and the determinants of trader margins and efficiency. Dillon and Dambro (2017) reviewed African studies on trader competitiveness and found that the segment of traders has efficient levels of competitive arbitrage, as entry and exit rates are high, there is co-movement of prices among markets, and trader margins are highly variable. A small number of studies used trader survey data. For example, Minten and Kyle (1999), for the Democratic Republic of the Congo, examined the determinants of trader margins as a function of transportation costs. They found that these costs, conditioned by road quality, explained food prices between farming regions, and that trader profit rates are higher on poorer roads. Fafchamps, Gabre-Madhin, and Minten (2005) found for Benin, Madagascar, and Malawi, in analyzing margins, costs, and value-added, that there are no significant increasing returns to scale among traders. From this they note the implication that policies to restrict entry into agricultural trade are not needed or useful, and governments should focus on technological and institutional innovations to upgrade markets. By contrast, for shrimp traders in Indonesia, Yi and Reardon (2015) find that larger traders have a cost advantage over small traders and are more efficient in allocating factors. They suggest that this is a possible explanation for the observed concentration occurring over time in the shrimp trader segment.

Other analytical studies examine the participation of small farmers in FVCs in the transitional stage (that is, not sales to modern enterprises or contract relations). An example is Bellemare and Barrett (2006), which analyzed the determinants, among which were transaction costs, of small pastoralists' decisions among autarchy or net sale or net purchase of animals in livestock markets in Ethiopia and Kenya. Barrett (2008) and Holloway et al. (2004) analyzed farmer participation in traditional and transitional grain and milk FVCs, controlling for threshold levels of assets (and thus the possibility of "poverty traps") keeping farmers out of markets, and controlling for farmer characteristics, studying the impacts of various shocks to the market (like drought and policies such as devaluation).

Second-Wave Literature: Meso and Business Strategy Analysis in the Product Differentiation Phase of the Product Cycle and Modern Stage of the FVC

TRENDS LEADING TO THE EMERGENCE OF THE SECOND WAVE

Three trends in the 1980s through the 1990s in developed countries and mainly the 1990s and beyond in developing countries drove the emergence of this second wave. We briefly summarize these trends:

1. Globalization of food trade and FDI, mainly in retail and processing and logistics from developed to developing regions occurred mainly from the 1980s through to the present.

2. Diets changed rapidly in the developing regions mainly from the 1980s through to the present, with diversification beyond foodgrains, increases in processed food consumption, and food purchases rising among rural households. There was a rise in demand for quality and safety of food.

3. With urbanization and globalization, there was an increase in the length of FVCs, and the shift into trade and internal commerce in products beyond bulk commodity grains and into perishables, and a large variety of processed products drove the implementation by food industry firms of coordination "institutions" ranging from contracts to private standards of quality and safety (Swinnen et al. 2015). These institutions were more widespread and complex than those studied in the first strand during the commoditization phase. The innovations mainly established in developed countries then transferred to developing countries in the 1990s via FDI and other means, thus "fast tracking" transformation of FVCs in developing countries (Reardon, Henson, and Berdegué 2007; Reardon et al. 2019).

Another factor besides competition and developed-developing country transfer of investments and techniques drove the jump from commodity supply chains to differentiated and coordinated value chains. Intra-developed country rivalry and transfer of techniques also spurred the process. A well-known example is the drive in Japan for quality differentiation and quality assurance (to shed the image and standing of a commodity manufacturer) in the 1960s. They pioneered techniques like total quality management. These were taken up by US manufacturers in the 1970s and the 1980s (Lillrank 1995) and studied by Michael Porter at Harvard Business School. Porter championed "competitive advantage" and the use of quality differentiation and assurance to move beyond the low margin and highly contested commodity markets. Porter was the founder of the second strand of FVC literature, discussed next.

Second-Wave Literature: Meso and Firm Analysis of Value Chains in the Differentiated Product and Modernization Stages

The second strand of FVC literature, often termed "food value chain governance" literature, emerged in the 1980s and 1990s along with globalization, with the transformation of the FVC in the United States and Europe from the transitional to the modern stage and with the shift along the product cycle from the commodity to the differentiated product stage. The latter was associated in the literature with "adding value" beyond commodity traits (mainly cost), and thus the "value" in value chain was introduced.

An important point is that while both the first and second waves can be said to have been started by two Harvard Business School professors, both in business management, the first wave tended to focus more on the structure and conduct of the whole supply chain, and on cost and efficiency, while the second strand, started by Porter (1985) has tended to focus on business management of quality differentiation and assurance, and by extension strategies of supply chain design by "lead" or "innovator" firms in the supply chain. These emphases were driven by the trends noted earlier in quality differentiation and food industry consolidation and competition. This wave of literature started to become important in developing regions starting in the 1990s and 2000s with the rise of large processors, supermarkets, and private standards and in general with the start of modernization in FVCs.

The second wave has two main strands. A first strand focuses on the institutions of "governance" present in FVCs. Kaplinsky and Morris (2000) inventory the following types of governance: (1) legislative governance (setting the rules such as standards for cost, quality, and volume); (2) judicial governance (monitoring compliance with requirements); and (3) executive governance (either sanctioning suppliers that do not meet the standards, or assisting suppliers, such as small farmers, to be able to meet the requirements). An example is "resource providing contracts" that provide credit and technical assistance to small farmers (Austin [1981]and Kuijpers and Swinnen [2016] in general, and, for cases in Slovakia and Poland, Gow and Swinnen [1998] and Dries and Swinnen [2004]).

Part of this first strand is analysis of private standards. That theme had been present in earlier work (such as Goldberg's [1968] work on industry standards for orange juice in the United States). But the global food industry greatly increased its use of private standards in the 1970s and 1980s in developed as well as developing countries for the portion of the food economy in interface with the world market as well as large food processors and supermarket chains focused on the domestic market (Reardon et al. 1999; Fulponi 2006; Swinnen 2007).

A second strand of this second wave analyzes the strategies of design choices made by businesses to gain and protect and build competitive advantage. Porter (1985) focused on this issue using case studies of firms choosing what and how to design and market a quality-differentiated product. The main choices concerned product traits, marketing strategies, and coordination mechanisms. Addressing a similar issue to that of Porter's work is that of Zilberman, Lu, and Reardon (2019), using economic models to examine the strategic choices of "innovator" businesses in terms of the design of value chains that will allow the innovator to create or capture a market for the innovation (product, technology, or application) and protect a competitive advantage. The work is integrative in comparison to earlier economic analyses that focused on specific aspects of value chain relations, such as transaction cost theory (Hobbs 1997) or contract theory (Goodhue 2011; Bolton and Dewatripont 2005).

Zilberman, Lu, and Reardon (2019) note that an innovator-cum-entrepreneur makes an integrated set of choices to implement the innovation in a supply chain. These include the following:

1. How much to produce of the product, given capital constraints.
2. Looking upstream, whether and how much a firm should grow its own intermediate input or buy it from farmers or other suppliers; looking downstream, whether and how much a firm should produce its own marketing services for processed output.
3. If the innovating processor decides to buy intermediate inputs (such as paddy for a rice mill) from farmers (or buy marketing services from distribution firms), whether to contract with the farmers (or distribution firms) or buy the intermediate input (or services) from them in a spot arrangement.

4. If the innovator, such as a processor, decides to grow its own intermediate input or marketing services, what technology to use.

5. If the innovator decides to contract, what design (terms) the contract should have.

6. The degree of monopsony and monopoly, as well as government regulation, change or condition the answers to the first five choices.

In the 1990s, as global FVCs formed and domestic and international chains stretched long and included actors of unequal economic power (such as vegetable farmers in Kenya selling to UK supermarkets), there arose a strand of literature focusing on the distribution of "rents" over segments of international chains. This was extended to domestic supply chains in developing countries in the 2000s. Here, a "rent" is a payment to an owned factor beyond the payment that would accrue under perfect competition (where the factor price is equated with the marginal value product from a unit of that factor). Although this arose as a "new" literature, it was in essence similar to the supply chain analyses of the 1960s in the United States that tested whether a given segment had monopoly or monopsony power arising from noncompetitive practices, looking for super-profits along the chain.

Such rents can be of different kinds, according to Kaplinsky and Morris (2000), including the following: (1) economic rents from differential productivity of factors plus barriers to entry (attained by exercise of market power); (2) technology rents, where the firm has command over scarce or innovative technologies; (3) human resource (skills) rents; (4) organizational rents (derived from internal organization of the firm, such as ability to combine economies of scale in procurement with close contact with clients in many markets, through chain store location, or from external organization, such as a farmers' cooperative or a buyers' club); (5) marketing rents (derived from marketing skills and organization but also from brands); (6) relational rents (derived from preferential relations with suppliers, such as in contract farming schemes or various "lock-in" arrangements, or with clients which in turn might be due to brands required by the buyer, such as a retailer, or payment of slotting fees [for example, for packaged rice or potato chips] or both); (7) resource (and we would add, infrastructural/locational) rents; (8) policy rents; and (9) financial rents (derived from better access to finance, which in turn might be derived from being in the formal sector, being near infrastructure, and other factors linked to other rents).

Rents can be correlated with "power" of the chain actor to ensure consequences along the chain (such as meeting of standards by suppliers) and to actively coordinate links (such as imposing standards via contracts and enforcing contracts with sanctions). This power typically emerges from an ability to set up barriers to competition (through some unique advantage that also generates rents) and derives from segment concentration. That power is sometimes vested in a "lead firm" or set of firms who set the rules (but are not necessarily the ones to implement and monitor the application of the rules by suppliers) (Swinnen et al. 2015). The lead firm may also issue contracts to resolve uncertainty of suppliers that impedes the needed investments to meet private standards and adopt requisite technologies, such as in the pork industry in the United States (Maples, Lusk, and Peel 2019).

Gereffi (1999) distinguished FVCs by the position in the chain of the lead firm (with "lead" conflated with "power" in the chain): (1) "buyer-driven FVCs" (such as the UK supermarket–driven chain sourcing from Kenya, described in Dolan and Humphrey [2000]); or (2) "supplier driven FVCs" where, say, a producers' organization sets the rules. The rise of the power of food industry firms due to concentration and their advantages drawn from branding and marketing have meant that buyer-driven FVCs have proliferated and supplier-driven FVCs have declined over the past several decades.

There is an emerging literature that combines lead-firm strategies and analysis of distributional impacts of these strategies. An example is the impacts of quality and sustainability certification in the coffee value chain. Minten et al. (2018) examine the distribution over segments in the coffee value chain in Ethiopia of the quality premium of certified coffee. To study the whole value chain, they combined datasets of small farmers, larger commercial coffee farms, cooperatives and traders, and coffee export transactions. They analyzed the impacts of certification of voluntary sustainability standards (VSS), in particular Fair Trade and Organic certifications, which are rapidly increasing in global value chains. They found that the transmission of export quality premiums to coffee producers is limited, with only less than a third of the premium passed on to farmers and totaling very little relative to farmers' overall incomes. Similar results of constraints to the benefits of certification were found in the coffee sector in Guatemala by de Janvry, McIntosh, and Sadoulet (2015).

Third-Wave Literature: The Microeconomics of Participation in and Effects of Modernizing and Product-Differentiating FVCs

This wave is a direct extension from meso to micro of the second wave and thus is motivated by the same trends. The transformation of FVCs from traditional to transitional to modern, and along the product cycle, presents opportunities for suppliers in the various segments of the chain. But they also present challenges. These impacts are studied in several strands of literature, emerging in the 1980s and 1990s but really taking off in the 2000s and especially the 2010s, along with the perception (we surmise born largely from the second wave of literature) of the importance and progress of the transformations as well as the confluence of sufficient survey data and econometric and RCT (randomized controlled trial) methods and computational power over the past several decades.

A first strand has used econometrics and farm survey data to study the determinants and impacts of participation in transformed FVCs. The cost and quality requirements implied by private standards as implemented in contracts and other enforcement mechanisms translate into a set of needed investments or "upgrading" measures by suppliers. This strand identifies the upgrading investments and empirically determines what kinds of suppliers can make the grade to participate or be included in the transforming FVC. This is akin to the earlier literature on "threshold investments" for technology adoption, such as during the Green Revolution (Feder, Just, and Zilberman 1985).

In this first strand, there has been a substantial literature on participation in and impacts of contract farming for some 30 years. This literature (114 articles) was critically reviewed as to findings and methods in Bellemare and Bloem (2018). They note that the literature does not conclusively answer whether contract farming improves farmer welfare because few articles are internally and externally valid. That is, the great majority are not rigorously causal, or apply to only a few crops, a particular area, or a single year. They recommend that future research go beyond just asking whether contracts make farmers better off, and explore the mechanisms by which contracts affect farmers, in particular whether and how they resolve market failures. Moreover, they note the importance of exploring further costs and not just the benefits of contract farming. They find there is too little research on the spillover effects of contract farming and recommend more work on spillovers. They note that internal validity, linked to the selection problem, is an important problem in many studies, and that researchers need either to undertake differences-in-differences design or conduct an RCT.

In addition, the contract farming literature has tended to find mixed results concerning whether smaller farmers are less likely to participate in transformed value chains compared with medium or larger farmers. The factor impeding participation at times is the transaction costs implied by farmers being in "hinterland" zones and difficult for buyers to access (Barrett et al. 2012). The constraint can also be nonland assets such as irrigation or education (Reardon et al. 2009). An extension of this is to test hypotheses concerning small farmers' entry into particular market channels, such as modern versus traditional FVCs (see, for example, Hernandez et al. 2007 for the case of traditional versus supermarket channels for tomatoes). A common hypothesis is that contract farming participation and/or sales to a large processor or supermarket is determined by farm size (such as in Indonesia by Simmons, Winters, and Patrick [2005] for seed rice and chicken) and by transaction costs faced by the farmer (such as in Peru by Escobal and Cavero [2012] for potatoes).

Furthermore, although much of the econometric literature examines effects on farmer welfare or incomes, some authors have studied other impacts, such as Rao, Brümmer, and Qaim (2012) on production technology and efficiency of vegetable farmers in Kenya of adoption of the supermarket channel. Finally, much of the econometric literature on contract farming and modern market channel participation (the first strand of the third wave) has framed the decision of the small farmer in terms of the price premium or average priced earned, usually relative to special costs incurred to "make the grade." But Michelson, Reardon, and Perez (2012) examined both price averages and price variability comparing small farmers' selling to the traditional market versus to Walmart in Nicaragua. They found that while Walmart's producer price was lower on average, so was its variance, and farmers preferred to sell to Walmart because of the lower risk.

A second, emerging strand is the use of field experiments such as randomized control trials to test for the effect of market channel and/or contract adoption by farmers. This literature is one way to address the internal validity problem noted by Bellemare and Bloem (2016). Two examples of this include the Casaburi and Macchiavello (2015) study of smallholder dairy markets in Kenya, and Saenger, Torero, and Qaim (2014) for such markets in Viet Nam. Other studies have attempted to overcome the challenge of endogeneity (selection bias) noted by Bellemare and Bloem (2018) using instrumental variable estimation. One example is Bellemare (2012), which uses contingent-valuation methods to control for unobserved heterogeneity among households in Madagascar. Indicators of willingness to pay (WTP) for contract farming are used as a vector of instrumental variables for participation in contract farming. Using that approach, Bellemare finds evidence of a positive farm income effect from participation in contract farming. This method allows for the control of factors omitted by other studies, such as subjective perceptions about contract farming and entrepreneurial ability.

A third, emerging strand is the study of the effects of modern channels such as supermarkets on the prices paid by consumers (compared with prices in traditional markets). An example is Atkin, Faber, and Gonzalez-Navarro (2018). They studied the impact on consumer welfare of the penetration of foreign supermarkets in Mexico. They find that penetration lowers prices (and thus overall cost of living) in particular for lower-income consumers, as well as benefits from new product variety, new store variety, and store amenities, beyond just price alone.

Conclusions and Assessment of Gaps in the Literature

This chapter laid out a heuristic conceptual framework to examine the structure, conduct, and performance of agrifood value chains from the perspective of stages of value chain transformation and commoditization and differentiation of products in the product cycle. We discussed evidence of trends in these transformations both in developed countries such as the United States and in western Europe as well as in the developing regions of Africa, Asia, and Latin America. We showed rapid transformation of value chains and a march along the product cycle. At each stage there were occasioned various "real world issues" such as those of competition and inclusion or exclusion of small farmers and cost of living for consumers. We showed that the literature on value chains developed in three waves that essentially tracked the actual transformation of the value chains and continuously oriented toward addressing the issues emerging at each stage.

We showed that the waves of literature have been particularly rich in and adapted to studying the meso-level changes in value chains, such as their length and structure, in both the commoditization and product differentiation phases, in the transitional and modernization stages. More recently, and fed by concerns about the welfare and inclusion impacts of the transformations, and armed with better survey data and computational and elicitation techniques than had characterized the literature before, say, the 1990s, the literature turned to the microeconomic determinants and consequences of the transformation. We discussed the econometric achievements of the literature concerning farmer participation determinants and effects as well as impacts on consumers.

That discussion revealed that there has been a relative dearth of surveys of and analysis of the off-farm components of value chains such as traders, logistics enterprises, and processors. It also revealed econometric challenges in identifying impacts of the transformations, as well as the recommendations of recent literature to broaden the applications, for more external validity of the tests and use of experimental and econometric methods such as differences-in-differences to improve the internal validity of the tests. The opportunities for and the need to greatly extend empirical work on the dynamics and impacts of value chain transformation stand as an urgent agenda for agricultural market development researchers in the decades to come.

References

Adjognon, S. G., L.S.O. Liverpool-Tasie, and T. Reardon. 2017. "Agricultural Input Credit in Sub-Saharan Africa: Telling Myth From Facts." *Food Policy* 67: 93–105.

Ahmadi-Esfahani, F. Z., and C. G. Locke. 1998. "Wholesale Food Markets with 'Chinese Characteristics.'" *Food Policy* 23 (1): 89–103.

Atkin, D., B. Faber, and M. Gonzalez-Navarro. 2018. "Retail Globalization and Household Welfare: Evidence from Mexico." *Journal of Political Economy* 126 (1): 1–73.

Austin, J. E. 1981. *Agroindustrial Project Analysis*. Baltimore: Johns Hopkins University Press.

Bardhan, P. K. 1980. "Interlocking Factor Markets and Agrarian Development: A Review of Issues." *Oxford Economic Papers* 32 (1): 79–98.

Barkema, A., M. Drabenstott, and K. Welch. 1991. "The Quiet Revolution in the U.S. Food Market." *Economic Review* (Kansas City Federal Reserve Bank) (May–June): 25–41.

Barrett, C. B. 2008. "Smallholder Market Participation: Concepts and Evidence from Eastern and Southern Africa." *Food Policy* 33: 299–317.

Barrett, C. B., M. E. Bachke, M. F. Bellemare, H. C. Michelson, and S. Narayanan. 2012. "Smallholder Participation in Contract Farming: Comparative Evidence from Five Countries." *World Development* 40 (4): 715–730.

Bellemare, M. F. 2012. "As You Sow, So Shall You Reap: The Welfare Impacts of Contract Farming." *World Development* 40 (7): 1418–1434.

Bellemare, M. F., and C. B. Barrett. 2006. "An Ordered Tobit Model of Market Participation: Evidence from Kenya and Ethiopia." *American Journal of Agricultural Economics* 88 (2): 324–337.

Bellemare, M. F., and J. R. Bloem. 2018. "Does Contract Farming Improve Welfare? A Review." *World Development* 112: 259–271.

Bennett, M. K. 1941. "International Contrasts in Food Consumption." *Geographical Review* 31 (3): 365–376.

Berdegué, J. A., F. Balsevich, L. Flores, and T. Reardon. 2005. "Central American Supermarkets' Private Standards of Quality and Safety in Procurement of Fresh Fruits and Vegetables." *Food Policy* 30 (3): 254–269.

Bolton, P., and M. Dewatripont. 2005. *Contract Theory*. Cambridge, MA: MIT Press.

Broutin, C., and N. Bricas. 2006. *Agroalimentaire et Lutte Contre la Pauvreté en Afrique Subsaharienne: Le Rôle des Micro et Petites Enterprises*. Paris: Gret.

Casaburi, L., and R. Macchiavello. 2015. "Loyalty, Exit, and Enforcement: Evidence from a Kenya Dairy Cooperative." *American Economic Review* 105 (5): 286–290.

Cochrane, W. W. 1979. *The Development of American Agriculture: A Historical Analysis*. Minneapolis: University of Minnesota Press.

COMTRADE. https://comtrade.un.org/.

Das Gupta, S., T. Reardon, B. Minten, and S. Singh. 2010. *The Transforming Potato Value Chain in India: From a Commercialized-Agriculture Zone (Agra) to Delhi*. Report of Value Chains Component of Asian Development Bank RETA (13th) International Food Policy Research Institute (IFPRI) Project on Policies for Ensuring Food Security in South and Southeast Asia. October.

Davis, J. H., and R. A. Goldberg. 1957. *A Concept of Agribusiness*. Boston: Division of Research, Graduate School of Business Administration, Harvard University.

de Figueirêdo, H. S., M. P. M. Meuwissen, and A. G. J. M. Oude Lansink. 2014. "Integrating Structure, Conduct and Performance into Value Chain Analysis." *Journal on Chain and Network Science* 14 (1): 21–30.

de Janvry, A., C. McIntosh, and E. Sadoulet. 2015. "Fair Trade and Free Entry: Can a Disequilibrium Market Serve As a Development Tool?" *Review of Economics and Statistics* 97 (July): 567–573.

Dillon, B., and C. Dambro. 2017. "How Competitive Are Crop Markets in Sub-Saharan Africa." *American Journal of Agricultural Economics* 99 (5): 1344–1361.

Dolan, C., and J. Humphrey. 2000. "Governance and Trade in Fresh Vegetables: The Impact of UK Supermarkets on the African Horticulture Industry." *Journal of Development Studies* 37 (2): 147–176.

Dolislager, M. D., D. Tschirley, and T. Reardon. 2015. "Consumption Patterns in Eastern and Southern Africa." Report to USAID. Michigan State University, Innovation Lab for Food Security Policy, East Lansing, US, May.

Dries, L., and J. Swinnen. 2004. "Foreign Direct Investment, Vertical Integration, and Local Suppliers: Evidence from the Polish Dairy Sector." *World Development* 32 (9): 1525–1544.

Echánove, F., and T. Reardon. 2006. "Wholesale Markets, Horticulture Products, and Supermarkets in Mexico." Staff Paper 2006-17, Michigan State University, East Lansing, US, April.

Engel, E. 1857. "Die Productions- und Consumtionsverhältnisse des Königreichs Sachsen." *Zeitschrift des sächsischen Statistischen Bureaus des Innern* 8 and 9: 1–54.

Escobal, J., and V. Agreda. 1997. *Analisis de la competitividad y la eficiencia en el Mercado mayorista de productos agropecuarios en Lima*. Lima, Peru: GRADE.

Escobal, J. A., and D. Cavero. 2012. "Transaction Costs, Institutional Arrangements and Inequality Outcomes: Potato Marketing by Small Producers in Rural Peru." *World Development* 40 (2): 329–341.

Fafchamps, M., E. Gabre-Madhin, and B. Minten. 2005. "Increasing Returns and Market Efficiency in Agricultural Trade." *Journal of Development Economics* 78 (2): 406–442.

FAOSTAT (Food and Agriculture Organization Corporate Statistical Database). www.fao.org/faostat/en/#home.

Farina, E. M. M. Q., ed. 1997. *Estudos de Caso em Agribusiness, Focalizando as Seguintes Empresas: Moinho Pacifico, Illycaffe, Cocamar, Sadia, Iochpe-Maxion, Norpac*. São Paulo: Pioneira.

Farina, E. M. M. Q., G. E. Gutman, P. J. Lavarello, R. Nunes, and T. Reardon. 2005. "Private and Public Milk Standards in Argentina and Brazil." *Food Policy* 30 (3): 302–315.

Feder, G., R. E. Just, and D. Zilberman. 1985. "Adoption of Agricultural Innovations in Developing Countries: A Survey." *Economic Development and Cultural Change* 33 (2): 255–298.

Fulponi, L. 2006. "Private Voluntary Standards in the Food System: The Perspective of Major Food Retailers in OECD Countries." *Food Policy* 31 (1): 1–13.

Gereffi, G. 1999. "International Trade and Industrial Upgrading in the Apparel Commodity Chain." *Journal of International Economics* 48: 37–70.

Goldberg, R. A. 1965. "Marketing Costs and Margins: Current Use in Agribusiness Market-Structure Analysis." *Journal of Farm Economics* 47 (5): 1352–1365.

———. 1968. *Agribusiness Coordination: A Systems Approach to the Wheat, Soybean and Florida Orange Economies*. Boston: Division of Research, Graduate School of Business Administration, Harvard University.

Goodhue, R. E. 2011. "Food Quality: The Design of Incentive Contracts." *Annual Review of Resource Economics* 3: 119–140.

Gow, H. R., and J. Swinnen. 1998. "Up- and Downstream Restructuring, Foreign Direct Investment, and Hold-Up Problems in Agricultural Transition." *European Review of Agricultural Economics* 25 (3): 331–350.

Henson, S., and T. Reardon. 2005. "Private Agri-Food Standards: Implications for Food Policy and the Agri-Food System." *Food Policy* 30 (3): 241–253.

Hernández, R., B. Belton, T. Reardon, C. Hu, X. Zhang, and A. Ahmed. 2018. "The 'Quiet Revolution' in the Aquaculture Value Chain in Bangladesh." *Aquaculture* 493: 456–468.

Hernández, R., T. Reardon, and J. A. Berdegué. 2007. "Supermarkets, Wholesalers, and Tomato Growers in Guatemala." *Agricultural Economics* 36 (3): 281–290.

Hobbs, J. E. 1997. "Measuring the Importance of Transaction Costs in Cattle Marketing." *American Journal of Agricultural Economics* 79 (4): 1083–1095.

Holloway, G., C. Nicholson, C. Delgado, S. Staal, and S. Ehui. 2004. "A Revised Tobit Procedure for Mitigating Bias in the Presence of Non-Zero Censoring with an Application to Milk-Market Participation in the Ethiopian Highlands." *Agricultural Economics* 31: 97–106.

Holtzman, J. S. 1986. "Rapid Reconnaissance Guidelines for Agricultural Marketing and Food System Research in Developing Countries." MSU International Development Papers, Working Paper 30. Department of Agricultural Economics, Michigan State University, East Lansing, MI.

Huang, J., X. Dong, Y. Wu, H. Zhi, X. Nui, Z. Huang, and S. Rozelle. 2007. *Regoverning Markets: The China Meso-Level Study*. Report. Beijing: Center for Chinese Agricultural Policy, Chinese Academy of Sciences, July.

Jaffee, S., R. Kopicki, P. Labaste, and I. Christie. 2003. *Modernizing Africa's Agro-Food Systems: Analytical Framework and Implications for Operations*. Africa Region Working Paper 44. Washington, DC: World Bank.

Jaffee, S., and J. Morton, eds. 1995. *Marketing Africa's High-Value Foods: Comparative Experiences of an Emergent Private Sector*. Dubuque, IA: Kendall Hunt.

Kaplinsky, R., and M. Morris. 2000. *A Handbook for Value Chain Research*. Ottawa: International Development Research Centre.

Kuijpers, R., and J. Swinnen. 2016. "Value Chains and Technology Transfer to Agriculture in Developing and Emerging Economies." *American Journal of Agricultural Economics* 98 (5): 1403–1418.

Lazzarini, S. L., F. R. Chaddad, and M. L. Cook. 2001. "Integrating Supply Chain and Network Analysis: The Study of Netchains." *Journal on Chain and Network Science* 1 (1): 7–22.

Lillrank, P. 1995. "The Transfer of Management Innovations from Japan." *Organization Studies* 16 (6): 971–989.

Liverpool-Tasie, S., T. Reardon, A. Sanou, W. Ogunleye, I. Ogunbayo, and B. T. Omonona. 2017. "The Transformation of Value Chains in Africa: Evidence from the First Large Survey of Maize Traders in Nigeria." Nigeria Agricultural Policy Project, Research Paper 91. December. Michigan State University, East Lansing, MI.

Loader, R. 1997. "Assessing Transaction Costs to Describe Supply Chain Relationships in Agri-Food Systems." *Supply Chain Management: An International Journal* 2 (1): 23–35.

Louw, A., D. Chikazunga, D. Jordaan, and E. Bienabe. 2007. "Restructuring Food Markets in South Africa: Dynamics in Context of the Tomato Sub Sector." Meso Report for the Regoverning Markets Program. University of Pretoria, South Africa. May.

Maples, J. G., J. L. Lusk, and D. S. Peel. 2019. "Technology and Evolving Supply Chains in the Beef and Pork Industries." *Food Policy* 83 (February): 346–354.

Michelson, H., T. Reardon, and F. Perez. 2012. "Small Farmers and Big Retail: Trade-Offs of Supplying Supermarkets in Nicaragua." *World Development* 40 (2): 342–354.

Minten, B., M. Dereje, E. Engida, and S. Tamru. 2018. "Tracking the Quality Premium of Certified Coffee: Evidence from Ethiopia." *World Development* 101: 119–132.

Minten, B., and S. Kyle. 1999. "The Effect of Distance and Road Quality on Food Collection, Marketing Margins, and Traders' Wages: Evidence from the Former Zaire." *Journal of Development Economics* 60 (2): 467–495.

Minten, B., T. Reardon, S. Das Gupta, D. Hu, and K. A. S. Murshid. 2016. "Wastage in Food Value Chains in Developing Countries: Evidence from the Potato Sector in Asia." In *Food Security in a Food Abundant World. Frontiers of Economics and Globalization* (Book 16), edited by A. Schmitz, P. L. Kennedy, and T. Schmitz, 225–238. Bingley, UK: Emerald Group Publishing.

Minten, B., T. Reardon, K. M. Singh, and R. Sutradhar. 2014. "The New and Changing Roles of Cold Storages in the Potato Supply Chain in Bihar." *Economic and Political Weekly*, December 27, 98–108.

Minten, B., E. Tamru, E. Engida, and K. Tadesse. 2016. "Feeding Africa's Cities: The Case of the Supply Chain of Teff to Addis Ababa." *Economic Development and Cultural Change* 64 (2): 265–297.

Minten, B., S. Tamiru, and D. Stifel. 2014. "Structural Transformation in Cereal Markets in Ethiopia." *Journal of Development Studies* 50 (5): 611–629.

Monke, E. A, and S. R. Pearson. 1989. *The Policy Analysis Matrix for Agricultural Development.* Ithaca, NY: Cornell University Press.

Nakasone, E., M. Torero, and B. Minten. 2014. "The Power of Information: The ICT Revolution in Agricultural Development." *Annual Review of Resource Economics* 6: 533–550.

Ortega, D. L., H. H. Wang, N. J. Olynk, L. Wu, and J. Bai. 2012. "Chinese Consumers' Demand for Food Safety Attributes: A Push for Government and Industry Regulations." *American Journal of Agricultural Economics* 94 (2): 489–495.

Pingali, P. 2006. "Westernization of Asian Diets and the Transformation of Food Systems: Implications for Research and Policy." *Food Policy* 32: 281–298.

Porter, M. E. 1985. *Competitive Advantage: Creating and Sustaining Superior Performance.* New York: Free Press.

Posri, W., and S. Chadbunchachai. 2006. "Consumer Attitudes towards and Willingness to Pay for Pesticide Residue Limit Compliant 'Safe' Vegetables in Northeast Thailand." *Journal of International Food & Agribusiness Marketing* 19 (1): 81–101.

Qanti, S. R., T. Reardon, and A. Iswariyadi. 2017. "Triangle of Linkages among Modernising Markets, Sprayer–Traders, and Mango-Farming Intensification in Indonesia." *Bulletin of Indonesian Economic Studies* 53 (2): 187–208.

Raikes, P., M. Friis-Jensen, and S. Ponte. 2000. "Global Commodity Chain Analysis and the French Filière Approach." *Economy and Society* 29 (3): 390–417.

Rao, E. J. O., B. Brümmer, and M. Qaim. 2012. "Farmer Participation in Supermarket Channels, Production Technology, and Efficiency: The Case of Vegetables in Kenya." *American Journal of Agricultural Economics* 94 (4): 891–912.

Reardon, T. 2015. "The Hidden Middle: The Quiet Revolution in the Midstream of Agrifood Value Chains in Developing Countries." *Oxford Review of Economic Policy* 31 (1): 45–63.

Reardon, T., C. Barrett, J. Berdegué, and J. Swinnen. 2009. "Agrifood Industry Transformation and Small Farmers in Developing Countries." *World Development* 37 (11): 1717–1727.

Reardon, T., and J. A. Berdegué. 2002. "The Rapid Rise of Supermarkets in Latin America: Challenges and Opportunities for Development." *Development Policy Review* 20 (4): 317–334.

Reardon, T., K. Z. Chen, B. Minten, and L. Adriano. 2012. *The Quiet Revolution in Staple Food Value Chains in Asia: Enter the Dragon, the Elephant, and the Tiger.* Manila: Asian Development Bank; Washington, DC: IFPRI.

Reardon, T., J.-M. Codron, L. Busch, J. Bingen, and C. Harris. 1999. "Global Change in Agrifood Grades and Standards: Agribusiness Strategic Responses in Developing Countries." *International Food and Agribusiness Management Review* 2 (3): 195–205.

Reardon, T., R. Echeverría, J. Berdegué, B. Minten, S. Liverpool-Tasie, D. Tschirley, and D. Zilberman. 2019. "Rapid Transformation of Food Systems in Developing Regions: Highlighting the Role of Agricultural Research and Innovations." *Agricultural Systems* 172: 47–59.

Reardon, T., S. Henson, and J. Berdegué. 2007. "'Proactive Fast-Tracking' Diffusion of Supermarkets in Developing Countries: Implications for Market Institutions and Trade." *Journal of Economic Geography* 7 (4): 1–33.

Reardon, T., C. Hu, D. Tschirley, M. Dolislager, and J. Snyder. 2015. "Urbanization, Diet Change, and Transformation of Food Value Chains in Asia." Report submitted to USAID from the Global Center for Food System Innovation, Michigan State University. June.

Reardon, T., B. Minten, K. Chen, S. Das Gupta, T. A. Dao, J. Wang, and K. A. S. Murshid. 2014. "Tied Output-Credit Markets Have Come Untied: The Fall of Traditional Agrifood Value Chain Finance in Asia." Paper for the Asian Development Bank, Manila, from Michigan State University and IFPRI. August.

Reardon, T., and C. P. Timmer. 2014. "Five Inter-Linked Transformations in the Asian Agrifood Economy: Food Security Implications." *Global Food Security* 3 (2): 108–117.

Reardon, T., C. P. Timmer, C. B. Barrett, and J. A. Berdegue. 2003. "The Rise of Supermarkets in Africa, Asia, and Latin America." *American Journal of Agricultural Economics* 85 (5): 1140–1146.

Reardon, T., D. Tschirley, B. Minten et al. 2015. "Transformation of African Agrifood Systems in the New Era of Rapid Urbanization and the Emergence of a Middle Class." In *Beyond a Middle Income Africa: Transforming African Economies for Sustained Growth with Rising Employment and Incomes*, edited by O. Badiane and T. Makombe, 62–74. ReSAKSS Annual Trends and Outlook Report 2014. Washington, DC: IFPRI.

Rello, F., and F. Saavedra. 2007. "Implicaciones estructurales de la liberalización en la agricultura y el desarrollo rural. El caso de México." Report to RuralStruc Project, mimeo, World Bank.

Ricardo, D. 1817. *Principles of Political Economy and Taxation*. London: John Murray.

Rubey, L. 1995. "The Impact of Policy Reform on Small-Scale Agribusiness: A Case Study of Maize Processing in Zimbabwe." *African Rural and Urban Studies* 2 (2–3): 93–119.

Saenger, C., M. Torero, and M. Qaim. 2014. "Impact of Third-Party Contract Enforcement in Agricultural Markets—A Field Experiment in Vietnam." *American Journal of Agricultural Economics* 96 (4): 1220–1238.

Senauer, B., D. Sahn, and H. Alderman. 1986. "The Effect of the Value of Time on Food Consumption Patterns in Developing Countries: Evidence from Sri Lanka." *American Journal of Agricultural Economics* 68 (4): 920–927.

Simmons, P., P. Winters, and I. Patrick. 2005. "An Analysis of Contract Farming in East Java, Bali, and Lombok, Indonesia." *Agricultural Economics* 33: 513–525.

Stifel, D., B. Minten, and B. Koru. 2016. "Economic Benefits of Rural Feeder Roads: Evidence from Ethiopia." *Journal of Development Studies* 52 (9): 1335–1356.

Swinnen, J., ed. 2007. *Global Supply Chains, Standards and the Poor.* Wallingford, UK: CABI Press.

Swinnen, J., K. Deconinck, T. Vandemoortele, and A. Vandeplas. 2015. *Quality Standards, Value Chains, and International Development: Economic and Political Theory.* Cambridge, UK: Cambridge University Press.

Tollens, E. 1997. *Wholesale Markets in African Cities: Diagnosis, Role, Advantages, and Elements for Further Study and Development.* FAO Food Supply and Distribution to Cities in French-Speaking Africa. Rome: FAO.

Tschirley, D., T. Reardon, M. Dolislager, and J. Snyder. 2015. "The Rise of a Middle Class in Urban and Rural East and Southern Africa: Implications for Food System Transformation." *Journal of International Development* 27 (5): 628–646.

Verhofstadt, E., and M. Maertens. 2013. "Processes of Modernization in Horticulture Food Value Chains in Rwanda." *Outlook on Agriculture* 42 (4): 273–383.

Vernon, R. 1966. "International Investment and International Trade in the Product Cycle." *Quarterly Journal of Economics* 80 (2): 190–207.

———. 1979. "The Product Cycle Hypothesis in a New International Environment." *Oxford Bulletin of Economics and Statistics* 41 (4): 255–267.

Wilkinson, J., and R. Rocha. 2009. "Agro-Industry Trends, Patterns and Development Impacts." In *Agro-Industries for Development*, edited by C. A. da Silva, D. Baker, A. W. Shepherd, C. Jenane, and S. Miranda-da-Cruz, 46–92. FAO/UNIDO/IFAD. Wallingford, UK: CAB International.

Williamson, O. E. 1979. "Transaction-Cost Economics: The Governance of Contractual Relations." *Journal of Law and Economics* 22 (2): 233–261.

———. 1981. "The Economics of Organization: The Transaction Cost Approach." *American Journal of Sociology* 87 (3): 548–577.

Yi, D., and T. Reardon. 2015. "Allocative Efficiency of Agrifood Traders: Shrimp Traders in Indonesia." *Bulletin of Indonesian Economic Studies* (3): 405–423.

Zhang, Y., X. Rao, and H. H. Wang. 2019. "Organization, Technology, and Management Innovations through Acquisition in China's Pork Value Chains: The Case of the Smithfield Acquisition by Shuanghui." *Food Policy* 83: 337–345.

Zilberman, D., L. Lu, and T. Reardon. 2019. "Innovation-Induced Food Supply Chain Design." *Food Policy* 83: 289–297.

Chapter 13

AGRICULTURAL DEVELOPMENT AND INTERNATIONAL TRADE

Kym Anderson and Will Martin

Traditionally societies and their governments have pursued agricultural development to ensure adequate food is available and affordable and incomes of farm households keep pace with those of nonfarm households. Today the farm sector is also expected to care for the natural environment, ensure the food it supplies is safe and nutritious, contribute to energy security, help reduce poverty and greenhouse gas emissions, and provide employment and investment opportunities for women as much as for men. Farm productivity growth can contribute to many of these goals and can be accelerated through more targeted investments in applied agricultural research and in rural infrastructure, education, and health. However, each society does not have to achieve these and their other goals in isolation and indeed will be less able to as climate changes add to the volatility of domestic production. Fortunately, each country's evolving consumption preferences and the wherewithal to satisfy them can be enhanced through trading more openly with other societies. This chapter shows how. It explains the contributions international trade openness can make—and has made—toward achieving these goals. In doing so, the chapter clarifies the role agricultural trade can play, in contrast to trying to remain self-sufficient in food. We draw out the implications for agricultural development prospects in various types of countries as the world economy grows.

Standard economic theory shows that opening to trade will generally raise real national income. The first demonstration of this, by David Ricardo in 1817, assumes differences in technology between countries and shows that gains from exploiting comparative advantage do not depend on absolute productivity levels but rather on countries' *relative* productivity in producing different goods. This means that both a poor and a rich country can—at the same time—benefit from being open to trade. The classic example assumes that only labor is used for production and focuses on a poor country that uses more labor to produce each good but still benefits from trade. The poor country can compete despite using more labor hours in producing its export than the rich country because it has a lower wage rate than the rich country. The rich country is similarly able to compete in its export commodity, despite having higher wage rates, because it uses labor more efficiently in its export than its import commodity and benefits by being able to import from the poor country.

Although 20th-century trade models accommodate factor endowments as well as productivity differences, they still conclude that both poor and rich countries can gain by trading with each other. A simple indicator of comparative cost advantage in farm products is national agricultural land endowment per worker. Agricultural land per person in the United States in 2005 through 2009 was slightly more than twice the world average, and Brazil's farm land endowment per person was nearly as high. At the other extreme, Japan and Korea had farm land endowments one-tenth of the world average. Little wonder that Brazil and the United States are large agricultural exporters, while Japan and Korea are large agricultural importers. These numbers alone are strongly suggestive of the high costs—to both importers and exporters—that would be associated with trying to be food self-sufficient. Yet some players in developing countries still advocate that policy goal.

The chapter begins by outlining how international trade in farm and other products has contributed to human welfare over the millennia. It then outlines how agricultural production and trade can be expected to change as economies grow. The evolution of agricultural trade patterns has been influenced nontrivially by policy choices of national governments, so they are summarized before turning to the evolution of actual patterns of trade in farm products over the past generation. For a long time agricultural policies had a strong anti-trade bias, but many reforms since the late 1980s have been trade liberalizing. We explain why further trade reform is even more important for developing countries now than in the past, before we conclude by pointing out what more could be done with trade-related policy reforms, together with complementary domestic policies, to further enhance sustainable agricultural development while reducing global income inequality, poverty, food insecurity, and hunger.

The Long History of Agricultural Trade, Briefly

Long-distance trade has contributed to agricultural development and global economic growth for millennia, but only in recent centuries via intercontinental trade in farmers' outputs. Its predominant contribution in earlier periods was through trade in crop seeds or cuttings, breeding animals, and farm production technologies. True, international trade in grain, olives, and wine flourished among Mediterranean countries and their neighbors during the days of the Roman Empire; and silk and spices were exported from China to Eurasia and the Mediterranean for many centuries along the Silk Road (Frankopan 2015). As well, the Hanseatic League facilitated maritime trade farther north into the Baltic states between the 13th and 16th centuries. Then, with Europeans settling in the Americas from the 16th century and opening sea lanes to Asia via South Africa, a few other high-valued unprocessed agricultural products began being transported back to Europe. They included cocoa, coffee, cotton, rubber, spices, sugar, tea, and tobacco. Large-scale plantation agriculture developed in Latin America and the Caribbean from the early 1500s, in North America from the early 1600s, and in South and Southeast Asia in the 19th century following European colonization. Once those countries became independent, smallholder agriculture replaced plantations in almost all cases, although there has been a recent resurgence in rubber, oil palm, cassava, and sugarcane plantations in newly opened economies at early stages of demographic and economic transition, such as Cambodia, Lao People's Democratic Republic, and Myanmar (Byerlee 2014).

The complementarity between knowledge of local growing conditions on the one hand and new crops and animals and associated technical know-how on the other (including through the domestic and international transmission of new crop varieties) led to very substantial output growth. That in turn supported population growth. So even though this exchange of farm inputs may not have accelerated GDP per capita or caused commodity prices to equalize across countries prior to 1800, Jones (2002: chapter 4) argues that it certainly improved agricultural output and national food security in many countries. Perhaps the most notable beneficiary country is China, where four major crops from the Americas (maize, peanuts, sweet potato, and white potato) were being cultivated within a century of Columbus's voyages. Being dryland crops, they did not stress the country's irrigation capability but instead encouraged the conversion of forests to arable land. Jones (2002: 55) notes that this globalizing impact of technical know-how and seed transfer contributed to the world's population increasing about 120 percent between 1500 and 1800, compared with only 18 percent between 1200 and 1500.

Britain's industrial revolution boosted intercontinental trade in farm products hugely thanks to the steam engine, which lowered the cost of transporting bulky products by rail, river, and sea. That lowering of trade costs encouraged agricultural development in the colonies of European powers especially, allowing one or two primary products to dominate the international exports of New World countries in the 19th century (Figure 13.1). Most notable were cotton from the United States and wool from Australia, for Britain's textile industry. But intranational trade in farm products also expanded greatly in all countries where transport infrastructure was developed, especially with railways as in India (Donaldson 2018).

Agricultural growth and trade were further helped by the development of refrigeration from the late 19th century and, in more recent decades, of air freight services that allowed fresh perishable products to be traded over long distances for the first time. That generated gains not only from greater national specialization in production of certain export crops but also through extending the seasonal availability of perishable foods for consumers. In recent decades it has even opened up the possibility for two consuming seasons per year for fresh annual temperate perishables, given the lowered cost of maintaining food safety and quality while transporting food between the Northern and Southern Hemispheres. Since 1800, declining costs of international commerce, and the increasing speed of ocean transport, led to the prices of farm and other products converging across countries, and hence to relative factor prices also converging. O'Rourke and Williamson (2002) as well as Findlay and O'Rourke (2007: 402–405) point to the huge declines in commodity price gaps between Europe and both America and Asia between about 1840 and World War I.

In addition to its contribution to that long-run decline in food (and factor) price gaps across continents, trade between countries has been hugely important in offsetting short-term seasonal shortfalls in food availability (both within and between years). As a consequence, famines are essentially a thing of the past, except where deliberately contrived by a country's leaders for local political purposes (Sen 1981; Ravallion 1987). With lower transport costs and the ever-rising incomes of consumers that raise the demand for variety in all things including foods, coupled with the emergence of modern supermarkets to satisfy those demands, one might expect the range of products available in food markets to have grown exponentially over the past two centuries. That indeed has happened in terms of the *number* of processed food items available in large affluent cities.

Despite that, just a dozen basic foods account today for 77 percent of the calorie intake and 72 percent of the protein consumed by the world's population, according to FAO Food Balance Sheets. They are three grains (wheat, rice, and maize), four meats (beef, mutton/lamb, pork, and poultry), two edible oils (soybean and palm oil), plus potatoes, milk, and sugar (FAO 2016).

FIGURE 13.1 Share of top-two goods in settler economies' exports, 1850–1913 (%)

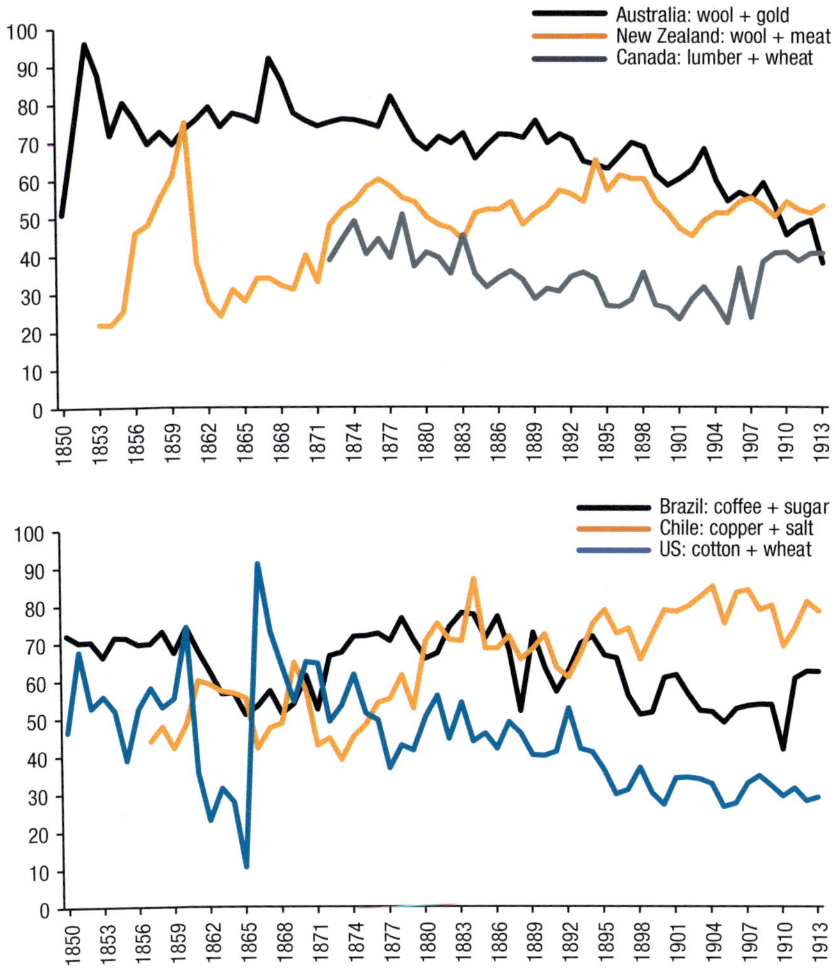

Source: Anderson (2017). Reproduced by permission of the publisher; this figure is not covered by the CC BY 4.0 license.

Those dozen basic farm products could not have become so dominant in the world without international trade in agricultural inputs/technologies or their products, given the small number of regions of the world from which those key species originated. Certainly, fresh fruits and vegetables plus a plethora of processed foods supplement our diets and provide important micronutrients, but a small number of them dominate (bananas, apples, oranges, and other citrus) and have become ubiquitous, thanks to international trade.

Drivers of Agricultural Production and Trade As Open Economies Grow

One of the best-known facts about growing economies is that their farm sector's shares of GDP and employment tend to fall. The reasons for those declines in a closed economy are well-known: low and falling income elasticities of demand for food (Engel 1857) plus rapid advances in farm production technologies mean that the domestic prices and quantities of farm relative to nonfarm products fall. It is less obvious that the farm sector of a small *open* economy—especially one with an abundance of farm land relative to other primary resources, labor, and capital—would have to face relative decline as its economy grows. The fact that it nonetheless almost always does has to do with the rising demand for nontradable goods and especially services as incomes rise. Being nontradable, enough of those products can be produced only by drawing mobile resources from sectors producing tradables. Thus agriculture's shares of national GDP and employment tend to fall with the expansion even of open, land-abundant economies (Anderson and Ponnusamy 2019) at a rate that may also be influenced by relative rates of factor accumulation and differences in sectoral rates of technical change (Martin and Warr 1993).

Agriculture's share of national *exports* depends on the country's comparative advantage, however, and so need not fall as the world economy expands. Indeed, the tradability of each sector's output increases as trade costs are lowered through investments in transport-related infrastructure. If a country's trade costs fall faster than the rest of the world's, and if its farm products gain more from that development than its nonfarm products, the country may strengthen its agricultural comparative advantage over time. According to the workhorse theory of comparative advantage developed in the 20th century, we should expect agricultural trade to occur between relatively lightly populated economies that are well-endowed with agricultural land and those that are densely populated with little agricultural land per worker. Leamer (1987) develops this model and relates it to paths of economic development. If the stock of natural resources is unchanged, rapid growth by one or more countries relative to others in their availability of produced capital (physical plus human skills and technological knowledge) per unit of available labor time would tend to cause those economies to strengthen their comparative advantage in nonprimary products. By contrast, a discovery of minerals or energy raw materials would strengthen that country's comparative advantage in mining and weaken its comparative advantage in agricultural and other tradable products, ceteris paribus. It would also boost national income and hence the demand for nontradables, which would cause mobile resources to move into the production of nontradable goods and services, further reducing farm and industrial production (Corden 1984). Conversely, a depletion or fall in the prices of minerals would strengthen the comparative advantage of agriculture and other sectors producing tradables and weaken the demand for nontradables.

At early stages of economic development, a country with high trade costs typically is agrarian, with most people working on the land and the farm sector producing the majority of GDP when estimated to include home-produced food. If that country has a relatively small stock of agricultural land (and other natural) resources per worker, labor rewards would be low. As trade costs fall or governmental trade restrictions are removed, the country would develop a comparative cost advantage in unskilled labor-intensive, standard-technology manufactures. Then, as the stock of industrial and human capital per worker grows, there would be a gradual move toward exporting manufactures that are relatively intensive in their use of physical capital, skills, and knowledge.

Natural resource–abundant economies, however, may attract migrants from more densely populated countries seeking to become farmers in a frontier region, thereby raising the settler economy's total if not per capita GDP. In such economies the agricultural sector's share of GDP would fall slower than in economies that are growing equally rapidly but are less land-abundant. And if resource-rich economies invest relatively more in capital (including new technologies) specific to primary production rather than to manufacturing, they would not develop a comparative advantage in manufacturing or services until a later stage of development, at which time their exports from those sectors would be relatively capital-intensive. This is all the more likely if new technologies developed for the primary sector become increasingly labor-saving as real wages rise—leading potentially to what are known as factor intensity reversals, whereby a primary industry in a high-wage country can retain competitiveness against a low-wage country by adopting capital-intensive new technologies. And agriculture's share of GDP would decline slower if farm productivity growth outpaced that of other sectors—as it did in high-income countries between 1970 and 2007, according to Herringdorf, Rogerson, and Valentinyi (2013).

Trade patterns are also affected by growth in domestic demands, insofar as preferences are nonhomothetic (Markusen 2013). Food has an income elasticity of demand of less than one, for example. Although this eventually slows the decline in comparative advantage in farm products in resource-poor emerging economies, the rate at which it does is initially slowed by consumers switching from staples to higher-valued foods, including intensively fed livestock (Bennett 1941). This theory of sectoral changes and evolving comparative advantages is consistent with global empirical evidence on changes in sectoral shares of GDP, employment, and exports (Anderson and Ponnusamy 2019); and it has been used successfully to explain the 20th-century flying geese pattern of comparative advantage and then disadvantage in unskilled labor-intensive manufactures as some rapidly growing economies expand their endowments of industrial capital per worker relative to the rest of the world—the classic example being clothing and textiles. The theory has also been used to explain the evolving trade patterns between Asia's resource-poor first- and second-generation industrializing economies and their resource-rich trading partners, and the prolonged importance of agriculture in some New World settler economies such as Australia (Anderson 2017). With the help of newer trade theories, it is also able to explain intra-industry trade and the contribution to productivity growth of opening up (Feenstra 2018).

But trade patterns are also affected by trade-related policy developments. We next summarize a century of such developments before turning to the evolution of actual patterns of trade in farm products.

Policy Developments Affecting Agricultural Trade

The globalization wave preceding World War I saw substantial liberalization of international trade, including in farm trade (Anderson 2016). Efforts to restore liberal trade following that war centered on international conferences. However, despite rhetoric in support of open markets, those meetings did not lead to renewed trade treaties with binding commitments to openness based on the most-favored-nation (MFN) principle used from 1860 by Britain and France. With no country willing or able to replace Britain as the hegemon, there was trade policy anarchy. When economic recession and low agricultural prices hit in the late 1920s, and the United States introduced the Smoot-Hawley tariff hikes of June 1930, governments elsewhere responded with beggar-thy-neighbor protectionist trade policies that together helped drive the world economy into depression. The volume of world trade shrank by one-quarter between 1929 and 1932, and its value fell by 40 percent. Over the entire interwar period both agricultural and other merchandise trade grew hardly at all. According to Federico (2005: 22–29), world exports of both agricultural and nonagricultural goods declined by 0.8 percent per year between 1925 and 1938, and real prices of farm products in international markets slumped following their highs during World War I.

The first attempts to reverse the growth in protection were discriminatory, benefiting colonies at the expense of other trading partners. By the end of the 1930s, protectionism was far more entrenched than in the late 19th century, when only nondiscriminatory tariffs had to be grappled with. Indeed, nontariff trade barriers were so rife as to make tariffs redundant and hence a return to MFN irrelevant unless and until "tariffication" of those barriers occurred. Out of the interwar trade policy experience, Britain and the United States were convinced that liberal world trade required a set of multilaterally agreed rules and binding commitments based on nondiscriminatory principles. After much negotiation, that led to the General Agreement on Tariffs and Trade (GATT). It was signed in 1947 by 23 trading countries—12 high-income and 11 developing—who at the time accounted for nearly two-thirds of the world's international trade (UNCTAD 2018). The GATT provided a forum to negotiate subsequent tariff reductions and changes in rules, plus a mechanism to help settle trade disputes. Eight so-called rounds of negotiations took place in the subsequent 46 years, as a result of which many tariffs on at least manufactured goods were progressively lowered in most high-income countries.

The last of those negotiations, the Uruguay Round (1986–1994), culminated in numerous agreements to further reduce trade barriers over the subsequent decade. One of those agreements involved, for the first time, a serious attempt to liberalize agricultural trade. Another agreement created the World Trade Organization in January 1995 to replace and extend the former GATT Treaty. WTO membership now involves 164 countries/customs territories that account for more than 97 percent of world trade. Global merchandise trade grew faster in the half century following the GATT's formation than in any other half century in history. However, agricultural trade grew less rapidly than trade in other goods, for three reasons. One is the fall in agriculture's share of global GDP (it is now only 3 percent, down from more than 50 percent not much earlier than 1900). A second contributor is the recent fragmentation of industrial production into ever more processes and the associated rapid expansion in the number of links in their global value chains (Baldwin 2016). The latter shows up in the global shares of exports to GDP, which rose during 1995 through 2010 from 66 percent to 105 percent for manufacturing while hardly changing (a rise from 53 percent to 58 percent) for agriculture.

The third key contributor to slow farm trade growth has to do with trade-restricting policies. The lack of strong GATT disciplines on agriculture's trade-related policies allowed two separate developments in farm policies between the 1950s and the 1980s: agricultural protection growth in high-income countries, and agricultural export taxation in low-income countries. In both Japan and the European Community in the 1950s, domestic prices exceeded international market prices for grains and livestock products, but by less than 40 percent. By the early 1980s, however, the difference was more than 80 percent for Japan and was around 40 percent for the EC, while still being close to zero for the agricultural-exporting rich countries of Australasia and North America. Nearly all of that assistance to Japanese and European farmers in that period was due to restrictions on imports of farm products. Assistance then rose markedly in the mid-1980s, particularly due to a North Atlantic food export subsidy "war" that was prompted by western Europe seeking to dispose abroad of support-induced output surpluses.

Meanwhile, developing countries had been heavily discriminating against their farmers. A major study of 18 developing countries by Krueger, Schiff, and Valdés (1988) shows that the depression of incentives facing developing-country farmers from the 1960s to the mid-1980s had been due partly to various forms of agricultural price and trade policies, including subsidies to food imports as well as taxes on farm exports and partly to developing countries' nonagricultural policies that hurt their farmers *indirectly*. The two key indirect ones were manufacturing protectionism (which attracts resources from agriculture to the industrial sector) and overvalued exchange rates (which attract resources to sectors producing nontradables, such as services).

A more recent World Bank (Anderson 2009) study covers 45 developing countries but also 13 European transition economies as well as 24 high-income countries. The results reveal that there have been substantial reductions in distortions to agricultural incentives in both high-income and developing countries since the mid-1980s (Figure 13.2). This is also clear from three decades of producer and consumer support estimates of the OECD (2019): the nominal rate of agricultural protection for the 36 OECD member countries (mostly high-income economies) fell from 48 percent in 1986–1988 to 29 percent in 2000–2002 and 11 percent in 2016–2018, while for 12 (mostly large) emerging economies, the nominal rate of agricultural protection rose on average from 0 percent in 2000–2002 to 5 percent in 2016–2018. Both these sources also reveal, however, that progress has not been uniform across countries and regions, and that the reform process therefore is far from complete. More specifically, many countries still have a wide dispersion in nominal rates of assistance (NRAs) for different farm industries and in particular have a strong anti-trade bias in the structure of assistance within their agricultural sector. Furthermore, some countries have "overshot" in the

FIGURE 13.2 Nominal rate of assistance to agriculture in high-income and developing countries, 1955–2014 (%)

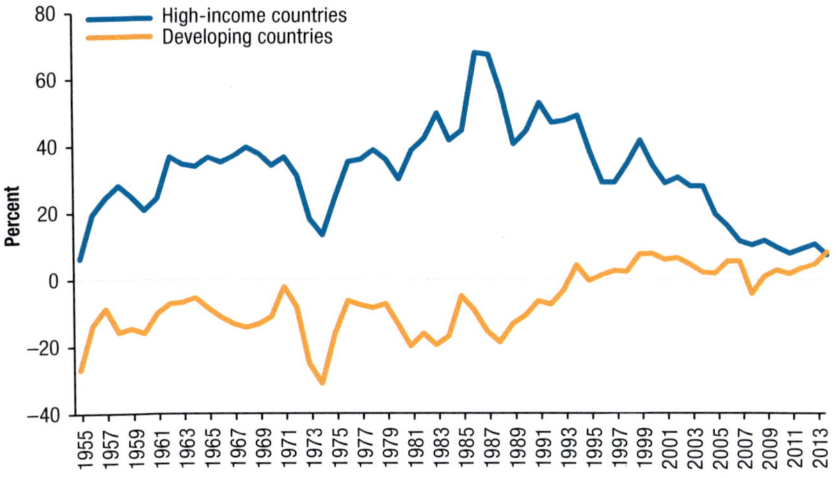

Source: Anderson and Nelgen (2013) to 2011 updated using nominal rates of protection from www.ag-incentives.org (accessed January 2019).

sense that they have moved from having an average relative rate of assistance to farmers (RRA) that was negative to one that is positive, rather than stopping at the welfare-maximizing rate of zero.[1] Moreover, the variance in rates of assistance across commodities within each country, and in aggregate rates across countries, remains substantial, which indicates that resources within the farm sector of each country and globally are not being put to their best use.

The developing-country average NRA conceals the fact that the exporting and import-competing subsectors of agriculture have very different NRAs. While the average NRA for exporters in developing countries has been negative throughout (but coming back from –50 percent in the 1960s and 1970s to almost 0 percent in 2000–2010), the NRA for import-competing farmers in developing countries has fluctuated around a trend rate that has risen from 10 percent and 30 percent (and it even reached 40 percent in the years of low international prices in the mid-1980s). This suggests that export-focused farmers in developing countries are still discriminated against in two respects: (1) by the anti-trade structure of assistance within their own agricultural

1 The NRA is the percentage by which gross returns to producers in a sector are raised because of government sectoral or trade policies. RRA is defined as $100 * [(100 + NRAag_t) / (100 + NRAnonag_t) - 1]$, where $NRAag_t$ and $NRAnonag_t$, respectively, are the NRAs for the tradable segments of the agricultural and nonagricultural sectors.

sectors, and (2) by the protection from imports that remains for farmers in high-income countries. Updated estimates from the agricultural incentives coalition (ag-incentives.org) suggest that by 2014 the NRA for developed and developing countries were essentially the same—a dramatic change from earlier decades (see Figure 13.2).

The improvement in farmers' incentives in developing countries is understated by the above NRA estimates, because those countries have also reduced their assistance to producers of nonagricultural tradable goods, most notably manufactures. The decline in the weighted average NRA for the latter was much greater than the increase in the average NRA for tradable agricultural sectors for the period to the mid-1980s, consistent with the finding of Krueger, Schiff, and Valdés (1988). For the period since the mid-1980s, changes in both sectors' NRAs have contributed to the improvement in farmer incentives. The relative rate of assistance for developing countries as a group went from −46 percent in the second half of the 1970s to just above 0 percent in the first decade of the current century. This increase (from a coefficient of 0.54 to 1.01) is equivalent to an almost doubling in the relative price of farm products, which is a huge change in the fortunes of developing-country farmers in just a generation (Anderson and Nelgen 2013).

The RRA estimates for Asia shed light on something that had perplexed agricultural trade analysts for some time, which is why self-sufficiency in farm products had fallen so little in China and India in recent decades despite their very strong growth in production and exports of manufactures (and of certain tradable services in the case of India): gradual removal of agricultural disincentives encouraged farm output growth even as Asia's industrial competitiveness was strengthening. Within the developing-country group, the spectrum of national RRA estimates remains wide, indicating great scope still for global economic welfare gains from further farm trade liberalization. Such reform not only would raise the mean level of real incomes but also would reduce the variance of international food prices by "thickening" international food markets. According to global economywide modeling results reported in Anderson (2009: chapter 13), liberalization of remaining trade barriers as of 2004 would have raised the share of farm production exported globally from 8 percent to 13 percent. Furthermore, such reform would reduce global income inequality and poverty, according to a follow-up study using numerous global and national economywide models all calibrated to 2004 and incorporating the same World Bank estimates of national price distortions as discussed earlier (Anderson, Cockburn, and Martin 2011).

The growth in agricultural protection to the mid-1980s added to a downward trend in real international farm product prices following the Korean War in the early 1950s. The developing countries' export taxes, including new ones introduced from around 1960, had a somewhat offsetting effect (Tyers and Anderson 1992). However, with strong global farm productivity growth resulting from agricultural R&D investments, supply outstripped demand growth and real international food prices trended downward to the mid-1980s. Those prices flatlined for the next twenty years, before rising dramatically for a combination of demand, supply, and policy reasons. One of the policy contributors was the emergence of biofuel subsidies and mandates in both the United States and the EU, just at a time when fossil fuel prices were rising for a decade from the early 2000s (Figure 13.3).

Although agricultural trade policies were officially all converted to tariffs following the Uruguay Round, many of the tariff bindings in developing countries are extremely high, allowing considerable flexibility to adjust protection rates below the limits imposed by these commitments. Many policymakers continue to seek to stabilize their domestic prices relative to world market prices by varying the protection or taxation rates applied on agricultural trade in the short run. This insulating behavior may make sense for individual countries, and frequently leads policymakers to think it is the solution to the problems created by commodity price volatility, but such thinking involves a fallacy of composition. This is most easily seen if we consider a case where each country's price is linked to the world price and each country responds in the same way to an increase in world prices. If a tightening of world market conditions raises world prices by $10, and all exporters offset this by applying a $10 export tax while all importing countries lower their import duties by $10 for the same reason, the combined impact is to raise world prices by $20 instead of $10. Thus domestic prices rise by the same $10 they would have risen in the absence of this collectively ineffective intervention (Martin and Anderson 2012; Ivanic and Martin 2014). In this environment only countries that insulate by more than the average can stabilize their prices using this policy (Anderson, Martin, and Ivanic 2017).

Another important form of agricultural trade policy response for staple foods—frequently observed in Africa south of the Sahara—focuses on shocks or perceived shocks to agricultural supply. This often sees export restrictions imposed, or state-traded imports brought in, when domestic output of maize or other key staples are expected to be below normal levels. Such policy responses tend to destabilize both domestic and international markets, and

FIGURE 13.3 Real international prices of food and fossil fuel (energy), 1960–2016 (2010 = 100, based on real 2005 US$)

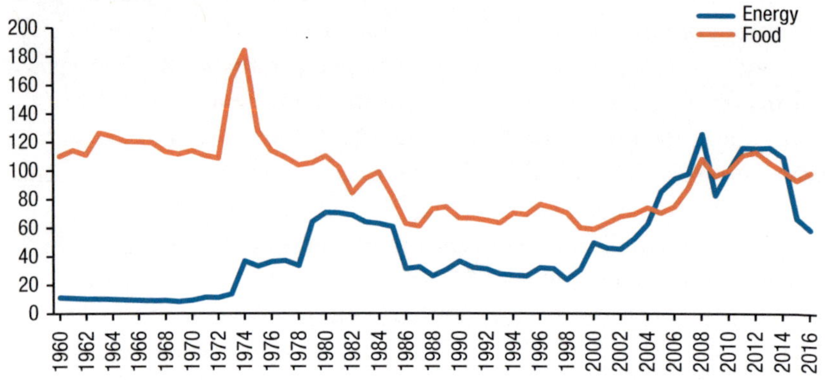

Source: Data from World Bank (2018b).

frequently reduce food security by focusing on the availability of food rather than on the more economically relevant ability of vulnerable groups to access food (Sen 1981).

Recent moves by the United States to a trade policy that seeks liberalization in trading partners by threatening increases in US protection echoes the approaches used for a short time against much smaller Latin American countries in the 1890s (Irwin 2017: chapter 6). Applied to major trading powers such as China, it appears retaliatory responses are bringing serious losses to both parties, including large losses to US agricultural exporters (Robinson and Thierfelder 2019). Hopefully policymakers on both sides will return to reform approaches that generate economic benefits by reducing own-country distortions or agreeing to reciprocal liberalization that generates benefits to both countries.

Agricultural Trade Patterns over the Past Quarter Century

Much of the discussion of global agricultural trade and associated policies over recent decades has focused on trade between high-income countries and on bulk commodities such as grains, beef, and oilseeds. That discussion does not seem to have caught up with more recent developments that have seen the importance of developing countries grow in world agricultural trade.

A key driver of the change in farm trade patterns since the early 1990s is the dramatic increase in economic growth rates of developing countries. For the first time since the beginning of the industrial revolution, developing countries have outperformed—by a large margin—the economic growth of the traditionally high-income countries. Between 1950 and 1993, per capita incomes in developing countries consistently fell relative to those in the high-income countries; but between 1993 and 2016 the growth rate of per capita income in developing countries exceeded that of the core high-income countries by an average of 2.5 percentage points per year. As a result, the share of developing countries in world GDP rose from 48 percent in 1993 to 65 percent in 2016. Of course, not all developing countries have grown rapidly since the early 1990s. Figure 13.4 shows that the growth rate of developing-country income in East Asia over this period was astoundingly high, at 6.84 percent per year. It was also relatively high in South Asia, at 4.40 percent per year. The developing countries of Europe and Central Asia and the Middle East and North Africa (MENA) also had growth rates above the high-income country average of 1.43 percent per year. Africa south of the Sahara was equal to the high-income country average, while the growth rate of per capita income in Latin America and the Caribbean was slightly below that average at 1.38 percent per year.

Another key change in the world economy has been a steady decline in the global population growth rate. This decline, from more than 2 percent per year in the mid-1960s, and close to 2 percent per year in 1990, to around 1 percent per year currently and a projected rate of around 0.5 percent per year in 2050, makes growth in per capita incomes relatively more important for food demand in the future than in the past, where global food demand growth was driven primarily by population. These developments have important implications for global food markets and food trade, because of two of the best-known regularities in economics: Engel's Law (Engel 1857; Houthakker 1957) and Bennett's Law (Bennett 1941). The former refers to the fact that the share of food in total spending declines as income rises, while the latter refers to the fact that consumers in middle-income countries transition as their incomes grow from diets dominated by starchy staple foods to diets that include animal products, vegetable oils, and fruits and vegetables. Together these laws mean that income growth in developing countries has a disproportionate impact on global food consumption. Their increase in the demand for animal products is particularly important for total food demand because of the large difference in the cost of producing livestock products relative to plant-based products.

FIGURE 13.4 Growth rates of per capita income at constant prices, by region, 1993–2016 (% per year)

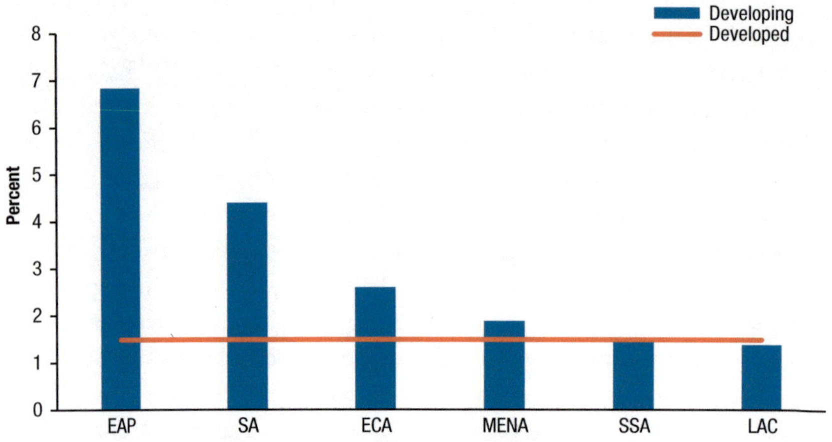

Source: World Bank (2018a), World Development Indicators, accessed April 2020.

Note: EAP = East Asia and Pacific; SA = South Asia; ECA = Central and Eastern Europe and Central Asia; LAC = Latin America and the Caribbean; MENA = Middle East and North Africa; and SSA = Africa south of the Sahara.

Figure 13.5 shows the relationship between income growth and calories consumed (as measured by cereal equivalents) directly by households and the calories required to produce the food bundles demanded at different levels of income. The figure shows how much larger is the energy requirement involved in the production of high-income diets, relative to those consumed by low-income people. It also shows that the total (direct plus indirect) calorie equivalent consumption increases for much longer as incomes grow than does just the direct consumption of calories. Notice that the growth of food consumption converges on zero as incomes rise in richer countries.

Yet another important development affecting agricultural trade since the early 2000s has been the biofuel policy responses by the United States, the EU, and others to climate change and energy security concerns. The sudden global increase in the demand for maize, oilseeds, and sugar by producers of biofuels in those countries reduced those rich countries' net exports of such farm products and thus raised their prices in international markets. That in turn boosted exports of them from developing countries.

On the supply side of the global food equation, the availability of agricultural land per person has declined over the period since the early 1990s at

FIGURE 13.5 Direct versus total (direct plus indirect equivalent) calorie consumption, 154 countries, 1992–2009

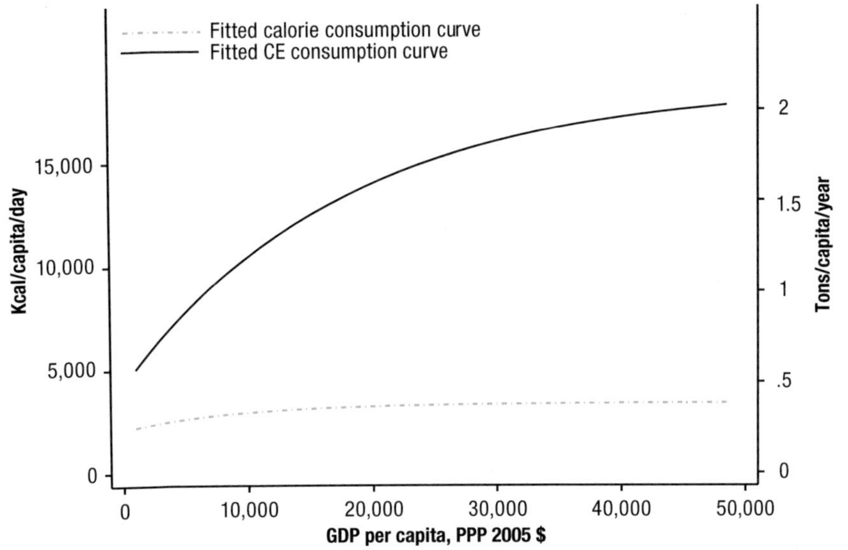

Source: Fukase and Martin (2016).
Note: PPP = purchasing power parity; CE = cereal equivalent; Kcal = kilocalories; GDP = gross domestic product.

different rates across regions.[2] Some countries—particularly in Latin America and Africa—have been able to bring new agricultural land into production, while others had little opportunity to do so. Sharp differences between countries in their population growth rates also influence the changes in agricultural land use per person: the declines in land area per person were quite small in Latin America and in East Asian developing countries but more than 15 percent in other regions and more than 30 percent in MENA (Figure 13.6).

Factor endowments are not the only supply-side determinant of farm trade. Agricultural research and development can also impact countries' abilities to export such products. Brazil has emerged as an agricultural export powerhouse in large measure because of rapid farm productivity improvements. The emergence of India as a large exporter of agricultural products, despite a relatively small land endowment per worker, also reflects improved farm

2 "Agricultural land" is defined, as in Fukase and Martin (2016), as arable land plus land in permanent crops and one-third of permanent pasture land.

FIGURE 13.6 Agricultural land per person, by region, 1992–1997 and 2008–2012 (hectares)

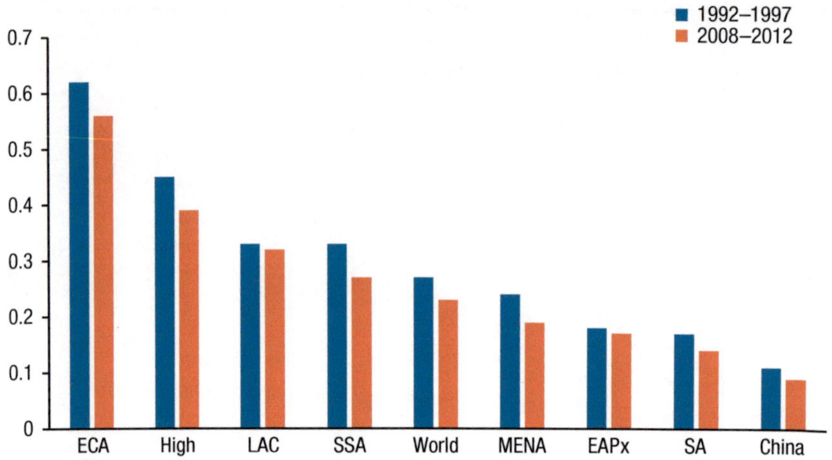

Source: Fukase and Martin (2020).

Note: Regions are based on those used by the World Bank, namely, high = high-income countries; EAPx = East Asia other than China; ECA = Eastern Europe and Central Asia; LAC = Latin America and the Caribbean; MENA = Middle East and North Africa; SA = South Asia; and SSA = Africa south of the Sahara.

productivity. These influences are reflected in the net trade positions of different regions. Figure 13.7 shows a dramatic expansion in the net agricultural exports of Latin America and the Caribbean as well as substantial increases in agricultural imports into China and MENA. Agricultural net exports from developing East Asia other than China also expanded substantially over this period. Net imports of agricultural products into high-income countries (which, as defined by the World Bank, include high-growth economies such as the Republic of Korea) increased from low levels in 1992 to around $57 billion in 2007, declined as prices in key exporting countries in the group responded to high world prices from 2008 but rose to around $60 billion in 2015 and 2016 as lower world prices reduced the incentive to produce.

These influences on agricultural production, consumption, and trade have brought substantial change to the pattern of global farm trade. At the beginning of this growth surge, in 1993, the high-income countries accounted for 56 percent of global agricultural exports and 61 percent of imports. By 2016 there had been an almost complete reversal in terms of importance as both importers and exporters: developing countries contributed 58 percent of those exports and accounted for 60 percent of imports (Table 13.1). That rising importance of developing countries in global farm trade is duplicated with

FIGURE 13.7 Net exports of agricultural products, by region, 1992–2016 (US$ billion)

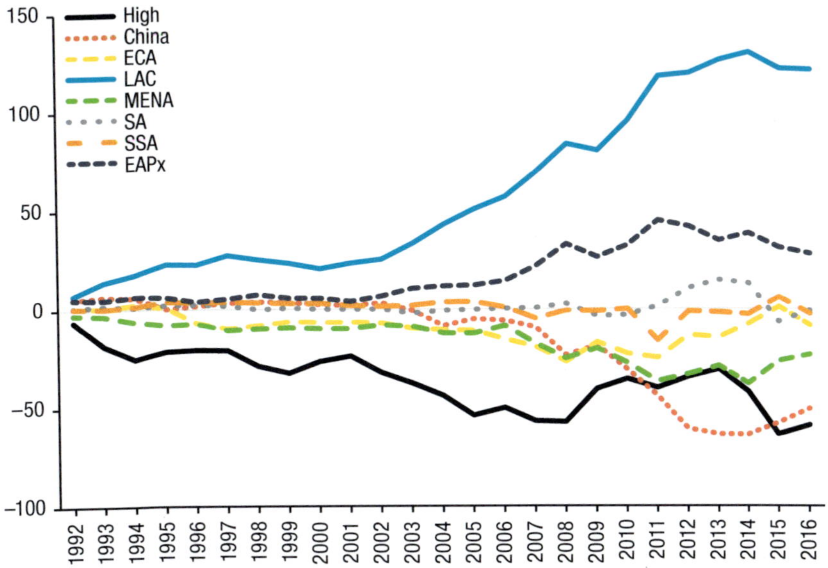

Source: UN COMTRADE data (accessed March 27, 2018, using wits.worldbank.org).

Note: Regions are based on those used by the World Bank, namely, high = high-income countries; EAPx = East Asia other than China; ECA= Europe and Central Asia; LAC = Latin America and the Caribbean; MENA = Middle East and North Africa; SA = South Asia; and SSA = Africa south of the Sahara.

TABLE 13.1 High-income and developing-country shares of world agricultural trade, 1993 and 2016 (%)

	High-income	Developing	World
1993			
Exports	56	44	100
Imports	61	39	100
2016			
Exports	42	58	100
Imports	40	60	100

Source: UN COMTRADE data (accessed July 21, 2018, using wits.worldbank.org).

Note: Intra-EU trade is excluded from EU and global trade.

TABLE 13.2 Share of agricultural products in total merchandise exports, major country groups, and world, 1960–2014 (%)

Country/country group	1960s	1970s	1980s	1990s	2000–2004	2014
Western Europe	17	14	13	12	9	11
Japan	8	3	2	1	1	1
United States and Canada	28	25	20	14	11	13
Australia and New Zealand	84	54	45	36	32	25
Subtotal: Above high-income	**22**	**16**	**13**	**12**	**9**	**11**
China	51	41	20	12	5	3
India	42	36	24	14	8	13
All developing countries:						
Upper-middle-income	—	41	24	15	10	10
Lower-middle-income	—	—	28	18	12	19
Low-income	60	39	29	22	19	—
World	**27**	**21**	**16**	**12**	**9**	**11**

Source: Anderson (2016: Table 4), based on UN COMTRADE data. Reproduced by permission of the publisher; this table is not covered by the CC BY 4.0 license.
Note: — = not available.

their trade in other goods (Baldwin 2016). To get a sense of how their comparative advantages have been changing, it is helpful to calculate two indexes of trade specialization. Over the past half-century, the share of farm products in national exports has been declining for each of the low-, lower-middle-, and upper-middle-income countries, as well as for the high-income group of countries, and so for the world as a whole—until recently when world food prices spiked upward (Table 13.2). A country is defined as having a "revealed" comparative advantage (disadvantage) in agriculture if the share of farm products in its exports exceeds (is less than) that share for the world. For the high-income group, and especially for western Europe, the index of "revealed" comparative advantage in farm products has risen and since the 1990s has averaged close to one. By contrast, that index for upper-middle-income countries has fallen from above two to below one as their manufacturing comparative advantage strengthened; while for low- and lower-middle-income countries, it has hovered around two for the past half century, declining only slightly (Table 13.3).

While imports are not captured in Table 13.3, they are in the index of farm trade specialization reported in Table 13.4, which shows the ratio of exports minus imports to exports plus imports. That index has been slightly negative over the past half-century for high-income countries (above zero for

TABLE 13.3 Indexes of revealed comparative advantage in agricultural and food products, by region, 1960–2014

Region	1960s	1970s	1980s	1990s	2000–2004	2014
Western Europe	0.6	0.7	0.8	1.0	1.1	1.0
Japan	0.3	0.2	0.1	0.1	0.1	0.1
United States and Canada	1.0	1.2	1.2	1.2	1.2	1.2
Australia and New Zealand	3.0	2.6	2.9	3.1	3.5	2.3
Subtotal: Above high-income	**0.8**	**0.8**	**0.9**	**1.0**	**1.0**	**1.0**
China	2.1	2.0	1.3	1.0	0.6	0.3
India	1.6	1.8	1.7	1.6	1.4	1.2
All developing countries:						
Upper-middle-income	—	2.0	1.6	1.3	1.2	0.9
Lower-middle-income	—	—	1.8	1.5	1.4	1.7
Low-income	2.2	1.9	1.8	1.9	2.1	—
World	**1.0**	**1.0**	**1.0**	**1.0**	**1.0**	**1.0**

Source: Compiled by Anderson (2016) from UN COMTRADE data. Reproduced by permission of the publisher; this table is not covered by the CC BY 4.0 license.

Note: — = not available. Revealed comparative advantage index is the share of agriculture and food in national exports as a ratio of that sector's share of global exports, hence 1 for the world. 1960s is 1961–1969, except for China which is 1965–1969.

TABLE 13.4 Indexes of trade specialization in agricultural and food products, by region, 1960–2014

Region	1960s	1970s	1980s	1990s	2000–2004	2014
Western Europe	–0.34	–0.22	–0.11	–0.05	–0.03	0.03
Japan	–0.72	–0.80	–0.84	–0.87	–0.84	–0.76
United States and Canada	0.11	0.24	0.25	0.21	0.08	0.16
Australia and New Zealand	0.73	0.71	0.68	0.65	0.62	0.47
Subtotal: Above high-income	**–0.18**	**–0.11**	**–0.05**	**–0.04**	**–0.04**	**–0.07**
China	0.10	0.07	0.08	0.10	–0.16	–0.40
India	–0.12	0.15	0.18	0.29	0.10	0.33
All developing countries:						
Upper-middle-income	—	0.50	0.32	0.17	0.10	–0.09
Lower-middle-income	—	—	0.30	0.14	0.09	–0.13
Low-income	—	0.25	0.17	0.14	0.07	—
World	**0.00**	**0.00**	**0.00**	**0.00**	**0.00**	**0.00**

Source: Compiled by Anderson (2016) from UN COMTRADE data. Reproduced by permission of the publisher; this table is not covered by the CC BY 4.0 license.

Note: — = not available. Agricultural trade specialization index is net exports as a ratio of the sum of exports and imports of agricultural and food products (so ranging between –1 and +1, positive for net exporters, and the world index is zero). 1960s is 1961–1969, except for China which is 1965–1969. The final column for China and India refers to 2010–2014; their index values in 2005–2009 were –0.22 and 0.25, respectively (based on FAOSTAT data).

Australasia and North America, below zero for western Europe and especially Japan), but it has declined from around 0.5 to −0.1 for middle-income countries and from 0.25 to below 0.1 for low-income countries. These changes in the relative importance and competitiveness of developing countries in world markets for agricultural products has major implications for future policy choices.

Why Further Trade Reform Remains Important for Developing Countries

Given the major increase in the share of developing countries in global farm trade over the past generation, those countries now have a far greater interest in policies affecting that trade. Both own-country and other-country policies are more important than ever in influencing farm incomes in food-exporting developing countries, in meeting food needs of food-importing countries, and in diversifying the available supply and affordability of safe and nutritious food in all countries (Martin and Laborde 2018). Those policy developments remain important because agricultural trade distortions in both high-income and developing countries still tend to be large, and together they account for two-thirds of the potential gains that could come from global merchandise trade liberalization (Anderson, Martin, and van der Mensbrugghe 2005; Laborde, Martin, and van der Mensbrugghe 2011). It is not just the income gains from further trade reform that matter for welfare in developing countries. Also important are the potential impacts of trade reform on poverty alleviation (since that is the key to solving food insecurity and many of the world's other problems), on malnutrition and hunger, on diet diversity and quality, and on domestic food price volatility.

Poverty Alleviation

Evidence surveyed by Ravallion (2006) suggests aggregate economic growth differences have been largely responsible for the differences in poverty alleviation across regions. Initiatives that boost economic growth are therefore likely to be helpful in the fight against poverty, and trade liberalization is one such initiative. But cuts to trade barriers also alter relative product prices domestically and in international markets, which in turn affect factor prices. Hence the net effect on poverty depends also on the way those price changes affect poor households' expenditure and their earnings net of remittances. If the consumer and producer price changes (whether due to own-country reforms and/or those of other countries) are pro-poor, then they will tend to reinforce any positive growth effects of trade reform on the poor. We know that reforms to the agricultural policies of high-income countries could provide a major source of developing country gains from trade reform. Such reform would boost the demand for unskilled labor and for farm products produced in poor countries. Since two-thirds of the world's poor live in rural areas (Castañeda et al. 2016), and since many poor rural households are net sellers of farm labor and/or food, one would expect such reforms to reduce the number in absolute poverty.

A set of analyses reported in Anderson, Cockburn, and Martin (2011), in which global and national economywide model results are carefully combined with household income and expenditure survey data for nearly a dozen developing countries, tests this hypothesis. It finds strong support for it in most of the country case studies considered. If full global trade reform were to be undertaken, that study concludes that it would reduce the number of people in extreme poverty by at least 26 million, and 87 million would be alleviated from $2 per day poverty (Anderson, Cockburn, and Martin 2011; Table 13.4). Bear in mind, too, that those estimates are from comparative static models, and so are underestimates because they do not include the poverty-reducing dynamic effects on economic growth of such reforms.

Malnutrition and Hunger

Hunger and undernutrition can be eased by trade not only in goods but also in agricultural technologies, in particular newly bred varieties of staple crops. The introduction of high-yielding dwarf wheat and rice varieties during the Green Revolution that began in Asia in the 1960s is a previous case in point, whereby producers and consumers shared the benefits in terms of higher farm profits and lower consumer prices for cereals.

A prospective case in point is the possibility of breeding crop varieties that are not only less costly to grow but are "nutriceuticals" in the sense they contain vitamin and mineral supplements. The most promising is so-called golden rice. Consumers in many poor countries suffer from chronic vitamin A deficiency that can lead to blindness, weakened immune systems, and increased morbidity and mortality for children as well as pregnant and lactating women. Golden rice has been genetically engineered to contain a higher level of beta-carotene in the endosperm of the grain and thereby provide a vitamin A supplement. By being cheaper and/or more nutritionally beneficial, golden rice would improve the health of poor people and thereby also boost their labor productivity. Anderson, Jackson, and Nielsen (2005) estimate that the latter economic benefit from this new biotechnology could be as much as ten times greater than just the traditional benefits of lower production costs—not to mention that poor people would live longer and healthier lives.

The benefits of such new biotechnology is being held back in many developing countries where genetically modified organisms (GMOs) have yet to be adopted. This is because the European Union and some other countries will not import from countries whose food may contain GMOs—even though there is no evidence that GM foods are a danger to human health (see, for example, EASAC 2013; UK, House of Commons 2015). The cost of that trade barrier to developing countries has been very considerable both through the loss of export opportunities and the dampening effect on adoption of improved technologies (Anderson 2010; Qaim 2016).

CRISPR approaches to breeding plants and livestock appear to offer additional potential for improving both yields and product quality (Zaidi et al. 2019; Godwin 2019). The fact that these technologies do not require transfers of genes between species and that gene-edited plants and animals are indistinguishable from those bred for the same attributes by conventional techniques may help promote community acceptance. The USDA has announced that it has no plans to regulate plants developed using CRISPR that could have been developed through traditional breeding techniques. However, the European Court of Justice has proposed regulating gene-edited crops under the same strict EU rules as apply to GM foods, and the US Food and Drug Administration has proposed mandatory premarket evaluation of food animals (Van Eenennaam, Wells, and Murray 2019). Clearly, the progress both of gene-editing technologies and of the regulatory responses to them are likely to have important influences on global food production and trade in the coming decades—hopefully with a more positive outcome than the current situation with respect to the regulatory rigidities applying to GMOs.

Dietary Diversity and Quality

Trade has considerable potential to improve food diversity and quality, particularly in countries that are small and have little agroecological diversity. This advantage of trade is likely to be exploited most by higher-income countries, where people have the spending power to diversify their diets. However, the process of dietary diversification begins at a relatively low level of income as consumers introduce vegetables, fruits, and oils into diets that were formerly focused heavily on basic staples (Bennett 1941). The link between openness to trade and food quality is more controversial. One would expect the higher incomes associated with trade to result in dietary improvements—assuming consumers are knowledgeable about what foods lead to better nutritional outcomes. But concerns have been raised about the role of trade, and globalization more generally, in creating nutritional problems, particularly those associated with obesity (for example, Hawkes, Chopra, and Friel 2009).

One strand of this literature (and related media discussion) focuses on Pacific Island countries. Contributions to this literature tend to claim that the precontact diet in these countries was a healthy mix of carbohydrates from root crops with proteins from tropical fish. The experiences of Easter Island and New Zealand (Flannery 1994) raise questions about the sustainability of such diets, particularly as populations grow. Concerns have been raised about the poor health outcomes associated with such imported foods as mutton flaps and turkey tails, and proposed solutions include banning imports of these foods if providing nutrition information is not enough to change diets. However, trade policy is an indirect and inefficient means of improving diets. It frequently leads to domestic firms producing a similar low-quality product and becoming a powerful lobby group against policies that encourage a move away from these products.

Because policymakers have multiple objectives—such as reducing malnutrition from both undernutrition and overconsumption of unhealthy foods, and avoiding hurting the poor—it seems likely that they will need multiple policy instruments to achieve these goals (Martin 2018). Just as in campaigns against smoking, these might include provision of information, attempts to deglamorize these products, excise taxation to discourage consumption, and perhaps "nudges" toward more nutritional foods (Just and Gabrielyan 2016). Whether imports "cause" nutritional problems, policy approaches that focus on trade policy, such as those advocated by Thow et al. (2011), not only stimulate production of undesired goods but also run into problems with WTO rules that distract policymakers away from a focus on nutrition and into unnecessary conflict over trade policy rules and commitments.

Domestic Food Price Volatility

Trade is often seen as a source of volatility to domestic agricultural markets. The reality is more typically the reverse. If the focus moves from a single, isolated market to one with many supplying and demanding regions linked by low-cost transport, the coefficients of variation of production and prices are likely to come down when trade is opened. Burgess and Donaldson (2010) found that connecting a district in India to the railway network resulted in a very sharp decline—almost the disappearance—of famines in that region. Inter-regional trade in this context was found to be particularly important because, as Donaldson (2018) explains, agricultural output volatility was large and internal transport costs extremely high prior to rail connectivity. Thus this work illustrates the positive role of trade in reducing the volatility of food prices and the risk of food insecurity. In his seminal study of the role of trade in famines in British India, Ravallion (1987) concluded that it had a modestly favorable impact on reducing the consumption impact of output shocks, an effect complemented by domestic storage. He found no evidence of "slump famines" in which the income decline associated with harvest failure reduced consumption enough to increase exports.

Trade and Climate Change

Food consumption by country and region is currently still strongly linked to production. Climate changes that alter the suitability of different regions for production of particular foods are therefore likely to increase the importance of agricultural trade for consumption smoothing (Hertel 2018). Increased output volatility is likely to further increase this importance on an annual basis (Baldos and Hertel 2015). Because the effects of climate change on crop and livestock productivity are expected to be very different across agroecological zones, it is important to assess these impacts at a fine level of detail. An important paper by Costinot, Donaldson, and Smith (2016) develops a new approach for this analysis using data for 1.7 million distinct fields around the world. Their conclusion is that trade would need to play a relatively minor role in adjusting for climate change.

A new paper by Gouel and Laborde (2019) builds on this analysis and raises concerns about key elasticities estimated in the Costinot, Donaldson, and Smith model and its assumed ability of agriculture to draw additional land under cultivation. Building more heavily on the econometric evidence about key parameters, and introducing an opportunity cost for additional land drawn into cultivation, they find that preventing trade adjustments in response to climate change would increase the cost of adjustment by more than 75 percent. That suggests openness to food trade will be far more important in the future, especially if current projections of increased frequency and severity of extreme weather events are vindicated.

What More Could Be Done to Continue the Trade-Related Policy Reform Process?

Open markets maximize the benefit that international trade can offer to boost global food security and ensure the world's agricultural resources are used sustainably. The decline in costs of trading internationally add to that prospect. Also, if global warming and extreme weather events are to become more damaging to food production in some regions, then all the more reason to have openness to international food markets and allow trade to buffer seasonal fluctuations in domestic production. The more countries that do so, the less volatile will be international food prices.

For those countries becoming more food import–dependent as their comparative advantage moves away from agriculture, slowing that process by raising food import barriers worsens rather than improves their national food security and nutrition, since it reduces economic access to food for the vast majority of households. By contrast, public investments to boost farm productivity—while achieving the same end of reducing import dependence—would enhance national economic growth and food security. Improving the efficiency of markets for all key factors of agricultural production (capital, labor, land, and water), and for inputs such as fertilizer and pump power, are additional ways to improve the sustainable use of the world's agricultural resources.

Developing countries still concerned that poor households would be too vulnerable if food markets were unrestricted have another option to consider. They can now invoke generic social safety net measures such as conditional targeted income supplements. Those measures can be made more affordable and more equitable if they are targeted at just the most vulnerable households. This option is far more practical now than just a few years ago, thanks to the information technology revolution that has reduced hugely the cost of administering such handouts, because they can be provided electronically as direct assistance to even remote households so long as they have access to electronic banking.

As for international efforts to reduce food price volatility, the most obvious option is for WTO member countries to agree collectively to desist from altering their food trade restrictions when international food prices spike. That would require binding not only import tariffs but also export taxes at zero or low levels, or reducing the extent of price insulation that contributes so strongly to world price volatility. Now may not seem a propitious time to think about further trade reform, with the US administration apparently thinking that industrial goals can be achieved through tariffs. But the genesis of today's global trading system was the US Reciprocal Trade Agreements Act of 1934, passed shortly after the damage from trade wars set off by the US Smoot-Hawley tariff of 1930 had become clear (Irwin 2008). And the prospects for agricultural trade reform in the industrial countries were surely gloomy when D. Gale Johnson (1973) wrote his seminal *World Agriculture in Disarray*, highlighting the need for global agricultural trade reform. As the costs of those policies became more obvious, and the potential for less costly alternatives more clear, it became possible to reach agreement on large-scale reforms of industrial-country agricultural policies by the end of the Uruguay Round in 1994. A similar challenge lies ahead for WTO members in dealing with the collective action problems remaining in both industrial and developing countries.

Finally, what about the well-being of net sellers of farm products who find it difficult to compete with imported like products? Protecting them from import competition is an extremely inefficient and inequitable way of assisting them. Partly this is because such a policy raises the prices of such products for all households who are net buyers of such goods. It is also inequitable in that it helps such farmers in proportion to their marketed output—and thus helps the biggest/least needy farmers most. Far more efficient would be to reduce any underinvestment in rural infrastructure (to lower transport and communication costs of getting farm products to market) and in agricultural R&D (to lower farmers' costs of production or raise the quality and thus price of their product). These messages are not new (see, for example, Johnson 1973), but—like the benefits from free trade—they are not intuitively obvious to noneconomists and so need to be repeatedly impressed upon every new generation if good trade-related policies are to prevail.

References

Anderson, K. 2009. *Distortions to Agricultural Incentives: A Global Perspective 1955–2007.* New York: Palgrave Macmillan; Washington DC: World Bank.

———. 2010. "Economic Impacts of Policies Affecting Biotechnology and Trade." *New Biotechnology* 27 (5): 558–564.

———. 2016. *Agricultural Trade, Policy Reforms, and Global Food Security.* London: Palgrave Macmillan.

———. 2017. "Sectoral Trends and Shocks in Australia's Economic Growth." *Australian Economic History Review* 57 (1): 2–21.

Anderson, K., J. Cockburn, and W. Martin. 2011. "Would Freeing Up World Trade Reduce Poverty and Inequality? The Vexed Role of Agricultural Distortions." *The World Economy* 34 (4): 487–515.

Anderson, K., L. A. Jackson, and C. P. Nielsen. 2005. "GM Rice Adoption: Implications for Welfare and Poverty Alleviation." *Journal of Economic Integration* 20 (4): 771–788.

Anderson, K., W. Martin, and M. Ivanic. 2017. "Food Price Changes, Domestic Price Insulation and Poverty (When All Policy Makers Want to be Above-Average)." In *Agriculture and Rural Development in a Transforming World*, edited by P. Pingali and G. Feder, 181–192. London: Routledge.

Anderson, K., W. Martin, and D. van der Mensbrugghe. 2005. "Market and Welfare Implications of Doha Reform Scenarios." In *Agricultural Trade Reform and the Doha Development Agenda*, edited by K. Anderson and W. Martin, 333–399. London: Palgrave Macmillan; Washington DC: World Bank.

Anderson, K., and S. Nelgen. 2013. "Updated National and Global Estimates of Distortions to Agricultural Incentives, 1955 to 2011." Database uploaded in June at www.worldbank.org/agdistortions.

Anderson, K., and S. Ponnusamy. 2019. "Structural Transformation to Manufacturing and Services: The Role of Trade." *Asian Development Review* 36 (2): 32–71.

Baldos, U., and T. W. Hertel. 2015. "The Role of International Trade in Managing Food Security Risks from Climate Change." *Food Security* 7 (2): 275–290.

Baldwin, R. E. 2016. *The Great Convergence: Information Technology and the New Globalization.* Cambridge, MA: Harvard University Press.

Bennett, M. 1941. "Wheat in National Diets." *Wheat Studies* 18 (2): 37–75.

Burgess, R., and D. Donaldson. 2010. "Can Openness Mitigate the Effects of Weather Shocks? Evidence from India's Famine Era." *American Economic Review* 100 (2): 449–453.

Byerlee, D. 2014. "The Fall and Rise Again of Plantations in Tropical Asia: History Repeated?" *Land* 3: 574–597.

Castañeda, A., D. Doan, D. Newhouse, M. C. Nguyen, H. Uematsu, J. P. Azevedo, and World Bank Data for Goals Group. 2016. *Who Are the Poor in the Developing World?* Policy Research Working Paper 7844, October. Washington, DC: World Bank.

Corden, W. M. 1984. "Booming Sector and Dutch Disease Economics: Survey and Consolidation." *Oxford Economic Papers* 36 (3): 359–380.

Costinot, A., D. Donaldson, and C. Smith. 2016. "Evolving Comparative Advantage and the Impact of Climate Change in Agricultural Markets: Evidence from 1.7 Million Fields around the World." *Journal of Political Economy* 124 (1): 205–248.

Donaldson, D. 2018. "Railroads of the Raj: Estimating the Impact of Transportation Infrastructure." *American Economic Review* 108 (4–5): 899–924.

EASAC (European Academies Science Advisory Council). 2013. *Planting the Future: Opportunities and Challenges for Using Crop Genetic Improvement Technologies for Sustainable Agriculture.* EASAC Policy Report 21. Halle, Germany: EASAC Secretariat, June.

Engel, E. 1857. “Die Productions und Consumtionsverhältnisse des Königreichs Sachsen.” *Zeitschrift des Statistischen Bureaus des Königlich Sächsischen Ministerium des Inneren* 8–9: 28–29.

FAO (Food and Agriculture Organization of the United Nations). 2016. *Food Balance Sheets.* Accessed April 4, 2016. http://faostat.fao.org/site/368/default.aspx#ancor.

Federico. G. 2005. *Feeding the World: An Economic History of Agriculture, 1800–2000.* Princeton, NJ: Princeton University Press.

Feenstra, R. C. 2018. “Alternative Sources of Gains from International Trade: Variety, Creative Destruction, and Markups.” *Journal of Economic Perspectives* 32 (2): 25–46.

Findlay, R., and K. H. O’Rourke. 2007. *Power and Plenty: Trade, War and the World Economy in the Second Millennium.* Princeton, NJ: Princeton University Press.

Flannery, T. 1994. *The Future Eaters.* New York: Grove Press.

Frankopan, P. 2015. *The Silk Roads: A New History of the World.* London: Bloomsbury.

Fukase, E., and W. Martin. 2016. “Who Will Feed China in the 21st Century? Income Growth and Food Demand and Supply in China.” *Journal of Agricultural Economics* 67 (1): 3–23.

———. 2020. “Economic Growth, Convergence, and World Food Demand and Supply.” *World Development* 132 (August): 1–12.

Godwin, I. D. 2019. *Good Enough to Eat? Next Generation GM Crops.* London: Royal Society of Chemistry.

Gouel, C., and D. Laborde. 2019. “The Crucial Role of International Trade in Adaptation to Climate Change.” Mimeo. IFPRI, updated August 2019 version of NBER Working Paper 25221, November 2018.

Hawkes, C., M. Chopra, and S. Friel. 2009. “Globalization, Trade, and the Nutrition Transition.” In *Globalization and Health: Pathways, Evidence and Policy*, edited by R. Labonte, T. Schrecker, C. Packer, and V. Runnels, 235–262. New York: Routledge.

Herrendorf, B., R. Rogerson, and A. Valentinyi. 2013. “Two Perspectives on Preferences and Structural Transformation.” *American Economic Review* 103 (7): 2752–2789.

Hertel, T. 2018. “Climate Change, Agricultural Trade and Global Food Security.” Background paper for *The State of Agricultural Commodity Markets 2018.* Rome: FAO.

Houthakker, H. 1957. “An International Comparison of Household Expenditure Patterns, Commemorating the Centenary of Engel’s Law.” *Econometrica* 25 (4): 532–551.

Irwin, D. 2008. *Trade Liberalization: Cordell Hull and the Case for Optimism.* New York: Council of Foreign Relations.

———. 2017. *Clashing over Commerce.* Chicago: University of Chicago Press.

Ivanic, M., and W. Martin. 2014. "Implications of Domestic Price Insulation for Global Food Price Behavior." *Journal of International Money and Finance* 42: 272–288.

Johnson, D. G. 1973. *World Agriculture in Disarray.* Revised edition. London: Macmillan, 1991.

Jones, E. 2002. *The Record of Global Economic Development.* Cheltenham, UK: Edward Elgar.

Just, D., and G. Gabrielyan. 2016. "Food and Consumer Behavior: Why the Details Matter." *Agricultural Economics* 47 (S1): 73–83.

Krueger, A. O., M. Schiff, and A. Valdés. 1988. "Agricultural Incentives in Developing Countries: Measuring the Effect of Sectoral and Economy-Wide Policies." *World Bank Economic Review* 2 (3): 255–272.

Laborde, D., W. Martin, and D. van der Mensbrugghe. 2011. "Potential Real Income Effects of Doha Reforms." In *Unfinished Business? The WTO's Doha Agenda,* edited by W. Martin and A. Mattoo, 255–275. London: Centre for Economic Policy Research; Washington DC: World Bank.

Leamer, E. E. 1987. "Paths of Development in the Three-Factor, n-Good General Equilibrium Model." *Journal of Political Economy* 95 (5): 961–999.

Markusen, J. R. 2013. "Putting Per-Capita Income Back into Trade Theory." *Journal of International Economics* 90 (2): 255–265.

Martin, W. 2018. "Food Trade Policy and the Dietary Transition." OCP Policy Center, Policy Brief PB-18/07. www.policycenter.ma/publications/food-trade-policy-and-dietary-transition.

Martin, W., and K. Anderson. 2012. "Export Restrictions and Price Insulation During Commodity Price Booms." *American Journal of Agricultural Economics* 94 (2): 422–427.

Martin, W., and D. Laborde. 2018. "The Free Flow of Goods and Food Security and Nutrition." In *Global Food Policy Report,* 20–29. Washington DC: International Food Policy Research Institute.

Martin, W., and P. G. Warr. 1993. "Explaining Agriculture's Relative Decline: A Supply Side Analysis for Indonesia." *World Bank Economic Review* 7 (3): 381–401.

OECD (Organisation for Economic Co-operation and Development). 2019. Producer and Consumer Support Estimates database. Accessed August 31, 2019. www.oecd.org.

O'Rourke, K. H., and J. G. Williamson. 2002. "When Did Globalisation Begin?" *European Review of Economic History* 6 (1): 23–50.

Qaim, M. 2016. *Genetically Modified Crops and Agricultural Development.* London: Palgrave Macmillan.

Ravallion, M. 1987. *Markets and Famines.* London: Oxford University Press.

———. 2006. "Looking beyond Averages in the Trade and Policy Debate." *World Development* 34 (8): 1374–1392.

Robinson, S., and K. Thierfelder. 2019. "Who's Winning the US-China Trade War? It's Not the United States or China." Peterson Institute for International Economics, Washington DC.

Sen, A. 1981. *Poverty and Famines.* London: Oxford University Press.

Thow, A., P. Heywood, J. Schultz, C. Quested, S. Jan, and S. Colagiuri. 2011. "Trade and the Nutrition Transition: Strengthening Policy for Health in the Pacific." *Ecology of Food and Nutrition* 50 (1): 18–42.

Tyers, R., and K. Anderson. 1992. *Disarray in World Food Markets: A Quantitative Assessment.* New York: Cambridge University Press.

UK, House of Commons. 2015. *Advanced Genetic Techniques for Crop Improvement: Regulation, Risk and Precaution.* Fifth Report of Session 2014–15, Science and Technology Committee, House of Commons, Parliament, London.

UN (United Nations). 2018. COMTRADE database. Accessed March 27, 2018. wits.worldbank .org.

UNCTAD (United Nations Conference on Trade and Development). 2018. UNCTADSTAT data on Merchandise Trade. Accessed April 5, 2018. http://unctadstat.unctad.org/wds/TableViewer/tableView.aspx?ReportId=10.1.

Van Eenennaam, A., K. Wells, and J. Murray. 2019. "Proposed U.S. Regulation of Gene-Edited Food Animals Is Not Fit for Purpose." *Science of Food* 3: 3.

World Bank. 2018a. World Development Indicators database. Accessed January 17, 2018. http://wdi.org.

———. 2018b. Pink Sheets database. Accessed February 27, 2018. www.worldbank.org/en/research/commodity-markets.

Zaidi, S., H. Vanderschuren, M. Qaim, M. Mahfouz, A. Kohli, S. Mansoor, and M. Tester. 2019. "New Plant Breeding Technologies for Food Security." *Science* 363 (6434): 1390–1391.

Chapter 14

THE POLITICAL ECONOMY OF AGRICULTURAL AND FOOD POLICIES

Johan Swinnen

Food and agriculture have been subject to heavy-handed government interventions throughout much of history and across the globe, both in developing and developed countries.[1] Political considerations are crucial to understand these policies since almost all agricultural and food policies have redistributive effects and are therefore subject to lobbying and pressure from interest groups and used by decision-makers to influence society for both economic and political reasons. Some policies, such as import tariffs or export taxes, have clear distributional objectives and reduce total welfare by introducing distortions in the economy. Other policies, such as food standards, land reforms, or public investments in agricultural research, may increase total welfare but at the same time also have distributional effects. These distributional effects will influence the preferences of different interest groups and thus trigger political action.

The inherent interlinkage between efficiency and equity issues in policy-making meant that for much of history, economics and politics were closely related disciplines, often written about by the same authors, as reflected in the works of the original architects of the economics discipline—Adam Smith, John Stuart Mill, David Ricardo, and so on. In the late 19th century the economics discipline started separating itself from the "political economy" framework. The revival (or return) of political economy started in the 1950s and 1960s and was referred to as "neoclassical political economy" or "new political economy," as economists started using their economic tools to analyze how incentives of political agents and constraints of political institutions influenced political decision-making. The start of this field is often associated with

1 This chapter is a revised version of J. Swinnen, "The Political Economy of Agricultural and Food Policies," chapter 21 in *Handbook of Agricultural Economics*, edited by G. Cramer, K. P. Paudel, and A. Schmitz, 381–398 (Oxon, UK: Routledge Publications, 2018). The material is reproduced with permission of the Licensor through PLSclear. For a more extensive discussion on all the issues discussed in this chapter and data illustrations, see also Swinnen (2018).

such publications as *An Economic Theory of Democracy* (Downs 1957), *The Logic of Collective Action* (Olson 1965), *The Calculus of Consent* (Buchanan and Tullock 1962), classic papers by Krueger (1974) and Bhagwati (1982) on "rent-seeking," and by Stigler (1971) and Becker (1983) on political economy theories of regulation.

These theories and insights have been applied to analyze food and agricultural policies, in particular to the puzzling question: *Why was agriculture subsidized in rich countries and taxed in poor countries?* The combination of an intriguing question, new theories to apply, and fascinating data induced a rich literature on the political economy of agricultural trade and distortions in the 1980s and the first part of the 1990s. Later, interest in the political economy of agricultural policies was sparked by a combination of fresh data, new theories, and novel intriguing questions. New political economy models focused on the role of institutions in economic policies (for example, Acemoglu 2003 and Persson and Tabellini 2000); better microfoundations for analyzing political economy decision-making (Grossman and Helpman 1994); and the role of (mass) media (McCluskey and Swinnen 2010; Strömberg 2004).

New datasets improved indicators on institutional and political variables. The World Bank's project on distortions to agricultural incentives provided novel and rich datasets (Anderson 2009, 2016). Studies focused on how major institutional and political reforms had affected agricultural policy. This includes the global shift from state-controlled to market-based governance of agricultural and food systems, not only in China and the former Soviet Union but also in Latin America and Africa (Swinnen and Rozelle 2006). Another question related to the impact of changes in international organizations and trade agreements, such as the Uruguay Round Agreement on Agriculture (URAA), the World Trade Organization (WTO), the North American Free Trade Agreement (NAFTA), the enlargement of the EU, and so on.

The turnaround in global agricultural and food markets in the second half of the 2000s induced new economic and political debates on agricultural and food policies. Instead of export subsidies and import tariffs, export barriers and price ceilings were introduced to prevent food prices from rising. The food crisis also drew attention to the failure of agricultural policies to stimulate investment and productivity growth. Another emerging issue was whether there is a shift from traditional trade barriers (such as import tariffs) to so-called nontariff measures. This was triggered by the rapid growth in public and private standards in global agrifood chains and a concern that with binding WTO constraints on tariffs, governments were looking for other instruments to protect their markets.

Political considerations are crucial to understanding these various agricultural and food policies of developing and developed countries. After a discussion on political models and coalitions, this chapter starts with the political economy of policies motivated primarily by redistributing income (or rents) between different groups in society, such as price and trade interventions in agricultural and food markets. The later sections focus on policies that may stimulate growth or reduce externalities but also redistribute income at the same time, such as standards and policies to reduce volatility as well as investments in public goods (research).

Political Coalitions in Agricultural and Food Policies

Political economy models of agricultural and food policy often consider producers, consumers, and taxpayers as the main groups with vested interests in policy outcomes. One (theoretical) reason is its didactic use—that is, to avoid unnecessary complications in deriving policy effects and identify equilibria. Another (empirical) reason is the absence of disaggregated information on policy impacts on various agents within (or outside) the value chain. It is, of course, well-known that in reality many more agents lobby governments to introduce or remove certain policies, including land owners, seed and agrochemical companies, rural banks, traders, food processors, retail companies, and so on. These agents may be differently affected by policies, depending on the nature of the policy—for example, whether the policy is targeted to a (raw) agricultural commodity or to a processed commodity, or whether farm subsidies affect land or other production factors.

As a consequence, these different agents have sometimes joined forces ("political coalitions") with farmers or with final consumers to influence policymakers in setting public policies. In other cases they have opposed each other on policy issues. For example, sugar processors and farmers may jointly lobby for sugar import tariffs or quotas but may oppose each other when governments consider regulating sugarcane or beet prices. In addition, the separation of interests may be less clear in reality than in models. For instance, poor farmers are affected both as producers and consumers of food, agribusinesses may own farms, and so on. These coalitions are also not static. There are several reasons why political coalitions may change: traditional power structures within value chains may change with some (sub)sectors growing and others declining with economic development, new technologies may bring new players into the value chains, new policy instruments may be introduced (or considered), and so on. As an illustration, consider changed coalitions because of "new players" that have emerged, or because the same players are interested in "new things."

New players have emerged for a variety of reasons. Technological advances, such as biotechnology and genetically manipulated (GM) crops, have created new vested interests—and changed those of others. In the 1970s there was no pro- or anti-GM lobby since there was no GM. Biofuels have emerged as an important factor in agricultural markets and food policy in response to rising oil prices and the search for renewable energy sources. Awareness of environmental issues and the lobbying of environmental organizations has increased. In countries such as those in the EU, environmental concerns have traditionally been less important in agricultural policy. This is different from the United States, where the 1930s Dust Bowl led to the introduction of a major conservation payment program.

The emergence of new policies, such as crop insurance subsidies, have brought new sectors, such as insurance companies, into the lobbying game for farm support programs. Growing concentration in retail and the emergence of preferred supplier systems have made the retail sector a more powerful sector in the value chain. This may benefit consumers, since for many agricultural policy issues consumer and retailer interests are aligned and therefore their political coalition may be reinforced by growing retail concentration. Vested interests may also change. In poor countries, for example, consumer interests are centered on having sufficient food at low prices. As incomes grow, however, consumers become more concerned with the quality of their food and with environmental and ethical standards of their food. With income growth and globalization, consumers are interested in local foods, while farm groups see it as a potential way of marketing and protecting their products. At the policy front, this has resulted in regulations on geographical indications (GI)—an issue that has created tensions in trade negotiations (Josling 2006; Huysmans and Swinnen 2019; Raimondi et al. 2020).

Finally, the spread of "global value chains" in recent decades is transforming not just the economics of production and exchange in the world but also the political economy of agricultural and food policies. Interest groups beyond borders have always played a role in agricultural and food policy. For example, in the 1990s the US company Monsanto actively lobbied the European governments to allow GMOs in Europe. However, with the growth of global value chains the distinction between "domestic interests" and "foreign interests" is no longer as obvious (Antràs 2015; Olper 2016). For example, if companies are sourcing inputs from foreign subsidiaries or contracting with foreign farms or companies for their raw materials, the policy interests of these (domestic) companies may be aligned with their (foreign) input suppliers. This integration of production and exchange in global value chains changes the incentives of agents in the value chains to lobby for or against import protection and integration in international trade agreements (Blanchard and Matschke 2015). For example, Blanchard, Brown, and Johnson (2017) and Gawande, Hoekman, and Cui (2015) show that protection is lower when the domestic content of foreign-produced final goods is higher and (vice versa) for foreign content of domestically produced goods. In other words, the integration of economies and companies in global value chains tends to dampen the incentives for protectionist policies.

Price and Trade Interventions

In the second half of the 20th century, there were major differences in agricultural and food policies between poor countries, where farmers were taxed, and rich countries that subsidized farmers (and taxed consumers). This difference was not only huge, it was also counterintuitive (Krueger, Schiff, and Valdés 1991). In countries where farmers were the majority of the population, they were losing out from agricultural policies that imposed a significant tax on them. In contrast, in countries where farmers were a small minority, farmers were subsidized, despite the fact that their numbers in the political arena had declined. This observation was referred to as "the development paradox." Political economy studies have since explained that the differences in agricultural policies between rich and poor countries captured in the development paradox are due to differences in political economy equilibria caused by the combination of structural economic differences, information costs, changes in governance structures, and so on (Anderson, Rausser, and Swinnen 2013).

Structural Change and Political Incentives

The structural changes that accompany economic development alter the costs and benefits of policies to various interest groups, and thus the incentives for political activities to be undertaken to influence governments. These, in turn, determine the government's political incentives and adjust the political economy equilibrium (Anderson 1995; Gardner 1987; Swinnen 1994). First, economic growth typically coincides with a rise in urban-rural income disparities, as growth in industry and services outpaces growth in the agricultural sector, whose specific assets make it slow to adjust. This income gap creates incentives for farmers and agricultural companies to demand—and politicians to supply—policies that redistribute income to reduce that income gap. Several mechanisms presented in the political economy literature explain these countercyclical policies. One is the "relative income hypothesis" of de Gorter and Swinnen (1993), driven by changes in marginal utility that in turn determine political incentives for governments to respond to interest groups. Another is the "loss aversion" argument, where political action is driven by interest groups who want to avoid losses coming from changing market conditions (Freund and Ozden 2008; Tovar 2009).

Second, in a poor economy most workers spend a large share of their income on food. They will therefore strongly oppose an increase in food prices through government interventions, such as import tariffs. Industrial capital will support worker opposition against food price increases because they are concerned about the inflationary effects on wages and their profits. In contrast, workers in rich countries generally spend a (much) smaller share of their income on food, and only a relatively small part of this is the cost of raw materials (agricultural products). This effect is reinforced by declining opposition from industry as the inflationary pressure on wages from agricultural protection declines.

Third, for a given per capita subsidy to farmers, it takes a much larger per capita tax on consumers (or workers in other sectors) when there are many farmers and fewer consumers (as in poor countries) than when there are few farmers and many consumers (workers in other sectors) as in rich countries. In other words, even though the share of farmers in the voting population declines, less opposition to protecting farmers arises when there are fewer of them. Swinnen (1994) shows that under plausible assumptions, the second of those two effects dominates. The combination of these factors causes a shift in the political economy equilibrium from taxing farmers to subsidizing farmers with economic growth.

Political Organization

Improvements in rural infrastructure with economic development also enhance farmers' ability to organize for political action. Olson (1965) explained that collective action by relatively large groups is difficult because of free-riding incentives, implying that in poor countries it is costly to politically organize farmers. Consumers are often concentrated in cities, where coordination and collective action are easier than in the rural areas. However, as the number of farmers declines and rural infrastructure improves, the cost of political organization for farmers decreases. In addition, the growth and concentration of agribusinesses and food-processing companies, which are often aligned with farm interests in lobbying for agricultural policies, strengthen pro-farm interests. In many countries the growth of agricultural protection has been associated with the growth of cooperative agribusiness and food-processing companies.[2]

Information Costs

Information plays a crucial role in political markets, organization, and policy design. Downs's (1957) "rationally ignorant voter" principle explains that it is rational for voters to be ignorant about certain policy issues if the costs of information are higher than the (potential) benefit of being informed. Rational ignorance, be it in the political arena ("voters") or in the economic arena ("consumers"), remains relevant despite reductions of information costs with the growth of mass media and social media (McCluskey and Swinnen 2004). People's opportunity costs and ideological differences between the information (media) source and the reader limit information consumption. The rationally ignorant voter argument implies that policies will be introduced that create concentrated benefits and dispersed costs (Stromberg 2004). This information effect reinforces the distributional effects caused by structural factors.

In addition, enhanced rural communication infrastructure, either through public investments (as in many high-income countries earlier in the 20th century) or through technological innovations and commercial distributions (as in the spread of mobile phone use in developing countries), reduces the relative costs of information in rural areas compared to urban areas. As a consequence, farmers are better informed about policies and they can use this enhanced information infrastructure to organize themselves better, improving the effectiveness of their lobbying activities. Several information-related aspects of economic development thus cause a shift in the political economy equilibrium from supporting consumers to supporting farmers (Olper and Swinnen 2013).

Political Reforms

There is a correlation between political regimes and economic development, with democratic regimes more prominent among richer countries than among poorer. Median-voter models predict that democracies tend to redistribute from the rich to the poor because the distribution of political power

2 Econometric studies by Gawande and Hoekman (2006) and López (2008) also show the influence of agribusiness and food companies' political contributions on US policies.

(measured by votes) is typically more equal than the distribution of income and wealth (Alesina and Rodrik 1994; McGuire and Olson 1996). As farmers are typically poorer, this suggests that farmers may benefit from democratization. Moreover, the same factors that make it difficult for farmers to organize politically in poor countries (such as their large number and geographic dispersion) render them potentially powerful in electoral settings (Bates and Block 2010; Varshney 1995).

It has been difficult to measure this empirically because of data and econometric constraints. Swinnen, Banerjee, and de Gorter (2001) use long-run data and find that changes in electoral rules that have disproportionately benefited agriculture (for example, extending voting rights to small farmers and tenants) have induced an increase in agricultural protectionism. Other electoral changes have not affected agricultural protection because they increased the voting power of both those in favor and those against protection. Olper, Fałkowski, and Swinnen (2014) analyze the impact of all democratic reforms in the world since the 1960s (which were concentrated in developing and emerging countries) and find that, on average, democratization has benefited farmers.

Policy Instrument Choice

The distortionary effects of government interventions are equally dependent on the choice of the instrument as on the level of the intervention. There is an extensive literature comparing the transfer efficiency and the distortions of various policy instruments in trade and agricultural policies (Alston and James 2002; Gardner 1983). The differences in distortionary effects of policies play an important role in policy discussions and trade negotiations.

Another stylized fact of agricultural and food policies is the "anti-trade bias"—that is, that import-competing sectors are protected by taxing imports, and exportable sectors are taxed more, thereby reducing exports (Anderson 2016). Trade-policy instruments are the most important agricultural and food policies to redistribute income between consumers and producers. In earlier history they were often the only policies but even today remain very important (Anderson, Ivanic, and Martin 2013). From 2007 through 2012 the anti-trade bias took on a particular version as many governments responded to rising food prices on world markets by restricting (sometimes outright banning) food exports, thereby exacerbating global price spikes (Anderson, Ivanic, and Martin 2014; Pinstrup-Andersen 2014).

There are several reasons why trade policies are used. First, import-competing sectors have lower comparative advantage than exporting sectors. Since benefits from market returns are lower in sectors with a comparative disadvantage, those sectors' incentives to seek income from government support are also relatively higher. In these (sub)sectors, returns to investment in lobbying activities dominate returns from market activities and so indirectly support an anti-trade bias. A second factor is the so-called revenue motive of public policy. Tariff revenues and export taxes increase government revenues and improve their terms of trade while export and import subsidies do the opposite. Third, distortions (deadweight costs) and budgetary costs of policy intervention typically increase with higher supply elasticities (Gardner 1983; de Gorter, Nielson, and Rausser 1992). Sectors with higher supply elasticities (typically exports) will be subsidized less because it is more costly to do so and causes more distortions (Becker 1983; Gardner 1987).

Fourth, policy instruments differ not only in deadweight costs but also in implementation costs. The most obvious explanation for the broad use of trade taxes (either import tariffs or export taxes) is that they are easiest and least costly to implement (Dixit 1996; Rodrik 1995). In many developing countries, tax-collection institutions are weakly developed and trade taxes (import tariffs or export taxes) are often an important—or the only substantive—source of tax revenue. In this case governments have greater incentives to use to assist farmers. Fifth, policy instruments also differ in their "transparency," the information available concerning policies, and their incidence. Politicians have an incentive to use policies that hide their costs or use policies that obfuscate the transfer itself (Magee, Brock, and Young 1989). This obfuscation perspective helps explain why nonbudget methods of redistribution, such as tariffs, are politically preferable to direct subsidies. Sixth, governments may prefer distortionary policies, such as tariffs, when they have imperfect information on their target group or the amount of transfer needed (Foster and Rausser 1993; Mitchell and Moro 2006). The total transfers—even with deadweight costs—may be lower than would be the case with direct (lump-sum) transfers when governments need to secure a minimum amount of political support. The combination of these political economy forces thus causes anti-trade bias.

These political factors can be constrained if the counterpressure is strong enough—for example, by integrating them into international trade agreements where (economic and political) costs and benefits can be weighed and compensated. In particular, the WTO has had a significant impact on the anti-trade bias and on policy instrument choice.

Policy Reforms in the Late 20th and Early 21st Century

In past decades there has been a change in the trend of agricultural protection and in policy instruments for several high-income countries. In the second half of the 20th century, for example, agricultural protection was high and coupled policies were the most important instrument in OECD countries, consistent with the strong anti-trade bias.[3] For instance, in the 1990s the average producer support estimate (PSE) of OECD countries was 30 percent to 35 percent, and the vast majority (more than 80 percent) of this support was in the form of coupled subsidies. Around the turn of the century, some important policy reforms in OECD countries brought down both total support and the amount of trade distortions. By the mid-2010s the average PSE was below 20 percent, and coupled support had fallen to around 50 percent (OECD 2019).

The implementation of GATT and the WTO have contributed importantly to policy reforms and to reducing the anti-trade bias in OECD countries (Swinnen, Olper, and Vandemoortele 2012). The collapse of Communist regimes and their distortive agricultural policies around the same period also contributed. In Eastern Europe and the former Soviet Union, economic and political liberalizations removed much of the heavy subsidies to farms that existed under the Communist regimes in the second half of the 20th century (Rozelle and Swinnen 2004). At the same time, developing countries have increased support to agriculture, primarily by reducing taxation of agricultural exports (Anderson 2009). The dominance of trade policies remains but the anti-trade bias has declined in recent decades, mainly owing to macroeconomic and trade policy reforms that reduced taxation of farmers and the anti-trade bias in developing countries. These political economy changes are due to economic growth, structural adjustments, and changes in information costs and governance structures, as explained earlier.

Anderson Rausser, and Swinnen (2013, p. 7) concluded that "since the 1980s, both the anti-agricultural bias in developing countries and the pro-agricultural bias in high-income countries have diminished and the two groups' average rates of assistance to agriculture have converged toward zero." This means that—rather that the *divergence* observed in the 1950s through the 1980s—there is now *convergence* in agricultural policies (see Chapter 13,

3 "Coupled PSE" includes all policy transfers (such as tariffs, price support, and subsidies) directly linked ("coupled") to agricultural production. These instruments are typically the most distortive. The second group of instruments ("decoupled") includes decoupled agricultural payments. These instruments are generally considered the least distortive.

in this volume). The EU and China illustrate this convergence well. A significant factor in the overall decline in OECD country support was the reforms of the EU's Common Agricultural Policy (CAP), which made up a substantial share of OECD farm subsidies. Both the level of EU subsidies and their distortions have significantly decreased since 1990. The 1994 Uruguay Round Agreement on Agriculture (URAA) of the GATT/WTO played a crucial role in this.[4] Further major changes in policy instruments from market interventions to direct payments increased the visibility of the transfers, as they occupied a large share of the EU's budget. This increased transparency of the transfers may have been an additional cause of reforms in the EU: over the past two decades, taxpayers have continuously pressured the EU's leaders to reduce agricultural subsidies.

China is another fascinating case of the political economy of agricultural and food policies. China is currently spending around US$200 billion per year on subsidies to farmers—much more than any other country in the world (OECD 2017). This is a huge turnaround from the situation in the 1950s and 1960s, when the Chinese countryside was characterized by hunger and poverty due to a policy system that caused enormous inefficiencies and imposed heavy taxes on farmers. Over this period the Chinese government has shifted from taxing farmers to heavily subsidizing agriculture, and now China spends more on agricultural subsidies than any other country in the world. Not surprisingly, this dramatic policy shift has coincided with a similar dramatic increase in incomes in China—as the development paradox principles would suggest. Economic growth in China, as in other countries, has coincided with an increase in the rural-urban income gap (Huang and Yang 2017). This growing inequality has induced the Chinese government to support farm incomes by a variety of market interventions and direct subsidies (Hejazi and Marchant 2017).

As in the EU in the 1970s and 1980s, these interventions created large market distortions which, in combination with WTO constraints, induced a reform of Chinese agricultural subsidy programs from market interventions to direct subsidies that are less market distorting (Huang, Wang, and Rozelle 2013). Hence albeit at different times and under vastly different political regimes, both China and the EU dramatically increased agricultural subsidies during times of rapid economic growth (in the EU after World War II and in

---

4 The URAA appears to have had less impact on US agricultural policies. Nonetheless, the US administration has attempted to introduce policy reforms with an eye toward ensuring that many US agricultural subsidies are classified as "green box" (that is, non–trade distorting) under the WTO agreement (Orden, Blandford, and Josling 2010).

China since 2000). Both first installed distortionary policy systems and later reformed their agricultural subsidy systems to less distortionary policy instruments; both capped their subsidy levels, after accession to the WTO.

Price Shocks and Political Economy of Aid and Food Policy

With a brief exception in the early 1970s, when prices moved up after the first oil crisis, for the past 50 years global agricultural markets were characterized by relatively stable and low prices. Most of the global agricultural and food policy discussions focused on the reduction of taxes on farmers in developing countries and the removal of policies that subsidized farmers in rich countries. This changed with dramatic increases in food prices in the 2000s. Urban consumers across the world protested, and governments reacted rapidly to the price spikes. Many governments—particularly in developing and emerging countries—intervened to reduce the effect of the global food price spikes (Barrett 2014; Naylor 2014; Pinstrup-Andersen 2014). Governments used price and trade policies to counter global price movements and to insulate the domestic market from the international price spikes (Demeke, Pangrazio, and Maetz 2009).[5] At the same time, food price spikes triggered media and policy attention to the broader issues of hunger and rural poverty.

Trading-Off Volatility and Distortions?

A key argument in favor of policy interventions to stabilize prices is that price volatility causes inefficiencies in demand and supply (FAO and OECD 2011; Prakash 2011; World Bank 2012). Unexpected price changes make it difficult for consumers and producers to make optimal decisions and reduces their confidence in the market and investment. However, others criticized government interventions to insulate domestic markets from global price fluctuations for (1) being ineffective, (2) causing distortions in the economy, and (3) reinforcing price fluctuations when food exporters reduced supply and food importers increased demand (Anderson, Ivanic, and Martin 2013; Ivanic and Martin 2014).

5 For example, India suspended rice exports, and Russia and Kazakhstan banned wheat exports (Martin and Anderson 2011). Countries with significant food stocks refused to release these on global markets and hoarded them for domestic purposes (Timmer 2010). At the same time, food-importing countries removed policies that would increase domestic prices, and many introduced price regulations and other interventions to restrict price increases (see chapters in Barrett 2014; Naylor 2014; Pinstrup-Andersen 2014).

From a political economy perspective, there is nothing surprising in the way that governments have reacted to changes in world market prices. Government interventions to counter market fluctuations are a key "stylized fact" of agricultural and food policies induced by the political economy mechanism caused by the relative income hypothesis or loss aversions explained earlier in the chapter. Hence, even without taking into account possible additional benefits for consumers or producers from price stability, political mechanisms will induce governments to respond to international price increases by policy interventions that limit price rises on domestic markets by export constraints and vice versa through import tariffs when prices fall on the international markets. To integrate benefits from price stability, Pieters and Swinnen (2016) derive a socially optimal distortions-volatility trade-off that takes into account both consumer and producer benefits from stability and production and consumption distortions caused by deviations from the world market price.[6] However, they find that many countries' policies during the past decade are far removed from the socially optimal distortions-volatility combinations. Hence political motives were very important.

Food Prices and Global Development Policy

The food crisis pushed food security and agricultural development from the bottom of the international development agenda toward the top of policymakers' priority list.[7] The price spikes of 2007 and 2008 led to urban protests and, in a number of cases, created political instability (Cohen and Garret 2010; Maystadt, Tanb, and Breisinger 2014). This captured the attention of global policymakers and donors. As soon as urban protests reached the streets, international organizations reacted much like local politicians and paid a disproportionate amount of attention to the problems of urban consumers. Global mass media played an important role in drawing reactions and policy attention from international organizations and global policymakers (Guariso, Squicciarini, and Swinnen 2014). The price spikes, and the ensuing urban

6 Their model is based on Barrett (1996); Bellemare, Barrett, and Just (2013); and Gouel and Jean (2015).

7 After the dramatic increase of food prices in 2006 and 2008, reports emphasized that *high* food prices have a devastating effect on developing countries and the world's poor. Before, most reports argued that *low* food prices were hurting developing countries' farmers and the poor (Swinnen, Squicciarini, and Vandemoortele 2011). Only a few studies initially pointed at the mixed effects of the high food prices on poverty and food security (for example, Aksoy and Hoekman 2010; Heady 2013; Verpoorten et al. 2013).

consumer unrest, created major "media events."[8] Local media reports were picked up by international mass media, which paid a disproportionate amount of attention to the problems of urban consumers, compared to the long-run hunger and poverty problems among the rural population.

Thus, while for many years experts pointed at the low level of investment in developing country agriculture as a source of poverty and food security, it was only after the "food crisis" that media attention increased and that policymakers worldwide put rural poverty and underinvestment in agriculture on their priority list and raised donor funding.[9] The "food crisis" acted as a catalyst of attention to long-standing issues related to food security and agricultural production, which were made particularly salient by the fact that urban consumers—whose voice is typically heard the most by mass media and policymakers—were hit the hardest by the spikes in food prices (Swinnen, Squicciarini, and Vandemoortele 2011). What is remarkable is that, despite the fact that rural malnutrition and poverty of farmers and low agricultural productivity in developing countries has been a major problem for a long time, it may have been an "urban (consumer) crisis" that helped to put poor farmers' situation on top of the agenda. Hence food price spikes may have succeeded where others have failed in the past: to put the problems of poor and hungry farmers on the policy agenda and to induce development policies and donor strategies to help them.

The Political Economy of Food Standards

Production and trade are increasingly regulated by stringent public and private standards on quality, safety, nutritional, environmental, and ethical as well as social aspects. An important critique is that standards are (nontariff)

8 Media attention is typically concentrated around events or shocks (Swinnen and Francken 2006). The agenda-setting effect of the media in international and aid policy has sometimes been referred to as the "CNN factor" (Hawkins 2002). A related factor is that the public at large is more interested in media reports concentrating on negative effects, the so-called bad news hypothesis (McCluskey and Swinnen 2004).

9 Between 2000 and 2005 the share of global overseas development aid (ODA) going to agriculture fell from 5 percent to 3.8 percent (OECD 2013), and the budget share in the UN system going to agriculture (FAO) fell from 20.1 percent to 15.5 percent (Global Policy Forum 2013). After the food crisis, donor funding reversed dramatically: between 2007 and 2011 the share going to agriculture (FAO) in the UN system increased from 15.2 percent to 22.2 percent, and the share of global development aid going to agriculture jumped from 3.7 percent to 6.5 percent (Global Policy Forum 2013; OECD 2013). Oxfam and global agricultural research centers, under the heading of the CGIAR, also saw their funding increase strongly (Guariso, Squicciarini, and Swinnen 2014).

trade barriers. As trade agreements such as WTO have reduced tariffs, countries may use standards to shield their domestic markets from foreign competition (Anderson, Damania, and Jackson 2004; Brenton and Manchin 2002; Fischer and Serra 2000). Convergence (or not) of standards is at the heart of recent trade negotiations such as CEFTA (Central America Free Trade Agreement), TTIP (Transatlantic Trade and Investment Partnership), and so on.

Standards affect trade.[10] However, the implicit comparison with tariffs in the trade debate is not entirely valid. In a small open economy the socially optimal tariff level is zero. A positive tariff level constrains trade, is harmful to social welfare, and is protectionist. But this is not necessarily the case for standards because this ignores the potential benefits of standards. If the standard reduces asymmetric information or externalities, there is no simple relationship between the trade effects of a standard and the social optimum (Beghin 2013; Marette 2014; Marette and Beghin 2010; Sheldon, 2012; Van Tongeren, Beghin, and Marette 2009). This result does not imply that there are no protectionist elements in standards setting. Standards can enhance aggregate welfare by reducing asymmetric information or negative externalities but can also create rents for specific interest groups. Because of the distributional effects of standards, interest groups have a vested interest in influencing governments' decisions on standards. When interest groups have differing lobbying strengths, the political equilibrium will generally differ from the social optimum.

The political equilibrium standard may be either too high or too low from a social welfare point of view. Influential lobby groups may push for both more stringent or less stringent standards depending on the relative magnitude of the price (demand) effect compared with the implementation cost (for producers) or the efficiency gain (for consumers) (Beghin, Maertens, and Swinnen 2015; Swinnen 2016). For example, if producers are more influential than consumers, overstandardization results when producers' profits increase with a higher standard and in understandardization otherwise. Higher profits for producers are more likely when the standard's price (demand) effect is large and when the implementation cost is small.

10 Only in very special circumstances do standards not affect trade: this is when the effect on domestic production exactly offsets the effect on consumption (Swinnen and Vandemoortele 2011).

Development and Pro-Standard and Anti-Standard Coalitions

This political economy can explain the empirically observed positive relationship between standards and economic development. First, and most obvious, higher income levels are typically associated with higher consumer preferences for high quality and safety standards as reflected in higher efficiency gains. Second, the quality of institutions for enforcement of contracts and public regulations is positively correlated with development. Better institutions imply better enforcement and control of standards. Although poor countries may have a cost advantage in the production of raw materials, the better institutions of rich countries lower the marginal increase in production costs caused by standards. Third, higher education and skills of producers, better public infrastructure, easier access to finance, and so on also lower implementation costs. Fourth is the different organization and structure of the media in rich and poor countries. Mass media is the main source of information for many people. The cost of media information is higher and government control of the media is stronger in poor countries. Therefore, the media structure and information provision is likely to induce a more pro-standard attitude in rich countries than in poor, as increased access to media increases attention to risks and negative implications of low standards (Curtis, McCluskey, and Swinnen 2008).

In combination these factors are likely to induce a shift of the political equilibrium from low standards to high standards with development. A pro-standard coalition of consumers and producers in rich countries results if consumers derive large efficiency gains from a standard, while producers incur only moderate increases in costs. In contrast, an anti-standard coalition may be present in poor countries if consumers are more concerned with low prices than with high quality (leading to small efficiency gains from a higher standard) while the implementation costs for producers may be large. Structural differences in information and media may reinforce the positive relationship between standards and development.

The Persistence of Standards: Dynamic Political Economics

Some of the most important political aspects of standards relate to their dynamic effects. Dynamic political economic aspects of standards can provide an explanation for different food standards in countries with similar levels of development, such as in the EU and the United States, and why such differences may persist.[11] Once adopted, countries will tend to stick to the status quo in standards because implementation costs depend on existing standards

11 See Swinnen et al. (2015) and Swinnen (2017) for more technical analysis and details.

that resulted from past investments. Differences in standards between countries may persist because of this and trade may enforce this. The reason is that producer or consumer preferences may change in a dynamic way once the standard is introduced.[12]

The standard will affect comparative advantages and will thus induce producers to support maintaining the standard to protect them from (cheaper) nonstandard imports. Hence, although standards may have been introduced because of consumer demands, their persistence in the long run results from (a coalition of consumer and) producer demands. Hysteresis in standards therefore can be driven by protectionist motives even if the initial standards were not introduced for protectionist reasons. With these forces in play, standards and regulations often persist over long periods of time and their protectionist effects and inefficiencies may increase over time. Regulatory differences among countries may cause major conflicts over time as vested interests and industries, which have invested in adhering to these standards, will lobby governments and international organizations to impose their own standards on foreign producers.

That said, there are many examples of standards and regulations that have changed over time when their use—or their vested interests—weakened (Vogel 2003). However, significant shocks to the political economy system may be required for such changes—that is, to move the political economy equilibrium to another equilibrium given the dynamic political and institutional constraints to overcome (Rausser, Swinnen, and Zusman 2012). Shocks may come from both internal and external sources. An internal source is when domestic crises affect food standards. For instance, the first wave of modern public food safety and quality regulations were induced in the late 19th century by public outrage of consumers over the use of cheap and sometimes poisonous ingredients in food production (Meloni and Swinnen 2013, 2015, 2017). In the early 21st century, major changes in public food standards in the EU followed food safety scandals in the late 1990s, with consumers demanding better protection and triggering new policies such as traceability through

12 The case that producers have different preferences and consumers have the same is analogous. For example, Paarlberg (2008) and Zilberman et al. (2013) argue that consumers on both sides of the Atlantic tend to dislike GM technology, but agribusiness lobbying has been much more pro-GM in the United States. In the longer run it may be that as consumers live in different GM-food environments in the United States and the EU, they develop different preferences. Consumer attitudes with respect to biotechnology are likely to be endogenous. In countries where GM products are available, consumer preferences may shift in favor of this technology, while inversely consumers may distrust GM technology more in countries where GM products have been banned.

value chains (McCluskey and Swinnen 2011). The introduction of various public regulations in China in the late 2000s followed the "milk scandal" where people died from consuming milk products with poisonous ingredients (Mo et al. 2012).

Another source of shocks is external. One example is the integration of countries with different standards through international agreements. This may either cause the removal of "inefficient standards" or the opposite: that inefficient standards are extended to other countries with international integration. In reality, both phenomena have been observed, often reflecting the bargaining power of the industries and countries where the (in-)efficient regulations were in place before integration.

GMOs, GIs, and Food Definitions

Three cases of public standards that have attracted wide attention in recent years and continue to do so are the cases of GMO regulations, GIs, and "food definitions." All of these cases represent interesting mixes of private and public interests and of changing political coalitions. Regarding GMOs, large agrochemical and seed companies as well as many NGOs have lobbied intensely to influence GM regulations around the world (Paarlberg 2001; Pinstrup-Andersen and Schioler 2003; Vigani and Olper 2014). While various groups tried to influence GM policy, Graff, Hochman, and Zilberman (2009) argue that the US agribusiness industry was lobbying more pro-GM than the EU agribusiness industry. Some important GM products are competing with traditional agribusiness products, especially GM traits that substitute for pesticides and insecticides. Because it threatened their traditional revenues, several large agropharmaceutical companies, such as Bayer and BASF, initially did not take a strong side in the debate—in contrast to, for example, US-based Monsanto, which lobbied strongly pro-GM. This created a very different political coalition in the United States than in the EU, contributing to a different regulatory outcome (Qaim 2009, 2016).

Second, globalization has increased the links between consumers and producers globally, but at the same time stimulated farmers to lobby for their "local products," seeking a coalition with consumers interested in local foods. At the policy front this has resulted in regulations on geographical indications (GI)—an issue that has created significant tensions in trade negotiations as the number of GIs has grown rapidly over the past 20 years, initially especially in the EU but now growing worldwide, and are an increasingly important item in trade negotiations (Josling 2006; Huysmans and Swinnen 2019; Raimondi et al. 2020). An interesting related issue is how organizations representing environmental interests will reinforce the "local products" coalition by pointing at the environmental costs of trade and global sourcing.

Third, another case of how standards may reduce information asymmetries and transaction costs but also protect vested interests, are the regulations that define specific foods. Historical cases include the definition of "wine" (first in France and later in the EU) and of "chocolate" in the late 19th and early 20th centuries, which had major implications for international trade in these food products a century ago and continues to affect trade and consumption patterns today (Meloni and Swinnen 2013, 2015, 2017). An interesting recent case is the definition of "meat" with technological advances and changing consumer preferences. As plant-based "meat" products have grown rapidly in recent years, US livestock farms have lobbied for regulation to prohibit companies from using words such as "meat," "burger," "sausage," and so on unless the product came from an animal. However, they face opposition from a coalition of new plant-based "meat" companies and large food companies that have invested in them.

The Political Economy of Public Investments and Compensation

There is relatively little research on the political economy of public goods and investments in agriculture. Two exceptions are the political economy of land reforms and of public investment in agricultural research.[13] Public investments in research are an important source of productivity growth (Alston and Pardey 2013; Alston 2017). Studies document high social rates of return to public agricultural research investments but also that there is significant underinvestment in research in both poor and rich countries (Huffman and Evenson 1992; Ruttan 1982). One political economy explanation of the underinvestment by governments is spillover effects (or externalities) in a policy environment where government research investments in one country affect other countries.[14]

13 For a review of the political economy of more general public investments in agriculture, see Mogues (2015). The discussion about investment in agricultural research in this chapter is useful to understand the analysis of agricultural research in Chapter 21 of this volume.

14 Studies have also argued that benefits of public investments in agricultural research are overestimated because of deadweight costs of taxation (Fox 1985), terms of trade effects (Edwards and Freebairn 1984), the effects on unemployment (Schmitz and Seckler 1970), the increase in the deadweight costs of existing commodity policies (Alston, Edwards and Freebairn 1988; Murphy, Furtan, and Schmitz 1993), or because they ignore private research and because of lags in the effects of research (Alston and Pardey 1996).

Research has both public and private good characteristics, as some of the benefits of research expenditures can be captured by specific groups, while other results spill over to other groups or countries (Cornes and Sandler 1986). This affects governments' incentives to invest in research. Spillover effects can thus induce free-riding behavior by governments. Governments in one country will invest less than optimal since they pay for all the costs while part of the benefits are reaped by other countries. Or, inversely, governments may think that they can reap (some of) the benefits from other countries' investments without having to bear the (full) costs of research investments (Huffman and Miranowski 1981; Khanna, Huffman, and Sandler 1994; Rose-Ackerman and Evenson 1985).

A different political economy explanation draws on the distributional effects of public investments (Baland and Kotwal 1998; de Gorter, Nielson, and Rausser 1992; de Gorter and Zilberman 1990; Rausser 1992). Although society as a whole may gain from public investments, different groups in society may have different preferences, depending on how it affects their personal welfare. They will prefer the government to choose their private optimum level of research and will negatively react to the government's choice if this diverges from their private optimum. If some groups oppose public investments because of income distribution effects, governments will underinvest in public goods as they balance the political costs and benefits of diverging from the social optimum.

For example, public investments in research have contributed to the dramatic increase in productivity of agriculture during the 20th century. This productivity increase contributed to the long-term decline in agricultural prices. Although this benefited food consumers, it put pressure on farm incomes (Alston 2017; Ruttan 1982). Coauthors de Gorter and Swinnen (1998) show that in general with unequal income distributional effects a government maximizing political support will underinvest in public research, both in rich countries and in poor countries. Gardner (1987) and Oehmke and Yao (1990) indeed find that underinvestment occurs if farmers gain relatively less from research.

Policy Interactions

So far this chapter has analyzed the political economy of various policies in isolation, meaning as if there were no other policies. However, in reality, many public policies exist simultaneously and may interact with each other. One can distinguish between "*economic* interaction effects" (EIEs), which arise if one policy affects the distributional and welfare effects of other policies, and "*political* interaction effects" (PIEs), which occur when one policy affects the political incentives of governments to introduce or change other policies.

One form of (positive) EIE is when combined reforms reinforce the (beneficial) impacts of separate policy reforms. For example, in the reform strategies in China and Eastern Europe in the 1990s, land reforms and privatization strategies provided new opportunities and better incentives for farmers, while at the same time distortionary price and market policies were reduced or removed. In these cases, both policy reforms combined to improve efficiency. An example of (negative) EIEs is the interaction between public agricultural research and commodity policies that regulate agricultural prices or production. Agricultural research increases productivity and may cause an increase in distortions of existing regulations. Under some conditions the research benefits may be outweighed by increased distortions (Alston, Edwards, and Freebairn 1988; Murphy, Furtan, and Schmitz 1993; Swinnen and de Gorter 1993).

An example of PIEs is the use of agricultural policies for compensation purposes. Compensation is an important element in the political economy of policy reform or public investment (Rausser, Swinnen, and Zusman 2012).[15] Reforms to a more efficient policy almost always imply gains for some groups and losses for others. Similarly, building a road may lead to major gains in rural development but may hurt those who have to move to allow the construction of the road. If the gains outweigh the losses, it is socially optimal to implement the reforms or make the investment since the gains of those who win are more than sufficient to compensate the losers. There are numerous empirical examples of "policy packages" that include compensation for certain groups. They are a traditional part of multiannual agricultural policy decision-making both in the EU and in the United States.

An important problem with compensation, however, is the credible implementation of such schemes. Those who lose from reform may oppose the reforms if they expect that (full) compensation will not take place. The latter may be the case when governments lack the credibility to effectively provide compensation when the reform effects emerge (Acemoglu and Robinson 2006; Swinnen and de Gorter 2002), when governments only offer partial

15 Trade policy reform and compensation have a long history in the economics literature, going back to the early analyses of Adam Smith and David Ricardo. A crucial element in the arguments on the optimality of free trade are that the gains of the winners of trade liberalization are more than sufficient to compensate the losers of reform, an issue that has clearly become highly relevant again in recent years with discussions on the gainers and losers from globalization.

compensation to mitigate political opposition sufficient to get the reforms through (Foster and Rausser 1993), when local institutions prevent the creation of effective compensation schemes (Swinnen 1997), or when there is uncertainty on the effect of the reforms—and thus on who will be the losers and gainers of the reforms (Fernandez and Rodrik 1991).

The inability of governments to credibly commit to compensate groups that are adversely affected is a prime cause of underinvestment in public goods or of failures to implement aggregate welfare improving policies more generally. An important question is therefore how to design mechanisms that constrain policymakers, to bring the discretionary political equilibrium closer to the social optimum. The creation of institutions that make policy reversal more difficult enhances the credibility of policymakers to commit to future compensation. Examples of such institutions are independent central banks for monetary policy or international trade agreements that impose constraints on government policies in agriculture and food. Another example is the role that the WTO has played in preventing the return to agricultural protectionism in recent years during periods of price fluctuations.

References

Acemoglu, D. 2003. "Why Not a Political Case Theorem? Social Conflict, Commitment, and Politics." *Journal of Comparative Economics* 31 (4): 620–652.

Acemoglu, D., and J. Robinson. 2006. *Economic Origins of Dictatorship and Democracy.* Cambridge: Cambridge University Press.

Aksoy, M. A., and B. Hoekman. 2010. *Food Prices and Rural Poverty.* Washington, DC: World Bank.

Alesina, A., and D. Rodrik. 1994. "Distributive Politics and Economic Growth." *Quarterly Journal of Economics* 109 (2): 465–490.

Alston, J. M. 2017. Fellows Address. "Reflections of Agricultural R&D, Productivity, and the Data Constraint: Unfinished Business, Unsettled Issues." *Proceedings of the American Journal of Agricultural Economics Conference*, Chicago, August.

Alston, J. M., G. W. Edwards, and J. W. Freebairn. 1988. "Market Distortions and Benefits from Research." *American Journal of Agricultural Economics* 70: 281–288.

Alston, J. M., and J. S. James. 2002. "The Incidence of Agricultural Policy." In *The Handbook of Agricultural Economics*, vol. 2, edited by B. Gardner and G. Rausser, 2073–2123. Amsterdam: Elsevier Science.

Alston, J. M., and P. Pardey. 1996. *Making Science Pay: The Economics of Agricultural R&D Policy.* Washington, DC: AEI Press.

———. 2013. "Agricultural R&D and Food Security of the Poor." *Economic Papers* 32 (3): 289–297.

Anderson, K. 1995. "Lobbying Incentives and the Pattern of Protection in Rich and Poor Countries." *Economic Development and Cultural Change* 43 (2): 401–423.

———. 2016. *Agriculture Trade, Policy Reforms, and Global Food Security.* Palgrave Studies in Agriculture Economics and Food Policy book series (AEFP). London: Palgrave Macmillan.

Anderson, K., ed. 2009. *Distortions to Agricultural Incentives: A Global Perspective, 1955–2007.* London: Palgrave Macmillan; Washington, DC: World Bank.

Anderson, K., R. Damania, and L. A. Jackson. 2004. "Trade, Standards, and the Political Economy of Genetically Modified Food." Discussion Paper 4526. Centre for Economic Policy Research.

Anderson, K., M. Ivanic, and W. Martin. 2014. "Food Price Spikes, Price Insulation, and Poverty." In *The Economics of Food Price Volatility,* edited by J-P Chavas, D. Hummels, and B. Wright, 311–344. Chicago: University of Chicago Press.

Anderson, K., G. C. Rausser, and J. Swinnen. 2013. "Political Economy of Public Policies: Insights from Distortions to Agricultural and Food Markets." *Journal of Economic Literature* 51 (2): 423–477.

Anderson, K., and J. Swinnen, eds. 2008. *Distortions to Agricultural Incentives in Europe's Transition Economies.* Washington, DC: World Bank.

Antràs, P. 2015. *Global Production: Firms, Contracts and Trade Structure.* Princeton, NJ: Princeton University Press.

Baland, J. M., and A. Kotwal. 1998. "The Political Economy of Underinvestment in Agriculture." *Journal of Development Economics* 55 (1): 233–247.

Barrett, C. B. 1996. "On Price Risk and the Inverse Farm Size-Productivity Relationship." *Journal of Development Economics* 51 (2): 193–215.

———. 2014. *Food Security and Sociopolitical Stability.* Oxford: Oxford University Press.

Bates, R. H., and S. Block. 2010. "Agricultural Trade Interventions in Africa." In *The Political Economy of Agricultural Price Distortions,* edited by K. Anderson, 304–331. Cambridge: Cambridge University Press.

Becker, G. S. 1983. "A Theory of Competition among Pressure Groups for Political Influence." *Quarterly Journal of Economics* 98 (August): 371–400.

Beghin, J. 2013. *Nontariff Measures with Market Imperfections: Trade and Welfare Implications.* Volume 12 of *Frontiers of Economics and Globalization.* Bingley, UK: Emerald.

Beghin, J., M. Maertens, and J. Swinnen. 2015. "Nontariff Measures and Standards in Trade and Global Value Chains." *Annual Review of Resource Economics* 7: 425–450.

Bellemare, M. F., C. B. Barrett, and D. R. Just. 2013. "The Welfare Impacts of Commodity Price Volatility: Evidence from Rural Ethiopia." *American Journal of Agricultural Economics* 95 (4): 877–899.

Bhagwati, J. N. 1982. "Directly Unproductive Profit Seeking Activities: A Welfare Theoretic Synthesis and Generalization." *Journal of Political Economy* 90: 988–1002.

Blanchard, E. J., C. P. Bown, and R. C. Johnson. 2017. *Global Supply Chains and Trade Policy.* NBER Working Paper 21883. Cambridge, MA: National Bureau of Economic Research.

Blanchard, E. J., and X. Matschke. 2015. "U.S. Multinationals and Preferential Market Access." *Review of Economics and Statistics* 97 (4): 839–854.

Brenton, P., and M. Manchin. 2003. "Making EU Trade Agreements Work: The Role of Rules of Origin." *The World Economy* 26 (5): 755–769.

Buchanan, J. M., and G. Tullock. 1962. *The Calculus of Consent.* Ann Arbor, MI: University of Michigan Press.

Cohen, M. J., and J. L. Garrett. 2010. "The Food Price Crisis and Urban Food (In)security." *Environment and Urbanization* 22: 467–482.

Cornes, R., and T. Sandler. 1986. *The Theory of Externalities, Public Goods, and Club Goods.* Cambridge: Cambridge University Press.

Curtis, K. R., J. J. McCluskey, and J. Swinnen. 2008. "Differences in Global Risk Perceptions of Biotechnology and the Political Economy of the Media." *International Journal of Global Environmental Issues* 8 (1/2): 77–89.

de Gorter, H., D. J. Nielson, and G. C. Rausser. 1992. "Productive and Predatory Public Policies: Research Expenditures and Producer Subsidies in Agriculture." *American Journal of Agricultural Economics* 74 (1): 27–37.

de Gorter, H., and J. Swinnen. 1998. "The Impact of Economic Development on Public Research and Commodity Policies in Agriculture." *Review of Development Economics* 2 (1): 41–60.

de Gorter, H., and D. Zilberman. 1990. "On the Political Economy of Public Good Inputs in Agriculture." *American Journal of Agricultural Economics* 72: 131–137.

Demeke, M., G. Pangrazio, and M. Maetz. 2009. "Country Responses to the Food Security Crisis: Nature and Preliminary Implications of the Policies Pursued." FAO Initiative on Soaring Food Prices. Rome: FAO.

Dixit, A. K. 1996. *The Making of Economic Policy: A Transaction Cost Politics Perspective.* Cambridge, MA: MIT Press.

Downs, A. 1957. *An Economic Theory of Democracy*. New York: Harper and Row.

Edwards, G. W., and J. W. Freebairn. 1984. "The Gains from Research into Tradable Commodities." *American Journal of Agricultural Economics* 66: 41–49.

FAO (Food and Agriculture Organization of the United Nations) and OECD (Organisation for Economic Co-operation and Development), eds. 2011. "Price Volatility in Food and Agricultural Markets: Policy Responses." Paris: OECD.

Fernandez, R., and D. Rodrik. 1991. "Resistance to Reform: Status Quo Bias and the Presence of Individual Specific Uncertainty." *American Economic Review* 81: 1146–1155.

Fischer, R., and P. Serra. 2000. "Standards and Protection." *Journal of International Economics* 52 (2): 377–400.

Foster, W. E., and G. C. Rausser. 1993. "Price-Distorting Compensation Serving the Consumer and Taxpayer Interest." *Public Choice* 77 (2): 275–291.

Fox, G. C. 1985. "Is the United States Really Underinvesting in Agricultural Research?" *American Journal of Agricultural Economics* 67: 806–812.

Freund, C., and C. Ozden. 2008. "Trade Policy and Loss Aversion." *American Economic Review* 98 (4): 1675–1691.

Gardner, B. L. 1983. "Efficient Redistribution through Commodity Markets." *American Journal of Agricultural Economics* 65 (2): 225–234.

———. 1987. "Causes of U.S. Farm Commodity Programs." *Journal of Political Economy* 95 (2): 290–310.

Gawande, K., and B. Hoekman. 2006. "Lobbying and Agricultural Trade Policy in the United States." *International Organization* 60: 527–561.

Gawande, K., B. Hoekman, and Y. Cui. 2015. "Global Supply Chains and Trade Policy Responses to the 2008 Crisis." *World Bank Economic Review* 29 (1): 102–128.

Global Policy Forum. 2013. *Global Policy Forum—Financing of the UN Programmes, Funds and Specialized Agencies*. Accessed June 2014. www.globalpolicy.org/un-finance/tables-and-charts-on-un-finance/the-financing-of-the-un-programmes-funds-and-specialized-agencies.html

Gouel, C., and S. Jean. 2015. "Optimal Food Price Stabilization in a Small Open Developing Country." *World Bank Economic Review* 29 (1): 72–101.

Graff, G. D., G. Hochman, and D. Zilberman. 2009. "The Political Economy of Agricultural Biotechnology Policies." *AgBioforum* 12 (1): 34–46.

Grossman, G. M., and E. Helpman. 1994. "Protection for Sale." *American Economic Review* 84 (4): 833–850.

Guariso A., M. P. Squicciarini, and J. Swinnen. 2014. "Food Price Shocks and the Political Economy of Global Agricultural and Development Policy." *Applied Economic Perspectives and Policy* 36 (3): 387–415.

Hawkins, V. 2002. "The Other Side of the CNN Factor: The Media and Conflict." *Journalism Studies* 3 (2): 225–240.

Headey, D. 2013. "The Impact of the Global Food Crisis on Self-Assessed Food Security." *World Bank Economic Review* 27 (1): 1–27.

Hejazi, M., and M. A. Marchant. 2017. "China's Evolving Agricultural Support Policies." *Choices* 32 (2): 1–7.

Huang, J., X. Wang, and S. Rozelle. 2013. "The Subsidization of Farming Households in China's Agriculture." *Food Policy* 41: 124–132.

Huang, J., and G. Yang. 2017. "Understanding Recent Challenges and New Food Policy in China, Global Food Security." *Global Food Security* 12: 119–126.

Huffman, W. E., and R. E. Evenson. 1992. "Contributions of Public and Private Science and Technology to U.S. Agricultural Productivity." *American Journal of Agricultural Economics* 74: 751–756.

Huffman, W. E., and J. A. Miranowski. 1981. "An Economic Analysis of Expenditures on Agricultural Experiment Station Research." *American Journal of Agricultural Economics* 63 (1): 104–118.

Huysmans, M., and J. Swinnen. 2019. "No Terroir in the Cold? A Note on the Geography of Geographical Indications." *Journal of Agricultural Economics* 70 (2): 550–559.

Ivanic, M., and W. Martin. 2014. "Implications of Domestic Price Insulation for Global Food Price Behaviour." *Journal of International Money and Finance* 42: 272–288.

Josling, T. 2006. "The War on Terroir: Geographical Indications as a Transatlantic Trade Conflict." *Journal of Agricultural Economics* 57: 337–363.

Khanna, J., W. E. Huffman, and T. Sandler. 1994. "Agricultural Research Expenditures in the United States: A Public Goods Perspective." *Review of Economics and Statistics* 76 (2): 267–277.

Krueger, A. O. 1974. "The Political Economy of the Rent-Seeking Society." *American Economic Review* 64 (3): 291–303.

Krueger, A. O., M. Schiff, and A. Valdés. 1991. *The Political Economy of Agricultural Pricing Policy.* Baltimore: Johns Hopkins University Press for the World Bank.

López, R. A. 2008. "Does 'Protection for Sale' Apply to the US Food Industries?" *Journal of Agricultural Economics* 9 (1): 25–40.

Magee, S. P., W. A. Brock, and L. Young. 1989. *Black Hole Tariffs and Endogenous Policy Theory.* Cambridge: Cambridge University Press.

Marette, S. 2014. "Non-Tariff Measures When Alternative Regulatory Tools Can Be Chosen." Mimeo. UMR Economie publique INRAE-AgroParis Tech.

Marette, S., and J. Beghin. 2010. "Are Standards Always Protectionist?" *Review of International Economics* 18 (1): 179–192.

Martin, W., and K. Anderson. 2011. "Export Restrictions and Price Insulation during Commodity Price Booms." Policy Research Paper 5645, World Bank, Washington, DC.

Maystadt, J. F., J. Tanb, and C. Breisinger. 2014. "Does Food Security Matter for Transition in Arab Countries?" *Food Policy* 46: 106–115.

McCluskey, J. J., and J. Swinnen. 2004. "Political Economy of the Media and Consumer Perceptions of Biotechnology." *American Journal of Agricultural Economics* 86: 1230–1237.

———. 2010. "Media Economics and the Political Economy of Information." In *The Oxford Handbook of Government and Business*, edited by D. Coen, W. Grant, and G. Wilson, 643–662. Oxford: Oxford University Press.

———. 2011. "Media and Food Risk Perceptions." *EMBO Journal* 12 (7): 467–486.

McGuire, M. C., and M. Olson Jr. 1996. "The Economics of Autocracy and Majority Rule: The Invisible Hand and the Use of Force." *Journal of Economic Literature* 34 (1): 72–96.

Meloni, G., and J. Swinnen. 2013. "The Political Economy of European Wine Regulations." *Journal of Wine Economics* 8 (3): 244–284.

———. 2015. "Chocolate Regulations." In *The Economics of Chocolate*, edited by M. P. Squicciarini and J. Swinnen, 268–306. Oxford: Oxford University Press.

———. 2017. "Standards, Tariffs and Trade: The Rise and Fall of the Greek-French Raisin Trade and the Definition of Wine." *Journal of World Trade* 51 (4): 711–740.

Mitchell, M., and A. Moro. 2006. "Persistent Distortionary Policies with Asymmetric Information." *American Economic Review* 96 (1): 387–393.

Mo, D., J. Huang, X. Jia, H. Luan, S. Rozelle, and J. Swinnen. 2012. "Checking into China's Cow Hotels: Have Policies Following the Milk Scandal Changed the Structure of the Dairy Sector?" *Journal of Dairy Science* 95 (5): 2282–2298.

Mogues, T. 2015. "Political Economy Determinants of Public Spending Allocations: A Review of Theories, and Implications for Agricultural Public Investment." *European Journal of Development Research* 27 (3): 452–473.

Murphy, J. A., W. H. Furtan, and A. Schmitz. 1993. "The Gains from Agricultural Research under Distorted Trade." *Journal of Public Economics* 51: 161–172.

Naylor, R. L., ed. 2014. *The Evolving Sphere of Food Security.* Oxford: Oxford University Press.

OECD. 2013. *Official Development Assistance.* www.oecd.org/dac/financing-sustainable-development/development-finance-standards/official-development-assistance.htm.

———. 2017. *Agricultural Policy Monitoring and Evaluation 2017: OECD Countries and Emerging Economies.* Paris: OECD.

———. 2019. *Agricultural Policy Monitoring and Evaluation 2019.* Paris: OECD.

Oehmke, J. F., and X. Yao. 1990. "A Policy Preference Function for Government Intervention in the US Wheat Market." *American Journal of Agricultural Economics* 72 (3): 631–640.

Olper, A. 2016. "The Political Economy of Trade-Related Regulatory Policy: Environment and Global Value Chain." *Bio-Based and Applied Economics* 5 (3): 287–324.

Olper, A., J. Fałkowski, and J. Swinnen. 2014. "Political Reforms and Public Policies: Evidence from Agricultural and Food Policy." *World Bank Economic Review* 28 (1): 21–47.

Olper, A., and J. Swinnen. 2013. "Mass Media and Public Policy for Agriculture." *World Bank Research Digest* 7 (3): 6.

Olson, M. 1965. *The Logic of Collective Action.* New Haven, CT: Yale University Press.

Orden, D., D. Blandford, and T. Josling. 2010. "Determinants of United States Farm Policies." In *The Political Economy of Agricultural Price Distortions,* edited by K. Anderson, 162–190. Cambridge: Cambridge University Press.

Paarlberg, R. 2001. *The Politics of Precaution: Genetically Modified Crops in Developing Countries.* Washington, DC: IFPRI.

———. 2008. *Starved for Science: How Biotechnology Is Being Kept out of Africa.* Cambridge, MA: Harvard University Press.

Persson, T., and G. Tabellini. 2000. *Political Economics: Explaining Economic Policy.* Cambridge, MA: MIT Press.

Pieters, H., and J. Swinnen. 2016. "Trading-Off Volatility and Distortions? Food Policy during Price Spikes." *Food Policy* (61): 27–39.

Pinstrup-Andersen, P., ed. 2014. *Food Price Policy in an Era of Market Instability: A Political Economy Analysis.* Oxford, UK: Oxford University Press.

Pinstrup-Andersen, P., and E. Schioler. 2003. *Seeds of Contention: World Hunger and the Global Controversy Over GM Crops.* Baltimore: IFPRI.

Prakash, A. 2011. "Why Volatility Matters." In *Safeguarding Food Security in Volatile Global Markets,* edited by A. Prakash, 1–27. Rome: FAO.

Qaim, M. 2009. "The Economics of Genetically Modified Crops." *Annual Review of Resource Economics* 1 (1): 665–694.

———. 2016. *Genetically Modified Crops and Agricultural Development*. New York: Springer.

Raimondi, V., C. Falco, D. Curzi, and A. Olper. 2020. "Trade Effects of Geographical Indication Policy: The EU Case." *Journal of Agricultural Economics*. 17 (2): 330–356.

Rausser, G. 1992. "Predatory versus Productive Government: The Case of U.S. Agricultural Policy." *Journal of Economic Perspectives* 6 (3): 133–157.

Rausser, G., J. Swinnen, and P. Zusman. 2012. *Political Power and Economic Policy: Theory, Analysis, and Empirical Applications*. Cambridge: Cambridge University Press.

Rodrik, D. 1995. "Political Economy of Trade Policy." In *Handbook of International Economics*, vol. 3, edited by G. M. Grossman and K. Rogoff, 1457–1494. Amsterdam: North-Holland.

Ronnen, U. 1991. "Minimum Quality Standards, Fixed Costs, and Competition." *RAND Journal of Economics* 22 (4): 490–504.

Rose-Ackerman, S., and R. Evenson. 1985. "The Political Economy of Agricultural Research and Extension: Grants, Votes, and Reapportionment." *American Journal of Agricultural Economics* 67 (1): 1–14.

Rozelle, S., and J. Swinnen. 2004. "Success and Failure of Reforms: Insights from Transition Agriculture." *Journal of Economic Literature* 42: 404–456.

Ruttan, V. W. 1982. *Agricultural Research Policy*. Minneapolis: University of Minnesota Press.

Schmitz, A., and D. Seckler. 1970. "Mechanized Agriculture and Social Welfare: The Case of the Tomato Harvester." *American Journal of Agricultural Economics* 52: 569–577.

Sheldon, I. 2012. "North-South Trade and Standards: What Can General Equilibrium Theory Tell Us?" *World Trade Review* 11 (3): 376–389.

Stigler, J. G. 1971. "Theory of Economic Regulation." *Bell Journal of Economics and Management Science* 2 (1): 3–21.

Strömberg, D. 2001. "Mass Media and Public Policy." *European Economic Review* 45 (4–6): 652–663.

———. 2004. "Mass Media Competition, Political Competition, and Public Policy." *Review of Economic Studies* 71 (1): 265–284.

Swinnen, J. 1994. "A Positive Theory of Agricultural Protection." *American Journal of Agricultural Economics* 76 (1): 1–14.

———. 1997. "Does Compensation for Disruptions Stimulate Reforms? The Case of Agricultural Reform in Central Europe." *European Review of Agricultural Economics* 24 (2): 249–266.

———. 2009. "The Growth of Agricultural Protection in Europe in the 19th and 20th Centuries." *World Economy* 32 (11): 1499–1537.

———. 2016. “Economics and Politics of Food Standards, Trade, and Development.” *Agricultural Economics* 47 (S1):7–19.

———. 2017. “Some Dynamic Aspects of Food Standards.” *American Journal of Agricultural Economics* 99 (2): 321–338.

———. 2018. *The Political Economy of Agricultural and Food Policies*. New York: Palgrave McMillan.

Swinnen, J., A. Banerjee, and H. de Gorter. 2001. “Economic Development, Institutional Change, and the Political Economy of Agricultural Protection: An Econometric Study of Belgium since the 19th Century.” *Agricultural Economics* 26 (1): 25–43.

Swinnen, J., K. Deconinck, T. Vandemoortele, and A. Vandeplas, eds. 2015. *Quality Standards, Value Chains and International Development*. New York: Cambridge University Press.

Swinnen, J., and H. de Gorter. 1993. “Why Small Groups and Low Income Sectors Obtain Subsidies: The ‘Altruistic’ Side of a ‘Self-Interested’ Government.” *Economics and Politics* 5 (3): 285–296.

———. 2002. “On Government Credibility, Compensation, and Under-Investment in Public Research.” *European Review of Agricultural Economics* 29 (4): 501–522.

Swinnen, J., and N. Francken. 2006. “Summits, Riots and Media Attention: The Political Economy of Information on Trade and Globalization.” *World Economy* 29: 637–654.

Swinnen, J., A. Olper, and T. Vandemoortele. 2012. “Impact of the WTO on Agricultural and Food Policies.” *World Economy* 35 (9): 1089–1101.

———. 2016. “The Political Economy of Policy Instrument Choice: Theory and Evidence from Agricultural and Food Policies.” *Theoretical Economics Letters* 6 (1): 106–117.

Swinnen, J., and S. Rozelle. 2006. *From Marx and Mao to the Market: The Economics and Politics of Agricultural Transition*. New York: Oxford University Press.

Swinnen, J., P. Squicciarini, and T. Vandemoortele. 2011. “The Food Crisis, Mass Media and the Political Economy of Policy Analysis and Communication.” *European Review of Agricultural Economics* 38 (3): 409–426.

Swinnen, J., and T. Vandemoortele. 2011. “Trade and the Political Economy of Food Standards.” *Journal of Agricultural Economics* 62 (2): 259–280.

Timmer, C. P. 2010. “Did Speculation Affect World Rice Prices?” In *The Rice Crisis: Markets, Policies and Food Security*, edited by D. Dawe, 29–60. Washington, DC: FAO and Earthscan.

Tovar, P. 2009. “The Effects of Loss Aversion on Trade Policy: Theory and Evidence.” *Journal of International Economics* 78: 154–167.

Van Tongeren, F., J. Beghin, and S. Marette. 2009. *A Cost-Benefit Framework for the Assessment of Non-Tariff Measures in Agro-Food Trade*. OECD Food, Agriculture and Fisheries Working Paper 21. Paris: OECD.

Varshney, A. 1995. *Democracy, Development, and the Countryside: Urban–Rural Struggles in India*. New York: Cambridge University Press.

Verpoorten, M., A. Arora, N. Stoop, and J. Swinnen. 2013. "Self-Reported Food Insecurity in Africa during the Food Price Crisis." *Food Policy* 39: 51–63.

Vigani, M., and A. Olper. 2014. "GM-Free Private Standards, Public Regulation of GM Products and Mass Media." *Environment and Development Economics* 19 (6): 743–768.

Vogel, D. 2003. "The Hare and the Tortoise Revisited: The New Politics of Consumer and Environmental Regulation in Europe." *British Journal of Political Science* 33: 557–580.

World Bank. 2012. *Responding to Higher and More Volatile World Food Prices*. Washington, DC.

Zilberman, D., S. Kaplan, E. Kim, G. Hochman, and G. Graff. 2013. "Continents Divided: Understanding Differences between Europe and North America in Acceptance of GM Crops." *GM Crops and Food* 4: 202–208.

Chapter 15

GENDER, HOUSEHOLD BEHAVIOR, AND RURAL DEVELOPMENT

Cheryl Doss and Agnes Quisumbing

Our understanding of decision-making within rural households has changed substantially since interest in intrahousehold decision-making emerged in the 1980s. Conventional wisdom, rooted in the unitary theory of the household, held that households are groups of individuals who have the same preferences and fully pool their resources (Becker 1981). Accumulating empirical evidence has shifted this concept of the household in which households decide "as one" to a "collective" model in which individual household members may have different preferences, may not completely pool resources, and may bargain over outcomes in both production and consumption (Haddad, Hoddinott, and Alderman 1997).

Much of the literature on gender and agriculture since the late 1990s has contributed to the empirical evidence supporting the collective models of the household. These have included studies showing differential propensities to spend out of income controlled by men or women (for example, Hoddinott and Haddad 1995); differential effects of men's and women's assets on consumption (Doss 2006), household expenditures, and schooling (Quisumbing and Maluccio 2003); incomplete sharing of risk within the household (Doss 2001; Goldstein and Udry 2008); and inefficiency of resource allocation on plots managed by men and women (Udry 1996). These findings have influenced the design of programs that targeted women, such as conditional cash-transfer programs, which have in turn generated a large pool of evidence that economic transfers controlled by women are associated with investments in child schooling, health, and nutrition (see the systematic review by Yoong, Rabinovich, and Diepeveen 2012).

Although this transition has provided important insights into how we understand rural households, it misses key elements of household dynamics. Even if household members do not completely pool resources, the fact that they form households and share ownership and control over some resources, work on family farms, produce output jointly, have and raise children together, and share in some consumption indicates that there are gains to jointness in gender and family dynamics (Fafchamps and Quisumbing 2008; Doss and Meinzen-Dick 2015). Understanding both individual roles within households and the levels of cooperation or jointness is essential to analysis of households, especially in rural areas where households engage in both production and consumption.

This chapter reviews recent conceptual and empirical developments regarding household behavior and gender norms in developing countries covering the following general topics: (1) What do the data tell us about gender gaps in control and ownership of resources? (2) What have we learned about jointness in household behavior? (3) What do the data tell us about the resources that men and women control, whether solely or jointly? (4) Why does it matter?

Household Behavior: Beyond Individual Threat Points to Cooperation

Early work on household decision-making (Becker 1981) treated the household as the decision-making unit.[1] Analytically, this assumes either that all household members have the same preferences, pool all resources, and agree on all decisions, or that one household member makes the decisions for everyone. Although these models have been enriched by including both production and consumption, as in the agricultural household model (for example, Singh, Squire, and Strauss 1986), which treats these decisions as recursive (households first decide how much to produce, then allocate what is produced across public and private household goods), most of them relied on the assumption of the unitary household. The women in development literature provided initial evidence that households did not necessarily make joint decisions and that men's and women's roles and responsibilities impacted production decisions.[2] While this literature initially focused on demonstrating that ignoring gender roles could result in inefficient outcomes for projects, it also had implications for understanding how gender roles and power dynamics shape household decisions. Studies of agricultural commercialization and nutrition drew attention to the possibility that potential losses of control of women's income would have detrimental effects on nutrition (von Braun and Kennedy 1994) and led to increased interest in resource allocation within households.

1 This section draws heavily on Doss and Meinzen-Dick (2015).

2 See the edited volume by Dwyer and Bruce (1988).

The increased attention to intrahousehold outcomes occurred in parallel with the development of the collective model of the household. All collective models of the household have two common features: they allow different decision-makers to have different preferences and do not assume a single household welfare index or utility function (Chiappori 1992). Much of this literature uses a game theoretic framework and begins with the assumption that the household will reach an efficient outcome. Thus the household could not produce more simply by reallocating labor or other resources, and goods and services could not be reallocated across household members to make at least one better off without making anyone worse off. Assumptions that preferences differ by gender allow tests of how men's and women's bargaining power affects outcomes.

In cooperative bargaining models, a subset of the collective models, each individual is defined as having an outside option or a "threat point." This is the amount of resources that they could access if they were not part of the household. Each individual must obtain at least their outside option within the household or they will leave. Depending on the context, "leaving" could involve divorce or desertion, or it could involve simply opting out of pooling resources and making joint decisions. The important policy insight from these models is that outside options affect household resource allocation. For example, women's wages, such as in public works programs, affect household resource allocation, even in households where women are not employed. Changes in laws governing men's and women's property rights and transfer programs, particularly those targeting specific household members, all change the outside options of individuals.

Studies have tested the collective model of the household in developing countries, using various proxies for women's bargaining power within the household including education, assets, and family background (for overviews of the literature, see Schultz 2003; Quisumbing 2003; Fafchamps and Quisumbing 2008). Although endogeneity problems often make it difficult to establish causal relationships, there is increasing evidence that women's bargaining power affects the outcomes of household decisions. Impact evaluations of conditional cash-transfer programs implemented in the late 1990s and beyond provided additional evidence that resources controlled by women had differential impacts on household decisions, particularly those related to child health and education (Yoong, Rabinovich, and Diepeveen 2012).

In contrast, noncooperative bargaining models do not assume that resources are pooled and explicitly allow for outcomes where potential gains from cooperation have not been realized. Individuals may act strategically; each individual makes separate but interrelated production and consumption decisions based on his or her own preferences and interests and expectations of what others will do. Many studies find outcomes that are consistent with noncooperative bargaining models. Jones (1983) found that women contributed less than optimal amounts of labor to household rice production, preferring to produce lower-value sorghum, since women felt that they were not adequately compensated within the household for their work on irrigated rice fields. Udry (1996) found that total household crop yields could have been increased by shifting fertilizer from men's fields to women's fields. McPeak and Doss (2006) found that male household heads among East African pastoralists located the household farther from town to limit women's milk marketing. Experimental games, in particular, have identified strategic behavior between spouses (Ashraf 2009), couples making choices that do not maximize surplus (Iversen et al. 2010), and that outcomes differ depending on the process through which income was acquired (Dasgupta and Mani 2013).

Bargaining models, however, typically formulate threat points as individual threat points—resources that the individual can control upon leaving the union—and most often focus on individually consumed goods (or leisure) rather than joint production or consumption of the public good. Yet the existence of public goods and economies of scale in household production are two reasons that households form. And the reality is that all over the world, men and women work together on family farms. Even if some plots may be solely managed by the man or the woman, there is joint production and management of at least some of the household's land, men's and women's labor are applied on each other's plots, and children are very often raised jointly by their parents.

The focus on individual control or ownership of resources may lead us to neglect the possibility that individuals within households may engage in some resource-pooling and risk-sharing, even if incomplete. A recent study on Malawi by Josephson (2016) tests the assumption that all household income is pooled, accounting for joint income, as well as income earned individually by men and women. Exploiting the variation in expenditure by different income earners resulting from exogenous variation in rainfall, she finds that household members partially insure one another for expenditure on essential goods (such as food, clothing, education, and healthcare) but do not insure one another for luxury goods, including cigarettes and alcohol, recreation, and housing and utilities. Interestingly, she finds that households in matrilineal areas of Malawi completely pool income and fully insure one another against income variation, while households in nonmatrilineal areas do not. This societal difference is not driven by the sex of the household head: female-headed households do not behave differently from those headed by men. She concludes that societal structures play an important role in intrahousehold income allocation.

Quisumbing, Kumar, and Behrman (2018) use panel data to examine whether shocks affect men's, women's, and jointly held assets differently in Bangladesh and Uganda. Across countries, covariate and idiosyncratic shocks have different effects on solely and jointly owned assets. Jointly held land and assets were better insured against food price increases in Bangladesh, unlike jointly held assets and wives' assets in Uganda. Similar to Josephson's findings on societal structures, Quisumbing, Kumar, and Behrman (2018) posit that differences in the institution of marriage and cultural concepts of joint and individual ownership affect the extent to which joint or individually owned assets are used to cope with shocks. In Bangladesh the results that show generally insignificant impacts on joint land and asset holdings—while individual assets are sacrificed at the margins—indicate that husbands and wives try to preserve the economic base of the household unit. In contrast, in Uganda husbands' assets appear better insured than wives' or even joint assets.

Most of the intrahousehold bargaining literature focuses on the relationships within a couple. This work has expanded to also consider the dynamics within polygynous households, especially regarding production efficiency (Akresh, Chen, and Moore 2016; Kazianga and Wahhaj 2013), cooperation (Barr et al. 2019), potential impacts of transfer programs on intimate partner violence (Heath, Hidrobo, and Roy 2020), and the dynamics between parents and children, especially in the context of migration (Cong and Silverstein 2011; Ambler 2016).

Research is beginning to explore the circumstances under which households cooperate and the circumstances under which there are gains to cooperation. Some of this research is inspired by work on collective action and natural resource management (see an extensive review in Doss and Meinzen-Dick 2015), by more detailed data that are better able to identify sole and joint asset ownership, and by findings from impact evaluations. Work by Barr et al. (2019) explores decision-making and cooperation within polygynous households; as cash-transfer programs expand to Africa, impact evaluations are also paying more attention to the implications of polygyny on the design and implementation of those programs (Heath, Hidrobo, and Roy 2020). We discuss emerging findings below.

What Do the Data Tell Us about Gender Gaps and Joint Ownership?

The shift to collective and bargaining models of the household resulted in a focus on the individual level. Slowly, large-sample household surveys have begun to collect more sex-disaggregated data. For agricultural household surveys this often includes information on ownership and management of plots within the household, information on the ownership of agricultural and other assets, and questions on who is involved in decision-making, regarding both production and consumption decisions. These data have facilitated both gender analyses and intrahousehold analyses, and have revealed the extent of gender gaps in the ownership and control of resources.

Gender Gaps

Although there are wide regional and country variations, most data sources show that there are persistent gender gaps in the ownership and control of land and assets (FAO 2011; World Bank 2011). Gender gaps have been documented across resources such as land (Lastarria-Cornhiel et al. 2014; Doss et al. 2015; Kieran et al. 2015; Deere and Leon 2003), assets (Deere et al. 2012; Doss et al. 2014), nonland inputs (Peterman, Behrman, and Quisumbing 2014), credit (Fletschner and Kenney 2014), and extension services (Ragasa 2015). In land, arguably one of the most important assets for rural households, analysis of nationally representative datasets in Africa south of the Sahara and Asia shows that men own a much larger proportion of household agricultural land than women do (Table 15.1), although a considerable proportion of land is also owned jointly.

Similarly, an analysis of housing and livestock data from surveys designed specifically to examine gendered assets ownership in Ecuador, Ghana, and Karnataka, India, finds that individual men are much more likely to own the principal residence as well as large and small livestock than women (Doss et al. 2011). The one exception is in Ecuador, where individual women are more likely than individual men to own their residence. This reflects the joint marital property regime in Ecuador, which attributes all property acquired during the marriage jointly to the husband and wife. Thus married men own their residence jointly with their wife, and fewer single heads of household who are men own a residence compared with those who are women. Only in Ecuador are individual women more likely than individual men to own any form of livestock and that is only for small livestock and poultry.

There is also evidence that gender gaps in schooling, historically favoring men, have begun to close. The World Bank (2011) estimates that gender gaps

TABLE 15.1 Share of household agricultural land area held by women, men, and jointly by both (%)

Country (date)	Definition of ownership	Women	Men	Joint
Ethiopia (2011–2012)	Registered	15	45	39
Malawi (2010–2011)	Owned	40	42	18
Niger (2011)	Owned	9	62	29
Nigeria (2010)	Right to sell (use as collateral)	4	87	9
Tanzania (2010–2011)	Owned	16	44	39
Uganda (2009–2010)	Owned	18	34	48
Bangladesh (2011–2012)	Documented	10	88	2
Timor-Leste (2007)	Land manager	12	88	n.a.
Tajikistan (2007)	Owner	14	86	n.a.
Viet Nam (2004)	Owner	72	15	13

Source: Doss et al. (2015); Kieran et al. (2015).
Note: n.a. = not applicable

in primary education have closed in almost all countries. In secondary education these gaps are closing rapidly and have reversed in many countries, especially in Latin America, the Caribbean, and East Asia—but it is now boys and young men who are disadvantaged. Among developing countries, girls now outnumber boys in secondary schools in 45 countries, and there are more young women than men in universities in 60 countries.

Joint Ownership

While the data reveal striking patterns of gender differences, they also reveal that property is often owned jointly, although the patterns differ widely across countries and assets. Table 15.1 shows a high proportion of agricultural land owned jointly; in Uganda 48 percent of agricultural land is owned jointly. In five of the eight countries in Africa and Asia for which data is available, more land is owned jointly than owned by women. Similarly, Table 15.2 shows that, in Ecuador, among households that owned their principal residence, 41 percent of the time it was owned jointly by the couple. Deere, Alvarado, and Twyman (2012) analyzed data from nationally representative surveys in Latin America and found wide variations in the extent of joint ownership.[3] Among households owning their home, the percentage of households

3 These data were all from nationally representative household surveys collected between 2000 and 2005. See Deere, Twyman, and Alvarado (2012) for details.

TABLE 15.2 Distribution of asset ownership, by form of ownership (%)

Asset	Country	Individual woman	Individual man	Couple	Household	Other
Principal residence	Ecuador	30	16	41	5	9
	Ghana	25	51	11	0	13
	Karnataka	23	64	4	0	9
Large livestock	Ecuador	10	48	27	12	2
	Ghana	8	84	1	2	5
	Karnataka	3	5	5	85	2
Small livestock	Ecuador	34	18	28	18	2
	Ghana	29	61	4	3	2
	Karnataka	3	5	6	84	2
Poultry	Ecuador	43	7	28	20	2
	Ghana	34	55	5	5	1
	Karnataka	10	7	7	75	0

Source: Based on Doss et al. (2011).

reporting joint ownership was very low in Central America (2.5 percent in Guatemala; 3 percent in Honduras; 6.3 percent in Nicaragua) and in Paraguay (3.5 percent), Chile (3.4 percent), and Mexico (3.3 percent). In other areas it was much higher: 15.9 percent in Panama, 40.7 percent in Argentina, and 41.3 percent in Ecuador.

For livestock, Table 15.2 indicates that it is common for livestock to be owned jointly in Ecuador and that in Karnataka the majority of respondents indicate that livestock is owned by everyone in the household. These data are all based on reports of ownership from one respondent in household surveys. Typically the respondent is the head of the household or the person who claims to be most knowledgeable. Yet evidence suggests that the responses to a variety of survey questions depend on who is providing the answers. These include questions about income (Fisher, Reimer, and Carr 2010), financial information (Fletschner and Mesbah 2011), and labor force participation (Badarsi et al. 2011). Data experiments conducted in Uganda, using an experimental design to identify the impacts of the choice of respondent(s) on patterns of asset ownership (Kilic and Moylan 2016), find differences in the reported patterns of asset ownership, depending on which household member was interviewed, whether multiple household members were interviewed, and whether respondents were asked about only their own assets or about those of all household members.

Analysis of the Bangladesh Integrated Household Survey finds that husbands and wives provide different answers about whether the household owns any agricultural land in 4.5 percent of households, and of those households that agree that land is owned, the spouses report the same answer for who owns most of the agricultural land in only 83 percent of households (Ambler et al. 2021). One reason for the discrepancy is that women are more likely than men to include others as owners. In other words, men will report that they own the land individually, while women report that it is owned jointly. Thus the analyses discussed earlier might suggest higher patterns of joint ownership if women were the ones responding.

A second challenge in analyzing data on joint ownership is that "joint" doesn't necessarily mean "equal." Evidence from both Uganda and South Africa finds that wives have fewer rights than their husbands over land that is reported as jointly owned (Jacobs and Kes 2015; Doss, Meinzen-Dick, and Bomuhangi 2013). Little quantitative data is available to analyze these issues. Even if surveys ask about joint ownership, they don't follow up with questions that would provide insights into what that means. Qualitative fieldwork may be useful for understanding the rights held by joint owners. The literature is still quite unclear on whether joint or individual ownership is better for women and what "better" might mean.[4] Women's property rights are embedded in the household and community; depending on the particular context, women may prefer individual or joint rights. Laws and norms identifying all property of married couples as being jointly owned will typically give married women claims to a larger amount of property than if they simply own some property individually. However, joint property rights may make it more difficult for women to leave abusive relationships, since it may be more difficult to retain their property.

Joint Decision-Making

Analyzing women's role in household decision-making has a relatively long history. The Demographic and Health Surveys have collected decision-making questions routinely for many years and these data are often used in analyses of outcomes related to the well-being of both women and children. Similar questions are now incorporated in many other large sample surveys. Many agricultural surveys have now incorporated questions about who decides with regard to choice of crops, use of inputs, and what to do with the output.

---

4 See Jackson (2003) and Agarwal (1994) for two different perspectives on this.

Decision-making within the household is used as an indicator of women's agency, women's empowerment, and women's bargaining power.

The DHS asks who usually makes decisions about six different areas, including contraception, expenditures, healthcare, and visits to family. The answers are typically coded as respondent, spouse, joint, or someone else. Most of the analyses using these data either combine the respondent and joint responses (Acharya et al. 2010) or treat them as hierarchical, with women making the decision alone as demonstrating the most decision-making power, and thus the most empowerment, followed by joint decision-making, with the woman not being involved as indicating the lowest level of empowerment. The questions may be framed in different ways. The DHS asks who usually makes decisions. Other surveys ask who has the final say. Other surveys simply ask if the woman participated in the decision. There is no consensus on which set of questions is best. However, these questions do not consider the processes of decision-making nor the issue of whether there is conflict or agreement within the household about what the decision should be. It is quite different to claim to be the one with the final say when you disagree with your spouse about the outcome than when you share the same preferences.[5] In addition, the reasons for why the decisions are made by a particular person may affect the outcome. If the man makes all of the agricultural decisions because he is running the farm while his wife is running another business, the implications may differ from a situation where she contributes much of her labor to the farm.

Surveys that allow options for reporting joint decision-making frequently find a relatively high proportion of decisions being made jointly. In a sample of couple households in Ecuador, Twyman, Useche, and Deere (2015) find that 63 percent of cultivation decisions and 50 percent of the input decisions are made jointly. They also find that the reporting differs depending on whether you interview a man or a woman. Similarly, in Tanzania, Anderson, Reynolds, and Gugerty (2017) find statistically significant differences in responses regarding wives' authority in making agricultural decisions, although the magnitudes of the differences are small. In Ecuador, Alwang, Larochelle, and Barrera (2017) randomized who was interviewed about agricultural decisions, interviewing only a man, only a woman, and interviewing both within a household (interviews were separate, but they knew the other would be interviewed). Men were more likely to claim that they were solely responsible; women claimed some responsibility for themselves and also were more likely to say that the responsibilities were joint.

5 These issues are discussed in much more detail in Donald et al. (2020).

Although households may make some decisions jointly, it is also important to recognize that household members may behave strategically with regard to resources and information. The experimental evidence on this is growing. Ashraf (2009) finds that the financial choices of married individuals vary depending on whether their choices are private or observable and whether spouses are encouraged to communicate as part of the savings choice experiment in the Philippines. In Uganda there is evidence of couples making choices that do not maximize surplus (Iversen et al. 2010); Dasgupta and Mani (2013) find that the process through which income was acquired affects outcomes. And finally, Ambler (2015) finds strategic behavior on remittance decisions in transnational households and the amounts of remittances vary based on the extent of the information between the migrant and the recipient.

Research on decision-making in polygynous households has lagged behind that focusing on spousal pairs. Yet the high incidence of polygyny in Africa south of the Sahara, with more than 40 percent of women in polygynous marriages (Barr et al. 2019), particularly in West Africa, demands that close attention be paid to this more complex household structure. Barr et al. (2019), using a series of public goods games in Kwara, Nigeria, find that in comparison to monogamous couples, polygynous husbands and wives are less cooperative with each other and that co-wives are least cooperative with each other. In contrast, in another study using observational data, Damon and McCarthy (2019) find that jointly managed plots in polygynous households in Malawi have higher yields and higher crop values than either men's plots in polygynous households or plots in monogamous households. Although the two studies do not study the same outcomes in the same context, it is clear that these findings are context-dependent and that it would be unwise to assume that polygynous households in different societies behave alike. Research on the institution of polygyny is an important area of future work, particularly in relation to the design and implementation of development programs. Heath, Hidrobo, and Roy (2020), for example, find that a cash-transfer program in Mali, where transfers were given to household heads, who were mostly male, significantly reduced intimate partner violence in polygamous households but had limited effects in monogamous households. They attribute the differential impact among polygamous households to reductions in men's stress and anxiety in polygamous households and larger reductions in disputes in polygamous relative to monogamous households. We have very little evidence on how other programs may differentially affect households with complex structures; this is a promising area of future work.

Why Understanding Household Behavior—and Jointness—Matters

Myths and assumptions regarding rural household behavior can be barriers to reducing poverty and achieving desired development outcomes. In this section we discuss the consequences of assuming that households act in a unitary manner, the unintended consequences of targeting women only, and the potential benefits to recognizing areas of joint and individual ownership and action in rural households.

Consequences of Assuming a Unitary Model of Household Behavior

In their 1997 book, Haddad, Hoddinott, and Alderman list four types of policy failures that may arise if policymakers use the unitary model of the household as their basis for policy prescriptions: (1) differences in the effect of public transfers, depending on the identity of the recipient; (2) the possibility that nonrecipients of the transfer may compensate for the transfer, thereby negating the intention of the transfer; (3) implications of targeting only one household member for information because of the assumption that information, like other resources, is shared within the household; and (4) most important, disabling policy instruments that could be brought to bear on development problems. The authors argue that the unitary model predicts that household behavior can be changed only by changes in prices and household incomes.

In contrast, collective models posit that a large range of policies can be used to affect household allocation outcomes, such as changes in access to common property resources, credit, public works schemes, and legal and institutional rights. Policymakers have used the findings from early research on intrahousehold issues by designing programs that aim to change household behavior, such as the conditional cash-transfer programs in Latin America (now all over the world), most of which designate the mother as the transfer recipient (see the edited volume on the Latin American experience by Adato and Hoddinott 2010 and the systematic review of programs transferring economic resources to women by Yoong, Rabinovich, and Diepeveen 2012).

The assumption that resources are completely pooled has been used to justify the lack of attention to closing gender gaps in resources. Yet, as gender gaps are documented, and as evidence accumulates about the nonpooling household, more estimates of the costs of gender gaps have become available. FAO (2011) estimated that if women had the same access to productive resources as men, they could increase yields on their farms by 20 percent to

30 percent, potentially raising total agricultural output in developing countries by 2.5 percent to 4 percent, which could reduce the number of hungry people in the world by 12 percent to 15 percent. Although these estimates are plausible and attracted attention to the potential costs of the gender gap in agriculture, these numbers were calculated using the estimated production functions for women assuming that they used the same levels of inputs as the men. They are not the result of programs that provide men and women with equal levels of inputs, such as a randomized controlled trial (Doss 2017). They do not account for the endogeneity of input application nor ways to feasibly increase input application by women.

More recently, estimates of the costs of gender gaps in access to resources have been further refined using data from the World Bank's Living Standard Measurement Study–Integrated Survey of Agriculture and Kitagawa-Oaxaca-Blinder decomposition analysis in six countries in Africa south of the Sahara (Kilic, Winters, and Carletto 2015).

These results are summarized in World Bank and ONE Campaign (2014) and reproduced in Table 15.3. The value of total crop output per hectare is compared across plots managed by men and women. Analyses from six countries (Ethiopia, Malawi, Niger, Nigeria [analyzed separately for northern and southern Nigeria], Tanzania, and Uganda) find statistically significant gender gaps in productivity for all but northern Nigeria and Tanzania when simply comparing the differences in value of output. Controlling for plot size and geographic factors, the gender differences generally increase rather than decrease. In Niger the gender gap increases from 19 percent to 66 percent. While the gap loses statistical significance in southern Nigeria, a gender gap of 46 percent results in northern Nigeria. Doss (2017) points out that one reason for these dramatic differences is that women, on average, have smaller holdings than men. Given the inverse relationship typically found between farm size and productivity, since women have smaller farms, they would be expected to have higher productivity per unit of land than men. For Malawi, Niger, Nigeria (both northern and southern), and Uganda, after accounting for the differences in farm size, the gender gap widens. In these countries, among smallholder farms, there is an inverse relationship between farm size and productivity; smaller farmers are more productive per unit.

There are fewer estimates of productivity gains in Asia, where men and women farm most land jointly, and therefore sex-disaggregated data on plot managers are rarely collected. A recent exception is a study from Bangladesh using nationally representative data from a 2012 survey that collected information on men's and women's empowerment using the Women's Empowerment

TABLE 15.3 Gender gaps in agricultural productivity, by country (%)

Country	Simple difference	Difference after accounting for plot size and regions
Ethiopia	23***	24***
Malawi	25***	25***
Niger	19***	66***
Nigeria, northern	4	46***
Nigeria, southern	24*	17
Tanzania	6	23***
Uganda	13***	33***

Source: World Bank and ONE Campaign (2014).

Note: * = difference significant at $p < 0.10$; *** = difference significant at $p < 0.01$.

in Agriculture Index (Alkire et al. 2013). Seymour (2017) found that reduced gender disparities within households (measured in terms of the empowerment gap between spouses) are associated with higher levels of technical efficiency both on plots that women jointly manage with their spouses as well as on those for which women did not report any involvement in agricultural decision-making. Aside from potential gains in agricultural productivity and incomes, closing gender gaps has potential returns in terms of investment in the next generation's education and health as well as more representative decision-making (World Bank 2011). This does not include the intrinsic gains from women's empowerment and gender equality in their own right.

The existence of social norms and gender-specific constraints affecting women is well documented, and so are their consequences both globally as well as in the agricultural sector (World Bank 2011; FAO 2011; Quisumbing et al. 2014). While national legislative structures are now in place in many countries, efforts to transform gender norms often continue to meet with reluctance or resistance, especially when they involve changing deeply embedded norms that are tied up with what it means to be men or women in different cultures. This issue is exemplified in discussions about women's land rights, where even when they are codified in statutory law, they are often not recognized in practice (Meinzen-Dick et al. 2017). Yet many development projects, particularly those implemented by civil society organizations, explicitly aim to transform gender roles and empower women.

Consequences of Targeting Women Only

To redress years of agricultural development programming targeted to men, many development organizations and governments have targeted women exclusively. One widespread example is the targeting of microfinance programs to women, arguing that such programs contribute to their empowerment. Another example is the targeting of transfers to women, as in conditional cash-transfer programs or food aid, as well as the targeting of nutrition behavior change communication (BCC) programs to mothers. Although these programs cite the findings from early work on intrahousehold allocation demonstrating the high returns in terms of child health and nutrition from targeting transfers and credit to women, as well as the hypothesized positive relationship between women's empowerment and child nutrition, much of this evidence comes from observational studies.

For example, the systematic review of interventions by Yoong, Rabinovich, and Diepeveen (2012) found that economic transfers controlled by women tended to improve child human capital outcomes, but only 15 studies using experimental or quasi-experimental techniques met the inclusion criteria. Another recent systematic review finds mixed effects of microcredit on women's empowerment, with small effect sizes. Moreover, the authors are doubtful about the methodological quality of the evidence base (Duvendack, Palmer-Jones, and Vaessen 2014). These mixed results are partly attributable to the heterogeneity of microcredit interventions, contexts, and target groups (Duvendack, Palmer-Jones, and Vaessen 2014: 75), which in turn come from differences in the way that women's empowerment is conceptualized, both in the interventions and in impact evaluations.

The synthesis of findings from gender-sensitive impact evaluations of eight agricultural development projects (Johnson et al. 2016) is also instructive about the limitations of designing and implementing any program or project that targets women in isolation from their households, families, and communities. First, there is no guarantee that such a transfer, even if designed to be controlled by women, will remain under their control—or that households will not undertake compensatory behavior. Of eight projects included, four increased women's assets. Three of these four distributed assets directly to women and took steps to ensure that women maintained control of the transferred assets. The steps taken—putting a women's name on the land title, supporting women to reclaim assets that were taken from them, influencing gender norms about asset ownership—varied but all had the effect of supporting the initial asset transfer. Although the CARE dairy value chain project did not distribute assets, households that received the value chain intervention were able to increase their stock of jointly held nonagricultural productive assets because dairy incomes were reinvested in other types of assets, diversifying the asset portfolio (Quisumbing et al. 2013).

However, men in participant households also increased their assets through the projects, and often more than women did, leading to a net increase in men's asset holdings relative to women's. An evaluation of BRAC's Targeting the Ultra Poor asset transfer program found that, relative to the control group, while women's ownership of program assets grew, men's sole ownership of all other assets grew (Das et al. 2013; Roy et al. 2015). CARE reported significant increases in household-level assets (goats, agricultural productive assets, nonagricultural productive assets) relative to a control group of households in the same villages that did not participate in the dairy value chain program. These increases were driven largely by increases in assets owned by men, although the increase in men's assets is not significantly different relative to the control group. Only in a homestead food production program in Burkina Faso, implemented by Helen Keller International (HKI), was there any evidence of women closing the gender asset gap. Women's value of agricultural assets in intervention villages increased, whereas men's decreased (van den Bold et al. 2015). The project had no impact on the area of land cultivated by either men or women, although qualitative work indicates that gender norms became more favorable toward women's landownership in treatment as compared with control areas.

Taken together, these results show that although it is possible to increase women's control and ownership of assets, it is not easy or automatic, even in projects that transfer assets to women. Given the gender norms that govern asset ownership, to the extent that project benefits get reinvested in assets, those assets are likely to be controlled primarily or exclusively by men. Explicit steps appear to be necessary to increase the chance that women will maintain and accumulate assets, including efforts to influence the norms around the acceptability of women having control of assets, either individually or jointly with others in the household. Some common barriers that projects face include: resistance or uncooperative attitudes of local government officials or even program staff toward women asset owners; deep-seated cultural norms that view particular types of assets as men's assets, such that even if a program targets women, men still maintain control of those assets and make major decisions regarding them; and possible dilution of the net effects of the asset transfer as proceeds are reinvested in assets owned by men (Johnson et al. 2016). Recognizing joint ownership—not only in how assets are measured but also in how asset transfers are designed in projects that target households—could broaden the scope for change.

In addition, targeting agricultural programming to women may inadvertently increase their workload. A recent systematic review of the impact of agricultural interventions on time use (Johnston et al. 2018) found that (1) women play a key role in agriculture, which is reflected in their time commitments to these activities, whether as farmers or as farmworkers; (2) women are important actors in the uptake and response to agricultural interventions; and (3) agricultural interventions tend to increase women's, men's, and children's time burdens. Consistent with the findings of the systematic review, the evaluations of asset-transfer projects reported that the transfer of assets had impacts on women's time. All projects that transferred livestock—BRAC, Land O'Lakes, and HKI—found that caring for livestock, especially improved or exotic breeds that tend to have greater nutrition and health needs, led to an increase in demand for women's time (Quisumbing et al. 2015). These new livelihood strategies increased demand for time of other household members as well, including men and children. The Land O'Lakes study reported that although the greatest increase in time spent on dairy was for women's time, labor increased for all household members, and men provided the largest amount of total labor for dairy production (Johnson et al. 2015).

To understand the impact of these increased time burdens, we need to know what household members were doing with their time *before* the projects. The CARE evaluation looked at changes in time spent by household members in a range of activities and found that women in program households spent more time on dairy and less on childcare (feeding and general care) than control households in the same communities (Quisumbing et al. 2013). This could be a cause for concern because the time women spend on childcare is a determinant of child nutritional outcomes (Herforth and Harris 2014). Findings from the BRAC evaluation state that some women complain about workloads associated with program assets and said that other family members had to help with care of livestock, especially cows (Das et al. 2013). Nonetheless, these women prefer their current situation to the previous situation of not having livestock and working outside the home.

Gains to Recognizing Jointness and Addressing Both Men and Women

Early work to include women in agricultural development projects typically targeted women as independent producers, growing particular crops on their own plots of land. Yet both women and men are involved in agricultural production within the same household as well as being involved in consumption decisions. Failing to recognize jointness in decision-making and control of productive resources may neglect gains to cooperation and to involving men as well as women. For example, most agricultural programs target extension advice about agriculture to men, and nutrition messages, as relevant, to women. For example, a HarvestPlus project that disseminated biofortified orange sweet potato (OSP) vines to farmers' groups gave nutrition messages about vitamin A to women but not their husbands. In examining adoption decisions within households, Gilligan et al. (2020) found that plots of land exclusively controlled by women are not more likely to contain OSP, but plots under joint control of men and women, in which a woman has primary control over decision-making, are significantly more likely to contain OSP. Plots exclusively controlled by men are the least likely to contain OSP.

This evidence indicates that women play an important role, and often a leading role, in the decision to adopt OSP, but that this decision is often jointly made with their husbands. Because of the jointness of these decisions, the current strategy of targeting only women with nutritional training may be missing an opportunity to create an awareness of the benefits of OSP among men. The evaluation of the REU project found no evidence of impact of fathers' knowledge of child-feeding practices in Uganda (de Brauw et al. 2010), but the contribution of nutrition messages received by women on the impact of the project on OSP adoption and dietary intakes of vitamin A appears to be relatively small (de Brauw et al. 2012). Nonetheless, in this setting our results suggest that engaging with adult men and women household members might be the best strategy to promote adoption.

In a study of the adoption of maize technologies in Ghana, Doss and Morris (2001) found that there were no significant differences in technology adoption between men and women farmers living in male-headed households. Women living in female-headed households were less likely to adopt the technologies. This suggests that women living in male-headed households had access to information or other resources through their households that women in female-headed households were not able to access.

In a study of social networks and the adoption of agricultural technologies in India, Magnan et al. (2013) found that men and women in the same households have very distinct networks of agricultural contacts. Although women's networks are as large as men's or, in the case of poor households, substantially larger, women's connections are more likely to be with poorer households that are less likely to adopt new technology. In contrast, poor men with smaller agricultural networks tend to be connected to wealthier and more progressive farmers who are more likely to be early technology adopters—either because being wealthy or progressive has a direct positive influence on adoption or because these factors attract extension assistance. Because of their wider reach, public extension services and private service providers could use women's social networks, particularly among poor households, to facilitate inclusive technology dissemination.

Conclusion and Ideas for Future Research

Many policies and interventions designed to address intrahousehold relations and promote gender equality have been influenced by the intrahousehold literature, particularly that based on bargaining models. These policies and interventions have targeted women, seeking to strengthen their incomes or bargaining power. Although this is an improvement over policies that were gender-blind, and assumed that gains from the project would be shared equally within the family, this framing implies a zero-sum game and can lead to opposition or backlash from men. By contrast, framing the policies or interventions as seeking to strengthen household welfare for shared gains can gain greater support from men as well as women. Children's welfare is often one of the household "public goods" that can be highlighted for cooperation. For example, a homestead garden project of Helen Keller International in Burkina Faso emphasized improving maternal and child nutrition but contributed to changes in attitudes that favored women's access to and control over land (van den Bold et al. 2015).

Similarly, acknowledging that households have joint interests, and designing policies and programs to strengthen cooperation and collective action among household members, may open new areas for policies and interventions. This could lead to designing and implementing agricultural development projects that (1) involve men as well as women in nutrition education activities; (2) target agricultural extension to both men and women within the same household; (3) discuss gender norms within the family and community to address women's decision-making power in agriculture and nutrition, time burden, mobility, as well as household resource allocation, domestic violence, and intrahousehold respect and communication; and (4) involve other household members, such as elders, in project activities (Quisumbing et al. 2017).

Simply shifting from a model of the household in which everyone is an independent actor to one in which there is both cooperation and conflict is an important first step. Household members do come together, to produce both agricultural and nonagricultural goods and use some for home consumption while selling others. We need to understand which resources are shared, and what "shared" means, in different social and cultural contexts, including contexts with more complex family structures such as extended families and polygynous households. We also need to better understand the decision-making processes; knowing when we need information from both spouses (in a monogamous union) and from co-wives (in a polygynous union) and finding ways to categorize a range of decision processes that involve different levels of input and final control from various household members. While we should not assume that they always obtain Pareto efficient outcomes, neither should we assume that there is no joint production.

References

Acharya, D. R., J. S. Bell, P. Simkhada, E. R. Van Teijlingen, and P. R. Regmi. 2010. "Women's Autonomy in Household Decision-Making: A Demographic Study in Nepal." *Reproductive Health* (1): 1–12.

Adato, M., and J. Hoddinott, eds. 2010. *Conditional Cash Transfers in Latin America*. Baltimore: Johns Hopkins University Press for IFPRI.

Agarwal, B. 1994. *A Field of One's Own: Gender and Land Rights in South Asia*. Cambridge: Cambridge University Press.

Akresh, R., J. J. Chen, and C. T. Moore. 2011. "Productive Efficiency and the Scope for Cooperation in Polygynous Households." *American Journal of Agricultural Economics* 94 (2): 395–401.

Alkire, S., R. Meinzen-Dick, A. Peterman, A. Quisumbing, G. Seymour, and A. Vas. 2013. "The Women's Empowerment in Agriculture Index." *World Development* 52: 71–91.

Alwang, J., C. Larochelle, and V. Barrera. 2017. "Farm Decision Making and Gender: Results from a Randomized Experiment in Ecuador." *World Development* 92: 117–129.

Ambler, K. 2015. "Don't Tell on Me: Experimental Evidence of Asymmetric Information in Transnational Households." *Journal of Development Economics* 113: 52–69.

———. 2016. "Bargaining with Grandma: The Impact of the South African Pension on Household Decision-Making." *Journal of Human Resources* 51 (4): 900–932.

Ambler, K., C. Doss, C. Kieran, and S. Pasarelli. 2021, forthcoming. "He Says, She Says: Spousal Disagreement in Survey Measures of Bargaining Power." *Economic Development and Cultural Change*. DOI: 10.1086/703082.

Anderson, C. L., T. W. Reynolds, and M. K. Gugerty. 2017. "Husband and Wife Perspectives on Farm Household Decision-Making Authority and Evidence on Intra-Household Accord in Rural Tanzania." *World Development* 90: 169–183.

Ashraf, N. 2009. "Spousal Control and Intra-Household Decision Making: An Experimental Study in the Philippines." *American Economic Review* (99): 1245–1277.

Bardasi, E., K. Beegle, A. Dillon, and P. Serneels. 2011. "Do Labor Statistics Depend on How and to Whom the Questions Are Asked? Results from a Survey Experiment in Tanzania." *World Bank Economic Review* 25 (3): 418–447.

Barr, A., M. Dekker, W. Janssens, B. Kebede, and B. Kramer. 2019. "Cooperation in Polygynous Households." *American Economic Journal: Applied Economics* 11 (2): 266–283.

Becker, G. 1981. *A Treatise on the Family*. Cambridge, MA: Harvard University Press.

Chiappori, P. A. 1992. "Collective Labor Supply and Welfare." *Journal of Political Economy* 100: 437–467.

Cong, Z., and M. Silverstein. 2011. "Intergenerational Exchange Between Parents and Migrant and Nonmigrant Sons in Rural China." *Journal of Marriage and Family* 73 (1): 93–104.

Damon, A. L., and A. S. McCarthy. 2019. "Partnerships and Production: Agriculture and Polygyny in Tanzanian Households." *Agricultural Economics* 50 (5): 527–542.

Das, N., R. Yasmin, J. Ara, M. Kamruzzaman, P. Davis, J. A. Behrman, S. Roy, and A. R. Quisumbing. 2013. *How Do Intrahousehold Dynamics Change When Assets Are Transferred to Women? Evidence from BRAC's Challenging the Frontiers of Poverty Reduction—Targeting the Ultra Poor Program in Bangladesh*. IFPRI Discussion Paper 1317. Washington, DC: IFPRI.

Dasgupta, U., and S. Mani. 2013. "Altruism in the Household: A Pilot Study." *Economic and Political Weekly* 48 (3): 17–19.

de Brauw, A., P. Eozenou, D. O. Gilligan et al. 2010. *The Impact of the HarvestPlus Reaching End Users Orange-Fleshed Sweet Potato Project in Mozambique and Uganda*. June 30. Washington, DC: IFPRI.

de Brauw, A., P. Eozenou, D. O. Gilligan, N. Kumar, and J. V. Meenakshi. 2012. "Biofortification, Crop Adoption, and Health Information: Impact Pathways in Mozambique and Uganda." November. Mimeo. IFPRI, Washington, DC.

Deere, C. D., G. E. Alvarado, and J. Twyman. 2012. "Gender Inequality in Asset Ownership in Latin America: Female Owners vs Household Heads." *Development and Change* 43 (2): 505–530.

Deere, C. D., and M. Leon. 2003. "The Gender Asset Gap: Land in Latin America." *World Development* 31 (6): 925–947.

Donald, A., G. Koolwal, J. Annan, K. Falb, and M. Goldstein. 2020. "Measuring Women's Agency." *Feminist Economics* 26 (3): 200–226.

Doss, C. 2001. "Is Risk Fully Pooled within the Household? Evidence from Ghana." *Economic Development and Cultural Change* 50 (1).

———. 2006. "The Effects of Intrahousehold Property Ownership on Expenditure Patterns in Ghana." *Journal of African Economies* 15 (1).

———. 2017. "Women and Agricultural Productivity: Reframing the Issues." *Development Policy Review* (February): 1–16.

Doss, C., C. D. Deere, A. D. Oduro et al. 2011. *The Gender Asset and Wealth Gaps: Evidence from Ecuador, Ghana, and Karnataka, India*. Bangalore: Indian Institute of Management Bangalore.

Doss, C., C. D. Deere, A. D. Oduro, and H. Swaminathan. 2014. "The Gender Asset and Wealth Gaps." *Development* 57 (3–4).

Doss, C., C. Kovarik, A. Peterman, A. Quisumbing, and M. van den Bold. 2015. "Gender Inequalities in Ownership and Control of Land in Africa: Myth and Reality." *Agricultural Economics* (46).

Doss, C. , and R. Meinzen-Dick. 2015. "Collective Action within the Household: Insights from Natural Resource Management." *World Development* 74 (October): 171–183.

Doss, C., R. Meinzen-Dick, and A. Bomuhangi. 2013. "Who Owns the Land? Perspectives from Rural Ugandans and Implications for Large-Scale Land Acquisitions." *Feminist Economics* 20 (1): 76–100.

Doss, C., and M. Morris. 2001. "How Does Gender Affect the Adoption of Agricultural Innovations? The Case of Improved Maize Technology in Ghana." *Agricultural Economics* 25: 27–39.

Duvendack, M., R. Palmer-Jones, and J. Vaessen. 2014. "Metaanalysis of the Impact of Microcredit on Women's Control over Household Decisions: Methodological Issues and Substantive Findings." *Journal of Development Effectiveness* 6 (2): 73–96.

Dwyer, D., and J. Bruce, eds. 1988. *A Home Divided: Women and Income in the Third World*. Stanford, CA: Stanford University Press.

Fafchamps, M., and A. R. Quisumbing. 2008. "Household Formation and Marriage Markets in Rural Areas." In *Handbook of Development Economics*, edited by T. P. Schultz and J. A. Strauss, 3187–3247. Volume 4. Amsterdam: North-Holland.

FAO (Food and Agriculture Organization of the United Nations). 2011. *The State of Food and Agriculture 2010–11: Women in Agriculture—Closing the Gender Gap for Development*. Rome.

Fisher, M., J. J. Reimer, and E. R. Carr. 2010. "Who Should Be Interviewed in Surveys of Household Income?" *World Development* 38 (7): 966–973.

Fletschner, D., and L. Kenney. 2014. "Rural Women's Access to Financial Services: Credit, Savings, and Insurance." In *Gender in Agriculture: Closing the Knowledge Gap*, edited by A. R. Quisumbing, R. Meinzen-Dick, T. Raney, A. Croppenstedt, J. A. Behrman, and A. Peterman, 187–208. Rome: FAO and Springer Science +Business Media, B.V.

Fletschner, D., and D. Mesbah. 2011. "Gender Disparity in Access to Information: Do Spouses Share What They Know?" *World Development* 39 (8): 1422–1433.

Gilligan, D. O., N. Kumar, S. McNiven, J. V. Meenakshi, and A. R. Quisumbing. 2020. "Bargaining Power, Decision Making, and Biofortification: The Role of Gender in Adoption of Orange Sweet Potato in Uganda." *Food Policy*. https://doi.org/10.1016/j.foodpol.2020.101909.

Goldstein, M., and C. Udry. 2008. "The Profits of Power: Land Rights and Agricultural Investment in Ghana." *Journal of Political Economy* 116 (6): 981–1022.

Haddad, L., J. Hoddinott, and H. Alderman. 1997. *Intrahousehold Resource Allocation in Developing Countries: Methods, Models, and Policy*. Baltimore: Johns Hopkins University Press.

Heath, R., M. Hidrobo, and S. Roy. 2020. "Cash Transfers, Polygamy, and Intimate Partner Violence: Experimental Evidence from Mali." *Journal of Development Economics* 143.

Herforth, A., and J. Harris. 2014. *Understanding and Applying Primary Pathways and Principles*. Brief 1. Improving Nutrition through Agriculture. Technical brief series. Arlington, VA: USAID/SPRING (Strengthening Partnerships, Results, and Innovations in Nutrition Globally).

Hoddinott, J., and L. Haddad. 1995. "Does Female Income Share Influence Household Expenditures? Evidence from Cote D'Ivoire." *Oxford Bulletin of Economics and Statistics* 57: 77–95.

Iversen, V., C. Jackson, B. Kebede, A. Munro, and A. Verschoor. 2010. "Do Spouses Realise Cooperative Gains? Experimental Evidence from Rural Uganda." *World Development* 39 (4): 569–578.

Jackson, C. 2003. "Gender Analysis of Land: Beyond Land Rights for Women?" *Journal of Agrarian Change* 3 (4): 453–480.

Jacobs, K., and A. Kes. 2015. "The Ambiguity of Joint Asset Ownership: Cautionary Tales from Uganda and South Africa." *Feminist Economics* 21 (3): 23–55.

Johnson, N., J. Njuki, E. Waithanji, M. Nhambeto, M. Rogers, and E. Hutchinson Kruger. 2015. "The Gendered Impacts of Agricultural Asset Transfer Projects: Lessons from the Manica Smallholder Dairy Development Program." *Gender Technology and Development* 19 (2): 145–180.

Johnson, N. L., C. Kovarik, R. Meinzen-Dick, J. Njuki, and A. R. Quisumbing. 2016. "Gender, Assets, and Agricultural Development: Lessons from Eight Projects." *World Development* 83: 295–311.

Johnston, D., S. Stevano, H. J. Malapit, E. Hull, and S. Kadiyala. 2018. "Review: Time Use as an Explanation for the Agri-Nutrition Disconnect? Evidence from Rural Areas in Low and Middle Income Countries." *Food Policy* (76): 8–18.

Jones, C. 1983. "The Mobilization of Women's Labor for Cash Crop Production: A Game Theoretic Approach." *American Journal of Agricultural Economics* (May): 1049–1054.

Josephson, A. 2016. "Share and Share Alike: The Impact of Rainfall on Gendered Income Allocation in Malawi." Working paper, Department of Agricultural Economics, Purdue University.

Kazianga, H., and Z. Wahhaj. 2013. "Gender, Social Norms, and Household Production in Burkina Faso." *Economic Development and Cultural Change* 61 (3): 539–576.

Kieran, C., K. Sproule, C. Doss, A. Quisumbing, and S. M. Kim. 2015. "Examining Gender Inequalities in Land Rights Indicators in Asia." *Agricultural Economics* 46 (supplement): 119–138.

Kieran, C., K. Sproule, A. Quisumbing, and C. Doss. 2017. "Gender Gaps in Landownership across and within Households in Four Asian Countries." *Land Economics* 93 (2): 342–370.

Kilic, T., and H. Moylan. 2016. *Methodological Experiment on Measuring Asset Ownership from a Gender Perspective (Mexa) Technical Report.* World Bank, Washington, DC.

Kilic, T., P. Winters, and C. Carletto. 2015. "Gender and Agriculture in Sub-Saharan Africa: Introduction to the Special Issue." *Agricultural Economics* 46: 281–284.

Lastarria-Cornhiel, S., J. A. Behrman, R. Meinzen-Dick, and A. R. Quisumbing. 2014. "Gender Equity and Land: Toward Secure and Effective Access for Rural Women." In *Gender in Agriculture: Closing the Knowledge Gap*, edited by A. R. Quisumbing, R. Meinzen-Dick, T. Raney, A. Croppenstedt, J. A. Behrman, and A. Peterman, 117–144. Rome: FAO and Springer Science +Business Media, B.V.

Magnan, N., D. J. Spielman, K. Gulati, and T. J. Lybbert. 2013. *Gender Dimensions of Social Networks and Technology Adoption: Evidence from a Field Experiment in Uttar Pradesh, India.* IFPRI Discussion Paper. Washington, DC: IFPRI.

McPeak, J. G., and C. R. Doss. 2006. "Are Household Production Decisions Cooperative? Evidence on Pastoral Migration and Milk Sales from Northern Kenya." *American Journal of Agricultural Economics* 88 (August): 525–541.

Meinzen-Dick, R., A. Quisumbing, C. Doss, and S. Theis. 2019. "Women's Land Rights As a Pathway to Poverty Reduction: Framework and Review of New Evidence." *Agricultural Systems* 172: 72–82.

Peterman, A., J. A. Behrman, and A. R. Quisumbing. 2014. "A Review of Empirical Evidence on Gender Differences in Nonland Agricultural Inputs, Technology, and Services in Developing Countries." In *Gender in Agriculture: Closing the Knowledge Gap*, edited by in A. R. Quisumbing, R. Meinzen-Dick, T. Raney, A. Croppenstedt, J. A. Behrman, and A. Peterman, 145–186. Rome: FAO and Springer Science +Business Media, B.V.

Quisumbing, A. R., ed. 2003. *Household Decisions, Gender, and Development: A Synthesis of Recent Research*. Washington, DC: IFPRI.

Quisumbing, A. R., N. Kumar, and J. A. Behrman. 2018. "Do Shocks Affect Men's and Women's Assets Differently? Evidence from Bangladesh and Uganda." *Development Policy Review* 36 (1): 3–34.

Quisumbing, A. R., and J. A. Maluccio. 2003. "Resources at Marriage and Intrahousehold Allocation: Evidence from Bangladesh, Ethiopia, Indonesia, and South Africa." *Oxford Bulletin of Economics and Statistics* 65 (3): 283–328.

Quisumbing, A. R., R. Meinzen-Dick, T. Raney, A. Croppenstedt, J. A. Behrman, and A. Peterman, eds. 2014. *Gender in Agriculture: Closing the Knowledge Gap*. Rome: FAO and Springer Science +Business Media, B.V.

Quisumbing, A. R., S. Roy, J. Njuki, K. Tanvin, and E. Waithanji. 2013. "Can Dairy Value-Chain Projects Change Gender Norms in Rural Bangladesh?: Impacts on Assets, Gender Norms, and Time Use." IFPRI Discussion Paper 1311. Washington, DC: IFPRI.

Quisumbing, A. R., D. Rubin, C. Manfre et al. 2015. "Gender, Assets, and Market-Oriented Agriculture: Learning from High-Value Crop and Livestock Projects in Africa and Asia." *Agriculture and Human Values* 32 (4): 705–725.

Quisumbing, A. R., K. Sproule, E. Martinez, and H. Malapit. 2017. *Gender, Women's Empowerment, and Nutrition: A Review, New Evidence, and Guidelines for Nutrition-Sensitive Agricultural Programming*. Report submitted to FAO, December 2017.

Ragasa, C. 2014. "Improving Gender Responsiveness of Agricultural Extension." 2014. In *Gender in Agriculture: Closing the Knowledge Gap*, edited by A. R. Quisumbing, R. Meinzen-Dick, T. Raney, A. Croppenstedt, J. A. Behrman, and A. Peterman, 411–430. Rome: FAO and Springer Science +Business Media, B.V.

Roy, S., J. Ara, N. Das, and A. R. Quisumbing. 2015. "'Flypaper Effects' in Transfers Targeted to Women: Evidence from BRAC's Targeting the Ultra-Poor Program in Bangladesh." *Journal of Development Economics* 117: 1–19.

Schultz, T. P. 2003. "Women's Role in the Agricultural Household: Bargaining and Human Capital Investments." In *Handbook of Agricultural Economics*, edited by B. L. Gardner and G. C. Rausser, volume 1A: *Agricultural Production*. Amsterdam: North-Holland.

Seymour, G. 2017. "Women's Empowerment in Agriculture: Implications for Technical Efficiency in Rural Bangladesh." *Agricultural Economics* 48 (4): 513–522.

Singh, I., L. Squire, and J. Strauss, eds. 1986. *Agricultural Household Models: Extensions, Applications, and Policy*. Baltimore: Johns Hopkins University Press.

Twyman, J., P. Useche, and C. D. Deere. 2015. "Gendered Perceptions of Land Ownership and Agricultural Decision-Making in Ecuador: Who Are the Farm Managers?" *Land Economics* 91 (3): 479–500.

Udry, C. 1996. "Gender, Agricultural Production and the Theory of the Household." *Journal of Political Economy* (104): 1010–1046.

van den Bold, M., A. Dillon, D. Olney, M. Ouedraogo, A. Pedehombga, and A. Quisumbing. 2015. "Can Integrated Agriculture-Nutrition Programs Change Gender Norms on Land and Asset Ownership? Evidence from Burkina Faso." *Journal of Development Studies* 51 (9): 1155–1174.

von Braun, J., and E. Kennedy, eds. 1994. *Agricultural Commercialization, Economic Development, and Nutrition*. Baltimore: Johns Hopkins University Press.

World Bank. 2011. *World Development Report 2012: Gender Equality and Development*. Washington, DC.

World Bank and ONE Campaign. 2014. *Leveling the Field: Improving Opportunities for Women Farmers in Africa*. Washington, DC: World Bank.

Yoong, J., L. Rabinovich, and S. Diepeveen. 2012. "The Impact of Economic Resource Transfers to Women versus Men: A Systematic Review." London: EPPI-Centre, Social Science Research Unit, Institute of Education, University of London.

Chapter 16

CREDIT FOR AGRICULTURAL DEVELOPMENT

Shahidur R. Khandker

Access to financial services is critical for agricultural development. By "access to financial services" we mean access to credit, savings, payments, and insurance. Such a broad measure of access is called "financial inclusion" in the recent development literature. Financial inclusion is also defined as a measure of an individual's and a business's access to use of financial services to save, borrow, make payments, and buy insurance to mitigate risk in production and consumption (Demirguc-Kunt and Klapper 2012; Demirguc-Kunt et al. 2015).[1] Access to financial services facilitates an individual's and a business's day-to-day transactions and helps them manage everything from investment plans to unexpected emergencies. "Financial access" means access to outlets where people save, borrow, pay bills, and buy insurance, in order to initiate and expand businesses, to invest in education or health, to manage risk, and to weather shocks. All these activities induced by financial inclusion lead to higher productivity, raising income, consumption, nutrition, and education, and thus achieving multiple Sustainable Development Goals (World Bank 2018).

For good reasons policymakers thus advocate financial inclusion as a vital factor for development. The Universal Financial Access (UFA) Initiative of the World Bank stresses the need for every individual and business to have at least a transaction account with a financial institution (Demirguc-Kunt and Klapper 2012; World Bank 2018).[2] A financial institution may be a bank, a microfinance institution, or a mobile financial service. A transaction account

1 "Financial inclusion" means that people have access to useful and affordable financial products and services that meet their day-to-day needs—transactions, payments, savings, credit, and insurance—delivered in a "responsible and sustainable way" (World Bank 2018).

2 As per the Global Financial Inclusion (coined as Global Findex) Survey (World Bank 2015c), "financial account" is defined as a transaction account at a bank or another type of financial institution, such as a credit union, cooperative, or microfinance institution. "Financial account" also includes a mobile money account that is limited to services that can be used without an account at a financial institution. In the Findex data the mobile money account is defined by whether the individual had used a mobile phone to pay bills or to send or receive money (Demirguc-Kunt et al. 2015).

with any such financial institution allows people to save money as well as to send and receive payments; thus, having such a transaction account is considered a gateway to accessing other financial services such as credit and insurance. This is the underlying philosophy of the World Bank Group's Universal Financial Access 2020 initiative emphasizing every adult must have a transaction account with a financial institution. Using the Global Findex data collected by the World Bank to measure financial inclusion, it is shown that 62 percent of adults worldwide reported having an account with a financial system in 2014, compared with 51 percent found in 2011 (Demirguc-Kunt et al. 2015). Studies also find that there is almost universal financial inclusion in OECD countries (94 percent) but limited coverage in the developing countries (54 percent), varying by level of economic and financial development.[3]

Like any sector, access to financial services is a necessity in agriculture. More important, smallholders who cultivate more than 80 percent of land in developing countries deserve to have equal access to different types of financial services. Farmers' access to financial services helps ensure better use of resources that enhances efficiency in the use of inputs such as fertilizer, pesticides, and irrigation in production. Access to financial services also helps purchase machinery and equipment to support high-value production and crop diversification. It also helps marketing and transportation of agricultural products (Koolwal and Khandker 2018; World Bank 2015a). Access to financial services thus is critical for transforming traditional subsistence into commercialized agriculture (meaning a higher degree of market-based production).

In developing countries, agriculture is a major source of livelihood for more than 80 percent of rural people (IFC 2013). Yet access to finance (that is, credit), especially for investment in agriculture, is very limited. In Africa less than 1 percent of commercial lending goes to the agriculture sector (IFC 2013). However, unlike other sectors, financial institutions are reluctant to extend financial services in agriculture for a variety of reasons such as production risk (due to drought and excessive rainfall), lack of physical collateral (lack of title to the cultivated land), and volatility of prices due to worldwide climate changes (World Bank 2015a).

Demand for financial services in agriculture varies by degree of commercialization, size of farm holding, and level of rural development including infrastructural development. Different financial instruments (such as savings,

3 The 2017 round of the Global Findex survey shows that 69 percent of adults were found to have an account with a financial institution with 94 percent in OECD countries and 63 percent in the developing world (Demirguc-Kunt et al. 2018). This means a 7 percent increase between 2014 and 2017.

credit, payments, and insurance) respond to different needs of actors engaged in agriculture, but the extent of financial inclusion or financial instruments available to farmers vary widely by the level of financial development, critical for agricultural development (for example, Ruete 2015). Agricultural development critically depends on financial development. However, reliable and appropriate financial instruments are only developed through financial innovations supported by government and international development agencies. A growing body of literature sheds light on the interdependence between agricultural development and financial development that is both inclusive and smallholders' friendly (for example, IFC 2013; World Bank 2015a).

Although an effective access to financial services demands access to efficient outlets for savings, borrowing, payments, and insurance, this chapter does not address farmers' needs of all categories of financial services and instruments and provisions available via a financial system. This chapter focuses only on access to credit with occasional references to other services such as savings and payments and the potential roles of governments in enhancing farmers' access to finance. Part of the fact is that access to finance is the most critical factor for agricultural development. Chapter 17 focuses on insurance to mitigate risk in agricultural production as a way to promote agricultural development.

This chapter is organized as follows. The first section discusses why agricultural credit is necessary for agricultural development. The current practices of agricultural finance with distinction between private and public sources of agricultural finance are outlined. The role of institutional finance in agricultural GDP of the developing world is examined. The chapter then highlights differential access to financial services relative to agricultural credit in selected developing countries. Successful agricultural finance practices in the developing world are discussed, to draw the conclusion that practices vary by countries even within the developing world. A review of impact studies showing how better access to financial services such as credit matters for raising agricultural productivity and hence, agricultural development, is presented. The conclusion discusses the challenges facing agricultural finance in meeting the needs of transforming traditional agriculture. It also presents a set of recommendations toward promoting agricultural development.

Why Credit

Agriculture is at a crossroads in the developing world. On the one hand, farm productivity has to be increased to support sustainable economic growth. On the other hand, food production has to be increased to feed a growing population, especially with a higher demand for high-valued diets such as meats, fish, and vegetables (Chapters 9–12 in this volume; FAO 2017). The global demand for food is expected to increase by 70 percent by 2050, and at least $80 billion annual investment is necessary to meet this food demand, almost double the current investment (World Bank 2015a). But agriculture is increasingly subject to volatility of climate changes (FAO 2016). Uncertainty of harvests, input availability, and income flows have created financial constraints on agricultural producers for raising investment in agriculture.

Ironically, low-income countries, which face high climate volatility that challenges agriculture and deepens poverty (Ahmed, Diffenbaugh, and Hertel 2009), depend the most on agriculture with the lowest productivity (Figure 16.1). That is, agricultural labor productivity is inversely related to a country's agricultural share of GDP. Such inverse relationship between productivity and agriculture share in GDP is prominent in the countries of Africa south of the Sahara (SSA) and South Asia (SA). But this is not the case in economies with greater agricultural diversification and higher private investment across Europe and Central Asia (ECA), East Asia and Pacific (EAP), and Latin America and the Caribbean (LAC). As 80 percent of the farmland in Africa south of the Sahara and Asia is managed by smallholders, the burden of raising food production and farm productivity for supporting sustainable growth via raising investment rests squarely on smallholders (FAO 2012).[4]

The investment decisions of smallholders are nonetheless conditioned by severe financial and agroclimate environment. Farmers have been described as constrained in credit markets, often characterized by asymmetry information, adverse selection, and moral hazard of lending (Stiglitz and Weiss 1981; Guirkinger and Boucher 2008; Kumar, Turvey, and Kropp 2013).[5] Financial institutions are reluctant to lend to agriculture, especially smallholders in agriculture, because of high default costs associated with high agricultural risk (due to drought, excessive rainfall, and pesticides) and high transaction costs of covering large geographical areas with poor infrastructures. For all such reasons, institutional credit is rationed; consequently, lack of credit becomes a major hurdle for undertaking investment in agriculture, let alone buying modern inputs and supporting crop diversification, key to producing high-value

4 While the interpretation of "smallholder" can vary across countries, FAO (2012) defines "smallholder" as working on up to 10 hectares.

5 In fact, studies from China and India show that binding credit constraints negatively affect food consumption and farm input applications, as well as health and educational outcomes (Kumar, Turvey, and Kropp 2013).

FIGURE 16.1 **Agricultural productivity and agricultural share of GDP, 2018**

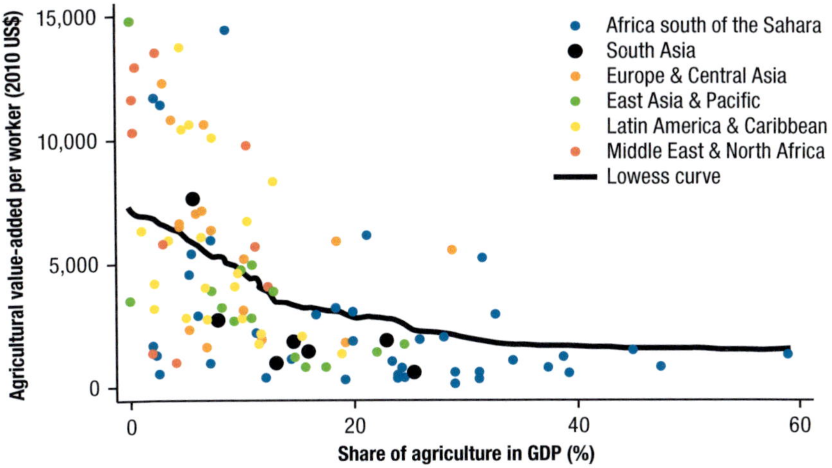

Source: Data obtained from World Bank (2019).

Note: Agriculture value-added per worker is a measure of agricultural productivity. Value-added in agriculture measures the output of the agricultural sector (ISIC divisions 1-5) less the value of intermediate inputs. Data are in constant 2010 US dollars. Regional labels and colors are used for different countries. Locally weighted regressions (bandwidth = 0.8) presented along with scatterplots. Agriculture includes crop, livestock, and fishery, but the WDI data includes forestry as well from 2017 on.

cash and food crops (for example, Feder et al. 1990). Recent empirical studies in developing countries confirm that credit constraints discourage farmers from borrowing and making investments, or encourage them to finance some investments purchases with cash from nonfarm activities and crop sales (for example, Adjognon, Liverpool-Tasie, and Reardon 2017).

Deteriorating weather and climate conditions make things only harder—farmers find the income cushion insufficient for confronting agricultural risk. Rather than investing in technology to confront production uncertainty, farmers in many rural settings are increasingly engaged in nonfarm activities to smooth income and consumption (for example, Pitt and Khandker 2002; Khandker and Koolwal 2016). But rural nonfarm activities are not independent of agricultural risk. In fact, the interlinkage between farm and nonfarm production underscores the importance of raising agricultural productivity (for example, McCullough 2017; ILO 2014).

Transforming subsistence agriculture to modern agriculture needs investment in agriculture. Investment means better access to agricultural finance (both short- and long-term), which can finance investment in modern technology by smallholders. With improved access to short-term credit, for instance, farmers can smooth income and consumption risks, raise productivity by purchasing inputs such as high-yielding varieties and modern fertilizers, and afford improved fodder for livestock and poultry. With access to long-term credit, on the other hand, farmers can purchase or lease production machinery such as tractors or invest in improved planting and irrigation technologies for crop diversification such as high value cash crop production. Improved access to credit therefore affects not only optimal input use and improved technology adoption by releasing credit constraints but also helps farmers cope with income shocks in an uncertain agroclimate environment. This in turn helps raise farm productivity and supports a country's sustainable growth agenda.

Private versus Public Sources of Agricultural Credit

Despite efforts of governments and multilateral agencies such as the World Bank to enhance access to credit and other financial services in developing countries, about 2 billion (some 38 percent of working-age adults) worldwide have no access to institutional finance and 73 percent of poor people are unbanked (World Bank 2018a). Many of the unbanked people are smallholder farmers and landless workers.

Farmers have very limited access to institutional credit for a number of reasons. Agricultural financial markets have historically been highly regulated in developing countries (for example, subsidizing interest and input prices). Government concessional lending with below-market interest rates and loan forgiveness policies create disincentives for lending of commercial banks and the government-directed agricultural banks (for example, Adams 1988; World Bank 2015a; Seibal 2000). Financial institutions, while successfully mobilizing savings from rural areas, are reluctant to lend in agriculture for a variety of reasons such as risk in production inherent in agriculture. They restrict lending to agriculture for uncertain market price fluctuations due to agroclimate shocks. Lack of sufficient collateral of farmers is another factor underlying limited lending to agriculture. These barriers to institutional lending are compounded by a limited network of banks, making it difficult and costly for them to reach out to farmers. Banks are also not fully equipped to deal with risk and uncertainly in agriculture.

Smallholders are often not able to afford loans at terms and conditions not suitable for borrowing. One limiting factor is covariate risk, meaning borrowing and repayment do not move in the same direction—that is, farmers would like to borrow during planting season when institutions are short of loanable funds and pay back during harvest period (when surplus of funds available are not demanded). The standard practice of fixed repayment schedules of commercial lending is thus not always compatible with the seasonality of crop production. Farmers interested in making investment in agriculture also need long-term credit, which is not always available from financial institutions such as credit unions and microfinance institutions (MFIs).

The limited access to formal finance (including banks, credit unions, MFIs, and mobile accounts) forces many farmers to rely on own/informal finance at high cost. As per the Findex data of 2014, for example, among 58 percent of individuals (who received payments from agriculture) who borrowed, only 11 percent of them borrowed from institutional sources (Table 16.1).[6] Access to formal credit varies by region—for example, some 20 percent in EAP, compared with only 7 percent in SSA. The bulk of the borrowing is reported for consumption smoothing and other purposes such as health and education—an average of 16 percent reported borrowing for farm and business purposes in the developing countries, compared with 19.8 percent in SSA and 16.7 percent in EAP. This suggests that less of borrowing goes to support agricultural production.[7] This of course does not mean borrowing for consumption smoothing or supporting health expenses is unproductive, especially for those food-insecure poor farmers in developing countries whose main production factor is labor. In fact, for enhancing and securing their labor input in farming, the poor farmers use borrowing funds, in times of shortages, for consumption to sustain their labor capacity (for example, Deaton 1997; Bocher, Alemu, and Kelbore 2017; Joshi, Roy, and Sonkar 2017; Pitt and Khandker 2002). The regional differences in farmers' use of borrowing for different purposes suggests the degree of direct use of funds for farming.

Because of limited access to formal finance, farmers in developing countries rely more on informal finance for different purposes. For example, among 58 percent of farmers who borrowed in the developing world, 47 percent borrowed from informal sources (Table 16.1). Informal finance has some advantages—it has lower transaction costs and is readily available on a

6 Farmers are defined as those who sell some or all of their output in the market, who comprise the middle of the landholding distribution. Subsistence farmers, which consume most of what they produce, are not included. Sampling weights are used to make the averages representiative at the country level.

7 This does not mean that all farmers need to borrow. In fact, there are many farmers who save and therefore need a safe outlet for savings on a regular basis (for example, Zeller and Sharma 2000). But the point is that like other sectors, better acess to institutional finance for both short- and long-term credit is essential for supporting agricultural development with diffusion of modern technology and crop diversification.

TABLE 16.1 Access to institutional finance by region, 2014

	Received agricultural payments				Did not receive agricultural payments			
Region	Has an account (%)	Borrowed in the past year* (%)	Borrowed for farm/ business (%)	Obser-vations	Has an account (%)	Borrowed in the past year* (%)	Borrowed for farm/ business (%)	Obser-vations
EAP	58.8	53.6 (19.6)	16.7	3,851	64.5	42.8 (14.3)	6.7	9,353
ECA	40.8	51.2 (14.9)	8.8	3,889	54.1	36.7 (12.5)	2.4	22,153
LAC	47.3	52.2 (19.2)	18.5	1,357	45.1	36.6 (12.6)	6.3	16,679
MENA	30.4	54.8 (6.6)	9.2	702	25.2	40.3 (6.8)	4.1	7,363
SA	41.6	50.3 (10.5)	13.0	2,375	40.3	38.8 (6.9)	5.3	6,757
SSA	29.3	65.0 (6.9)	19.8	11,749	30.3	45.7 (5.5)	7.4	22,295
Developing world	38.3	58.4 (11.4)	16.4	23,923	43.7	40.2 (10.0)	5.4	84,600

Source: World Bank (2015c).

Note: * Figures in parentheses show percentage of borrowing from formal sources.

1. Having an account indicates either institutional financial account, or mobile money account, or both.

2. EAP = East Asia and Pacific, ECA = Europe and Central Asia, LAC = Latin America and Caribbean, MENA = Middle East and North Africa, SA = South Asia, SSA = Africa south of the Sahara. Developing world excludes OECD countries.

3. Sampling weights used. About 21.81 percent of individuals affirmatively responded to the question "if received agricultural payments in last 12 months."

short-term notice. But informal financing often at short-terms is insufficient to enable farmers to access modern technologies, fertilizer, irrigation, and other inputs (World Bank 2015a). Informal finance also has limits in leveraging resources or pooling risks, as informal finance is provided by village-level lenders who are normally associated with the same covariate risk that affects farmers. These limits are costly to small farmers in diversifying agriculture, undertaking investments, and assuming risks for growing high-valued crops, meats, and vegetables. Agricultural-led growth needs long-term investment in farming and agribusinesses for marketing high-valued crops and farm products. The resources for such investments would come only from financial institutions such as banks.

But commercial banks lend a small share of portfolio to agriculture compared with agriculture's share of GDP. In 2015 agriculture accounted for 27.5 percent of GDP and received only 5.1 percent of commercial lending in SSA (Figure 16.2). Consequently, agricultural lending of commercial banks accounted for 7 percent of agricultural GDP in 2000 and 9 percent in 2015 in SSA (Figure 16.3). In contrast, in EAP with more developed agriculture, agricultural credit of commercial banks accounted for 13 percent in 2000 and 47 percent of agricultural GDP in 2015.

FIGURE 16.2 Agricultural credit and agricultural GDP share, 2000 and 2015

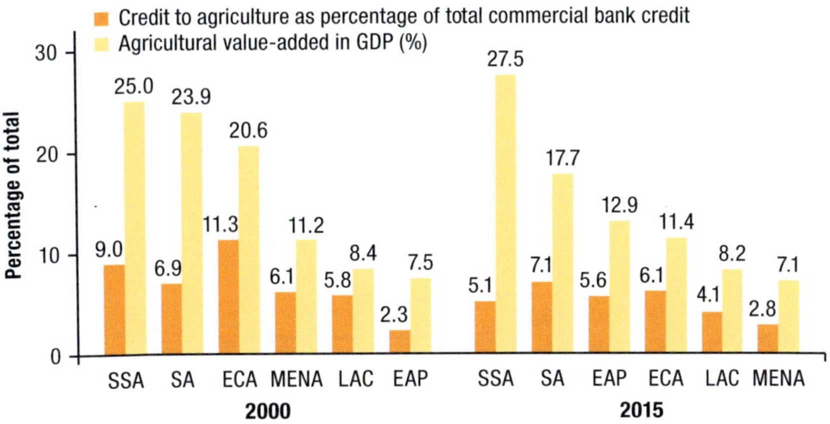

Source: Data from World Bank (2019).

Note: Sample size 127. EAP = East Asia and Pacific, ECA = Europe and Central Asia, LAC = Latin America and Caribbean, MENA = Middle East and North Africa, SA = South Asia, SSA = Africa south of the Sahara.

FIGURE 16.3 Agricultural credit as share of agricultural GDP, 2000 and 2015

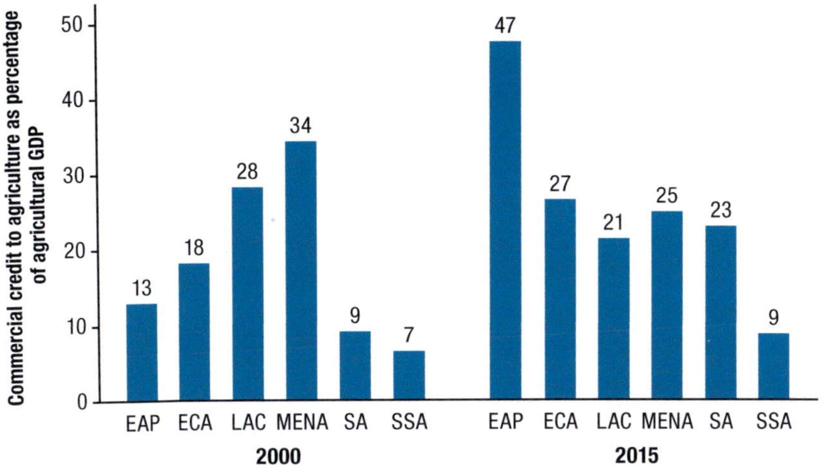

Source: Data from World Bank (2019).

Note: Sample size 117. EAP = East Asia and Pacific, ECA = Europe and Central Asia, LAC = Latin America and Caribbean, MENA = Middle East and North Africa, SA = South Asia, SSA = Africa south of the Sahara.

FIGURE 16.4 Agricultural GDP and commercial bank branches, 2018

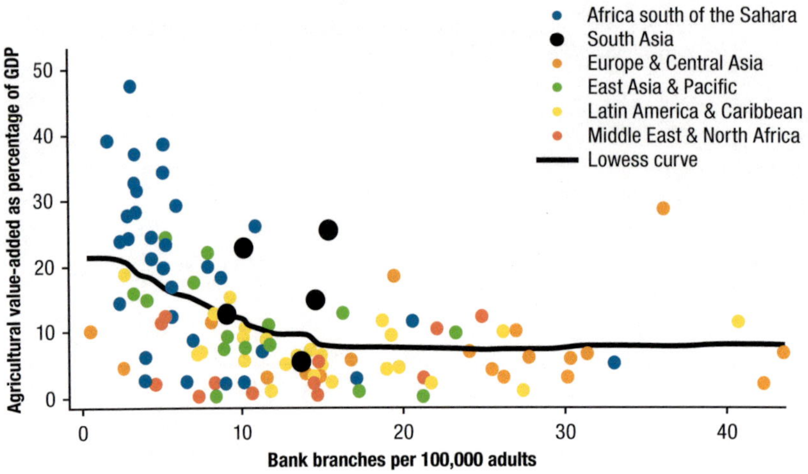

Source: World Bank (2019).

Commercial banks played a bigger role in agricultural development of EAP relative to SSA and SA regions. Limited commercial lending to agriculture is a direct outcome of an inverse relationship between the extent of per capita commercial bank branches and share of agricultural GDP in a country's total GDP (Figure 16.4). We see fewer commercial bank branches per capita with high share of agriculture GDP in total GDP. Of course, commercial banks are not the only provider of institutional credit. There are other sources of institutional credit such as MFIs, agricultural development banks, agricultural cooperatives, and registered credit unions and/or rotating societies for credit and savings (RoSCAS). Yet all these institutions together do not account for more than 11 percent of farm-sector lending.[8]

However, mobile financial services (MFS) are increasingly providing financial services such as payments and transfer of remittance. But MFS hardly extends credit and therefore has not much impact on agricultural investment. Consequently, the role of financial institutions (banks, microfinance institutions, and mobile networks) in agriculture is limited, especially in SSA and SA, where the need for institutional credit is the highest for promoting agricultural investment and productivity.

---

8 See discussion below with reference to Table 16.1.

Current State of Agricultural Finance in Developing Countries

Borrowing is only one measure of access to financial services. As indicated earlier, a broad measure of financial access is defined by financial inclusion, which is defined as a measure of an individual's and a business's access to and use of financial services to save, borrow, make payments, and manage risk in production and consumption (Demirguc-Kunt and Klapper 2012; Demirguc-Kunt et al. 2015). Using the Global Findex data of 2014, we can present the degree of financial inclusion among farmers and nonfarmers in both developed and developing countries.[9]

Financial inclusion among farmers (those who received income from agriculture over the past 12 months) in the developing world (excluding OECD countries) was 38.3 percent, meaning only 38 percent of farmers had an account with a financial system in 2014 (see Table 16.1). Interestingly, among the 38 percent of those who have a financial account, only 11 percent borrowed from institutional sources. This means some 27 percent of farmers in the developing world use financial services for other purposes such as savings, payments, and remittances. Financial inclusion in agriculture varies by region—it is highest among farmers in EAP (58.8 percent) and lowest in SSA (29.3 percent). Borrowing from financial institutions is the highest in EAP and LAC regions and the lowest in MENA and SSA regions. Such variations are due to varying financial practices and policies. Consider financial inclusion in a few selected countries to understand why financial inclusion matters for agricultural lending for investment. Among the selected seven countries, Thailand from EAP perhaps has the best practices of financial services—some 87 percent of Thai farmers have an account with a financial institution, which is close to the average of OECD countries (94 percent), higher than the EAP average (58.8 percent), and even higher than the average of the developing world (38.3 percent) (Table 16.2).

In contrast, Viet Nam has only 28.9 percent financial inclusion among farmers, much lower than the EAP average but close to the average of the SSA region (29.3 percent). In SA, India has 60.6 percent coverage, higher than the SA average of 41.6 percent, while Bangladesh's coverage of 31.4 percent

9 In its 2014 round the Global Financial Inclusion Survey (also called Global Findex survey) asked individuals whether they received any payments for the sale of agricultural products. "Farmers" are defined as those who received payments from the sale of agricultural products. Consequently, the definition of being a farmer is restricted to those involved in market-based agriculture rather than in agriculture purely for own consumption (Koolwal and Khandker 2018).

TABLE 16.2 Access to institutional finance in seven developing countries (%), 2014

	Farmers[a]					Nonfarmers				
Country	Has account	Borrowed	From formal institutions	Borrowed for ag/bus[b]	Observations	Has account	Borrowed	From formal institutions	Borrowed for ag/bus	Observations
Bangladesh	31.4	56.4	8.4	3.8	222	30.9	45.8	10.3	4.0	778
Ethiopia	22.8	49.0	9.3	17.1	366	21.1	39.9	6.1	7.8	638
India	60.6	64.8	13.6	20.1	607	51.1	41.3	5.3	7.0	2,393
Peru	28.8	37.3	15.5	12.6	102	29.0	26.3	11.6	6.5	898
Thailand	87.4	55.5	21.6	17.7	356	72.7	47.3	11.8	9.9	644
Uganda	48.9	87.8	18.7	27.5	680	34.7	59.5	9.4	11.2	320
Viet Nam	28.9	53.3	24.8	13.3	273	31.8	44.3	16.1	4.8	727
Total	48.6	64.0	16.1	18.8	2,606	41.8	41.4	9.0	6.9	6,398

Source: Data from World Bank (2015c).

Note: a. "Farmers" are defined as those who received agricultural payments in the past 12 months (weighted sample).

b. "Borrow for ag/bus" refers to the proportion of respondents who report that they borrowed for agricultural or business purposes.

is lower than the SA average. Uganda is an outlier in SSA, as its financial inclusion for farmers is 48.9 percent, compared with the SSA average of 29.3 percent and 22.8 percent average in Ethiopia, the lowest among the seven countries considered. Uganda also has the highest rate of borrowing among the seven countries—18.7 percent of farmers borrowed from financial institutions and 27.5 percent of the borrowing was for agricultural purposes. Bangladesh has the lowest (3.8 percent) and Uganda has the highest coverage (27.5 percent) of institutional borrowing for agriculture.

Low rates of financial inclusion in agriculture rest on the following factors: (1) the transaction costs of rural financial intermediation are high due to poor infrastructures and need of small loans and savings to cover a wide geographical areas; (2) high default costs of lending due to problems of loan enforcement and frequent loan forgiveness policies of the government; and (3) reluctance of banks and other financial institutions to lend money to farmers because of perceived high risks associated with agriculture in terms of weather uncertainty and production risks (World Bank 2015a). Rather than facilitating rural financial intermediation, governments often create obstacles such as interest caps and concessional lending for banks to operate efficiently for agricultural lending.

Examples of Agricultural Credit and Microcredit Programs

Because of the importance for extending credit to smallholders, governments with the help of the World Bank and other agencies are supporting banks and microfinance institutions (MFIs) to extend credit to agriculture and operate efficiently in rural settings. The agencies' credit access includes extending agricultural credit directly to local banks, which then provide loans to farmers and rural entrepreneurs. Prior to the 1990s, the World Bank and other multilateral institutions managed the disbursement of agricultural finance directly through project implementation units (World Bank 2003). However, in recent years banks provide lines of credit to financial institutions to extend credit. Examples of recent World Bank–supported credit access programs include the Rural Finance Project in Viet Nam, the SAGARPA Program in Mexico, the Financial Services for the Poorest Project in Bangladesh, and programs to provide local financial institutions technology upgrades and training in other countries (for example, Strengthening India's Rural Credit Cooperatives Project and Ghana Rural Financial Services Project). In 2014 the World Bank financed a nongovernment microfinance facility called the

Bandhan to become a rural bank to support rural lending, including agricultural lending (World Bank 2015b). Commercial bank lending to agriculture has somewhat improved in recent years for such initiatives but is not enough to enhance agricultural investment to support agricultural development.

Similarly, microfinance institutions in developing countries have no doubt made some headway in reaching small and marginal farmers with finance of donors to resolve supply-side credit constraints (for example, Khandker, Khalily, and Samad 2017). MFIs have nonetheless been promoting rural nonfarm activities in developing countries to generate additional income and employment for marginal farmers. However, MFIs have limited capacity to extend farm lending because they typically lack required licenses to operate as a bank like the Bandhan Bank of India or Grameen Bank of Bangladesh that can mobilize savings. Yet many of these microfinance institutions refrained from offering a wide array of financial products to suit the needs of farmers and small producers. As a result, microfinance institutions have tended to operate on a small scale, offering local, demand-driven options, such as group-liability for short-term lending, to better reach clients and improve their own profitability and sustainability.

Nonetheless, microfinance has been lending increasingly to support agriculture, given that many small and marginal farmers are willing to participate in MFI programs to support income earning activities, including farming. MFI share of agriculture is rising over time, based on the demand for credit in agricultural production to meet high demand for food and other staples. Microfinance institutions are also encouraged to extend seasonal credit, sometimes with funds of commercial banks to satisfy their government-set target to reach farmers in their loan portfolio (Khandker, Samad, and Badruddoza 2018).

The limit of higher MFI lending to agriculture, even if it is successful in many countries in reaching marginal and small farmers, is that they have failed to support long-term investment in agriculture necessary for raising agricultural productivity. Take the cases of two countries—Bangladesh and Thailand—with the deliberate policies to expand both microfinance and bank lending to agriculture. Bangladesh's agricultural development banks and commercial banks, including nationalized commercial banks (NCBs), have been lending increasingly in rural areas.[10] Lending to smaller farmers is a relatively recent phenomenon, handled by microfinance agencies such as

10 Bangladesh's banking sector is dominated by four state-owned NCBs, which control approximately half the assets in the banking sector.

Grameen Bank, the Association for Social Advancement (ASA), and BRAC. Palli Karma-Sahayak Foundation (PKSF), Bangladesh's wholesale microfinance lending facility, recently started a Seasonal Loans and Agricultural Lending program, nearly doubling the funds it directs toward crop agriculture (Khandker, Samad, and Badruddoza 2018). Yet its agricultural portfolio is less than 15 percent of total portfolio.

With such recent efforts, banks (both commercial and agricultural development banks) lent 7.2 percent of their total lending to agriculture, and microfinance institutions, including the famous Grameen Bank, lent about 39.5 percent of their lending to agriculture. However, they together lent only 11.2 percent of total institutional credit disbursed in 2015. Bangladesh's wholesale microfinance agency's total annual portfolio was about US$2 billion of which agricultural lending was no more than 12 percent in 2015. In the same year, while agricultural share of GDP was about 16 percent in Bangladesh, agricultural lending was about 11 percent of total institutional lending. While total institutional lending accounts for 42 percent of the GDP, agricultural credit provided by institutions explains 31 percent of the agricultural GDP (Khandker, Samad, and Badruddoza 2018). Although MFIs are increasingly financially solvent, agricultural development banks are highly subsidized.

In contrast, Thailand's specialized agricultural bank, known as the Bank for Agriculture and Agricultural Cooperatives (BAAC), provides half of agricultural loans, while the microfinance facility known as Thai Village Fund (TVF), the largest MFI in the world, provides 29.2 percent of its portfolio for agricultural purposes (Haughton, Khandker, and Torero 2018). Together they accounted for 10.5 percent of all loans (which is 13.5 percent) that went to agriculture in 2013. BAAC accounted for 6.6 percent (which was 51 percent of the total BAAC loan advanced), while TVF accounted for only 3.9 percent (which was 29 percent of the total TVF loan advanced) of total loans received by agriculture. While agriculture accounts for 8.4 percent of GDP, institutional credit in Thailand explained some 81 percent of total credit received by agriculture of which 30 percent hailed from microfinance institutions. So the bulk of agricultural loans comes from the BAAC. Although both institutions are government-owned and -supported, TVF is highly subsidized and the BAAC is not (Haughton, Khandker, and Torero 2018).

One common parameter for both countries is, however, that the agricultural portfolio of financial institutions is low—11.1 percent in Bangladesh and 13.5 percent in Thailand, although agricultural GDP in Thailand is almost 10 times Bangladesh's agricultural GDP. But Thailand's agriculture is funded more by an agricultural development bank (BAAC) rather than microfinance institutions such as TVF, while the opposite is the case for Bangladesh. MFI lending to agriculture in Bangladesh is largely short-term, such as seasonal crop loans or annual general loans. Seasonal or annual loans are important for crop production as well as income and consumption smoothing. But long-term loans are necessary for promoting large investment to commercialize and diversify crops in agriculture. It is no wonder that agricultural productivity is much higher in Thailand (largely because of diversification from cereal production to rubber and sugar production) than Bangladesh (predominantly cereal production). In 2014 agricultural value-added per worker (in constant 2010 US$) was US$715 in Bangladesh compared with US$2,133 in Thailand.[11] Higher productivity in Thailand must be an outcome of modern technology diffusion for high-value crop production and agro-based processing with private investment in agriculture and agribusinesses (Haughton, Khandker, and Torero 2018).

Mobile financial services (MFS) are the latest innovative way of extending financial services to agriculture. Mobile phones are ubiquitous in many countries and increasingly in developing countries. In 2017 there were more than 1.5 billion cellular subscriptions in South Asia and 764 million in Africa south of the Sahara (World Bank 2018b). Mobile money is a prominent innovation emerging from widespread mobile phone use, and it enables owners to deposit, transfer, and withdraw funds without physically going to a bank or owning a bank account. While MFS have transaction fees, they reduce costs for some transactions, including those geographically disparate and those where opportunity cost of holding cash may be high, as in high-crime cities (Economides and Jeziorski 2015).

Mobile financial services are more easily accessible than traditional financial institutions and are also associated with lower costs of managing financial instruments (Suri 2017). As of 2017, the mobile money industry processes a billion dollars a day with 690 million registered accounts globally (GSMA 2018). In particular, MFS could play an important role in addressing the fixed costs of managing financial instruments, which often constrain the ability of formal banks to provide financial services to the poor in general and smallholders in particular. However, as the data shows, there is no overwhelming evidence of achieving this end using this means by commercial banks

11 In contrast, rice yield per hectare in the same year was 46,188 kilograms per hectare in Bangladesh and 29,118 kilograms per hectare in Thailand. This suggests that short-term credit must play a more important role in producing rice (a short-term cereal crop) in Bangladesh than in Thailand.

in extending credit services to smallholders, although MFS has successfully helped achieve higher financial inclusion by providing financial services other than credit.

Mobile phones are in fact widespread in emerging and developing economies, although there is significant variation in the use of mobile financial services across countries and sectors, as well as extending financial services. Using the Findex data of 2014, we see that mobile financial services accounted for 11.7 percent of financial accounts across the developing world with the highest (15.9 percent) in SSA, explaining 54 percent of financial accounts, and lowest in ECA (0.5 percent), explaining only 0.5 percent of financial accounts (Table 16.3). While banks and MFIs need physical infrastructure to support financial inclusion, MFS does not find this as a barrier. However, MFS is not reaching farming communities equally. While 79 percent of financial accounts are facilitated by mobile network in Uganda, MFS is nonexistent in Ethiopia; 22.8 percent of financial inclusion in Ethiopia is facilitated by banks and microfinance institutions, compared to only 10.2 percent in Uganda. Branch networks are therefore the single most effective way of reaching farmers in a more developed agricultural economy. For instance, in Thailand, while 87 percent of individuals in agriculture have an account with a financial institution that is entirely facilitated by banks and MFIs, only 1.6 percent of financial inclusion is facilitated via mobile financial services.

A mobile financial account often provides financial services such as payments and transfer of remittance, but it is not yet capable of extending credit or mobilizing savings. In Uganda, for example, although financial inclusion is high (48.9 percent), mainly facilitated by mobile phone network, the extent of borrowing from financial institutions to support agriculture is negligible. Thus, while 87.8 percent of the Ugandans (who received payment from agriculture) have borrowed, only 18.7 percent of them borrow from financial institutions, leaving the remaining 69 percent to borrow from informal sources (see Table 16.2). Moreover, only 27.5 percent of the loans are meant for agriculture. This shows that even if financial inclusion is high in Uganda mostly because of having a transaction account with mobile financial services, the extent of borrowing from institutional services is highly limited and thus farmers rely on noninstitutional sources.

Financial inclusion via mobile financial services is nonetheless worth supporting. Mobile finance can reduce reliance on cash transactions, and time burdens for those living in more geographically isolated areas with less access to transport, financial institutions, and infrastructure. In Kenya, Uganda, and other countries, mobile money accounts are increasingly being developed to

TABLE 16.3 Access to mobile account, 2014 (%)

Region	Farmers[a]			Nonfarmers		
	Any financial account	Mobile account	(%)	Any financial account	Mobile account	(%)
By region[b]						
EAP	58.8	5.2	8.8	64.5	3.2	5.0
ECA	40.8	0.2	0.5	54.1	0.5	0.9
LAC	47.3	5.4	11.4	45.1	1.8	4.0
MENA	30.4	1.7	5.6	25.2	0.7	2.8
SA	41.6	3.1	7.5	40.3	1.6	4.0
SSA	29.3	15.9	54.3	30.3	9.4	31.0
Total	38.3	11.7	30.5	43.7	4.6	10.5
By seven developing countries[c]						
Bangladesh	31.4	1.9	6.1	30.9	2.9	9.4
Ethiopia	22.8	0.0	0.0	21.1	0.0	0.0
India	60.6	3.7	6.1	51.1	2.0	3.9
Peru	28.8	0.0	0.0	29.0	0.0	0.0
Thailand	87.4	1.4	1.6	72.7	1.3	1.8
Uganda	48.9	38.7	79.1	34.7	27.1	78.1
Viet Nam	28.9	0.4	1.4	31.8	0.5	1.6
Total	48.6	11.0	22.6	41.8	2.6	6.2

Source: Data from World Bank (2015c).

Note: a. "Farmers" are defined as those who "received agricultural payments in the past 12 months" (weighted sample).

b. EAP = East Asia and Pacific, ECA = Europe and Central Asia, LAC = Latin America and the Caribbean, MENA = Middle East and North Africa, SA = South Asia, SSA= Africa south of the Sahara.

c. "Developing world" excludes OECD countries.

link farmers, agents, and buyers (for example, Kikulwe, Fischer, and Qaim 2014). Mobile-based financial services are emerging to draw a greater share of the population into formal financial services (not necessarily for borrowing but for payment services and remitting money), particularly in Africa south of the Sahara and other regions. But mobile financial services are not yet found to be an effective mechanism for extending credit to small farmers in areas where institutional credit is highly due but not currently available.

But this does not mean that MFS cannot offer financial services that can extend credit and mobilize savings using the concept of branchless banking. In principle, mobile money can serve as a facilitator to bring financial services to the doorsteps of smallholders, where users can display some financial track record through financial transactions. Using such records, financial institutions can develop credit scores of potential customers (via reduction of information asymmetries), upon which they can extend credit through reducing transaction costs as well as the costs of managing these financial accounts. However, evidence so far does not lend support to rising expected lending portfolios for farmers via this method of banking. Thus, further innovations are warranted in promoting linkages of mobile financial services with agricultural finance and how such linkage can be facilitated by the government and international development agencies to promote digitation in agrifinance.

Impact of Agricultural Credit on Smallholders' Income and Productivity

Estimating the effect of credit or financial access is necessary for determining whether institutional credit matters, how it matters, and what it means to policymaking. More specifically, estimating credit effect is a way to understand the mechanisms in which provision of credit through financial institutions is a prudent policy for boosting agricultural investment and productivity for supporting the government agenda of growing sustainably and attaining food security.

Evidence of Potential Role of Financial Inclusion in Agriculture

Agricultural productivity (measured by agricultural valued-added per worker) at the country level is found to have the strongest positive correlation with access to financial institutions (either measured by account ownership, commercial bank branches, or agricultural credit as a share of total institutional credit). Figure 16.5 shows that a country's share of agricultural GDP is positively associated with the share of agricultural credit in total credit disbursed, meaning that the higher the share of lending to agriculture via financial institutions, the higher is the share of agricultural GDP. Similarly, as Figure 16.6 shows, a country's agricultural productivity is positively associated with the country's number of commercial bank branches. This suggests that availability of banks increases the chances of borrowing for agricultural investment and the induced effects on agricultural productivity.[12]

Figure 16.7 shows the trend in financial account ownership against agricultural value-added per worker as a measure of agricultural productivity. Thus

12 Availability of a bank branch also increases savings mobilization from rural areas, which in turn can help smooth farmers' income and consumption, thereby ensuring a higher liquidity in absorbing risks in production.

FIGURE 16.5 Agricultural GDP and agricultural borrowing, 2014

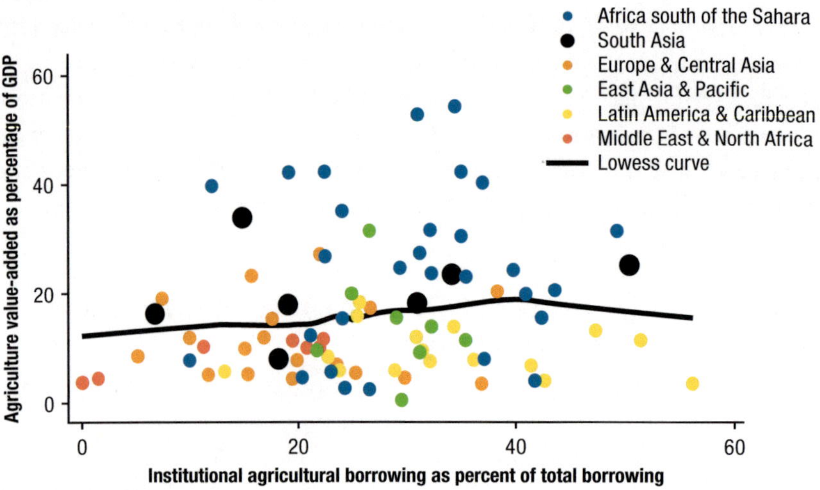

Source: World Bank (2015c) and World Bank (2019).

Note: Agricultural borrowing is measured in the following manner: (1) From FINDEX data, we first selected all the individuals who received agricultural payments in the past 12 months. (2) Among those individuals, we observed whether the individual borrowed (of any kind) in the past 12 months. (3) Those who mentioned the purpose of borrowing as farm/business are considered to have agricultural borrowing, which implies the percentage of sampled individuals who borrowed for agricultural/business purposes who also received agricultural payment in the past 12 months.

FIGURE 16.6 Growth in agricultural productivity by growth in bank branches, 2018

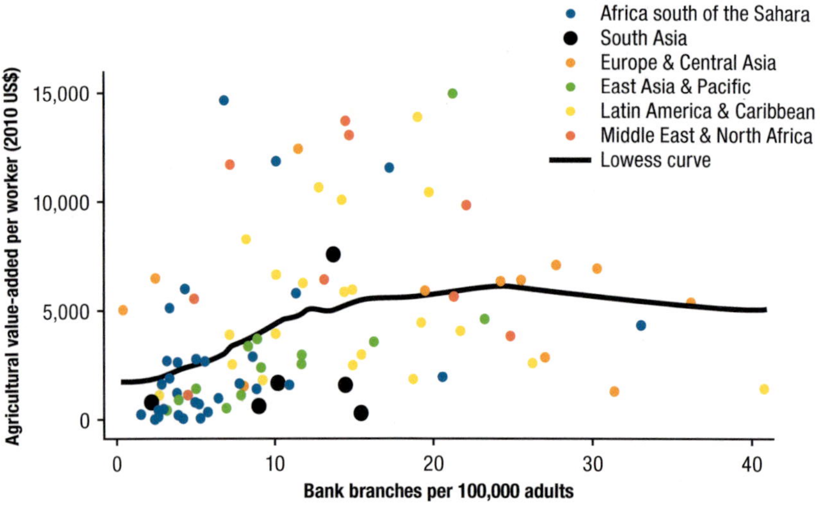

Source: Global Findex data; World Bank (2019).

FIGURE 16.7 Financial account ownership against agricultural productivity (country agriculture value-added per worker), 2014

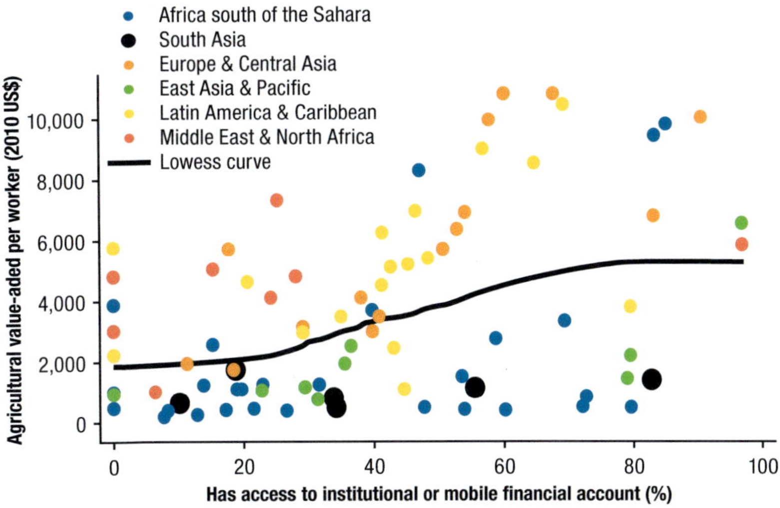

Source: World Bank (2015c; 2019).

financial inclusion (measured broadly in terms of having an account with a financial institution or mobile financial service) is higher among countries with a higher index of agricultural productivity than countries with lower agricultural productivity. That is, increasing the level of financial inclusion in agriculture can lead to raising agricultural productivity, investment, food production, and food security. These relations are simple correlations but do not necessarily mean causality, meaning that there is a potential role of financial services in raising agricultural productivity, but it is not clear if financial inclusion actually increases productivity or if higher productivity raises demand for financial services. Disentangling this relationship is a causality question—what would have happened if credit access were not improved? Identification of the causal effect of credit is a challenge that involves understanding how improved access and use of specific services affects productivity.

Evidence of the Causal Effect of Agricultural Credit

Measuring the causal effect of financial service is a major goal of any study on credit policy for agricultural development. However, establishing causality—that is, identifying the net effect of financial access—is difficult because agricultural development outcomes can be affected by the same factors that affect farmers' decisions to borrow from financial institutions. This is the classic problem of endogeneity that needs to be resolved satisfactorily to estimate the causal effect of credit.

In theory, credit access does not seem to affect agricultural productivity unless farmers are liquidity constrained (for example, Feder et al. 1990; Karlan et al. 2014). When farmers are liquidity constrained, they are under a constrained decision-making regime in which credit access does directly affect agricultural decision-making processes such as input use and farming practices. Even with a credit-constrained regime, it does not necessarily follow without further assumptions that credit induces an effect on agricultural outcomes such as productivity, as borrowing decisions are not determined independently of agricultural productivity. The endogeneity issue involves estimating the demand for credit as part of a farmer's investment decision and its consequential impact on agricultural productivity. Various methods (for example, randomized controlled trial, propensity score matching, and instrumental variable) can help resolve the problem of endogeneity in estimating credit effects (for details, see Khandker, Koolwal, and Samad 2010).

Studies using alternate methods have shown that financial inclusion has positive effects on income, productivity, consumption, education, health, gender equality, and other indicators of well-being (for example, Cull, Ehrbeck, and Holle 2014; Aker et al. 2014; Klapper, El-Zoghbi, and Hess 2016). In particular, financial inclusion can help mitigate risks in production through enhanced investment and reduced costs of easing credit constraints in production. A large body of literature suggests that higher access to and use of financial services by farmers can help increase modern use of inputs and technology, and investment in agriculture, leading to higher farm productivity and income and hence improved household welfare (for example, Feder et al. 1990; Cull, Ehrbeck, and Holle 2014; Karlan et al. 2014).

Assuming credit access is measured by bank branch density, several studies have found that access to commercial and agricultural banks has created viable alternatives to moneylenders for households with somewhat greater assets and collateral. For example, Binswanger and Khandker (1995) found that better access to farm credit provided by banks spurred fertilizer use and investment in agriculture; however, it is less successful in generating viable institutions to generate agricultural employment. But bank density increases agricultural investment and productivity in India via investment in pump set, a very productive tool for raising agricultural output. Rural bank branches increase nonfarm employment and rural wages (Binswanger, Khandker, and Rosenzweig 1993). Rural bank branch expansion, savings mobilization, and credit disbursement are all found to increase farm and nonfarm output significantly and hence reduce rural poverty consequently (for example, Burgess and Pande 2005; Prina 2015; Brune et al. 2016).

Recent efforts on identifying the effect of farm credit include micro-data studies that attempt to identify the extent of credit constraint faced by farmers and its induced effect on productivity. This is because of the fact that credit effect depends on the extent of credit constraint encountered by farmers (for example, Feder et al. 1990). Studies from Peru show that credit constraints lower the value of agricultural output substantially for poor households; however, for those households without credit constraints, productivity is independent of such endowments as land and liquidity (Boucher, Guirkinger, and Trivelli 2009; Guirkinger and Boucher 2008). Some other studies also confirm that credit constraints affect agricultural output and productivity (Feder et al. 1990; Sial and Carter 1996), farm profit (Carter 1989; Foltz 2004), and farm investment (Carter and Olinto 2003). The presence of credit constraint among MFI borrowers also matters in raising farm income—the effect of borrowing on farm income is higher for credit-constrained than noncredit-constrained borrowers (Khandker, Khalily, and Samad 2016).

Farmers face risks such as unpredictable weather and crop price variation, affecting how they choose to borrow and invest. Thus smoothing consumption across seasons is an important motivation for microfinance participation in Bangladesh (Pitt and Khandker 2002; Khandker 2012). The enhanced ability to smooth consumption permits households to choose riskier but higher-yielding contracts in agricultural markets (Pitt 2000). Although microfinance has traditionally been targeted toward the nonfarm sector, because of the linkages across farm and nonfarm sectors and funds are fungible, it nonetheless promotes farm income growth (Khandker, Khalily, and Samad 2016).

Several studies examined the effect on household consumption and income risk by credit when credit is obtained through mobile money accounts. Access to mobile money services (for example, remittances) reduces transaction costs of money transfer as well as consumption risks, thereby increasing consumption and reducing poverty (Jack and Suri 2014, 2016). Among the most successful mobile money products, M-PESA in Kenya, launched in 2007, now has virtually universal coverage in the country. Suri (2017) reviewed the impacts of M-PESA on responses to a range of outcomes and finds users spend more on medical expenses after a health shock while also increasing expenses on food and maintaining education expenditures. Moreover, mobile money contributed to a decline in poverty rate by 2 percentage points as a result of better access to mobile money services and thus increased household consumption and savings.

Several other studies, using experimental methods, have evaluated the impacts of rural farmers' improved access to credit. In Morocco, Crepón and others (2011) found that the effects of rural expansion of microfinance institutions depend on activities that households are initially involved in, particularly self-employed agriculture—a large increase in sales and profits leads to substantial increases in expenditure and employment. Karlan et al. (2014) used an experimental design of a combination of cash grants to test the relevance of imperfect credit and an incomplete insurance market to agricultural investment and productivity in Nigeria. They found that an incomplete insurance market, which protects farm income from perceived agricultural risk, is more binding than a liquidity (that is, credit) constraint for promoting investment in agriculture.[13]

Banerjee, Karlan, and Zinman (2015) reviewed randomized controlled trials measuring the impact of MFI access across countries; the review revealed that microfinance access improved business activity, ranging from investments, business size, and profits, by at least a marginal extent across the six countries examined. However, the extent and type of impacts varied across countries. In India, small business investment and profits of preexisting businesses increased, but consumption did not significantly increase (Banerjee et al. 2015). In Ethiopia, Tarozzi, Desai, and Johnson (2015) find no treatment effect for the majority of outcomes. A joint liability microcredit program for women had a positive impact of access to group loans on female entrepreneurship and household food consumption but not on total working hours or income in the household in Mongolia (Attanasio et al. 2015). Angelucci, Karlan, and Zinman (2015), however, do not find transformative impacts on many outcomes in a clustered randomized trial in Mexico.

Although microfinance institutions have scope to improve credit access to agriculturalists and rural people, the evidence is not definitive on whether microfinance alleviates poverty and what exactly are its impacts on business activity. In investigating the question of external validity—concerns the studies may be too different from each other to generalizable their results—Meager (2019) aggregates evidence from seven randomized controlled trials and finds negligible impact of microfinance access on household business and consumption variables, and that this impact is unlikely to be transformative.

13 This of course does not mean that liquidity (and hence, credit) is not a binding constraint for smallholders in agriculture. This study observes that when an insurance market is missing in a risky environment, lack of insurance is perhaps more binding than a liquidiy constraint for raising farm investment and productivity.

Indeed, these experimental studies are not conclusive. Methods also matter in evaluating microfinance. Experimental methods are usually applied to evaluate a short-term program (usually less than 18 months of intervention) such as PROGRESA in Mexico—a conditional cash transfer (Angrist et al. 2002; de Janvry et al. 2006; Schultz 2004). In contrast, microfinance or institutional finance for that matter, is not a cash transfer, and accrued benefits from self-employed activities supported under microfinance require a certain minimum length of exposure (for example, a minimum of three years of exposure means an experimental method is inappropriate). Therefore, nonexperimental methods, in contrast to experimental methods, are used to explore the impact of certain length of program exposure. Many nonexperimental studies of microfinance are found to have substantive positive contribution of microfinance in enhancing income, consumption, and poverty reduction (for example, Islam 2011; Imai, Arun, and Annim 2010; Imai et al. 2012; Pitt and Khandker 1998; Khandker 2005: Khandker, Khalily, and Samad 2016). It is thus important to distinguish between short-term versus long-term impacts of microfinance and draw complementarity between these alternative methods. Further studies are necessary to reconcile the findings drawn from alternative methods of impact evaluation regarding roles of microfinance in development (Wydick 2016). It is important to carry out more studies to demonstrate how a microfinance facility works and helps smallholders and poor farmers to avert the short-term and long-term consequences of exposures to climate and other risks increasingly associated with smallholder farming.

Enhancing profitability of smallholder agriculture is the prime goal associated with the agenda of growing sustainably and attaining food security. Smallholders are not, however, a homogeneous group; some smallholder farmers have the potential to increase farm profits and undertake commercial activities in the agricultural sector, while others should be supported in exiting agriculture and seeking nonfarm opportunities. For smallholder farmers with profit potential, their ability to be successful is apparently hampered by challenges including limited financial inclusion. By overcoming this and other challenges, such as climate change and price shocks, smallholders can move from subsistence to commercially oriented agricultural systems and increase their profits (Fan et al. 2013). However, a major knowledge gap exists as to whether and how an improvement in financial access can help increase profitability of smallholders. We do not know if providing credit alone is sufficient to improve profitability. It is indeed possible that other innovative financial products are necessary that provide additional services using new technologies so that actors in the agri–value chain can make better profit-enhancing investments.

Challenges and Recommendations to Support Innovations in Agricultural Credit

Agricultural investment is absolutely necessary to raise food production and farm productivity, especially in smallholder agriculture, to support a country's agenda of growing sustainably and attaining food security. Access to finance is thus critical to promote investment in agriculture and hence agricultural growth and productivity. However, farming in developing countries, where agricultural growth is the driver of growth and development, is managed largely by smallholders. Smallholders in agriculture need funds to invest but for a variety of reasons are constrained in accessing finance, especially institutional finance. Although access to financial services such as savings, credit, payments, and insurance is limited in general in rural areas, smallholders are limited further in accessing finance for reasons such as high transaction costs of small loans and savings, high risk associated with farming caused by drought and excessive rains, and lack of appropriate policies facilitating coverage of the financial services provided by a country's financial institutions.

Evidence suggests that institutional credit plays an important role in enhancing investment and productivity in smallholder agriculture. Better access to institutional credit is desirable because it is less expensive compared to own or informal finance to buy fertilizer, irrigation, and other productivity-enhancing inputs to raise farm productivity. Credit also helps resolve cash flow problems of crop production and mitigate risks in farming due to uncertainty of weather patterns and climate changes. Provision of weather-indexed insurance helps farmers also mitigate risks, but credit at easy terms with lower transaction costs to support investments can mitigate agricultural risk effectively. Institutional credit, especially long-term, encourages private investment in marketing and agro-processing, and thus helps produce high-valued crops, meat, and vegetables. Diversification, modernization, and commercialization of agriculture are the hallmarks of agricultural development for attaining sustainable growth and food security.

However, financial institutions, especially agricultural development banks and commercial banks, are not capable of serving the smallholder agriculture sustainably and effectively. Banks have been asked to support subsidized credit in many low-income countries but have failed to support profitably government credit schemes. In the process, however, many of them have failed miserably not only to support agricultural credit extension but also to become self-sustainable institutions. Agricultural finance institutions must learn how to serve farmers as per demand and deliver products profitably to resolve both short-term cash flow problems and long-term investment needs of smallholders. Institutions must be developed to address idiosyncratic and systematic risks that characterize agriculture in a highly volatile agroclimate environment. Government policies must be facilitating rather than dictating institutional lending to enhance private investment in agriculture. Governments must refrain from directing misguided credit policies and programs, which are found to stifle growth and sustainability of financial institutions.

A variety of bilateral and nongovernmental organization–funded programs have recently evolved to provide various options to the challenging task of extending credit and other financial services to farmers, especially among smallholders in Africa and Asia (for example, Kloeppinger-Todd and Sharma 2010). For example, to address limited collateral among smaller farmers, techniques have been developed to tailor lending strategies to the agricultural supply chain. Thus farmers can borrow against output stored in licensed warehouses, or producers and processors can make binding contracts for crop output after which processors repay the producer's loan to the bank. Other alternatives being used include direct product-distribution channels (for example, mobile banking), electronic point-of-sale devices (for example, those run by the Uganda Microfinance Union in rural areas), and partnerships with market-facilitating institutions. Also, as is discussed in Chapter 17, agricultural insurance (for example, weather-indexed insurance) is an emerging policy area for helping farmers to manage production and investment risks (Cai 2012; Giné and Yang 2009).

Government can help develop alternative institutions with appropriate incentive structures for agricultural development; a combination of bank finance, microfinance, and mobile finance is perhaps necessary for successful agro-based lending and other financial services. One group of finance may not fit the various needs of farmers. Examples of Thailand's BAAC and India's Bandhan Bank are perhaps instructive for other countries to follow (World Bank 2015b).[14] They show how an agricultural development bank with a mandate to support agriculture or a nongovernment microfinance facility in a rural setting—both of them relying entirely on donor funds and not permitted or encouraged to mobilize savings—can be turned into a viable financial

14 Another example is Indonesia's Bank Rakayat Indonesia (BRI), which had gone through reforms for several years before it became a self-sustainble agricultural development bank (Seibel 2000).

institution that is now allowed to mobilize savings and able to meet the banking needs of farmers and other small rural producers. In the absence of the large-scale government subsidies/donor funding of the past, policymakers may facilitate the institutions to explore ways of expanding financial access to smallholders effectively and sustainably.

Governments may support innovations in how to develop the facilitating role of financial institutions in enhancing smallholders' access to finance. They need to provide a sound macroeconomic and regulatory framework, a prudent monitoring framework through central banks to protect savers, and fund innovations in financial product design and diversification via experimentation. Learning from microfinance institutions and informal finance to use social collateral may be a way to mitigate risk in rural lending and should be part of innovation in agricultural finance. Institutional innovation may also promote linkages between formal financial institutions and microfinance institutions, rotating savings and credit associations (RoSCAS), and village banks. In short, governments should support innovations and institutional development but never subsidize interest rates on lending to support agricultural finance. After all, government support is to be directed toward lowering transaction costs of banking for the poor smallholders to access financial services such as credit. This is perhaps the only way for any government to support agricultural finance that is institutionally viable as well as effective in enhancing agricultural investment in support of agricultural development.

References

Adams, D. W. 1988. "The Conundrum of Successful Credit Projects in Floundering Rural Financial Markets." *Economic Development and Cultural Change* 36 (2): 355–367.

Adjognon, S. G., L.S.O. Liverpool-Tasie, and T. Reardon. 2017. "Agricultural Input Credit in Sub-Saharan Africa: Telling Myth from Facts." *Food Policy* 67: 93–105.

Ahmed, S. A., N. S. Diffenbaugh, and T. W. Hertel. 2009. "Climate Volatility Deepens Poverty Vulnerability in Developing Countries." *Environment Research Letters* 4 (034004).

Aker, J., R. Boumnijel, A. McClelland, and N. Tierney. 2016. "Payment Mechanisms and Anti-Poverty Programs: Evidence from a Mobile Money Cash Transfer Experiment in Niger." *Economic Development and Cultural Change* 65 (1): 1–37.

Angelucci, M., D. Karlan, and J. Zinman. 2015. "Microcredit Impacts: Evidence from a Randomized Microcredit Program Placement Experiment by Compartamos Banco." *American Economic Journal: Applied Economics* 7 (1): 151–182.

Angrist, J., E. Bettinger, E. Bloom, E. King, and M. Kramer. 2002. "Vouchers for Private Schooling in Colombia: Evidence from a Randomized Natural Experiment." *American Economic Review* 92 (5): 1535–1558.

Attanasio, O., B. Augsburg, R. De Haas, E. Fitzsimons, and H. Harmgart. 2015. "The Impacts of Microfinance: Evidence from Joint-Liability Lending in Mongolia." *American Economic Journal: Applied Economics* 7 (1): 90–122.

Banerjee, A., E. Duflo, R. Glennerster, and C. Kinnan. 2015. "The Miracle of Microfinance? Evidence from a Randomized Evaluation." *American Economic Journal: Applied Economics* 7 (1): 22–53.

Banerjee, A., D. Karlan, and J. Zinman. 2015. "Six Randomized Evaluations of Microcredit: Introduction and Further Steps." *American Economic Journal: Applied Economics* 7 (1): 1–21.

Binswanger, H., and S. R. Khandker. 1995. "The Impact of Formal Finance on the Rural Economy of India." *Journal of Development Studies* 32: 234–262.

Binswanger, H., S. Khandker, and M. Rosenzweig. 1993. "How Infrastructure and Financial Institutions Affect Agricultural Output and Investment in India." *Journal of Development Economics* 41: 337–466.

Bocher, T., B. Alemu, and Z. Kelbore. 2017. "Does Access to Credit Improve Household Welfare? Evidence from Ethiopia Using Endogenous Regime Switching Regression." *African Journal of Economic and Management Studies* 8 (1): 51–65.

Boucher, S., C. Guirkinger, and C. Trivelli. 2009. "Direct Elicitation of Credit Constraints: Conceptual and Practical Issues with an Application to Peruvian Agriculture." *Economic Development and Cultural Change* 57 (4): 609–640.

Brune, L., X. Giné, J. Goldberg, and D. Yang. 2016. "Facilitating Savings for Agriculture: Field Experimental Evidence from Malawi." *Economic Development and Cultural Change* 64 (2): 187–220.

Burgess, R., and R. Pande. 2005. "Do Rural Banks Matter? Evidence from the Indian Social Banking Experiment." *American Economic Review* 95 (3): 780–795.

Cai, J. 2012. "Social Networks and the Decision to Insure: Evidence from Randomized Experiments in China." Working Paper, University of California, Berkeley.

Carter, M. R. 1989. "The Impact of Credit on Peasant Productivity and Differentiation in Nicaragua." *Journal of Development Economics* 103: 13–36.

Carter, M. R., and P. Olinto. 2003. "Getting Institutions Right for Whom? Credit Constraints and the Impact of Property Rights on the Quantity and Composition of Investment." *American Journal of Agricultural Economics* 85 (1): 173–186.

Crepón, B., F. Devoto, E. Duflo, and W. Pariente. 2011. "Impact of Microfinance in Rural Areas of Morocco: Evidence from a Randomized Evaluation." MIT Working Paper, Massachusetts Institute of Technology, Cambridge, MA.

Cull, R., T. Ehrbeck, and N. Holle. 2014. "Financial Inclusion and Development: Recent Impact Evidence." *CGAP Focus Note 92.* World Bank, Washington, DC.

Deaton, A. 1997. *The Analysis of Household Surveys: A Microeconometric Approach to Development Policy.* Baltimore: Johns Hopkins University for the World Bank.

de Janvry, A., F. Finan, E. Sadoulet, and R. Vakis. 2006. "Can Conditional Cash Transfer Programs Serve as Safety Nets in Keeping Children at School and from Working When Exposed to Shocks?" *Journal of Development Economics* 79 (2): 349–373.

Demirguc-Kunt, A., and L. Klapper. 2012. "Measuring Financial Inclusion: The Global Findex Database." Policy Research Working Paper 6025. World Bank, Washington, DC.

Demirguc-Kunt, A., L. Klapper, D. Singer, S. Ansar, and J. Hess. 2018. *The Global Findex Database 2017: Measuring Financial Inclusion and the Fintech Revolution.* Washington, DC: World Bank.

Demirguc-Kunt, A., et al. 2015. "The Global Findex Data Base 2014: Measuring Financial Inclusion around the World." Policy Research Working Paper 7255. World Bank, Washington, DC.

Economides, N., and P. Jeziorski. 2015. "Mobile Money in Tanzania." Working Paper. New York University, New York.

Fan, S., J. Brzeska, M. Keyzer, and A. Halsema. 2013. *From Subsistence to Profit: Transforming Smallholder Farms.* Food Policy Report. Washington, DC: International Food Policy Research Institute.

FAO (Food and Agricultural Organization of the United Nations). 2012. "Smallholders and Family Farmers." Sustainability Pathways, Smallholders. Rome.

———. 2016. *Climate Change and Food Security: Risks and Responses.* Rome.

———. 2017. *The State of Food Security and Nutrition in the World 2017: Building Resilience for Peace and Food Security.* Rome.

Feder, G., L. J. Lau, J. Y. Lin, and X. Luo. 1990. "The Relation between Credit and Productivity in Chinese Agriculture: A Model of Disequilibrium." *American Journal of Agricultural Economics* 72 (5): 1151–1157.

Foltz, J. 2004. "Credit Market Access and Profitability in Tunisian Agriculture." *Agricultural Economics* 30: 229–240.

Giné, X., and D. Yang. 2009. "Insurance, Credit, and Technology Adoption: Field Experimental Evidence from Malawi." *Journal of Development Economics* 89: 1–11.

GSMA (Group System Mobile Association). 2018. *2017 State of the Industry Report on Mobile Money*. London.

Guirkinger, C., and S. Boucher. 2008. "Credit Constraints and Productivity in Peruvian Agriculture." *Agricultural Economics* 39: 295–308.

Haughton, J., S. Khandker, and M. Torero. 2018. "Agricultural Finance in Thailand." Mimeo. Washington, DC.

IFC (International Finance Corporation). 2013. "IFC and Agri-Finance: Creating Opportunity Where It's Needed Most." Washington, DC.

ILO (International Labour Organization). 2014. "Economic Diversification of the Rural Economy." Decent Work in the Rural Economy Policy Guidance Notes. Geneva.

Imai, K. S., T. Arun, and S. K. Annim. 2010. "Microfinance and Household Poverty Reduction: New Evidence from India." *World Development* 38 (12): 1760–1774.

Imai, K. S., R. Gaiha, G. Thapa, and S. K. Annim. 2012. "Microfinance and Poverty Reduction: A Macro Perspective." *World Development* 40 (8): 1675–1689.

Islam, A. 2011. "Medium and Long-Term Participation in Microcredit: An Evaluation Using a New Panel Dataset from Bangladesh." *American Journal of Agricultural Economics* 93 (3): 847–866.

Jack, W., and T. Suri. 2014. "Risk Sharing and Transactions Costs: Evidence from Kenya's Mobile Money Revolution." *American Economic Review* 104 (1): 183–223.

———. 2016. "The Long-Run Poverty and Gender Impacts of Mobile Money." *Science* 354 (6317): 1288–1292.

Joshi, P. K., D. Roy, and V. K. Sonkar. 2017. "Elements of Agriculture Value Chain Financing: A Review." In *Financing Agriculture Value Chains in India: Challenges and Opportunities*, edited by G. Mani, P. K. Joshi, and M. V. Ashok. Part 1, *Agriculture Value Chain Financing: Theoretical Framework*, Chapter 2, 15–32. Singapore: Springer.

Karlan, D., R. Osei, I. Osei-Akoto, and C. Udry. 2014. "Agricultural Decisions after Relaxing Credit and Risk Constraints." *Quarterly Journal of Economics* 129 (2): 597–652.

Khandker, S. 2005. "Microfinance and Poverty: Evidence Using Panel Data from Bangladesh." *World Economic Review* 19 (2): 263–286.

———. 2012. "Seasonality of Income and Poverty in Bangladesh." *Journal of Developmental Economics* 97: 244–256.

Khandker, S., B. Khalily, and H. Samad. 2016. *Beyond Ending Poverty: The Dynamics of Microfinance in Bangladesh*. Washington, DC: World Bank.

Khandker, S., and G. Koolwal. 2016. "How Has Microcredit Supported Agriculture? Evidence Using Panel Data from Bangladesh." *Agricultural Economics* 47 (2): 157–168.

Khandker, S., G. Koolwal, and H. Samad. 2010. *Handbook on Impact Evaluation: Quantitative Methods and Practices*. Washington, DC: World Bank.

Khandker, S., H. Samad, and S. Badruddoza. 2018. "Agricultural Finance in Bangladesh." Mimeo. Washington, DC.

Kikulwe, E., E. Fischer, and M. Qaim. 2014. "Mobile Money, Smallholder Farmers, and Household Welfare in Kenya." *PLoS ONE* 9 (10): e109804.

Klapper, L., M. El-Zoghbi, and J. Hess. 2016. "Achieving the Sustainable Development Goals: The Role of Financial Inclusion." Consultative Group to Assist the Poorest (CGAP), World Bank, Washington, DC.

Kloeppinger-Todd, R., and M. Sharma. 2010. *Innovations in Rural and Agricultural Finance*. Washington, DC: IFPRI and World Bank.

Koolwal, G., and S. Khandker. 2018. "Financial Inclusion in Agriculture across Countries: An Overview and Descriptive Analysis using the Global Findex." Mimeo. Washington, DC.

Kumar, C., C. Turvey, and J. Kropp. 2013. "The Impact of Credit Constraints on Farm Households: Survey Results from India and China." *Applied Economic Perspectives and Policy* 35(3): 508–527.

McCullough, E. B. 2017. "Labor Productivity and Employment Gaps in Sub-Saharan Africa." *Food Policy* 67: 133–152.

Meager, R. 2019. "Understanding the Average Impact of Microcredit Expansions: A Bayesian Hierarchical Analysis of Seven Randomized Experiments." *American Economic Journal: Applied Economics* 11 (1): 57–91.

Pitt, M. 2000. "The Effect of Nonagricultural Self-Employment Credit on Contractual Relations and Employment in Agriculture: The Case of Microfinance Programs in Bangladesh." *Bangladesh Development Studies* (June–September): 15–48.

Pitt, M., and S. Khandker. 1998. "The Impact of Group-Based Credit Programs on Poor Households in Bangladesh: Does Gender of Program Participants Matter?" *Journal of Political Economy* 106: 958–996.

———. 2002. "Credit Programs for the Poor and Seasonality in Rural Bangladesh." *Journal of Development Studies* 39 (2): 1–24.

Prina, S. 2015. "Banking the Poor Via Savings Accounts: Evidence from a Field Experiment." *Journal of Development Economics* 115 (C): 16–31.

Ruete, M. 2015. "Financing for Agriculture: How to Boost Opportunities in Developing Countries." Investment in Agriculture Policy Brief 3, International Institute for Sustainable Development (IISD).

Schultz, T. P. 2004. "School Subsidies for the Poor: Evaluating the Mexican PROGRESA Poverty Program." *Journal of Development Economics* 74 (1): 199–250.

Seibal, H. D. 2000. "Agricultural Development Banks: Close Them or Reform Them." *Finance and Development* 37 (2).

Sial, M. H., and M. R. Carter. 1996. "Financial Market Efficiency in an Agrarian Economy: Microeconometric Analysis of the Pakistani Punjab." *Journal of Development Studies* 32 (5): 771–798.

Stiglitz, J., and A. Weiss. 1981. "Credit Rationing in Markets with Imperfect Information." *American Economic Review* 71 (3): 393–410.

Suri, T. 2017. "Mobile Money." *Annual Review of Economics* 9: 497–520.

Tarozzi, A., J. Desai, and K. Johnson. 2015. "The Impacts of Microcredit: Evidence from Ethiopia." *American Economic Journal: Applied Economics* 7 (1): 54–89.

World Bank. 2003. "Review of the Bank's Rural Finance Experience." Working Paper, Operations Evaluation Department. World Bank, Washington, DC.

———. 2015a. "Agricultural Finance." Washington, DC. www.worldbank.org/en/topic/financialsector/brief/agriculture-finance.

———. 2015b. "Bandhan Becomes India's Youngest Bank." World Bank News, August 23.

———. 2015c. Databank: Global Financial Inclusion (database). https://databank.worldbank.org/reports.aspx?source=1228

———. 2018a. "UFA2020 Overview: Universal Financial Access Initiative 2020." www.worldbank.org/en/topic/financialinclusion/brief/achieving-universal-financial-access-by-2020.

———. 2018b. "Financial Inclusion: An Overview." www.worldbank.org/en/topic/financialinclusion/overview.

———. 2019. "World Development Indicators." https://data.worldbank.org/indicator/IT.CEL.SETS.

Wydick, B. 2016. "Microfinance on the Margin: Why Recent Impact Studies May Understate Average Treatment Effects." *Journal of Development Effectiveness* 8: 257–265.

Zeller, M., and M. Sharma, 2000. "Many Borrow, More Save, and All Insure: Implications for Food and Microfinance Policy." *Food Policy* 25: 143–167.

Chapter 17

AGRICULTURAL INSURANCE FOR DEVELOPMENT: PAST, PRESENT, AND FUTURE

Miguel Robles

Agricultural production is a risky activity subject to several contingencies that make farming incomes unstable and unpredictable from year to year. Bakst, Sewell, and Wright (2016) report that the Economic Research Service of the US Department of Agriculture (USDA) identifies five types of farming risk: (1) human and personal risk (such as human health), (2) institutional risk (regarding governmental action), (3) financial risk (such as access to capital), (4) price or market risk, and (5) production risk (such as weather and pests). The HIV/AIDS pandemic in African countries in the early 2000s is an example of a human risk that affected agricultural output and labor supply and demand. Land expropriation, land reforms, and unanticipated changes in the agricultural tax system are examples of institutional risk that can depress investment in the agricultural sector. Nonavailability of funds in the formal or informal financial sector or high interest rates reflecting liquidity scarcity at the beginning of the crop season can limit area planted in a year. Uncertain crop prices at harvest time and input prices during the crop calendar can have severe impacts on farm income and become a social problem when affecting many farmers.

Agricultural production is directly tied to weather variables such as rainfall, temperature, humidity, and wind. When extreme weather conditions occur, agricultural production typically suffers and, in some cases, may be lost completely. Year after year, unexpected weather extremes are a constant threat in several regions of the world, with devastating effects on agricultural production and rural livelihoods. Some examples of risky events happening within a short period of time globally are prolonged droughts in the Horn of Africa and the midwestern part of the United States, extensive floods in the Philippines and northern India, abnormally low temperatures in Japan and the United States, and heat waves in Australia and Europe. Also, extreme weather events allow pests and diseases to flourish, with potential large impacts over vast regions.

According to Hazell and Hess (2016), risks can be characterized according to the following elements:

- *Covariance.* The degree to which they are correlated across households within a community or region, ranging from independent (affecting one person) to highly covariate (affecting everyone at the same time).
- *Frequency.* How often they occur.
- *Types and severity of losses incurred.* Shortfalls in seasonal production and income, damage to assets, and loss of life.

Table 17.1 expands on each of these elements and provides examples. The 2002 drought in Ethiopia that affected most of the country and led to 12.5 million people requiring food aid is an example of a highly covariate risk that generally occurs with low frequency but with catastrophic impacts within affected regions. Deaths and illnesses of people in individual households or livestock are risks that are weakly if at all covariate but that occur with high frequency. In a year it is not possible to predict which individual household will be impacted, but the proportion of total households affected each year is often predictable. Other risks are moderately covariate and occur with moderate frequency. These include losses in production and income or damage to assets due to less severe drought, excess rainfall, or market and price risks. When lost farm output at the local level is considerable and the degree of economic integration with other regions is low, then the rural nonfarm economy might suffer as well. This is because the local demand for other goods and services might also decrease substantially and widespread defaults on loans might happen as well, undermining the local rural financial systems.

Arias and Covarrubias (2006) classify agricultural risks as natural hazards and human-made risks and as risks that are within or beyond the control of the farmer (for instance, natural catastrophes or international financial crises). An example of a controllable risk is variations in river water flows by building proper irrigation and dam systems. Some risks are controllable to a certain extent, such as the effects of some plagues and diseases. This is an important distinction when studying crop insurance and designing policies, as access to insurance instruments should complement rather than substitute for best agricultural practices, especially those that help reduce controllable risks. Kyle et al. (2016) present evidence that introducing to Indian farmers a new rice variety that reduces downside risk has the potential to increase agricultural productivity in normal years. Insurance mechanisms should encourage this type of investment and be designed to take care of residual or noncontrollable risks.

TABLE 17.1 Types of risk and loss—and local capacity to cope

Type of risk		Type of loss			
Degree of covariance	Frequency	Life	Assets	Seasonal production/Income	Examples
High	Low	Widespread loss of life and injuries from catastrophic weather events such as hurricanes, floods, or severe drought. Little or no capacity to cope locally; recovery is difficult and slow.	Widespread loss of homes and productive assets from catastrophic weather events. Little or no capacity to cope locally; recovery is difficult and slow.	Impacts of catastrophic weather events on regional production and income can be severe, with limited local coping capacity. Recovery can be slow if lives and assets are also lost.	Catastrophes such as tsunami, severe drought, flood, hurricane, or earthquake.
Medium	Medium	Some loss of life and widspread health problems can arise from seasonal malnutrition. Moderate capacity to cope with the effects of the shock locally; recovery occurs.	Widespread loss of animals from drought or contagious diseases. Moderate capacity to cope locally and slow recovery; some people fall into poverty traps.	Loss of income from poor market prices; regional production and income impacts can be widespread owing to shrinkage of the rural nonfarm economy. Moderate capacity to cope locally and quick recovery if assets are not lost as well; some people fall into poverty traps.	Less-severe drought or excess rainfall in critical periods, new pest outbreaks and animal diseases.
Low	High to medium	Deaths, accidents, and illnesses that affect a predictable share of the population each year. Some local capacity to pool these risks, but recovery from losses can be slow for the households involved.	Loss, damage, or disease of a predictable share of the total stock of homes or productive assets each year. Good local capacity to pool these risks, but recovery from losses can be slow for the households involved.	Low yields for some farmers due to a variety of localized weather and pest problems. Good local capacity to cope with these risks; recovery is usually quick.	Localized weather and pest problems (e.g., frost in a particular valley, pest outbreak in certain fields).

Source: WFP and IFAD (2010).

The severity of agricultural risks can be very high with catastrophic consequences. Hazell and Hess (2016) document the extent of events and losses due to natural disasters by region. They show that more than 7,000 natural disasters occurred between 1995 and 2015 worldwide, affecting a total of 4.3 billion people with damages estimated at US$2.3 trillion. In Africa, 1,145 natural disasters—including droughts, extreme temperatures, floods, storms, wildfire, earthquakes, tsunamis, mass movements, volcanic activity, and landslides—occurred in this period with 308 million people affected. In Asia, 2,977 natural disasters affected 3.8 billion people. In Latin America and the Caribbean, 1,268 natural disasters occurred with total damage of US$158 billion and 146 million people affected.

There are different options to manage agricultural risks, agricultural insurance being one of them. Actions taken before the risk materializes are known as ex-ante measures and actions taken after the fact as ex-post. Three main approaches can be used: risk reduction (ex-ante); risk mitigation (ex-ante); and risk coping (ex-post).

1. *Risk reduction* can occur in several ways: investments in hazard-resistant technology, such as irrigation systems and pest-resistant seed varieties, and through the diversification of income sources including off-farm employment and migration away from hazardous areas.

2. *Risk-mitigation* activities include crop insurance and saving. While saving is a risk-retention strategy, crop insurance is a risk-transfer strategy. In the former, the amount of savings must be large enough to be prepared for worst case scenarios either through own savings or by taking savings from others (credit). In the latter, only a fraction (typically a small fraction) of the potential losses are paid as a premium to secure the right to be compensated if a risk materializes; if the risk doesn't materialize, the premium is lost.

3. *Risk-coping* strategies when uninsured shocks hit include selling productive assets such as land and livestock, cutting back on consumption, and reducing investments in education among others.

Traditional Crop Insurance in Developed Economies

A key feature of a good agricultural insurance product is that it will compensate the farmers for losses that come after any negative event for the risks insured. Also, the insurance product should not induce inefficient production or behavioral decisions (to avoid moral hazard) and should let the insurance company discriminate between high-risk and low-risk clients (to avoid adverse selection). This is the case for any insurance market including agricultural insurance. Ideally when a farmer acquires agricultural insurance, he or she pays a premium. If a negative event occurs, for a risk that has been insured, which induces a loss in his agricultural income, he expects a compensation from the insurance provider equal to the size of the loss. In this sense agricultural insurance follows a similar logic to standard car or fire insurance. The insured customer pays a premium to the insurance company expecting to be compensated when a negative event coming from an insured risk triggers a loss. Standard insurance contracts are also denominated "indemnity" insurance since the client is "indemnified" by the insurance provider after verifying the size of the loss that is attributed to an insured risk. The size of the indemnity is in line with the size of the evaluated loss.

In practice, traditional agricultural insurance consists of multiple-peril or single-peril insurance instruments that require an insurance adjustor to physically verify assets (ex-ante) and losses (ex-post) on an individual farm basis. Under single-peril insurance, indemnities are paid for losses incurred for a single risk, such as hail. Under multiple-peril insurance, indemnities are paid for losses incurred from a broad range of risks. The World Bank (2005) provides an overview of agricultural insurance programs in the United States and Canada. These countries have been able to implement substantial programs aimed to reduce yield and revenue risk for agricultural producers. Although these programs offer a variety of risk-management products for farmers, in all cases programs require sizable levels of government support.

In the United States multiple-peril yield and revenue insurance products are offered through the Federal Crop Insurance Program (FCIP), a public-private partnership between the federal government and various private-sector insurance companies. A comprehensive review of how the FCIP expanded during the 1990s and first decade of the 2000s can be found in Glauber (2013). A key milestone is that under the Federal Crop Insurance Act of 1980, the Federal Crop Insurance Corporation was encouraged to privatize delivery functions to the maximum extent possible and to promote a rapid expansion. Currently the program is delivered entirely by private crop insurance companies and independent insurance agents (Glauber 2016).

By 2015 area insured under the program totaled almost 300 million acres, accounting for more than 85 percent of potentially insurable area and total liability (coverage in force) topped US$100 billion. Crop insurance is viewed as a key piece of the federal farm safety net, and its annual costs have grown significantly since 2000. Estimated annual costs are projected at $8.9 billion over fiscal years 2016–2025, making it the largest single farm program in the farm safety net. By contrast, price and income support programs are estimated to cost $5.6 billion annually, while conservation programs are estimated at $5.8 billion over the same period. In the case of US federal crop insurance, producers are charged premiums that reflect the underlying actuarial risks of crop production (and revenue), and the same rates apply to all companies (no price competition). The government provides separate reimbursements for administrative and operating expenses. Also, insurance companies must take all eligible participants regardless of risk profile, not being allowed to turn down customers or adjust rates.

In Canada, since 2003, there are two main schemes: production insurance and income stabilization. The production insurance (PI) scheme offers producers a variety of multiple-peril production or production value loss products like many of those offered in the United States. One major distinction, however, is that the Canadian program is marketed, delivered, and serviced entirely and jointly by federal and provincial government entities, and Canadian producers are not allowed to separately insure different parcels but rather must insure together all parcels of a given crop type. Combined, the federal and provincial governments cover approximately 66 percent of program costs, including administrative costs. Beginning in 2004, the Canadian Agricultural Income Stabilization (CAIS) scheme replaced and integrated former income-stabilization programs. The program generates a payment when a producer's current year production margin falls below that producer's reference margin (based on an average of the program's previous five-year margins, less the highest and lowest).

It is important to emphasize that most, if not all, traditional multiple-peril crop insurance programs in developed countries, as well as in developing countries, require substantial government support. Goodwin and Smith (1995) concluded that purely private markets for traditional crop insurance are not feasible because agricultural insurance provision is too costly. This would make commercial unsubsidized premiums high enough to discourage the demand for insurance among farmers. They argue that large informational problems (moral hazard and adverse selection) are the reason why the provision of crop insurance is too costly.[1]

---

1 Moral hazard occurs when insurance clients take riskier actions once insured while insurance companies cannot perfectly observe such actions, leading to higher indemnity payments than

One can ask why other insurance markets, such as car and home insurance, seem to work fairly well. One hypothesis is that information problems can be managed at a lower cost when insuring assets (cars, homes, equipment) than income (crop yields, unemployment), although we are not aware of rigorous research supporting this claim. In fact, Goodwin and Smith (1995) identify perils such as fire and hail for which private crop insurance do exist, although participation is small. One reason why such private markets are possible would be because measurement of such risks appears to be more straightforward, lowering monitoring and loss assessment costs. An open question for future research is whether new remote-sensing technologies could lower those costs, increasing the chance for commercially viable private crop insurance markets.

Insurance Opportunities for Small-Scale Farmers: Index-Based Insurance

In high-income countries farmers have access to and commonly use insurance and other financial products (for example, futures and options and credit) to protect themselves from shocks. Indemnity insurance products are available, although largely subsidized and supported by public funds. Access to these types of formal financial risk management products is essentially nonexistent in developing countries. Without formal insurance mechanisms, small-scale farmers in developing countries employ ex-ante risk-mitigation strategies such as relying on low-risk/low-yield production techniques, which can have negative livelihood consequences in the long run. Rosenzweig and Binswanger (1993) found that smaller and poorer farmers in a semi-arid region in India sacrificed 27 percent of their expected income to reduce risk. Similar results have been analyzed by others (Carter 1997; Morduch 1995).

In contrast, using a natural experiment among Chinese tobacco farmers, Cai (2016) shows that provision of agricultural insurance increases the insured crop production by 16 percent. Cole, Giné, and Vickery (2017) find that insurance provision induces farmers to invest more in higher-return but rainfall-sensitive cash crops. Jensen and Barrett (2017) document that when shocks such as droughts do occur, households might rely on detrimental

otherwise without insurance. "Adverse selection" describes a situation where a farmer's demand for insurance is positively correlated with his likelihood of crop (or income) loss while the insurance company doesn't have perfect information on that likelihood of loss. Farmers with high risk of crop loss will likely purchase insurance in larger amounts than farmers with low risk of loss, leading to higher aggregate indemnity payments that make insurance companies raise their premiums. This would discourage low-risk farmers even more from buying insurance, which ends up limiting the insurance market.

coping strategies that include selling off productive capital, skipping meals, and withdrawing children from school (Hoddinott 2006; Janzen and Carter 2019). In the case of pastoralists, droughts can decrease the herd size to a critical level, leading to a poverty trap (Chantarat et al. 2017). The threat of shocks (that is, risk), the shocks themselves (that is, adverse realized outcomes), and the strategies used to cope with them play a crucial role in the long-term well-being of rural households in developing countries (Rosenzweig and Binswanger 1993; Morduch 1995; Zimmerman and Carter 2003; Dercon 2004; Barnett, Barrett, and Skees 2008; Barrett and Carter 2013).

Ceballos and Robles (2014) identify several reasons why agricultural indemnity insurance has failed to expand successfully in developing countries. First, given that farms in developing countries are significantly smaller than in countries like the United States and Canada, traditional crop insurance products would have higher administrative costs as a percentage of total premiums. A portion of these costs are related to marketing and servicing (loss adjustment) insurance policies. Another portion is related to the lack of farm-level data and cost-effective mechanisms for controlling moral hazard. These costs are even larger if rural infrastructure is inadequate (low-quality roads). Moreover, the lack of formal financial service networks and legal records may add to the cost of premium collection and compensation disbursements.

Second, indemnity insurance is prone to significant information asymmetry problems such as adverse selection (whereby only the most at-risk farmers purchase insurance) and moral hazard (whereby an insured farmer may not exert optimal effort to reduce risk or mitigate its impact). Both problems generally result in an increased cost that would make purely private markets non-viable even in developed countries. Third, developing countries have more limited fiscal resources to support and subsidize indemnity products than do developed countries; and since larger segments of the population are engaged in farming, the scarcity of fiscal resources is even more acute. More important, the opportunity cost of those limited fiscal resources may be significantly greater than in a developed country. Fourth, developing countries have far less access to global crop reinsurance markets than do developed countries. Reinsurance contracts typically involve high transaction costs related to due diligence. For a global reinsurer to be willing to enter a market, the enabling environment must foster confidence in contract enforcement and institutional regulations. An enabling environment is, in fact, a prerequisite for effective and efficient insurance markets, and these components are largely missing in developing countries.

Index insurance is promoted as a low-cost alternative for conventional insurance products (Alderman and Haque 2007; Barnett, Barrett, and Skees 2008; Hazell et al. 2010; Mahul and Stutley 2010). With index insurance products, payments are based on an independent measure regarded as highly correlated with farm-level yield or revenue outcomes. Unlike traditional crop insurance that attempts to measure individual farm yield loss or revenue loss, index insurance makes use of variables exogenous to the individual policyholder—such as temperature or rainfall measurement—but have a strong correlation to farm-level losses. The index can also be an average in an outcome related to loss over a small area, such as average crop yield or livestock mortality rate. They can be estimated using statistical sampling (such as random crop cuttings) or average pasture damage based on satellite observations.

As an example of a typical weather index insurance product, consider the one presented by Ceballos, Kramer, and Robles (2018) to insure Indian wheat farmers against excessive unseasonal rains. They use the maximum accumulated rain (in millimeters) over any two consecutive days during February through April as the index that could trigger a payment. And that happens when the index takes a value equal to or higher than a number called the index strike. The higher the index value, the higher the compensation per acre to the farmer up to a maximum that reflects the cost of production per acre. The index value is computed using data from a local reference weather station. It is relevant to discuss the following aspects that are common to most agricultural index insurance products.

The Index

The selected index must be highly correlated with yield loss or revenue loss for farms across a geographic area. Ideally one would like to use data on historical yields and rainfall to establish a model (mathematical, statistical, or agronomical) to help select the best index. Unfortunately, in many rural contexts in developing countries, such data are not available and/or the data quality is very low. Consultation with farmers and local experts can help identify the best index selection. The index must satisfy other properties that would make clients trust the product.

First, the index must be observable and easily measured by an independent party. Typically the national meteorological agency is the best candidate to measure weather data as it has the expertise and reputation to do it as part of its regular activities. Based on daily weather records the independent party can periodically (daily) compute the index and make the information publicly available through, for example, a website or local media. Second, the index must be calculated objectively following well-defined protocols known to all parties involved, including farmers. Such protocols should include the data-collection process, the specific formula for index value calculation, and how to handle contingencies such as temporary weather station malfunctions.

Third, the index must not be subject to tampering and human manipulation. This implies setting supervision protocols and restricting access to weather stations. In a 2015 weather index program implemented in Uruguay, automatic weather stations were installed in police stations to ensure proper supervision. Having algorithms that can quickly detect abnormal changes in rainfall or index values can help detect tampering and/or malfunctioning. Finally, the index should be reportable in a timely manner. When insured farmers suffer a crop loss, the faster the insurance payment arrives the better. Working with automatic weather stations or satellite data can significantly reduce the time to report the index and determine payment amounts and that can quickly reach the affected farmers.

The Payment Schedule

Payment amounts are estimated ex-ante, based on the best available prediction on what could be the yield loss for every possible value of the index and considering partial and full crop loss. In general, the payment schedule is structured as a function that goes from index to payment, and in principle this function could take different shapes depending on how the estimated crop loss is related to the index. In practice, most index products take a linear payment function to compensate for ex-ante estimated partial crop loss up to a point beyond which full crop loss is assumed.

Main Advantages of Index Insurance

There are three main reasons why index-based insurance has been regarded as having great potential to reach smallholder farmers in developing countries.

1. *Lower administrative costs.* Given that payments are based only on publicly observed data, inspections of individual farms are not required, thus reducing drastically loss verification costs. The cost of sending an agent to assess yield loss at a remote small farm can be quite large relative to the insurance compensation, making premiums too high to be financially viable. With index insurance this cost is eliminated and replaced by the cost of generating, collecting, and reporting index data.

2. *Asymmetric information problems are minimized.* Two classical problems in the insurance literature are moral hazard and adverse selection.

Deductibles, copayments, or other partial payments for loss are commonly used by traditional indemnity insurance providers to mitigate these problems, although the evidence for agricultural insurance in particular is that informational problems are high enough to prevent the existence of purely private all-risk crop insurance markets. However, index insurance is not subject to moral hazard as the likelihood of making payments does not depend on the client's behavior but only on the index value. Clients have no incentive to take extra risk as they will not be protected for the additional risk. Similarly, adverse selection problems are much lower with index insurance because the likelihood of insurance payments is not affected by the type (high or low risk) of farmer buying insurance but only by the probability that the index will trigger payments. High risk farmers might select themselves in buying more insurance compared to low risk farmers, but this will not be an "adverse" selection as it will not have an impact on the average insurance payments. Also, it is very unlikely that farmers will have better information than the insurer regarding the probability distribution of the index value. Because index insurance is not subject to asymmetric information problems, there is less need for deductibles and copayments and few restrictions need to be placed on the amount of coverage a farmer decides to purchase.

3. *Quick payment disbursement.* Index insurance compensations can be quickly determined based solely on index readings and the preestablished payment schedule. Therefore payments can be disbursed quickly to farmers not long after the end of the coverage period (in some cases even before). New technologies are helping to make the disbursement compensations even faster. With automatic weather stations that can report weather variables online, the index can be computed and reported at a very high frequency (daily) and compensations calculated immediately. Also, with more farmers accessing mobile banking, they will see payments in their accounts as soon as the insurer and corresponding bank authorize them.

Index Insurance Challenges

Although index insurance has some important advantages that make it a promising product among smallholder farmers in developing countries, it presents some disadvantages that prevent an agricultural microinsurance revolution based on index-based products.

BASIS RISK

"Basis risk" refers to the potential mismatch between index-triggered payouts and actual losses. It occurs when an insured farmer has a loss and does not receive an insurance payment sufficient to cover the loss that resulted due to the insured risk. This is known as "downside basis risk." The opposite situation, referred to as "upside basis risk," takes place when a farmer has a loss (or no loss at all) and receives a compensation that exceeds the loss amount. Clarke (2016) formally defined basis risk as the unconditional probability of both experiencing a critical loss and not receiving an insurance payout (downside basis risk) or as the unconditional probability of both not experiencing a critical loss and receiving an insurance payout (upside basis risk). The worst situation would be one in which an insured farmer has total or close-to-total crop loss and the observed index does not trigger any payment.

According to Ceballos (2016), a broad characterization of the different components of basis risk can be thought of as (1) spatial basis risk and (2) design risk. "Spatial basis risk" is related to the geographic variation between the index being measured at the farmer's plot and the index being measured somewhere else—for example, at a distant weather station. Since the relevant rainfall that explains crop yields is the rainfall at the farmer's plot, the larger the distance between the reference weather station and the farm of the insured farmer the lower the expected correlation between the index (computed with data from the reference weather station) and crop yields. "Design basis risk" is related to the design of the insurance product, including the choice of the weather variable(s) and the specific index as well as the functional form of the payment schedule. Even if there were a weather station at every farm (no spatial basis risk), the correlation between crop yield loss and the index can be far from perfect. Also, how specifically one links the index to payments (for example, linearly) could be another source of basis risk. The more data and knowledge from experts, the higher the chances to select a better index and payment schedule to maximize the correlation with farmer losses.

In practice, there is a third source of basis risk that is related to idiosyncratic farmer characteristics. In the design of an index product one collects the best data and expert knowledge available to design a product targeting the typical or average farmer in a region. However, farmers could be heterogeneous in several dimensions, including farming practices and plot characteristics (such as soil quality) that make them depart from the "average" regional farmer. Therefore the index product might not perfectly fit the risk profile of any farmer other than the "average" one.

An index product with too much basis risk is an insurance product that offers poor insurance coverage as payments do not come when most needed (high losses) and might come when not needed (no loss). It is expected that above a certain level of basis risk a farmer would not demand the index product (see Clarke 2016 and Hill, Robles, and Ceballos 2016 for a formal discussion). Also, basis risk might introduce behavioral aspects that would make index-based products even less appealing. For example, Carter, Elabed, and Serfilippi (2015) show that in the presence of basis risk, farmers might become excessively sensitive to insured basis risk in a way that is not consistent with standard expected utility theory.

LACK OF INFRASTRUCTURE AND INFORMATION

To design index products, historical data are needed to correctly estimate probabilities of triggering payments and to price the product. In developing countries, especially in rural areas, this could be a serious limitation as such historical data are nonexistent or of poor quality. Data limitations have been emphasized by Miranda and Farrin (2012) and others as a serious constraint to the expansion of index insurance programs beyond small-scale pilots, and they suggest that governments and donors should take responsibility for collecting and maintaining index data, given their public-good nature. Similarly, a dense network of weather stations is needed as one would like to limit spatial basis risk. Again, in developing countries these networks are often not very dense. One way to overcome these limitations is by using satellite information to proxy for weather information and/or other variables that can inform about crop yields (greenness index, area leaf index, and so on).

COMPLEXITY

Index products are not necessarily easy to understand for potential clients, especially as the level of education among smallholder farmers in developing countries is unfortunately still low. As farmers might not easily understand an insurance product that pays based on readings from a distant weather station, or even worse based on satellite information, they might be reluctant to purchase it. Several studies reviewed in the next section show how lack of product understanding explains low insurance demand. While private insurers might invest in marketing their products, Hazell and Hess (2016) suggest that their educational efforts will be far from socially optimum levels and that government and donor interventions will be required. Also, designing and pricing index products is not straightforward and requires professionals with relevant expertise. In many instances these professionals might not be available locally or among the staff of local microfinance institutions that would be natural candidates to design and offer index products to smallholder farmers. This might limit the supply.

The Current State of Agricultural Index Insurance in Developing Countries

Hazell and Hess (2016) estimate that the total number of insured smallholders in 2014 in the developing world is 198 million, with approximately 650,000 in Africa, 3.3 million in Latin America and the Caribbean, and about 194 million in Asia, including 160 million in China and 33 million in India.[2] Other estimates are provided by the regional landscape studies conducted by the Micro Insurance Centre, as part of the Microinsurance Network's World Map of Microinsurance (Microinsurance Network n.d.). A total of 1.1 million Africans were identified as being covered by agricultural microinsurance as of the end of 2014 (Micro Insurance Center 2016). Since 2014, index-based agricultural microinsurance has become more popular in Latin America and the Caribbean; the number of people covered by agricultural insurance has increased by 129 percent, from 35,000 in 2014 to 80,000 in 2017 (A2F Consulting 2018). In Asia and Oceania the number of people covered by agricultural microinsurance in 2012 is estimated at 23.8 million, and agricultural products were registering the highest growth of all microinsurance product types. These numbers do not include social agricultural microinsurance schemes, which are mainly delivered as products tied to crop loans in these countries. China's PICC agricultural insurance scheme is the largest scheme and covers more than 100 million individuals. India's National Agricultural Insurance Scheme and Pakistan's Crop Loan Insurance are also large programs with an outreach of 16.3 million and 10.5 million respectively.

The most comprehensive study assessing the extent of agricultural insurance around the world was conducted by Mahul and Stutley (2010) based on a survey of 65 countries and showing the situation through 2007. Hazell and Hess (2016) summarize very well the most salient facts of this study, among them:

2 See Hazell and Hess (2016), Appendix 1, for a detailed list of agricultural insurance programs in the developing world.

- Lower-middle-income and low-income countries accounted for only 7.5 percent of the total agricultural insurance premiums (including premium subsidies) of US$15.1 billion.
- Market penetration remains small, even in rich countries. The total insurance premium collected (including subsidies) amounted to 0.9 percent of agricultural gross domestic product, ranging from virtually zero in low-income countries to 2.3 percent in high-income countries (5 percent in North America).
- 82 percent of countries offered both crop and livestock insurance, but crop insurance accounted for 90 percent of the premium.
- Multiple-peril crop insurance was available in 65 percent of the countries but was most popular in the middle-income countries. Named peril insurance was even more widely available (in 69 percent of countries) and was even available in half of the low-income countries.
- Index-based crop insurance is available, mainly at a pilot stage and with very low aggregate premium volume, in one in three surveyed countries. Area-yield insurance was available in 15 percent of the countries, and weather-index insurance in 22 percent of countries. Index-based insurance had also penetrated the low-income countries; 17 percent had area-yield insurance and 33 percent had weather-index insurance.

Although index-based insurance has been regarded as having great potential to reach smallholder farmers in developing countries, attracting a lot of attention from donors, international organizations, governments, researchers, and microfinance institutions, most programs have repeatedly experienced low take-up, due to among other reasons basis risk (low-quality products), lack of trust in the insurance company, lack of understanding of the product, liquidity constraints, or crowding out of insurance by implicit public guarantees. Currently India is the country with the largest partially subsidized agricultural index insurance program with more than nine million farmers purchasing index products, although this can be explained to a large extent by the fact that agricultural insurance is mandatory for those who want access to agricultural credit.

Some of the most successful cited index-based insurance programs in the developing world are the following (see Greatrex et al. 2015 for a full description of these programs):

- The Indian National Agricultural Insurance Scheme (NAIS) started in 1999 and over time has been complemented by the modified NAIS (mNAIS) and the Weather-Based Crop Insurance Scheme (WBCIS). These programs insure cereals, millets, pulses, oilseeds, and horticultural products. In 2013 the programs provided coverage to 16.79 million farmers under NAIS, 3 million under mNAIS, and 13.62 million under WBCIS. Considering that the number of farmers in 2011 is estimated at 118.7 million, insurance penetration was at most 28 percent at the country level.[3] The programs are state-subsidized.
- The Agriculture and Climate Risk Enterprise (ACRE, formerly Kilimo Salama) started in 2009 and now operates in Kenya, Rwanda, and Tanzania. In 2013 it provided coverage to more than 187,466 farmers, 60 percent in Kenya and 40 percent in Rwanda, for the following crops: maize, beans, wheat, sorghum, coffee, and potatoes. The program has strong links to aggregators and mobile technology and offers a wide range of products, mostly linked to credit or inputs.
- R4 Rural Resilience Initiative (formerly HARITA) started in 2009 in Ethiopia and now operates in Senegal as well. In 2014 it provided coverage to 24,133 farmers in 82 villages in Ethiopia and 1,989 in Senegal to insure teff, beans, maize, wheat, barley, sorghum, and millet. This is a farmer-led, integrated risk management project in which farmers can pay premiums with labor employed in risk-reduction projects. Satellite rainfall indexes are used in the design of products.
- Index-Based Livestock Insurance Project in Mongolia was created in 2006. In 2014 it provided insurance for livestock (camels, cattle, sheep, goats, and horses) of approximately 15,000 herders. This is a public-private partnership with innovative risk layering.
- Index-based livestock insurance started in 2010 in Kenya and provides protection against drought to pastoralists with camels, cattle, sheep, and goats (Chantarat et al. 2013). Approximately 3,000 contracts have been sold up to 2016. This is an innovative cattle mortality index-based insurance based on satellite data and a normalized difference vegetation index (NDVI).

3 See India, Ministry of Agriculture (2018).

Index Insurance Take-Up: Empirical Evidence

Most agricultural index insurance pilot programs in developing countries show that the overall demand for index products is still very low even in the presence of large premium subsidies. Among multiple studies, in India, Gine, Townsend, and Vickery (2008) report a lower than 5 percent take-up rate in Andhra Pradesh, and Cole et al. (2013) report that in 2006 one contract was estimated to be sold for every 13.2 households in Indian villages where rainfall insurance contracts were available and households that buy insurance generally purchase just one or two policies, hedging only a modest fraction of monsoon agricultural income. Hill, Robles, and Ceballos (2016) report overall take-up rates of 6.8 percent and 4.0 percent during 2011 and 2012, respectively, in Madhya Pradesh. A more encouraging study is Karlan et al. (2011), reporting a 40 percent to 50 percent take-up in Ghana at fair price plus a 50 percent loading, although there is an income effect as farmers were offered grants to access insurance products. In sum, uptake to this date has been generally disappointing, and the empirical and academic literature on demand for index insurance has mostly focused on household and contract characteristics that vary within a given project (Jensen and Barrett 2017). This approach is useful for conducting econometric analysis to study marginal impacts in the demand for insurance.

Basis risk is a usual suspect to explain low demand for index products. The higher a product's basis risk, the less hedging it provides to a farmer (Miranda and Farrin 2012; Binswanger-Mkhize 2012; Clarke 2016). Studies that quantify basis risk find that it can be considerable. Jensen, Barrett, and Mude (2016) find that in the Index-Based Livestock Insurance (IBLI) program in northern Kenya, policyholders are left with an average of 69 percent of their original risk; in other words 69 percent of all potential losses are not compensated by the insurance product. This is surprisingly high as the IBLI product is regarded as close to a best-case scenario since the index was selected based on a regression model using a long panel of household data to expressly minimize basis risk.

Although IBLI reduces exposure to covariate risk by an average of 63 percent, the presence of idiosyncratic risk is high, accounting for a large overall basis risk. Clarke et al. (2012) find in the Weather Based Crop Insurance Scheme (WBCIS) in India that the relationship between claim payments and yields appears to be rather weak, with low average claim payments in the event of extreme yield losses (in the event of a zero yield, the average WBCIS claim payment is only 12 percent of the sum insured). They estimate that conditional on total crop loss (average crop yield of zero), there is a one-in-three chance of receiving no claim payment from the WBCIS. Other studies estimate by how much exogenous variations in basis risk (proxied by distance to weather stations) reduces demand for index products. Hill, Robles, and Ceballos (2016) find that doubling the distance to a reference weather station decreases demand by 18 percent. Mobarak and Rosenzweig (2012) estimate that for every kilometer increase in the (perceived) distance of the weather station for a farmer without any informal risk protection, there is a drop-off in demand for formal index insurance of 6.4 percent. Karlan et al. (2014) and Jensen, Barrett, and Mude (2016) also study the impact of basis risk on demand, although they don't have access to exogenous variations to basis risk.

Lack of product understanding is also regarded as an important driver for low take-up of index insurance. Index-based products might not be easy to understand, which is made even more problematic by the low levels of formal education among small farmers in developing countries. For example, Hill, Robles, and Ceballos (2016) report that on average farmers with 3.5 acres of land have less than 4.5 years of formal education in a selected sample of relatively better-off small famers in Madhya Pradesh, India. In Karlan et al. (2014) only 15 percent of farmers in the selected sample are literate; these are farmers that on average have less than 8 acres of land. Similarly, Cole et al. (2013) show that 67 percent of household heads in Andhra Pradesh and 42 percent in Gujarat have at most primary school education.

Also, many smallholder farmers have no experience with formal measurements of weather variables and, for example, cannot relate to rainfall measured in millimeters per day. Satellite data, normalized difference vegetation index, and leaf area index are all unknown concepts for most of them. And experience with any type of formal insurance contract or product is also not common. All these elements make the understanding of index products difficult. Hence promoting the demand for index-based products requires educating farmers on the general concepts of insurance and the specifics of index insurance products. Experimental studies have found that improved understanding of the product is an important demand driver. Hill, Robles, and Ceballos (2016) find that the impact that increased investments in training have on demand for weather-index products is significant in the short term. Households that receive more intense training are 5 percentage points more likely to purchase the insurance product in the season immediately following the training. Gaurav, Cole, and Tobacman (2011) find that a two-day educational program, involving games that simulate rainfall insurance, did increase rainfall insurance demand by 5 percentage points.

Another challenge to demand is trust. Index insurance pilots often are implemented in regions where insurance companies have little or no presence and where farmers have no previous experience with insurance products in general. In these regions there is little legal recourse in reclaiming insurance payments. For index insurance, the expected payout is difficult to know because the relationship between weather and loss is imperfect. And when the insurance provider is better informed on risk, farmers must rely on the insurance company in setting a fair price. Relations of trust with the insurance provider and its agents are thus very important for uptake. Lack of trust resulting in low willingness to pay for index insurance has been confirmed experimentally. Cole et al. (2013) show that endorsement of the insurance product from a trusted third party increased uptake by 40 percent compared with farmers who heard of no endorsement. In the context of an insurance for sows and a large randomized natural field experiment conducted in southwestern China, Cai et al. (2009) provide several pieces of evidence suggesting that trust, or lack thereof, for government-sponsored insurance products is a significant barrier for farmers' willingness to participate in the insurance program.

Cai et al. (2015) find that in China trust is established by experiencing payouts to oneself or by witnessing payouts to members of your social network. Payouts were the main instrument in building trust, suggesting that trust can be increased by initial subsidies to boost demand and increases the likelihood of observing actual payouts, or by increasing the frequency of payouts by insuring smaller high-frequency losses in addition to larger low-frequency losses (Carter 2009). Dercon, Gunning, and Zeitlin (2016) have developed a theoretical model of insurance demand under limited trust and using field and laboratory-experimental data from a randomized controlled trial introducing a composite health insurance product to tea farmers in Kenya. They find evidence that limited trust is an important barrier to the adoption of insurance, particularly among poor and risk-averse households.

Hellmuth et al. (2009) document that (1) in an Indian contract farming case study involving PepsiCo, farmers trust the program because of its timely settlement of claims; (2) Oxfam's long-standing presence in Ethiopia, combined with high levels of respect and trust for its local partner NGO, the Relief Society of Tigray (REST), made farmers willing to explore weather-index insurance; and (3) feedback from Thai farmers suggested that the primary motivation for purchasing insurance was trust in the Bank for Agriculture and Agricultural Cooperatives (BAAC), an institution that has long-term relationships with the farmers. They conclude that the lesson is clear: identifying partner organizations that farmers already trust is critical to successfully scale up index insurance programs. Government regulation and providing legal recourse to farmers is also important in building trust on insurance schemes because premiums are generally paid at the onset of insurance contracts, leaving farmers with the risk associated with contractual nonperformance. Karlan et al. (2014) conclude that trust is a key issue that can be tackled through product design (increasing states of the world with payouts), proper linkage with trusted institutions, and proper regulation.

Demand for index insurance is price sensitive. Jensen, Barrett, and Mude (2017) report that published estimates of price elasticities range from −0.35 to −1.16, and often demand remains very low even when premiums are heavily subsidized. Cole et al. (2013) found insurance demand in the Indian states of Andhra Pradesh and Gujarat to be highly price elastic, with a 10 percent reduction in the price of insurance associated with a 10 percent to 12 percent increase in insurance take-up. Karlan et al. (2014) find that 40 percent to 50 percent of Ghanaian farmers purchased insurance at actuarially fair prices, but take-up rates dropped to 10 to 20 percent when charged double the actuarially fair price. Mobarak and Rosenzweig (2012) find the price elasticity of insurance demand to be −0.44, and Hill, Robles, and Ceballos (2016) estimate the price elasticity as −0.58 and show that price elasticity is sensitive to different levels of basis risk. They find that households located less than 5 kilometers from a weather station have a sensitivity to price almost 10 times higher than that of those located more than 12 kilometers from a weather station.

Standard insurance models predict that in the absence of basis risk, demand for insurance is increasing with the degree of risk aversion. However, Clarke (2016) provides a theoretical framework predicting that in the presence of basis risk and premiums above the actuarially fair price there is a hump-shaped relationship between demand and risk aversion: insurance demand is initially increasing with risk aversion before decreasing such that, for very risk-averse farmers, purchasing insurance makes them worse off. Hill, Robles, and Ceballos (2016) provide indicative evidence of the predicted hump-shaped demand for index insurance.

Liquidity constraints have been shown to be important to demand rates across a variety of products and populations, which suggests limitations in the use of commercial index insurance as a tool for helping the poorest subpopulations (Jensen, Barrett, and Mude 2016). Note that farmers would have to buy agricultural insurance at the start of the growing season, when there are many competing uses for limited funds increasing the opportunity cost of insurance. Giné, Townsend, and Vickery (2008) provide empirical evidence that credit constraints reduce the probability of take-up by 30 percent. Cole et al. (2013) find suggestive experimental and nonexperimental evidence that liquidity constraints reduce demand. After randomly assigning certain households with high cash rewards (enough to buy one policy), they observe that take-up increases by 140 percent with a magnified effect among poor households, which are likely to have less access to credit markets.

Jensen, Mude, and Barrett (2018) find suggestive evidence that reducing liquidity constraints is associated with an increase in the demand for index-based livestock insurance (IBLI) in northern Kenya. Two studies remove credit constraints by directly postponing premium payments until harvest time. Liu et al. (2016) finds that for a government-managed livestock mortality insurance in China, delaying premium payment increases take-up from 5 percent to 16 percent. Casaburi and Willis (2018) show that in the context of a randomized controlled trial in Kenya, take-up goes from 5 percent to 72 percent among sugarcane producers. However, insurance companies will not be willing to postpone premiums until harvest time unless new mechanisms to enforce premium payments are available. In Casaburi and Willis (2018), default on premiums is less of a concern since the product is interlinked with a contract farming scheme.

Most of the empirical work studying the demand for index insurance has the standard expected utility theory as the theoretical framework. However, recent work based on behavioral economics provides additional explanations for low index insurance take-up. Carter, Elabed, and Serfilippi (2015) find that substantial numbers of farmers depart from expected utility behavior in ways that predict excess sensitivity to uncovered basis risk in index-based insurance contracts. Experiments by Elabed and Carter (2015) find that farmers perceive index-insured risk as a compound lottery with uncertainty around the insured risk and uncertainty around how well the index will reflect their losses (basis risk). They find that 60 percent of their sample from southern Mali are compound risk averse and that in the presence of moderate basis risk, the levels of compound risk aversion observed in their sample could cut demand for index insurance in half, amplifying the already detrimental impact of basis risk on demand.

In a dynamic setting, an interesting piece of evidence shown by Cole, Stein, and Tobacman (2014) is that demand is highly sensitive to payouts being made in a household's village in the most recent year, with the effect being stronger when more individuals in a village receive payouts, and the effect remains positive over multiple seasons, but the estimated size decreases over time. As pointed out by the authors, these results stand in contrast to standard rational models, in which the realization of recent insurance outcomes should not affect forward-looking insurance decisions.

Challenges and Recommendations to Scale Up Agricultural Insurance in Developing Countries

The main challenge of agricultural insurance in developing countries is how to create sustainable agricultural microinsurance markets. To meet this challenge, the following three elements are needed: (1) designing high-quality products that are commercially viable with minimum possible government support, (2) existence of a large and sustained demand, and (3) presence of a competitive supply mainly from the private sector. Given limited resources in developing countries and many other sectors requiring attention, promoting strong private-sector participation in agricultural microinsurance markets complemented with well-targeted government support is the most realistic option.

The evidence with traditional all-risk indemnity insurance programs in developed countries is that they require massive government support because of high costs due to information asymmetries (moral hazard and adverse selection), let alone the political economy behind disguised transfers from tax payers to farmers as premium subsidies. The small-scale nature of farming in developing countries would add costs to such traditional insurance schemes, making them nonviable unless new technologies or approaches make them feasible at affordable rates to either small-scale producers or governments willing to provide support within their limited budgets. In a broader context agricultural microinsurance markets should provide incentives to adopt best agricultural practices rather than substitute for them, complement other efficient risk management strategies such as risk-reduction investments including newly available seed varieties that have tolerance to abiotic shocks, such as flood, drought, and extreme temperatures, and when possible become an ex-ante system for government disaster risk management.

High-Quality Insurance Products Commercially Viable

A high-quality agricultural insurance product is one that compensates farmers when and only when they suffer a crop or income loss and the amount of the compensation is in direct proportion to the size of the loss. It is known that traditional indemnity insurance products are in this sense high-quality products but unfortunately, because of information asymmetries and monitoring and loss assessment costs, they are not commercially viable. Index-based products, however, have low monitoring and loss assessment costs and minimize if not eliminate information asymmetries, but they introduce basis risk, the imperfect relationship between the indemnity payments and the actual insured losses. When basis risk is too high, the insurance product is more like a lottery ticket. Most efforts are being placed in trying to reduce basis risk for index-based products.

Recent innovations in index insurance design are worth noting (Jensen and Barrett 2017). The ACRE product in Kenya provides a menu of coverage options, each developed to cover a specific agricultural phase, allowing clients greater flexibility in the risk that they choose to insure. The index-based livestock insurance (IBLI) product in Kenya used historical household-level data and econometric methods to statistically minimize expected basis risk. Hill and Robles (2010) have proposed an approach that is based on a variety of weather contracts in a region and then allow individual farmers the flexibility to form their own portfolios of weather contracts that best match their own crop mix and locational characteristics.

The following actions will help in designing better index products (see Hazell and Hess 2016 and Carter et al. 2017):

- Increase the number and dispersion of weather stations to better capture the spatial diversity of farming conditions in a region. Technological advances are rapidly reducing the cost of adding automatic weather stations, and in some countries private firms offer weather station services for a fee (for example, India). Although additional weather stations could add to the cost of developing and marketing index contracts, and as new weather stations come without site-specific historical records, the calculation of "synthetic" datasets based on the triangulation of existing historical weather data will be required.

- Invest in agrometeorological research and crop-weather modeling to identify weather indexes that minimize basis risk for as many households as possible in a region given the available weather data.

- Insure groups of farmers who can pool basis risk among themselves through local institutions. This idea is based on the theoretical work of Dercon et al. (2014) that has been explored in a series of insurance games by McIntosh, Povel, and Sadoulet (2019).

- Limit the insurance to the kinds of low-frequency, high-impact weather risks that affect most people in a region at the same time. In this case individual losses would be more highly correlated with the insured weather station event (Giné, Townsend, and Vickery 2007).

- In the absence of weather stations and any local weather data at all, area-based yield insurance might be a viable alternative. Mongolia has pioneered a livestock insurance program in which the index is a county-level livestock mortality rate measured through an annual livestock census (Hellmuth et al. 2009).
- Complement the primary index-based contract with a secondary contract that can be index-based over a broader area or damage assessment–based (Carter et al. 2014), minimizing as much as possible downside basis risk by adding an audit rule (see Flatnes and Carter 2015) that, for example, is triggered when more than a certain fraction of farmers after not being compensated request a sample of crop-cutting exercises.

There has been a lot of recent innovation in developing indexes that can be assessed remotely with satellites at a high resolution (for example, 5 hectares), such as cloud cover, vegetative cover, or soil moisture content for a chosen region during critical agricultural periods.

A more recent and innovative approach—picture-based insurance (PBI) aiming at developing high-quality insurance products—has been proposed by Ceballos, Kramer, and Robles (2019). It is based on taking regular geo-referenced pictures using smartphones as a mechanism to deliver plot-level assessments of damage at low cost. This approach takes a different direction in the sense that it moves away from a standard index-based product and goes back to the idea of traditional indemnity insurance but using new technologies to reduce loss assessment verification costs. The aim of PBI is to combine key advantages of both index-based insurance (timely compensations without expensive loss assessments) and indemnity insurance (minimum basis risk and an easy to understand product).

The product takes advantage of recent advances in image processing for near-surface remote sensing (see Richardson et al. 2018), aims to apply machine learning in developing algorithms that can capture accurately losses from a time-series of images, and expects that direct participation of farmers in the insurance process will increase understanding of and trust in the product. While PBI is being tested at a small scale with wheat farmers in India, interestingly Kramer et al. (2017) find indicative evidence that farmers are willing to pay more for PBI than for a comparable weather index–based product. The main challenge of this approach would be to overcome possible moral hazard and adverse selection issues and to enforce that insured farmers comply with picture taking according to certain protocol. One option to be explored is to link PBI with agro-advisory services based on the same time-series of images. In any case, PBI is a work in progress.

Large and Sustained Demand

So far, the only forms of index insurance to be adopted at scale have been area-yield insurance in both China and India, which are heavily subsidized, and in the Indian case, are compulsory for farmers who take agricultural loans from state banks. Everywhere else in developing countries the demand is very low, limiting the interest of private providers to participate in these markets. To have a larger demand, the first step is to have a high-quality product and more generally a value proposition to farmers. As explained earlier, a high-quality index product is one with no or small basis risk. But besides having a high-quality product, there are other options that can help in inducing a higher demand.

A major challenge to demand is client education and trust. Here there is a role for government intervention by providing insurance literacy programs among potential clients. Also, the government has a critical role in providing a sound regulatory environment to provide standards for consumer protection and improving trust. This should include standard insurance regulations, such as minimum capital-to-liability holdings requirements for insurers and reinsurers, clear index certification processes, a process for speedy and accessible disputed settlement resolutions and regulations, allowing for unconventional insurance agents such as NGOs or microfinance institutions (Jensen and Barrett 2017), which are more familiar to small farmers.

Developing quality standards and metrics for measuring and disseminating information on product quality is highly recommended (Carter et al. 2017). Assuming a latent untapped demand for high-quality insurance exists, if consumers could distinguish between two equally priced products of differing quality, market forces could be relied on to reward investments in quality (Jensen and Barrett 2017). Providing smart subsidies is another possibility (Hazell and Hess 2016). The evidence shows that agricultural microinsurance is unlikely to ever scale up quickly without increased levels of public support by governments and donors. Provision of subsidies is more obvious when insuring disaster relief agencies, given difficulties in recovering costs from poor beneficiaries. More generally, if subsidies are not used carefully, they might encourage farmers to take on too much risk and environmentally damaging practices, increasing their dependence on future subsidized assistance. For using subsidies in smart ways that avoid creating disincentive problems, or becoming a financial burden on the state, the following principles should be applied:

- Subsidies will be less distorting if made directly to the insurer to offset administration and development costs rather than subsidizing the premium rates paid by farmers.
- If premium rates are to be subsidized, it is better to do this on an ad-valorem—that is, per farmer basis—rather than on a risk-premium proportional basis, to benefit smaller and poorer buyers who buy smaller amounts of insurance.
- The final subsidized net premium for the farmer should not be lower than the pure risk, so that farmers will not have an incentive to gamble the system or plant the wrong crops in the wrong areas.
- There should be an explicit exit strategy or strategy for long-term financing.

Expanding index insurance beyond farmers may be desirable, as it has the potential to insure many other types of rural people who are engaged in non-agricultural activities that are dependent directly or indirectly on local agriculture—for example, agricultural traders and processors, landless workers, and village shopkeepers.

Recently Casaburi and Willis (2018) show that postponing premium payments until harvest time can go a large way in boosting insurance take-up. As they rightly point out, the gains from insurance arise from the transfer of income across states of nature, and by requiring premiums to be paid upfront, insurance products will also transfer income across time with no value-added. Credit and savings are financial instruments much better suited to transfer income across time. The reason why premiums are required to be paid upfront by insurance companies is to avoid defaults on premium payments. Therefore innovations that would allow postponement of premium payments while at the same time minimizing defaults are required.

Carter et al. (2017) argue in favor of offering index-based insurance to a mesolevel institution—such as a bank that might offer agricultural loans, an administrative entity, or a producer organization—as a potentially attractive option. In the case of banks, they worry about large correlated shocks that might affect their entire portfolio of agricultural loans in a given region. This implies that basis risk is less of a problem for such a mesolevel institution. Farrin and Miranda (2015) show in the context of a heterogeneous agent model that when banks lay first claim on indemnities there is a reduction in credit default, which can decrease interest rates and expand credit access among farmers. De Janvry, Dequiedt, and Sadoulet (2014) also support the idea of insuring farmer groups, especially when farmers are members of groups with common interests.

Competitive Supply Mainly from the Private Sector

Index insurance needs insurance companies willing and capable to sell the product in rural communities. In most cases available insurance companies and brokers have a well-established presence in urban centers but are typically unfamiliar with the rural customer base. Before index insurance can be offered, sales agents need to be trained in the new product, and since farmers may not trust insurance agents from outside their region, an entirely new sales force may need to be recruited and trained (Jensen, Barrett, and Mude 2017). If the insurance is linked to other products (for example, fertilizer, seed, or credit), additional legal arrangements also need to be developed.

Availability of reinsurers for specialized agricultural insurance products in developing countries is limited, and typically they offer very high rates that end up being reflected in high premiums. Giné, Townsend, and Vickery (2007) emphasize the importance of having reinsurance mechanisms to diversify regionally covariate weather shocks, and Carter (2013) points out that in the absence of adequate high-quality data, uncertainty-averse reinsurers place a large premium penalty on uncertainty. Regional or domestic public reinsurance could be put in place to incentivize a more competitive behavior among reinsurers. For example, the African Risk Capacity, a specialized agency of the African Union, provides alternative risk pooling and transfer mechanisms that may directly or indirectly improve the reinsurance market for index insurance.

To address the problem of collecting premiums and making payouts in a timely and cost-effective manner, some insurers are taking advantage of mobile phone and mobile banking technologies. A good example is the ACRE program in East Africa, which enables farmers to pay their insurance premiums and receive payouts via the M-PESA mobile banking system (Hazell and Hess 2016). On the supply side of microinsurance products there is also a first-mover problem: the high initial investment costs in research and development of index insurance products might not be recouped given the ease with which competitors can replicate such products if they prove profitable to sell. Therefore, there may be an important public role in funding the development of index insurance products.

References

A2F Consulting. 2018. *The Landscape of Microinsurance in Latin America and the Caribbean 2017.* Final Report: The World Map of Microinsurance. Luxembourg: Microinsurance Network; Munich: Munich Re Foundation.

Alderman, H., and T. Haque. 2007. *Insurance against Covariate Shocks: The Role of Index-Based Insurance in Social Protection in Low-Income Countries of Africa.* No. 95. Washington, DC: World Bank Publications.

Arias, D., and K. Covarrubias. 2006. "Agricultural Insurance in Mesoamerica: An Opportunity to Deepen Rural Financial Markets." Inter-American Development Bank—Economic and Sector Study Series RE2-06-006. Washington, DC: Inter-American Development Bank.

Bakst, D., J. Sewell, and B. Wright. 2016. "Addressing Risk in Agriculture." Heritage Foundation Special Report 189. Washington, DC: Heritage Foundation.

Barnett, B. J., C. B. Barrett, and J.R. Skees. 2008. "Poverty Traps and Index-Based Risk Transfer Products." *World Development* 36 (10): 1766–1785.

Barrett, C. B., and M. R. Carter. 2013. "The Economics of Poverty Traps and Persistent Poverty: Empirical and Policy Implications." *Journal of Development Studies* 49 (7): 976–990.

Binswanger-Mkhize, H. P. 2012. "Is There Too Much Hype about Index-Based Agricultural Insurance?" *Journal of Development Studies* 48 (2): 187–200.

Cai, H., Y. Chen, H. Fang, and L.-A. Zhou. 2009. "Microinsurance, Trust and Economic Development: Evidence from a Randomized Natural Field Experiment." PIER Working Paper 09-034. Penn Institute for Economic Research, Philadelphia.

———. 2015. "The Effect of Microinsurance on Economic Activities: Evidence from a Randomized Field Experiment." *Review of Economics and Statistics* 97 (2): 287–300.

Cai, J. 2016. "The Impact of Insurance Provision on Household Production and Financial Decisions." *American Economic Journal: Economic Policy* 8 (2): 44–88.

Cai, J., A. de Janvry, and E. Sadoulet. 2015. "Social Networks and the Decision to Insure." *American Economic Journal: Applied Economics* 7 (2): 81–108.

———. 2016. *Subsidy Policies and Insurance Demand.* NBER Working Paper 22702. Cambridge, MA: National Bureau of Economic Research.

Carter, M. R. 1997. "Environment, Technology, and the Social Articulation of Risk in West African Agriculture." *Economic Development and Cultural Change* 45 (3): 557–590.

———. 2009. "Intelligent Design of Index Insurance Contracts for Smallholder Farmers and Pastoralists." In *Innovations in Insuring the Poor,* edited by R. Vargas-Hill and M. Torrero. Washington, DC: International Food Policy Research Institute (IFPRI).

———. 2013. "Sharing the Risk and the Uncertainty: Public-Private Reinsurance Partnerships for Viable Agricultural Insurance Markets." I4 Index Insurance Innovation Initiative Brief 2013-1. Feed the Future and UC Davis.

Carter, M. R., A. de Janvry, E. Sadoulet, and A. Sarris. 2014. "Index-Based Weather Insurance for Developing Countries: A Review of Evidence and a Set of Propositions for Up-Scaling." Foundation pour les études et recherches sur le développement international Working Paper 111. FERDI, France.

———. 2017. "Index Insurance for Developing Country Agriculture: A Reassessment." *Annual Review of Resource Economics* 9 (1): 421–438.

Carter, M. R., G. Elabed, and E. Serfilippi. 2015. "Behavioral Economic Insights on Index Insurance Design." *Agricultural Finance Review* 75 (1): 8–18.

Casaburi, L., and J. Willis. 2018. "Time versus State in Insurance: Experimental Evidence from Contract Farming in Kenya." *American Economic Review* 108 (12): 3778–3813.

Ceballos, F. 2016. *Estimating Spatial Basis Risk in Rainfall Index Insurance: Methodology and Application to Excess Rainfall Insurance in Uruguay*. IFPRI Discussion Paper 1595. Washington, DC: IFPRI.

Ceballos, F., B. Kramer, and M. Robles. 2019. "The Feasibility of Picture-Based Insurance (PBI): Smartphone Pictures for Affordable Crop Insurance." *Development Engineering* 4: 100042.

Ceballos, F., and M. Robles. 2014. *Weather Risks and Insurance Opportunities for the Rural Poor.* 2020 Conference Brief 10. May 17–19, Addis Ababa, Ethiopia. Washington, DC: IFPRI.

Chantarat, S., A. G. Mude, C. B. Barrett, and M. R Carter. 2013. "Designing Index-Based Livestock Insurance for Managing Asset Risk in Northern Kenya." *Journal of Risk and Insurance* 80: 205–237.

Chantarat, S., A. G. Mude, C. B. Barrett, and C. G. Turvey. 2017. "Welfare Impacts of Index Insurance in the Presence of a Poverty Trap." *World Development* 94 (June): 119–138.

Clarke, D. J. 2016. "A Theory of Rational Demand for Index Insurance." *American Economic Journal: Microeconomics* 8 (1): 283–306.

Clarke, D. J., O. Mahul, K. N. Rao, and N. Verma. 2012. *Weather Based Crop Insurance in India.* Policy Research Working Paper WPS 5985. Washington, DC: World Bank.

Cole, S., X. Giné, J. Tobacman, P. Topalova, R. Townsend, and J. Vickery. 2013. "Barriers to Household Risk Management: Evidence from India." *American Economic Journal: Applied Economics* 5 (1): 104–135.

Cole, S., X. Giné, and J. Vickery. 2017. "How Does Risk Management Influence Production Decisions? Evidence from a Field Experiment." *Review of Financial Studies* 30 (6): 1935–1970.

Cole, S., D. Stein, and J. Tobacman. 2014. "Dynamics of Demand for Index Insurance: Evidence from a Long-Run Field Experiment." *American Economic Review* 104 (5): 284–290.

de Janvry, A., V. Dequiedt, and E. Sadoulet. 2014. "The Demand for Insurance against Common Shocks." *Journal of Development Economics* 106: 227–238.

Dercon, S. 2004. "Growth and Shocks: Evidence from Rural Ethiopia." *Journal of Development Economics* 74 (2): 309–329

Dercon, S., J. W. Gunning, and A. Zeitlin. 2016. "The Demand for Insurance under Limited Trust: Evidence from a Field Experiment in Kenya." Working paper presented at the Poverty and Applied Microeconomics Seminar Series of the World Bank. www.worldbank.org/en/events/2016/09/21/demand-for-insurance-under-limited-trust.

Dercon, S., R. Hill, D. Clarke, I. Outes-Leon, and A. Taffesse. 2014. "Offering Rainfall Insurance to Informal Insurance Groups: Evidence from a Field Experiment in Ethiopia." *Journal of Development Economics* 106 (January): 132–143.

Elabed, G., and M. R. Carter. 2015. "Compound-Risk Aversion, Ambiguity and the Willingness to Pay for Microinsurance." *Journal of Economic Behavior and Organization* 118: 150–166.

Farrin, K., and M. Miranda. 2015. "A Heterogeneous Agent Model of Credit-Linked Index Insurance and Farm Technology Adoption." *Journal of Development Economics* 116: 199–211.

Flatnes, J. E., and M. Carter. 2015. "Fail-Safe Index Insurance without the Cost: A Satellite Based Conditional Audit Approach." Working paper, University of California–Davis.

Gaurav, S., S. Cole, and J. Tobacman. 2011. "Marketing Complex Financial Products in Emerging Markets: Evidence from Rainfall Insurance in India." *Journal of Marketing Research* 48 (special interdisciplinary issue 2011: Consumer Financial Decision Making): S150–S162.

Giné, X., R. Townsend, and J. Vickery. 2007. "Statistical Analysis of Rainfall Insurance Payouts in Southern India." *American Journal of Agricultural Economics* 89 (5): 1248–1254.

———. 2008. "Patterns of Rainfall Insurance Participation in Rural India." *World Bank Economic Review* 22 (3): 539–566.

Glauber, J. W. 2013. "The Growth of the Federal Crop Insurance Program, 1990–2011." *American Journal of Agricultural Economics* 95 (2): 482–488.

———. 2016. "Crop Insurance and Private Sector Delivery Reassessing the Public-Private Partnership." Taxpayers for Common Sense. October.

Goodwin, B., and V. H. Smith. 1995. *The Economics of Crop Insurance and Disaster Aid.* Washington, DC: AEI Press.

Greatrex, H., J. Hansen, S. Garvin, R. Diro, S. Blakeley, M. Le Guen, K. Rao, and D. Osgood. 2015. *Scaling Up Index Insurance for Smallholder Farmers: Recent Evidence and Insights.* CCAFS Report 14. Copenhagen: CGIAR Research Program on Climate Change, Agriculture and Food Security.

Hazell, P., J. Anderson, N. Balzer, A. Hastrup Clemmensen, U. Hess, and F. Rispoli. 2010. "The Potential for Scale and Sustainability in Weather Index Insurance for Agriculture and Rural Livelihoods." International Fund for Agricultural Development and World Food Programme, Rome.

Hazell, P., and U. Hess. 2016. *Innovations and Emerging Trends in Agricultural Insurance*. Bonn: GIZ.

Hellmuth, M. E., D. E. Osgood, U. Hess, A. Moorhead, and H. Bhojwani, eds. 2009. *Index Insurance and Climate Risk: Prospects for Development and Disaster Management*. Climate and Society No. 2. International Research Institute for Climate and Society (IRI), Columbia University, New York.

Hill, R. V., and M. Robles. 2010. *Flexible Insurance for Heterogeneous Farmers: Results from a Small-Scale Pilot in Ethiopia*. IFPRI Discussion Paper 1092. Washington, DC: IFPRI.

Hill, R. V., M. Robles, and F. Ceballos. 2016. "Demand for a Simple Weather Insurance Product in India: Theory and Evidence." *American Journal of Agricultural Economics* 98: 1250–1270.

Hoddinott, J. 2006. "Shocks and Their Consequences across and within Households in Rural Zimbabwe." *Journal of Development Studies* 42: 301–321.

India, Ministry of Agriculture. 2018. *Agricultural Statistics at a Glance 2017*. Farmers Welfare Department of Agriculture, Cooperation and Farmers Welfare Directorate of Economics and Statistics.

Janzen, S. A., and M. R. Carter. 2019. "After the Drought: The Impact of Microinsurance on Consumption Smoothing and Asset Protection." *American Journal of Agricultural Economics* 101 (3): 651–671.

Jensen, N., and C. Barrett. 2017. "Agricultural Index Insurance for Development." *Applied Economic Perspectives and Policy* 39 (2): 199–219.

Jensen, N., C. Barrett, and A. Mude. 2016. "Index Insurance Quality and Basis Risk: Evidence from Northern Kenya." *American Journal of Agricultural Economics* 98 (5): 1450–1469.

———. 2017. "Index Insurance and Cash Transfers: A Comparative Analysis from Northern Kenya." *Journal of Development Economics* 129: 14–28.

Jensen, N., A. Mude, and C. Barrett. 2018. "How Basis Risk and Spatiotemporal Adverse Selection Influence Demand for Index Insurance: Evidence from Northern Kenya." *Food Policy* 74: 172–198.

Karlan, D., E. Kutsoati, M. McMillan, and C. Udry. 2011. "Crop Price Indemnified Loans for Farmers: A Pilot Experiment in Rural Ghana." *Journal of Risk and Insurance* 78: 37–55.

Karlan, D., R. Osei, I. Osei-Akoto, and C. Udry. 2014. "Agricultural Decisions after Relaxing Credit and Risk Constraints." *Quarterly Journal of Economics* 129 (2): 597–652.

Kramer, B., F. Ceballos, M. Krupoff, M. S. Toor, A. Mishra, S. Karekar, and M. Robles. 2017. "Picture-Based Insurance: Is It Sustainable? Effects on Willingness to Pay, Adverse Selection, and Moral Hazard." Project Note 2. Washington, DC: IFPRI.

Kyle, E., A. de Janvry, E. Sadoulet, and M. H. Dar. 2016. "Technological Innovations, Downside Risk, and the Modernization of Agriculture." *American Economic Review* 106 (6): 1537–1561.

Liu, Y., K. Chen, R. Hill, and C. Xiao. 2016. *Delayed Premium Payment, Insurance Adoption, and Household Investment in Rural China*. IFPRI Discussion Paper 1306. Washington, DC: IFPRI.

Mahul, O., and C. J. Stutley. 2010. *Government Support to Agricultural Insurance: Challenges and Options for Developing Countries*. Washington, DC: World Bank.

McIntosh, C., F. Povel, and E. Sadoulet. 2019. "Utility, Risk and Demand for Incomplete Insurance: Lab Experiments with Guatemalan Co-Operatives." *Economic Journal* 129 (August): 2581–2607.

Micro Insurance Center. 2016. *The Landscape of Microinsurance Africa 2015: The World Map of Microinsurance*. Luxembourg: Microinsurance Network; Bonn: Munich ReFoundation.

Microinsurance Network (n.d.). World Map of Microinsurance. http://worldmapofmicroinsurance.org/.

Miranda, M. J., and K. Farrin. 2012. "Index Insurance for Developing Countries." *Applied Economic Perspectives and Policy* 34 (3): 391–427.

Mobarak, A. M., and M. Rosenzweig. 2012. "Selling Formal Insurance to the Informally Insured." Working paper 1007, Economic Growth Center, Yale University, New Haven, CT.

Morduch, J. 1995. "Income Smoothing and Consumption Smoothing." *Journal of Economic Perspectives* 9 (3): 103–114.

Richardson, A. D., K. Hufkens, T. Milliman et al. 2018. "Tracking Vegetation Phenology across Diverse North American Biomes Using PhenoCam Imagery." *Scientific Data* 5: 180028.

Rosenzweig, M., and H. Binswanger. 1993. "Wealth, Weather Risk and the Composition and Profitability of Agricultural Investments." *Economic Journal* 103 (416): 56–78.

World Bank. 2005. *Managing Agricultural Production Risk Innovations in Developing Countries*. Washington, DC: World Bank Agriculture and Rural Development Department.

WFP (World Food Program) and IFAD (International Fund for Agricultural Development). 2010. *The Potential for Scale and Sustainability in Weather Index Insurance for Agriculture and Rural Livelihoods*. Rome.

Zimmerman, F. J., and M. R. Carter. 2003. "Asset Smoothing, Consumption Smoothing and the Reproduction of Inequality under Risk and Subsistence Constraints." *Journal of Development Economics* 71 (2): 233–260.

Chapter 18

NATURAL RESOURCE MANAGEMENT AND RESOURCE RIGHTS FOR AGRICULTURE

Frank Place, Ruth Meinzen-Dick, and Hosaena Ghebru

For the foreseeable future, agriculture in developing countries will be overwhelmingly based on farming that is dependent on natural resources to produce edible plants, provide feed for livestock, or to raise fish. The abundance of natural resources affects total agricultural production but the health, state, or quality of natural resources affects both total production and productivity of agriculture. Productivity growth in agriculture is of paramount importance for feeding growing populations in a world with limited natural resources (FAO 2012b). However, achieving high productivity is not easily accomplished in smallholder-dominant countries as evidenced by high yield gaps—the difference between actual and potential yields—in much of the developing world (FAO 2011) and productivity gaps between agriculture and other sectors (Gollin, Lagakos, and Waugh 2014). Obtaining high productivity in a way that sustainably uses natural resources is even more elusive globally.

The concept of environmental sustainability has two important dimensions. The first is whether the productivity of the current land under agriculture can be maintained or increased over time through the prevention of soil degradation, overuse of water, and other destructive practices (for example, salinization). The second is whether this productivity can be maintained without creating adverse effects on offsite natural resources. Negative externalities are manifested in flows of soil, nutrients, and wastes associated with agriculture that move on surfaces or below surfaces or atmospheric borne effects such as with greenhouse gases or silt. They are also manifested by expansion of agriculture into new land areas, which is destructive to forests, wetlands, and other natural habitats, and by negative changes in biodiversity (loss of wild species, agrobiodiversity, or increase of pests). It is apparent that farming involves or touches many types of natural resources that are both near and far from farms. This of course involves many stakeholders and poses great challenges for natural resource governance and property rights systems.

This chapter begins with a discussion of the importance of natural resources for agriculture and the links between agriculture and surrounding natural resources. It presents evidence on trends in natural resources use in agriculture and the drivers behind those changes. Also described are the many technological and institutional natural resource management (NRM) options that have proven to be successful, at least in some settings, as well as the main factors that restrain their wider use. The chapter focuses on tenure and governance challenges in managing natural resources for the benefit of users and wider society. Some recent innovations in tenure and governance of natural resources are discussed as well as an exploration of how well they perform in terms of achieving efficiency, equity, and sustainability objectives. The chapter concludes with key messages as well as a summary of the major development challenges and knowledge gaps.

Natural Resources and Agriculture

Soil and water are essential natural resources for agriculture. Soils have physical, chemical, and biological attributes—all of which contribute to soil health and subsequent productivity of agriculture (Sanchez 2019). Although all three properties can be degraded quickly through erosion, they have different timescales for their enrichment. Soil chemical properties can be changed most quickly through additions of nutrients, while some soil physical properties take decades to rebuild. Soils themselves have complex relationships with other natural resources; soils determine the degree to which other vegetation, such as trees and grasses, thrive while the integration of certain types of plants or their derivatives can enrich soils. Soil processes occur at a small scale (for example, in situ effects of no tillage) and at a large scale (for instance, soils are key to global carbon and nitrogen cycles and some of the other planetary boundaries) (Rockström et al. 2009).

Water for agriculture is largely from rainfall, but techniques exist to collect and move water both within farms (for example, water harvesting ponds or drip irrigation) and across larger land areas (for instance, canal irrigation schemes or pumping from aquifers) (Comprehensive Assessment of Water Management in Agriculture 2007). Managing water for agriculture is particularly important when rainfall is erratic and can in some cases extend the growing season or number of growing seasons within a calendar year. In addition, water flows across (and under) landscapes are affected by land use including farm management practices. Water for agriculture is therefore immediately recognizable as a landscape or catchment-level issue, and its management almost always involves externalities.

Important natural resources for farming go beyond soil and water. On farms the strategic use of agrobiodiversity, including edible or nonedible plants, can improve soil and water conservation, reduce exposure to pests and disease, or reduce wind damage to crops. They also serve as feed for livestock and beekeeping. Trees can provide for microclimate benefits through effects on ground temperatures and improving soil structures (Lasco et al. 2014). Leguminous plants can provide organic nutrients to nourish crops in the short term and to improve soil biology in the medium to long term (Peoples et al. 2009). These same plants can be important when found on nonagricultural lands to stem soil erosion, improve water infiltration and groundwater replenishment, provide needed feed for livestock, and host pollinator species. Livestock can be viewed as a natural resource that provides an important function of converting biomass into a high-quality manure. Forests and woodlands are sources of food (for example, wild fruits) and help to regulate water services in agriculture areas.

As described, the management of natural resources on farms not only has obvious direct effects on the farms. In addition, there are possible significant externalities from the management of natural resources and other farm practices within agriculture. Important negative externalities include sediment flow into rivers, chemical leaching into water supplies, deforestation, overgrazing, reduction in biodiversity, and greenhouse gas emissions. However, some types of practices may have positive externalities such as carbon sequestration, soil erosion control, and evapotranspiration effects (Kremen and Miles 2012). Reducing agriculture's negative environmental effects and enhancing its positive externalities is much discussed in global policy fora (for example, Food and Land Use Coalition 2019). The direction and level of externalities are highly influenced by the security of rights and management systems of local users over natural resources.

Trends of Natural Resources in Agriculture

The area under agriculture in developing countries has been increasing rapidly, by 230 million hectares between the early 1960s and mid-2000s (FAO 2012b). The trend in agricultural land area up to the mid-2010s varies by region and country, slightly declining in such places as China and India (where peaks were around 1990), slowing in such densely populated countries as Kenya and Rwanda, but still expanding in others with lower rural population densities (for example, the Democratic Republic of the Congo, Mozambique, and Zambia). It is expected that a further increase of 107 million hectares will be needed by 2050 to feed the increased population of the current developing countries (FAO 2012b).

Recent studies demonstrate that there has been significant degradation of land and soil quality. Le, Nkonya, and Mirzabaev (2016) find that globally 25 percent of cultivated land is degrading (as measured by declining plant productivity) over the past 25 years as has more than 30 percent of land used for livestock grazing. One key factor behind these trends is the continuous use of land with little fallowing and insufficient use of biological or mineral inputs, resulting in nutrient mining (Henao and Banaante 2006). Another is poor conservation of soils, resulting in soil erosion of the more productive top layers (Tamene and Le 2015).

Most of agriculture is based on rainfed conditions with irrigation used on only 16 percent to 20 percent of arable land (FAO 2011). China and India account for more than half of the irrigated area (54 percent) in developing countries (FAO 2011). In Africa south of the Sahara, only 3 percent of cultivated land is irrigated (Calzadilla 2013). There had been significant expansion of irrigation in developing countries in the latter half of the 1900s, but the pace has slowed since 2000. Some water resources are becoming depleted—for example, groundwater in West Asia, North Africa, and India, and glacier reservoirs in the Himalayas and Andes—and climate change may reduce rains in other areas (for example, southern Africa). Furthermore, continued population growth increases demand for domestic water use and competes with water supplies for agriculture. Water quality is also poor and threatened in many cases. It was estimated that 21 percent of surface water in China was unfit even for agriculture (2030 Water Resources Group 2009).

Forest, tree, and other vegetative cover is important for conserving soils on landscapes and for regulating watershed hydrology, and has been identified as terrestrial sources of rainfall through evapotranspiration (Ellison et al. 2017). Deforestation continues (about 130 million hectares of forests between 2000 and 2010), is slowing, or is reversing in many important countries such as China and India (FAO 2012a) but increased overall in the tropics in 2017 (Weisse and Goldman 2018). The effects of global deforestation on loss of carbon sequestration is partly offset by increasing tree cover on agricultural land (Zomer et al. 2016).

Drivers of Availability and Degradation of Natural Resources

Among the anthropogenic factors affecting trends in natural resource availability and quality are urban and rural population growth, development of markets for agriculture and other natural resource–based products, and policies that affect the ways in which people use resources. Population growth, whether urban or rural, raises the demand for food, historically resulting in the expansion of agricultural land. This has negative consequences on such natural habitats as forests, shrublands, and wetlands, which are converted to agriculture. Conversion to agriculture is the proximate cause of about 80 percent of deforestation (Kissinger, Herold, and De Sy 2012). The other effect of increasing rural population density is to reduce the median farm size, most noticeably in Africa (AGRA 2013). This can lead to either better or worse natural resource management on farms, depending on such factors as market conditions and property rights systems.

Market development will affect incentives to manage natural resources on farms or in landscapes. These could be positive—for example, by increasing the returns to on-farm investments in soil fertility management, water management, and other similar practices through better access to productivity-enhancing inputs as well as increasing the ratio of output to input prices. These could be negative as well, creating incentives to overextract or mine resources to seize market opportunities. The direction of this effect depends on clarity and security of resource rights, especially long-term rights. Market development also includes markets for the resources or the ecosystem services they may provide (Kroeger and Casey 2007). Land markets are subjected to a variety of legal and nonlegal barriers, markets for water are rarely formalized in rural areas, and the valuation and recognition of environmental services is only operational in a few pilot cases. Markets for watershed protection services, carbon sequestration, or other ecosystem services have emerged and are growing but are still uncommon and minor contributors to income streams of rural resource users (Salzman et al. 2018). Finally, property rights and governance institutions play a key role in shaping the ways in which natural resources are exploited, managed, and conserved, which is the main topic discussed in this chapter.

Natural Resource Management Practices and Institutional Approaches in Agriculture

This section does not attempt to provide a comprehensive review of natural resource practices used by farmers and communities or to explore all the links between the management of natural resources and agriculture. The purpose is rather to introduce a range of practices and to identify important constraints farmers or communities face in adopting them to better understand the role of tenure factors in NRM.

Farmers in developing countries use a variety of natural resource management practices on their farms. The specific practices that are applied vary considerably depending on the local context (for example, soil type, aridity, level of market development for inputs and outputs) and household characteristics (for instance, farm size, available family labor for agriculture) (see, for example, Vanlauwe et al. 2015). Most soils in the tropics have deficiencies in physical structure (for instance, high sand content), chemical composition (high acidity or low phosphorus), or poor biological activity (low soil carbon and microbial function) (Sanchez 2019). In these cases the conventional approach of relying only on mineral fertilizer to generate consistently high yields will not work (Marenya and Barrett 2009), and other soil management strategies are necessary.

Some dimensions of soil health can be affected by NRM practices in the short- to medium-term and thus be attractive for farmers to invest in. This includes the planting of trees and shrubs to help break hard pans and improve water retention, the use of minimum tillage to improve soil biological function, and the application of mulch or conservation structures to reduce top soil degradation. Soil conservation practices are varied and include terracing and ridging, which themselves can be formed through altering the soil through labor, applying materials like stones, or planting grass strips or other plants. Soil biology is improved mainly through application of organic matter and farmers use a variety of methods to achieve this, including use of animal manure or compost, incorporation of crop residues, natural or planted fallowing, and application of green mulch from woody or herbaceous plants often grown in situ. Acidic soils require an offsetting chemical agent like lime to be applied. These practices are also referred to as sustainable land management practices (SLM).

Water is another essential natural resource for agriculture and its effective management can lead to significant effects on production. There are some practices that simply encourage rain to accumulate in areas where farmers intend to plant seed. These include planting basins, zai pits, half-moons, or any other type of excavated area where water can accumulate and infiltrate. Soil conservation structures will also act to manage water and prevent it from moving too fast or far on the surface. Runoff may be stored in an on-farm source such as a water pan, farm pond, or container. Rooftop water harvesting may be used for domestic water supply and homestead gardens. Farmers construct channels that can be used to move water onto or off of fields. Water can also be moved via hoses above or below ground (for example, drip irrigation).

Beyond such on-farm water control, watershed management is one of the clearest examples of landscape-level NRM. This may involve technical interventions such as building check dams to slow water flows or increase recharge, terracing, planting trees, or buffer strips along waterways to protect against pollution, but generally also include some forms of restrictions on behavior, such as excluding animals from grazing in certain areas or limiting water withdrawals. Kerr (2007: 91) notes that "technical interventions are likely to be fruitless without subsequent management." This management depends on institutional arrangements—rules that are set, adapted, and enforced.

Most irrigation draws water from an off-farm source of water such as a reservoir, canal, river, pond, or aquifer. These require investment and coordination above the level of the individual farm to construct, operate, and maintain the system as well as to regulate the timing and amounts of water flows and withdrawals across users. The importance of these institutional factors for irrigation use has often been noted, as in the experimental study in Burney and Naylor's (2012) research in Benin. Farmers also often engage in negotiations and contracting to bring water resources onto their farms from surface or groundwater. These can take varied and complex forms across and even within villages, as illustrated by Michler and Wu (2014) in Bangladesh.

To manage flows of biodiversity harmful to agriculture, there has been significant investment in training on integrated pest and disease management that rely on the application of natural resource–based solutions to reduce threats. This includes providing for a diverse habitat to promote natural predators, to act as barriers (to disease encroachment), or to attract the population of pollinators that are essential for a number of plants (Karp et al. 2018). Managing biodiversity often requires coordination across farms and with mosaics of cultivated and uncultivated lands (including forests and other types of land use). Symbiotic relationships over space and time can also be managed through intercrops or rotations of plants that are complementary in resource use or other ways.

All pastoralists and most agropastoralists depend on pastures for at least some of the fodder for livestock production. In most developing countries, these are not private pastures but shared rangelands, which are often dry or mountainous areas with high spatial and temporal variability (IUCN 2011). Without careful management these lands are subject to degradation through overgrazing or undergrazing, encroachment, or weather and climate changes.

Paths created by cattle can become significant sources of erosion. Customary systems for rangeland management have often included livestock mobility to access fodder and water resources as well as to prevent undergrazing or overgrazing. Interventions for rangeland improvement may include planting grasses, shrubs, or trees to increase fodder availability, provide shelter, and reduce erosion; limiting grazing (whether by rotating areas to encourage natural regeneration or limiting stocking numbers); and developing or protecting watering points. As with watersheds, improving rangeland conditions depends on effective institutions to accompany any technical interventions (McCarthy 2004).

Forests provide inputs and ecosystem services that are critical for agriculture and rural livelihoods by providing fencing and housing materials, firewood, leaf litter, and other soil amendments to support cultivation, fodder for livestock, and medicinal plants as well as timber and nontimber products to diversify farmers' livelihoods. Beyond their contribution to local livelihoods, the role of forests in carbon sequestration, regulating temperature, rainfall and water flows, and biodiversity mean that not only forest communities or nearby farmers depend on them, but also other stakeholders—from downstream cities (for water regulation) to the global community (for biodiversity and carbon functions). Forests may cover entire landscapes, as in the Amazon, or occur in patches interspersed with cultivated lands and rangelands. Effective management of forests requires regulating cutting and withdrawals as well as selective planting or protecting regrowth. But because of the many different functions and interests in forests, coordinating forest management requires addressing many different stakeholders.

Constraints to Adoption of Natural Resources Management Practices in Agriculture

A body of research, largely comprised of observational case studies, has found that many NRM practices entail some characteristics or constraints that limit their wide use, such as demands on labor, opportunity costs of land, and uncertainty of long-run returns to investment. While we explore tenure- and governance-related factors here, other factors are featured elsewhere, such as Stevenson and Vlek (2018), covering a range of NRM practices developed by CGIAR; Bationo et al. (2012) on benefits from integrated soil fertility management (ISFM); Yamano, Otsuka, and Place (2011) on various integrative practices in East Africa; Giller et al. (2009) and Corbeels et al. (2014) on adoption challenges for conservation agriculture; and Farrow et al. (2019) on grain legumes in Africa. Furthermore, rural transformation processes continue to shift the incentives for undertaking NRM practices. For example, development of input markets and rising rural wages may favor labor-saving technology like herbicides (Haggblade et al. 2017) over labor-intensive land management practices (Jayne et al. 2019).

Constraints to adoption of on-farm NRM practices have been investigated in better designed cause-effect studies in recent years. A study by Michler et al. (2019) on the use of minimum tillage in Zimbabwe found that in normal rainfall years the technology did not pay off and in fact led to economic losses for a number of farmers. Jack et al. (2015) found that dis-adoption of a new technology (planted *Faidherbia albida*) was common in Zambia due to uncertain information on benefits. Jack and Jayachandran (2018) further investigated the effectiveness of payments for environmental service programs related to tree planting and conservation and found costs of conservation as well as enrollment costs to be important factors in self-selection and participation in the programs. The important conclusion of this body of research is that there are many potential constraints to adoption of natural resource management practices and one or more of them appear to be present in any given developing country contexts. While tenure or governance of resources are among those constraints, the importance of other factors should not be overlooked.

Natural Resources Management Practices and Implications for Property Rights and Collective Action

Although a full discussion of each of these landscape-level natural resource topics is beyond the scope of this chapter, they share some important features. First, they depend on a careful assessment of the resource conditions, threats, and interrelationships across a large area, to identify points of most effective interventions. Second, effective management usually combines tangible and technical activities, such as building a check dam or planting specific trees, with behavioral or institutional interventions such as rules regarding resource extraction and provision (Anderies and Janssen 2013). Third, there needs to be consideration of the human as well as the biophysical environment: Who are the users and how will they use the resources, and what types of user organizations and government agencies will participate in the development and implementation of the programs?

Figure 18.1 illustrates the need for institutional arrangements to address natural resource management. The lower horizontal scale shows the time scale. Many aspects of soil structure and biological properties require considerable time to rebuild. Agroforestry practices that can improve soils, water flows, and carbon storage also take time for the returns to be realized. The upper

FIGURE 18.1 Key institutions for natural resource management

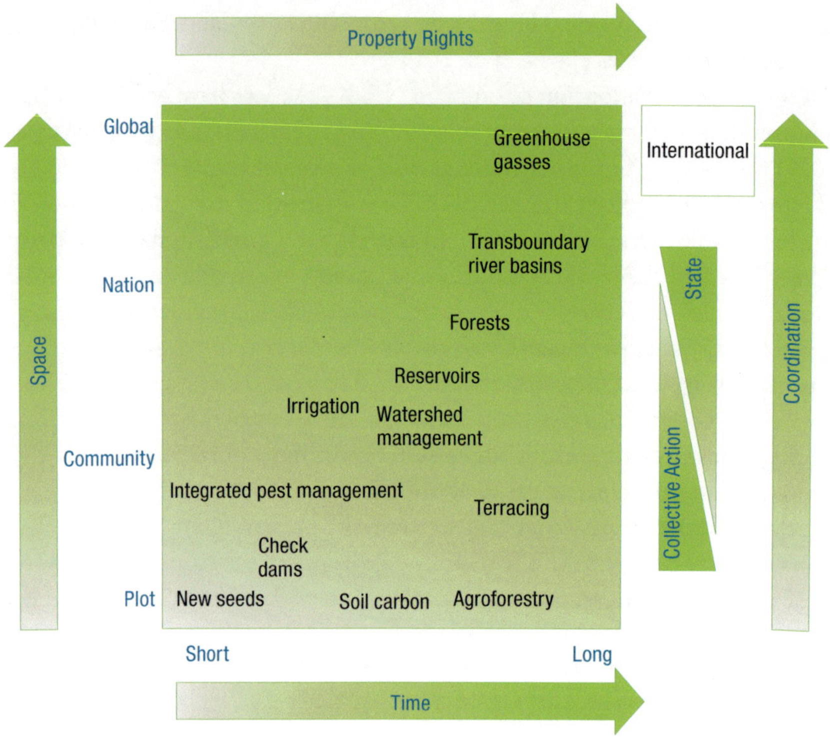

Source: Meinzen-Dick, Markelova, and Moore (2010).

Note: Placement of particular NRM practices is illustrative and depends on the particular characteristics of each practice in each place as well as the size of farms, communities, and nations. NRM = natural resource management.

horizontal scale indicates the corresponding institution (property rights) to deal with long time horizons between investment and returns. Farmers without secure land tenure may have less incentive to invest in long-term soil improvements or may not even be authorized to plant trees. However, in some cases individuals make land investments with the purpose of securing long-term rights to the land, especially by planting trees or careful husbandry of existing land (Bruce and Migot-Adholla 1994; Besley 1995; Otsuka and Place 2001; Platteau 2008).

The vertical axes in Figure 18.1 show the spatial scale and corresponding institutions for coordination. Whereas planting a tree can be done on an individual farm, practices such as integrated pest management (IPM) are more successful when coordination leads to concerted actions by many farmers.

Most irrigation systems, watershed management, or forests go beyond the boundaries of an individual farm and therefore require some form of coordination of investment (for example, in planting) and extraction. Whether long-term property rights are also important depends on the period of time to positive net benefits. Farmers with weak long-term tenure security may still have incentive for taking short-term measures to reduce pests or apply chemical fertilizer but will not be inclined to invest in structures for soil or water conservation that do not pay off for several years. Ultimately, this framework can be extended to the global scale to address such issues as greenhouse gas emissions or sequestration or widespread loss of biodiversity. Although resource management at the plot, farm, and community level (for example, planting trees or limiting deforestation and hunting) can contribute to addressing these global resource problems, above the level of the nation-state, achieving coordination proves challenging, as discussed in Chapter 19 in this volume.

Tenure and Governance Challenges in Managing Resources

This section defines concepts and identifies challenges in property rights and coordination of management of natural resources.

Property Rights

The term "property rights" refers to "claims, entitlements and related obligations among people regarding the use and disposition of a scarce resource" (Furubotn and Pejovich 1972). Although rights to land are the most prominent form of property rights in agriculture and rural areas, rights to water, trees, forests, fisheries, and other aquatic resources are also important, and these may be distinct from the rights to the land on which they are found. According to UN-Habitat (2008), "land tenure" is defined as "the way land is held or owned by individuals and groups, or the set of relationships legally or customarily defined among people with respect to land. In other words, tenure reflects relationships between people and land directly, and between individuals and groups of people in their dealings in land." This definition can be extended to other resources as well.

Most natural resource tenure is characterized by the concurrent existence of multiple legal orders, referred to as "legal pluralism" (Knight 2010; Meinzen-Dick and Pradhan 2002). The most prominent of these are statutory and customary law, but international, religious, and even regulations developed by particular programs may also be important. Customary tenure may be or may not be recognized and enforced by the formal legal system. In situations where one system does not recognize the other, the coexistence of customary and statutory tenure could be a source of confusion and may lead to greater tenure insecurity (Knight 2010; USAID 2013). UN-Habitat (2008) presents a range of different tenure types as a continuum from informal to formal rights, with various sets of rights and degrees of security, responsibility, and enforcement. Tenure types and strength and weakness of tenure vary across countries and even across different regions within a country. In particular, different ethnic groups may have different customary laws and rules governing uses of resources. Ultimately, tenure claims are only as strong as the institution that backs them up. This means that in areas where the state is strong, statutory tenure is very important, but where the state is weak (often in more remote areas), customary tenure and resource management rules hold greater sway.

Property rights over natural resources are not limited to "ownership" but are better understood as distinct bundles of rights. There are many ways of classifying the bundles of rights, but Schlager and Ostrom (1992) identify five key bundles that are relevant for resource management.

- Access (for example, to walk across a field)
- Withdrawal (for example, to draw water or harvest plants)
- Management (for example, to plant a crop or tree)
- Exclusion (for example, to prevent others from accessing the field)
- Alienation (for example, to rent out, sell, or give away the rights)

The first two (access and withdrawal) are considered use rights, and the latter three (management, exclusion, and alienation) are control or decision-making rights. Galik and Jagger (2015) argue for a sixth bundle—alteration—to reflect the right to transform the resource, often by making improvements. "Ownership" often implies having all of these rights, but different bundles of rights are often held by different individuals, groups, or institutions (for example, government agencies) over the same resource. For instance, a farmer may have the right to plant a crop on a piece of land, but anyone can cross that land to get water; pastoralists may have the right to graze their herds on that land in the fallow season; family elders may have the right to allocate or reallocate that land; and the state may claim ultimate "ownership" of the resource. How these bundles of rights are distributed will affect who has authority and incentive to manage the resources as well as the distribution of benefits in society.

The holders of these rights can be individuals, households, companies, groups, or the state. The first three are generally referred to as private property; rights held by various forms of groups or collectives are common property, and rights held by the state are public property. Although most farms (especially smallholder farms) are private property, forests, rangelands, and water resources are often common or public property. Even farmers with private holdings may rely on common property for critical inputs to their livelihoods (for example, fuel, animal fodder) and ecosystem services.

The NRM practices discussed earlier can generally be mapped into three different types of management domains: (1) private property management without much externality (for example, improved soil nutrient management); (2) private property management with significant externality (for instance, pest management, soil conservation); and (3) common property management (for example, forest and rangelands). In common property management, communal rights to the resources and coordination mechanisms are important for successful resource management. An example of common property management is the case of forests (see Figure 18.1), where the establishment of governance mechanisms including the establishment of access and ownership rights among various stakeholders is important. The case of private property management with externalities lies between the two other cases. Private property requires clarity and secure rights for the holder, while the presence of externalities clearly indicates the need for coordination.

Place, Roth, and Hazell (1994) identify three dimensions of tenure security: robustness, duration, and assurance of rights. "Robustness" refers to how many of the rights are held. "Duration" is the length of time for which a right is valid. "Assurance" is the certainty with which the rights are held and the extent to which the rights are enforceable. The individual's subjective perception of their own tenure situation is identified as an important factor that influences their behavior (Van Gelder 2010). Investment in land as well as sustainable management of natural resources is based on the perception households and individuals have regarding tenure security in the future, when the return of a current investment in natural resource management is realized in the future.

Increased land tenure security can lead to three other reinforcing channels for increasing investment in land. First, greater tenure security can increase demand for land-related investment by reducing land disputes and freeing up resources for investment that have been spent to defend land rights. Second, many formalized tenure systems facilitate land transfers and enable investors to benefit from their investment even if they will not be able to use the land. Third, tenure security can enhance access to credit for investment and inputs (Piza and Moura 2011) and more so where agricultural credit markets are functional (Deininger 2003). In addition to economic benefits, secured land tenure promotes social stability by reducing land conflicts and social exclusion and improving shelter conditions (UN-Habitat 2008; Deininger 2003).

Land tenure (in)security can be measured at farm plot, individual, household, group, or community levels. At the community or group level, tenure security is often measured by indicators such as whether collective land rights are recognized by the government, whether community rights can be held in perpetuity, and whether the law requires the consent of communities before the government or outsiders may acquire their land (Wily et al. 2016). Marginalized ethnic groups often experience less tenure security, especially where the state does not recognize their resource rights. This particularly applies to pastoralists, forest-dwelling communities, or other Indigenous peoples. At the household level, a sense of tenure insecurity may come from the perception that someone may challenge their land rights and eventually make them lose their rights to the land. This threat arises either due to (1) hazard of expropriation by the government (Jacoby, Li, and Rozelle 2002), or (2) encroachment or eviction by other individuals.

Underlying drivers of insecurity may result from foreign interest in land or be related to local socioeconomic trends, such as urban expansion, rural population growth, commercialization of agriculture and other factors, conditioned by the degree of protection the government offers against such risks (Ghebru and Girmachew 2017, 2019; Ghebru and Lambrecht 2017). The individual level is also very important because even if a households' rights to tenure are secure, not all individuals may be secure. In particular, women who depend on men for their access to land face additional layers of tenure insecurity—for example, in cases of widowhood or divorce as well as youth who depend on their elders for land, these individuals may lack full decision-making rights (see Chapter 15 in this volume; Ghebru and Girmachew 2018). At the plot level, tenure security may vary based on location, tenure status, and time and mode of acquisition. More remote lands are often less secure than those close to the household, and land in other villages may be even less secure. Thus it is common to find that households operating multiple plots of land will have different levels of tenure security on each of them. Where state law is strong, land with formal titles or other certification giving statutory recognition may be stronger than undocumented land under customary law. Common property is often less secure than individual property.

The bundles of rights and implicit security of tenure associated with land are affected by the mode of acquisition. There are modes of acquisition for which long-term rights to land are usually conferred including purchase, inheritance, and gift. Land allocated by formal or traditional authorities represents a dominant form of land acquisition in many developing countries where lack of proper documentation of such land implies an adverse effect on natural resource management. Other modes of acquisition such as renting and sharecropping are mainly for short-term rights to land. This distinction has important consequences for NRM because renting and other short-term acquisitions do not afford sufficient time for the tenant to reap benefits from many of the NRM practices noted in Figure 18.1. Moreover, landlords may restrict tenants from undertaking certain visible long-term investments that could lead to contested rights beyond the agreed-upon termination date of the contract. Thus it is not surprising that studies find lower levels of long-term investment on land under short-term tenure arrangements (Place 2009; Deininger and Jin 2006).

Information and knowledge about rights, legal restrictions, and the various types of threats and protection opportunities matter (Deininger et al. 2008; Ghebru and Girmachew 2018). The government may be the source of the risk or the source of protection against the risk, depending on the setting or the formal land rights recognized by government and held by the individual claiming rights in a piece of land. With conflicting claims over land, the increase in the (in)security of one party may imply a reduction in the (in)security of another party or parties. The strength of (in)security can depend on traditional rights (customs and social norms), modes of land acquisition, legal protection (laws and law enforcement), duration of possession, social networks, political connections and power structure, the degree of scarcity (competition) and value of the land, as well as individual and group abilities.

From an empirical perspective, the types of tenure security weaknesses that pose the greatest challenges differ across and within countries. However, a common empirical finding is that women are less likely to own land and have fewer rights to land than men (see Chapter 15 in this volume; Doss et al. 2015; Kieran et al. 2015). Although a far cry from the myth that women own just 1 percent to 2 percent of the land, the differences between men and women can often be large, as in the case of Niger and Nigeria. Women also may face more challenges in protecting their limited rights in family, customary, or formal conflict resolution mechanisms. This has important implications for the bargaining power and well-being of women themselves, their households, and the condition of resources. Individuals/households with more secure tenure have been shown to invest more in terracing and bunding for long-term soil conservation, tree planting, and fallowing of land (Ali et al. 2014; Deininger et al. 2008; Holden, Deininger, and Ghebru 2011; Goldstein et al. 2018), and more so in the case of women (Meinzen-Dick et al. 2019; Goldstein et al. 2018).

Coordination for Resource Governance

As noted, many natural resource management practices involve positive or negative spatial externalities. Thus coordination is important for adoption. While an individual farmer may be able to construct on-farm water storage or tap groundwater in a well, most surface irrigation requires coordination of multiple farmers and other stakeholders to build and convey the water. As many individual farmers take water from wells, coordination among users becomes important to manage the underlying aquifer sustainably. Similarly, finding the optimal place to put check dams or terracing in watersheds and ensuring that they are maintained requires coordination across multiple stakeholders in the watershed. Forest management requires rules for planting, harvesting, and managing the trees and other resources. Mobility of herds has been a key strategy for sustainable management of rangelands, often over a combination of private and collectively held lands, requiring coordination on what animals can be grazed in what areas at what times (McCarthy 2004).

At higher levels, the state has an important role to play in the coordination of NRM (see the right-hand vertical axis in Figure 18.1), but at the most local (community and group) levels, the state has less effective reach and collective action among users is more important. However, these are not mutually exclusive: government agents may operate at the local level, and federations of user groups can extend collective action to the national level, as exemplified the Federation of Community Forest Users (FECOFUN), which links more than 14,000 community forest user groups in Nepal (McLain and Lawry 2015). Thus the rights and responsibilities of each type of institution, and the relationships between state and collective action, are important for NRM outcomes. Although the specific nature of the coordination differs depending on the resource and context, studies of the commons—shared resources such as water, forests, rangelands, and fisheries—have identified several broad patterns. After systematically comparing numerous case studies of common-pool resources (CPRs) characterized by subtractability (in which one person's use of the resource makes less available to others) and the high costs of excluding others from using the resource, Ostrom (1990) identified a set of design principles that characterize many long-enduring CPRs. These have been subsequently validated and updated by Cox, Arnold, and Villamayor Tomás (2010), and include the following:

1. Clearly defined boundaries of the resource itself and the group of authorized users.
2. Congruence between appropriation and provision rules and local conditions, so that the rules make sense in the local context and ensure that the benefits obtained are proportional to the amount of inputs required.
3. Collective-choice arrangements that allow most people affected by the rules to participate in setting or modifying the rules.
4. Monitoring of the resource.
5. Graduated sanctions for breaking the rules.
6. Conflict-resolution mechanisms.
7. Minimal recognition of rights to organize.
8. Nested enterprises so that all the governance activities are organized at multiple levels.

Security of tenure on common property depends on external recognition of the group's rights to the land as well as effective internal governance arrangements that will ensure that resources are effectively managed. The external recognition of collective rights is even more rare than for private lands. In a review of 64 countries, RRI (2015) report that half of the countries recognize less than 5 percent of the country's land area as collectively owned or controlled by Indigenous peoples and local communities, even though the total land area used collectively is many times larger. Reforms that have been implemented to secure tenure for individual and collective resources, as well as efforts to strengthen collective management of common pool resources within and across property boundaries, are examined next.

Policy and Institutional Innovations in Tenure Security and Governance

Over the past few decades, significant changes that affect land tenure systems were witnessed in many parts of the developing world. These include demographic growth, urbanization, migration, livelihood diversification, monetization of the economy and commodification of land, greater integration in the global economy, and cultural change (see Chapters 9–13 as well as Chapters 19 and 20 in this volume). Such changes have especially major implications for the "traditional" or "customary" land tenure system (Cotula and Neves 2007; Ghebru and Lambrecht 2017; Linkow 2016).

Individual Land Rights

Even though there is a consensus on the importance of tenure security, the mechanisms to secure land rights have been debatable, with countries following different approaches. Many countries, particularly in Asia and parts of Africa, undertook redistributive land reforms to compensate for colonial and postindependence land-ownership inequalities and regressive land-use policies (Sikor and Muller 2009). Such "land to the tiller" programs were intended to remove large landlords and inefficiencies in tenancy arrangements and to provide for a strong smallholder agricultural sector. In other countries land titling and registration programs without redistribution were introduced as one means to increase tenure security, facilitate land markets, and improve access to credit using land as collateral (Deininger and Binswanger 1999).

Although formal land registration is taken as a means to reduce disputes, it might be cause for uncertainty in situations where the wealthy and educated groups use it to capture land from smallholder farmers, women, or communities who have limited or no access to information and are not able to cover the cost of registration. In some situations formal registration fails to recognize the rights of secondary right holders such as women, migrants, and herders (Platteau 2000 cited in Quan and Toulmin 2004; Meinzen-Dick and Mwangi 2007). Moreover, experience of countries has shown that land titling programs were time-consuming, expensive, and not accessible for poor farmers. For instance, only about 1 percent of land in Africa was registered under formal systems in the early 2000s (Easterly 2008).

In recognition of the limitation of land titling programs, alternative approaches have been developed to improve tenure security. Among individual titling and registration programs, some (such as Ethiopia or Rwanda) have made explicit provision for inclusion of women's names as joint owners of household lands. Another approach has been to give legal recognition to customary tenure systems that could be done through direct contact with individual holders of customary rights or nationwide recognition of customary land rights (Van den Brink et al. 2006 cited in Lawry et al. 2017; Palmer et al. 2009). This approach is practiced in some African countries, such as Burkina

Faso, Ghana, Mali, Mozambique, Niger, and Uganda, that have legally recognized customary tenure (Palmer et al. 2009). In Ghana, Customary Land Secretariats were formed to manage land in peri-urban areas and make arrangements for community members who lost agricultural land because of urban expansion (UN-Habitat 2008). In Uganda the 1998 Land Act recognizes land acquired by households through allocation by customary authorities. In Mozambique the Land Law gave recognition to customary tenure by protecting land rights of communities through a community land delimitation program (Palmer et al. 2009; USAID 2009).

Another approach involves providing intermediate forms of tenure, such as a land certificate, right of occupancy, formalization of leasing or rental transactions, and so on. This type of recognition of land rights is implemented in countries like Brazil, Colombia, Ethiopia, Mexico, Tanzania, and Trinidad and Tobago (Palmer et al. 2009). For instance, in Ethiopia low-cost land certification (for example, not based on formal demarcation) was implemented that provides land-use rights in rural areas (Palmer et al. 2009). Several studies have analyzed the effect of the certification program on Ethiopian households and found benefits in the form of perceived tenure security, investment in land improvements, and in promoting land rentals (Bezabih, Holden, and Mannberg 2016; Deininger et al. 2008; Ghebru and Holden 2013; Holden, Deininger, and Ghebru 2011). In Trinidad and Tobago, Certificates of Comfort are issued to people living in informal settlements, which enable households to use and invest in the land (UN-Habitat 2008). Two recent studies using experimental approaches provided certificates of land rights following boundary demarcation and examined investment behavior in a short period that followed. Goldstein et al. (2018) found significant enhancement of perceived security and land investment in Benin, while Huntington, Haflett, and Ewing (2018) found that the certificates do not affect planting of trees in Zambia.

Despite the various efforts made to improve land governance in many developing countries, tenure insecurity in the form of land disputes among families/neighbors, land expropriation by the government, land grabbing by the elite and foreign investors, and expansion of the informal land market is still a threat to the rural poor and urban dwellers. Empirical studies have found mixed results in the impact of land titling and registration on investment, access to credit, land rights of vulnerable groups (female, migrants, economically poor), and agricultural productivity. In a systematic review of the impact of tenure reforms in developing countries, Lawry et al. (2017) found no evidence for a credit mechanism. In another systematic review, Higgins et al. (2018) report that only two of seven studies found a link between improved tenure security and credit. However, both of these systematic reviews found stronger evidence of land tenure recognition programs on investment in the land, particularly in long-term improvements in the natural resources.

In their systematic review of women's land rights on poverty reduction, Meinzen-Dick et al. (2019) found a similar pattern: greater tenure security for women led to increased investments in tree planting as well as terracing and bunding to reduce soil erosion. However, not all titling or registration programs strengthen women's land rights; without special efforts to ensure that women's names are recorded as sole land owners or jointly with their spouses, the effect of the registration may be to reduce the rights women have under customary tenure (Lastarria-Corhiel 1997). Lawry et al. (2017) also found substantial productivity and income gains from land tenure recognition, but these were greater in Asia and Latin America than in Africa. Underestimating the contribution of customary tenure systems to farmers' tenure security, lower level of household income to make investment in agriculture, and lack of complementary public investments (in infrastructure, provision of input and market access, training for farmers) to support agricultural investment are mentioned as possible causes for the weaker linkage in Africa.

Beyond land registration, recognition of land rights by the community, availability of an institution to enforce land rights, duration of land rights, clear definition of land rights and boundary demarcation of land are important contributors to security of land tenure (Abdulai and Antwi 2005). Moreover, the insecurity of "informal" or customary rights is overestimated. Williamson and Kerekes (2011) found that the role of informal institutions has been underestimated, and the efforts to secure property rights through formal means cannot be effective without considering informal practices in the society.

Communal Rights

Formalization of individual rights is not the only form of tenure reforms that is important for natural resource management. Many countries have or are engaging in reforms to strengthen collective land and resource rights, notably for forests and rangelands. More than 73 percent of forests remain formally under state ownership (RRI 2014), with government forestry agencies relying on regulation and policing of forest use, often with adversarial relations with communities (Lawry et al. 2012). In recent years there have been notable initiatives to devolve rights to forests, in particular, from the state to communities. Progress has been most noted in Latin America (especially Bolivia, Brazil,

Colombia, Honduras, and Peru), with both long-term community or group concessions and titling of Indigenous peoples' domains (Lawry et al. 2012). In Asia the state has devolved a more limited set of forest resource rights to communities, notably in China, India, Indonesia, Nepal, and the Philippines.

However, these co-management arrangements between the state and communities often limit use of high-value forest resources (especially timber). Often it is only degraded forest lands that are transferred to communities or forest user groups, with requirements to use particular practices to restore degraded forest lands. Forest tenure reform is most limited in Africa, with restrictions even on rights to trees on private land. Lawry et al. (2012) note numerous successes of cases that strengthen community rights to forests and trees, but "tenuous use rights and weak benefit-sharing models only go so far toward providing the security and financial incentives needed to invest in forest improvements and protection at landscape-scales" (Lawry et al. 2012: viii).

For example, with unclear benefit-sharing arrangements, restoration through planting and management of high-value species will be inhibited, although enclosure methods may still take place (Otsuka and Place 2001). Furthermore, even when countries adopt policies to devolve rights, these are often not implemented because of opposition from elites or forest agencies; even if they are implemented, communities may require more than secure tenure: strengthening cooperatives, training, and other resources may be required to sustainably exploit forest resources (Larson et al. 2016). However, reviewing cases from Guatemala, Mexico, Namibia, and Nepal, Gnych et al. (2018) found that devolution of forest rights that gives communities secure tenure has catalyzed investment in collectively held forests and other natural resources, with environmental and social returns as well as profits. But this also requires the emergence of legitimate community-level governance organizations for benefit sharing within the communities and capacity to engage with external state and market actors.

As noted, pastoralists rely on mobility across large areas to sustainably exploit rangelands and have historic claims to use rights over large areas as well as complex arrangements to access areas in extreme years, which are not easily accommodated in formal tenure systems. Indeed, the formalization process and recognition of individual tenure has often been a threat to pastoralists' tenure: "As a landscape is progressively surveyed, demarcated and allocated, pastoralists' mobility may be obstructed and their practices of repeatedly renegotiating access rights to resources become less effective, essentially depriving them of those rights" (Davies et al. 2016: 20). Securing tenure for pastoralists therefore requires a different way of thinking about the land:

instead of exclusive, individual holdings, accommodating overlapping collective rights, recognizing customary institutions, and building capacity of pastoralists, other communities, and government to negotiate and resolve conflicts (Davies et al. 2016). For example, Tanzania's Village Land Use Planning process was often missing pastoralists because they moved across village boundaries. A new Joint Village Land Use Planning process across village boundaries where pastoralists move allowed for better representation of pastoralists and led to issuing of group certificates of customary rights of occupancy (CCROs) (ILC 2014).

Participatory land-use planning approaches that bring together various stakeholders—including diverse local resource users, local government, and line agencies—to organize resource uses and practices are also used in watershed management. UN-Habitat (2016) identifies ways that tenure-sensitive land-use planning can help strengthen the resource rights of local resource users, especially those with more insecure tenure. Although the participation of stakeholders in such planning processes is likely to increase compliance with plans for resource use, plans alone are not sufficient. Some form of institutions are needed for ongoing activities such as setting and enforcing operational rules and managing conflicts. Many watershed, irrigation, or forest programs have responded to this by setting up formalized user groups, but effective coordination is more likely if programs identify and build on existing bases of cooperation, either through customary (for example, tribal) institutions or forms of forest or water users' groups.

Because of the wide variability of social and ecological systems over space and time, there is no single formula that works for all conditions. Rather, many countries and programs seek to establish adaptive collaborative management (or "adaptive co-management") arrangements that involve sharing of rights and responsibilities by government and civil society, combined with an adaptive learning process that draws on both local and external scientific knowledge to deal with complex systems (Armitage, Berkes, and Doubleday 2007). Instead of blueprint approaches to establish steady state conditions, these call for flexibility to create resilience of linked human and natural resource systems.

Conclusion

With population continuing to grow while additional land available for agriculture is shrinking, there is no doubt that maintaining or improving the status of natural resources already in use for agricultural resources is critical for future food security. There are numerous management practices that can safeguard or enrich the natural resource base, but their application in developing countries needs to be greatly expanded. Constraints to wider use are manifested at various levels. Price incentives may be weak, especially in remote areas, and dampen enthusiasm for undertaking any type of investment in agriculture. Uptake of nontraditional NRM practices may face basic informational constraints (for example, for conservation agriculture). Market constraints may limit availability of seed of leguminous plants or equipment for some land or water management improvements. Some farm-level practices are land consuming (for instance, rotations, fallows) while others are labor intensive (for example, composting, terracing) and may therefore be attractive only to households with available resources. Landscape-level practices are inhibited by lack of functional mechanisms for coordination. Finally, weak tenure security has a significant impact on the willingness of farmers and communities to invest in NRM.

There are many mechanisms at the disposal of governments to improve incentives, but a key one will be the improvement of tenure security over resources. Recent institutional innovations have proven that tenure security enhancement can be made efficiently and equitably without expensive demarcation of rights at the household or plot level. It is not a given, however, that legislation or regulation will enhance tenure security for the groups who are in most need. There is need for investment in awareness creation and capacity development to operationalize reforms. A challenge facing policymakers is that there is no single tenure policy or institutional blueprint that will prove to be the best option in all countries or even all locations within a country. From an agricultural perspective, the natural resource and tenure priorities will continue to differ for rainfed versus irrigated areas, for sedentary versus pastoral areas, for key cash/export crop areas versus staple food areas, and for peri-urban versus remote areas. In addition, sound stewardship of natural resources requires attention to their management at different spatial scales, from the plot to global scale, and the voice of future generations needs to be heard. This poses a tremendous coordination challenge for which mechanisms are still elusive. Finally, demographic, economic, and other forces will undoubtedly require modifications to today's tenure and coordination institutions.

Learning what works under which national and local contexts will continue to be a priority of researchers in support of property rights reforms and sustainable governance of resources. Multiple methods contribute to this learning. Detailed case studies provide nuanced insights. Systematic comparisons of multiple case studies (for example, Ostrom 1990; Cox, Arnold, and Villamayor Tomás 2010) and qualitative comparative analysis (QCA) help to identify the conditions that shape outcomes. Although it is generally difficult to randomize natural resource conditions, there have been some randomized controlled trials and other quasi-experimental methods to assess the impact of interventions, especially in tenure security (for example, Deininger et al. 2008), and systematic reviews (for example, Lawry et al. 2017; Higgins et al. 2018) build the evidence base across multiple interventions in different contexts. Given the multiple factors affecting natural resource management, no single approach can provide definitive answers. Rather, it is important to triangulate among different methods, perspectives, resources, and regions to improve our understanding of what contributes to (or undermines) sustainable natural resource management.

References

Abdulai, R., and A. Antwi. 2005. "Traditional Landholding Institutions and Individual Ownership of Land Rights in Sub-Saharan Africa." *World Review of Science, Technology and Sustainable Development* 2: 302–319.

AGRA (Alliance for a Green Revolution in Africa). 2013. *African Agriculture Status Report 2013*. Nairobi.

Ali, D., K. Deininger, and M. Goldstein. 2014. "Environmental and Gender Impacts of Land Tenure Regularization in Africa: Pilot Evidence from Rwanda." *Journal of Development Economics* 110: 262–275.

Anderies, J. M., and M. A. Janssen. 2013. *Sustaining the Commons.* Tempe: Arizona State University.

Armitage, D., F. Berkes, and N. Doubleday. 2007. *Adaptive Co-Management: Collaboration, Learning and Multi-Level Governance.* Vancouver, Canada: UBC Press.

Bationo, A., B. Waswa, J. Kihara, I. Adolwa, B. Vanlauwe, and K. Saidou, eds. 2012. *Lessons Learned from Long-Term Soil Fertility Management Experiments in Africa.* Dordrecht, Netherlands: Springer Press.

Besley, T. 1995. "Property Rights and Investment Incentives: Theory and Evidence from Ghana." *Journal of Political Economy* 103 (5): 903–937.

Bezabih, M., S. Holden, and A. Mannberg. 2016. "The Role of Land Certification in Reducing Gaps in Productivity between Male- and Female-Owned Farms in Rural Ethiopia." *Journal of Development Studies* 52 (3): 360–376.

Bruce, J. W., and S. E. Migot-Adholla, eds. 1994. *Searching for Land Tenure Security in Africa.* Dubuque, IA: Kendall/Hunt Publishing Co.

Burney, J., and R. Naylor. 2012. "Smallholder Irrigation as a Poverty Alleviation Tool in Sub-Saharan Africa." *World Development* 40 (1): 110–123.

Calzadilla, A., T. Zhu, K. Rehdanz, R. S. J. Tol, and C. Ringler. 2013. "Economywide Impacts of Climate Change on Agriculture in Sub-Saharan Africa." *Ecological Economics* 93 (September): 150–165.

Comprehensive Assessment of Water Management in Agriculture. 2007. *Water for Food, Water for Life: A Comprehensive Assessment of Water Management in Agriculture.* London: Earthscan; Colombo, Sri Lanka: International Water Management Institute.

Corbeels, M., J. de Graaff, T. H. Ndah, E. Penot, F. Baudron, K. Naudin, and L. Rusinamhodzi. 2014. "Understanding the Impact and Adoption of Conservation Agriculture in Africa: A Multi-Scale Analysis." *Agriculture Ecosystem and Environment* 187: 155–170.

Cotula, L., and B. Neves. 2007. "The Drivers of Change." In *Changes in "Customary" Land Tenure Systems in Africa*, edited by L. Cotula, 15–31. London: International Institute for Environment and Development.

Cox, M., G. Arnold, and S. Villamayor Tomás. 2010. "A Review of Design Principles for Community-Based Natural Resource Management." *Ecology and Society* 15 (4): 38.

Davies, J., P. Herrera, J. Ruiz-Mirazo, J. Mohamed-Katerere, I. Hannam, and E. Nuesri. 2016. *Improving Governance of Pastoral Lands Implementing the Voluntary Guidelines on the Responsible Governance of Tenure of Land, Fisheries and Forests in the Context of National Food Security.* Governance of Tenure Technical Guide 6. Rome: FAO.

Deininger, K. 2003. *Land Policies for Growth and Poverty Reduction.* Washington, DC: World Bank.

Deininger, K., D. A. Ali, S. Holden, and J. Zevenbergen. 2008. "Rural Land Certification in Ethiopia: Process, Initial Impact, and Implications for Other African Countries." *World Development* 36 (10): 1786–1812.

Deininger, K., and H. Binswanger. 1999. "The Evolution of the World Bank's Land Policy: Principles, Experience, and Future Challenges." *World Bank Research Observer* 14 (2): 247–276.

Deininger, K., and S. Jin. 2006. "Tenure Security and Land Related Investment: Evidence from Ethiopia." *European Economic Review* 50 (5): 1245–1277.

Doss, C., C. Kovarik, A. Peterman, A. Quisumbing, and M. van den Bold. 2015. "Gender Inequities in Ownership and Control of Land in Africa: Myth and Reality." *Agricultural Economics* 46 (3): 403–434.

Easterly, W. 2008. "Design and Reform of Institutions in LDCs and Transition Economics: Institutions Top Down or Bottom Up?" *American Economic Review* 98 (2): 95–99.

Ellison, D., C. Morris, B. Locatelli et al. 2017. "Trees, Forests and Water: Cool Insights for a Hot World." *Global Environmental Change* 43: 51–61.

FAO (Food and Agriculture Organization of the United Nations). 2011. *The State of the World's Land and Water Resources for Food and Agriculture (SOLAW)—Managing Systems at Risk.* Rome: FAO; London: Earthscan.

———. 2012a. *State of the World's Forests.* Rome.

———. 2012b. *World Agriculture towards 2030/2050: The 2012 Revision*, by N. Alexandratos and J. Bruinsma. ESA Working Paper 12-03, June. Rome.

Farrow, A., E. Ronner, G. van den Brand et al. 2019. "From Best Fit Technologies to Best Fit Scaling: Incorporating and Evaluating Factors Affecting the Adoption of Grain Legumes in Sub-Saharan Africa." *Experimental Agriculture* 55: 226–251.

Food and Land Use Coalition. 2019. *Growing Better: Ten Critical Transitions to Transform Food and Land Use.* Global Consultation Report of the Food and Land Use Coalition.

Furubotn, E. G., and S. Pejovich. 1972. "Property Rights and Economic Theory: A Survey of Recent Literature." *Journal of Economic Literature* 10 (4): 1137–1162.

Galik, C. S., and P. Jagger. 2015. "Bundles, Duties, and Rights: A Revised Framework for Analysis of Natural Resource Property Rights Regimes." *Land Economics* 91 (1): 76–90.

Ghebru, H., and F. Girmachew. 2017. *Scrutinizing the Status Quo: Rural Transformation and Land Tenure Security in Nigeria.* IFPRI-NSSP Working Paper 43. Washington, DC: IFPRI.

———. 2019. *Perceived Tenure (In)security in the Era of Rural Transformation: Gender-Disaggregated Analysis from Mozambique.* IFPRI Discussion Paper 1799. Washington, DC: IFPRI.

Ghebru, H., and S. Holden. 2013. *Links between Tenure Security and Food Security: Evidence from Ethiopia.* IFPRI Discussion Paper 1288. Washington, DC: IFPRI.

Ghebru, H., and I. Lambrecht. 2017. "Drivers of Perceived Land Tenure (In)Security: Empirical Evidence from Ghana." *Land Use Policy* 66: 293–303.

Giller, K., E. Witter, M. Corbeels, and P. Tittonell. 2009. "Conservation Agriculture and Smallholder Farming in Africa: The Heretics' View." *Field Crops Research* 114: 23–34.

Gnych, S., S. Lawry, I. Monterroso, and A. Adhikary. 2018. "Common Benefits: Is Community Tenure Facilitating Investment in the Commons for Inclusive and Sustainable Development?" Paper presented at the "2018 World Bank Conference on Land and Poverty," Washington, DC, March 19–23.

Goldstein, M., K. Hounkebedji, F. Kondylis, M. O'Sullivan, and H. Selod. 2018. "Formalization without Certification? Experimental Evidence on Property Rights and Investment." *Journal of Development Economics* 132: 57–74.

Gollin, D., D. Lagakos, and M. Waugh. 2014. "The Agricultural Productivity Gap." *Quarterly Journal of Economics* 129 (2): 939–993.

Haggblade, S., B. Minten, C. Pray, T. Reardon, and D. Zilberman. 2017. "The Herbicide Revolution in Developing Countries: Patterns, Causes, and Implications." *European Journal of Development Research* 29 (3): 533–559.

Henao, J., and C. Banaante. 2006. *Agricultural Production and Soil Nutrient Mining in Africa: Implications for Resource Conservation and Policy Development: Summary*. Muscle Shoals, AL: International Fertilizer Development Center.

Higgins, D., T. Balint, H. Liversage, and P. Winters. 2018. "Investigating the Impacts of Increased Rural Land Tenure Security: A Systematic Review of the Evidence." *Journal of Rural Studies* 61: 34–62.

Holden, S., K. Deininger, and H. Ghebru. 2011. "Tenure Insecurity, Gender, Low-Cost Land Certification, and Land Rental Market Participation in Ethiopia." *Journal of Development Studies* 47 (1): 31–47.

Huntington, H., A. Haflett, and B. Ewing. 2018. "The Impact of Interventions to Promote Climate Change Adaptation: Does Stronger Tenure Security Increase Farmer Investment in Sustainable Agroforestry?" Paper presented at the "2018 World Bank Conference on Land and Poverty," Washington, DC, March 19–23.

ILC (International Land Coalition). 2014. *Participatory Rangeland Resource Mapping in Tanzania*. Rome.

IUCN. 2011. *Supporting Sustainable Pastoral Livelihoods: A Global Perspective on Minimum Standards and Good Practices*. Nairobi.

Jack, B., P. Oliva, C. Severen, E. Walker, and S. Bell. 2015. "Technology Adoption under Uncertainty: Take-Up and Subsequent Investment in Zambia." NBER Working Paper 21414. National Bureau of Economic Research, Cambridge, MA.

Jack, K., and S. Jayachandran. 2018. "Self-Selection into Payments for Ecosystem Services Programs." *Proceedings of the National Academy of Sciences* 116 (12): 5326–5333.

Jacoby, H., G. Li, and S. Rozelle. 2002. "Hazards of Expropriation: Tenure Insecurity and Investment in Rural China." *American Economic Review* 92 (5): 1420–1447.

Jayne, T., S. Snapp, F. Place, and N. Sitko. 2019. "Sustainable Agricultural Intensification in an Era of Rural Transformation in Africa." *Global Food Security* 20: 105–113.

Karp, D. S., R. Chaplin-Kramer, T. D. Meehan et al. 2018. "Crop Pests and Predators Exhibit Inconsistent Responses to Surrounding Landscape Composition." *Proceedings of the National Academy of Sciences (PNAS)* 115 (33): E7863-E7870.

Kerr, J. 2007. "Watershed Management: Lessons from Common Property Theory." *International Journal of the Commons* 1 (1): 89–110.

Kieran, C., K. Sproule, C. Doss, A. Quisumbing, and S. M. Kim. 2015. "Examining Gender Inequalities in Land Rights Indicators in Asia." *Agricultural Economics* 46: 119–138.

Kissinger, G., M. Herold, and V. De Sy. 2012. "Drivers of Deforestation and Forest Degradation: A Synthesis Report for REDD+ Policymakers." Lexeme Consulting, Vancouver.

Knight, S. R. 2010. *Statutory Recognition of Customary Land Rights in Africa*. Rome: FAO.

Kremin, C., and A. Miles. 2012. "Ecosystem Services in Biologically Diversified versus Conventional Farming Systems Benefits, Externalities, and Trade-Offs." *Ecology and Society* 17 (4): 40.

Kroeger, T., and F. Casey. 2007. "An Assessment of Market-Based Approaches to Providing Ecosystem Services on Agricultural Lands." *Ecological Economics* 64 (2): 321–332.

Larson, A. M., I. Monterroso, M. R. Banjade, and E. Mwangi. 2016. "Community Rights to Forests in the Tropics: Progress and Retreat on Tenure Reforms." In *Comparative Property Law: Global Perspectives*, edited by M. Graziadei and L. Smith, 435–457. Cheltenham, UK: Edward Elgar.

Lasco, R., R. J. Delfino, D. Catacutan, E. Simelton, and D. Wilson. 2014. "Climate Risk Adaptation by Smallholder Farmers: The Roles of Trees and Agroforestry." *Current Opinion in Environmental Sustainability* 6: 83–88.

Lastarria-Cornhiel, S. 1997. "Impact of Privatization on Gender and Property Rights in Africa." *World Development* 25 (8): 1317–1333.

Lawry, S., R. McLain, B. Swallow, K. Biedenweg, and M. Matt Sommerville. 2012. *Devolution of Forest Rights and Sustainable Management Volume 1: A Review of Policies and Programs in 16 Developing Countries*. Washington, DC: Tetra Tech ARD.

Lawry, S., C. Samii, R. Hall, A. Leopold, D. Hornby, and F. Mtero. 2017. "The Impact of Land Property Rights Interventions on Investment and Agricultural Productivity in Developing Countries: A Systematic Review." *Journal of Development Effectiveness* 9 (1): 61–81.

Le, Q. B., E. Nkonya, and A. Mirzabaev. 2016. "Biomass Productivity-Based Mapping of Global Land Degradation Hotspots." In *Economics of Land Degradation and Improvement: A Global Assessment for Sustainable Development*, edited by E. Nkonya, A. Mirzabaev, and J. von Braun, 55–84. Cham, Switzerland: Springer Open.

Linkow, B. 2016. "Causes and Consequences of Perceived Land Tenure Insecurity: Survey Evidence from Burkina Faso." *Land Economics* 92 (2): 308–327.

Marenya, P., and C. Barrett. 2009. "Soil Quality and Fertilizer Use Rates among Smallholder Farmers in Western Kenya." *Agricultural Economics* 40 (5): 561–572.

McCarthy, N. 2004. *Managing Resources in Erratic Environments: An Analysis of Pastoralist Systems in Ethiopia, Niger, and Burkina Faso.* Research Report 135. Washington, DC: International Food Policy Research Institute (IFPRI).

McLain, R. J., and S. Lawry. 2015. "Good Governance: A Key Element of Sustainable Nontimber Forest Product Harvesting Systems." In *Ecological Sustainability and Non-Timber Forest Products,* edited by C. M. Shackleton, A. K. Pandey, and T. Ticktin, 235–259. London: Routledge.

Meinzen-Dick, R., H. Markelova, and K. Moore. 2010. *The Role of Collective Action and Property Rights in Climate Change Strategies.* CAPRi Policy Brief 7. Washington, DC: IFPRI.

Meinzen-Dick, R., and E. Mwangi. 2007. "Cutting the Web of Interests: Pitfalls of Formalizing Property Rights." *Land Use Policy* 26: 36–43.

Meinzen-Dick, R. S., and R. Pradhan. 2002. *Legal Pluralism and Dynamic Property Rights.* CAPRi Working Paper 22. Washington, DC: IFPRI.

Meinzen-Dick, R. S., A. R. Quisumbing, C. R. Doss, and S. Theis. 2019. "Women's Land Rights As a Pathway to Poverty Reduction: Framework and Review of Available Evidence." *Agricultural Systems* 172: 72–82.

Michler, J., K. Baylis, M. Arends-Kuenning, and K. Mazvimavi. 2019. "Conservation Agriculture and Climate Resilience." *Journal of Environmental Economics and Management* 93: 148–169.

Michler, J., and S. Wu. 2014. "The Effects of Governance on Relational and Formal Contracts: Theory and Evidence from Groundwater Irrigation Markets." Paper presented at 4th International Rice Conference, Bangkok.

Ostrom, E. 1990. *Governing the Commons: The Evolution of Institutions for Collective Action.* Cambridge: Cambridge University Press.

Otsuka, K., and F. Place. 2001. *Land Tenure and Natural Resource Management: A Comparative Study of Agrarian Communities in Asia and Africa.* Baltimore: Johns Hopkins University Press for IFPRI.

Palmer, D., S. Fricska, B. Wehrmann et al. 2009. *Towards Improved Land Governance.* Land Tenure Working Paper 11. Rome: FAO and United Nations Human Settlements Program.

Peoples, M., J. Brockwell, D. Herridge et al. 2009. "The Contributions of Nitrogen-Fixing Crop Legumes to the Productivity of Agricultural Systems." *Symbiosis* 48 (1): 1–17.

Piza, C., and M. Moura. 2011. "How Does Land Title Affect Access to Credit? Empirical Evidence from an Emerging Economy." Working Paper Series 2211, University of Sussex Business School, Brighton.

Place, F. 2009. "Land Tenure and Agricultural Productivity in Africa: A Comparative Analysis of the Economics Literature and Recent Policy Strategies and Reforms." *World Development* 37: 1326–1336.

Place, F., M. Roth, and P. Hazell. 1994. "Land Tenure Security and Agricultural Performance in Africa: Overview of Research Methodology." In *Searching for Land Tenure Security in Africa*, edited by J. W. Bruce and S. Migot-Adholla, 15–40. Washington, DC: World Bank.

Platteau, J.-P. 2008. "The Evolutionary Theory of Land Rights as Applied to Sub-Saharan Africa: A Critical Assessment." *Development and Change* 27 (1): 29–86.

Quan, J., and C. Toulmin. 2004. "Formalizing and Securing Land Rights in Africa." In *Land in Africa: Market Asset or Secure Livelihood, Proceedings and Summary of Conclusions from the Land in Africa Conference*, edited by J. Quan, S. Fei Tan, and C. Toulmin, 133–162. London: IIED.

Rockström, J., W. Steffen, K. Noone et al. 2009. "Planetary Boundaries: Exploring the Safe Operating Space for Humanity." *Ecology and Society* 14 (2): 32.

RRI (Rights and Resources Initiative). 2014. *What Future for Reform? Progress and Slowdown in Forest Tenure Reform since 2002*. Washington, DC.

———. 2015. *Who Owns the World's Land? A Global Baseline of Formally Recognized Indigenous and Community Land Rights*. Washington, DC.

Salzman, J., G. Bennett, N. Carroll, A. Goldstein, and M. Jenkins. 2018. "The Global Status and Trends of Payments for Ecosystem Services." *Nature Sustainability* 1: 136–144.

Sanchez, P. 2019. *Properties and Management of Soils in the Tropics*. Cambridge: Cambridge University Press.

Schlager, E., and E. Ostrom. 1992. "Property-Rights Regimes and Natural Resources: A Conceptual Analysis." *Land Economics* 68 (3): 249–262.

Sikor, T., and D. Muller. 2009. "The Limits of State-Led Land Reform: An Introduction." *World Development* 37 (8): 1307–1316.

Stevenson, J. R., and P. Vlek. 2018. *Assessing the Adoption and Diffusion of Natural Resource Management Practices: Synthesis of a New Set of Empirical Studies*. Rome: Independent Science and Partnership Council (ISPC).

Tamene, L., and Q. B. Le. 2015. "Estimating Soil Erosion in Sub-Saharan Africa Based on Landscape Similarity Mapping and Using the Revised Universal Soil Loss Equation (RUSLE)." *Nutrient Cycling in Agroecosystems* 102: 17–31.

2030 Water Resources Group. 2009. *Charting Our Water Future: Economic Frameworks to Inform Decision-Making*. McKinsey.

UN-Habitat. 2008. *Secure Land Rights for All*. Nairobi: United Nations Human Settlements Programme.

———. 2016. *Tenure Responsive Land Use Planning: A Guide for Country Level Implementation*. Nairobi.

USAID (US Agency for International Development). 2009. *Integrating Customary Land Tenure into Statutory Land Law: A Review of Experience from Seven Sub-Saharan African Countries and the Kyrgyz Republic*. Property Rights and Resource Governance Project.

———. 2013. *The Future of Customary Tenure: Options for Policymakers*. USAID Issue Brief.

Van Gelder, L. J. 2010. "What Tenure Security? The Case of a Tripartite View." *Land Use Policy* 27: 449–456.

Vanlauwe, B., K. Descheemaeker, K. Giller et al. 2015. "Integrated Soil Fertility Management in Sub-Saharan Africa: Unravelling Local Adaptation." *Soil* 1: 491–508.

Weisse, M., and E. D. Goldman. 2018. "2017 Was the Second-Worst Year on Record for Tropical Tree Cover Loss." Blog, World Resources Institute, June 26.

Williamson, R. C., and C. B. Kerekes. 2011. "Securing Private Property: Formal versus Informal Institutions." *Journal of Law and Economics* 54: 537–572.

Wily, L. A., N. Tagliarino, A. Vidal, C. Salcedo-La Vina, S. Ibrahim, and B. Almeida. 2016. Indicators of the Legal Security of Indigenous and Community Lands. Data file from LandMark: The Global Platform of Indigenous and Community Lands. www.landmarkmap.org.

Yamano, T., K. Otsuka, and F. Place. 2011. *Emerging Development of Agriculture in East Africa: Markets, Soils and Innovations*. Dordrecht, Netherlands: Springer Press.

Zomer, R., H. Neufeldt, J. Xu et al. 2016. "Global Tree Cover and Biomass Carbon on Agricultural Land: The Contribution of Agroforestry to Global and National Carbon Budgets." *Nature Scientific Reports* 6: 29987.

PART IV

Emerging Challenges and Opportunities in Agricultural Development

Chapter 19

CLIMATE CHANGE AND AGRICULTURAL DEVELOPMENT

Mark W. Rosegrant, Keith Wiebe, Timothy B. Sulser, Daniel Mason-D'Croz, and Dirk Willenbockel

Climate change will be a major driver of change in the agricultural sector in the coming decades, along with changes in population, income, urbanization, dietary preferences, and technology.[1] Agriculture is unique among economic sectors in its dependence on temperature, precipitation, and other climate variables, and is thus unique in its sensitivity to changes in those variables. Farmers around the world have long been accustomed to dealing with the vagaries of weather, but climate change is now occurring on a larger scale and will bring bigger challenges in terms of what farmers produce, where and how they produce it, and what we eat.

Throughout the entire history of agriculture over the past 10,000 years, including the period of rapid growth and intensification during the Green Revolution over the past half century, global mean temperatures have remained within a range of about 1°C from current levels (Schellnhuber, Rahmstorf, and Winkelmann 2016). The 2016 Paris climate accord set a target of keeping temperatures well below 2°C above preindustrial levels and to pursue efforts to limit the increase to 1.5°C. But most climate change scenarios show future temperatures rising well above these levels and well beyond historical experience. This chapter explores the implications of these changes. The first part examines alternative scenarios for climate change. The next section presents the latest findings on climate change impacts on agriculture and food security. Options to adapt to climate change and its impacts are considered as well as mitigation strategies.

1 The research on which this chapter is based was supported by funding from the CGIAR Research Program on Policies, Institutions, and Markets (PIM), the CGIAR Research Program on Climate Change, Agriculture and Food Security (CCAFS), the Bill & Melinda Gates Foundation, and the United States Agency for International Development. Analytical support from Nicola Cenacchi, Richard Robertson, and Shahnila Dunston is also gratefully acknowledged.

FIGURE 19.1 The IMPACT system of models

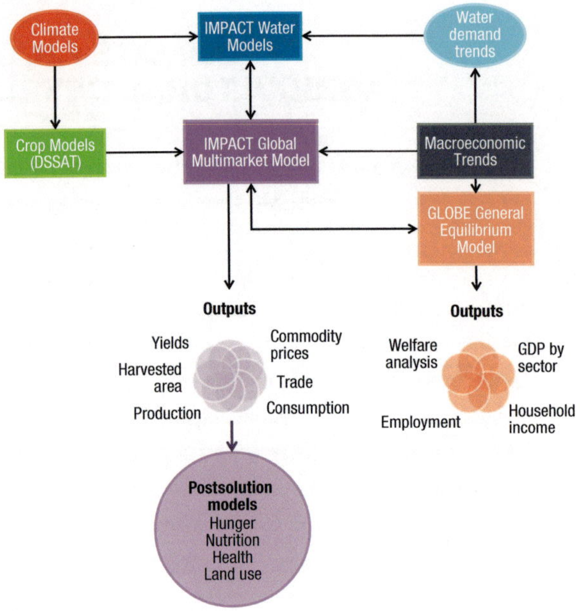

Alternative Scenarios for Climate Change

Analyzing climate change and its impacts requires a wide range of data and analytical tools. These may include general circulation models (GCMs) that generate simulations of future climates based on assumptions about future growth in greenhouse gas emissions; hydrology models that simulate water flows and storage; crop models that simulate crop growth; and economic models that simulate interactions between agricultural production, consumption, prices, and trade.

Methods and Scenarios

IFPRI's long-term projections of food and agriculture are based on the International Model for Policy Analysis of Agricultural Commodities and Trade (IMPACT), a multimarket model that simulates the operation of national and international markets, solving for production, demand, and prices that equate supply and demand across the globe (Rosegrant and IMPACT Development Team 2012; Robinson et al. 2015). The core model is linked to a number of "modules" that include climate models, water models (hydrology, water basin management, and water stress models), global

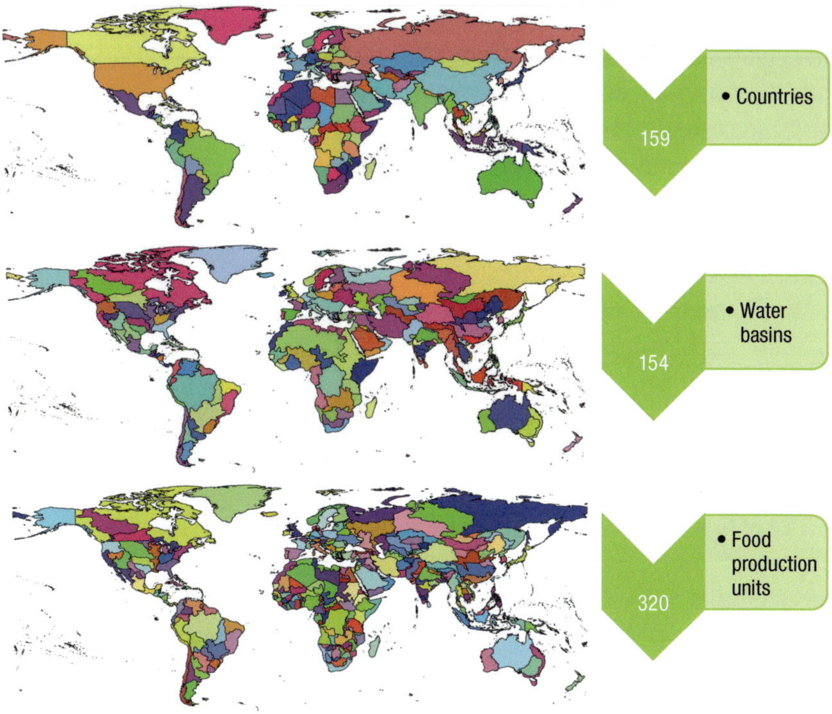

Source: Reproduced from Robinson et al. (2015).

gridded crop simulation models (Robertson 2017) such as the Decision Support System for Agrotechnology Transfer (DSSAT) (Jones et al. 2003; Hoogenboom et al. 2017), value chain models (for example, sugar, oils, livestock), land use (pixel-level land use, cropping patterns by regions), nutrition and health models, and welfare analysis (Figure 19.1).

The core multimarket model focuses on national and global markets including 159 countries. Agricultural production is specified by models of land supply, allocation of land to irrigated and rainfed crops, and determination of yields. Production is modeled at a subnational level, including 320 regions called food production units (FPUs). FPUs are defined to link to the water models and correspond to water basins within national boundaries—154 basins (that is, Nile, Amazon, and so forth) and 159 countries. The multimarket model simulates 62 agricultural commodity markets, representing the bulk of food and cash crops. A range of methodological innovations and improvements were incorporated to allow quantitative assessments of the potential trade-offs between competing and complementary multidimensional future scenarios.

The IMPACT partial equilibrium modeling framework was extended with a global computable general equilibrium model (GLOBE) (Willenbockel et al. 2018) and several linked postsolution models to evaluate the scenario effects of various investment requirements, along with evolving land-use changes, greenhouse gas (GHG) emissions, biodiversity, water quality, and micronutrient availability as well as dietary diversity. Another key modeling innovation was a new methodology to estimate research costs under reference and alternative investment scenarios (Mason-D'Croz et al. 2019). The IMPACT model is continually being upgraded and enhanced, through close collaboration between IFPRI and the other 14 CGIAR centers through the Global Futures & Strategic Foresight project, and with other leading global modeling groups through the Agricultural Model Intercomparison and Improvement Project (AgMIP) (Rosenzweig et al. 2013; Rosenzweig et al. 2018).

To facilitate analysis and comparison involving multiple models, standard sets of assumptions about different climate change trajectories, called "representative concentration pathways," have been developed as part of the Intergovernmental Panel on Climate Change (IPCC) process (Moss et al. 2010). These range from relatively slow rates of change in Representative Concentration Pathway (RCP) 2.6 (expressed in units of radiative forcing, in this case 2.6 watts per square meter in 2100), to relatively rapid change in RCP 8.5, with intermediate cases RCP 4.5 and 6.0 (Figure 19.2).

To estimate impacts of future climate change, we need to distinguish those impacts from the effects of other major drivers of change, including population, income, and technology. As with climate change, the global modeling community has also developed a set of standardized assumptions about changes in these other drivers to facilitate analysis and comparison involving multiple models. These "Shared Socioeconomic Pathways" (O'Neill et al. 2014; O'Neill et al. 2017) include a business-as-usual case (SSP2), a case with faster income growth and slower population growth (SSP1), a case with slower income growth and faster population growth (SSP3), and two other cases characterized by different assumptions about the nature of technological change and the degree of economic inequality (Figure 19.3).

Global and Regional Patterns of Climate Change

Climate change is characterized by multiple dimensions, including rising temperatures, changing precipitation patterns, changing frequency and intensity of extreme weather events, changing patterns of pests and diseases affecting crops and livestock as well as humans, sea-level rise, and glacial melting. The

FIGURE 19.2 Climate change as characterized in different Representative Concentration Pathways (RCP), 2000–2100

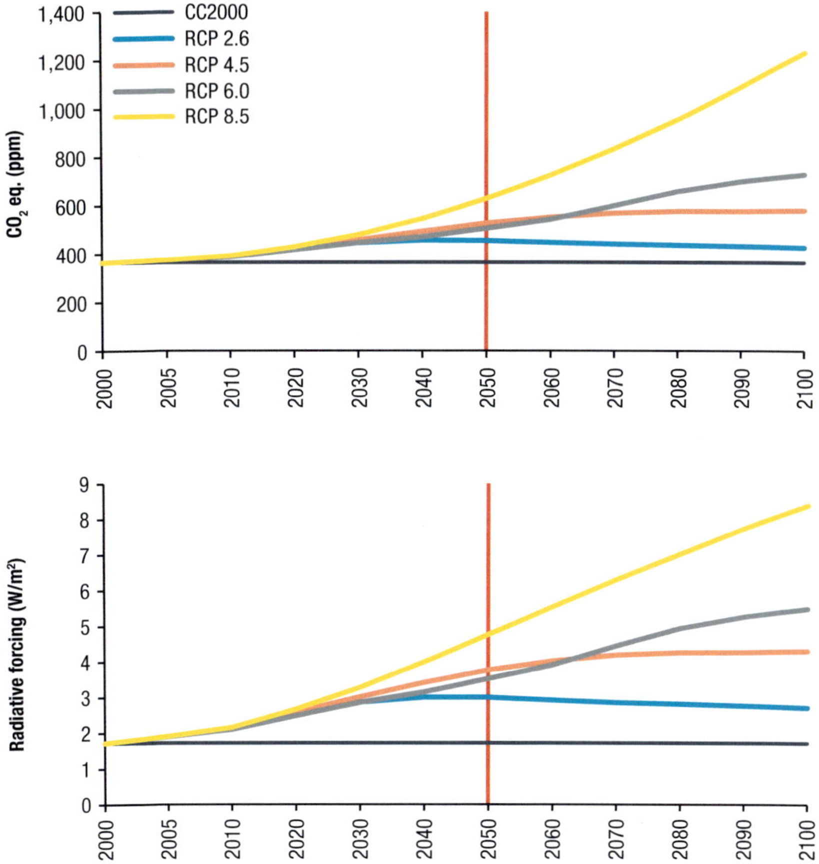

Source: Compilation by Robinson et al. (2015) from IIASA (2015).

first four are global in scope, while the last two are concentrated in coastal areas and particular river basins, respectively. Of the first four, current economic modeling capacities limit our primary focus to the first two: temperature and precipitation. General circulation models vary in the details of their projections about changes in temperature (Figure 19.4), but they generally agree that increases will be largest (up to 4°C or more by 2050) at higher latitudes, especially in the Northern Hemisphere. But lower latitudes, with higher temperatures to begin with, will feel the effects of heat stress with even smaller increases.

FIGURE 19.3 Socioeconomic drivers of change, 1980–2100

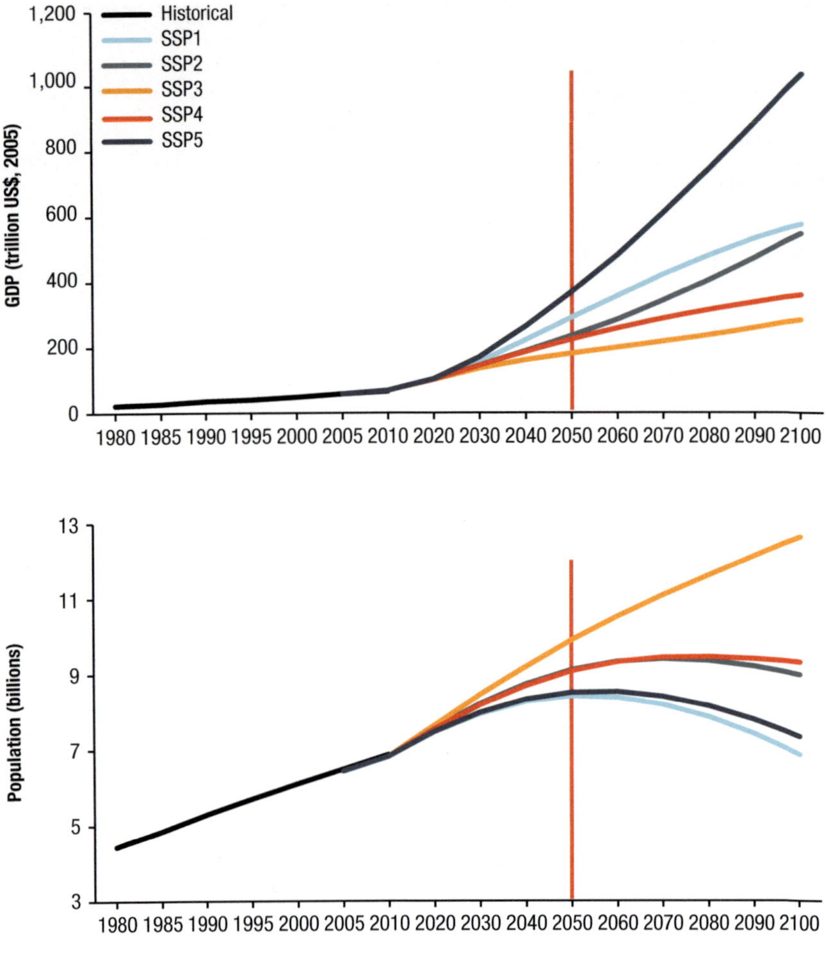

Source: Compilation by Robinson et al. (2015) from IIASA (2013).

Note: SSP = Shared Socioeconomic Pathway; GDP = gross domestic product.

Models vary more widely in their projections of changes in precipitation (Figure 19.5). There is general agreement that North Africa and southern Europe will become drier and that high northern latitudes will be wetter, but there are significant differences across models for most other regions. This highlights the uncertainty inherent in projections about climate change and its impacts.

Climate Change Impacts on Agriculture

Climate change affects agriculture most directly through the impact of changing temperatures and precipitation patterns on the growth of crops and livestock. Modeling tools such as the Decision Support System for Agrotechnology Transfer (DSSAT) system of crop models can analyze how future changes in these variables will affect the growth of particular crops and varieties, given levels of other variables such as soil quality and management practices. Figure 19.6, for example, shows how climate change would affect yields of rainfed maize, according to the HadGEM general circulation model and the DSSAT maize model and assuming climate changes rapidly as characterized by representative concentration pathway 8.5. Yield losses of 25 percent or more (indicated in red) are widespread, including in major producing areas in Asia, Brazil, Europe, and the United States. Impacts are mixed in Africa, and positive in a few areas such as northern China and western Canada.

It is essential to note that the projections in Figure 19.6 show only the direct biophysical impacts of changes in temperature and precipitation, holding constant such factors as technology and management practices. But these other factors are also changing over time, as a result of changes in demand (and thus prices) driven by changes in population, income, and dietary preferences, and also as a result of changes in prices driven by the biophysical impacts of climate change itself. Yield shocks of the magnitude indicated in Figure 19.6, were they to be realized, would trigger price increases for maize that would in turn induce changes in technology and management that would ripple through the agricultural economy. In fact, these changes are already occurring, for example as farmers adjust planting times and experiment with different crop varieties, and as researchers develop new varieties that are more tolerant of higher temperatures. Analyzing these indirect effects to determine the impacts of climate change more completely requires the use of global economic models in combination with climate and crop models.

Recent studies using multiple climate, crop, and economic models have explored this question under a range of climate and socioeconomic scenarios. Nelson et al. (2014) found that climate change would reduce yields of maize, rice, wheat, and soybeans by an average of 11 percent worldwide by 2050, relative to the hypothetical reference case of no climate change in 2050, and increase real prices by 20 percent over the reference case. Using a similar set of models, Wiebe et al. (2015) found yield reductions of 5 percent to 7 percent in 2050 for the same four crops plus sugar, attributing the difference to a broader range of socioeconomic and climate scenarios considered, with model

FIGURE 19.4 Changes in maximum temperature in 2050 compared to 2000 (°C)

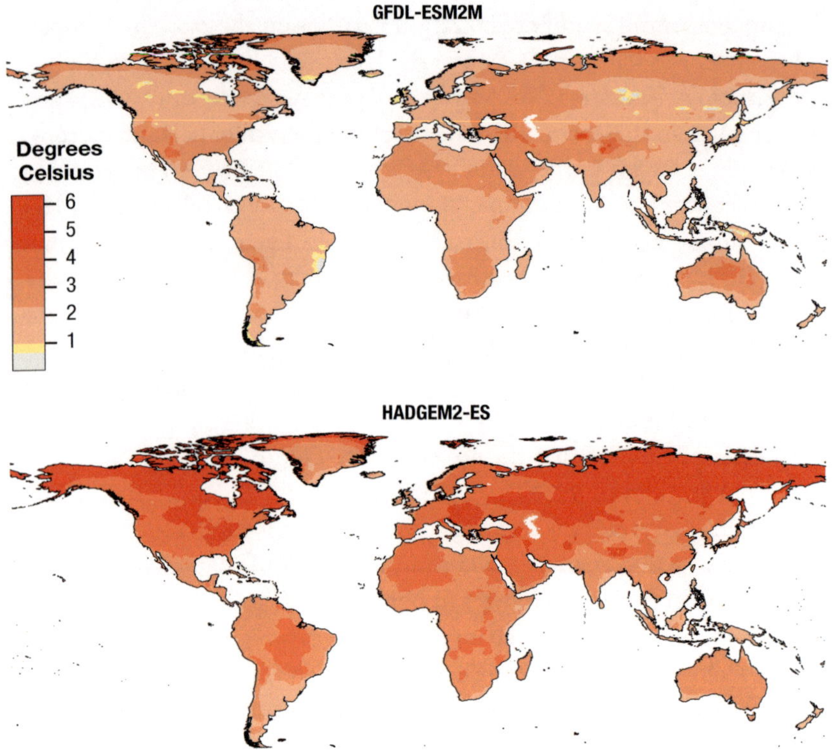

improvements allowing greater flexibility in responding to climate change, and the fact that sugar yields were found to respond more favorably to climate change than those of the other four crops considered. These yield losses in turn triggered real price increases for these commodities of 10 percent to 15 percent over 2050 levels in the absence of climate change and increases in area harvested of around 4 percent. While these increases in prices and area may not sound large over several decades, they are nevertheless double the increases that would be expected over that period in the absence of climate change, with important implications for access to food and environmental quality, respectively.

These impacts need to be seen in the context of broader changes driven by other drivers such as population, income, and technology. The latest baseline projections from IMPACT (IFPRI 2019) indicate that global food production will grow by about 60 percent over 2010 levels by 2050 in the context

IPLS-CM5A-LR

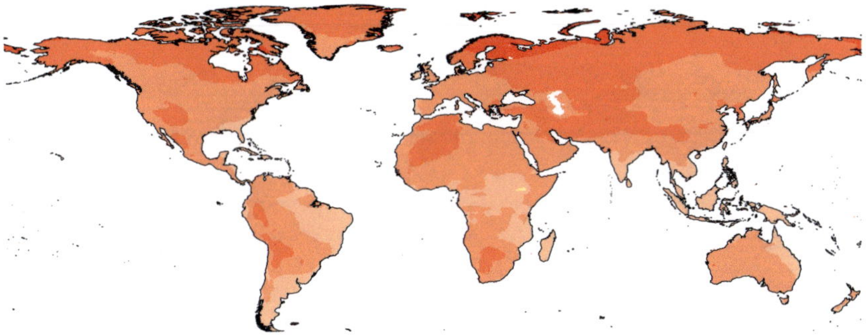

MIROC-ESM

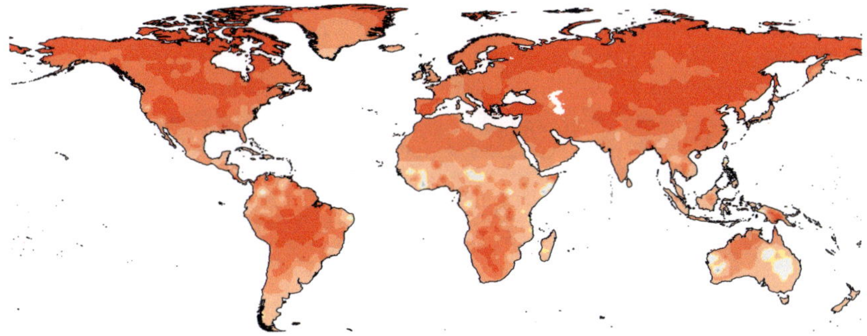

Source: Robinson et al. (2015), according to four general circulation models using RCP 8.5.

Note: Geophysical Fluid Dynamics Laboratory-Earth System Model version 2M (GFDL-ESM2M); Hadley Centre Global Environment Model version 2-Earth System (HADGEM2-ES); Institut Pierre-Simon Laplace Coupled Model, version 5A, low resolution (IPSL-CM5A-LR); Model for Interdisciplinary Research on Climate-Earth System Model (MIROC-ESM).

of climate change—10 percentage points less than would be the case without climate change. Production will grow more rapidly in developing countries, particularly in Africa. Even with population growth and climate change, per capita consumption is projected to increase by 10 percent globally to more than 3,000 kilocalories per day. But differences in access to food within and between countries mean that nearly 500 million people will remain at risk of hunger. In Africa south of the Sahara an additional 46 million people are projected to be at risk of hunger in 2050 as a result of climate change—30 percent more than would be at risk in the absence of climate change.

Despite the impacts of climate change, meat production is projected to grow by 65 percent globally by 2050, and by 76 percent in developing countries, although per capita consumption levels in developing countries will remain less than half of those in developed countries. Production of fruits and vegetables, pulses, and oilseeds will grow even more rapidly, by more

FIGURE 19.5 Changes in annual precipitation in 2050 compared to 2000 (millimeters)

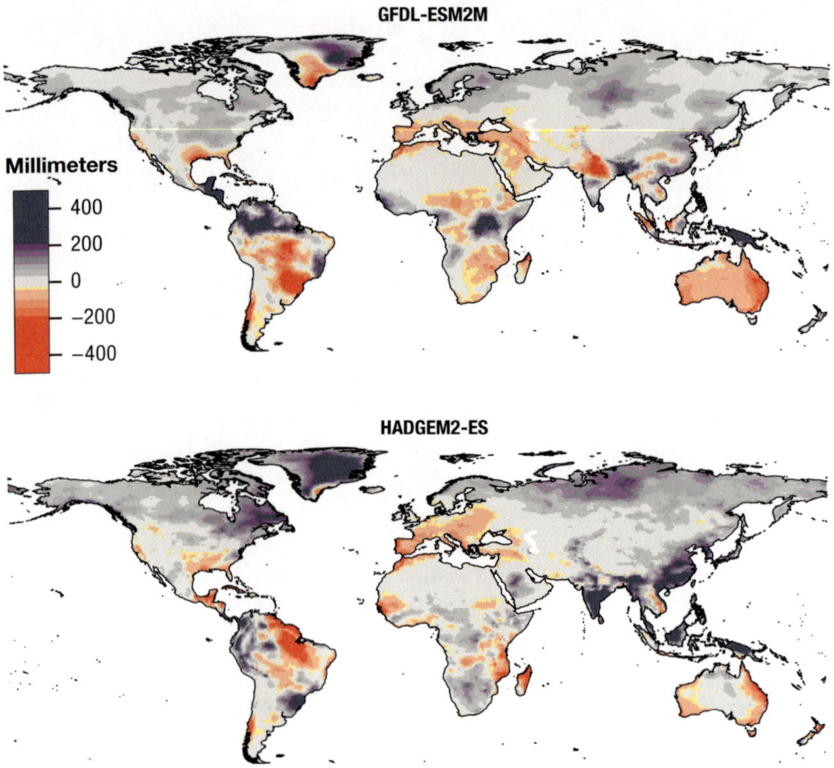

than 80 percent globally and more than doubling in some regions. Per capita consumption of fruits and vegetables in developing countries is projected to surpass that of developed countries by 2050, with important benefits for nutrition and health. Production of cereals and roots and tubers will grow more slowly, by around 40 percent globally but roughly doubling in Africa south of the Sahara. Developing countries as a group will become larger net importers of food from developed countries.

In addition to the indicators presented here, IMPACT allows us to explore changes in prices, land and water use, greenhouse gas emissions, and other socioeconomic and environmental indicators. Prices, for example, are projected to rise by about 50 percent for most food commodity groups by 2050 when the impacts of climate change are considered—about double the increase projected in the absence of climate change.

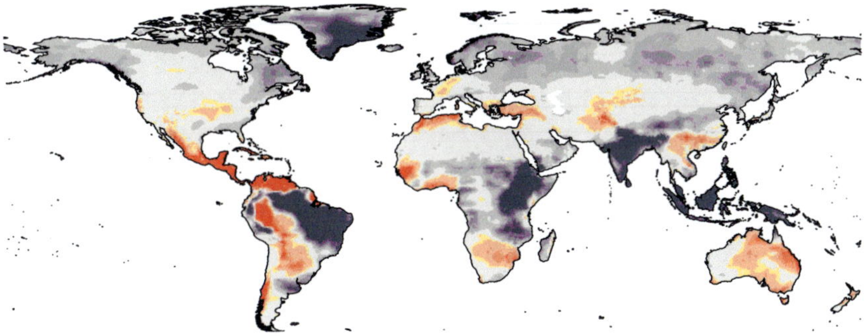

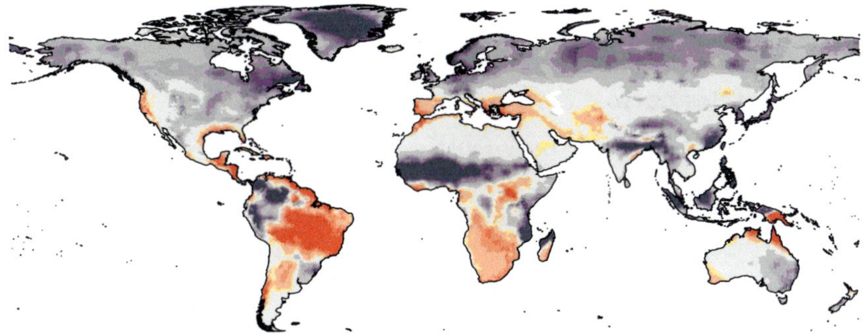

Source: Robinson et al. (2015), according to four general circulation models using RCP 8.5.

Note: Geophysical Fluid Dynamics Laboratory-Earth System Model version 2M (GFDL-ESM2M); Hadley Centre Global Environment Model version 2-Earth System (HADGEM2-ES); Institut Pierre-Simon Laplace Coupled Model, version 5A, low resolution (IPSL-CM5A-LR); Model for Interdisciplinary Research on Climate-Earth System Model (MIROC-ESM).

FIGURE 19.6 Climate change impacts on rainfed maize yields

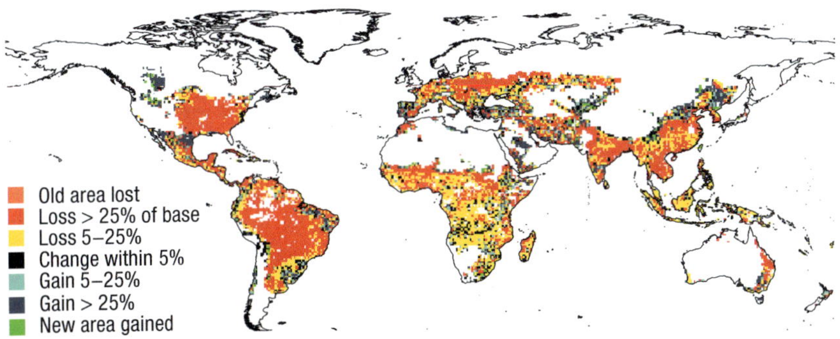

Source: Authors (based on HadGEM, DSSAT, and RCP 8.5).

Investments for Climate Change Adaptation

This section draws on work by Rosegrant et al. (2017), which has also been elaborated in recent work by Chan et al. (2019), Enahoro et al. (2019), Petsakos et al. (2019), and Frija et al. (2020). We apply the IMPACT modeling system linked to the GLOBE general equilibrium model to assess whether increased investments in the agricultural sector and rural infrastructure can effectively adapt the sector to climate change, improving the longer-term results on a range of food security, income, and environmental outcomes under climate change to better performance than under a future with no climate change. We assess the impacts of alternative agricultural R&D investment options as well as the role of complementary investments in irrigation and water resource management, soil management, and infrastructure.

Reference Scenario

The reference scenario used in this analysis, employed for comparison with alternative investment portfolios, assumes middle-of-the-road changes in population and income and rapid climate change. These assumptions are based on the IPCC's Shared Socioeconomic Pathway 2 (SSP2), in which the global population reaches 9.2 billion by 2050 with an average income of US$25,000 per person, and Representative Concentration Pathway 8.5 (RCP 8.5), as modeled by the HadGEM general circulation model (GCM).[2] Under RCP 8.5, by mid-century, the global mean surface temperature is projected to increase by 2.0°C relative to levels in 1986 through 2005 (Stocker et al. 2013).

Agricultural Research

Improvements in agricultural productivity in the reference scenario are represented by exogenous growth rates for each commodity and country, based on historical trends and expert opinion about future changes. We have developed an R&D investment-yield model to assess the investment required to achieve projected growth in agricultural productivity. Investments in research take time to bear fruit, as innovative ideas can take years to be developed and diffused widely. To capture these lags, the investment-yield estimation model is based on the perpetual inventory method, where research investments

---

2 Two additional variants of the baseline scenario were also run to explore sensitivity to climate change assumptions, one using a different GCM (REF_IPSL) and one assuming no climate change (REF_NoCC).

contribute to the stock of knowledge over time. Knowledge decays as older technologies become obsolete or irrelevant. Productivity grows if the stock of knowledge grows at a faster rate than it decays. The lag structure in the perpetual inventory method used here follows a gamma distribution in which R&D investments reach peak impact ten years after the initial investment and then decline over time, reaching zero ten years after peak impact.

With regionally differentiated research elasticities and decay rates, these imputed lag structures would vary by region according to existing R&D capacity and the potential trajectories for each region. Research capacity itself also varies significantly by region. To reflect these differences, we use elasticities of productivity with respect to research investments from the literature and incorporate spillover effects to represent each region's capacity to access and apply outside knowledge.

Irrigation and Water Resource Management

Water availability, including rainfall, streamflows, and evaporation, is determined in a hydrological model that downscales precipitation and temperature from climate scenarios generated by the GCM. Water supply and demand for each sector are determined in a simulation model that allocates water across irrigation, livestock, domestic, and industrial use. Water supply and demand are solved in 154 river basins globally and are linked annually to the IMPACT economic model (Robinson et al. 2015). Two of the key drivers in this model are assumptions on trends in irrigation expansion and water use efficiency (WUE). As with our assumptions on agricultural productivity, the reference assumptions used for these drivers are based on historical trends combined with expert opinion about future pathways. Total harvested area expands by about 18 percent in the projection period from 2010 to 2050. Irrigated area grows at a faster rate than rainfed area. Expansion of irrigation requires investments in water infrastructure such as dams, canals, and other conveyance systems. While the largest expansion in irrigated area is projected in Asia, the largest investments will be needed in Africa south of the Sahara due to the higher costs of expanding irrigation.

Infrastructure

The economic growth assumed in the reference scenario also includes investments in new infrastructure and in maintenance of existing electrification, roads, and other items.

Alternative Investment Scenarios

Three sets of alternative investment scenarios were analyzed, each of which increases investment in one of the areas described in the previous section. A fourth comprehensive scenario combines elements from the first three.

1. Enhanced productivity through increased investments in agricultural R&D.
2. Improved water resource management.
3. Improved marketing efficiency through increased investment in infrastructure.
4. A comprehensive scenario combining select elements of 1–3.

R&D for enhanced productivity. We model boosts to agricultural productivity through increased investment in R&D. These gains were first expressed as potential changes in absolute yield levels and then translated into differential yield growth rates used in the IMPACT modeling framework. The final endogenous yields and output growth generated by the scenarios are functions of interactions between these growth rates and projected changes in prices, demand, and other factors. Increased yields are combined with increased research efficiency to accelerate development, dissemination, and adoption of new technologies. Sources of improved research efficiency include more effective breeding techniques, brought about by advances in genomics, bioinformatics, and high-throughput gene sequencing, and more effective regulatory and intellectual property rights systems that reduce the lag times from discovery to deployment of new crop varieties.

Improved water resource management. Three alternative scenarios focus on investments and improvements in agricultural water resource management that affect crops and livestock directly through changes in water availability, and livestock indirectly through changes in feed prices. They include (1) accelerated investments in irrigation expansion; (2) the combination of irrigation expansion with improved water use efficiency on all irrigated cropland; and (3) improvements in rainwater harvesting and soil water-holding capacity. The expansion scenario simulates an expansion of irrigated areas in developing countries by 2030, relative to the reference scenario, with changes thereafter following the growth rates in the reference scenario. In the water use efficiency scenario, irrigation efficiencies are assumed to increase by 15 percent by 2030 and remain constant thereafter. The water management scenario simulates the benefits of technologies such as no-till agriculture and water harvesting that increase the water-holding capacity of soil or otherwise make precipitation more readily available to plants (that is, effective precipitation). Improvements vary by region due to varying levels of application, with a maximum increase in effective precipitation of 5 percent to 15 percent by 2045. Projected increases are largest in Africa south of the Sahara, where adoption of improved water-saving technologies is currently lower than in South Asia and other regions.

Improved infrastructure and market access. The Reduced Marketing Margins scenario assumes a mix of infrastructure improvements throughout the economies of developing countries, focusing primarily on transportation (road building, road maintenance, and railroads) and increased rural electrification. These improvements enhance productivity along the value chain, increase the speed of moving commodities to markets, and improve storage capacity—all of which boost market efficiency by better matching supply and demand over time. We represent these improvements as a reduction in the cost of moving goods from the farm to market. In IMPACT this is done by adjusting the price wedges between producer and consumer prices, reducing the margin from producer prices to consumer prices by 1 percentage point per year in all regions between 2015 and 2030.

Comprehensive investment portfolio. The final scenario considers the potential outcomes of a comprehensive investment portfolio that combines the investments of three other scenarios, including productivity enhancement, water management, and infrastructure investments.

Impacts on Poverty, Agriculture, Food Security, and the Environment to 2050

The scenarios vary widely in cost and generate a wide range of impacts. In some cases, impacts on different objectives are complementary, while others have significant trade-offs. The productivity enhancement scenarios generally offer moderate improvements in income, agricultural supply, and food security, with little impact on environmental improvement by 2030, but larger improvements by 2050, at relatively low cost. The water management scenario reduces water use and shows small improvements in income, supply, and food security. The infrastructure scenario increases income, supply, and food security, but at the cost of increased conversion of forestland and added GHG emissions. Such variable outcomes highlight the importance of a mixed portfolio of investments combining productivity enhancement with improved water resource management and market access. The comprehensive scenario achieves significant improvements in all outcome areas, particularly by 2050.

Income

In the reference scenario, global average incomes in 2050 increase by more than 150 percent compared to 2010, driven primarily by faster growth in the developing world. However, climate change slows income growth in all regions, the most in developing countries. The largest impacts are in South Asia and Africa south of the Sahara, where incomes in 2050 are 3 percent lower than they would be in the absence of climate change, compared to a reduction of 0.25 percent in developed countries. This translates into a negative impact of US$4.6 trillion on the developing world; Asia is hit hardest, although negative effects are felt around the globe (in trillion US$ in 2050: East Asia and Pacific −1.9; South Asia −1.4; Africa south of the Sahara −0.5; Middle East and North Africa −0.3; Latin America and the Caribbean −0.2).

In the enhanced agricultural productivity scenarios, as in the reference case, average global per capita income is projected to increase by about 2.5-fold between 2010 and 2050. This rate of increase is broadly similar across all of these scenarios because the agricultural sector is relatively small compared to the global economy. The average increase in income across developing countries is larger—closer to a fourfold increase compared to 2010—due to the larger share of agriculture in developing economies. In general, climate change results in per capita income about 3 percent lower in 2050, relative to levels in the absence of climate change, but incomes grow more rapidly under all productivity-enhancement scenarios. The highest increases are in the scenarios where CGIAR investments are supplemented by those from national governments or through increases in CGIAR R&D system efficiency. Increases are projected to be largest in South Asia (especially Afghanistan, Nepal, and Pakistan) and Africa south of the Sahara (especially Benin, Ghana, and Nigeria).

The three water scenarios all have a positive but small (less than 1 percent) impact on developing world incomes. In general, we see relatively small gains from expanding irrigation alone. This suggests that it will take increased investments in water use efficiency to fully realize the benefits of expanding irrigation. The improved infrastructure scenario increases the efficiency of transportation and processing sectors, thus reducing the cost of getting raw commodities from farm to table. These gains lead to increases in income of about 1 percent in developing countries. The gains vary by region, with the largest observed in Latin America and the Caribbean and Africa south of the Sahara, though smaller per capita gains across Asia add up when population size is factored in.

The comprehensive scenario sees significant increases in income, which would make achieving the CGIAR System Level Outcomes (SLOs) and the United Nations Sustainable Development Goals (SDGs) much easier. Relative to the reference in 2050, average incomes across developing countries rise by just over 4 percent in 2030 and by nearly 6 percent in 2050, adding more than US$9 trillion to the global economy. The largest improvements are in South Asia (with an increase of 9 percent), followed by Africa south of the Sahara (7 percent) and East Asia and Pacific (5 percent). (Due to complexity and overlap in the combination of scenarios, increases in income for the comprehensive scenario are less than the additive income gains from each of the component scenarios individually.)

Yield, Production, and Area

Globally, climate change will compound pressure on agriculture.[3] Although global average yields for the majority of crops are estimated to increase between 2010 and 2050 regardless of climate effects, climate change results in generally slower growth compared to the no climate change (NoCC) scenario. As a result, aggregate yields in 2050 are reduced under climate change for all commodity groups, although this is not necessarily the case for each single crop. Global average yields across the cereals group are estimated to decline by 6 percent to 9 percent under climate change by 2050. Overall, yield losses across all crops due to climate change are largest in South Asia, followed by Africa south of the Sahara and Latin America and the Caribbean.

Production for all major commodity groups increases between 2010 and 2050 across both developed and developing regions. As with yields, however, production growth is slower under climate change conditions across developing countries, while in developed countries effects are more variable. This leads to lower agricultural production under climate change across most regions, especially in Africa south of the Sahara and South Asia. Cereals production is estimated to be hit especially hard in South Asia (for example, in Pakistan and India) and Latin America and the Caribbean (especially in Central American countries). But agricultural production may benefit from climate change in higher latitudes, especially across the Former Soviet Union (FSU) (for example, in Armenia, Belarus, and Kyrgyzstan). Under the reference scenario, area harvested for all crops is projected to grow about 18 percent

3 Climate change impacts on livestock systems are currently modeled only through feed systems. Direct effects on livestock health, water use, and other components of the livestock production systems are currently under development for a future version of IMPACT.

to 20 percent between 2010 and 2050 (or about 200 million additional harvested hectares), which combines both intensification and extensification of agricultural production.

The comprehensive scenario generates large increases in productivity across the developing world, greatly improving agricultural competitiveness. The combination of investments in productivity, irrigation expansion, water use efficiencies, and reductions in marketing inefficiencies increases the developing world's total agricultural production by about 9.8 percent and 11.5 percent in 2030 and 2050, respectively, relative to the reference scenario. All targeted regions benefit; the largest increases in total agricultural production are projected in Africa south of the Sahara (20 percent and 25 percent in 2030 and 2050), with South Asia and Middle East and North Africa increasing between 15 percent and 18 percent. In response to these increases, the developed world lowers its agricultural production by 3 percent to 5 percent, with much of this decline (2 percent to 4 percent) coming from reductions in cropland area.

Prices and Trade

In general, climate change drives world commodity prices higher. In 2050 the average aggregated world crop commodities prices are projected to be between 12 percent and 18 percent higher under climate change than under the NoCC reference scenario. The price impacts on specific commodity groups vary depending on crop adaptability and specific demands on producers. Maize, groundnut, and potato see the largest price increases; barley, lentils, and meat are less affected (and could see price decreases).

In IMPACT, prices influence commodity supply and demand (and vice versa), while trade links national production and demand to world markets. Developing countries increase their net agricultural imports from the developed world between 2010 and 2050 under the reference scenario. Net imports for cereals and meat rise 2.6-fold and 6-fold from 86.6 million metric tons and 3.6 million metric tons, respectively, between 2010 and 2050. Imports of pulses and oilseeds rise 3.5- to 3.8-fold both from about 3 million metric tons, respectively. The developing world will shift from being net exporters of fruits and vegetables and roots and tubers to being moderate importers.

Increases in yields and production drive down world prices in 2050 across all of the alternative investment scenarios, relative to the reference with climate change. Increasing production through improved water resource management pushes prices down slightly (less than 1 percent on average across the projection period). The largest declines are observed for heavily irrigated crops such as rice, sugarcane, and cotton. Improved infrastructure shrinks price wedges as marketing efficiency increases, allowing producers to capture a greater share of the final consumer price. This leads world prices and consumer prices to decline by 4 percent to 5 percent in tandem, while producer prices increase by around 8 percent. South Asia and Africa south of the Sahara see larger producer price increases than the developing-country average, with maize farmers, in particular, benefiting from higher producer prices (12 percent increase).

In the comprehensive scenario, large increases in production push down prices globally, with reductions of nearly 20 percent in 2030. All targeted commodities see their world prices decline by more than 10 percent, with the largest declines occurring for roots, tubers, and pulses. In the comprehensive scenario the developing world improves its agricultural terms of trade, becoming a net exporter, as opposed to a net importer in the reference scenario (that is, under climate change without additional investments). This transition varies regionally; exporting regions such as Latin America and the Caribbean become larger exporters, while the importing regions South Asia and Africa south of the Sahara import less.

Food Security and Nutrition

In the reference scenario without climate change, both the number of undernourished children ages zero to five and the global risk of hunger fall between 2010 and 2050 due to rising food production. The South Asia region sees the largest reduction in population at risk of hunger by 2050: about 170 million people, with the number of affected falling from about 16 percent to about 4 percent of the population. The trend for Africa south of the Sahara is also worth noting. In 2010 the estimated number of people at risk of hunger is comparable between South Asia and Africa south of the Sahara, but in Africa south of the Sahara by 2050 the population at risk of hunger falls by only about 60 million. Under climate change these improvements are still significant but less pronounced. Across developing countries, the number of undernourished children increases by 3 percent to 5 percent in 2050 due to climate change compared to the reference without climate change. Climate change hits Africa south of the Sahara particularly hard. Its share of population at risk of hunger in 2050 rises from 8.6 percent under the NoCC scenario to 10–11 percent under climate change. This corresponds to an increase in population at risk of hunger of between 30 million and 46 million people in Africa south of the Sahara due to climate change (with East Africa, Malawi, and Tanzania among the most affected), against an increase between 5 million and 9 million in South Asia.

Higher food supplies under the enhanced productivity scenarios raise the availability of dietary energy (kilocalories) per capita across each region. This reduces the population at risk of hunger by 20 percent in developing countries (and by 30 percent in Africa south of the Sahara) and the number of malnourished children by about 7 percent in developing countries. The water resource management scenarios' small changes in prices and income lead to insignificant changes in overall welfare. Nevertheless, South Asia gets a large boost from increased water use efficiency—hunger falls by 9 percent in 2030 compared to the reference. Improved soil and water management, meanwhile, contributes more to reducing hunger in Africa south of the Sahara. When marketing costs are reduced, consumers' purchasing power rises as commodity prices fall and they gain increased income. These factors, together with increased food supply, boost consumption and reduce hunger, with the at-risk population falling by 6 percent in Africa south of the Sahara in 2050.

In the comprehensive scenario, commodity prices fall steeply while income sharply rises, leading to significant increases in kilocalorie availability across the developing world (from 4 percent in Latin America and the Caribbean to 10 percent in Africa south of the Sahara relative to the reference in 2050). These consumption increases are several percentage points higher than in any other of the investment scenarios considered, bringing the developing world average consumption to over 3,200 kilocalories per person per day. South Asia and Africa south of the Sahara see the largest increases in kilocalorie availability, 9.1 percent and 10.0 percent, respectively, bringing both regions to an average food supply near 3,000 kilocalories per person per day.

This dramatically reduces the population at risk of hunger, which falls almost one-quarter in developing countries by 2050 (compared to the reference scenario with climate change), while the absolute number at risk of hunger falls from 823 million in 2010 to 361 million in 2050. Figure 19.7 shows the potential of the comprehensive scenario for making progress toward SDG2 (for hunger). Most regions achieve the SDG2 target under the comprehensive scenario except for Africa south of the Sahara. In addition, although climate change is a drag on achieving SDG2, investments of the type modeled under the comprehensive scenario show there is potential to mitigate those effects.

Climate Change Mitigation

As shown in Figure 19.7, climate change has significant impacts on agriculture. Mitigation includes measures that reduce the amount of emissions

FIGURE 19.7 Prevalence of hunger, in millions of people and as a share of the total population (%)

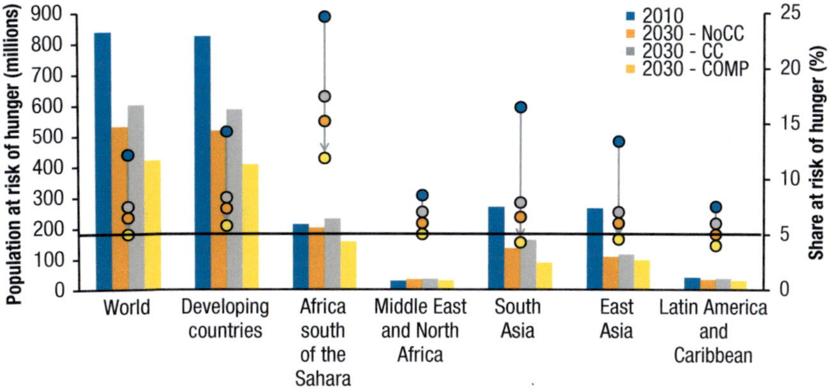

Source: Rosegrant et al. (2017).

Note: NoCC assumes a constant 2005 climate; CC reflects a climate future using RCP 8.5 and the HGEM Climate Model. The bars represent the number of people at risk of hunger in each region (left axis). The bubbles represent the share of the region's total population at risk of hunger (right axis). The gray lines reflect the change in the share at risk of hunger over time and across scenarios. The solid black line represents a target threshold of 5 percent of the population at risk of hunger.

(abatement) or enhance the absorption capacity of greenhouse gases (sequestration). The total global potential for mitigation depends on emissions levels, technology availability, enforcement, and incentives. In many situations the efficiency of agriculture can be improved at a low cost; however, when low-cost incentives are unavailable, policy development is important. The following is a short summary of key points.

Greenhouse Gas Emissions from Agriculture

- Agricultural emissions account for a significant amount of GHG emissions and could increase emissions substantially with growth in food demand if it is a sector of high economic importance in developing countries, with growth expected in the coming decades due to population growth and shifts in diets among other factors. Agriculture not only contributes to emissions from agricultural activities in the narrow sense but is also a driver of land-use change, which is usually treated separately in greenhouse gas accounting.

- Within the agricultural sector, emissions from fertilizer application, livestock and manure management, and rice cultivation are the major emission sources.

Mitigation Potential and Options

- Of developing regions, Africa has the lowest economic potential, contributing only 3.4 percent to the total potential reductions at carbon prices of US$30 per tCO_2eq. Similar results can be found for Brazil and India.
- China and South and Southeast Asia, on the other hand, have a higher potential, contributing together more than 40 percent of reductions at carbon prices of US$30 per tCO_2eq.
- Based on these results, rice cultivation mitigation strategies have the highest economic potential in developing countries.

Framework Conditions for Realizing the Mitigation Potential

- Agriculture in developing countries can play its role in the mitigation of greenhouse gases, but incentives to date are not conducive to investing in mitigation. At the same time, a major challenge will lie in aligning growing demand for agricultural products with sustainable and emission-saving development paths.
- The carbon market for the agricultural sector is underdeveloped. This is in part for good reason, as costs of verification and monitoring and transaction costs are rather high. However, it could be stimulated through different rules of access and operational rules in carbon trading as well as capacity building and advances in measurement and monitoring (Wollenberg et al. 2016; Godoy and Saes 2015; Smith et al. 2007; Rosegrant et al. 2008).
- Policies focused on mitigating greenhouse gas emissions, if carefully designed, can help create sustainable new income streams for farmers by increasing the profitability of environmentally sustainable practices (Rosegrant et al. 2008; Smith et al. 2008; Smith et al. 2013).

Emissions Trends

Climate change is the result of an increase in the concentration of greenhouse gases like carbon dioxide (CO_2), nitrous oxide (N_2O), and methane (CH_4). Rising greenhouse gas emissions are associated with economic activities including energy, industry, transport, and patterns of land use, including agricultural production and deforestation. As shown in Figure 19.8, agriculture—together with related emissions from land-use change and forestry

FIGURE 19.8 Global greenhouse gas emissions by economic sector, 2010 (%)

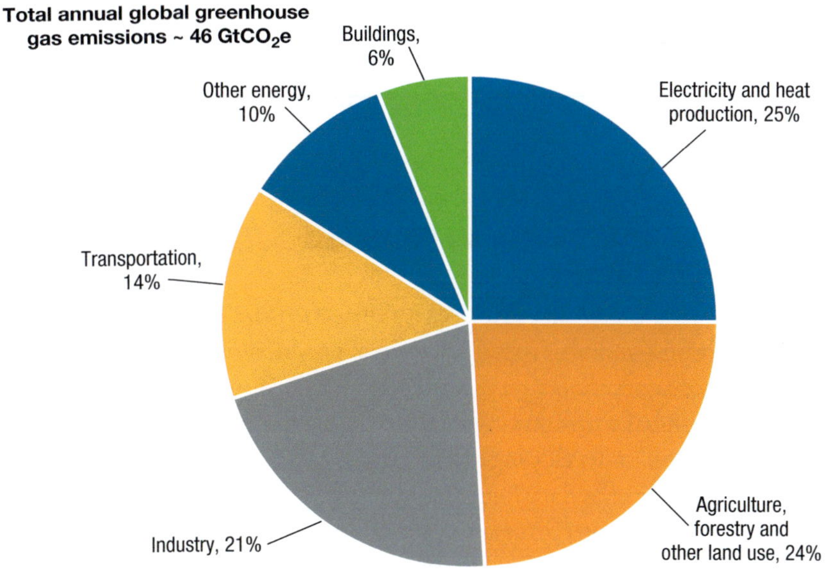

Source: IPCC (2014).

(LUCF)—create nearly one-quarter of global greenhouse gas emissions (IPCC 2014).

According to CCAFS (n.d.), direct agricultural emissions were about 5.38 $GtCO_2$ equivalent in 2012, which is 11.7 percent of total greenhouse gas emissions.[4] The vast majority of agricultural emissions from this sector are methane and nitrous oxide, making the agricultural sector the largest producer of non-CO_2 emissions. Although agricultural lands also generate very large CO_2 fluxes both to and from the atmosphere via photosynthesis and respiration, this flux is nearly balanced on existing agricultural lands. Significant carbon releases, however, result from the conversion of forested land, which is accounted for under the LUCF category. Concerning food production specifically, estimates of the amount of total emissions in this sector that are due to land conversion for agricultural intensification are difficult to make; however,

4 One million metric tons (MMt) of methane (CH_4) emissions equals 21 million metric tons of carbon dioxide (CO_2) emissions (1 MMt CH_4 = 21 MMt CO_2); similarly, 1 MMt N_2O = 320 MMt CO_2. This indicates that the global warming potential of methane and nitrous oxide are higher than carbon dioxide because they exist longer in the atmosphere. Yet because of their significantly smaller concentrations, the actual radiative forcing of CH_4 and N_2O are one-third and one-tenth of CO_2, respectively.

one estimate attributes 9 percent of total global emissions—one half of LUCF emissions—due to the expansion into forests for feedcrops and livestock production (Steinfeld et al. 2006). Finally, other agricultural activities related to GHG emissions are accounted for in other sectors, such as the upstream manufacture of equipment, fertilizers, and pesticides, the on-farm use of fuels, and the transport of agricultural products.

Emissions and Impacts Related to Food Production

The remainder of this chapter focuses on emissions and impacts related to food production—mainly crop and livestock production—and their related mitigation and adaptation strategies. Non-OECD countries emit nearly 75 percent of global emissions (WRI 2008). As a result, the theoretical potential for mitigation in developing countries is greater in the agricultural sector than in industrialized nations. Asian countries account for 37 percent of the world total emissions from agricultural production, with Latin America and Europe a distant second and third place, with 16 percent and 12 percent, respectively (WRI 2008). In Asia, China accounts for more than 18 percent of the total, while Brazil alone is responsible for nearly 10 percent of agricultural emissions in Latin America (WRI 2008).

Emissions from agriculture come from four principal sectors: agricultural soils, livestock and manure management, rice cultivation, and the burning of agricultural residues and savanna for land clearing. Figure 19.9 presents the share of emissions from each of these sectors. The largest shares of emissions originate from agricultural soils, with most of this coming from fertilizer applications, and enteric fermentation (CH_4) associated with livestock production. Each of these accounts for about 39 percent of agricultural GHG emissions. Flooded rice fields are the third largest source of agricultural emissions, contributing nearly 10 percent in the form of methane that results from anaerobic decomposition of organic matter.

Options and Potential for Mitigation in Agriculture

"Mitigation" is defined as any anthropogenic intervention that can either reduce the sources of GHG emissions (abatement) or enhance their sinks (sequestration). Following this, there are two categories of mitigation methods in agriculture: carbon sequestration into soils and on-farm emissions reductions. While not as large as the savings potential from reducing the consumption of fossil fuels, the total potential savings from various agronomic and

FIGURE 19.9 Greenhouse gas emissions from agriculture by source, 2012 (%)

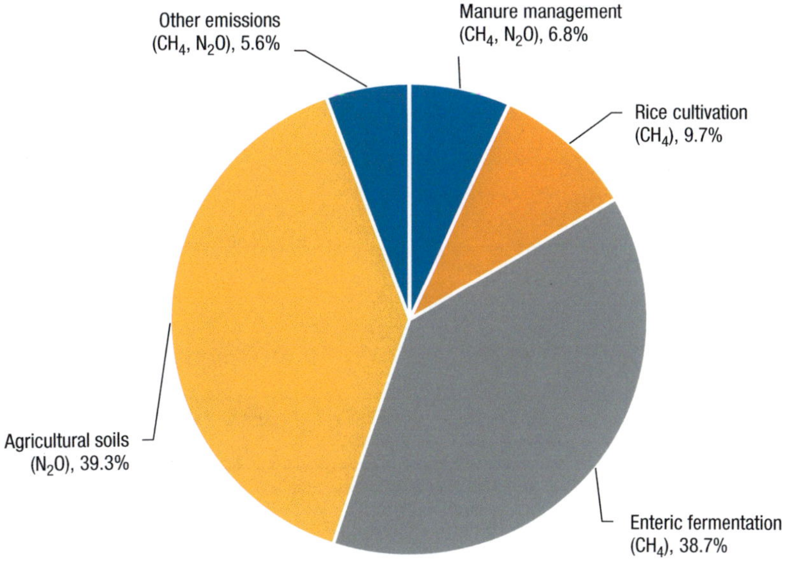

Source: CCAFS (n.d.) with supporting data and information from FAO (2015), Gerber et al. (2013), US EPA (2012).
Note: "Other emissions" includes burning of crop residues and savanna land, among others.

land-use activities are still substantial and can be achievable at a competitive cost. The technical mitigation potential is the theoretical amount of emissions that can be reduced given the full application of current technologies without considering the costs of implementation.

A better measure of the potential for mitigation is the economic potential. Many estimates have been made for the economic potential for GHG mitigation in agriculture, and Figure 19.10 provides a summary of the central tendency of the available estimates for the different sources of emissions from agriculture. The highest potential is soil carbon sequestration at 1.2 $GtCO_2$ equivalent. Soil carbon sequestration can be achieved through conservation tillage, integrated soil fertility management, restoring cultivated organic soils and degraded lands, retaining crop residues, and growing high-residue crops. A drawback of soil carbon sequestration is that renewed plowing following periods of low or zero tillage will release carbon to the atmosphere.

Dietary change also has significant potential, through reduction of consumption of animal-sourced food in favor of plant-based diets. A combination of meat-based or GHG emissions–based taxes, education, school feeding programs, and long-term lifestyle change can be used to achieve diet-based savings.

FIGURE 19.10 Potential for agricultural greenhouse gas emission reduction

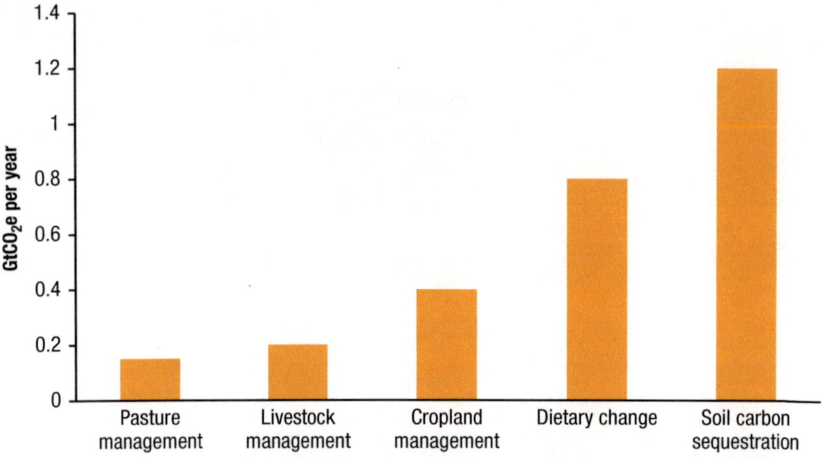

Source: Synthesized from Wollenberg et al. (2016); Smith et al. (2008); Smith et al. (2013); Del Grosso and Cavigelli (2012); Springmann et al. (2017); Havlík et al. (2014); Stehfest et al. (2013).

Note: The potential for reduction is substantial, at carbon tax of $20 per metric ton CO_2 equivalent.

Improved cropland management is another key source of GHG emission savings, through improved nitrogen-use efficiency through precision agriculture, slow release fertilizer, nitrogen-use efficient new crop varieties, stabilized N sources (polymer-coated urea and nitrification inhibitors), improved rice management (midseason drainage of rice paddies or alternate wetting and drying), and improved water management to reduce fertilizer runoff. Removal of subsidies on fertilizer, water, and energy would provide a major incentive to reduction in GHG emissions in cropland agriculture by reducing the use of these inputs. Other sources of reduction in emissions include livestock management through optimizing animal feed mixtures and feed additives; improving manure management systems, enhancing reproductive efficiency, and breeding for reduced methane emissions; and pasture management through improved grasses and pasture management and use of legumes.

Although significant potential exists for mitigation in agriculture, there are also barriers to mitigation, especially in developing countries. These barriers include the lack of a price on carbon that would incentivize the reduction of GHG emissions, weak property rights, political economy that prevents subsidy reductions, higher production costs for sustainable practices, and a lack of access to inputs and technical assistance. As a result, policy interventions are needed to create pro-poor mitigation strategies and to maximize synergies with sustainable rural development and adaption. Perhaps the most important step is to establish a value for carbon, through the implementation of carbon taxes and development of payment programs for carbon-reducing agricultural practice and technologies.

Policies should also look beyond purely project-based mechanisms—for example, on programmatic approaches or on crediting of sustainable policies and measures with positive mitigation effects. International capacity building and advisory services can enable countries to improve their ability to participate in and benefit from the carbon market. Global sharing of innovative technologies for efficient use of land resources and low emission management practices will also be needed. Finally, further investment in advanced measurement and monitoring can dramatically reduce costs of verification. Measurement and monitoring techniques have been improving rapidly, thanks to a growing body of field measurements and the use of statistics and computer modeling, remote sensing, global positional systems, and geographic information systems, so that changes in stocks of carbon can now be estimated more accurately at lower cost.

Climate change mitigation policies, if carefully designed, can create a new development strategy that encourages the creation of new value in pro-poor investments by increasing profitability of environmentally sustainable practices. To achieve this dual goal, it will be necessary to streamline the measurement and enforcement of offsets, financial flows, and carbon credits for investors. It is important to enhance global financial facilities and governance to simplify rules and increase funding flows for mitigation in developing countries. Climate policies and increased integration of agriculture into carbon markets will still face substantial barriers. In many regions, nonclimate policies related to macroeconomics, agriculture, and the environment have a larger impact on agricultural mitigation than climate policies, and reforms will be necessary to stimulate climate change mitigation.

Conclusion and Policy Implications

Population and income will continue to drive growth in demand through midcentury. Food and nutrition security are projected to improve, but climate change will slow this progress. Impacts are uncertain and vary by location, time, and scale. The results presented here may be conservative in that they focus on the impacts of changes in mean temperature and precipitation, and exclude variability, extreme events, sea-level rise, and changing patterns of pests and diseases. These are critical areas for further research.

It is essential to distinguish final impacts of climate change from the initial and direct biophysical impacts on crop growth. Final impacts depend on complex feedback interactions, including—critically—choices made by individuals, businesses, and governments. Key among these are increased investments in agricultural R&D and infrastructure, which we find can offset climate change impacts on agriculture and food security, at least through midcentury. These findings reinforce what we already know to be sound measures to improve productivity and resilience. The menu of management, technology, and investment options for climate adaptation and mitigation is essentially the same that has been developed for agricultural productivity growth. And essentially the same constraints need to be overcome with (and without) climate change—risks, uncertainty, imperfect markets, lack of credit, and insurance. So what difference does climate change policy make?

As noted in Rosegrant and Sombilla (2018), climate change policy is good agricultural policy—with a twist. As shown in this chapter, both agricultural growth and climate adaptation require increased investments in agricultural research and development. Under climate change there should be a shift in R&D investment on the margin to breeding for nitrogen use efficiency, drought tolerance, and livestock efficiency and greenhouse gas reduction. Increased irrigation investments are another priority for both growth and climate adaptation. In some regions climate change will make large dams more valuable to handle increased variability in precipitation and runoff, but in more cases greater emphasis should be given to small-scale irrigation for flexibility. Removal of subsidies for fertilizer, water, and energy, as well as putting the financial savings into productivity-enhancing investments, will boost agricultural growth and simultaneously reduce GHG emissions. Promotion of healthy diets has even higher benefits under climate change because it can also reduce GHG emissions. Finally, the increased variability in production over time due to climate change can increase the benefits from removal of agricultural trade and macroeconomic distortions. Open trade becomes even more important because climate change will increase the reliance of many developing countries on food imports.

References

CCAFS (CGIAR Research Program on Climate Change, Agriculture and Food Security). n.d. CCAFS Big Facts. Theme 2. Food Emissions: Direct Agricultural Emissions. Accessed January 2020. https://ccafs.cgiar.org/bigfacts/#theme=food-emissions&subtheme=direct-agriculture.

Chan, C. Y., N. Tran, S. Pethiyagoda, C. C. Crissman, T. B. Sulser, and M. J. Phillips. 2019. "Prospects and Challenges of Fish for Food Security in Africa." *Global Food Security* 20: 17–25.

Del Grosso, S. J., and M. J. Cavigelli. 2012. "Climate Stabilization Wedges Revisited: Can Agricultural Production and Greenhouse Gas Reduction Goals Be Accomplished?" *Frontiers in Ecology and the Environment* 10: 571–578.

Enahoro, D., D. Mason-D'Croz, M. Mul, K. M. Rich, T. P. Robinson, P. Thornton, and S. S. Staal. 2019. "Supporting Sustainable Expansion of Livestock Production in South Asia and Sub-Saharan Africa: Scenario Analysis of Investment Options." *Global Food Security* 20: 114–121.

FAO (Food and Agriculture Organization of the United Nations). 2015. FAOSTAT database. Accessed September 2017. http://faostat3.fao.org/faostat-gateway/go/to/home /.

Frija, A., A. Chebil, K. A. Mottaleb, D. Mason-D'Croz, and B. Dhehibi. 2020. "Agricultural Growth and Sex-Disaggregated Employment in Africa: Future Perspectives under Different Investment Scenarios." *Global Food Security* 24: 100353.

Gerber, P. J., H. Steinfeld, B. Henderson, A. Mottet, C. Opio, J. Dijkman, A. Falcucci, and G. Tempio. 2013. *Tackling Climate Change Through Livestock—A Global Assessment of Emissions and Mitigation Opportunities*. Rome: FAO.

Godoy, S. G. Marques de, and M. S. M. Saes. 2015. "Cap-and-Trade and Project-Based Framework: How Do Carbon Markets Work for Greenhouse Emissions Reduction?" *Ambiente & Sociedade* 18 (1): 135–154.

Havlík, P., H. Valin, M. Herrero et al. 2014. "Climate Change Mitigation through Livestock System Transitions." *Proceedings of the National Academy of Sciences of the United States of America* 111: 3709–3714.

Hoogenboom, G., C. H. Porter, V. Shelia et al. 2017. Decision Support System for Agrotechnology Transfer (DSSAT) Version 4.7. https://DSSAT.net. DSSAT Foundation, Gainesville, FL.

IFPRI (International Food Policy Research Institute). 2019. *2019 Global Food Policy Report*. Washington, DC: IFPRI.

IIASA (International Institute for Applied Systems Analysis). 2013. SSP Database version 1.0. Accessed May 10, 2015. https://tntcat.iiasa.ac.at/SspDb.

———. 2015. RCP Database version 2.0.5. Accessed May 10, 2015. www.iiasa.ac.at/web-apps/tnt/RcpDb.

IPCC (Intergovernmental Panel on Climate Change). 2014. *Climate Change 2014: Synthesis Report*. Contribution of Working Groups I, II and III to the Fifth Assessment Report of the Intergovernmental Panel on Climate Change. Core Writing Team, R. K. Pachauri and L. A. Meyer, eds. Geneva: IPCC.

Jones, J. W., G. Hoogenboom, C. H. Porter et al. 2003. "DSSAT Cropping System Model." *European Journal of Agronomy* 18: 235–265.

Mason-D'Croz, D., T. Sulser, K. Wiebe et al. 2019. "Agricultural Investments and Hunger in Africa Modeling Potential Contributions to SDG2–Zero Hunger." *World Development* 116: 38–53.

Moss, R. H., J. A. Edmonds, K. A. Hibbard et al. 2010. "The Next Generation of Scenarios for Climate Change Research and Assessment." *Nature* 463: 747–756.

Nelson, G. C., H. Valin, R. D. Sands et al. 2014. "Climate Change Effects on Agriculture: Economic Responses to Biophysical Shocks." *Proceedings of the National Academy of Sciences* 111 (9): 3274–3279.

O'Neill, B. C., E. Kriegler, K. L. Ebi et al. 2017. "The Roads Ahead: Narratives for Shared Socioeconomic Pathways Describing World Futures in the 21st Century." *Global Environmental Change* 42: 69–80.

O'Neill, B. C., E. Kriegler, K. Riahi et al. 2014. "A New Scenario Framework for Climate Change Research: The Concept of Shared Socioeconomic Pathways." *Climatic Change* 122: 387–400.

Petsakos, A., S. D. Prager, C. E. Gonzalez et al. 2019. "Understanding the Consequences of Changes in the Production Frontiers for Roots, Tubers and Bananas." *Global Food Security* 20: 180–188.

Robertson, R. D. 2017. "Mink: Details of a Global Gridded Crop Modeling System." Washington, DC: IFPRI.

Robinson, S., D. Mason-D'Croz, S. Islam et al. 2015. *The International Model for Policy Analysis of Agricultural Commodities and Trade (IMPACT): Model Description, Version 3*. IFPRI Discussion Paper 1483. Washington, DC: IFPRI.

Rosegrant, M., G. Yohe, I. Burton, S. Huq, M. Ewing, and R. Valmonte-Santos. 2008. "Climate Change and Agriculture: Threats and Opportunities." Commissioned paper submitted to GTZ, Germany, July.

Rosegrant, M. W., and the IMPACT Development Team. 2012. *International Model for Policy Analysis of Agricultural Commodities and Trade (IMPACT): Model Description*. Washington, DC: IFPRI.

Rosegrant, M. W., and M. A. Sombilla. 2018. *The Future of Philippine Agriculture under a Changing Climate: Policies, Investments and Scenarios*. Singapore: Yusof Ishak Institute (ISEAS).

Rosegrant, M. W., T. B. Sulser, D. Mason-D'Croz et al. 2017. *Quantitative Foresight Modeling to Inform the CGIAR Research Portfolio*. Washington, DC: IFPRI.

Rosenzweig, C., J. W. Jones, J. L. Hatfield et al. 2013. "The Agricultural Model Intercomparison and Improvement Project (AgMIP): Protocols and Pilot Studies." *Agricultural and Forest Meteorology* 170: 166–182.

Rosenzweig, C., A. Ruane, J. Antle et al. 2018. "Coordinating AgMIP Data and Models across Global and Regional Scales for 1.5°C and 2.0°C Assessments." *Philosophical Transactions of the Royal Society A* 376 (2119).

Schellnhuber, H. J., S. Rahmstorf, and R. Winkelmann. 2016. "Why the Right Climate Target Was Agreed in Paris." *Nature Climate Change* 6: 649–653.

Smith, P., H. Haberl, A. Popp, K.-H. Erb et al. 2013. "How Much Land-Based Greenhouse Gas Mitigation Can Be Achieved without Compromising Food Security and Environmental Goals?" *Global Change Biology* 19 (8): 2285–2302.

Smith, P., D. Martino, Z. Cai et al. 2007. "Agriculture." In *Climate Change 2007: Mitigation*, edited by B. Metz, O. R. Davidson, P. R. Bosch et al., 498–540. Contribution of Working Group III to the Fourth Assessment Report of the Intergovernmental Panel on Climate Change. New York: Cambridge University Press.

Smith, P., D. Martino, Z. Cai et al. 2008. "Policy and Technological Constraints to Implementation of Greenhouse Gas Mitigation Options in Agriculture." *Agriculture, Ecosystems & Environment* 118 (1–4): 6–28.

Springmann, M., D. Mason-D'Croz, S. Robinson, K. Wiebe, C. J. Godfray, M. Rayner, and P. Scarborough. 2017. "Mitigation Potential and Global Health Impacts from Emissions Pricing of Food Commodities." *Nature Climate Change* 7: 69–74.

Stehfest, E., M. Berg, G. Woltjer, S. Msangi, and H. Westhoek. 2013. "Options to Reduce the Environmental Effects of Livestock Production—Comparison of Two Economic Models." *Agricultural Systems* 114: 38–53.

Steinfeld, H., P. Gerber, T. D. Wassenaar, V. Castel, M. Rosales, and C. de Haan. 2016. *Livestock's Long Shadow: Environmental Issues and Options*. Rome: FAO.

Stocker, T. F., D. Qin, G.-K. Plattner et al. 2013. "Technical Summary." In *Climate Change 2013: The Physical Science Basis,* edited by T. F. Stocker et al., 1–84. Contribution of Working Group I to the Fifth Assessment Report of the Intergovernmental Panel on Climate Change. Cambridge, UK: Cambridge University Press.

US EPA (United States Environmental Protection Agency). 2012. *Global Anthropogenic Non-CO_2 Greenhouse Gas Emissions: 1990–2030*. EPA 430R-12-006. Washington, DC.

Wiebe, K., H. Lotze-Campen, R. Sands et al. 2015. "Climate Change Impacts on Agriculture in 2050 under a Range of Plausible Socioeconomic and Emissions Scenarios." *Environmental Research Letters* 10: 085010.

Willenbockel, D., S. Robinson, D. Mason-d'Croz, M. W. Rosegrant, T. Sulser, S. Dunston, and N. Cenacchi. 2018. *Dynamic Computable General Equilibrium Simulations in Support of Quantitative Foresight Modeling to Inform the CGIAR Research Portfolio: Linking the IMPACT and GLOBE Models*. IFPRI Discussion Paper 1738. Washington, DC: IFPRI.

Wollenberg, E., M. Richards, P. Smith et al. 2016. "Reducing Emissions from Agriculture to Meet the 2 degree C Target." *Global Change Biology* 22 (12): 3859–3864.

WRI (World Resources Institute). 2008. Climate Analysis Indicators Toolkit (CAIT). Accessed September 2017. http://cait.wri.org/.

Chapter 20

THE ROLE OF WATER IN SUPPORTING FOOD SECURITY: WHERE WE ARE AND WHERE WE NEED TO GO

Claudia Ringler, Nicostrato Perez, and Hua Xie

Although global annual water availability is largely stable, with small increases as a result of accelerated water cycles under climate change (Oki and Kanae 2006), the demand on water resources has grown substantially over the past 50 years, due to population growth, agricultural and economic growth, and urbanization (WWAP 2016; Bates et al. 2008). This has led to increasing competition across water-using sectors and contributed to severe degradation of water and related ecosystems and biodiversity loss in parts of the globe and, in some cases, outmigration of humans. Higher temperatures, less certain precipitation patterns, as well as shorter, more concentrated rains together with prolonged dry seasons are putting further pressure on available water supplies (Bates et al. 2008; Fishman, Jain, and Kishore 2013; WWAP 2016). All of these developments have put access by farmers to water for food production at risk.

Globally, most crops and livestock are produced using soil moisture that comes from precipitation (also known as rainfed agriculture) (FAO 2011). Rainfed agriculture accounts for approximately 95 percent of the harvested crop area in Africa south of the Sahara, 86 percent in Latin America, two-thirds in the Middle East and North Africa region, half in Asia, and 85 percent of the harvested crop area in high-income countries (Ringler 2017). Conversely, Asia features the world's largest irrigated areas, followed by the Middle East and North Africa regions, and there has been little investment in irrigation in Africa south of the Sahara until recently.

While most crop area is rainfed, the contribution of irrigation to food security has been essential, generating 40 percent of global food production on less than a third of the world's harvested land. For this, irrigation requires large volumes of freshwater. This is reflected in the agriculture sector accounting for approximately 70 percent of global water withdrawals, largely for irrigation but also for livestock watering and aquaculture as well as for more than 80 percent of consumptive water use of withdrawn water (FAO AQUASTAT

2019; Ringler 2017; WWAP 2019). Irrigated agriculture supports food production in dry seasons, in areas that receive too little rainfall to grow food, and increasingly supplements production in areas with less predictable rainfall. Irrigated yields are generally 30 percent to 60 percent higher than those of rainfed crops, as irrigation supports higher-yielding seeds through improved water control. As a result of growing water variability and climate change, the role of irrigation as a key climate change adaptation strategy is growing in importance (Rosegrant, Ringler, and Zhu 2009; Ringler 2017).

The call for more sustainable food systems and an overall healthier planet, coupled with acute water shortages in cities around the globe, are putting pressure on irrigation water use. At the same time, there are demands for accelerated irrigation development to fuel agricultural and rural growth in areas particularly affected by climate variability and change as well as in those regions where irrigation development has lagged, such as in Africa south of the Sahara (Malabo Montpellier Panel 2018; Vicuña, McPhee, and Garreaud 2012; Olmstead 2013). Given growing pressures on water resources, there is agreement that agricultural water development and management needs to be dramatically improved to meet increased food demands and nutrition goals while also improving the environment. The next section describes the role of irrigation for food security, including through modeling a series of irrigation expansion scenarios. An overview of the key challenges associated with increased irrigation development follows, along with a series of promising solutions. The final section concludes with policy recommendations.

The Role of Irrigation for Food Security

Climate change exacerbates the adverse effects of growing water scarcity on agricultural productivity through increasing temperatures and changing patterns of precipitation and consequent increased variability in the supply of water (Bates et al. 2008). To assess the role of irrigation expansion for future food security, we use IFPRI's International Model for Policy Analysis of Agricultural Commodities and Trade (IMPACT), a multimarket model that simulates the operation of national and international food markets subject to water and other constraints (Robinson et al. 2015; see Chapter 19 in this volume for additional details). IMPACT takes into account both biophysical factors, such as climate and water availability, and economic factors, such as food prices and food demand, and can thus assess the contribution of changes in irrigation development for key food security parameters under climate change. The climate change scenario chosen for this analysis is the HGEM Global

Circulation Model with a Representative Concentration Pathway (RCP) 8.5 scenario and a midrange Shared Socioeconomic Pathway scenario (SSP2) (see Chapter 19).

Without climate change (NoCC), average global yields of cereals are projected to increase from 3.3 metric tons per hectare to 4.3 metric tons per hectare between 2015 and 2050. With climate change (the HGEM scenario), global yields are projected to be 9 percent lower, on average, in 2050. Yield declines due to climate change are projected to be largest in the North America region, albeit from high levels, followed by the Latin America and Caribbean region and the South Asia region. Yield declines translate into a projected reduction of global cereal production of 8 percent, with an 18 percent decline in the group of developed countries and a 2 percent decline in the Global South. Production shortfalls put pressure on food prices and dampen demand for food (Figure 20.1).

Irrigation can counteract production declines and associated food price increases under climate change through ensuring that enough water is available during the cropping season, through increased cropping intensity, by expanding the cropping season into the dry season, for example, and through supporting higher yields through more optimal water provision on irrigated lands. However, irrigated area has been expanding only slowly as a result of high investment costs for large-scale irrigation—compared to long-term secular declines in global food prices—due to growing pressure to use water in other sectors and due to growing environmental concerns. As a result, growth in irrigated area is estimated at around 2 million hectares of harvested area a year globally over the next several decades, with growth concentrated in the group of developing countries.

To assess the role of irrigation for agriculture and food security, we simulate three irrigation expansion scenarios under climate change with varying rates of irrigation expansion. In IRREXP_30, we accelerate irrigated area development over baseline expansion to increase irrigated area by 30 percent by 2030 from 2015 levels. In IRREXP_50, global irrigated harvested area grows by 50 percent by 2030 from 2015 levels; and in IRREXP_100, area increases by 100 percent, but over a longer timeframe, from 2015 to 2045. Figure 20.2 presents changes in total harvested areas as a result of irrigation investment and Figure 20.1 presents associated changes in average crop and in average cereal prices. Under moderately accelerated irrigation development, global irrigated harvested area is estimated to increase by 101 million hectares by 2050, including the conversion of 47 million hectares of rainfed lands. If investments would generate a 50 percent increase in irrigated

FIGURE 20.1 Projected world prices of food commodities under alternative irrigation expansion scenarios, 2050 (index 2015 = 1.00)

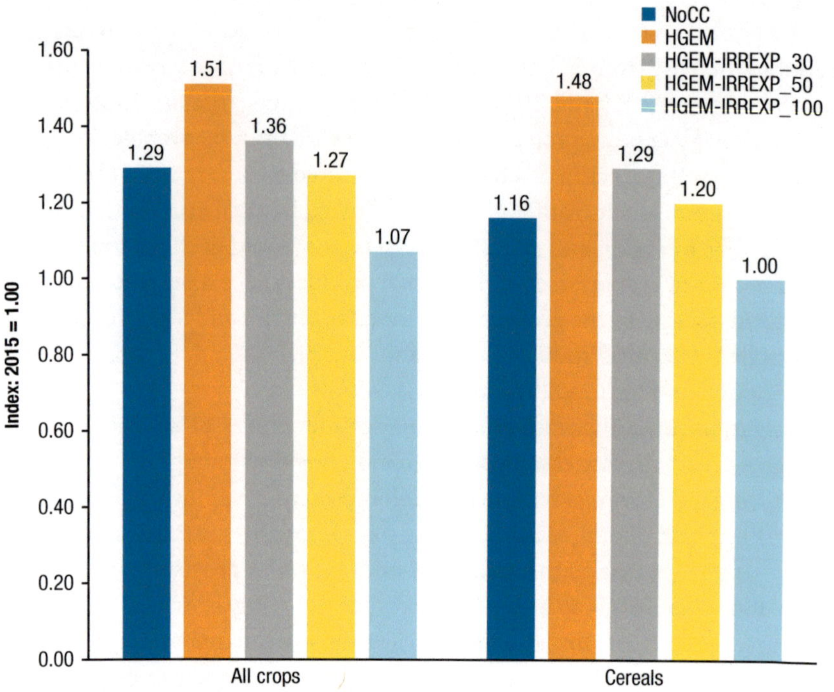

Source: Authors' estimates from IMPACT simulations.

Note: NoCC = no climate change; HGEM = climate change scenario based on the Hadley Centre Global Environmental Model; IRREXP_30 = 30 percent irrigation expansion; IRREXP_50 = 50 percent irrigation expansion; IRREXP_100 = 100 percent irrigation expansion.

area, an additional 165 million hectares of irrigated harvested area would be added, including through the conversion of 73 million hectares of rainfed harvested area. Under the most rapid irrigation development scenario (HGEM_IRREXP_100), 315 million hectares of irrigated harvested area would be added, including 100 million hectares from conversion of rainfed area.

Increases in irrigated area support substantial increases in food production. Production of all crops is projected to increase by 2.1 percentage points to 7.7 percentage points in 2050 as a result of irrigation expansion; with larger increases in the group of developing countries. As Figure 20.1 shows, an increase of irrigated areas by 50 percent over baseline rates by 2030 can close to reverse food price increases due to climate change. Even larger investments, at 100 percent over baseline increases, can reduce food prices to levels below

FIGURE 20.2 Changes in harvested irrigated and rainfed areas, alternative irrigation expansion scenarios, 2050

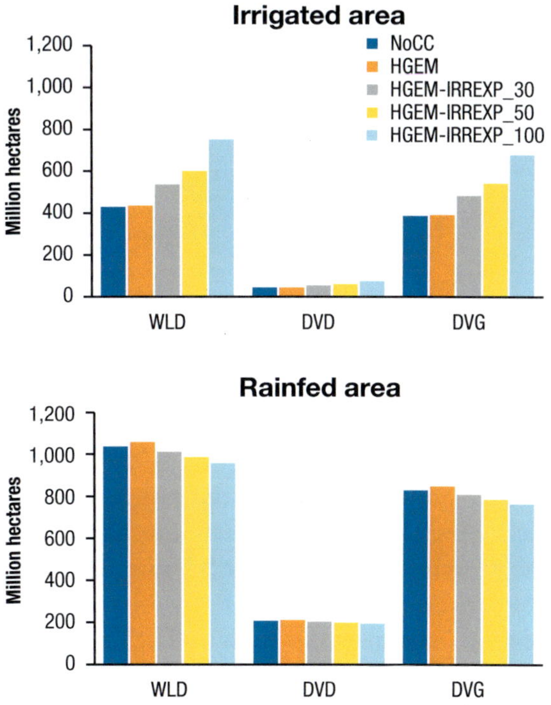

Source: Authors' estimates from IMPACT simulations.

Note: WLD = world; DVD = developed countries; DVG = developing countries; NoCC = no climate change; HGEM = climate change scenario based on the Hadley Centre Global Environmental Model; IRREXP_30 = 30 percent irrigation expansion; IRREXP_50 = 50 percent irrigation expansion; IRREXP_100 = 100 percent irrigation expansion.

a NoCC scenario. This suggests that irrigation can make effective contributions to address adverse impacts of climate change.

Global Challenges of Agricultural Water Management

Water Depletion, Including Groundwater Depletion

With population and economic growth, water scarcity has been on the rise in many areas around the world and is now often considered to be the single largest water problem (Jury and Vaux 2005). Using water withdrawals as a share of water availability as a water scarcity indicator, Figure 20.3 shows the

FIGURE 20.3 **Water withdrawals as a share of internal renewable water resources at country level, 2010**

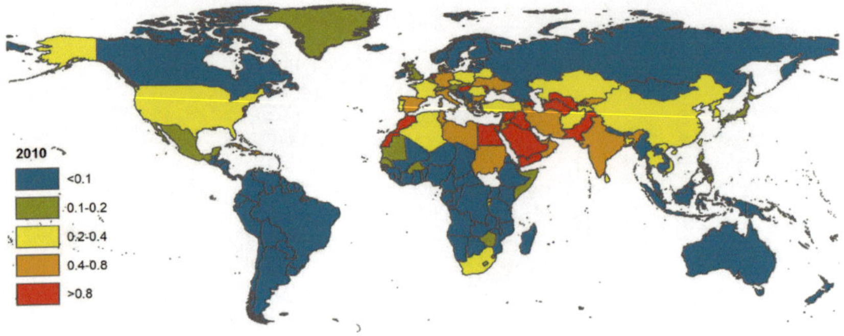

Source: IFPRI IMPACT (2019).

present water scarcity situation at the country level. A country is considered to be under low water stress if its water withdrawal-to-availability ratio is below 0.2, under moderate water stress if the ratio is between 0.2 and 0.4, and under severe water stress if the ratio is greater than 0.4 (Alcamo and Henrichs 2002). Following this definition, water stress levels are largest in the Middle East and North Africa region, Pakistan, and selected Central Asian countries, and are also large for selected countries in Europe.

Ringler et al. (2016) estimate that in 2010 one-third of the global population lived in river basins with severe risks of water scarcity (see also Oki and Kanae 2006) and that the share of the population at severe risk was growing dramatically to affect more than half (52 percent) of the global population by 2050. In a different estimate, Mekonnen and Hoekstra (2016) find that two-thirds of the world's population currently live in areas that experience water scarcity for at least one month a year—focusing attention on the seasonality of water scarcity. However, these water scarcity indictors mask important subnational variations in water availability and access; highly water-scarce countries, such as in the North Africa region, might have adjusted to lower water withdrawal levels, complicating cross-country comparison, and water access by the poorest can be lowest in some of the most water-rich countries, such as the Democratic Republic of the Congo (Mehta et al. 2019). As such, the water scarcity indicator—and indeed any water-related indicator—would need to be supplemented by additional information for water management investments.

Agricultural water was traditionally supplied by surface systems through river diversions, or through harnessing seasonal floods (Sojka, Bjorneberg, and Entry 2002). However, groundwater is the largest distributed store of liquid freshwater in the world (Shiklomanov 1993; Gleeson et al. 2015). With technological innovations, groundwater use has been increasing dramatically over the last 50 years, and groundwater depletion and contamination with pollutants and seawater in coastal areas has become a further sign of the growing risks and lack of sustainability of current agricultural water management practices. The most rapid increase in groundwater withdrawals for agriculture has occurred in Asia, where the world's largest irrigated areas are located (Shah et al. 2007). In parts of Asia, large, centralized pumping systems were used in the 1950s and 1960s, often to reduce waterlogging and salinization, such as in the Salinity Control and Reclamation Project (SCARP) of Pakistan (World Bank 1986). With the advent of cheap drilling technologies and more affordable, individual pump sets since the 1970s, combined with a secular decline in investments in public surface systems in Asia, private groundwater-sourced irrigation has rapidly increased and in many instances overtaken large-scale public infrastructure in growth and extent.

The most well-known case of smallholder-led groundwater irrigation is India. According to Mukherji and Das (2014), by 2011–2012 approximately 39 million hectares of the total of 65 million hectares of net irrigated area in India were irrigated with groundwater. Much of this development in India has been due to (1) affordable, individual pump technology; (2) cheap drilling technology; (3) inflexible and unreliable canal irrigation—sometimes coupled with the availability of high water tables due to runoff and seepage and percolation from canal systems; and (4) in several states, access to free electricity for pumping (Mukherji and Das 2014). Smallholder- or farmer-driven developments in South Asia have since been replicated in parts of Southeast Asia, Latin America, and Africa south of the Sahara.

Today, groundwater plays a major role in irrigation and food production globally. More than one-third of the world's 301 million hectares of area equipped for irrigation relies on groundwater. In addition, global consumptive use of groundwater in irrigation is estimated to account for 43 percent of total consumptive irrigation water use (Siebert et al. 2010). Among the top 15 groundwater-using countries in the world, 11 are in Asia. India is the largest groundwater user in the world, abstracting 251 cubic kilometers of groundwater annually, of which 223 cubic kilometers or 89 percent is estimated to be used for irrigation (Figure 20.4). China and the United States are the second and third largest users of groundwater with most of the water extracted used

FIGURE 20.4 Groundwater extraction by sector in the top 15 groundwater-using countries, 2010 (cubic kilometers)

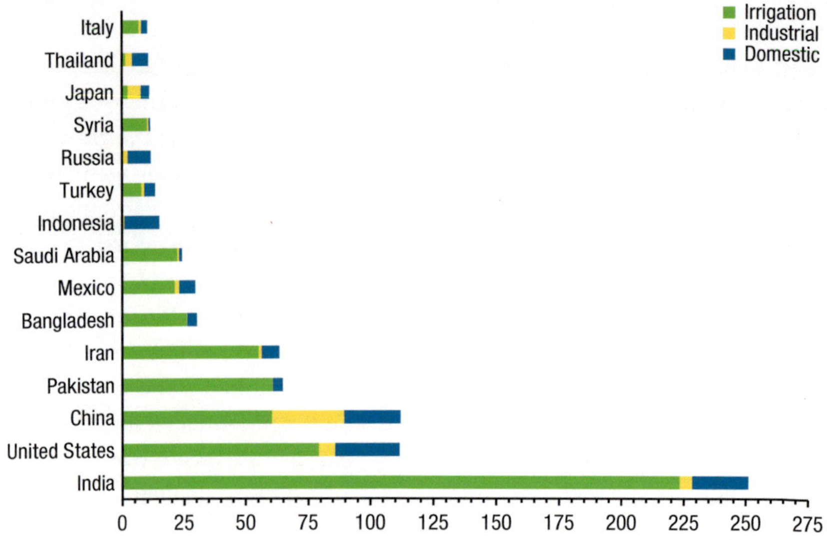

Source: Groundwater extraction statistics in Margat and van der Gun (2013).

for irrigation. Pakistan and Iran are the fourth and fifth largest groundwater users, abstracting an estimated 65 cubic kilometers and 63 cubic kilometers in 2010, again largely for irrigation (Margat and van der Gun 2013).

Intensive groundwater irrigation has led to groundwater depletion in many arid and semiarid agricultural regions, where groundwater is extracted at rates exceeding groundwater recharge (Wada et al. 2010; Wada et al. 2012; Gleeson et al. 2012; Döll et al. 2014). Groundwater tables have been declining at alarming rates in some of the world's major aquifers (Rodell, Velicogna, and Famiglietti 2009; Wada et al. 2010; Jiménez et al. 2011; Gleeson et al. 2012; Cao et al. 2013; Long et al. 2016; Tang et al. 2017). Many of those depleted aquifers overlap with the world's most important breadbaskets (Villholth et al. 2016). Sustained groundwater overdraft puts future irrigated food production at risk and leads to undesirable environmental consequences, including land subsidence and seawater intrusion, which can have profound social and environmental impacts (Giordano 2009; Kløve et al. 2011; Medellín-Azuara et al. 2015).

Water Pollution

Water pollution from the overuse of agricultural chemicals is a growing, but largely unaddressed, threat to public health and water-related ecosystems.

FIGURE 20.5 Estimated nutrient-loading intensity from agricultural land in the base year (2000)

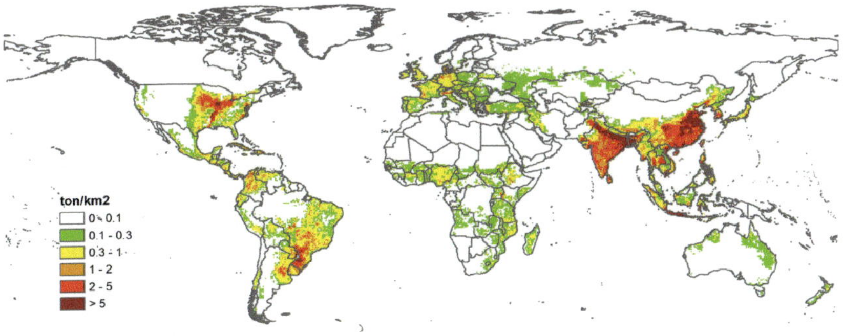

Source: Xie and Ringler (2017).

Major pollutants from agricultural sources that lead to water quality deterioration include nutrients, pesticides, and a family of emerging contaminants such as hormones and feed additives (Mateo-Sagasta et al. 2017). While most water pollution studies are carried out at small scales due to data constraints and the role of local biophysical environments in shaping pollution levels, a few macronutrient pollution studies have been implemented at larger scales, because sources and impacts are relatively well-known and data are more readily available. Nitrogen pollution, which has been clearly linked to agricultural practices, is harmful for ecosystems, causing algal blooms and dead zones in coastal areas, such as the hypoxia in the Gulf of Mexico (Good and Beatty 2011). Nitrates in drinking water can decrease the ability of blood to carry oxygen, which can be fatal for infants.

Xie and Ringler (2017) assessed the likely future development of global nitrogen (N) and phosphorous (P) loadings from agricultural production under alternative socioeconomic, climate, and technological change scenarios. They estimate global N loadings from agriculture around the year 2000 at 46 million metric tons per year and P loadings at 2.7 million metric tons per year. An example of the modeled distribution of N loadings across the globe is presented in Figure 20.5. Global N and P loadings are projected to rapidly increase under all climate and socioeconomic scenarios with N loadings growing to between 62 million metric tons and 81 million metric tons. Patterns of change are similar for projections of P loadings but with lower overall projected growth, and a total range of 2.8 million metric tons to 3.4 million metric tons per year. Moreover, growth in pollution is projected to be fastest in the group of low-income developing countries, with a projected increase of up to 118 percent for N loadings and up to 47 percent for P loadings. While nutrient-loading levels are currently largest in the Asia region (Figure 20.5), rates of growth are fastest in Africa.

It is much more challenging to assess the impacts of micropollutants, such as pesticides and emerging contaminants, as data on sources of pollution and their impacts are lacking, transformation processes of individual components and mixtures of various components are not always known, and no data are collected on most of these pollutants. Moreover, in many countries only some micropollutants are regulated and many are not treated. It is thus difficult to monitor and model the share of pesticide components and other micropollutants that reach water bodies and to design broad mitigation measures (Schwarzenbach et al. 2010).

Water pollution can occur at recommended levels of agrochemical applications, as only a share of chemicals are taken up by plants. Pollution levels can be larger if farms are near water bodies or if large rainfall events wash chemicals into water bodies. At the same time overuse of chemicals occurs and is linked to farmers' risk aversion, fertilizer subsidies (often for nitrogen, which can lead to unbalanced use of application levels), underinvestment in extension services and related overreliance on agrodealers for advice on chemicals, and farmers' lack of knowledge regarding pesticide use when new plant diseases occur (Borin et al. 2005; Jin, Bluemling, and Mol 2015; Good and Beatty 2011).

Water Variability

Water variability, or hydrologic variability, is a natural phenomenon driven by climate variability. When manifesting in extreme hydroclimatic events such as floods and droughts, water variability can damage crops (Ray et al. 2015), destroy livelihoods, and adversely affect economic growth (Brown and Lall 2006; Thurlow, Zhu, and Diao 2012). Owing to agriculture's strong reliance on and direct exposure to weather, management of water variability is a key concern for food production. According to Lesk, Rowhani, and Ramankutty (2016), globally, droughts and extreme heat alone reduced cereal production by 9 percent to 10 percent during 1964 through 2007. Moreover, agricultural production in developing countries is more vulnerable to weather shocks, due to the sector's lower capital investment and knowledge intensity and associated coping capacity. In recent years, almost one-quarter of weather-related damages has been in the agricultural sector in developing countries (FAO 2015).

Figure 20.6 displays mean annual runoff and the coefficient of variation of annual runoff at a 0.5-degree resolution. Annual runoff depth is a

FIGURE 20.6 **Mean annual runoff and coefficient of variation (CV) of annual runoff**

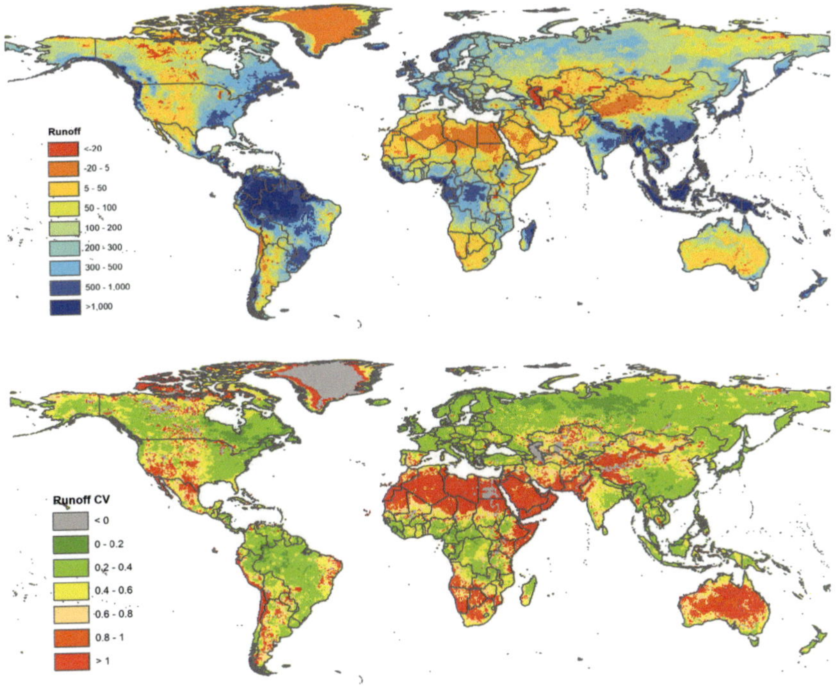

Source: IMPACT Global Hydrological Simulation for the 1951–2000 period.

Note: In open-water bodies grid cells where actual evapotranspiration exceeds precipitation runoff values are negative. Runoff is in mm per year.

measure of water availability. The coefficient of variation (CV) is the standard deviation of annual runoff divided by mean annual runoff and is a measure of interannual water variability. Strong water variability exists in many regions of the world, especially arid and semiarid regions such as southern Africa, the Middle East and North Africa, the southwestern United States, and southern Latin America. Poorer countries generally exhibit higher water variability (Grey and Sadoff 2007). While floods can sweep away crops and properties, and pollute domestic water sources, droughts can cause crop losses through reducing both harvested areas and crop yields (Lesk, Rowhani, and Ramankutty 2016). In much of Africa south of the Sahara, where agriculture remains largely rainfed, interannual climate variability significantly affects agricultural GDP and total GDP. Increasing water variability associated with climate change is expected to increase pressure on food production. Existing and growing uncertainties regarding precipitation are adversely affecting investments in agricultural productivity, particularly in Africa south of the Sahara where investment in irrigation infrastructure is especially low (Cooper et al. 2008).

Food production and prices can be affected by water variability and climatic shocks in significant ways, both at the local and global levels. Climate change projections generally only assess the mean impact of changes in temperatures and precipitation and do not consider changes in the variability of climate. To assess the role of irrigation in addressing water variability, we simulate the impact of climate variability and change on maize production and net trade in maize, the key staple crop in southern Africa. Southern Africa is of particular interest to the analysis of climate variability due to its overall water scarcity and high variability of water availability (see Figure 20.6; Conway et al. 2015). We then simulate the potential of expanded irrigation in southern Africa to reduce the adverse impact of climate variability and change, projecting a 50 percent increase in irrigated area in the region (using the scenario introduced earlier in the chapter for the southern African region only).

Figure 20.7 presents the cumulative distribution functions (CDF) for production and net trade of maize under climate variability and change, compared to a NoCC scenario as well as the benefits from irrigation expansion for maize production and trade in southern Africa. The probability that maize production falls below 35 million metric tons increases from 13.2 percent under the NoCC scenario to 20.0 percent under the HGEM scenario. Conversely, if irrigated area is expanded in southern Africa, the probability for maize production to decline below 35 million metric tons declines to 6.9 percent.

In terms of maize trade, under the HGEM climate change scenario, the probability and range of maize imports in southern Africa declines from 16 million metric tons to 31 million metric tons under the NoCC scenario to 9 million metric tons to 23 million metric tons as a result of climate change, which puts pressure on maize prices, resulting in a decline in demand for maize. With substantial expansion of irrigated area, the region would see a further shift to the right with a range of yet lower net imports of 5 million metric tons to 21 million metric tons. In this case, net imports would decline due to the rapid expansion in irrigated area in the southern African region, which would reduce the pressure on maize prices and also reduce the need to import maize.

The key response to water variability has been investment in water storage. Historically, most investment has been in reservoirs. Such reservoirs are increasingly being built as multipurpose systems, storing water for irrigation,

FIGURE 20.7 **Variability in maize production and net trade, plotted as cumulative distribution functions, southern Africa, 2050**

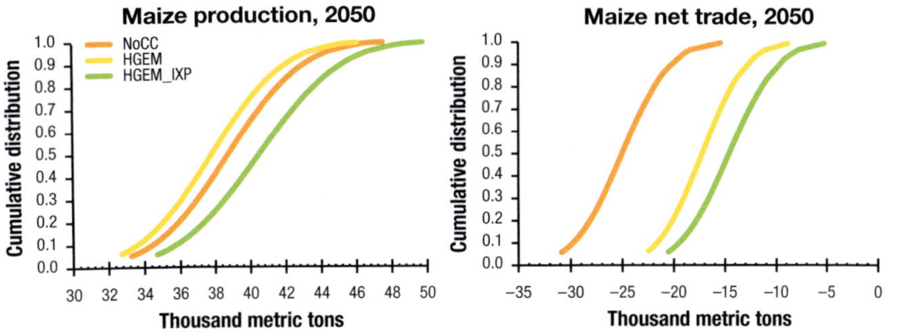

Source: Authors' estimates from IMPACT simulations.

Note: NoCC = no climate change; HGEM = climate change from Hadley Centre Global Environmental Model; IXP = irrigation expansion.

municipal-industrial uses, hydroelectricity production and flood control, and sometimes reservoir fisheries, and to ensure timely releases for environmental flow purposes. According to the International Energy Agency (IEA 2016), power production from reservoirs accounted for more than 85 percent of global renewable electricity generation in 2015. However, such reservoirs, or gray storage, often affect the migration patterns and spawning patterns of fish; change hydrological regimes, temperature, and flows of rivers; and can disrupt nutrient and sediment flows for fish and other aquatic resources.

A more holistic consideration of the storage continuum is now increasingly considered. This continuum ranges from natural water storage underground, in the soil and in water bodies to gray infrastructure, and includes small and large reservoirs (McCartney and Smakhtin 2010). A relatively recent measure to enhance the natural storage capacity of groundwater for both reduced flooding and increased availability of water during the dry season, for irrigation and other purposes, is managed aquifer recharge (MAR). MAR is the intentional recharge of water to aquifers for subsequent recovery or environmental benefit. Dillon et al. (2019) estimate that MAR has accelerated at a rate of 5 percent per year and has reached an estimated 10 cubic kilometers per year, equivalent to more than 2 percent of groundwater extraction in countries reporting MAR. India and the United States, large extractors of groundwater, also report the largest MAR capacity.

There are measures to directly reduce water variability in irrigation systems, for example, through improved irrigation-scheduling tools that can reduce excess water application. These tools, including soil moisture and yield sensors, are increasingly linked to large databases that trigger irrigation applications or other agricultural management interventions. Improved climate forecasting tools, drought-monitoring bulletins, and apps that provide real-time and near-future information on precipitation, temperatures, droughts, and crop prices are further tools that help farmers manage climate, water, and associated food production risks (Rodrigues et al. 2016).

Conclusion and Policy Recommendations

Water scarcity is a reality today for many people on the globe, and this scarcity is particularly felt in the agriculture sector, which is often the user of last resort. Driven by population and economic growth, the demand for water is expected to substantially increase in nearly all developing countries and many developed countries over the coming decades. While domestic and industrial water demands continue to grow faster than irrigation water demands, agricultural water use will remain the largest consumptive use sector out to 2050. Climate change adds an additional layer of complexities to the water and food challenges in the future, and considerable uncertainties exist in climate model projections and associated projections of water availability and resulting gaps between demands and supplies. However, there is little doubt that the adverse effects of climate change on water most directly affect the agriculture sector, where growing more food to meet the caloric and nutrition demands of an increasing population will require that more water is used for food production unless significant water savings (that is, crop per drop) can be achieved. Increasing water productivity can be achieved both in the water and irrigation sectors but also through a host of other measures.

Policies and strategies in the irrigation sector that can help support water security for food production in developing countries include removal of restrictions to the import of advanced irrigation technologies, and increased investment in extension services and other knowledge management platforms geared toward the use of often complex irrigation technologies as well as for associated knowledge areas in integrated pest management, agrochemical applications, and climate information services. Knowledge platforms on advanced irrigation technologies can also directly address water pollution concerns. Drip irrigation systems can apply agrochemicals with the irrigation water in more precise doses than traditional application mechanisms can provide—that is, through fertigation—and can furthermore enhance the quality of horticultural products. Various precision-agricultural technologies, such as soil moisture sensors or simple wetting-front detectors, can provide guidance to farmers on when irrigation can improve crop yields and when not (for example, Stirzaker 2003). To achieve savings through direct irrigation interventions, it will be essential to further advance measurements of irrigation water use through remote sensing technologies, combined with field validation. Several research and private consultancy groups continue to advance measurements of evapotranspiration from space (Bastiaanssen, Molden, and Makin 2000).

Importantly, investments in advanced irrigation technologies, such as drip or sprinklers, without proper assessment of water resources and flows, and without institutions that put a cap on withdrawals for agriculture, might well increase total irrigation water consumption, which is also called the "paradox of irrigation efficiency" (Rosegrant, Ringler, and Zhu 2009; Ringler 2017; Grafton et al. 2018). In rainfed environments, key measures to improve water-use efficiency when irrigation is not available include conservation agriculture, improved weeding, reduced tillage, and application of mulches. Governments need to strengthen investments in crop breeding for increased plant transpiration efficiency. Investment in agricultural research in this and other areas continue to fall short of research investments in other sectors like health. Moreover, the development of drought- and heat-stress-tolerant varieties of key food crops as well as submergence-tolerant varieties for coastal ecoregions can further improve the productivity and efficiency of irrigation applications.

In addition to investment in technologies, growing reliance on, and increased competition over shared water resources requires the strengthening of water management and allocation institutions. Such institutions include river basin organizations at the basin or watershed level, water user associations at the level of irrigation systems, and community rules or regulations for using shared aquifer resources. Mekonnen, Channa, and Ringler (2015) found that the existence of effective water user organizations improved agricultural productivity of farmers at the tail end of irrigation canals. At the community level, pilot studies in India have shown that collective groundwater management facilitated by experimental games can be effective in improving local groundwater governance (Meinzen-Dick et al. 2018). These are now being scaled up to several thousand communities across multiple Indian states.

Many of these technologies and institutions can also help remediate key challenges to irrigation expansion: water pollution and groundwater depletion. However, this requires increased investment in these often invisible environmental impacts. Unless serious investments and actions are taken toward addressing agricultural water pollution and groundwater depletion, clean water will be increasingly difficult to find in the breadbaskets of Asia and the rapidly growing farming centers of Africa south of the Sahara. Solutions exist and include enhanced nutrient-use efficiency, phasing out fertilizer subsidies, no-till or reduced tillage, crop rotations and other conservation measures, and closing the nutrient cycle, but both enforcement of regulation and economic incentives to use fertilizers and pesticides more judiciously are currently lacking (Good and Beatty 2011; IFPRI and Veolia 2015; Sun et al. 2012; Huang et al. 2015).

References

Alcamo, J., and T. Henrichs. 2002. "Critical Regions: A Model-Based Estimation of World Water Resources Sensitive to Global Changes." *Aquatic Sciences* 64 (4): 352–362.

Bastiaanssen, W. G. M., D. J. Molden, and I. W. Makin. 2000. "Remote Sensing for Irrigated Agriculture: Examples from Research and Possible Applications." *Agricultural Water Management* 46 (2): 137–155.

Bates, B. C., Z. W. Kundzewicz, S. Wu, and J. P. Palutikof, eds. 2008. "Climate Change and Water." Technical Paper of the Intergovernmental Panel on Climate Change, IPCC Secretariat, Geneva.

Borin, M., M. Vianello, F. Morari, and G. Zanin. 2005. "Effectiveness of Buffer Strips in Removing Pollutants in Runoff from a Cultivated Field in North-East Italy." *Agriculture, Ecosystems & Environment* 105 (1–2): 101–114.

Brown, C., and U. Lall. 2006. "Water and Economic Development: The Role of Variability and a Framework for Resilience." *Natural Resources Forum* 30: 306–317.

Cao, G., C. Zheng, B. R. Scanlon, J. Liu, and W. Li. 2013. "Use of Flow Modeling to Assess Sustainability of Groundwater Resources in the North China Plain." *Water Resources Research* 49 (1): 159–175.

Conway, D., E. Archer van Garderen, D. Deryng et al. 2015. "Climate and Southern Africa's Water-Energy-Food Nexus." *Nature Climate Change* 5 (9): 837–846.

Cooper, P. J. M., J. Dimes, K. P. C. Rao, B. Shapiro, B. Shiferaw, and S. Twomlow. 2008. "Coping Better with Current Climatic Variability in the Rain-Fed Farming Systems of Sub-Saharan Africa: An Essential First Step in Adapting to Future Climate Change?" *Agricultural Ecosystems and Environment* 126 (1–2): 24–35.

Dillon, P., P. Stuyfzand, T. Grischek et al. 2019. "Sixty Years of Global Progress in Managed Aquifer Recharge." *Journal of Hydrogeology* 27: 1.

Döll, P., H. Müller Schmied, C. Schuh, F. T. Portmann, and A. Eicker. 2014. "Global-Scale Assessment of Groundwater Depletion and Related Groundwater Abstractions: Combining Hydrological Modeling with Information from Well Observations and GRACE Satellites." *Water Resources Research* 50: 1–23.

FAO (Food and Agriculture Organization of the United Nations). 2011. *The State of the World's Land and Water Resources for Food and Agriculture (SOLAW): Managing Systems at Risk*. Rome.

———. 2015. *Impact of Natural Hazards and Disasters on Agriculture and Food Security and Nutrition*. Rome.

FAO AQUASTAT. 2019. FAO AQUASTAT Main Database. Accessed September 26, 2019. www.fao.org/nr/water/aquastat/data/query/index.html?lang=en.

Fishman, R., M. Jain, and A. Kishore. 2013. "Patterns of Migration, Water Scarcity and Caste in Rural Northern Gujarat." London: IGC.

Giordano, M. 2009. "Global Groundwater? Issues and Solutions." *Annual Review of Environmental Resources* 34: 153–178.

Gleeson, T., K. M. Befus, S. Jasechko, E. Luijendijk, and M. B. Cardenas. 2015. "The Global Volume and Distribution of Modern Groundwater." *National Geoscience Advance* (November): 1–15.

Gleeson, T., Y. Wada, M. F. P. Bierkens, and L. P. H. van Beek. 2012. "Water Balance of Global Aquifers Revealed by Groundwater Footprint." *Nature* 488 (7410): 197–200.

Good, A. G., and P. H. Beatty. 2011. "Fertilizing Nature: A Tragedy of Excess in the Commons." *PLoS Biology* 9 (8): p.e1001124.

Grafton, R. Q., J. Williams, C. J. Perry et al. 2018. "The Paradox of Irrigation Efficiency." *Science* 361 (6404): 748–750.

Grey, D., and C. W. Sadoff. 2007. "Sink or Swim? Water Security for Growth and Development." *Water Policy* 9 (6): 545–571.

Huang, J., Z. Huang, X. Jia, R. Hu, and C. Xiang. 2015. "Long-Term Reduction of Nitrogen Fertilizer Use Through Knowledge Training in Rice Production in China." *Agricultural Systems* 135: 105–111.

IEA (International Energy Agency). 2016. Electricity Data Browser. Accessed October 2016. www.eia.gov/electricity/data/browser/.

IFPRI (International Food Policy Research Institute) and Veolia. 2015. "The Murky Future of Global Water Quality. New Global Study Projects Rapid Deterioration in Water Quality." White paper. IFPRI, Washington, DC.

Jiménez, C., et al. 2011. "Global Intercomparison of 12 Land Surface Heat Flux Estimates." *Journal of Geophysical Research* 116 (D2): 1–27.

Jin, S., B. Bluemling, and A. P. J. Mol. 2015. "Information, Trust and Pesticide Overuse: Interaction between Retailers and Cotton Farmers in China." *NJAS-Wageningen Journal of Life Sciences* 72–73: 23–32.

Jury, W. A., and H. Vaux. 2005. "The Role of Science in Solving the World's Emerging Water Problems." *Proceedings of the National Academy of Sciences* 102 (44): 15715–15720.

Kløve, B., et al. 2011. "Groundwater Dependent Ecosystems. Part II. Ecosystem Services and Management in Europe under Risk of Climate Change and Land Use Intensification." *Environmental Science Policy* 14 (7): 782–793.

Lesk, C., P. Rowhani, and N. Ramankutty. 2016. "Influence of Extreme Weather Disasters on Global Crop Production." *Nature* 529 (7584): 84–87.

Long, D., X. Chen, B. R. Scanlon et al. 2016. "Have GRACE Satellites Overestimated Groundwater Depletion in the Northwest India Aquifer?" *Science Rep.* 6 (April): 24398.

Malabo Montpellier Panel. 2018. *Water-Wise: Smart Irrigation Strategies for Africa.* Dakar: IFPRI.

Margat, J. F., and J. van der Gun. 2013. *Groundwater around the World: A Geographic Synopsis.* New York: CRC Press.

Mateo-Sagasta, J., S. M. Zadeh, H. Turral, and J. Burke. 2017. "Water Pollution from Agriculture: A Global Review." Food and Agriculture Organization of the United Nations and the International Water Management Institute, Rome.

McCartney, M., and V. Smakhtin. 2010. *Water Storage in an Era of Climate Change: Addressing the Challenge of Increasing Rainfall Variability.* Blue paper, IWMI Reports 212430. Colombo, Sri Lanka: International Water Management Institute.

Medellín-Azuara, J., D. MacEwan, R. E. Howitt et al. 2015. "Hydro-Economic Analysis of Groundwater Pumping for Irrigated Agriculture in California's Central Valley, USA." *Journal of Hydrogeology* 23 (6): 1205–1216.

Mehta, L., T. Oweis, C. Ringler, and S. Varghese. 2019. *Water for Food Security, Nutrition and Social Justice.* New York: Routledge.

Meinzen-Dick, R. S., M. Janssen, S. Kandikuppa, R. Chaturvedi, K. Rao, and S. Theis. 2018. "Playing Games to Save Water: Collective Action Games for Groundwater Management in Andhra Pradesh, India." *World Development* 107 (July 2018): 40–53.

Mekonnen, D., H. Channa, and C. Ringler. 2015. "The Impact of Water Users' Associations on the Productivity of Irrigated Agriculture in Pakistani Punjab." *Water International* 40 (5–6): 733–747. Special Issue: Sustainability in the Water-Energy-Food Nexus.

Mekonnen, M. M., and A. Y. Hoekstra. 2016. "Four Billion People Facing Severe Water Scarcity." *Science Advances* 2 (2): p.e1500323.

Mukherji, A., and A. Das. 2014. "The Political Economy of Metering Agricultural Tube Wells in West Bengal, India." *Water International* 39 (5): 671–685.

Oki, T., and S. Kanae. 2006. "Global Hydrological Cycles and World Water Resources." *Science* 313 (5790): 1068–1072.

Olmstead, S. M. 2013. "Climate Change Adaptation and Water Resource Management: A Review of the Literature." *Energy Economics* 46.

Ray, D. K., J. S. Gerber, G. K. MacDonald, and P. C. West. 2015. "Climate Variation Explains a Third of Global Crop Yield Variability." *Nature Communications* (6): 5989.

Ringler, C. 2017. *Investment in Irrigation for Global Food Security*. IFPRI Policy Note. Washington, DC: IFPRI.

Ringler, C., T. Zhu, S. Gruber et al. 2016. "Role of Water Security for Agricultural and Economic Development—Concepts and Global Scenarios." In *Handbook on Water Security*, edited by C. Pahl-Wostl, J. Gupta, and A. Bhaduri, 183–200. Cheltenham, UK: Edward Elgar Publishing Ltd.

Robinson, S., D. Mason d'Croz, S. Islam et al. 2015. *The International Model for Policy Analysis of Agricultural Commodities and Trade (IMPACT): Model Description for Version 3*. IFPRI Discussion Paper 1483. Washington, DC: IFPRI.

Rodell, M., I. Velicogna, and J. S. Famiglietti. 2009. "Satellite-Based Estimates of Groundwater Depletion in India."*Nature* 460 (7258): 999–1002.

Rodrigues, J., J. Thurlow, W. Landman, C. Ringler, R. Robertson, and T. Zhu. 2016. *The Economic Value of Seasonal Forecasts: Stochastic Economy-Wide Analysis for East Africa*. IFPRI Discussion Paper 1546. Washington, DC: IFPRI.

Rosegrant, M. W., C. Ringler, and T. Zhu. 2009. "Water for Agriculture: Maintaining Food Security under Growing Scarcity." *Annual Review of Environmental Resources* 34: 205–222.

Schwarzenbach, R. P., T. Egli, T. B. Hofstetter, U. von Gunten, and B. Wehrli. 2010. "Global Water Pollution and Human Health." *Annual Review of Environment and Resources* 35: 109–136.

Shah, T., J. Burke, K. Villholth, M. Angelica, and E. Custodio et al. 2007. "Groundwater: A Global Assessment of Scale and Significance." In *Water for Food, Water for Life: A Comprehensive Assessment of Water Management in Agriculture*, edited by D. Molden, 395–423. London and Colombo, Sri Lanka: Earthscan.

Shiklomanov, I. A. 1993. "World Fresh Water Resources." In *Water in Crisis: A Guide to the World's Fresh Water Resources*, 13–24. Oxford: Oxford University Press.

Siebert, S., J. Burke, J. M. Faures, K. Frenken, J. Hoogeveen, P. Döll, and F. T. Portmann. 2010. "Groundwater Use for Irrigation—A Global Inventory." *Hydrology Earth Systems Science* 14 (10): 1863–1880.

Sojka, R. E., D. L. Bjorneberg, and J. A. Entry. 2002. "Irrigation: An Historical Perspective." *Encyclopedia of Soil Science* (1): 745–749.

Stirzaker, R. J. 2003. "When to Turn the Water Off: Scheduling Micro-Irrigation with a Wetting Front Detector." *Irrigation Science* 22 (3–4): 177–185.

Sun, B., L. Zhang, L. Yang, F. Zhang, D. Norse, and Z. Zhu. 2012. "Agricultural Non-point Source Pollution in China: Causes and Mitigation Measures." *Ambio* 41 (4): 370–379.

Tang, Y., M. Hooshyar, T. Zhu, C. Ringler, A. Y. Sun, D. Long, and D. Wang. 2017. "Reconstructing Annual Groundwater Storage Changes in a Large-Scale Irrigation Region Using GRACE Data and Budyko Model." *Journal of Hydrology* 551: 397–406.

Thurlow, J., T. Zhu, and X. Diao. 2012. "Current Climate Variability and Future Climate Change: Estimated Growth and Poverty Impacts for Zambia." *Review of Developmental Economics* 16 (3): 394–411.

Vicuña, S., J. McPhee, and R. Garreaud. 2012. "Agriculture Vulnerability to Climate Change in a Snowmelt-Driven Basin in Semiarid Chile." *Journal of Water Resources, Planning, and Management* 138 (5): 431–441.

Villholth, K. G., A. Sood, N. Liyanage, T. Zhu, and Y. Wada. 2016. "Global Food Production Share from Sustainable and Unsustainable Groundwater Use." In *The 43rd International Association of Hydrogeologists Congress "Groundwater and Society: 60 years of IAH."* Montpellier, France: IAH.

Wada, Y., L. P. H. van Beek, and M. F. P. Bierkens. 2012. "Nonsustainable Groundwater Sustaining Irrigation: A Global Assessment." *Water Resources Research* 48 (November).

Wada, Y., L. P. H. van Beek, C. M. van Kempen, J. W. T. M. Reckman, S. Vasak, and M. F. P. Bierkens. 2010. "Global Depletion of Groundwater Resources." *Geophysical Research Letters* 37 (20): 1–5.

World Bank. 1986. *Pakistan—Salinity Control and Reclamation (SCARP) Transition Pilot Project.* Washington, DC.

WWAP (UNESCO World Water Assessment Programme). 2016. *The United Nations World Water Development Report 2016: Water and Jobs.* Paris: UNESCO.

———. 2019. *The United Nations World Water Development Report 2019: Leaving No One Behind.* Paris: UNESCO.

Xie, H., and C. Ringler. 2017. "Agricultural Nutrient Loadings to the Freshwater Environment: The Role of Climate Change and Socioeconomic Change." *Environmental Research Letters* (12): 104008.

Chapter 21

FUTURE OF AGRICULTURAL RESEARCH

David Zilberman

Increased application of scientific knowledge has been a major contributor to feeding a growing population at higher per capita food levels while decreasing use of land and labor per unit of food (Goldewijk 2001).[1] However, growth in agricultural productivity varied significantly by location and crop, as was emphasized by Ritchie and Roser (2017) and in Chapters 3 through 7 of this book. This heterogeneity is associated to a large extent with varying application of scientific knowledge and advanced technology (Evenson and Gollin 2003). The coming decades will further challenge our agricultural system to meet the needs of a growing population and shifting demands of consumers, as well as to address the implications of climate change.

This chapter addresses some of the basic questions regarding agricultural research systems, especially in the context of developing countries, raised in the preceding chapters. Increased urbanization and urban food demand, as well as increasing per capita income, will lead to greater demand for high-value products, including fruits, vegetables, and animal-sourced foods (Chapters 9 and 10). Modern food value chains have emerged in response to such structural changes in food demand (Chapters 11 and 12). However, undernutrition, food insecurity, and poverty are persistent issues in many developing countries (Chapter 10). Building new equitable and viable agrifood systems will be one of the major issues of agricultural development in the coming decades. This chapter asks what agricultural research should be done, who should do it, and who should pay for it. A major challenging global issue in agricultural development is how to cope with climate change (Chapter 19), which accompanies deteriorating stock of natural resources (Chapter 18) and growing water scarcity (Chapter 20). Here too, the questions arise as to who should do what kind of agricultural research and who should shoulder the cost.

1 I would like to thank Joel Ferguson, Ben Gordon, and Jed Silver for excellent assistance.

To answer these questions, this chapter analyzes the innovation supply chains, relying on multiple bodies of literature. Some of the major and specific challenges identified in the previous chapters, which ought to be addressed by agricultural research, are discussed. Finally, the chapter explores the policy implications of this analysis. One major conclusion is that agricultural research can adequately enhance capacity to feed humans as well as address social welfare and climate change considerations by actively investing in research excellence around the world. This includes developing effective mechanisms of technology transfer from the lab to the field and the factory and from developed to developing countries, as well as relying on sound regulatory frameworks that allow scientific results to flourish while maintaining adequate levels of safety. It is also of paramount importance that sufficient public and private sources of funding are allocated to agricultural research and development (R&D) efforts, managed in ways that allow these efforts to be most effective.

Innovation Supply Chains

Agricultural research is part of an evolving system of general scientific knowledge, product development and food supply chains, and political economy forces that affect allocation of resources and funding to agricultural research. Furthermore, agriculture itself is going through a major transformation, and research is both a contributor to and an outcome of this transformation. Previous chapters have discussed induced innovation as well as political economy of agricultural policy, including in relation to agricultural research (Chapters 1, 2, and 14). This section focuses on an emerging topic related to the innovation process—that is, innovation supply chains.

To understand how innovations are carried out, it is important to understand the *innovation supply chains* connecting major players in agricultural research. Most innovations start with a new idea and subsequently there are several steps to bring the idea to an implementable product or practice. These steps can be defined as the research and development process, which may be different for different types of innovation. While many important innovations originated with practitioners, the share of innovations initiated with scientific discovery is increasing. Once a discovery is made, it needs to be verified and quantified both in terms of efficacy and safety, and then it needs to be scaled up and, if it is embodied in products, an industrial and commercialization process needs to be established. The next stage is dissemination of information so that it can reach the final users. For simplicity, I treat research that leads to discoveries as basic research.[2]

I refer to upscaling and testing as applied research, and commercialization can be viewed as development of the innovation. Dissemination can be carried out by private firms through marketing efforts as well as by public extension, and at times both. The R&D and dissemination activities may be done by different organizations, and the innovation supply chain consists of activities that lead from discovery to implementation in terms of product or practice. Table 21.1 illustrates the different stages and types of the innovation supply chain.

There may be different configurations of the innovation supply chain. One configuration that has become prominent in developed countries, especially the United States, is the educational-industrial complex, where innovations originate at research institutions, like universities. Table 21.1 illustrates two typical patterns under the educational-industrial complex, in one case where a technology transfers from a university to a company and another in which it goes from university through a start-up to a company. These innovations are typically patented, and the rights to develop them are sold to private companies or start-ups. These start-ups may become major corporations themselves or be acquired by large corporations. Graff, Heiman, and Zilberman (2002) documented this process of technology transfer, which served as a crucial element of the emerging medical and agricultural biotechnology sectors. Similar processes of interaction between universities and the private sector were important for innovations like the tomato harvester (Rasmussen 1968). One of the most important elements of the educational-industrial complex is the involvement of university researchers with the commercialization of their ideas.

The intrinsic knowledge held by innovators is often essential for commercialization, and the intense interaction between universities and the private sector has contributed to the overall performance of technological advances. These developments are consistent with the recent findings of Etzkowitz et al. (2000) that universities have played an increasingly important role in supporting private industry and government in an increasingly knowledge-based economy. The literature emphasizes that the innovation supply chain is not linear, and new innovations may emerge from experience in the field as well

2 Agricultural sciences are a form of applied research and use findings in other fields. From my perspective this discovery research is basic research.

TABLE 21.1 The innovation process

<table>
<tr><th rowspan="2">Stages of the innovation process</th><th colspan="2">Educational-industrial complex</th><th colspan="2">International innovation systems</th></tr>
<tr><th>Arrangement 1</th><th>Arrangement 2</th><th>Arrangement 1</th><th>Arrangement 2</th></tr>
<tr><td>Research</td><td rowspan="2">University</td><td rowspan="2">University</td><td rowspan="2">CGIAR* centers</td><td rowspan="2">CGIAR centers</td></tr>
<tr><td>Discovery</td></tr>
<tr><td>Development</td><td rowspan="2">Start-up</td><td rowspan="4">Major corporations</td><td rowspan="2">National Agriculture Research System</td><td rowspan="3">National Agriculture Research System</td></tr>
<tr><td>Patent (if applicable) and approval</td></tr>
<tr><td>Production</td><td rowspan="2">Major corporations</td><td>Private firms</td></tr>
<tr><td>Dissemination</td><td>Private firms/ extension</td><td>Extension</td></tr>
<tr><td>Adoption</td><td>Farmers</td><td>Farmers</td><td>Farmers</td><td>Farmers</td></tr>
</table>

Source: Based on Graff, Heiman, and Zilberman (2002).
Note: *Formally known as the Consultative Group on International Agricultural Research.

as commercialization of products. Interaction between researchers at different institutions is becoming more fluid. Structures akin to the educational-industrial complex are emerging in developing countries, where private firms are assuming a greater role in the application and introduction of technologies. Pray (2001) provides examples from Brazil, China, and India, where new forms of private-public partnerships are emerging to carry out basic research in plant breeding. Fuglie et al. (2012) emphasize the growing importance of private-sector research in the development and introduction of chemicals and mechanical solutions to agricultural problems in developing countries. This research has benefited from important transfers of knowledge through foreign direct investment as well as other forms of interaction with advanced research institutions.

Another structure quite prevalent in developing countries is an international/national innovation system where CGIAR, working with researchers at universities in developing and developed countries, conducts some of the fundamental research, such as developing a new variety or pest control strategy for a specific location, and then this technology may be transferred to the National Agriculture Research System (NARS) for further adaptive research (see Table 21.1). Sometimes the NARS may release their technologies through extension directly to farmers, and at other times they transfer it to private firms with their own marketing efforts and extension.

The public-sector research and outreach system based on the CGIAR and NARS can provide a base to supply food in developing countries. But the agrifood sectors of the developing world would benefit from the emergence of private-sector research, development, and marketing. Indeed, Pray and Fuglie (2015) show that private agricultural R&D has been growing faster than public R&D in recent decades, and there is a complementarity between public and private research. Graff and Zilberman (2001) suggest that public intellectual property should be managed to optimize net social benefit, and they suggest establishing a clearinghouse that would obtain access to intellectual property for applications that are not likely to be pursued by the private sector but could significantly benefit society, especially the poor in developing countries and specialty crops generating low volumes of income. However, Spielman and Ma (2016) find that intellectual property protection, either biologically through the use of hybrids or legally through enforcement of patents and other means, allows domestic seed industries to evolve in developing countries and contribute to the reduction of the gap in productivities between developing and developed countries.

The nature of the innovation supply chain varies by the nature of the technology. When it comes to technologies that are embodied in products, the private sector plays a much greater role. The large literature on technology diffusion (Feder, Just, and Zilberman 1985) emphasizes the importance of economic factors, like pricing, risk considerations, lack of credit, and regulations in introducing technologies. It argues that the diffusion of technologies is gradual due to heterogeneity among potential adopters and that processes of learning-by-doing and learning-by-using increase the attractiveness of new technologies and expand the range of adopters. Adoption may be hindered by uncertainty, and hence marketing efforts, like demonstrations, sampling, and money-back guarantees, are important for inducing adoption (Zilberman, Zhao, and Heiman 2012). Furthermore, implementation of innovations may require the establishment of particular supply chains—that is, the *food value chain* discussed in Chapter 12.

The design of the supply chain determines the capacity to disseminate the new technology, its price, and marketing strategy. One of the key roles of supply chains is provision of credit as well as insurance, as lack of credit and risk are major obstacles for adoption of new technology (Zilberman, Lu, and Reardon 2017). For example, growing flowers in Kenya can be viewed as an innovation, which may require research with respect to varieties and practices, but the main challenge is the establishment of a value chain, which includes nurseries for genetic material, farms to grow flowers, and then systems of processing and shipping. In the case of embodied technologies with increasing economies of scale in production, such as tractors and sprayers, many early adopters will lease the technology from a technology provider, and over time users will own the technology (Du et al. 2016).

The public sector may play a bigger role in the innovation supply chain for disembodied technologies, such as improved cultural practices, which are basically characterized by public goods. Public-sector extension agents are especially important in disseminating technologies that are relatively easy to implement and that do not require much human capital. However, when implementation of agricultural practices requires significant human capital, it may lead to the emergence of specialized consultants and information services (Wolf 1995). There is a growing sector of specialized consultants for irrigation, waste management, and pest control, and farmers' choices and views are affected by these sources of information (Lichtenberg and Zimmerman 1999).

Reshaping Agricultural Research to Tackle Future Challenges

The evolution of agriculture provides the background to assess some of the challenges faced by agricultural research and how they can be addressed. These challenges are multidimensional. On the one hand, agriculture needs to produce more food, estimated at 60 percent to 70 percent more by 2050 than in 2005 (FAO 2009). On the other hand, agriculture needs to address challenges of climate change as well as heterogeneity among regions and even among farmers in the same community in terms of productivity and capacity and the need to develop value-added agriculture. These issues have been identified as key challenges highlighted in previous thematic chapters. I address each of these issues with the consideration that agricultural research must be smallholder-friendly for poverty reduction and equitable development. In addition, I touch upon inherent issues about agricultural research, such as use of microlevel data and funding for agricultural research.

Increase Productivity

Although agricultural productivity gains have been remarkable during the past 150 years, the rate of increase has been declining more recently (Alston, Beddow, and Pardey 2009). Alston, Beddow, and Pardey (2009) argue that much of this decline is due to reduced public investment in agricultural research and the major emphasis of research on issues unrelated to productivity. Given increasing population, decreasing room for expansion of agricultural land, and climate change, continued emphasis on the enhancement of productivity is obviously warranted. According to Chapter 19, as long as investment in agricultural research increases and emphasizes input use efficiency and effective use of water and other resources, overall agricultural production is likely to meet the growing demand at modestly higher food prices. This analysis, however, does not explicitly consider some of the emerging breakthroughs in the life sciences, including agricultural biotechnology developments such as gene editing. These technologies can improve yields, reduce pesticide use, and enhance product quality. There is evidence that the use of Bt corn can even reduce postharvest risk, which has become a major concern (Wu 2006).

Investment in research to enhance productivity would allow countries to expand food production and prevent deforestation for expansion of agricultural land and deterioration of soil fertility. Improved crop and water management and technology, such as enhanced water harvesting, conservation tillage, and small-scale precision farming that optimizes application of water and other inputs within the field, are important means to improve yields in both irrigated and rainfed settings as are continued investments in high-efficiency irrigation in irrigated settings (Chapter 20). And here, the ability to take advantage of these advanced technologies depends, to a large extent, on the political will.

The recent experience with genetically modified crops suggests how regulations can hinder a technology reaching its potential. While Klümper and Qaim (2014) and Barrows, Sexton, and Zilberman (2014) show that adoption of genetically modified crops has enhanced yields, increased farmers' income, and decreased use of pesticides in the few countries and for the few crops where it has been adopted, it is far from reaching its potential. Regulations, in particular, hamper the ability of developing countries, especially in Asia and Africa (Herring and Paarlberg 2016). The case of golden rice (Wesseler and Zilberman 2014) demonstrates the immense monetary and human health loss due to regulatory delay; and the bans on technologies developed in and for Africa show the large cost of regulation in terms of malnutrition (Wesseler et al. 2017). Furthermore, costly and uncertain regulations of new technologies tend to reduce investment in these technologies in university research, and most important, in the private sector, which commercializes new knowledge throughout the innovation supply chain. From the viewpoint of political economy (discussed in Chapter 14), equilibrium with undesirable regulations should be broken by providing scientifically correct information about genetically modified crops.

Genetic modification was one of the first major applications of modern molecular and cellular biology in agriculture. With more recent breakthroughs, such as gene editing, the potential for new innovation is immense.

The extent of the societal gain from these technologies depends on the development of a regulatory framework that (1) balances the benefits of these technologies with sound risk considerations and (2) uses adaptive management where some experimentation is allowed with new technologies before large-scale use (Bradford et al. 2005). Furthermore, the capacity to take advantage of these promising new tools depends on an expanded network of research centers that study how and when new technology packages can be applied effectively throughout the world. Foley et al. (2011) suggest that the major discrepancy in terms of productivity between regions points to a large potential to increase yield and efficiency as technologies are introduced.

However, introduction of new varieties and production methods is not feasible without applied research that allows for adaptation and an effective supply chain for dissemination of these technologies. This further emphasizes the important role of a wide network of research stations associated with private-sector firms and extension distributed throughout the world. The ability to introduce more productive systems will benefit from the transformation of agricultural systems, introduction of advanced supply chains, and improved transportation and information technologies as well as demographic changes in agriculture.

Address Heterogeneity

The differences between regions in terms of productivity are among the most notable features of agrifood systems around the world (Chapters 3–8). No less important is the heterogeneity of productivity among farmers even in the same areas (Chapter 11). Needless to say, agricultural productivity and welfare or poverty are intimately related, so that in order to eradicate poverty, enhancing agricultural productivity in left-out areas is indispensable. Different regions have different research and agrifood systems. Advanced countries, to some extent, have strong research universities, effective agribusiness firms, robust infrastructure, and safety nets. Middle-income countries have many of the same features, while low-income countries lack these features, in part or whole.

Low-income countries, though, are generally most vulnerable to climate change, and their production practices and yields lag substantially. While these countries will continue to be dependent on CGIAR centers in the foreseeable future, policies that are focused on developing stronger national research centers and extension to adapt and develop technologies are crucial (Chapter 11). The work of Evenson and Kislev (1975) emphasizes the importance of excellence in research in being able to identify the best suited cultivars. Thus each region should pursue establishing centers of agricultural research excellence. Sanchez (2010) emphasizes the complementarity between agricultural practices and inputs, and thus research centers should have expertise in multiple aspects of agricultural production (for example, soil, water, fertilizers, seeds, and so on). In addition to establishing strong NARS, the decentralized extension system, which can deal with technology diffusion in vastly heterogeneous production environments, must be developed (Chapter 11). Furthermore, the public sector should not operate in isolation but rather develop links and alliances with the private sector that must grow even in the least developed countries in order to take advantage of new knowledge and opportunities to establish mechanisms for dissemination of new technologies.

The issue of heterogeneity is particularly relevant for natural resource management or NRM (Chapter 18). One clear finding from the existing studies is that the performance of NRM technologies, individually or in combination, varies considerably across different contexts (for example, by soil type), and their adoption also varies considerably as a consequence of this heterogeneity as well as socioeconomic variation (for instance, land rights, wage rates, and degree of market development). Therefore, research can probably have the largest impact by better understanding which of the existing NRM practices are likely to be most suitable for different contexts. Another clear finding is that desirable NRM practices or rules are insufficiently applied if measured by physical indicators of sustainability such as land degradation or water availability. So agricultural R&D could usefully reallocate funds away from pure technological research into understanding of institutional and socioeconomic factors, how practices can be adapted to fit better into local contexts in the short run, and how to improve the institutional environment in the longer term.

Develop Value-Added Agriculture

Rural poverty and low income have been major problems in both developed and developing countries, stemming from inelastic demand for agricultural commodities and technological change. Agricultural policies should aim to ameliorate these problems by transforming the rural economy (Chapters 11 and 12), but migration away from agriculture and larger farm operations have contributed to addressing them (Schultz 1964; Gardner 1992). In fact, the economic potential and viability of small subsistence farms in developing countries is limited (Hazell 2005). An increase in farm size is a key step for increasing farm incomes, and the development of labor-saving technology, such as mechanization, is a prerequisite for farm size expansion. Foster and Rosenzweig (2017) provide evidence from India showing that labor market frictions explain the persistence of small farms, but that this inhibits optimal machine use that takes advantage of economies of scale. Preservation of small farms is inefficient and exacerbates poverty. The seminal paper by Kislev and Peterson (1982) suggests that farmers in developed countries were pulled by higher urban incomes, which led to increases in farm size that helped remaining farmers to increase their income.

Urbanization and the development of local towns will be facilitated by the development of such technology (Chapters 9 and 11). Another avenue is to increase value-added of agricultural products both on the farm and in rural areas (Thirtle, Lin, and Piesse 2003). Value-added agriculture can be achieved by increasing production of specialty crops (for example, fruits, vegetables, flowers) and transition from basic commodities to differentiated products (for instance, corn or potato with unique nutritional or taste characteristics). The bifurcated nature of agriculture, where a small share of higher-income consumers has significant willingness to pay for uniquely produced agricultural products (for example, organic, exotic crops), can benefit the rural sector. As income grows, emphasis on differentiated, high-quality products should become a major priority for the agricultural sector (Chapter 11). However, development of value-added agriculture requires additional knowledge, sophisticated supply chains, and supportive policies. Also needed are innovative credit and insurance programs to facilitate the adoption of potentially profitable but possibly risky production of high-value products (Chapters 16 and 17).

Chapter 10 has clearly shown that many nutritious foods are very expensive in poor countries, which is likely to result in undernutrition. Thus the reduction of prices of nutrient-dense foods is an urgent issue, and diversifying agricultural R&D into nonstaple foods is the way to go. Also indispensable are related investments in extension and veterinary services, as many high-value products are new to farmers and the population of livestock is expected to increase. Also important for the development of value-added activities is the development of food-processing technologies and transport systems of fresh farm products. For example, agricultural research on solar-powered post-harvest drying and refrigeration technologies could have a high payoff.

Value-added agriculture can benefit from the expansion of the bioeconomy. This is the growing part of the economy that uses advances in life sciences and utilizes living organisms to produce goods and services. Besides food products, the bioeconomy produces fuels, chemicals, and medical products (Wesseler and von Braun 2017). It is a key to transitioning from relying on nonrenewable resources to a sustainable economy. Agricultural research should be expanded to include production of multiple products from agricultural feedstocks, as expansion of the range of products produced by agriculture will increase the well-being of the rural sector (Zilberman et al. 2013). While there has been significant debate about diversion of agricultural resources away from food to produce biofuels, and it seems that in the short run there have been some negative effects of biofuels on poverty, with the right policies, biofuels and poverty reduction are not substitutes but can complement one another (Kline et al. 2017). In other words, agricultural research to boost productivity and promote utilization of agricultural resources can contribute to both supply of ample food and enhancement of farmers' income.

Adapt to Climate Change

Although the scientific understanding of climate change and ability to predict its impacts are improving, they are still shrouded in uncertainty. Nevertheless, climate change will have a significant impacts on agrifood systems that will require, and already have prompted, major adaptations. Climate change is likely to result in (1) migration from regions closer to the tropics to regions closer to the poles, (2) rising sea levels and loss of coastal lands, (3) increased snow and ice melt as well as changes in supply of surface water, and (4) increased probability of extreme events with increased likelihood of fires, floods, and droughts.

The responses to climate change ought to include innovation and adoption of new technologies, changes in land use and trade patterns, development of insurance schemes, and investment in risk-reducing capacity (Zilberman, Zhao, and Heiman 2012). Climate change may require significant relocation. It will provide new opportunities for agricultural settlements in northern regions (for example, Canada, Russia) and may result in losses in opportunities closer to the equator. Thus climate change may result in climate migrants and refugees; and the ability to deal with this challenge is a major policy issue (Olmstead and Rhode 2011). Since social adaptation to migration is difficult and costly, research that allows for adaptation of agricultural and food systems to changes in climate will have significant value. Adaptation will benefit from the use of modern biotechnology tools to accelerate development of new varieties and the capacity to develop agronomical practices to use them effectively. Examples are development of flood-, drought-, disease- and pest-resistant, heat-tolerant, and water-saving varieties. For example, plant breeding can improve plant biomass per unit of water through transpiration rates and efficiency of biomass growth per unit of transpiration.

In the face of increasing climate risk, efficient insurance contracts will become critically important (Chapter 17). The development of improved varieties with a variety of resistant traits will contribute to the stabilization of crop yield, which will reduce the risk of production. This will increase the demand for credit and helps justify calls for institutional finance (Chapter 16). If crop insurance markets fail, stabilization of crop yield will aid in consumption smoothing.

Nitrous oxide and methane are major greenhouse gases emitted from agriculture. Thus, also needed is agricultural research on the development of nitrogen-efficient varieties that would allow lower use of fertilizer and reduced nitrous oxide emissions from agriculture, and farming practices to reduce methane emission from paddy fields. Breeding cattle for lower methane emissions will also contribute to the mitigation. Spatial changes in climate and water resource distribution will require significant investment in infrastructure, including dams, waterways, processing facilities, and so on. Since timing of responses to climate change is essential, it is important not only to develop better technical decision tools but also to improve political decision-making processes that do not impose unnecessary constraints. Finally, also important will be research that improves predictions of extreme climate events, designs mechanisms to cope with them, and develops guidance for policies, infrastructure design, and farming systems that are more resilient to possible changes.

Reardon and Zilberman (2018) also suggest that climate change may affect the rural sector through its impact on food value chains. Disruption of supply chains as a result of climate events may harm farmers dependent on them. Agribusiness firms may elect to invest in farming activities in regions more resilient to climate change. Thus awareness of these links and research and decision-making that takes them into account will improve adaptive capacity to climate change. Because of the uncertainty regarding climate change and its impacts, climate policies and research need to be designed within a larger context of agrifood growth and development. Combatting climate change does not mean reducing the effort to address food security. Policy design should emphasize "no regrets" policies that address climate challenges while addressing other crucial agricultural issues in an efficient and equitable manner (Lipper et al. 2018).

Use Microlevel Data

The recent expansion of microlevel data has led to better understanding and measurement of the internal workings of biological, physical, and social systems. In the case of physics and chemistry, the understanding of the atom and the development of electricity allowed transition from mechanical systems to electronic systems and now systems that manipulate nanodata. The discovery of DNA enabled plant biologists to move from selective breeding to biotechnology. In social and managerial sciences, we see the increased use of big microlevel behavioral data matched with biophysical and climatic data. This refined knowledge allows for greater disaggregation, enabling deeper analysis of microbehavior that addresses heterogeneity. The challenge in all fields is to develop a method that takes advantage of these new data and opportunities. Both precision agriculture and genetic engineering are based on the availability of new data. The availability of these new data will be key to more precise and effective supply chains. Einav and Levin (2014) suggest that the big data revolution will allow more refined understanding of the impact of policies and technologies, and help develop policies that adjust to heterogeneity and randomness. Mullainathan and Spiess (2017) suggest that machine learning would complement and enable more effective estimation and identify links that have not been used by traditional methods.

However, the challenge is to be able to make sense of the new available data and to reconcile microsystems and macrosystems. In biology there is a challenge to understand the functions of microelements in a bigger system and to realize the relationships between ecosystems and different organisms within them. In economics and management there is a challenge to integrate microrelationships and macrorelationships, and to use them in harmony. Furthermore, linked models at different degrees of aggregation are required to understand and develop policy related to climate change. We have seen this emerge through the integrated use of programming, econometrics, and general equilibrium for the case of biofuels (see Khanna and Zilberman 2012). But this is a modest beginning and the new data, as well as fresh approaches in econometrics and artificial intelligence, provide novel opportunities for policy and management analysis.

While data science is providing new opportunities, the challenges we face are immense. How do we incorporate narratives in empirical analysis for policy assessment? How do we take advantage of human ingenuity in interpreting emerging data sources? How do we increase efficiency to users and practitioners in using looming data-intensive technologies? Finally, how do we prevent abuse of expanding data sources?

Expand Resource Availability for and Effectiveness of Agricultural Research

The literature on the political economy of research (Rausser, Swinnen, and Zusman 2011) suggests that the funding and direction of public research represent political choices and aim to meet the objectives of governments, donors, multilateral organizations, NGOs, and environmental groups as well as the private sector. The development of the research agenda in the Global North to some extent affects the research that is done in the Global South. For example, donor countries' growing concerns about climate change and its implications in terms of migration and food security are likely to reemphasize adaption to climate change and prevention in the agricultural research agenda.

Similarly, issues of gender and sustainability are likely to be promoted by donors. The European objections to GMOs (genetically modified organisms) may reduce the weight in research done on GMOs in developing countries. Private actors may support activities that will enhance the reliability and quality of their supply and enhance the demand for their products. Research administrators have to delicately balance the different objectives of their major supporters and obtain funds without strings attached in order to pursue the most appropriate research agenda. (Defining this concept is its own topic of research.) One of the major challenges of research management is to obtain and allocate resources to basic research. The seminal paper by Rosenberg (1990) suggests that the value of this research and its spillovers may, from a social perspective, exceed the returns of targeted research efforts.

Conclusion and Summary

Agricultural research is challenged to address multiple societal issues, including sufficient and healthy foods, improved protection of the environment, and growing income for the agrifood sector. Its future, to a large extent, will depend on major policy decisions, both in the private and public sectors. A clear understanding of the major features of the agricultural sector, and agricultural technology in particular, is critical in designing policies to allow agricultural research to be a source of value and benefit for society.

Agricultural research is an essential element of the innovation supply chain, and its impact is hampered by not taking advantage of and establishing systems that may complement research within agricultural supply chains. In particular, establishing effective links between public and private research is important. The analysis suggests that public research and private research, in many cases, are not substitutes but rather complements. Basic knowledge produced by the public sector provides a foundation to much of the commercialization efforts of the private sector. While the educational-industrial complex will play a growing and important role in developed countries, expanding this model elsewhere can enhance research productivity. In this model, public research results are transferred to the private sector and then expanded and commercialized, and researchers are actually involved in expansion and commercialization processes.

For this research model to be effective, it is important to develop strong support for basic research at universities, develop a legal framework for technology transfer, and develop a culture that encourages collaboration between academics and practitioners. To be socially effective, the educational-industrial complex requires efficient management and utilization of intellectual property. This includes high thresholds for patenting as well as nonexclusive transfer of rights for process innovations and mechanisms to ensure access to patents needed for products to address the needs of the poor and developing countries. Support for research is provided not only in terms of direct support but also tax and other incentives to researchers and others to pursue work on public goods research.

While development of private-sector research capacity in developing countries is desirable and important, the public sector will continue to play a major role in development of agricultural innovations for the poor in developing countries. Sufficient investment in these capabilities is important and probably will continue to rely on internal funds as well as foreign aid. Strengthening the NARS and extension system is important as well, and these public-sector activities should be encouraged to improve, but not constrain, the evolution of stronger private-sector research, especially technology delivery. Appropriate division of efforts between the public and private sectors is needed to effectively introduce and disseminate new technologies. Furthermore, given the human capital and other constraints on the capacities of the private sector in developing countries, foreign direct investment in agricultural research, within limits, may contribute significantly to enhancing productivity.

CGIAR will continue to have a major role in providing the foundation for the development of agricultural technologies in developing countries. Development of research and education excellence in developing countries can contribute to the human capital and knowledge base needed to develop new technologies and apply them to specific conditions. That may suggest investment in establishing centers of excellence in research, development, and extension throughout developing countries. Some of them may be an outgrowth of the existing CGIAR centers. In addition, the example of the educational-industrial complex suggests that involvement of CGIAR centers and researchers in developing countries in further development and commercialization of technologies can accelerate the agricultural development process.

Another important dimension of policy is regulation. While regulations are essential to prevent mishaps, reduce risk, and control externalities, excessive regulation can be a significant hindrance to new technologies, both directly by preventing their use and indirectly by discouraging investment in new research. As argued earlier, the developing world has not gained much of the potential benefits from the introduction of new agricultural biotechnology tools because of excessive regulation. Furthermore, scarce resources have been diverted away from agricultural research to regulatory compliance. Introduction of more sound regulations is also important in developed countries, especially Europe, because we are at risk of losing some of the potential gains from modern biotechnology in agriculture. Agricultural research must recognize the evolution of agricultural and agrifood systems, and a growing priority should be given to investigate improved productivity of agrifood systems beyond the farm-gate. While studying crop and livestock systems will continue to be important elements of research, increased focus on storage, transportation, processing, and even utilization of food will be of much importance. Future research on these aspects should be on technological parameters but also emphasize the social and political implications of agrifood systems and how to design such systems to enhance social welfare and equity.

Agrifood systems will continue transforming. The economic viability of subsistence farmers is limited, and agricultural research can help some smallholders transition to a more viable scale of operation, add new value-added activities, and usher in agriculture to be a main contributor to and beneficiary of a growing bioeconomy. Agricultural research is likely to play a major role in guiding agriculture through adaptation to climate change, changes in land use, adoption of new technologies, and relocation, as well as provision of new sources of income from mitigation efforts. Finally, the ongoing information and biotechnology revolutions are likely to transform agriculture in significant ways. The challenge as researchers is to help make these transformations beneficial to the poor, consumers, society as a whole, and the environment.

References

Alston, J. M., J. M. Beddow, and P. G. Pardey. 2009. "Agricultural Research, Productivity, and Food Prices in the Long Run." *Science* 325 (5945): 1209–1210.

Barrows, G., S. Sexton, and D. Zilberman. 2014. "Agricultural Biotechnology: The Promise and Prospects of Genetically Modified Crops." *Journal of Economic Perspectives* 28 (1): 99–119.

Bradford, K. J., A. Van Deynze, N. Gutterson, W. Parrott, and S. H. Strauss. 2005. "Regulating Transgenic Crops Sensibly: Lessons from Plant Breeding, Biotechnology and Genomics." *Nature Biotechnology* 23 (4): 439–444.

Du, X., L. Lu, T. Reardon, and D. Zilberman. 2016. "Economics of Agricultural Supply Chain Design: A Portfolio Selection Approach." *American Journal of Agricultural Economics* 98 (5): 1377–1388.

Einav, L., and J. Levin. 2014. "The Data Revolution and Economic Analysis." *Innovation Policy and the Economy* 14 (1): 1–24.

Etzkowitz, H., A. Webster, C. Gebhardt, and B. R. C. Terra. 2000. "The Future of the University and the University of the Future: Evolution of Ivory Tower to Entrepreneurial Paradigm." *Research Policy* 29 (2): 313–330.

Evenson, R. E., and D. Gollin, eds. 2003. *Crop Variety Improvement and Its Eon Productivity: The Impact of International Agricultural Research.* Wallingford, UK: CABI Publishing.

Evenson, R. E., and Y. Kislev. 1975. *Agricultural Research and Productivity.* New Haven, CT: Yale University Press.

FAO (Food and Agriculture Organization of the United Nations). 2009. "How to Feed the World in 2050." In *Rome: High-Level Expert Forum.* Rome.

Feder, G., R. E. Just, and D. Zilberman. 1985. "Adoption of Agricultural Innovations in Developing Countries: A Survey." *Economic Development and Cultural Change* 33 (2): 255–298.

Foley, J. A., N. Ramankutty, K. A. Brauman et al. 2011. "Solutions for a Cultivated Planet." *Nature* 478 (7369): 337–342.

Foster, A. D., and M. R. Rosenzweig. 2017. "Are There Too Many Farms in the World? Labor-Market Transaction Costs, Machine Capacities and Optimal Farm Size." NBER Working Paper 23909. Cambridge, MA: National Bureau of Economic Research.

Fuglie, K., P. Heisey, J. King, C. E. Pray, and D. Schimmelpfennig. 2012. "The Contribution of Private Industry to Agricultural Innovation." *Science* 338 (6110): 1031–1032.

Gardner, B. L. 1992. "Price Supports and Optimal Spending on Agricultural Research." Working Paper 92–20. Department of Agricultural Economics, University of Maryland, College Park.

Goldewijk, K. K. 2001. "Estimating Global Land Use Change over the Past 300 Years: The HYDE Database." *Global Biogeochemical Cycles* 15 (2): 417–433.

Graff, G., A. Heiman, and D. Zilberman. 2002. "University Research and Offices of Technology Transfer." *California Management Review* 45 (1): 88–115.

Graff, G., and D. Zilberman. 2001. "An Intellectual Property Clearinghouse for Agricultural Biotechnology." *Nature Biotechnology* 19 (12): 1179–1180.

Hayami, Y., and V. W. Ruttan. 1971. *Agricultural Development: An International Perspective.* Baltimore: Johns Hopkins University Press.

Hazell, P. B. 2005. "Is There a Future for Small Farms?" *Agricultural Economics* 32 (s1): 93–101.

Herring, R., and R. Paarlberg. 2016. "The Political Economy of Biotechnology." *Annual Review of Resource Economics* 8: 397–416.

Janssen, E., and J. Swinnen. 2019. "Technology Adoption and Value Chains in Developing Countries: Evidence from Dairy in India." *Food Policy* 83: 327–336.

Khanna, M., and D. Zilberman. 2012. "Modeling the Land-Use and Greenhouse-Gas Implications of Biofuels." *Climate Change Economics* 3 (3): 1–15.

Kislev, Y., and W. Peterson. 1982. "Prices, Technology, and Farm Size." *Journal of Political Economy* 90 (3): 578–595.

Kline, K. L., S. Msangi, V. H. Dale et al. 2017. "Reconciling Food Security and Bioenergy: Priorities for Action." *GCB Bioenergy* 9 (3): 557–576.

Klümper, W., and M. Qaim. 2014. "A Meta-Analysis of the Impacts of Genetically Modified Crops." *PloS ONE* 9 (11): p.e 111629.

Lichtenberg, E. 2002. "Agriculture and the Environment." In *Handbook of Agricultural Economics*, vol. 2, edited by B. Gardner and G. Rausser, 1249–1313. Amsterdam: Elsevier Science.

Lichtenberg, E., and R. Zimmerman. 1999. "Information and Farmers' Attitudes about Pesticides, Water Quality, and Related Environmental Effects." *Agriculture, Ecosystems and Environment* 73 (3): 227–236.

Lipper, L., N. McCarthy, D. Zilberman, S. Asfaw, and G. Branca. 2018. *Climate Smart Agriculture: Building Resilience to Climate Change.* Rome: FAO; Dordrecht, Netherlands: Springer.

Mullainathan, S., and J. Spiess. 2017. "Machine Learning: An Applied Econometric Approach." *Journal of Economic Perspectives* 31 (2): 87–106.

Olmstead, A. L., and P. W. Rhode. 2011. "Responding to Climatic Challenges: Lessons from US Agricultural Development." In *The Economics of Climate Change: Adaptations Past and Present,* edited by G. Libecap and R. Steckel, 169–194. Chicago: University of Chicago Press.

Pardey, P. G., J. M. Alston, C. Chan-Kang et al. 2018. "The Shifting Structure of Agricultural R&D: Worldwide Investment Patterns and Payoffs." In *From Agriscience to Agribusiness*, edited by N. Kalaitzandonakes, E.G.Carayannis, E. Grigoroudis, and S. Rozakis, 13–39. Cham, Switzerland: Springer.

Pardey, P. G., J. M. Alston, and R. Piggott, eds. 2006. *Agricultural R&D in the Developing World.* Washington, DC: IFPRI.

Pray, C. E. 2001. "Public-Private Sector Linkages in Research and Development: Biotechnology and the Seed Industry in Brazil, China and India." *American Journal of Agricultural Economics* 83 (3): 742–747.

Pray, C. E., and K. O. Fuglie. 2015. "Agricultural Research by the Private Sector." *Annual Review of Resource Economics* 7 (1): 399–424.

Rasmussen, W. D. 1968. "Advances in American Agriculture: The Mechanical Tomato Harvester as a Case Study." *Technology and Culture* 9 (4): 531–543.

Rausser, G. C., J. Swinnen, and P. Zusman. 2011. *Political Power and Economic Policy: Theory, Analysis, and Empirical Applications.* Cambridge: Cambridge University Press.

Reardon, T., and D. Zilberman. 2018. "Climate Smart Food Supply Chains in Developing Countries in an Era of Rapid Dual Change in Agrifood Systems and the Climate." In *Climate Smart Agriculture,* edited by L. Lipper, N. McCarthy, D. Zilberman, S. Asfaw, and G. Branca, 335–351. Cham, Switzerland: Springer.

Ritchie, H., and M. Roser. 2017 (revised 2019). "Yields and Land Use in Agriculture." Our World In Data. Accessed 2019. https://ourworldindata.org/yields-and-land-use-in-agriculture.

Rosenberg, Nathan. 1990. "Why Do Firms Do Basic Research (with Their Own Money)?" *Research Policy* 19 (2): 165–174.

Sanchez, P. A. 2010. "Tripling Crop Yields in Tropical Africa." *Nature Geoscience* 3 (5): 299–300.

Schultz, T. W. 1964. *Transforming Traditional Agriculture.* New Haven, CT: Yale University Press.

Spielman, D. J., and X. Ma. 2016. "Private Sector Incentives and the Diffusion of Agricultural Technology: Evidence from Developing Countries." *Journal of Development Studies* 52 (5): 696–671.

Thaler, R. H. 2015. *Misbehaving: The Making of Behavioral Economics.* New York: W. W. Norton & Company.

Thirtle, C., L. Lin, and J. Piesse. 2003. "The Impact of Research-Led Agricultural Productivity Growth on Poverty Reduction in Africa, Asia and Latin America." *World Development* 31 (12): 1959–1975.

Wesseler, J., R. D. Smart, J. Thomson, and D. Zilberman. 2017. "Foregone Benefits of Important Food Crop Improvements in Sub-Saharan Africa." *PloS ONE* 12 (7): p.e0181353.

Wesseler, J., and J. von Braun. 2017. "Measuring the Bioeconomy: Economics and Policies." *Annual Review of Resource Economics* 9 (2017): 275–298.

Wesseler, J., and D. Zilberman. 2014. "The Economic Power of the Golden Rice Opposition." *Environment and Development Economics* 19 (6): 724–742.

Wolf, S. 1995. "Cropping Systems and Conservation Policy: The Roles of Agrichemical Dealers and Independent Crop Consultants." *Journal of Soil and Water Conservation* 50 (3): 263–270.

Wu, F. 2006. "Mycotoxin Reduction in Bt Corn: Potential Economic, Health, and Regulatory Impacts." *Transgenic Research* 15 (3): 277–289.

Zilberman, D., G. Hochman, D. Rajagopal, S. Sexton, and G. Timilsina. 2012. "The Impact of Biofuels on Commodity Food Prices: Assessment of Findings." *American Journal of Agricultural Economics* 95 (2): 275–281.

Zilberman, D., E. Kim, S. Kirschner, S. Kaplan, and J. Reeves. 2013. "Technology and the Future Bioeconomy." *Agricultural Economics* 44 (s1): 95–102.

Zilberman, D., L. Lu, and T. Reardon. 2019. "Innovation-Induced Food Supply Chain Design." *Food Policy* 83: 289–297.

Zilberman, D., J. Zhao, and A. Heiman. 2012. "Adoption versus Adaptation, with Emphasis on Climate Change." *Annual Review of Resource Economics* 4 (1): 27–53.

Chapter 22

RESHAPING AGRIFOOD SYSTEMS TO ACHIEVE MULTIPLE DEVELOPMENT GOALS

Shenggen Fan and Keijiro Otsuka

Today, agrifood systems are undergoing remarkable changes, reflected in the modernization of food value chains and rural transformation responding to urbanization, income growth, and expansion of international trade. At the same time, agrifood systems are expected to contribute to a wide range of development goals, reaching beyond agricultural productivity growth and food security. Thus this book has examined key issues using new lenses reflecting the rapidly changing world, such as nutrition, household decision-making behavior (related not only to gender but also to credit and insurance), natural resource management (including land, forests, and water), climate change, reprioritization of agricultural research, and political economy. Major findings from regional chapters (Chapters 3–7) were summarized in Chapter 8, while this chapter aims to synthesize the thematic chapters (Chapters 9–21). We highlight priority areas for action to reshape agrifood systems for achieving multiple development goals, including reductions in poverty and malnutrition, without harming the environment.

Agricultural Development and Transformation of Agrifood Systems

Despite the different pathways regional agrifood systems have taken over the past decades, rapid urbanization is occurring in almost all developing countries. This is driving structural change as workers move from rural to urban areas and leave agriculture for other sectors with higher labor productivity and incomes. Urbanization and urban income growth generate increased demand for agricultural products, while also shifting the composition of products demanded. This affects agrifood systems, as urban households consume a lower share of cereals, fats, and oils but higher shares of meat, dairy, fish, fruits, and vegetables as well as processed foods. While the latter products are generally nutritious, their prices are relatively high in developing countries, reflecting inadequate investment in research and development (Chapter 10).

Furthermore, governments in developing countries have a role to play in diversification of agricultural production and consumption, by investing, for example, in community health/nutrition initiatives.

Urbanization has also led to an increase in the share of poor people living in urban areas, likely driven by an inflow of poorer migrants or an overcrowding of low-income urban areas (Chapter 9). At the same time, transformation of the rural economy occurred in some regions, marked by increased productivity in staple crop farming, introduction of more profitable high-value agricultural products, better employment opportunities in rural nonfarm sectors, and migration to urban areas (Chapter 11). Since the farming environment is heterogeneous across major regions and even within a community, and since high-value products are usually new to farmers, the role of applied research and extension in disseminating environment-specific technological and managerial information to farmers has become critically important. Innovation is key to the development of agro-processing industries, so investment in human capital of entrepreneurs and farmers is indispensable (Chapter 11). To the extent that innovation by private individual initiatives is imitated, collective action plays an important role in internalizing the benefits of innovation.

As agrifood systems transform, food value chains transform as well. Traditional food value chains are fragmented and spatially short, with a low urban share and a high share of staple crops and grains. As economies continue to develop, food value chains become spatially longer, with a rising urban share, growth in the midstream of the value chain, and more diversified products. In the modern era, food value chains are spatially long, with a high urban share, expanded transportation, and development of peri-urban production. Growing urban market demand, especially for nongrain products, is transmitted to rural areas through rural-urban supply chains. The rapid development of nongrain and processed food markets in both urban and rural areas creates profitable new opportunities for actors along the value chain, including farmers, wholesalers, and processors (Chapter 12). Additional modern trends emerge, including disintermediation and multinationalization of the food value chain. This is reflected in expanding international trade of agricultural products, which is also supported by decreased distortion of agricultural markets (Chapter 13).

This transforming context of agrifood systems and value chains has significant implications for and is simultaneously affected by household decision-making behavior, particularly in relation to gender dynamics. Women have less access to agricultural land and other assets in developing countries, as men traditionally own or control land more often than women do (Chapter 15).

In addition, women tend to spend more resources on health and nutrition as well as education of children. Thus women's constrained access to productive resources may lead to underinvestment in these areas, even though the rate of return to such investment increases with the transformation of agrifood systems.

Although access to credit and insurance is critical for households to smooth consumption over time under risky production environments, to purchase modern inputs, and to invest in improving long-term productivity, many households still lack access to institutional finance. An important new financial institution, mobile banking, is particularly common in East Africa. While certain transactions, such as payments and remittance transfers, can be conducted via mobile banking, extension of credit cannot. To improve access to credit and insurance, institutional innovations—for example, bank finance and microfinance, the use of satellite imagery for monitoring crop damage, and index insurance programs—will be key (Chapters 16 and 17).

Understanding household behaviors will be crucial to facilitate gender equity. Moreover, to ensure inclusiveness of modern agrifood systems that leaves no one behind, investments in child schooling, nutrition, and health as well as in farm equipment and nonfarm businesses will be needed. Yet important aspects of household decision-making are not always properly accounted for in the macroeconomic literature; these include gender dynamics, liquidity or credit constraints (mismatches between the timing of income and expenditures), and risk (for which insurance is an important policy instrument).

With the population continuing to grow, while additional available land for agriculture is declining, there is no doubt that maintaining or improving natural resources already in use for agriculture—for example, land, water, and trees—is critical for food security. There are numerous management practices that can safeguard or enrich the natural resource base, but their application in developing countries needs to be greatly expanded. There are many mechanisms to improve incentives to adopt desirable management practices, but a key one will be the improvement of tenure securities over resources (Chapter 18). It is important to note, however, that there is no single tenure policy that will prove to be the best option in all countries, all locations within a country, or even within a community because of environmental, cultural, and institutional heterogeneities.

Agriculture continues to contribute to climate change, and climate change impacts agrifood systems. Climate impacts involve many dimensions, including changes in precipitation patterns, changing frequency and intensity of extreme weather events, changing patterns of pests and diseases affecting crops, livestock, and humans, rising sea levels, and glacial melting. As a result, projection models indicate that, for example, by 2050 climate change will reduce yields of maize, rice, wheat, and soybeans by an average of 11 percent or so worldwide and increase real prices by 20 percent, relative to the hypothetical reference case of no climate change (Chapter 19). Also worrisome is the growing scarcity of irrigation water for agriculture, which is affected by climatic volatility, urbanization and industrialization, and diet shift toward water-intensive foods (Chapter 20). Continuing the excessive use of water in agriculture will not be a sustainable option. Strengthening property rights on environmental resources, accelerating investment in irrigation, and improving water management are necessary to preserve and improve the stock of water resources.

Using new scientific possibilities, agricultural research must address issues arising from global-level transformations, including (1) raising farm productivity to feed the growing population under declining natural resources (Chapters 18, 19, and 20); (2) dealing with heterogeneity of natural and cultural environments by strengthening national agricultural research and extension systems (Chapters 11 and 18); (3) developing value-added agriculture to raise farmers' incomes and to supply healthy foods to consumers (Chapters 10, 11, and 12); and (4) facilitating adaptation to and mitigation of climate change (Chapter 19). A clear understanding of the major features of agricultural technology is critical in designing policies to allow agricultural research to be a source of value and benefit for both farmers and consumers, while addressing global challenges like climate change and degradation and depletion of natural resources. In order to achieve these goals, it is particularly important to establish effective linkages between public and private research and among international agricultural research systems represented by CGIAR, national agricultural research systems, and decentralized extension systems. We expect that the ongoing information and biotechnology revolutions are likely to transform agriculture in significant ways, and our challenge is to make this transformation beneficial to the poor, consumers, society as a whole, and the environment.

Transforming the Agrifood System for Achieving the UN Sustainable Development Goals

To engage in a changing global context and transforming agrifood systems, an integrated food systems approach is needed. The approach will need to consider traditional issues that remain important, emerging challenges in formulating policies, and a way forward for agrifood systems, many of which were discussed in the previous chapters. Improving agricultural productivity, employment, and livelihoods of smallholders, which are among the issues that have been the focus in agricultural development in past decades, will still be crucial to address challenges that face agrifood systems. At the same time, health and diets, the environment, and climate change have grown in prominence within the food systems approach, with a recognition that these issues are inseparably related to food production and global agrifood systems (Willett et al. 2019; IPCC 2018, 2019a, 2019b). If consumers' dietary habits change, the food production system must change. Consequently, greenhouse gas (GHG) emissions will change, as will the ecosystem, leading to changes in climate. The challenge for modern agrifood systems is how to mitigate climate change and preserve or improve ecosystems, while delivering healthy foods to consumers. To cover these multiple facets and contribute to multiple development goals, the approach should seek to reshape agrifood systems to be nutrition- and health-driven, productive and efficient, environmentally sustainable and climate-smart, inclusive, and business-friendly.

A nutrition- and health-driven system that uses nutrition as goals and promotes healthy diets is important, considering the persistence of and increases in multiple forms of malnutrition and diet-related diseases related to food systems. The food system needs to be productive and efficient by producing more food with fewer resources and reducing food waste and loss to efficiently meet current and future needs. The system must be environmentally sustainable and climate-resilient to minimize negative environmental impact, including but not limited to land degradation, deforestation, and overuse of species. The global food system needs to be inclusive, especially of smallholder farmers, women, and youth, who often lack access to assets and markets and face risk of exclusion from modern food value chains. Last, the food system should be business-friendly, with an enabling environment for well-functioning markets, partnerships, and a vibrant private sector. To reshape agrifood systems in light of the growing urgency and consensus in the global community toward the food systems approach, we propose the following priority areas for action based on the results of diverse analyses in the preceding chapters.

Empower Consumers with Knowledge

To improve nutrition, knowledge and understanding of the links between nutrition and health must reach consumers (Chapter 10). Understanding various barriers to consumer knowledge and addressing them with proper education and knowledge dissemination is essential to ensure enhanced nutritional outcomes (Willett et al. 2019). It is equally important that consumers are equipped with nutrition knowledge to better leverage social protection programs, such as cash transfers (de la Paz 2016).

While many strategies for consumer education are available, some successful experiences can be helpful in providing best practices and lessons learned. One such experience was in Bangladesh, where nutrition training sessions and cooking contests were held to test their impact on nutrition knowledge and subsequent dietary changes toward healthier consumption (Kramer 2017). And according to a randomized controlled trial study in Malawi, provision of information on child nutrition to mothers improved not only child nutrition but also health and food expenditures (Fitzsimons et al. 2016). These interventions were low-cost and successful in improving knowledge and education on nutrition. However, the study found that complementary interventions are needed to further change consumer behavior toward healthier diets.

In addition, many other potential avenues of education—such as nontechnical publications, public informal debates, blogs, webinars, media, targeted community training programs, nutritional curriculum development, research grants, and fellowship promotion—are available (Fan, Yosef, and Pandya-Lorch 2018). Targeting education to women and children will be particularly important in these campaigns. These two population subsets are those most affected by malnutrition and should therefore be well educated on the importance of good nutrition and how to attain it (Chapters 10 and 15). Women often hold considerable influence in the nutrition and health outcomes of a household and should be inclusively engaged with education. Food system regulations, both mandatory and voluntary, are an important complement to new technologies and consumer awareness for nutrition and health. The nutritional content and safety of foods is often unknown by the consumer, especially after processing or if food is purchased and consumed away from home. Labeling and quality standards based on nutrient testing and other food safety measures are therefore needed to improve consumers' awareness so that they can make good nutritional choices that will benefit their health. Information and communications technologies have also improved surveillance of food products through adulteration testing and traceability.

Transform Farmers into Entrepreneurs

The policies pursued under the proposed agrifood systems approach need to be inclusive of social, economic, and environmental issues and of all sections of the population, especially the vulnerable and marginalized. Particularly important will be investments in human capital of subsistence-oriented smallholder farmers (Chapter 11). Investment in their human capital improves household income as well as development of rural nonfarm sectors. Schooling and health, in particular, affect farming efficiency, nonfarm employment, and migration.

In addition, human capital investment, including the entrepreneurial training of farmers, is key to stimulating new innovations, such as the introduction of nutritious high-value products by farmers themselves (Chapter 11). Particularly important is the ability to "deal with disequilibria" (Schultz 1975)—that is, the ability to learn new technologies, products, and production methods, and to analyze the changing structure of market demands. Farmers cannot be made significantly better-off if they passively follow the instruction given by marketing agents. Unless farmers are transformed from subsistence-oriented to entrepreneurial, so that they can bargain with marketing agents (for example, agro-processors, traders, and supermarkets), an inclusive modern agrifood system can never be constructed. To support more efficient value chains linking rural and urban areas, institutional innovation such as the formation of producer cooperatives is often required (Chapter 11; Hashino and Otsuka 2016). A good example is the coffee unions of Ethiopia that provide marketing and input-supply services and connect producers to export markets (da Silva and Fan 2017). Producer cooperatives also play a key role in the introduction of new high-value products to rural areas, as these reduce the transaction costs between buyers and a large number of smallholder farmers (Chapter 11).

Incentivize the Private Sector

Private sector leadership is critical as urbanization and market growth around the world provide profitable business opportunities. Business interests have the power to improve nutritional outcomes by boosting demand for nutritious products through appropriate marketing and pricing. While there is great potential in business-friendly agrifood systems, it is by now widely recognized in the literature that the managerial inability of enterprise managers is a major constraint on the growth of enterprises and, consequently, local industries in developing countries (Bloom et al. 2012; Sonobe and Otsuka 2011, 2014).

Since new valuable information spills over, the social benefit of introducing fresh ideas is larger than the private benefit. Herein lies opportunity for government and aid agencies to invest, through training, in the managerial human capital of enterprise managers, who can appropriately respond to changing demand and create improved marketing systems. Using randomized controlled trials, several studies demonstrated that management training improves enterprise performance significantly (Bloom et al. 2013; Mano et al. 2012; Higuchi, Mhede, and Sonobe 2019). Establishment of industrial clusters or industrial parks, for example, can also help create hubs for vibrant development of agro-processing industries (Chapter 11). Improvements in infrastructure, transportation, and communications systems along the supply chain will be important for reducing food loss and waste and creating a food system that is responsive to changing consumer demands across rural and urban areas (Fan 2016).

The public sector can also support private stakeholders by setting quality standards, maintaining market institutions, and introducing tax incentives or other structures to effectively encourage the supply of healthy foods as well as fostering innovation and providing technical support. If markets for foods high in micronutrients and free of undetectable contaminants are to function and thrive, sellers must have the ability to signal quality. Low- and variable-quality products will remain in the market until new technologies and institutions permit quality standards to be set and enforced (Masters, Nene, and Bell 2017). Therefore, governments and trade associations need to be equipped and capable of regulating business practices and production methods that affect nutrition and health (Fan, Yosef, and Pandya-Lorch 2018). We should eventually aim to construct a food system that has strong links with private sector parties along the supply chain and facilitates partnerships between private sector actors and public institutions, international organizations, and civil society organizations. Adequate regulatory mechanisms to mitigate market shocks along with information and communications infrastructure can contribute to a well-functioning food system to engage the private sector (Fan 2016; IFPRI 2016).

Reprioritize Research and Development

Relatively few resources have been invested in agricultural research and development (Chapter 21). Probably the best indicator of underinvestment in research on nutrition-rich, high-value products in developing countries is how exceedingly high-priced these products are (Chapter 10). This is notable given the remarkable increases in demand for nutritious foods and increasing interest in the agriculture-nutrition nexus. It is true that nutrition-driven and nutrition-sensitive technologies, such as biofortification, have been developed. They can potentially play an important role by increasing the nutrition density in crops (HarvestPlus 2017). To strengthen linkages between agriculture, nutrition, and health, it is essential to invest in agricultural research and development, particularly for nutritious foods.

There also are potential technology options and cultivation practices that can substantially mitigate and adapt to climate change (Chapters 19 and 20; IPCC 2018, 2019a; Willett et al. 2019). For mitigation, resource-saving technology is needed, whereas for adaptation, technologies resistant to harsh climates are needed (Chapter 21). Also needed is the development of technology to use renewable resources, including biomass. In particular, the following are important technological innovations:

1. Technology or crop varieties (including fruits, vegetables, and beans) resistant to drought, heat, submergence, and salinity.
2. Input-saving technologies and management practices, including water, nitrogen fertilizer, and feed.
3. Technology to monitor activities leading to GHG emissions and to assess their impacts.
4. Cost-effective bioenergy, whose production does not compete with food production.

These technologies are global public goods, and global institutions such as the CGIAR must undertake research with the aim of mitigating and adapting to climate change.[1] Yet because of the environmental heterogeneity, adaptive research by national agricultural research systems and decentralized local extension systems are also an integral part of the innovation system. If technology development is the ultimate solution for climate change and the achievement of sustainable development, we must recognize that proper pricing of resources is the key to inducing the desired resource-saving technological changes. For example, restricting the use of fossil fuels, which will lead to increased energy prices, will stimulate the development of energy-saving technology in general and bioenergy in particular. Bioenergy deployment offers significant potential for climate change mitigation, which requires extensive use of agricultural residues and second-generation biofuels extracted from planted fast-growing tree species (IPCC 2018, 2019a). To avoid the risk of reducing food production, such "renewable biomass" whose production does not conflict with food production should be used. Although often unnoticed, technology to monitor GHG emissions and assess their impacts is a particularly important global public good (Chapter 19). Development of such

1 This does not imply that CGIAR has not undertaken relevant research (see, for example, Rosenstock et al. [2016] for the analysis of methods for measuring GHG emission).

technology will help implement appropriate policies to reduce GHG emissions, such as carbon tax, which is in line with the Coase Theorem (Coase 1960) that indicates that lower transaction costs tend to lead to socially improved resource allocations.

We will be able to transform agriculture to help achieve the UN Sustainable Development Goals only if the global community is willing to invest sufficiently to successfully develop a new set of global public goods. Investment in research and development is particularly necessary to fill extensive gaps in the evidence base regarding the effectiveness of actions for shifting diets in more healthy and sustainable directions. Research and development will also need to shift more toward improvements in health and environmental sustainability, and to focus on achieving multiple wins. Rigorous estimation methods that are being largely used in applied microeconomics should be leveraged and further strengthened. Particular attention should be paid to knowledge gaps for low- and middle-income countries, where there is a dearth of evidence on the linkages among agriculture, nutrition, health, and ecosystems (Willett et al. 2019).

Internalize Environment and Climate Effects

Achieving high productivity in agriculture and maintaining its growth in a way that does not degrade natural resources such as water, soil, and natural vegetation are major challenges to achieving sustainable agriculture (Chapters 18 and 20). We must note that management of water, soils, application of manure, conversion of forestland and woodland to farmland, and raising ruminants all affect the emission and sink of GHGs, which give rise to global externalities or climate change. Thus we must consider policies to reduce global negative externalities and to promote positive ones by improving the management of natural resources and the global climate. For example, irrigation water is often a local or regional common-pool resource and its use is free, which is bound to result in socially excessive use. A fundamental cause of excessive use of water is the absence of appropriate pricing that reflects its scarcity value. Thus, to reduce water use, a low-cost method of measuring its use is needed to apply volumetric water pricing to individual farmers or their group (for instance, a water users' association, which in turn charges water fees to individual farmers).

In addition, incentives to preserve and improve the stock of natural resources hinge on security of land rights, because they affect the probability that future benefits accrue to those who invest in the future. There is a host of empirical evidence that devolving forest property rights from the state to communities contributes to the sustainable management of forest resources (Chapter 18). The establishment of clear land rights, be they individual or communal, is also essential for sustainable management of farmland and other natural resources. Policy reforms, especially granting individual land rights to women, will be crucial for inclusive growth of agriculture (Fan 2016; Chapters 15 and 18).

Furthermore, agriculture must contribute to massive reductions in GHG emissions or even to sequestration of greenhouse gases. Major potential activities to reduce GHG emissions include reduction in deforestation, increase in forest cover and agroforestry, improvement of agricultural and livestock management, and introduction of sustainable bioenergy (Chapters 18–21; IPCC 2019a). The payment for ecosystem services (Farley 2012; Bennett and Gosnell 2015)—that is, the "price" paid for suppliers of ecosystem services as compensation for their continued provision of such services—is an application of the Coase Theorem to the environmental issue. Since forests and other ecosystems provide global externality in terms of sink of CO_2 and preservation of biodiversity, it is reasonable for the global community to pay landowners for the amount of global benefits.

Cost-effective methods to reduce emissions of nitrous oxide could include imposing taxes on nitrogen fertilizer and biomass burning and development of nitrogen-efficient crop varieties. Taxes could also be placed selectively on purchased feed, as the use of improved feed and dietary additives and improved breeds with higher productivity and lower methane emissions can contribute to reducing GHG emissions. The promotion of precision agriculture, particularly microdosing, can be an important complement to increase crop yields with less fertilizer (ICRISAT 2015). Global institutions, such as the Global Environmental Facility, will also need to be part of building a global consensus on simple practical rules on the amount of taxation and payments for ecosystem services, which are needed to achieve sustainable world agriculture.

Make Key Stakeholders Accountable

It must be clearly recognized that agrifood systems encompass the entire range of actors and their interlinked value-adding activities throughout the world—from food production, processing, and consumption to the broader economic, social, and natural environments in which these diverse food systems are embedded. Although preceding chapters did not explicitly discuss the issues of intragovernmental and intergovernmental collective actions and governance, it is obvious that in order to transform and reshape agrifood systems for global benefits, stakeholders will be required to undertake unprecedented collective action based on a solid understanding of the political economy in the agrifood system at the local, national, and global levels. This will require effective governance mechanisms at all levels under an integrated approach that cuts across political, sectoral, and geographical boundaries. To facilitate such an approach, we must establish formal and informal collaborative interactions and structures among governing groups, including various UN bodies and specialized agencies, to focus on cross-sectoral issues like healthy and sustainable diets. Relevant sectors and ministries at the national level, such as ministries of health, agriculture, and environment, will need to coordinate their efforts.

Given the deteriorating global environment and increasing emission of greenhouse gases by developing countries, now amounting to more than half of total global emissions (IPCC 2018), both developed and developing countries must contribute to the mitigation of climate change. However, considering that large developed economies are largely responsible for climate change, they should be held accountable and ought to consider how they can compensate developing countries for loss and damage due to climate change. Enhanced monitoring ability and government accountability are important to track our progress, and we must leverage the data revolution and big data analytics for quality, evidence-based evaluation. Backed by high-quality, timely data, stakeholders will need to track progress toward multiple development goals and use data to guide policy action. For example, tools like the Women's Empowerment in Agriculture Index, which measures women's empowerment, agency, and inclusion in agriculture, will play a critical role in identifying current gaps and constraints.

To sum up, in order to achieve multiple goals of sustainable, healthy, and inclusive agricultural development all over the world, national, regional, and global collaborative efforts are clearly called for to reshape global agrifood systems in line with the recommendations proposed in this chapter.

References

Bennett, D. E., and H. Gosnell. 2015. "Integrating Multiple Perspectives on Payments for Ecosystem Services through a Social-Ecological Systems Framework." *Ecological Economics* 116: 172–181.

Bloom, N., B. Eifert, A. Mahajan, D. McKenzie, and J. Roberts. 2013. "Does Management Matter? Evidence from India." *Quarterly Journal of Economics* 128 (1): 1–51.

Bloom, N., C. Genakos, R. Sadun, and J. Van Reenen. 2012. "Management Practices across Firms and Countries." *Academy of Management Perspectives* 26 (1): 12–33.

Coase, R. H. 1960. "The Problem of Social Cost." *Journal of Law and Economics* 3: 1–44.

da Silva, J. G., and S. Fan. 2017. "Smallholders and Urbanization: Strengthening Rural–Urban Linkages to End Hunger and Malnutrition." In *2017 Global Food Policy Report*, 14–23. Washington, DC: International Food Policy Research Institute (IFPRI).

de la Paz, J. 2016. "Channeling Social Protection Programs for Improved Nutrition in Bangladesh." IFPRI blog, July 5, IFPRI, Washington, DC.

Fan, S. 2016. "Food Policy in 2015–2016: Reshaping the Global Food System for Sustainable Development." In *2016 Global Food Policy Report*, 1–11. Washington, DC: IFPRI.

Fan, S., S. Yosef, and R. Pandya-Lorch. 2019. *Seizing the Momentum to Reshape Agriculture for Nutrition*. Wallingford, UK: CABI.

Farley, J. 2012. "Ecosystem Services: The Economics Debate." *Ecosystem Services* 1 (1): 40–49.

Fitzsimons, E., B. Malde, A. Mesnard, and M. Vera-Hernández. 2016. "Nutrition, Information, and Household Behavior: Experimental Evidence from Malawi." *Journal of Development Economics* 122: 113–126.

HarvestPlus. 2017. *Driving Impact: 2016 Annual Report*. Washington, DC: IFPRI.

Hashino, T., and K. Otsuka. 2016. *Industrial Districts in History and the Developing World*. Dordrecht, Netherlands: Springer.

Higuchi, Y., E. P. Mhede, and T. Sonobe. 2019. "Short- and Medium-Run Impacts of Management Training: An Experiment in Tanzania." *World Development* 114: 220–236.

ICRISAT (International Crops Research Institute for Semi-Arid Tropics). 2015. *Fertilizer Microdosing Increases Agriculture Productivity*. Project Policy Brief 2. Niamey, Niger

IFPRI (International Food Policy Research Institute). 2016. "Accelerating the Contribution That Nutrition's Underlying Drivers Make to Nutrition Improvements." In *Global Nutrition Report 2016: From Promise to Impact: Ending Malnutrition by 2030*, 60–75. Washington, DC.

IPCC (Intergovernmental Panel on Climate Change). 2018. *Global Warming of 1.5°C*. An IPCC Special Report on the impacts of global warming of 1.5°C above pre-industrial levels and related global greenhouse gas emission pathways, in the context of strengthening the global response to the threat of climate change, sustainable development, and efforts to eradicate poverty.

———. 2019a. *Climate Change and Land*. An IPCC Special Report on climate change, desertification, land degradation, sustainable land management, food security, and greenhouse gas fluxes in terrestrial ecosystems.

———. 2019b. *IPCC Special Report on the Ocean and Cryosphere in a Changing Climate.*

Kramer, B. 2017. *Cooking Contests for Healthier Recipes: Impacts on Nutrition Knowledge and Behaviors in Bangladesh.* IFPRI Discussion Paper 1661. Washington, DC: IFPRI.

Mano, Y., A. Iddrisu, Y. Yoshino, and T. Sonobe. 2012."How Can Micro and Small Enterprises in Sub-Saharan Africa Become More Productive? The Impacts of Experimental Basic Managerial Training." *World Development* 40 (3): 458–468.

Masters, W. A, M. D. Nene, and W. Bell. 2017. "Nutrient Composition of Premixed and Packed Complementary Food for Sale in Low- and Middle-Income Countries: Lack of Standards Threatens Infant Growth." *Maternal & Child Nutrition* 13 (4): 12421–12431.

Rosenstock, T. S., M. C. Rufino, K. Butterbach-Bahl, E. Wollenberg, and M. Richards, eds. 2016. *Methods for Measuring Greenhouse Gas Balances and Evaluation Mitigation Options in Smallholder Agriculture.* Dordrecht, Netherlands: Springer.

Schultz, T. W. 1975. "The Value of Ability to Deal with Disequilibria." *Journal of Economic Literature* 13 (3): 827–846.

Sonobe, T., and K. Otsuka. 2011. *Cluster-Based Industrial Development: A Comparative Study of Asia and Africa.* Hampshire, UK: Palgrave Macmillan.

———. 2014. *Cluster-Based Industrial Development: Kaizen Management for MSE Growth in Developing Countries.* Hampshire, UK: Palgrave Macmillan.

Willett, W., et al. 2019. "Our Food in the Anthropocene: The EAT-*Lancet* Commission on Healthy Diets from Sustainable Food Systems." *Lancet* 393 (10170): 447–492.

CONTRIBUTORS

Keijiro Otsuka is a professor of Development Economics at the Graduate School of Economics, Kobe University, Japan, and a chief senior researcher at the Institute of Developing Economies, Tokyo.

Shenggen Fan is chair professor, College of Economics Management, and dean of the Academy of Global Food Economics and Policy, China Agricultural University, Beijing, and former director general of the International Food Policy Research Institute (IFPRI), Washington, DC.

Akhter U. Ahmed is a senior research fellow in the Poverty, Health, and Nutrition Division of IFPRI and IFPRI representative for Dhaka, Bangladesh.

Kym Anderson is a professor of economics at the University of Adelaide and at the Australian National University's Crawford School of Public Policy, Canberra, Australia.

Ousmane Badiane is the founder and executive chairperson for AKADEMIYA2063, Kigali, Rwanda.

Saule Burkitbayeva is a researcher at the LICOS Centre for Institutions and Economic Performance of the University of Leuven, Belgium.

Xinshen Diao is the deputy division director and a senior research fellow in the Development Strategy and Governance Division of IFPRI, Washington, DC.

Eugenio Díaz-Bonilla is a senior research fellow and head of the Latin America and Caribbean program of IFPRI, Washington, DC, and former executive director of the Inter-American Development Bank.

Paul Dorosh is the director of the Development Strategy and Governance Division of IFPRI, Washington, DC.

Cheryl Doss is an associate professor and senior departmental lecturer in the Department of International Development, Oxford University, UK.

Ruben G. Echeverría is director general emeritus of the International Center for Tropical Agriculture (CIAT), Cali, Colombia, and a senior research fellow at IFPRI, Washington, DC.

Hosaena Ghebru is a research fellow in the Development Strategy and Governance Division of IFPRI, Washington, DC.

Derek Headey is a senior research fellow in the Poverty, Health, and Nutrition Division of IFPRI, Washington, DC.

Jikun Huang is a professor at the School of Advanced Agricultural Sciences and director of the China Center for Agricultural Policy at Peking University, Beijing, China.

Thomas Jayne is University Foundation Professor of Agricultural, Food, and Resource Economics and founding co-director of the Alliance for African Partnership at Michigan State University, East Lansing, US.

P. K. Joshi is the former director for South Asia, IFPRI, New Delhi.

Shahidur R. Khandker is a development economist with research experience at the World Bank and other development organizations.

Anjani Kumar is a senior research fellow in the South Asia Office of IFPRI, New Delhi, India.

William Liefert is a senior agricultural economist with the Market and Trade Economics Division of the Economics Research Service, US Department of Agriculture, Washington, DC.

Will Martin is a senior research fellow in the Markets, Trade, and Institutions Division of IFPRI, Washington, DC.

Daniel Mason-D'Croz is a senior research scientist with the Commonwealth Scientific and Industrial Research Organisation, Canberra.

William A. Masters is a professor in the Department of Economics and the Friedman School of Nutrition Science and Policy at Tufts University, Boston.

Ruth Meinzen-Dick is a senior research fellow in the Environment and Production Technology Division of IFPRI, Washington, DC, and co-leader of the CGIAR Research Program on Policies, Institutions, and Markets (PIM) flagship on governance of natural resources.

Bart Minten is a senior research fellow in the Development Strategy and Governance Division of IFPRI and the program leader of the Myanmar Agricultural Policy Support Activity, Yangon.

Nicostrato Perez is a research fellow in the Environment and Production Technology Division of IFPRI, Washington, DC.

Frank Place is the director of the CGIAR Research Program on Policies, Institutions, and Markets (PIM), which is led by IFPRI, Washington, DC.

Agnes Quisumbing is a senior research fellow in the Poverty, Health, and Nutrition Division of IFPRI, Washington, DC.

Thomas Reardon is a professor in the Department of Agricultural, Food, and Resource Economics at Michigan State University, East Lansing, US.

Claudia Ringler is the deputy division director in the Environment and Production Technology Division of IFPRI, Washington, DC.

Miguel Robles is a professor in the Department of Economics and Finance at the Universidad del Pacífico, Lima.

Mark W. Rosegrant is a research fellow emeritus at IFPRI, Washington, DC, and a nonresident fellow of the Chicago Council on Global Affairs.

Timothy B. Sulser is a senior scientist in the Environment and Production Technology Division of IFPRI, Washington, DC.

Johan Swinnen is the director general of IFPRI, Washington, DC.

Hiroyuki Takeshima is a senior research fellow in the Development Strategy and Governance Division of IFPRI, Washington, DC.

James Thurlow is a senior research fellow in the Development Strategy and Governance Division of IFPRI, Washington, DC.

Keith Wiebe is a senior research fellow in the Environment and Production Technology Division of IFPRI, Washington, DC.

Dirk Willenbockel is a research fellow at the Institute of Development Studies, University of Sussex, UK.

Hua Xie is a research fellow in the Environment and Production Technology Division of IFPRI, Washington, DC.

Futoshi Yamauchi is a senior research fellow in the Markets, Trade, and Institutions Division of IFPRI, Washington, DC.

Xiaobo Zhang is a distinguished chair professor of economics at the National School of Development, Peking University, Beijing, and a senior research fellow in the Development Strategy and Governance Division of IFPRI, Washington, DC.

David Zilberman is a professor and holds the Robinson Chair in the Department of Agricultural and Resource Economics at the University of California–Berkeley, US.

INDEX

Page numbers for entries occurring in figures are followed by an *f*; those for entries in notes, by an *n*; and those for entries in tables, by a *t*.

Made in the USA
Monee, IL
13 January 2026

41512650R00457